# Sustainable Environmental Technologies—Volume I

# Sustainable Environmental Technologies—Volume I

Topic Editors

**Dibyendu Sarkar**
**Rupali Datta**
**Prafulla Kumar Sahoo**
**Mohammad Mahmudur Rahman**

MDPI

Basel • Beijing • Wuhan • Barcelona • Belgrade • Novi Sad • Cluj • Manchester

*Topic Editors*
Dibyendu Sarkar
Stevens Institute of Technology
Hoboken
USA

Rupali Datta
Michigan Technological University
Houghton
USA

Prafulla Kumar Sahoo
Central University of Punjab
Bathinda
India

Mohammad Mahmudur Rahman
The University of Newcastle (UON)
Callaghan
Australia

*Editorial Office*
MDPI AG
Grosspeteranlage 5
4052 Basel, Switzerland

This is a reprint of the Topic, published open access by the journals *Applied Sciences* (ISSN 2076-3417), *Energies* (ISSN 1996-1073), *Environments* (ISSN 2076-3298), *Toxics* (ISSN 2305-6304) and *Water* (ISSN 2073-4441), freely accessible at: https://www.mdpi.com/topics/Sustainable_Environmental_Technologies.

For citation purposes, cite each article independently as indicated on the article page online and as indicated below:

Lastname, A.A.; Lastname, B.B. Article Title. *Journal Name* **Year**, *Volume Number*, Page Range.

**Volume I**
**ISBN 978-3-7258-3039-8 (Hbk)**
**ISBN 978-3-7258-3040-4 (PDF)**
**https://doi.org/10.3390/books978-3-7258-3040-4**

**Set**
**ISBN 978-3-7258-3037-4 (Hbk)**
**ISBN 978-3-7258-3038-1 (PDF)**

# Contents

# About the Editors

**Dibyendu Sarkar**

Dr. Dibyendu "Dibs" Sarkar, Ph.D., is a professor of Environmental Engineering at the Department of Civil, Environmental and Ocean Engineering, Stevens Institute of Technology, Hoboken, U.S.A. He is also the Director of the Stevens Center for Sustainability. Dibs earned his Ph.D. in geochemistry from the University of Tennessee and worked as a postdoctoral researcher in soil and water chemistry at the University of Florida. He has 25 years of experience as an environmental science and engineering faculty member and has authored 3 books, 20 book chapters, more than 170 journal articles, and more than 280 technical abstracts and conference proceedings. Dibs is an elected Fellow of the Geological Society of America and Soil Science Society of America. He is the founding Editor-in-Chief of *Current Pollution Reports*, a Q1 environmental science journal, and serves as a Section Editor-in-Chief, Technical Editor, and Associate Editor of several other journals. His research interests include environmental quality and remediation, soil and water chemistry, green technology development, and risk assessment.

**Rupali Datta**

Dr. Rupali Datta, Ph.D., is a professor of Biological Sciences at Michigan Technological University (MTU), Michigan, U.S.A. She has a Ph.D. in Life Sciences from the University of Hyderabad, India. She completed her postdoctoral studies at the University of Niigata in Japan, the University of Florida at Gainesville, and the University of Texas at San Antonio (UTSA). Dr. Datta's primary research lies in developing green remediation technologies for environmental pollutants through the application of molecular biology and microbiology, plant biochemistry, genetics, phytoremediation, bioremediation, and plant–microbe interactions. Her current research is focused on the study of the bioavailability of metal and organic contaminants in aquatic media and the potential of using plants to remediate contaminated sites, as mediated by microbes. She has authored more than 300 publications in high-quality journals.

**Prafulla Kumar Sahoo**

Dr. Prafulla Kumar Sahoo, Ph.D., is an Assistant Professor in the Department of Environmental Science and Technology, Central University of Punjab, India. Dr. Sahoo received his Ph.D. degree from IIT Kharagpur. He has more than 13 years of research experience (including Ph.D. and postdoctoral) in the field of environmental geochemistry, especially for acid mine drainage, groundwater contamination, and soil and sediment geochemistry. Much of his research is focused on societal and environmental issues, including emerging toxic contaminants (such as As, U, and F) in groundwater, soils and paddy rice. The importance of his research and potential at the international level has been recognized through his selection as a BK21-postdoctoral fellow in South Korea and Young Scientist Fellow at the prestigious Brazilian Academy of Science. So far, he has authored over 100 journal articles and 10 book chapters indexed in the Scopus database. He currently serves on the Editorial Board of *Current Pollution Reports* (Springer, a Q1 environmental science journal), in addition to serving as an Associate Editor for *Soil Pollution & Remediation of Frontiers in Soil Science*.

**Mohammad Mahmudur Rahman**

Dr. Mohammad Mahmudur Rahman, Ph.D., is an Associate Professor in the Global Centre for Environmental Remediation (GCER), College of Engineering, Science and Environment, University of Newcastle (UON), Australia. He is a globally recognized expert in arsenic-related research. He is also recognized as one of Australia's leading Environmental Scientists by Researcher.com. His work primarily examines the impacts of metals and metalloids on the environment and their potential risks for human health. His research encompasses varied topics, including chemistry, toxicity, bioavailability, environmental nanomaterials, environmental remediation, environmental analytical chemistry, phytoremediation, risk analysis and human health effects of arsenic, and the speciation of metals/metalloids using hyphenated techniques. Dr. Rahman's most notable achievements include publishing 242 publications in international peer-reviewed high-impact journals and refereed book chapters, underscoring his outstanding research leadership. In addition to his research, he serves as an Associate Editor for the Elsevier journal *Groundwater for Sustainable Development* and is a member of the Editorial Boards for several journals, including *Soil and Environmental Health*, *Water* (MDPI), *Journal of Sustainable Agriculture and Environment*, and *Frontiers in Nanotechnology: Environmental Nanotechnology*.

*applied sciences*

Editorial

# Sustainable Environmental Technologies: Recent Development, Opportunities, and Key Challenges

**Prafulla Kumar Sahoo [1,*], Rupali Datta [2], Mohammad Mahmudur Rahman [3] and Dibyendu Sarkar [4,*]**

1 Department of Environmental Science and Technology, Central University of Punjab, Bathinda 151401, India
2 Department of Biological Sciences, Michigan Technological University, Houghton, MI 49931, USA; rupdatta@mtu.edu
3 Global Centre for Environmental Remediation (GCER), College of Engineering, Science and Environment, The University of Newcastle, Callaghan Campus, Callaghan, NSW 2308, Australia; mahmud.rahman@newcastle.edu.au
4 Department of Civil, Environmental and Ocean Engineering, Stevens Institute of Technology, Hoboken, NJ 07030, USA
* Correspondence: pk.sahoo@cup.edu.in (P.K.S.); dsarkar@stevens.edu (D.S.)

## 1. Introduction

The ongoing increase in global population, industrialization, urbanization, and intensive agricultural practices has resulted in a wide range of environmental challenges including increased waste generation, rising greenhouse gas emissions, the uncontrolled release of emerging and toxic pollutants, degraded soil, water, and air quality, the depletion of natural resources, and the escalating impact of climate change [1]. In response to these growing concerns, the development of sustainable environmental technologies has emerged as an indisputable priority since the beginning of the twenty-first century [1–4]. These technologies, especially in areas such as pollution prevention and remediation, waste-to-energy conversion, renewable/clean energy, greenhouse gas reduction, resource recovery, and the conservation of natural resources, play a pivotal role in advancing sustainable development [4–6]. The primary goals of these technologies are to focus on innovation, develop new technological solutions through sustainable processes, and significantly address the most pressing environmental issues facing our society [2,7,8]. Additionally, they aim at promoting a transition to a more sustainable and resilient planet for future generations. Various organizations/companies/industries are being urged to adopt sustainable technology in order to lower their carbon footprints and track their progress towards achieving sustainable development goals. While sustainable environmental technologies have evolved in various fields in recent decades, there is still limited knowledge regarding their full potential and their ability to address environmental problems effectively.

This Topic project was a special collection of the latest multi-disciplinary and comprehensive research/review papers as well as innovative research models, published recently (2022–2023) in four MDPI journals (*Water*, *Environment*, *Toxins*, and *Applied Science*). The collection explores diverse facets of sustainable environmental technologies (Figure 1) and aims at summarizing these technologies, highlighting the latest developments and future directions in the field. The goal is to enhance the understanding of their application in combating environmental problems and fostering a greener future. It also highlights some exciting developments of frontier technologies with sustainability potential, while discussing the current challenges and perspectives of future research and innovations required for sustainable technologies to be successfully adapted and effective. This special collection accepted 65 articles including 55 research papers and 10 reviews for publication, covering research areas such as environmental remediation, sustainable materials, waste to energy, resource recovery, renewable/clean energy, and the role of AI and ML in sustainability. These papers cover 162 primary and partner institutions across the world (from 40 countries and six continents—including Asia, Africa, North America, South America, Europe, and

**Citation:** Sahoo, P.K.; Datta, R.; Rahman, M.M.; Sarkar, D. Sustainable Environmental Technologies: Recent Development, Opportunities, and Key Challenges. *Appl. Sci.* **2024**, *14*, 10956. https://doi.org/10.3390/app142310956

Received: 28 September 2024
Revised: 19 November 2024
Accepted: 20 November 2024
Published: 26 November 2024

Australia—), showcasing the global diversity of the subject covered here. An overview of the themes of sustainable environmental technology, the focus of this Topic project, is given below:

**Figure 1.** Major research themes in the subject of sustainable environmental technology.

## 2. Renewable/Clean Energy Technologies

Renewable energy technologies such as solar, hydro, and wind power are leading the way in providing alternatives to fossil fuels [4,9]. These technologies aim to reduce carbon, energy consumption, and greenhouse gas emissions. The growing interest in solar energy has led to the development of several photovoltaic (PV) technologies, with silicon-based PV modules becoming increasingly popular because of their exceptional long-term stability, high efficiency, and durability. Pandey et al. [10] reviewed the recent advances in solar PV systems and their potential application towards more sustainable energy solutions. They found photovoltaic thermal, building-integrated PV systems, desalination, and concentrated photovoltaic applications of solar PV to be the most technically sound and feasible solutions at the user end for future energy challenges. Despite their benefits, these PV modules generally have an operational lifespan of 25 to 30 years; thus, the number of PV modules that are approaching the end of their useful life has gradually increased, which poses a significant risk to the environment if they are disposed of incorrectly. To address this issue, Ko et al. [11] examined the processes for separating silicon photovoltaic modules and compiled efforts to produce modules that are easier to recycle. Their study findings provide valuable guidance for creating effective waste management strategies aimed at minimizing the environmental impact of PV modules at the end of their life cycle.

Furthermore, green technologies, including green hydrogen electric vehicles and natural gas, have gained increasing attention as sustainable and clean energy alternatives in recent years. According to Zhang et al. [12], natural gas has responded more favorably to the substitution impact in G7 nations as they move away from conventional energy sources and toward clean and renewable ones, and this is more significant in countries like the UK, France, Italy, and Japan. This trend underscores the importance of emphasizing natural gas's role as a key component of the energy transition and energy conservation efforts. Green hydrogen, produced through the process of hydrolysis, is another promising energy resource that is gaining momentum. It has diverse potential applications in various sectors, such as transportation, industry, and power generation [13]. Recently, several ground-breaking research studies have further advanced this field. Falcone et al. [14] reviewed green hydrogen technology in the context of sustainable development goals. They have demonstrated the potential of hydrogen as a clean energy source and how it can be used to achieve both social and environmental objectives. Yu et al. [15] discussed various low-carbon hydrogen generation techniques and green, blue, and aqua hydrogen, exploring their benefits and the differences in each approach while highlighting how each reduces carbon emissions. The article also highlighted the process involved in producing green hydrogen from renewable resources, blue hydrogen from carbon capture and storage, and aqua hydrogen using nuclear power and electrolysis. In another study, Abate et al. [16] indicated that microalga–bacterium consortia have significant potential for advancements in biotechnology related to renewable energy generation, particularly in the area of biohydrogen. These findings demonstrate how these techniques could promote sustainable energy transitions and contribute to a low-carbon hydrogen economy.

## 3. Waste-To-Energy Conversion

Waste management technology is currently undergoing a significant transformation, revolutionizing how we manage different types of waste [17]. Recent innovations in this field are enabling the conversion of waste into energy, producing sustainable resources, and replacing traditional, non-renewable materials with energy derived from plant and animal sources [17–19]. As a result, biomass waste is increasingly being converted into valuable products such as ethanol, biofuels, bioenergy, and biogas, all of which are key contributors to a more sustainable future. For example, Anacleto et al. [20] conducted a detailed analysis of biogas production, specifically methane ($CH_4$), from textile waste. Their study found that using biological pretreatments, instead of chemical methods, could boost methane production by up to 360%, while being more environmentally friendly and cost-effective. This method not only generates clean energy but also reduces environmental impact by facilitating the reuse of textile wastewater. In another study, Sharma et al. [19] investigated the potential of lignocellulosic biomass like rice straw and other agricultural residues as potential raw material for high-value-added products like bioethanol and biodiesel. The result indicated that the use of these biomasses can replace petrochemical products, positioning them as essential components of sustainable and renewable energy production. For instance, rice straw can be gasified using fermentation techniques to produce biogas or methane. Furthermore, rice straw could be used for renewable biohydrogen production, where it undergoes fast pyrolysis to produce bio-oil, which is then reformed with steam to produce hydrogen [18]. Additionally, Okolie et al. [21] experimented with the potential of converting human and animal waste into biofuels, sustainable materials, and value-added chemicals. These technologies could significantly reduce waste deposition as well as greenhouse gas emissions into the atmosphere. The study demonstrates considerable advancements in converting human and animal waste into biofuels and value-added materials through thermochemical and biological conversion pathways. It also noted that the USA, China, and England are the leading nations in research on resource recovery from human or animal waste4. Remediation of Emerging/Toxic Contaminants

With the increasing presence of emerging/chemical contaminants in the environment, the use of sustainable, low-cost materials plays a major role for the cleanup of contaminated sites and the treatment of wastewater or contaminated water bodies [6,22–24]. Several studies are currently underway to address this formidable pollution crisis. Ganie et al. [23] studied the role of carbon nanotubes as a nano-remediation approach for water decontamination. Their results indicated that carbon nanotubes are regarded as effective adsorbents for the removal of both organic and inorganic pollutants such as heavy metals, harmful organic contaminants including Dichlorodiphenyltrichloroethane (DDT), pesticides, chlorinated solvents, and perfluoroalkyl and polyfluoroalkyl substances (PFAS). The high adsorption capacity of carbon nanotubes is due to their large surface area and hollow layered structure, making them a promising alternative to conventional methods. In contrast to many conventional techniques, nano-remediation techniques offer long-term sustainable solutions by reducing the cost of cleaning up polluted areas and producing less waste. In addition, biochar, a low-cost sustainable material, has shown enormous potential as an engineered material in environmental remediation. Al Masud et al. [24] reviewed the diverse applications of biochar in remediating contaminated soil, water, and air. The review highlighted biochar's sustainability and eco-friendly characteristics from lab-scale to field-scale experiments, demonstrating its effectiveness to in mitigate contaminants, improving soil quality and restoring ecosystems. Similarly, Sachdeva et al. [25] investigated the use of biochar in immobilizing potentially hazardous elements (PTEs) in agricultural soils. Their study demonstrated that biochar effectively reduced the mobility and bioavailability of heavy metals such as Pb, Cd, Cu, and Ni in soils, leading to a lower uptake of these metals by plants. This enhanced biochar adsorption potential is due to its high surface area, porosity, pH, aromatic structure, and several functional groups, which mostly rely on the feedstock type and pyrolysis temperature. Thus, this study shows that agriculture waste-based biochar can be an effective option for soil remediation and sustainable agriculture. Biochar has also been effectively applied in the treatment of polluted water and wastewater. Thakur et al. [22] reviewed the possible application of various feedstock-based raw and modified biochar in the remediation of geogenic pollutants such as fluoride ($F^-$) and uranium (U) from aqueous solutions. The data showed that cellulose- and hemicellulose-rich feedstocks, including agricultural residues, grasses, softwood, and manure-based biochars, were successful in removing U and $F^-$ at low-to-medium pyrolysis temperatures. While metal oxide, hydroxide, and alkali modification promoted F-adsorption, acidic and magnetic modification favored U adsorption.

Similarly, Neulls et al. [26] investigated the effectiveness of plant-based waste materials as adsorbents for removing nitrogen-based chemicals from water. The results demonstrated that in comparison to animal biomass and chitosan, vegetal biomass and pineapple crowns were more effective in eliminating organic pollutants from water bodies. Therefore, the use of these plant-based biomasses is a practical and affordable solution to lessen the amount of waste that ends up in sanitary landfills and dumps while also addressing environmental pollution. The performance of the denitrifying bacterial strain *Pseudomonas* sp. in a laboratory-scale sequencing batch reactor was examined in another study conducted by Chen et al. [27]. The findings imply that microorganisms can improve total nitrogen (TN) elimination by bioaugmentation. A metal–organic framework (MOF) was developed by Sun et al. [28] as an effective sorbent for environmentally friendly oil–water separation. Similar to this, an MOF was used by Karami et al. [29] to remove methyl orange from water, showing the versatility of MOFs in water treatment. Phytoremediation, a process that uses plants to remove contaminants from the environment without creating secondary pollution, is another low-cost, sustainable, and effective option compared to conventional procedures based on chemical extraction. In another study conducted by Park and Son [30], the phytotoxicity and accumulation of antibiotics in two different types of plants were examined. The researchers found that water lettuce had a higher accumulation of antibiotics in its tissues compared to parrot feather plants, indicating its great potential for removing xenobiotics from contaminated environments. These plants possess a natural ability to adapt and thrive in harsh, polluted conditions, making them valuable tools for environmental cleanup. Overall, these studies offer insightful information on the technical developments in the production of sustainable materials for the removal of pollutants from water and soils. The integration of plant-based solutions, microbial bioaugmentation, and MOFs highlights the growing potential of sustainable, low-cost methods for addressing environmental contamination.

## 4. AI and ML Application in Environmental Sustainability

Artificial intelligence (AI) and machine learning (ML) models are increasingly playing a pivotal role in addressing key challenges related to the management of natural resources, improving the accuracy of contaminant detection and optimizing remediation strategies [31–33]. AI is also being used for the timely monitoring of natural resources, including water, soil, energy, minerals, coal, oil, natural gas, biodiversity, and forests, as well as for decision assistance, in order to increase environmental sustainability. Pandey et al. [31] demonstrate an in-depth review on the use of AI and ML approaches and big data in the management of natural resources. This study explained the data-driven methods for decision-making, emphasizing how AI and ML can help policymakers assess the current status of natural resources, create evidence-based policies, and identify areas that require improvement. By utilizing these technologies, both public and private sectors can better manage natural resources, ensuring their responsible extraction and long-term preservation for future generations. In another study, Rao et al. [32] examined the nexus between natural resources and environmental sustainability for the period 1990–2020 in 19 Asian countries through the ML approach. This study mainly focused on the impacts of the Industrial Revolution, and digitalization revealed interesting results. This research indicated that when natural resource use is between 0.1 and 9.87, $CO_2$ emission decreases, but when the range is 0–0.1 or above 9.87, the impact on $CO_2$ emission is positive. The analysis further revealed that urbanization, energy use, and economic complexity make the greatest contributions in Asian $CO_2$ predictions. This finding suggested that governments in Asian countries should use these AI/ML techniques to effectively manage their resources. Similarly, AI has demonstrated effective use for the discovery of mineral resources [33], as well as the management of water bodies, energy, transportation, and biodiversity [34]. Similarly, Zhao et al. [35] used a multi-indicator weighted robustness analysis to monitor planktonic communities in marine plankton ecosystems. This network technology exposed unseen topological flaws and provided a new direction for future research. The findings of this work can help with marine conservation and the sustainable management of these sources. Additionally, AI approaches are revolutionizing agriculture for sustainable and precision farming. These methods make food production more environmentally friendly and sustainable by utilizing resources more wisely. Overall, these studies advance scientific knowledge and innovative ideas that could aid in highlighting the importance and dynamic nature of sustainable environmental technology, which is not just transforming the environment and industries, but also laying the foundation for a more sustainable and eco-friendly future. The continued advancement and widespread use of these technologies will be essential to solving contemporary environmental problems on a worldwide scale.

## 5. Key Challenges and Future Prospects for Advancing Sustainable Technologies

Although sustainable technologies have proven to be an effective and promising solution for addressing a range of various environmental issues, they also face some key challenges that hinder their widespread adoption and advancement. These challenges include high production costs, technological limitations, infrastructure and logistics barriers, and research and regulatory hurdles [5]. There are also several challenges to improve efficiency and effectiveness in innovation and turn sustainability into a reality/practice for field-based applications One such issue is evident in the renewable energy sector. Despite the fact that a significant number of patents have been filed on various green energy technologies such as solar, wind, tidal, and hydrogen energy, all of which have the potential to lower greenhouse gas emissions, some of these emerging technologies are still at the research or development stage within some energy companies. As a result, despite the promise of these technologies, they are not yet ready for large-scale commercial deployment, limiting their ability to deliver technological innovations for broader market applications. To overcome these challenges and support the further development of these technologies, the following future directions are essential: (1) Advance research infrastructure and significant investment should be made across multiple sectors to foster innovation, increase efficiency, minimize costs, and create cutting-edge applications in sustainable technologies [3]. (2) In the energy sector, funding low-carbon projects is urgently needed, and interest-free or low-interest allowances must be established in order to improve the use of contemporary technologies [9]. Additionally, breakthrough research is needed to upgrade the energy system and introduce advanced batteries and hydrogen fuel cells that can produce less waste and fewer toxic fumes, with the potential to overcome conventional technologies. (3) In the transportation sector, the rise in electric vehicles (EVs) alongside policy measures can be a major focus to create a more sustainable future by reducing greenhouse gas emissions. Also, innovations in smart grid technologies are required for a further contribution to environmental sustainability. (4) In newly developed cities or buildings, there is an emerging need to use the Internet of Things (IoT) to optimize energy usage and minimize waste. This smart technology will also optimize resource usage and provide easy solutions for water quality and quantity management, air quality monitoring, and urban planning. (5) Furthermore, to address the real demands of mitigating climate change, green and low-carbon efforts as well as the idea of the "dual-carbon" goal should be taken into consideration in conjunction with the initiative of energy saving and carbon emission reduction. Additionally, to strengthen the renewable energy market system and resource conservation, each country's SDG targets must include environmental tax and financial subsidy policies [2]. (6) Furthermore, emphasis should be placed on strong collaboration and partnership between various stakeholders including governments, financial institutions, and civil society to finance renewable energy projects, sustainable infrastructure, and initiatives to address climate change and drive significant positive changes towards achieve sustainable development goals [9]. (7) Governments worldwide should also prioritize climate change action plans to regulate greenhouse gas emissions and effect the transition to a low-carbon economy. They should also implement methods like carbon pricing in order to incentivize the reduction in environmental impact. (8) Moreover, the integration of artificial intelligence (AI), machine learning (ML), and big data into sustainability research will bring new opportunities to optimize resource usage, manage natural resources in real time, mitigate climate change, and drive innovation in the energy sector [36]. By integrating vast amounts of data from diverse sources, including sensors, Internet of Things (IoT) devices, and digital platforms, these technologies can find novel solutions for urgent environmental issues and advance the understanding required to comprehend the extent of their use in addressing those issues and pinpoint areas for improvement.

Other challenges include ecological destruction, land use, and habitat impact. For example, land use and habitats will be affected by any large-scale renewable energy projects, which emphasizes the importance of ecological impact assessments and sustainable practices to minimize disturbance and preserve biodiversity. Moreover, the lack of a scientific framework for the systematic and integrated assessment of technology innovation at various phases of the life cycles of technological advances may result in an unclear picture of the full range of their sustainability performance. In order to undertake a systematic sustainability assessment of such technologies at different stages, urgent research is needed to develop such frameworks. Addressing these challenges is essential to effectively incorporate and optimize the capabilities of environmentally friendly technology, which would be useful to researchers, decision makers, and organizations in guiding technological advancements for sustainable environmental practices and keeping the planet in good shape for future generations.

## 6. Opportunities in Sustainable Environmental Technologies

Environmental technologies offer numerous opportunities for innovation and technological advancements across various sectors, creating potential for new enterprises or research centers that provide employment opportunities. By enabling businesses to create sustainable materials and services in response to the rising consumer demand for sustainability, these technologies promote corporate innovation and encourage collaboration among various stakeholders from governments, industries, and international organizations.

Moreover, environmentally friendly technologies play a major role in the battle against climate change, as they help reduce carbon emissions and resource consumption, thus mitigating environmental degradation. From a social perspective, these technologies contribute to improved living standards by ensuring access to cleaner water and air, as well as creating a healthier environment for agriculture. They also provide a means to achieve national and global sustainability goals, advancing progress toward a more sustainable future. In essence, the widespread adoption of environmental technologies holds significant potential to not only stimulate economic growth but also pave the way for a fairer and more sustainable global community.

## 7. Conclusions

This Editorial, prepared for the Topic project 'Sustainable Environmental Technologies', provides a comprehensive overview of the current trends and recent technological advancements in the field of environmental sustainability. It highlights the key challenges faced in this area, along with potential solutions and strategies for moving forward. As demonstrated throughout this paper, the utilization of sustainable environmental technologies, driven by innovation, can significantly contribute to mitigating environmental degradation, reducing greenhouse gas emissions, enhancing energy security, and offering promising solutions to resource depletion and waste reduction. Moreover, the integration of artificial intelligence and machine learning models has proven effective in predicting and optimizing resource conservation and contributing to the shaping of a more sustainable future. However, many challenges related to such technology remain, including cost barriers, the lack of a scientific framework for the systematic assessment of technology innovation, and the potentially negative environmental impacts of certain technologies, that constitute significant hurdles to overcome. It is crucial for researchers, policymakers, industry leaders, governments, and research institutions to collaborate in a unified effort to address these challenges. A collective commitment to advancing technological innovation through high-quality research, systematic assessment, and methodological rigor is critical for overcoming the current barriers and ensuring practical and impactful innovations. Such collaborative efforts are key to unlocking the full potential of sustainable technologies and safeguarding the health of our planet for future generations. Furthermore, the success of these technologies towards long-term environmental goals will require the establishment of supportive policy frameworks, incentives for investment, and global cooperation to accelerate their development and implementation. These combined efforts will be vital in overcoming the existing challenges and enabling a successful transition toward environmental sustainability.

**Acknowledgments:** We would like to express our sincere gratitude to all authors whose valuable works were published in this Topic project. PKS sincerely acknowledges the core research grant (CRG/2021/002567) of the DST SERB New Delhi (Government of India). MMR would like to acknowledge the Global Centre for Environmental Remediation (GCER) located in The University of Newcastle, Australia.

**Conflicts of Interest:** The authors declare no conflict of interest.

## References

1. Huang, Y. Technology innovation and sustainability: Challenges and research needs. *Clean Technol. Environ. Policy* **2021**, *23*, 1663–1664. [CrossRef] [PubMed]
2. Bashir, M.F.; Sharif, A.; Staniewski, M.W.; Ma, B.; Zhao, W. Environmental taxes, energy transition and sustainable environmental technologies: A comparative OECD region climate change analysis. *J. Environ. Manag.* **2024**, *370*, 122304. [CrossRef] [PubMed]
3. Rehman, M.A.; Hasan, M.; Pata, U.K.; Chen, X.H.; Kontoleon, A. Exploring the Nexus of technology, environmental policy stringency, and political globalization: Pathways to achieving sustainability. *Energy Econ.* **2024**, 107979. [CrossRef]
4. Sharif, A.; Bashir, U.; Mehmood, S.; Cheong, C.W.; Bashir, M.F. Exploring the impact of green technology, renewable energy and globalization towards environmental sustainability in the top ecological impacted countries. *Geosci. Front.* **2024**, *15*, 101895. [CrossRef]

5. Sebbagh, T.; Şahin, M.E.; Beldjaatit, C. Green hydrogen revolution for a sustainable energy future. *Clean Technol. Environ. Policy* **2024**, 1–24. [CrossRef]
6. Ezeorba, T.P.C.; Okeke, E.S.; Nwankwo, C.E.; Emencheta, S.C.; Enochoghene, A.E.; Okeke, V.C.; Ozougwu, V.E. Emerging eco-friendly technologies for remediation of Per-and poly-fluoroalkyl substances (PFAS) in water and wastewater: A pathway to environmental sustainability. *Chemosphere* **2024**, *364*, 143168. [CrossRef]
7. Feng, Q.; Usman, M.; Saqib, N.; Mentel, U. Modelling the contribution of green technologies, renewable energy, economic complexity, and human capital in environmental sustainability: Evidence from BRICS countries. *Gondwana Res.* **2024**, *132*, 168–181. [CrossRef]
8. Huang, Y. Reinforcing sustainability assessment and reshaping technology innovation for highly sustainable manufacturing in the post-COVID-19 era. *Smart Sustain. Manuf. Syst.* **2020**, *4*, 341–345. [CrossRef]
9. Doğan, B.; Chu, L.K.; Khalfaoui, R.; Ghosh, S.; Shahbaz, M. Strategy towards sustainable energy transition: The effect of policy uncertainty, environmental technology and natural resources rent in the OECD nations. *Resour. Policy* **2024**, *98*, 105333. [CrossRef]
10. Pandey, A.K.; Tyagi, V.V.; Jeyraj, A.; Selvaraj, L.; Rahim, N.A.; Tyagi, S.K. Recent advances in solar photovoltaic systems for emerging trends and advanced applications. *Renew. Sustain. Energy Rev.* **2026**, *53*, 859–884. [CrossRef]
11. Ko, J.; Kim, K.; Sohn, J.; Jang, H.; Lee, H.; Kim, D.; Kang, Y. Review on Separation Processes of End-of-Life Silicon Photovoltaic Modules. *Energies* **2023**, *16*, 4327. [CrossRef]
12. Zhang, X.; Yu, G.; Ibrahim, R.L.; Sherzod Uralovich, K. Greening the E7 environment: How can renewable and nuclear energy moderate financial development, natural resources, and digitalization towards the target? *Int. J. Sustain. Dev. World Ecol.* **2024**, *31*, 447–465. [CrossRef]
13. Hassan, Q.; Algburi, S.; Sameen, A.Z.; Salman, H.M.; Jaszczur, M. Green hydrogen: A pathway to a sustainable energy future. *Int. J. Hydrogen Energy* **2024**, *50*, 310–333. [CrossRef]
14. Falcone, P.M.; Hiete, M.; Sapio, A. Hydrogen economy and sustainable development goals: Review and policy insights. *Curr. Opin. Green Sustain. Chem.* **2021**, *31*, 100506. [CrossRef]
15. Yu, M.; Wang, K.; Vredenburg, H. Insights into low-carbon hydrogen production methods: Green, blue and aqua hydrogen. *Int. J. Hydrogen Energy* **2021**, *46*, 21261–21273. [CrossRef]
16. Abate, R.; Oon, Y.S.; Oon, Y.L.; Bi, Y. Microalgae-bacteria nexus for environmental remediation and renewable energy resources: Advances, mechanisms and biotechnological applications. *Heliyon* **2024**, *10*, e31170. [CrossRef]
17. Chicaiza-Ortiz, C.; Peñafiel-Arcos, P.; Herrera-Feijoo, R.J.; Ma, W.; Logroño, W.; Tian, H.; Yuan, W. Waste-to-Energy technologies for municipal solid waste management: Bibliometric review, life cycle assessment, and energy potential case study. *J. Clean. Prod.* **2024**, *480*, 143993. [CrossRef]
18. Zheng, H.; Wang, Y.; Feng, X.; Li, S.; Leong, Y.K.; Chang, J.S. Renewable biohydrogen production from straw biomass–Recent advances in pretreatment/hydrolysis technologies and future development. *Int. J. Hydrogen Energy* **2022**, *47*, 37359–37373. [CrossRef]
19. Sharma, S.; Tsai, M.L.; Sharma, V.; Sun, P.P.; Nargotra, P.; Bajaj, B.K.; Dong, C.D. Environment friendly pretreatment approaches for the bioconversion of lignocellulosic biomass into biofuels and value-added products. *Environments* **2023**, *10*, 6. [CrossRef]
20. Anacleto, T.; Kozlowsky-Suzuki, B.; Wilson, A.; Enrich-Prast, A. Comprehensive Meta-Analysis of Pathways to Increase Biogas Production in the Textile Industry. *Energies* **2022**, *15*, 5574. [CrossRef]
21. Okolie, J.; Jimoh, T.; Akande, O.; Okoye, P.; Ogbaga, C.; Adeleke, A.; Ikubanni, P.; Güleç, F.; Amenaghawon, A. Pathways for the Valorization of Animal and Human Waste to Biofuels, Sustainable Materials, and Value-Added Chemicals. *Environments* **2023**, *10*, 46. [CrossRef]
22. Thakur, A.; Kumar, R.; Sahoo, P. Uranium and Fluoride Removal from Aqueous Solution Using Biochar: A Critical Review for Understanding the Role of Feedstock Types, Mechanisms, and Modification Methods. *Water* **2022**, *14*, 4063. [CrossRef]
23. Ganie, A.S.; Bano, S.; Khan, N.; Sultana, S.; Rehman, Z.; Rahman, M.M.; Khan, M.Z. Nanoremediation technologies for sustainable remediation of contaminated environments: Recent advances and challenges. *Chemosphere* **2021**, *275*, 130065. [CrossRef] [PubMed]
24. Al Masud, M.A.; Shin, W.S.; Sarker, A.; Septian, A.; Das, K.; Deepo, D.M.; Malafaia, G. A critical review of sustainable application of biochar for green remediation: Research uncertainty and future directions. *Sci. Total Environ.* **2023**, *904*, 166813. [CrossRef] [PubMed]
25. Sachdeva, S.; Kumar, R.; Sahoo, P.K.; Nadda, A.K. Recent advances in biochar amendments for immobilization of heavy metals in an agricultural ecosystem: A systematic review. *Environ. Pollut.* **2023**, *319*, 120937. [CrossRef]
26. Neulls, T.; Gouveia, P.; da Silva Pereira, C.; Souza, C.; Chaves, F.; Souza, I.; Tavarez, R.; Aliança, A.; Gonçalves, M.; Gomes, W.; et al. Comparative Study of Two Organic Wastes as Adsorbents in the Treatment of Water Rich in Nitrogen Compounds. *Water* **2023**, *15*, 876. [CrossRef]
27. Chen, T.; Yang, X.; Sun, Q.; Hu, A.; Qin, D.; Li, J.; Wang, Y.; Yu, C. Changes in Wastewater Treatment Performance and the Microbial Community during the Bioaugmentation of a Denitrifying Pseudomonas Strain in the Low Carbon–Nitrogen Ratio Sequencing Batch Reactor. *Water* **2022**, *14*, 540. [CrossRef]
28. Sun, Y.; Ke, Z.; Shen, C.; Sun, R.; Wei, Q.; Yin, Z.; Yang, W. Fabrication of Carbon Aerogels Derived from Metal-Organic Frameworks/Carbon Nanotubes/Cotton Composites as an Efficient Sorbent for Sustainable Oil–Water Separation. *Appl. Sci.* **2022**, *12*, 7285. [CrossRef]

29. Karami, A.; Shomal, R.; Sabouni, R.; Al-Sayah, M.; Aidan, A. Parametric Study of Methyl Orange Removal Using Metal–Organic Frameworks Based on Factorial Experimental Design Analysis. *Energies* **2022**, *15*, 4642. [CrossRef]
30. Park, Y.; Son, J. Phytotoxicity and Accumulation of Antibiotics in Water Lettuce (*Pistia stratiotes*) and Parrot Feather (*Myriophyllum aquaticum*) Plants under Hydroponic Culture Conditions. *Appl. Sci.* **2022**, *12*, 630. [CrossRef]
31. Pandey, D.K.; Hunjra, A.I.; Bhaskar, R.; Al-Faryan, M.A.S. Artificial intelligence, machine learning and big data in natural resources management: A comprehensive bibliometric review of literature spanning 1975–2022. *Resour. Policy* **2023**, *86*, 104250. [CrossRef]
32. Rao, A.; Talan, A.; Abbas, S.; Dev, D.; Taghizadeh-Hesary, F. The role of natural resources in the management of environmental sustainability: Machine learning approach. *Resour. Policy* **2023**, *82*, 103548. [CrossRef]
33. Yang, F.; Zuo, R.; Kreuzer, O.P. Artificial intelligence for mineral exploration: A review and perspectives on future directions from data science. *Earth-Sci. Rev.* **2024**, *258*, 104941. [CrossRef]
34. Nti, E.K.; Cobbina, S.J.; Attafuah, E.E.; Opoku, E.; Gyan, M.A. Environmental sustainability technologies in biodiversity, energy, transportation and water management using artificial intelligence: A systematic review. *Sustain. Futures* **2022**, *4*, 100068. [CrossRef]
35. Zhao, D.; Zhang, T.; Chen, T.; He, Q.; Huang, D. Multi-Indicator Weighted Robustness Analysis of Planktonic Community Systems under Different Destructive Factors. *Appl. Sci.* **2023**, *13*, 8742. [CrossRef]
36. Yoshikuni, A.C.; Dwivedi, R.; dos Santos, M.Q.L.; Liu, F.; Yoshikun, M.M. Sustainable Environmental Performance: A Cross-Country Fuzzy Set Qualitative Comparative Analysis Empirical Study of Big Data Analytics and Contextual Factors. *J. Clean. Prod.* **2024**, *481*, 144040. [CrossRef]

*applied sciences*

*Article*

# Phytotoxicity and Accumulation of Antibiotics in Water Lettuce (*Pistia stratiotes*) and Parrot Feather (*Myriophyllum aquaticum*) Plants under Hydroponic Culture Conditions

**Young-Jae Park [1] and Jae-Gwon Son [2,*]**

1 Horse Industry Complex Center, Jeonju Kijeon College, Jeonju 54989, Korea; song352@naver.com
2 Department of Rural Construction Engineering, Jeonbuk National University, Jeonju 54906, Korea
* Correspondence: sjg@jbnu.ac.kr

**Abstract:** The aim of this research was to investigate the accumulation of antibiotics in two kinds of plants. The hydroponic culture solution included a mixture of antibiotics, including three commonly used antibiotics in South Korea: norfloxacin (NOR), sulfamethazine (SMZ), and tetracyclines (TC). None of these antibiotics were detected in the shoots of water lettuce plants, only in the roots. However, in parrot feather plants, antibiotics were detected in both the shoots and the roots, with higher amounts detected in the shoots than in the roots. SMZ and TC were most likely to be detected in the roots and shoots of water lettuce and parrot feather plants, and about one-third of the NOR administered was later detected in the plants. The BCF (bioconcentration factor) of antibiotics ranged from 0.24 to 0.78, while that of NOR was much lower, ranging from 0.24 to 0.38. The SMZ (0.59–0.64) and TET (0.72–0.78) exhibited higher uptake accumulation in the water lettuce tissues compared with the parrot feather plants.

**Keywords:** antibiotics; parrot feather; phytoremediation; water lettuce; water plants

**Citation:** Park, Y.-J.; Son, J.-G. Phytotoxicity and Accumulation of Antibiotics in Water Lettuce (*Pistia stratiotes*) and Parrot Feather (*Myriophyllum aquaticum*) Plants under Hydroponic Culture Conditions. *Appl. Sci.* **2022**, *12*, 630. https://doi.org/10.3390/app12020630

Academic Editors: Dibyendu Sarkar, Rupali Datta, Prafulla Kumar Sahoo and Mohammad Mahmudur Rahman

Received: 13 December 2021
Accepted: 7 January 2022
Published: 10 January 2022



## 1. Introduction

Phytoremediation is a technique that utilizes plants to remove or transform toxic chemicals incorporated in soils, ground water, and surface water. Various plants are used for removal or degradation of inorganic and organic contaminants, such as pesticides, antibiotics, heavy metals, and nutrients.

Phytoremediation techniques are used extensively because they have a number of advantages. Phytoremediation is low cost compared with conventional techniques and is suitable for treating very large land areas [1]. Furthermore, these techniques may be more effective than conventional techniques based on chemical extraction because the use of plants for environmental cleanup does not cause secondary pollution. However, phytoremediation also has limitations. A significant disadvantage of phytoremediation by aquatic plants is their shallow root systems, which limit the depth of soil treatment by the rhizosphere. Another disadvantage is the slow rate of environmental cleanup, which may take more than 10 years [2]. Phytoremediation effectiveness decreases during winter, and plant diseases or pests may occur in the condition of plant damage [3].

For decades, a wide variety of antibiotics has been extensively applied worldwide for preventing or treating human, animal, and plant infections and as feed additives for animals to prevent or treat diseases and promote growth [4]. Consumption of veterinary antibiotics has been increasing in both industrialized and developing countries. However, in Korea, the total production of antibiotics decreased significantly from 1500 tons in 2007 to 820 tons in 2013. This dramatic reduction in use of veterinary antibiotics was due to restrictions on their use for feed supplementation. Tetracyclines, sulfonamides, and penicillin are the most commonly used antibiotic types, with usage of 165, 60, and 160 tons, respectively, in 2014. These antibiotics are commonly found in both aquatic and agricultural environments, associated with the discharge of wastewater to the environment. Despite the low concentrations usually found in water ecosystems (from nanograms to micrograms per liter), continuous discharge of antibiotics and their degradation products through different pathways makes them 'pseudo-persistent' in the environment [5].

Macrophytes accumulate pollutants by incorporating them into the structures of their cells. In addition, these plants have a natural ability to absorb and metabolize xenobiotics while adapting to harsh conditions in a contaminated environment [6]. Some studies have used aqueous solution amended with different antibiotics to address plant uptake. The water plants *Myriophyllum aquaticum* (parrot feather) and *Pistia stratiotes* (water lettuce) were previously used for the study of phytoremediation of tetracyclines and oxytetracyclines in aqueous media [7]. Comparable levels of antibiotic removal by filtered root exudates from these two species a suggested involvement of root-secreted enzymes/metabolites in degrading/transforming the antibiotics [7].

The aim of the present study was to investigate the accumulation of antibiotics in two kinds of plants and to survey the phytotoxicity of antibiotics. The study was performed under hydroponic culture conditions to avoid potentially confounding effects of adsorption of antibiotics on soil particles [8]. The hydroponic culture solution contained a mixture of three commonly used antibiotics in South Korea: norfloxacin, sulfamethazine, and tetracyclines.

## 2. Materials and Methods

### 2.1. Antibiotics and Chemicals

Target antibiotic compounds were selected based on frequency of detection in the environment and physicochemical differences. In this study, tetracycline of the tetracycline type and sulfamethazine of the sulfonamide type, which are widely used in South Korea, were first selected as target antibiotics. In addition, norfloxacin, a quinolone type, was frequently detected in sewage wastewater and was additionally selected. We examined tetracyclines (TC), sulfamethazine (SMZ), and norfloxacin (NOR), which belong to three different antibiotic classes. All antibiotic standards and some internal standards were obtained from Sigma-Aldrich (St. Louis, MO, USA). All organic solvents used were HPLC grade and purchased from Merck Corporation (Darmstadt, Germany). Stock solutions and internal standards were prepared at 100 mg/L in methanol and stored in amber glass vials at 4 °C.

### 2.2. Experimental Design

Selection of plants is important for achieving significant phytoremediation in aquatic ecosystems. The plants must be able to tolerate pollution and be easy to grow and fast growing. Prior to the experiments, 2 different types of water plants that have been used in wetland systems were cultured for 1 month in modified Hoagland's nutrient solution for acclimatization at a nearby plant nursery. We then obtained the plants and washed them to remove sediment particles. Identical plants were assigned to 9 glass containers, each containing 10 individuals arranged in parallel rows. We covered the surface of each container with silver foil to prevent the photodegradation of the antibiotics and to mitigate the effects of sunlight on the root systems. Throughout the study, hydroponic culture systems were supplied with 5 L Hoagland's nutrient solution fortified with the test antibiotics, the pH was adjusted to 6.5. The experiments took place in a greenhouse environment (12 h photoperiod, 30/25 $\pm$ 2 °C day/night), and all plants were treated for 30 days with the following concentrations of TC, SMZ, or NOR in nutrient solution: 0, 1, 5, and 10 mg/L. The Hoagland's nutrient solution was prepared as described by Dordio et al. [9]. The antibiotics in wastewater were frequently detected and ranged from ng/L up to lower µg/L (norfloxacin: 2.775 µg/L; sulfamethazine: 1.280 µg/L; and tetracycline: 0.412 µg/L; respectively) [10,11]. However, in this study, antibiotics were treated at relatively high concentrations to search for antibiotic hyperaccumulater water plants. We measured the following phytotoxicity parameters: leaf chlorophyll and fresh and dry weights of the water plants. At the end of the treatment period, samples of roots and shoots were obtained from the same pot. Roots were carefully washed in tap water. Plant materials were dried for 2 days at 60 °C and then frozen until analyzed. Control and treated plant materials were dried in separate drying ovens to prevent cross-contamination. The distributions of antibiotics in the hydroponic culture solution and water plants were analyzed.

### 2.3. Assay of Phytotoxicity Parameters of Water Plants

Chlorophyll contents (Chl a, Chl b and total Chl) of water plants were measured according to the method described by Huang et al. (2004) [12]. Plant leaves (0.2 g/3 parallel individuals in each container) were cut into 0.2 cm segments and treated in 80% (*v/v*) acetone for 24 h in the dark. Absorbance of the solutions was measured with a spectrophotometer (DR 5000, HACH, Loveland, CO, USA) at 663 nm and 645 nm. After 30 days, the plants were harvested and washed with tap water and distilled water. Fresh weights of the plants were determined for each treatment, and dry weights of shoots and roots were determined after drying at 65 °C for 48 h.

### 2.4. Antibiotic Analysis in Water and Plant Samples

Liquid chromatography in combination with tandem mass spectrometry (LC–MS/MS) was used to analyze the antibiotics in the plant and water samples. Prior to extraction, the water was filtered through Whatman no. 1 filter paper to eliminate the suspended matter and then filtered through a 0.45 μm mixed cellulose membrane. The pH of water samples was adjusted to 4.0 with 1 mol/L $H_2SO_4$. The antibiotics were extracted in 250 mL of filtered water. The plant samples were freeze-dried and ground to 80 mesh. A mixed solution (5 mL) containing phosphate buffer, acetonitrile, and the plant samples was added to each centrifuge tube. The tubes were shaken in a shaker at 250 rpm for 30 min, sonicated for 10 min, and then centrifuged at 12,000× *g* for 10 min.

The extraction process was performed 3 times for each sample. The samples were extracted using Oasis HLB extraction cartridges (500 mg, 6 mL, Waters, Milford, MA, USA) in the following manner. Each cartridge was sequentially pre-conditioned with 6.0 mL acetone, 6.0 mL methanol, and 6.0 mL 0.1% formic acid solution. The samples were passed through an extraction cartridge at a rate of 3 mL/min. Subsequently, the cartridge was rinsed with 5 mL 0.1% formic acid solution containing 5 mmol/L ammonium acetate and then eluted with 5.0 mL methanol. Finally, the target fraction was re-dissolved in 10% methanol solution to a final volume of 1.0 mL for LC–MS/MS (ABI 3200 Q TRAP) analysis. The antibiotics were analyzed using a LC–MS/MS system with an Inertsil® ODS-SP column for separation. The chromatographic mobile phase consisted of methanol (A) and 0.1% formic acid solution dissolved with 5 mmol/L ammonium acetate (B). The column temperature was set to 40 °C. The target antibiotics were detected by electrospray tandem mass spectrometry in positive-ion mode. Data acquisition was performed by multiple reaction monitoring (MRM), recording two MRM sessions per compound. The optimal conditions are summarized in Table 1. The correlation coefficients ($R^2$) of the calibration curves exceeded 0.99 for all analytes. Recoveries of the antibiotics in the plant and water samples were determined at different concentrations in triplicate and calculated as percentages of the measured concentration, relative to the spiked concentration. Limits of quantification (LOQ) of the antibiotics were calculated with signal–noise ratios of 10. Retention time, correlation coefficient, LOQ, and recovery of the samples are summarized in Table 2.

**Table 1.** Optimum LC–MS/MS parameters of precursor and product ions and collision energy (CE).

| Antibiotics | Precursor Ion (m/z) | Product Ion (m/z) | Collision Energy (eV) |
|---|---|---|---|
| NOR | 152 | 110.0 | 16 |
| TC | 445 | 410.0 | 25 |
| SMZ | 279 | 186.1 | 24 |

NOR—norfloxacin; SMZ—sulfamethazine; TC—tetracycline.

**Table 2.** Retention time, correlation coefficient, recovery, and limit of quantification (LOQ) of samples.

| Antibiotics | Retention Time | $R^2$ | LOQ (ng/g) | Recovery (%) | |
|---|---|---|---|---|---|
| | | | | Plant | Water |
| NOR | 3.52 | 0.99 | 2.59 | 80.1 ± 2.8 | 91.2 ± 3.2 |
| TC | 7.49 | 0.99 | 5.17 | 74.1 ± 3.6 | 84.6 ± 2.0 |
| SMZ | 7.68 | 0.99 | 1.54 | 79.9 ± 2.2 | 88.7 ± 3.2 |

### *2.5. Bioconcentration Factor*

To evaluate the abilities of the two water plant species to extract and accumulate antibiotics into plant tissues, the bioconcentration factor (BCF) was determined for roots, according to the method described by Zayed et al. [13], as follows:

$$\mathrm{BCF} = \frac{\text{Pollutant concentration in plant tissue at harvest (mg/kg)}}{\text{Initial concentration in the external growth medium }\left(\frac{\mathrm{mg}}{\mathrm{L}}\right)}$$

## 3. Results and Discussion

### *3.1. Effects of Antibiotics on Growth of Water Plants*

The objective of this study was to identify water plant species capable of efficient removal of antibiotics from aquatic environments without adverse effects on their growth and morphology. During the 30-day experimental period using 1–5 ppm treatment concentrations, we noticed no visible signs of detrimental effects or yellow leaves in either species. The growth of plants in the hydroponic culture solution was good, as demonstrated by significantly greater biomass yield at the end than at the beginning of the experiment (Figure 1). By the 10th day of the experiment, the containers were completely covered by plants. However, at 10 ppm treatment concentrations, plant growth was inhibited by 40–50%. In particular, we observed chlorosis and early abscission of leaves. Neither water lettuce nor parrot feather plants showed difference in fresh plant weight according to type of antibiotic used. The fresh weight of plants decreased in a concentration-dependent manner as antibiotic concentration increased (Figure 2).

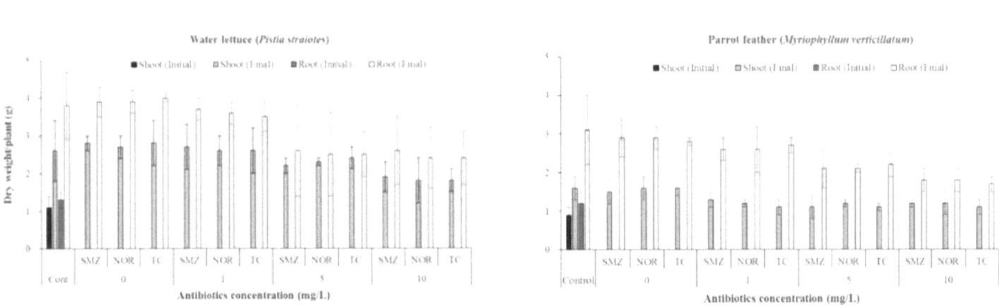


**Figure 1.** Dry weights of water lettuce and parrot feather under various treatments at the end of the experimental period.

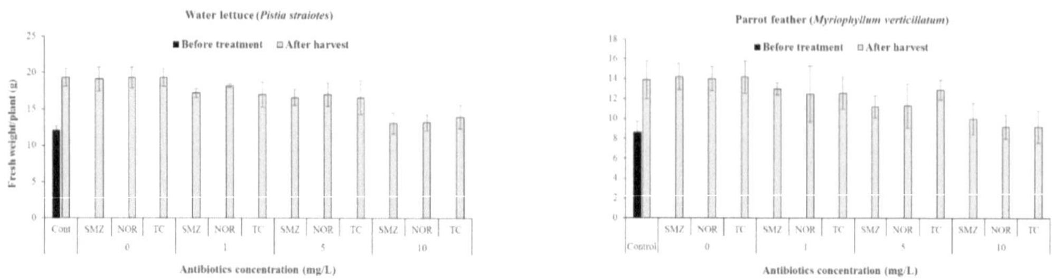


**Figure 2.** Fresh weights of water lettuce and parrot feather plants under various treatments at the end of the experimental period.

### 3.2. *Effects of Antibiotics on Chlorophyll Content*

Plant chlorophyll content is an important parameter for evaluating photosynthetic activity, and such measurements can be used as indicators of pollutant-induced plant stress [12]. Plant total chlorophyll, chlorophyll a, and chlorophyll b were all affected by antibiotic treatment. At the end of the treatment period, inhibition rates of total chlorophyll content of the 2 water plants for control, 1 ppm, 5 ppm, and 10 ppm treatments were 3.1–5.9%, 13.3–16.8%, 33.9–44.2%, and 41.5–45.2%, respectively (Tables 3 and 4). Antibiotics administered at concentrations greater than 5 mg/L seriously disrupted the normal chlorophyll content of water lettuce and parrot feather plants. The inhibition rate of chlorophyll b was higher than that of chlorophyll after antibiotic treatment. Plant total chlorophyll, chlorophyll a, and chlorophyll b did not significantly differ between the water lettuce and parrot feather plants. The inhibition rates of chlorophyll in water plants increased as antibiotic concentrations increased, possibly because organic pollutants block electron transport flow from photosystem II to photosystem I, thereby inhibiting plant biosynthesis of chlorophyll b [11].

### 3.3. *Antibiotic Residues in Aqueous Solution*

Antibiotic concentrations in hydroponic culture solution under various treatment concentrations during the experiment period are shown in Figures 3 and 4. In water lettuce plants, SMZ and TC were lower than the limit of detection in aqueous solution from 21 days after treatment. However, 60% of the treated amount of NOR remained in aqueous solution. The water solubilities of treated antibiotics were SMZ 1500 mg /L $H_2O$ (25 °C), TC 231 mg/L $H_2O$ (25 °C), and NOR 0.28 mg/mL $H_2O$ (25 °C). The high residual amount of NOR in aqueous solution is presumably due to water solubility. It was previously reported that absorption, abiotic transformation, and biotic transformation are major processes that antibiotics undergo in aquatic environments [14]. Hydrolysis and biodegradation also comprise an important degradation pathway for organic pollutants in aquatic ecosystem. In the present study, treated water plants showed greater removal efficiency of antibiotics compared with the control group (data not shown). We conclude that water plants play an important role in reduction in antibiotic contamination in aquatic environments, especially SMZ and TC. Parrot feather plants showed similar trends to water lettuce plants.

**Table 3.** Effects of antibiotics on concentrations of Chl a, Chl b, and total Chl in leaf tissues of water lettuce (*Pistia straiotes*). Values are the mean ± S.D. of three parallel individuals in each container, with three measurements from each sample.

| Item | DAT | Inhibition Rates of Chlorophyll (%) | | | | | | | | | | | |
|---|---|---|---|---|---|---|---|---|---|---|---|---|---|
| | | 0 | | | 1 mg/L | | | 5 mg/L | | | 10 mg/L | | |
| | | SMZ | NOR | TC | SMZ | NOR | TC | SMZ | NOR | TC | SMZ | NOR | TC |
| Chl a | 1 | 0 | 0 | 0 | 0 | 0 | 0 | 0 | 0 | 0 | 0 | 0 | 0 |
| | 3 | 0 | 0 | 0 | 0 | 0 | 0 | 0 | −3.1 ± 0.1 | −1.8 ± 0.2 | 0 | −10 | −9 |
| | 7 | 0 | 0 | 0 | −6.9 ± 0.6 | −3.0 ± 0.1 | −1.2 ± 0.1 | −6.5 ± 0.3 | −10.5 ± 0.6 | −12.9 ± 0.3 | −21.3 ± 3.1 | −24.4 ± 3.1 | −32.9 ± 1.4 |
| | 14 | −3.8 ± 0.2 | −2.2 ± 0.1 | −1.6 ± 0.2 | −8.5 ± 0.4 | −3.3 ± 0.2 | −6.5 ± 0.2 | −15.5 ± 1.1 | −15.3 ± 1.5 | −10.3 ± 1.1 | −30.0 ± 2.0 | −32.5 ± 2.9 | −39.8 ± 2.0 |
| | 21 | −3.9 ± 0.5 | −3.8 ± 0.4 | −3.7 ± 0.3 | −13.1 ± 0.9 | −8.2 ± 0.6 | −6.0 ± 0.6 | −25.6 ± 5.6 | −16.3 ± 1.1 | −16.6 ± 2.0 | −35.3 ± 3.1 | −30.1 ± 6.9 | −42.2 ± 1.3 |
| | 30 | −4.6 ± 0.3 | −2.6 ± 0.3 | −3.2 ± 0.3 | −15.3 ± 1.2 | −10.1 ± 1.9 | −14.9 ± 1.9 | −31.3 ± 6.2 | −26.1 ± 5.6 | −31.6 ± 2.3 | −42.9 ± 3.9 | −43.6 ± 3.4 | −50.4 ± 3.5 |
| Chl b | 1 | 0 | 0 | 0 | 0 | 0 | 0 | 0 | 0 | 0 | 0 | 0 | 0 |
| | 3 | 0 | 0 | 0 | −1.9 ± 0.3 | −3.0 ± 0.3 | −3.4 ± 0.3 | −2.3 ± 0.1 | −3.8 ± 0.9 | −3.2 ± 0.2 | −4.2 ± 0.2 | −5.2 ± 0.3 | −5.3 ± 0.4 |
| | 7 | 0 | 0 | 0 | −9.6 ± 0.6 | −8.9 ± 1.2 | −9.3 ± 0.3 | −14.9 ± 2.5 | −19.8 ± 1.1 | −16.5 ± 1.5 | −20.3 ± 1.4 | −16.4 ± 0.9 | −14.4 ± 1.0 |
| | 14 | −1.2 ± 0.1 | 0 | 0 | −11.4 ± 0.2 | −15.2 ± 1.9 | −16.3 ± 1.0 | −32.4 ± 3.5 | −25.3 ± 2.5 | −29.9 ± 3.2 | −33.4 ± 1.2 | −31.6 ± 1.2 | −22.1 ± 0.6 |
| | 21 | −2.3 ± 0.1 | −2.2 ± 0.1 | −3.8 ± 0.1 | −25.6 ± 0.9 | −21.3 ± 3.6 | −18.8 ± 1.9 | −36.0 ± 5.2 | −29.1 ± 3.3 | −33.3 ± 3.0 | −36.8 ± 2.2 | −32.0 ± 2.9 | −30.9 ± 2.9 |
| | 30 | −3.7 ± 0.2 | −3.7 ± 0.4 | −4.3 ± 0.3 | −26.2 ± 0.5 | −22.5 ± 4.1 | −21.6 ± 3.5 | −41.3 ± 5.1 | −32.5 ± 3.6 | −39.8 ± 4.2 | −36.9 ± 2.5 | −34.1 ± 3.3 | −33.2 ± 3.6 |
| Tot Chl | 1 | 0 | 0 | 0 | 0 | 0 | 0 | 0 | 0 | 0 | 0 | 0 | 0 |
| | 3 | 0 | 0 | 0 | 0 | 0 | 0 | −2.2 ± 0.1 | −3.9 ± 0.6 | −3.3 ± 1.0 | −3.3 ± 0.6 | −4.3 ± 0.1 | −4.1 ± 0.2 |
| | 7 | 0 | 0 | 0 | −5.4 ± 0.4 | −6.1 ± 0.3 | −5.1 ± 0.1 | −5.6 ± 0.1 | −12.1 ± 0.1 | −9.8 ± 0.3 | −18.2 ± 0.9 | −26.3 ± 0.9 | −29.6 ± 0.9 |
| | 14 | 0 | −1.3 ± 0.2 | −1.1 ± 0.1 | −9.3 ± 0.2 | −12.9 ± 0.3 | −11.1 ± 0.6 | −24.6 ± 0.6 | −21.3 ± 3.3 | −20.2 ± 1.3 | −26.1 ± 1.5 | −30.9 ± 0.6 | −28.2 ± 0.3 |
| | 21 | −3.4 ± 0.2 | −3.2 ± 0.2 | −4.3 ± 0.2 | −13.9 ± 1.3 | −13.3 ± 0.9 | −12.9 ± 0.2 | −29.1 ± 0.9 | −25.5 ± 3.6 | −26.6 ± 1.3 | −41.6 ± 1.9 | −38.2 ± 1.5 | −33.6 ± 1.9 |
| | 30 | −3.6 ± 0.5 | −3.1 ± 0.1 | −5.2 ± 0.5 | −16.8 ± 1.0 | −15.5 ± 1.9 | −16.3 ± 0.5 | −36.1 ± 0.2 | −35.6 ± 5.2 | −39.5 ± 3.0 | −42.2 ± 1.8 | −42.6 ± 2.9 | −45.2 ± 2.1 |

DAT—days after treatment.

**Table 4.** Effects of antibiotics on concentrations of Chl a, Chl b, and total Chl in leaf tissues of parrot feather (*Myriophyllum verticillatum*). Values are the mean ± S.D. of three parallel individuals in each container, with three measurements from each sample.

| Item | DAT | Inhibition Rates of Chlorophyll (%) | | | | | | | | | | | |
|---|---|---|---|---|---|---|---|---|---|---|---|---|---|
| | | 0 | | | 1 mg/L | | | 5 mg/L | | | 10 mg/L | | |
| | | SMZ | NOR | TC | SMZ | NOR | TC | SMZ | NOR | TC | SMZ | NOR | TC |
| Chl a | 1 | 0 | 0 | 0 | 0 | 0 | 0 | 0 | 0 | 0 | 0 | 0 | 0 |
| | 3 | 0 | 0 | 0 | 0 | 0 | 0 | 0 | −1.0 ± 0.2 | −1.1 ± 0.1 | −1.9 ± 0.3 | −3.7 ± 0.1 | −3.1 ± 0.1 |
| | 7 | 0 | 0 | 0 | −2.1 ± 0.1 | −2.3 ± 0.4 | −1.6 ± 0.2 | −2.9 ± 0.1 | −9.3 ± 0.4 | −3.2 ± 0.1 | −5.5 ± 1.1 | −20.5 ± 1.1 | −26.1 ± 2.9 |
| | 14 | −1.2 ± 0.1 | −1.6 ± 0.1 | −1.1 ± 0.1 | −3.9 ± 0.2 | −5.1 ± 0.1 | −5.3 ± 0.3 | −9.3 ± 0.6 | −11.4 ± 0.9 | −9.8 ± 0.2 | −11.4 ± 1.3 | −24.4 ± 1.9 | −26.9 ± 3.9 |
| | 21 | −3.0 ± 0.3 | −3.1 ± 0.1 | −3.2 ± 0.1 | −7.4 ± 0.3 | −11.6 ± 0.9 | −7.2 ± 0.1 | −13.2 ± 2.3 | −19.1 ± 1.3 | −19.3 ± 0.6 | −18.9 ± 2.6 | −29.6 ± 2.6 | −31.2 ± 2.5 |
| | 30 | −4.1 ± 0.9 | −3.2 ± 0.2 | −4.3 ± 0.2 | −9.5 ± 0.9 | −15.9 ± 1.2 | −13.3 ± 0.6 | −19.9 ± 1.9 | −26.0 ± 3.4 | −26.4 ± 1.9 | −29.6 ± 2.9 | −32.5 ± 2.5 | −34.9 ± 3.4 |
| Chl b | 1 | 0 | 0 | 0 | 0 | 0 | 0 | 0 | 0 | 0 | 0 | 0 | 0 |
| | 3 | 0 | 0 | 0 | 0 | 0 | −2.2 ± 0.1 | −1.1 ± 0.2 | −2.1 ± 0.2 | −2.1 ± 0.1 | −6.5 ± 0.1 | −4.1 ± 0.9 | −6.6 ± 0.2 |
| | 7 | 0 | 0 | 0 | −2.1 ± 0.6 | −3.6 ± 0.4 | −6.0 ± 0.5 | −3.2 ± 0.5 | −12.9 ± 0.5 | −15.9 ± 2.2 | −16.2 ± 0.3 | −16.0 ± 1.1 | −10.1 ± 1.8 |
| | 14 | 0 | 0 | 0 | −6.3 ± 0.2 | −10.1 ± 1.2 | −10.1 ± 0.2 | −19.6 ± 1.1 | −21.4 ± 0.9 | −19.1 ± 2.1 | −29.9 ± 0.9 | −30.9 ± 1.9 | −20.9 ± 1.1 |
| | 21 | 0 | −2.5 ± 0.2 | −4.6 ± 0.6 | −9.1 ± 0.9 | −19.9 ± 3.0 | −13.6 ± 1.2 | −30.2 ± 1.9 | −41.2 ± 2.9 | −32.2 ± 1.5 | −29.1 ± 1.1 | −39.2 ± 2.1 | −41.2 ± 2.0 |
| | 30 | 0 | −1.3 ± 0.1 | −2.1 ± 0.1 | −19.9 ± 0.5 | −26.3 ± 2.9 | −20.1 ± 2.1 | −36.6 ± 3.2 | −45.9 ± 4.1 | −29.1 ± 2.9 | −42.0 ± 1.9 | −39.8 ± 2.5 | −41.0 ± 2.9 |
| Tot Chl | 1 | 0 | 0 | 0 | 0 | 0 | 0 | 0 | 0 | 0 | 0 | 0 | 0 |
| | 3 | 0 | 0 | 0 | 0 | 0 | 0 | −2.9 ± 0.1 | −3.1 ± 0.2 | −3.0 ± 0.1 | −3.1 ± 0.1 | −3.6 ± 0.6 | −4.9 ± 0.4 |
| | 7 | 0 | 0 | 0 | −5.0 ± 0.3 | −6.0 ± 0.1 | −5.9 ± 0.3 | −5.2 ± 0.3 | −12.9 ± 0.9 | −8.1 ± 0.6 | −17.1 ± 0.3 | −25.4 ± 1.8 | −23.5 ± 0.2 |
| | 14 | 0 | −1.1 ± 0.1 | −1.5 ± 0.2 | −9.2 ± 0.1 | −12.1 ± 0.2 | −12.2 ± 0.5 | −23.1 ± 0.2 | −30.1 ± 1.5 | −21.4 ± 0.6 | −23.2 ± 2.1 | −31.2 ± 2.9 | −29.7 ± 3.2 |
| | 21 | −3.9 ± 0.6 | −3.3 ± 0.5 | −4.4 ± 0.3 | −13.1 ± 0.6 | −12.6 ± 0.9 | −13.0 ± 0.5 | −29.9 ± 1.5 | −33.1 ± 2.6 | −25.6 ± 3.5 | −40.9 ± 3.3 | −37.1 ± 2.2 | −30.3 ± 2.1 |
| | 30 | −3.3 ± 0.2 | −3.9 ± 0.2 | −5.9 ± 0.2 | −14.1 ± 0.5 | −13.3 ± 2.1 | −16.1 ± 0.9 | −44.2 ± 1.3 | −33.9 ± 3.8 | −41.1 ± 2.9 | −41.5 ± 2.9 | −42.5 ± 4.2 | −42.9 ± 4.3 |

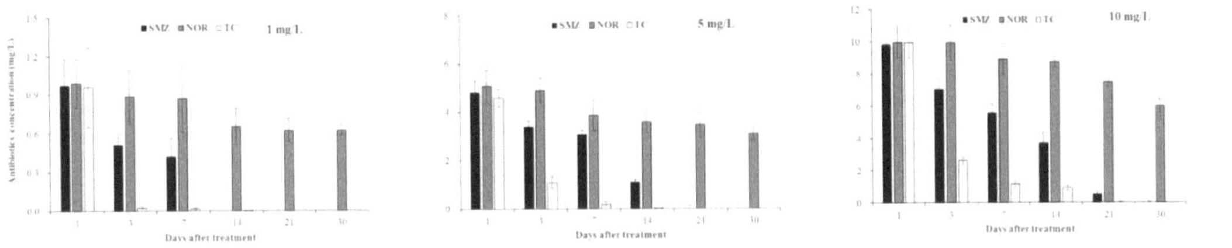


**Figure 3.** Selected antibiotic concentrations in hydroponic culture solution (water lettuce) under various treatments at the end of the experimental period.

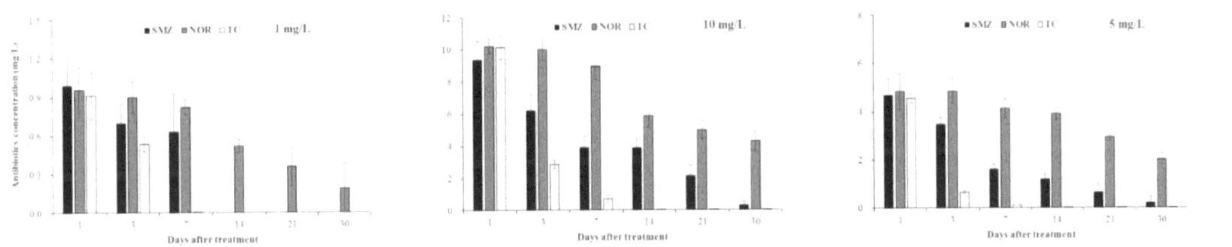


**Figure 4.** Selected antibiotic concentrations in hydroponic culture solution (parrot feather) under various under various treatments at the end of the experimental period.

### 3.4. Uptake Translocation of Antibiotics into Water Lettuce and Parrot Feather

In the harvest stage of the plant, we analyzed the SMZ, NOR, and TC contents of water lettuce and parrot feather. The results clearly indicated that the three antibiotics were taken up by both plants. Further, the uptake translocation of antibiotics was positively correlated with antibiotic concentration, a finding that is similar to those of previous studies [15,16]. None of the three antibiotics were detected in the shoots of the water lettuce plants, only in the root. However, in the parrot feather plants, antibiotics were detected in both the shoots and the roots, with higher amounts detected in the shoots than in the roots. Most SMZ and TC were detected in the roots and shoots of the water lettuce and parrot feather plants, and about one-third of the treated amount of NOR was detected (Figure 5). The low NOR uptake by plants is associated with water solubility. In this study, we found that the uptake translocation mechanisms of parrot feather differed from those of water lettuce. The antibiotics were taken up and transported via mass flow and active transport [15]. In general, there is a close correlation between antibiotic uptake translocation into plants and the antibiotic octanol/water partition coefficient (log $K_{OW}$). There is an optimal hydrophobicity value for antibiotic uptake translocation, and that organic compounds that are either highly polar (log $K_{OW} < 1$) or highly lipophilic (log $K_{OW} > 4$) will not be significantly taken up by plants [15,17]. In this study, log $K_{OW}$ values for SMZ, NOR, and TC were below 0.5. Further, we conducted experiments in a plastic film house environment at a temperature of $30/25 \pm 2$ °C day/night, and the volume of evapotranspiration water in all treatments was greater than 160–170 mL/day. Therefore, we hypothesized that antibiotics were taken up through mass flow and passive absorption. SMZ, NOR, and TC accumulated at different levels in water plants, likely because of their different solubilities, affinities, and absorption mechanisms. Specifically, the solubility value of SMZ was much higher than those of NOR and TC. Dettenmaier et al. [18] indicated that highly water soluble organic compounds are most likely to be taken up by plants. The three antibiotics were detected in shoots and roots, but there were different distributions for different antibiotics. The plant root was the primary point of contact with antibiotic solution; thus, most antibiotics accumulated in the root system via biological and physical–chemical uptake. Removal of antibiotics from hydroponic culture solution takes place through a variety of mechanisms, namely hydrolysis, biodegradation, photodegradation, and plant uptake. In hydrosphere ecosystems, further research is needed to investigate the mechanism of phytoremediation of antibiotics by aquatic plants.

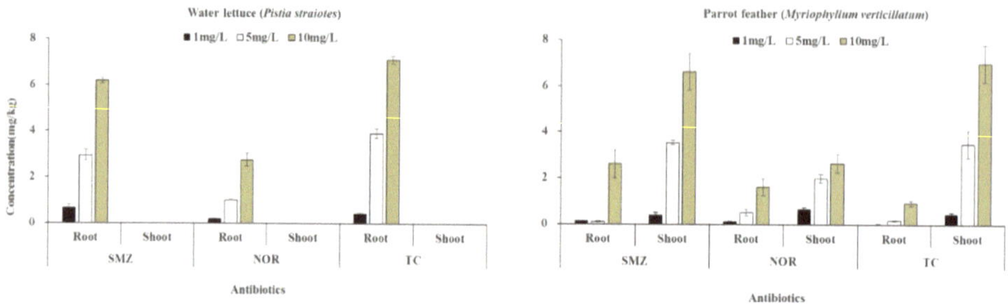


**Figure 5.** Selected antibiotic concentrations (on a dry weight basis) in water lettuce and parrot feather plants under various treatments at the end of the experimental period. Values are the mean ± S.D. of three parallel individuals in each container.

### 3.5. Bioconcentration Factor

Plants having both phytostabilization and pollution-tolerant capacities are potentially useful for phytoremediation purposes. The mean BCF values for antibiotics in water lettuce and parrot feather tissues are shown in Figure 6. Uptake translocation of antibiotics varied among treatments. In water lettuce tissues, BCF of antibiotics ranged from 0.24 to 0.78, while that of NOR was much lower, ranging from 0.24 to 0.38. In addition, SMZ (0.59–0.64) and TET (0.72–0.78) showed higher uptake accumulation into water lettuce tissues. The BCF values of water lettuce and parrot feather showed similar trends but were higher for parrot feather (Figure 6). The differences in uptake accumulation of antibiotics in water plants are governed by antibiotic concentrations, crop species, and environmental conditions. SMZ and TC showed 60–80% removal, in the range of 1–10 mg/L of treatment concentration, but NOR removed only 20–30%. Parrot feather was slightly more effective in this respect than water lettuce. BCF values greater than 1 are generally believed to indicate the potential success of a plant species for phytoremediation. The BCF values of inorganic metal ions are often higher than 1 [19]; however, in this study, BCF values of antibiotics, which are toxic organic substances, were less than 1. The BCF values of antibiotics were similar to those measured in previous studies [18]. In this study, the metabolites of the three antibiotics could not be identified due to various research limitations. We present data for estimating metabolites of three antibiotics using previous research data. Biotransformation was shown to be the main removal process for the parent antibiotics from the aqueous phase and abiotic transformation, mineralization, and sorption to the suspended solids [20]. *Microbacterium* sp. strain 4N2-2 produces four metabolites (8-hydroxynorfloxacin, 6-defluoro-6-hydroxynorfloxacin, desethylene-norfloxacin, and N-acetylnorfloxacin) from norfloxacin [21], *Streptomyces* sp. strain AL-16012 produces five metabolites (chlorotetracycline, doxytetracycline, meclotetracycine, minotetracycline, and rolitetracycline) from tetracycline [20], and ammonia-oxidizing archaeon strain produces two metabolites (acetyl + hydroxy (AcOH-sulfonamide) and N4-formyl-sulfonamide) from sulfamethazine [22] through glycosylation, decarboxylation, hydroxylation, oxidative defluorination, desethylation, and *N*-acetylation. These metabolites, of course, may be subjected to further degradation by other environmental microorganisms. Three antibiotic types—quinolones, sulfonamides, and tetracyclines—which are thought to be genetically toxic to fishes, have been reported to bioaccumulate in fish tissues, as well as in aquatic environments, such as rivers and surface water. The adverse effects of these antibiotics are known to cause damage to developmental, cardiovascular, and metabolic systems, as well as in altering antioxidant and immune responses, in fishes. Therefore, there are continuous concerns about the toxicity of antibiotics in fishes and management strategies are needed [20–22].

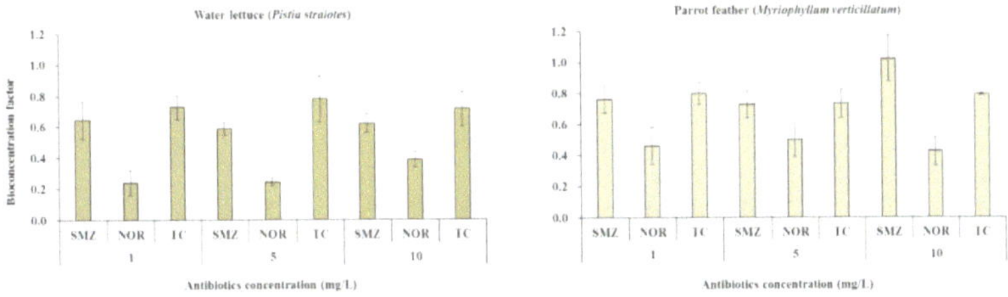


**Figure 6.** Bioconcentration factor (BCF) values of antibiotics in water lettuce and parrot feather plants (mean ± SD, n = 3).

**Author Contributions:** Y.-J.P. and J.-G.S. performed the experiments and wrote the article. All authors have read and agreed to the published version of the manuscript.

**Funding:** This paper was supported by research funds of Jeonbuk National University in 2019.

**Institutional Review Board Statement:** Not applicable.

**Informed Consent Statement:** Not applicable.

**Data Availability Statement:** The data used in this study can be available from the authors on reasonable request.

**Conflicts of Interest:** The authors declare no conflict of interest.

## References

1. Garbisu, C.; Alkorta, I. Phytoextraction: A cost-effective plant-based technology for the removal of metals from the environment. *Bioresour. Technol.* **2001**, *77*, 229–236. [CrossRef]
2. Paz-Alberto, A.M. Phytoremediation: A green technology to remove environmental pollutants. *Am. J. Clim. Chang.* **2013**, *2*, 71–86. [CrossRef]
3. Milena, M.; Anna, W.; Elbieta, S. Phytoremediation techniques in wastewater treatment. *Environ. Biotechnol.* **2015**, *11*, 10–13.
4. Du, L.F.; Liu, W.K. Occurrence, fate, and ecotoxicity of antibiotics in agroecosystems: A review. *Agron. Sustain. Dev.* **2012**, *32*, 309–327. [CrossRef]
5. Grassi, M.; Luigi, R.; Anna, F. Endocrine disruptors compounds, pharmaceuticals and personal care products in urban wastewater: Implications for agricultural reuse and their removal by adsorption process. *Environ. Sci. Pollut. Res.* **2013**, *20*, 3618–3628. [CrossRef] [PubMed]
6. Tasho, R.P.; Cho, J.Y. Veterinary antibiotics in animal waste, its distribution in soil and uptake by plants: A review. *Sci. Total Environ.* **2016**, *563*, 366–376. [CrossRef]
7. Gujarathi, N.P.; Bryan, J.H.; James, C.L. Phytoremediation potential of Myriophyllum aquaticum and Pistia stratiotes to modify antibiotic growth promoters, tetracycline, and oxytetracycline, in aqueous wastewater systems. *Int. J. Phytoremediat.* **2015**, *7*, 99–112. [CrossRef] [PubMed]
8. Hillis, D.G.; Fletcher, J.; Solomon, K.R.; Sibley, P.K. Effects of ten antibiotics on seed germination and root elongation in three plant species. *Arch. Environ. Contam. Toxicol.* **2011**, *60*, 220–232. [CrossRef]
9. Dordio, A.V.; Duarte, C.; Barreiros, M.; Carvalho, A.; Pinto, A.; Costa, C.T. Toxicity and removal efficiency of pharmaceutical metabolite clofibric acid by Typha spp. *Bioresour. Technol.* **2009**, *100*, 1156–1161. [CrossRef] [PubMed]
10. Ghosh, G.C.; Hanamoto, S.; Yamashita, N.; Huang, X.; Tanaka, H. Antibiotics removal in biological sewage treatment plants. *Pollution* **2016**, *2*, 131–139.
11. Jathwa, A.I.; Muna, Y.A.A. Detection of tetracycline, doxycycline, chlortetracycline, and oxytetracycline antibiotics in Nineveha drug wastewater. *Al-Nahrain J. Eng. Sci.* **2012**, *15*, 215–221.
12. Huang, X.D.; El Alawi, Y.; Penrose, D.M.; Glick, B.R.; Greenberg, B.M. Responses of three grass species to creosote during phytoremediation. *Environ. Pollut.* **2004**, *130*, 453–463. [CrossRef]
13. Zayed, A.; Gowthaman, S.; Terry, N. Phytoaccumulation of trace elements by wetlands plants: I. *Duckweed. J. Environ. Qual.* **1998**, *27*, 715–721. [CrossRef]
14. Huang, C.; Renew, J.; Sivieby, K. Assessment of potential antibiotic contaminants in water and preliminary occurrence analysis. *Water Res.* **2001**, *120*, 30–40.
15. Boonsaner, M.; Hawker, D.W. Accumulation of oxytetracycline and norfloxacin from saline soil by soybeans. *Sci. Total Environ.* **2010**, *408*, 1731–1737. [CrossRef] [PubMed]
16. Michelini, L.; Meggio, F.; La Rocca, N.; Ferro, S.; Ghisi, R. Accumulation and effects of sulfadimethoxine in Salix fragilis L. plants: A preliminary study to phytoremediation purposes. *Int. J. Phytoremediation* **2012**, *14*, 388–402. [CrossRef]

17. Herklotz, P.A.; Gurung, P.; Heuvel, B.V.; Kinney, C.A. Uptake of human pharmaceuticals by plants grown under hydroponic conditions. *Chemosphere* **2010**, *78*, 1416–1421. [CrossRef]
18. Dettenmaier, E.M.; Doucette, W.J.; Bugbee, B. Chemical hydrophobicity and uptake by plant roots. *Environ. Sci. Technol.* **2009**, *43*, 324–329. [CrossRef] [PubMed]
19. Ashfaq, N.; Riffat, N.M.; Muhamamd, A.; Nasrullah, K.; Muhammad, F.S. Hyperaccumulators of heavy metals of industrial areas of Islamabad and Rawalpindi. *Pak. J. Bot.* **2011**, *43*, 1925–1933.
20. Ramesh, P.P.; Chu, L.L.; Kim, T.S.; Sohng, J.K. Bioconversion of tetracycline antibiotics to novel glucoside derivatives by single-vessel multienzymatic glycosylation. *J. Microbiol. Biotechnol.* **2018**, *28*, 298–304.
21. Kim, D.W.; Thomas, M.H.; Kim, B.S.; Laura, K.S.; Kellie, A.W.; John, B.S. Modification of norfloxacin by a *Microbacterium* sp. strain isolated from a wastewater treatment plant. *Appl. Environ. Microbiol.* **2011**, *77*, 6100–6108. [CrossRef]
22. Tzeng, T.W.; Liu, Y.T.; Deng, Y.; Hsieh, Y.C.; Tan, C.C.; Wang, S.L.; Huang, S.T.; Tzou, Y.M. Removal of sulfamethazine antibiotics using cow manure-based carbon adsorbents. *Int. J. Environ. Sci. Technol.* **2016**, *13*, 973–984. [CrossRef]

*applied sciences*

MDPI

*Article*

# Hydrogen Energy Demand Growth Prediction and Assessment (2021–2050) Using a System Thinking and System Dynamics Approach

**Talal Yusaf [1,*], Mohamd Laimon [2], Waleed Alrefae [3], Kumaran Kadirgama [4], Hayder A. Dhahad [5], Devarajan Ramasamy [6], Mohd Kamal Kamarulzaman [7] and Belal Yousif [8]**

1 School of Engineering and Technology, Central Queensland University, Brisbane, QLD 4008, Australia
2 Engineering Faculty, Al-Hussein Bin Talal University, Ma'an 71111, Jordan; u1093176@umail.usq.edu.au
3 Mechanical Engineering Technology Department, The Public Authority of Applied Education and Training (PAAET), Adailiyah P.O. Box 23167, Kuwait; waleed_alrefae@hotmail.com
4 Faculty of Mechanical and Automotive Engineering Technology, Universiti Malaysia Pahang, Pekan 26600, Pahang, Malaysia; kumaran@ump.edu.my
5 Mechanical Engineering Department, University of Technology, Baghdad P.O. Box 18310, Iraq; Hader.A.Dhahad@uotechnology.edu.iq
6 College of Engineering, Universiti Malaysia Pahang, Pekan 26600, Pahang, Malaysia; deva@ump.edu.my
7 Automotive Engineering Centre, Universiti Malaysia Pahang, Pekan 26600, Pahang, Malaysia; kamalkz@hotmail.com
8 School of Engineering, The University Southern Queensland, Toowoomba, QLD 4350, Australia; belal.yousif@usq.edu.au
* Correspondence: t.yusaf@cqu.edu.au

**Abstract:** Adoption of hydrogen energy as an alternative to fossil fuels could be a major step towards decarbonising and fulfilling the needs of the energy sector. Hydrogen can be an ideal alternative for many fields compared with other alternatives. However, there are many potential environmental challenges that are not limited to production and distribution systems, but they also focus on how hydrogen is used through fuel cells and combustion pathways. The use of hydrogen has received little attention in research and policy, which may explain the widely claimed belief that nothing but water is released as a by-product when hydrogen energy is used. We adopt systems thinking and system dynamics approaches to construct a conceptual model for hydrogen energy, with a special focus on the pathways of hydrogen use, to assess the potential unintended consequences, and possible interventions; to highlight the possible growth of hydrogen energy by 2050. The results indicate that the combustion pathway may increase the risk of the adoption of hydrogen as a combustion fuel, as it produces $NO_x$, which is a key air pollutant that causes environmental deterioration, which may limit the application of a combustion pathway if no intervention is made. The results indicate that the potential range of global hydrogen demand is rising, ranging from 73 to 158 Mt in 2030, 73 to 300 Mt in 2040, and 73 to 568 Mt in 2050, depending on the scenario presented.

**Keywords:** hydrogen energy; hydrogen demand; hydrogen fuel cell pathway; hydrogen combustion pathway; emissions; nitrogen oxides; renewable alternative fuel; systems thinking; system dynamics

**Citation:** Yusaf, T.; Laimon, M.; Alrefae, W.; Kadirgama, K.; Dhahad, H.A.; Ramasamy, D.; Kamarulzaman, M.K.; Yousif, B. Hydrogen Energy Demand Growth Prediction and Assessment (2021–2050) Using a System Thinking and System Dynamics Approach. *Appl. Sci.* **2022**, *12*, 781. https://doi.org/10.3390/app12020781

Academic Editor: Andreas Sumper

Received: 28 December 2021
Accepted: 11 January 2022
Published: 13 January 2022



## 1. Introduction

Hydrogen is a promising alternative to fossil fuels in many applications. Some of these applications are the domain of hydrogen energy, and no other alternatives can compete with hydrogen, such as heavy and long-distance transport (e.g., heavy-duty trucks, ships, and planes), as well as energy-intensive manufacturing sectors (e.g., ferrous and non-ferrous metals, petroleum refining, chemicals, and cement). With the current technology, these sectors are difficult to electrify, and other alternatives (e.g., bio energy) cannot cope with demand. Hence, decarbonisation of these sectors will remain a challenge. Therefore, adoption of hydrogen as a replacement for a fossil fuel could be a major step towards decarbonising and fulfilling the needs of the energy sector.

Hydrogen is an abundant (the most abundant element in the universe), efficient, reliable, clean, non-toxic, and renewable fuel, and it has a very high energy content (about three times as much as gasoline and natural gas, and almost seven times that of coal) [1–3]. Hydrogen's entire process has four stages: (1) resources, (2) systems, (3) pathways of use, and (4) end use, as shown in Figure 1. Hydrogen can be produced by diverse energy resources, such as RE and fossil fuels. If it is produced by RE, it is called green hydrogen, if it is produced by fossil fuels with Carbon Capture Storage (CCS), it is called blue hydrogen, and if it is produced by fossil fuels without CCS, it is called grey hydrogen [4]. Three systems are involved: production systems, storage systems, and delivery systems. Production systems include many routes; the most prevalent are: Steam Methane Reforming (SMR), which is the process in which methane extracted from natural gas is heated by steam to generate hydrogen [5]; coal gasification, which is the process in which Carbon Dioxide ($CO_2$) is generated by adding air to the coal through combustion. Carbon dioxide then reacts with the rest of the carbon in the coal to form carbon monoxide, and in the final process, carbon monoxide reacts with steam, generating hydrogen [6]; electrolysis of water, which is the process of using electricity to decompose water into oxygen and hydrogen [7]. Storage systems also include many routes, of which the most common storage systems are: as a compressed gas in high-pressure gas cylinders; as a liquid in cryogenic tanks; as a solid by reacting with metals or chemical compounds or being stored in an alternative chemical form. Delivery systems are through three main methods: gaseous pipelines; pressurized tube trailers; cryogenic liquid tankers [8]. Hydrogen energy can be used through two key pathways: fuel cells and combustion [9]. Fuel cells produce electricity, while combustion is used in boilers and engines [10]. End use includes a large number of: chemicals (e.g., methanol, and ammonia), industry (refinery, steel, and cement), residential and commercial (e.g., blending into natural gas pipelines), mobility (e.g., fuel cells electric vehicle), power (e.g., power generation by turbine and storage as a backup power in fuel cells or tank storage), and space applications (e.g., rocket fuel).

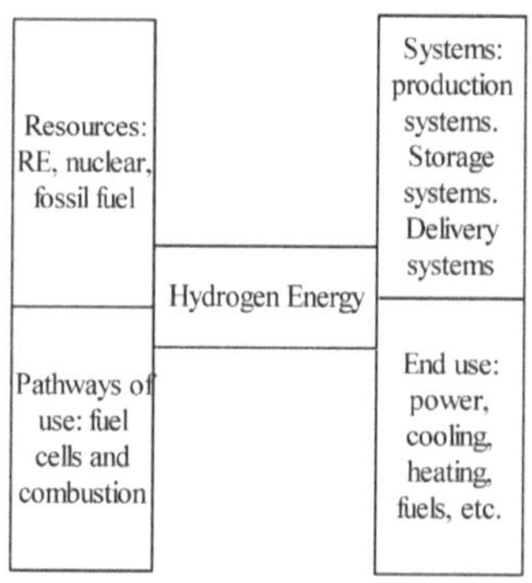


**Figure 1.** Hydrogen entire process stages.

Previous efforts have focused, mainly, on hydrogen production resources and systems [11–17]. For example, Dincer et al. [11] assessed innovative hydrogen production methods under various criteria. Hosseini et al. [12] highlighted the state-of-the-art hydrogen production technologies using renewable and sustainable energy resources. Acar et al. [18] provided a review on sustainability of hydrogen production systems. Niaz et al. [14] focused on various hydrogen producing and storing methods for creating a hydrogen economy. When considering the impact of hydrogen on the environment, the focus rests on issues related to the leakage of hydrogen [19–24]. For example, Li et al. [19] and Shen et al. [20] investigated the safety of hydrogen fuel cells by a numerical simulation of hydrogen leakage from fuel cell ship and vehicle, while Gye et al. [21] assessed the risk of an urban hydrogen refuelling station. The pathways of hydrogen use have received little attention in research or policy, which may be explained by the common belief that nothing but water is released as a by-product when hydrogen energy is used, a belief that is only true for the fuel cells pathway. Burning of hydrogen in the combustion pathway can produce Nitrogen Oxides ($NO_x$), which have something in common between hydrogen and fossil fuel when burned at high temperatures [25–27]. $NO_x$ is a harmful air pollutant that can be formed in all fuel-air mixed in high flames (above 1300 °C); less than 750 °C no $NO_x$ is formed [26,28]. This may pose a risk to the future of hydrogen energy as a combustion fuel. On the other hand, previous studies have employed different approaches to explore the future of hydrogen energy demand, such as the Geographic Information System (GIS) method [29–31], the Optimal Power Flow (OPF) technique [32], and the mathematical model [33].

However, there is a need to understand the dynamics underlying interactions among the components of any complex system, which is the domain of systems thinking and system dynamics approaches, as applications of these approaches in the hydrogen energy sector have been largely absent. Further, there is a need to focus on fields and/or sectors in which only hydrogen can play a vital role compared with other alternatives, which is missing in many previous studies, and has been taken into consideration in this paper. Exploring the future of hydrogen energy demand and highlighting the risks that may affect the sector are of paramount importance to understand the transition to a hydrogen economy.

The aim of this study is twofold: (1) using a systems thinking approach to build a conceptual model for hydrogen energy, with a special focus on the pathways of hydrogen use, to assess the potential consequences that could lead to unintended impacts, and possible interventions; (2) using a system dynamics approach to highlight the possible demand growth of hydrogen energy until 2050. The demand meant here is the global demand, and it will be linked to a case study of a country that may be a leader in hydrogen production in the foreseen future [34–36]. Australia's path will be compared with the global path, hence potential opportunities need to be seized. Australia is one of the world's largest exporters of coal and natural gas, and by 2032, with excessive fossil fuel extraction, coal will be the only fossil fuel resource that Australia totally relies on [37]; hence, Australia should think about an export alternative of energy, and hydrogen would be a suitable alternative, as the current infrastructure used for gas can also be used for hydrogen [38–40].

## 2. Hydrogen Recent Development

Hydrogen, today, is gaining unprecedented traction. This is clearly seen through: the increasing demand, which has grown more than threefold since 1975; the mega projects being implemented; the increasing of policies that directly support investment in hydrogen technologies; the mandates and policy incentives that directly support hydrogen; the global spending on hydrogen energy research [41,42].

## 3. Hydrogen Economy

Nearly every day, there is a new update on the use of hydrogen in transportation and energy. The potential for Hydrogen to X (H2X) to easily merge with present transportation and energy systems, while dramatically lowering air pollution (zero-carbon emission at point of use) [2,43], is fascinating. Simultaneously, human reliance on fossil fuels is responsible for anything from oil spills to ozone threats to climate change [44,45]. The political plan likewise prioritizes the guarantee of price stability and energy supply [46]. These factors are driving the world's transition far from a fossil fuels economy and toward a greener hydrogen economy. The Hydrogen Economy or Hydrogen Energy System is a theoretical concept of a system in which hydrogen is the primary energy source. In 1970, John Bockris was the earliest one to introduce the phrase "hydrogen economy" at the General Motors (GM) Technical Center presentation and during a hydrogen meeting in Miami [47].

The main objective of a hydrogen economy is to produce hydrogen primarily from readily available energy sources in order to replace fossil fuels currently used in transportation, industry, domestic, and commercial applications. The hydrogen economy has been presented as a highly refined and long-term solution to the world's interrelated difficulties, such as (i) world environment concern, (ii) natural resource depletion, (iii) food shortages and malnutrition in third-world countries, and (iv) the world's growing population. Despite the fact that the issues connected with the fossil fuel economy are serious and overwhelming, successful growth of the hydrogen economy will result in significant benefits for the environmental, energy supplies, economic, and end users [48]. Besides the benefits, rapid transition from fossil fuel energy to a hydrogen energy system has been slowed by substantial socioeconomic, technological, and scientific challenges. The extraordinarily low density of hydrogen makes storage a crucial issue during the transfer [49]. Although chemical and refinery industries commonly utilize hydrogen, its costs for production, storage, and delivery are too expensive and undesirable for the majority of energy applications [50].

Furthermore, the great advantages of the hydrogen economy are so exciting that policymakers from all over the world are spending heavily in improving the energy system's prospects [14]. The European Commission's High Level Group on Hydrogen and Fuel Cell Technologies proposed, in 2003, that the European Union achieve a hydrogen-based economy by 2050 and anticipates that, by 2040, 35% of new manufactured automobiles would be powered by zero carbon hydrogen [51]. The Energy Efficiency and Renewable Energy, Fossil Energy, Nuclear Energy, and Science Offices of the US Department of Energy, on the other hand, recommended that transition to hydrogen-powered fuel cell automobiles take place around 2020 [52].

The hydrogen economy has been intensively explored, according to the research; various roadmap reports, case studies, and reviews have been documented, summarizing the main achievements and the function of hydrogen's potential in the energy sector [53–55]. For the purpose of completeness, researchers demonstrates an increasing attention in exploring the social consequences, policy issues, environmental impact, and techno-economic of hydrogen economy [18]. As a result, a number of national and international organizations have been established to educate the policymakers, companies, and public in order to develop a foundation and promote preparations for the hydrogen economy age. Several examples are shown in Table 1.

**Table 1.** National and international hydrogen organization.

| Location | Organization |
| --- | --- |
| Africa | *The Hydrogen South Africa (HySA)* |
| Australia | *The Australian Association for Hydrogen Energy (AAHE)* |
| Canada | *Canadian hydrogen and fuel cell industry* |
| Europe | *Hydrogen Europe* |
| Global | *Hydrogen Council* |
| Spain | *Hydrogen Foundation* |
| Global | *IEA Hydrogen* |
| Global | *International Association for Hydrogen Energy (IAHE)* |
| Global/USA | *The Fuel Cell and Hydrogen Energy Association (FCHEA)* |
| UNIDO | *The International Centre for Hydrogen Energy Technologies (ICHET)* |
| USA | *National Hydrogen Association (NHA)* |
| USA | *The U.S. DOE Hydrogen and Fuel Cells Program* |
| WA-Australia | *Renewable Hydrogen Council* |

The hydrogen economy is not presented as a substitute to existing energy vectors in the world scale but rather, as a compliment in the aspect of a smart and low-carbon strategy for certain purposes and areas [53]. As a result, the transformation to a worldwide "hydrogen economy" is unlikely to occur in the near future [44]. Recent publically funded

R&D initiatives, cost-effectiveness, policies, and the development of a hydrogen system, on the other hand, will undoubtedly contribute to making this dream a reality [56,57].

Renewable hydrogen (i.e., Green Hydrogen), production from renewable energy resources, as well as the development of all corners of square-shaped Hydrogen Square (HydS) (shown in Figure 2) seems to be a promising way to comply with a 100% renewable electric grid and expanding to a 100% renewable energy supply and, hence, a hydrogen economy [58,59]. The literature shows consensus on green hydrogen has the potential to decarbonise a large range of industrial applications or serve as a clean energy carrier for storing renewable energy, utilised in a variety of applications [59,60]. However, parallel development in the four corners of the HydS is essential for the green pathways to become technoeconomically feasible.

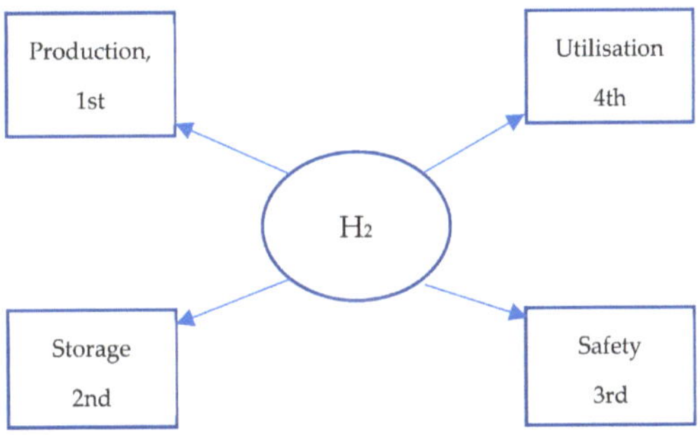


**Figure 2.** Hydrogen Square (HydS): four corners model.

Hydrogen economy seems achievable when a 100% renewable electric grid complied and expanded to a 100% renewable energy supply though the development of all corners of HydS, production from renewable energy resources, and renewable hydrogen (i.e., Green Hydrogen) [61,62]. Green hydrogen has the ability to decarbonise a wide range of economic uses or act as a clean energy carrier for storing renewable energy to be be used in a lot of areas [63]. Nevertheless, for green pathways to become technoeconomically viable, simultaneous growth at the four corners of the HydS is required.

## 4. Exergy Destruction

Exergy is the amount of work that can be done with a given amount of energy. Exergy is the highest theoretical useable energy that can be extracted from an energy resource or a substance that is subjected to an almost reversible technique from its starting condition until it is returned to its natural condition. Exergy is determined by a total of parameters that can be equal to zero (in an equilibrium state with the surroundings), and it is reliant on the relative state of a system and its ambient conditions [64]. Exergy is an indicator that measures how much a system deviates from equilibrium with its surroundings. Exergy is a term that encompasses both the qualitative and quantitative aspects of energy. Any irreversible occurrence results in exergy losses, resulting in process exergy destruction (loss of useful work) or elevated energy consumption from whichever source the energy was acquired from. The Gouy–Stodola theorem connects exergy destruction to total entropy production in a system:

$$\dot{E} = T_0 \dot{S}_{gen} \quad (1)$$

where $\dot{E}$ is the exergy destruction, commonly known as loss of useful work, $T_0$ is the environment's reference temperature, and $\dot{S}_{gen}$ is the system's total entropy production.

Limited numbers of scientists and researchers are reported in prediction and modelling of hydrogen energy demands. A hybrid energy system for electricity and hydrogen generation is theoretically conceived and implemented to an island, Bozcaada, by Kalinci et al. [65]. Exergy, energy, and techniques are used to undertake a thermodynamic evaluation. The Hybrid Optimization Model for Electric Renewable (HOMER) software examines the theoretical structure from the perspectives of energy load distributions and the economy. The study's analytical scenarios are based on the stand-alone island model. The converter, hydrogen tank, polymer electrolyte membrane fuel cell (PEMFC), electrolyser, wind turbines, and photovoltaic (PV) array are all included in the model. The magnitudes and locations of system inefficiencies are determined using exergy and energy assessments. The studies are mostly performed for the primary equipment under dynamically changing environmental and operational circumstances. There are also hourly distributions of varying exergy and energy rates shown. As a result, the PV array's daily average exergy and energy efficiencies are 14.26% and 13.31%, respectively. In addition, the efficiency values for the electrolyser equipment is calculated as exergy at 60.26% and energy at 59.68%, while the wind turbine is determined as exergy at 50.12% and energy at 46%, respectively.

## 5. Research Method

### 5.1. Developing Systems Thinking and System Dynamics Models for Hydrogen Energy

Systems thinking and system dynamics approaches are used to deal with complexity in systems. Systems thinking represents qualitative modelling, where the end goal is to develop a conceptual model that presents the dynamic interaction between system components. System dynamics represents quantitative modelling, where the end goal is to build a simulation model to simulate the dynamic relationships between the components. The steps of systems thinking are: problem articulation and formulating dynamic hypotheses (conceptual model), while the steps of systems dynamics are: formulating a simulation model, validating/testing, policy design, and evaluation, as shown in Figure 3. Understanding, analysing, and improving the performance of complex systems are the aims of using systems thinking and system dynamics approaches.

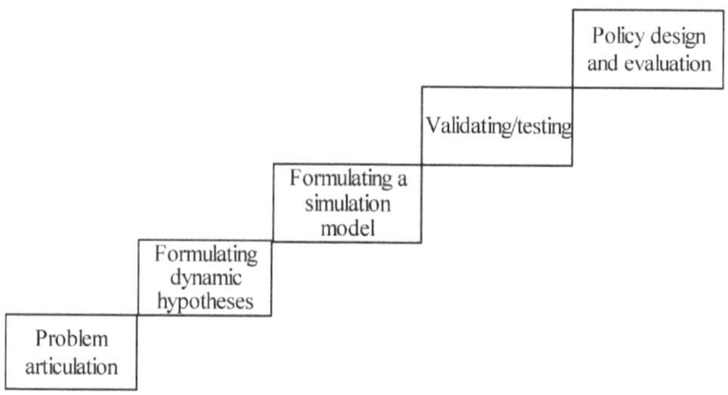


**Figure 3.** The steps for systems thinking and systems dynamics.

A conceptual model for hydrogen energy was constructed using a Causal Loop Diagram (CLD). A CLD consists of variables and arrows that present the causal relationships between the variables. The arrows within a CLD link variables together to form either reinforcing (positive) denoted by R, or balancing (negative) denoted by B feedback loops. R feedback loops create exponential growth or decline over time, while B feedback loops act to stabilize system behaviour over time. A system dynamics model for hydrogen demand was constructed in order to calculate the hydrogen demand for different sectors where hydrogen is dominant. Systems dynamics consist of stocks, flows, and auxiliary variables. The stock is a variable that accumulates or depletes over time through flow into or out of the stock, which leads to feedback that can cause changes, either accelerating (reinforcing feedback) or balancing out (balancing feedback). Reinforcing feedback (positive) arises when a stock causes changes, leading to further growth; balancing feedback (negative) arises when a stock causes changes leading to slower growth. A stock responds to changes by its flows, while the flows are controlled by stock and auxiliary variables. Figure 4 represents an example for feedback loops and stock-flow model.

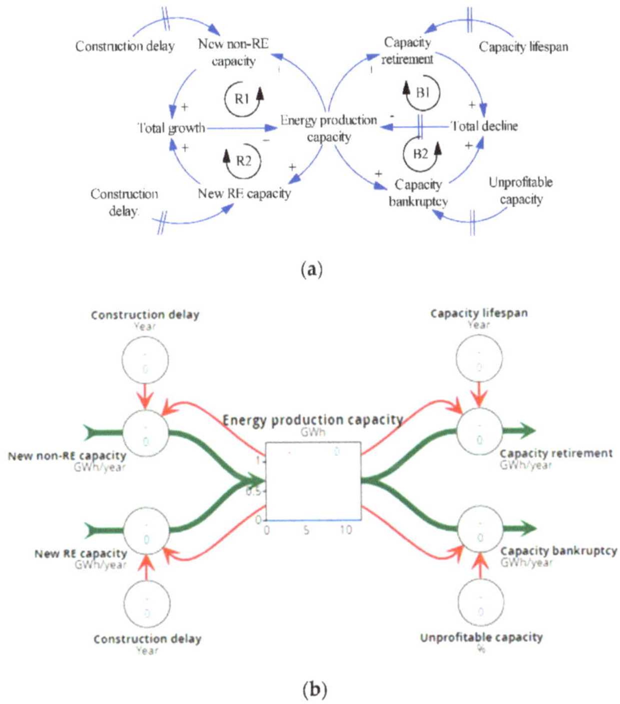


**Figure 4.** Feedback loops (**a**) and stock-flow model (**b**) [37].

## 5.2. Leverage Points and System Archetypes

CLD is a powerful tool that goes beyond understanding of a dynamic and complex system; it can also be used to uncover leverage points (e.g., intervention strategies) that lead to improve system performance. Leverage points are not easy to identify, but System Archetypes (SAs) may help uncover them. SAs are highly aggregated situations in isolation, derived from the main model. There are many archetypes, but the most common are *Limits to Growth* and *Fixes that Fail* archetypes. The *Limits to Growth* archetype represents situations in which growth driven by R loops is followed by stagnation or decline driven by B loops. In that case, leverage points are latent in B loops; thus, limiting factors that change the behaviour must be identified and changed. The *Fixes that Fail* archetype relates to fixes where unintended, and often harmful, consequences resulted from well intended actions. Thus, a long-term fix should be maintained, and short-term quick fixes should be avoided.

# 6. Results

## 6.1. The Conceptual Model of Hydrogen Energy

The CLD of hydrogen energy is shown in Figure 5, which contains seven reinforcing loops (R1 to R7) and three balancing loops (B1 to B3). The CLD highlights the interrelationships among the main components of the variables. The variables involved and the key message conveyed by each loop are summarized in Table 2 and as in Appendix A (Table A1).

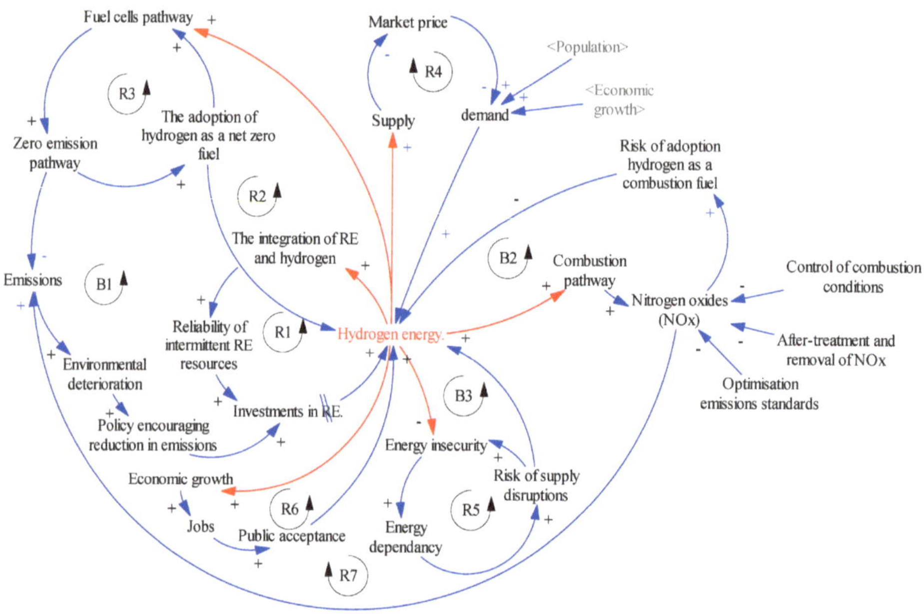


**Figure 5.** The CLD of hydrogen energy. A positive polarity (+) means that the two variables change in the same direction, while a negative polarity (-) means that they move in the opposite direction. Double bars (| |) indicate a time delay between the two variables. R represents Reinforcing loops, while B represents Balancing loops.

**Table 2.** The feedback loops within Figure 5.

| Loop Name | Feedback Loop | Variables Involved | Key Message |
|---|---|---|---|
| Reinforcing | R1 | Hydrogen energy → + The integration of RE and hydrogen → + Reliability of intermittent RE resources → + Investments in RE → + Hydrogen energy | Growth of hydrogen and RE |
| | R2 | Hydrogen energy → + Fuel cells pathway → + Zero emission pathway → + Hydrogen energy → + The adoption of hydrogen as a net zero fuel → + Hydrogen energy | Emissions mitigation |
| | R3 | Fuel cells pathway → + Zero emission pathway → + The adoption of hydrogen as a net zero fuel → + Fuel cells pathway | Emissions mitigation |
| | R4 | Hydrogen energy → + Supply → − Market price → − demand → + Hydrogen energy | Supply-demand |
| | R5 | Energy insecurity → + Energy dependancy → + Risk of supply disruptions → + Energy insecurity | Energy insecurity |
| | R6 | Hydrogen energy → + Economic growth → + Jobs → + Public acceptance → + Hydrogen energy | Economic growth |
| | R7 | Emissions → + Environmental deterioration → + Policy encouraging reduction in emissions → + Investments in RE → + Hydrogen energy → + Combustion pathway → + Nitrogen oxides ($NO_x$) → + Emissions | Emissions |
| Balancing | B1 | Hydrogen energy → + Fuel cells pathway → + Zero emission pathway → − Emissions → + Environmental deterioration → + Policy encouraging reduction in emissions → + Investments in RE → + Hydrogen energy | Emissions mitigation |
| | B2 | Hydrogen energy → + Combustion pathway → + Nitrogen oxides ($NO_x$) → + Risk of adoption hydrogen as a combustion fuel → − Hydrogen energy | Risk of adoption hydrogen as a combustion fuel |
| | B3 | Hydrogen energy → − Energy insecurity → + Energy dependency → + Risk of supply disruptions → + Hydrogen energy | Energy insecurity |

### 6.2. System Archetypes

The first SA seen in the conceptual model is 'Limits to Growth'. This archetype consists of a reinforcing (R1) and a balancing loop (B2), as illustrated in Figure 6a. Hydrogen energy increases the reliability of intermittent RE resources (wind and solar power), and this, in turn, will increase investments in RE. Reliability of supply is one of the most crucial challenges of RE intermittent resources. Using hydrogen energy as a backup power in RE intermittent resources will enhance flexibility and solve the uncertainty of RE. On the other hand, hydrogen as a combustion fuel may increase $NO_x$, which, in turn, may affect the adoption of hydrogen as a combustion fuel. Flame temperature is the limiting factor that may affect hydrogen growth as a combustion fuel. Reducing or controlling the limiting factor may minimize $NO_x$ emissions.

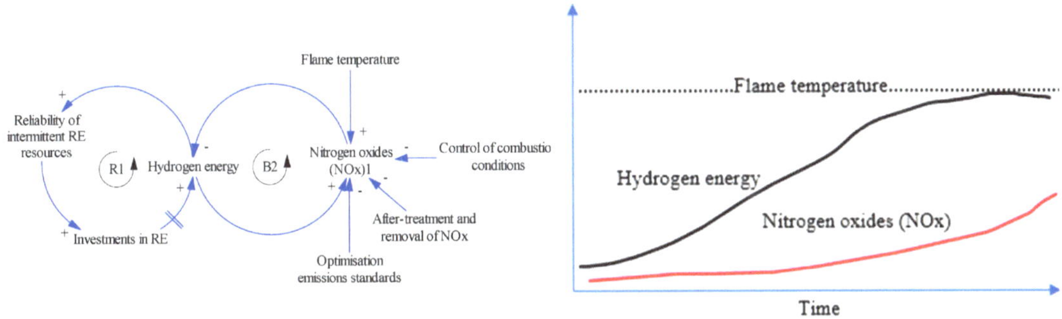


(a) (b)

**Figure 6.** Structure (**a**) and behavioural graph (**b**) of the Limits to Growth.

The second SA is 'Fixes that Fail'. This archetype consists of reinforcing (R7) and a balancing loop (B1), as illustrated in Figure 7a. To mitigate emissions, the government is investing in hydrogen energy (loop B1, Figure 6a). While investing in hydrogen is important to mitigate $CO_2$ emissions, hydrogen as a combustion fuel may lead to further emissions ($NO_x$), which will, in turn, increase the net of emissions (represented by loop R7).

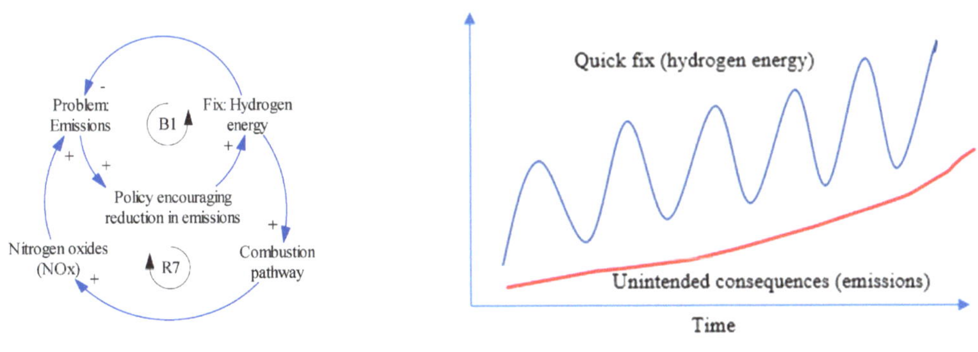


(a) (b)

**Figure 7.** Structure (**a**) and behavioural graph (**b**) of the Fixes that Fail archetype.

### 6.3. The System Dynamics Model of Hydrogen Demand

The system dynamics model of hydrogen demand is shown in Figure 8. In regard to the global hydrogen demand, four scenarios have been taken into consideration until 2050:

(1) a no-growth scenario (current production); (2) a base case scenario that represents the current trend (current growth), and no dramatic changes are assumed, and the base year is 2019, while 2020 has been excluded, due to drops in demand amid the global pandemic (COVID 19); (3) average growth, based on average growth over the last ten years; (4) best case scenario assumes that hydrogen will be the fuel of choice for: grid electrical supply as a back-up capacity (20% of total energy demand); heavy and long-distance transport (20% of total energy demand); energy-intensive manufacturing sector (20% of total energy demand); ammonia production (41 MtH in 2019). These sectors have been selected, as hydrogen can play a vital role in them. The same sectors have been targeted in the case study, considering the current production of these sectors in the Australian context until 2050. The only difference is that the Australian gas sector as an important export has been added to the model to be replaced by hydrogen in the future. The data have been extracted from the "Australian Energy Statistics"—Australian Energy Update report 2019–2020 [66]. It is updated each year and consists of detailed historical energy consumption, production, and trade statistics and balances. The unit used in the report for all energy sources is PJ, however the total hydrogen demand calculated in the model is converted from PJ to KWh to Mt. In this study, all the different forms of energy are put into the same units, and one-to-one conversion rate is used. This is used by many sources and it is the scientific way to compare and summarize energies [67]. Figure 9 shows the potential global demand for hydrogen for the four scenarios. Figure 10 shows the potential hydrogen demand in the Australian context compared to the global demand for the four scenarios.

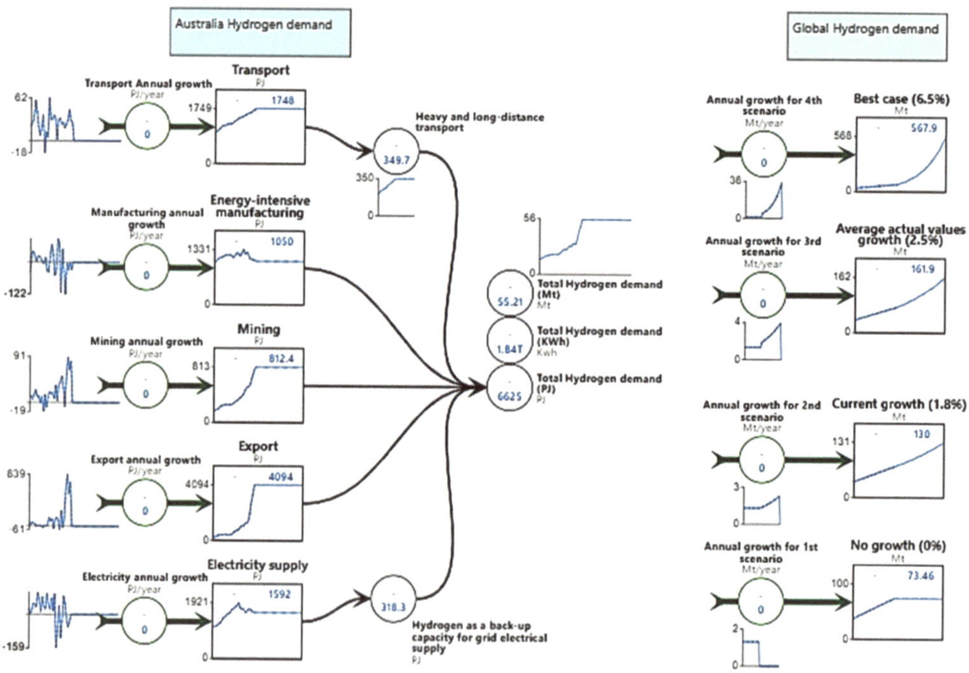


**Figure 8.** The system dynamics model of hydrogen demand.

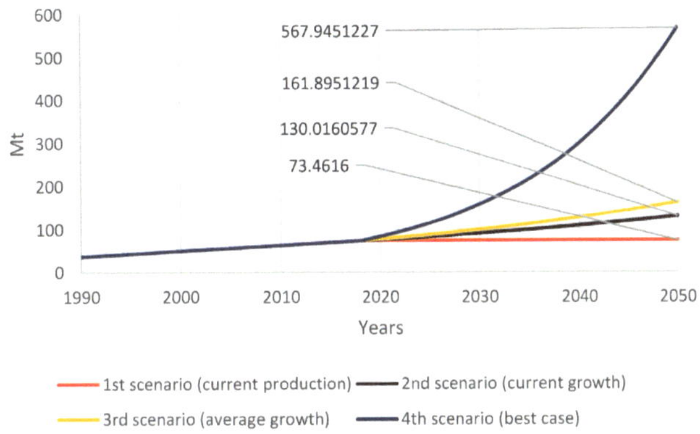


**Figure 9.** The potential global demand for hydrogen for the four scenarios.

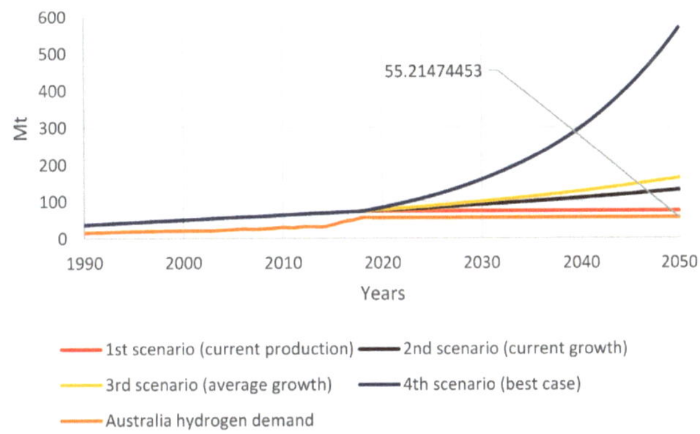


**Figure 10.** The potential hydrogen demand in the Australian context compared to the global demand for the four scenarios.

Model Validating/Testing

The validating of system dynamics models generally involves structural and behavioural tests. Structural tests assess whether the model's structure represents the real system. Behavioural tests assess whether the model provides a sensible output behaviour. Regarding the structural tests, dependency and unit consistency test, laws of conservation and accumulation test, and negative stock test have been applied. Dependency and check unit consistency was performed using the "dependency tracking" feature in the software used (Silico) [68] to check the relationship between parameters and track their units. The laws of conservation and accumulation test implies that the value of the stock must equal the sum of inflows minus the sum of outflows. The negative stock test implies that the stock cannot go below zero.

In relation to the behavioural tests, the model should include a number of real/actual values (historical time-periods). The current study used a historical time series consisting of 30 years from 1990–2019.

## 7. Discussion

The findings indicate that hydrogen energy is important for supporting energy security and renewable energy, zero emission pathway, and economic growth. Hydrogen is an ideal option for an energy-intensive manufacturing sector, grid electrical supply, heavy and long-distance transport, gas networks, and a key ingredient for producing chemicals such as ammonia. Ammonia is an effective hydrogen storage, and it can be used by extracting its stored hydrogen or used directly as fuel. Ammonia can be produced from many different types of energy sources, including renewables, fossil fuels and surplus energy [69]. Recently, Ammonia and hydrogen production from algae is gaining attention as high total energy efficiency was reported to be achieved [70]. However, there are many potential environmental challenges that are not limited to production and distribution systems, but they also relate to how the hydrogen is used through fuel cells and combustion pathways, a concept which has received little attention in research and policy. Combustion pathway may increase the risk of adoption of hydrogen as a combustion fuel (loop B2), as it produces $NO_x$, a key air pollutant that causes environmental deterioration (loop R7), which may limit the application of the combustion pathway [71].

Through the CLD we have constructed, two SAs were revealed, namely, limits to growth and fixes that fail—Figures 6 and 7. Both SAs indicate that the combustion pathway may undermine the use of hydrogen as a combustion fuel if no intervention is made. These interventions are pointed to in the CLD, namely, control of combustion conditions, after-treatment and removal, and optimization emissions standards. These interventions can reduce $NO_x$, but they have downsides too: control of combustion conditions may affect performance; after-treatment and removal may increase cost; optimization emissions standards require extensive testing of performance of new engineered boilers, and engines before deployment to be approved by the new standards [26]. Considering the aforementioned is of paramount importance, as it may have an effect on hydrogen demand in the future.

The findings indicate that the range of global hydrogen demand is from 73 to 158 Mt in 2030, 73 to 300 in 2040, and 73 to 568 Mt in 2050, depending on the scenario presented. Revealing and defining these scenarios will help countries to leverage global energy opportunities. One of these countries is Australia, which is planning to become a global player in hydrogen energy. According to our findings, hydrogen demand could reach 21 Mt for domestic use, if Australia adopted hydrogen in its fields (100% of heavy and long-distance transport, 100% of energy-intensive manufacturing, 100% of mining, and 20% of grid electrical supply as a backup capacity); 34 Mt for export, if replacing gas exports with hydrogen in the future. Australia is one of the world's biggest exporters of coal and natural gas, and it is looking at replacing fossil fuel exports with an alternative; this alternative could not be better than hydrogen. It is expected that, by 2030 and 2040, hydrogen will be one of the preferred technologies [72,73].

## 8. Conclusions

Adoption of hydrogen energy as a future fuel requires a comprehensive understanding of its components and interactions, and intervention scenarios that may change system behaviour. This is the domain of systems thinking and system dynamics approaches. We used systems thinking and system dynamics approaches to examine the behaviour and trends of hydrogen energy through different possible scenarios. A systems thinking approach was used to build a conceptual model for hydrogen energy, with a special focus on the pathways of hydrogen use, to assess the potential consequences that could lead to unintended impacts and possible interventions. A system dynamics approach was used to highlight the growth of the possible global demand of hydrogen energy until 2050, and it has been linked to Australia's potential hydrogen demand in order to leverage global energy opportunities.

Findings indicate that hydrogen global demand is rising and ranges from 73 Mt (pessimistic case) to 568 Mt (optimistic case) by 2050. For countries seeking to be leaders in hydrogen production like Australia, this is motivating news, as adopting hydrogen for some sectors (heavy and long-distance transport, energy-intensive manufacturing, mining, and grid electrical supply) and replacing fossil fuel (mainly gas) with hydrogen, for export, means that the Australian hydrogen demand could reach 55 Mt.

This study provides insights and information on the future of hydrogen energy demand and highlights the risks that may affect the sector, which may assist policy decision makers to understand and test their options for the transition to a hydrogen economy, which may enable more effective decisions and/or policy changes to obtain much better outcomes and avoid undesirable ones. Regarding limitations of the work, energy demand growth has been taken as an exogenous variable, although other studies may take it as an endogenous variable. Further, future studies can implement the model in other countries, as the proposed model can be used in any country. There are significant benefits for the transition to a hydrogen economy. However, the supply side includes many variables, which will be a focus of the next paper.

**Author Contributions:** Conceptualization, T.Y. and W.A.; methodology, K.K. and D.R.; software, M.L.; validation, H.A.D. and M.K.K.; formal analysis, B.Y. All authors have read and agreed to the published version of the manuscript.

**Funding:** This research received no external funding.

**Institutional Review Board Statement:** Not applicable.

**Informed Consent Statement:** Not applicable.

**Data Availability Statement:** Not applicable.

**Conflicts of Interest:** The authors declare no conflict of interest.

## Appendix A

**Table A1.** Parameters used in the simulation model. The model produces behaviours over time graphs from 1990 to 2050. The software used is Silico (https://silico.app/).

| **Australia Hydrogen Demand** | | |
|---|---|---|
| **Variable Name** | **Units** | **Parameter Value** |
| Transport Annual growth | PJ/year | Real/actual values (historical values) from 1990 to 2019. After 2019 no growth is assumed |
| Transport | PJ | Accumulation of transport Annual growth |
| Heavy and long-distance transport | | "Transport"*0.2 |
| Manufacturing annual growth | PJ/year | Real/actual values (historical values) from 1990 to 2019. After 2019 no growth is assumed |
| Energy-intensive manufacturing | PJ | Accumulation of manufacturing annual growth |
| Mining annual growth | PJ/year | Real/actual values (historical values) from 1990 to 2019. After 2019 no growth is assumed |
| Mining | PJ | Accumulation of mining annual growth |
| Export annual growth | PJ/year | Real/actual values (historical values) from 1990 to 2019. After 2019 no growth is assumed |
| Export | PJ | Accumulation of export annual growth |
| Electricity annual growth | PJ/year | Real/actual values (historical values) from 1990 to 2019. After 2019 no growth is assumed |
| Electricity supply | PJ | Accumulation of electricity annual growth |
| Hydrogen as a back-up capacity for grid electrical supply | PJ | "Electricity supply"*0.2 |
| Total Hydrogen demand (PJ) | PJ | "Export" + "Heavy and long-distance transport" + "Hydrogen as a back-up capacity for grid electrical supply" + "Energy-intensive manufacturing" + "Mining" |
| Total Hydrogen demand (KWh) | KWh | "Total Hydrogen demand (PJ)"*277800000 |
| Total Hydrogen demand (Mt) | Mt | ("Total Hydrogen demand (KWh)"/33.33)/1000000000 |
| **Global Hydrogen demand** | | |
| Annual growth for 1st scenario | Mt/year | Real/actual values (historical values) from 1990 to 2019. After 2019 no growth is assumed |
| No growth (0%) | Mt | Accumulation of annual growth for 1st scenario |
| Annual growth for 2nd scenario | Mt/year | Real/actual values (historical values) from 1990 to 2019. After 2019 1.8% growth is assumed |
| Current growth (1.8%) | Mt | Accumulation of annual growth for 2nd scenario |
| Annual growth for 3rd scenario | Mt/year | Real/actual values (historical values) from 1990 to 2019. After 2019 2.5% growth is assumed |
| Average actual values growth (2.5%) | Mt | Accumulation of annual growth for 3rd scenario |
| Annual growth for 4th scenario | Mt/year | Real/actual values (historical values) from 1990 to 2019. After 2019 6.5% growth is assumed |
| Best case (6.5%) | Mt | Accumulation of annual growth for 4th scenario |

## References


1. Zohuri, B. Hydrogen energy. In *Cryogenics and Liquid Hydrogen Storage: Challenges and Solutions for a Cleaner Future*; Springer: New York, NY, USA, 2019; pp. 121–139.
2. Abe, J.O.; Popoola, A.; Ajenifuja, E.; Popoola, O. Hydrogen energy, economy and storage: Review and recommendation. *Int. J. Hydrogen Energy* **2019**, *44*, 15072–15086. [CrossRef]
3. Weger, L.B.; Leitao, J.; Lawrence, M.G. Expected impacts on greenhouse gas and air pollutant emissions due to a possible transition towards a hydrogen economy in German road transport. *Int. J. Hydrogen Energy* **2021**, *46*, 5875–5890. [CrossRef]
4. Atilhan, S.; Park, S.; El-Halwagi, M.M.; Atilhan, M.; Moore, M.; Nielsen, R.B. Green hydrogen as an alternative fuel for the shipping industry. *Curr. Opin. Chem. Eng.* **2021**, *31*, 100668. [CrossRef]
5. Yang, J.-H.; Yoon, Y.; Ryu, M.; An, S.-K.; Shin, J.; Lee, C.-J. Integrated hydrogen liquefaction process with steam methane reforming by using liquefied natural gas cooling system. *Appl. Energy* **2019**, *255*, 113840. [CrossRef]
6. Li, J.; Cheng, W. Comparative life cycle energy consumption, carbon emissions and economic costs of hydrogen production from coke oven gas and coal gasification. *Int. J. Hydrogen Energy* **2020**, *45*, 27979–27993. [CrossRef]
7. Lamy, C.; Millet, P. A critical review on the definitions used to calculate the energy efficiency coefficients of water electrolysis cells working under near ambient temperature conditions. *J. Power Sources* **2020**, *447*, 227350. [CrossRef]
8. Tarhan, C.; Çil, M.A. A study on hydrogen, the clean energy of the future: Hydrogen storage methods. *J. Energy Storage* **2021**, *40*, 102676. [CrossRef]
9. Vinoth Kanna, I.; Paturu, P. A study of hydrogen as an alternative fuel. *Int. J. Ambient Energy* **2020**, *41*, 1433–1436. [CrossRef]
10. Hosseini, S.E.; Butler, B. An overview of development and challenges in hydrogen powered vehicles. *Int. J. Green Energy* **2020**, *17*, 13–37. [CrossRef]
11. Dincer, I.; Acar, C. Innovation in hydrogen production. *Int. J. Hydrogen Energy* **2017**, *42*, 14843–14864. [CrossRef]
12. Hosseini, S.E.; Wahid, M.A. Hydrogen production from renewable and sustainable energy resources: Promising green energy carrier for clean development. *Renew. Sustain. Energy Rev.* **2016**, *57*, 850–866. [CrossRef]
13. Acar, C.; Beskese, A.; Temur, G.T. Sustainability analysis of different hydrogen production options using hesitant fuzzy AHP. *Int. J. Hydrogen Energy* **2018**, *43*, 18059–18076. [CrossRef]
14. Niaz, S.; Manzoor, T.; Pandith, A.H. Hydrogen storage: Materials, methods and perspectives. *Renew. Sustain. Energy Rev.* **2015**, *50*, 457–469. [CrossRef]
15. Zhang, F.; Zhao, P.; Niu, M.; Maddy, J. The survey of key technologies in hydrogen energy storage. *Int. J. Hydrogen Energy* **2016**, *41*, 14535–14552. [CrossRef]
16. Ren, J.; Musyoka, N.M.; Langmi, H.W.; Mathe, M.; Liao, S. Current research trends and perspectives on materials-based hydrogen storage solutions: A critical review. *Int. J. Hydrogen Energy* **2017**, *42*, 289–311. [CrossRef]
17. Nagpal, M.; Kakkar, R. An evolving energy solution: Intermediate hydrogen storage. *Int. J. Hydrogen Energy* **2018**, *43*, 12168–12188. [CrossRef]
18. Acar, C.; Dincer, I. Review and evaluation of hydrogen production options for better environment. *J. Clean. Prod.* **2019**, *218*, 835–849. [CrossRef]
19. Li, F.; Yuan, Y.; Yan, X.; Malekian, R.; Li, Z. A study on a numerical simulation of the leakage and diffusion of hydrogen in a fuel cell ship. *Renew. Sustain. Energy Rev.* **2018**, *97*, 177–185. [CrossRef]
20. Shen, Y.; Zheng, T.; Lv, H.; Zhou, W.; Zhang, C. Numerical Simulation of Hydrogen Leakage from Fuel Cell Vehicle in an Outdoor Parking Garage. *World Electr. Veh. J.* **2021**, *12*, 118. [CrossRef]
21. Gye, H.-R.; Seo, S.-K.; Bach, Q.-V.; Ha, D.; Lee, C.-J. Quantitative risk assessment of an urban hydrogen refueling station. *Int. J. Hydrogen Energy* **2019**, *44*, 1288–1298. [CrossRef]
22. Chen, J.; Gong, Y.; Jiang, T.-H.; Pan, A.-X.; Wang, S.-H.; Yang, Z.-G. Failure analysis on abnormal leakage of TP321 stainless steel pipe of medium temperature shifting gas in hydrogen production system. *Eng. Fail. Anal.* **2021**, *125*, 105413. [CrossRef]
23. Moradi, R.; Groth, K.M. Hydrogen storage and delivery: Review of the state of the art technologies and risk and reliability analysis. *Int. J. Hydrogen Energy* **2019**, *44*, 12254–12269. [CrossRef]
24. Li, W.; Kou, H.; Zeng, X.; Cui, Y.; Chen, H.; Wang, F. Numerical simulations on the leakage and diffusion of tritium. *Fusion Eng. Des.* **2020**, *159*, 111749. [CrossRef]
25. Ingenito, A.; Agresta, A.; Andriani, R.; Gamma, F. NOx reduction strategies for high speed hydrogen fuelled vehicles. *Int. J. Hydrogen Energy* **2015**, *40*, 5186–5196. [CrossRef]
26. Lewis, A.C. Optimising air quality co-benefits in a hydrogen economy: A case for hydrogen-specific standards for NOx emissions. *Environ. Sci. Atmos.* **2021**, *1*, 201–207. [CrossRef]
27. Wang, D.; Ji, C.; Wang, S.; Yang, J.; Wang, Z. Numerical study of the premixed ammonia-hydrogen combustion under engine-relevant conditions. *Int. J. Hydrogen Energy* **2021**, *46*, 2667–2683. [CrossRef]
28. Yu, Q.; Hendrix, Y.; Lorencik, S.; Brouwers, H. Field study of NOx degradation by a mineral-based air purifying paint. *Build. Environ.* **2018**, *142*, 70–82. [CrossRef]
29. Ni, J.; Johnson, N.; Ogden, J.M.; Yang, C.; Johnson, J. Estimating hydrogen demand distribution using geographic information systems (GIS). In Proceedings of the National Hydrogen Association (NHA) Annual Hydrogen Conference, Entitled "Partnering for the Global Hydrogen Future", Washington, DC, USA, 29 March–1 April 2005.

30. Melendez, M.; Milbrandt, A. *Regional Consumer Hydrogen Demand and Optimal Hydrogen Refueling Station Siting*; National Renewable Energy Lab. (NREL): Golden, CO, USA, 2008.
31. Rahmouni, S.; Settou, N.; Negrou, B.; Gouareh, A. GIS-based method for future prospect of hydrogen demand in the Algerian road transport sector. *Int. J. Hydrogen Energy* **2016**, *41*, 2128–2143. [CrossRef]
32. Carr, S.; Premier, G.C.; Guwy, A.J.; Dinsdale, R.M.; Maddy, J. Hydrogen storage and demand to increase wind power onto electricity distribution networks. *Int. J. Hydrogen Energy* **2014**, *39*, 10195–10207. [CrossRef]
33. Kim, J.; Lee, Y.; Moon, I. Optimization of a hydrogen supply chain under demand uncertainty. *Int. J. Hydrogen Energy* **2008**, *33*, 4715–4729. [CrossRef]
34. Burdon, R.; Palmer, G.; Chakraborty, S. *National Hydrogen Strategysubmission*; Energy Transition Hub: Canberra, Australia, 2019.
35. Feitz, A.J.; Coghlan, R.; Tenthorey, E. *Prospective Hydrogen Production Regions of Australia*; Geoscience: Symonston, Australia, 2019.
36. Low, D.; Rodger, A.; Gallagher, B.; Sharma, P. Is hydrogen Asia's new energy commodity? Opportunities and challenges for Australia. *APPEA J.* **2020**, *60*, 468–472. [CrossRef]
37. Laimon, M.; Mai, T.; Goh, S.; Yusaf, T. Energy Sector Development: System Dynamics Analysis. *Appl. Sci.* **2020**, *10*, 134. [CrossRef]
38. Witkowski, A.; Rusin, A.; Majkut, M.; Stolecka, K. Analysis of compression and transport of the methane/hydrogen mixture in existing natural gas pipelines. *Int. J. Press. Vessel. Pip.* **2018**, *166*, 24–34. [CrossRef]
39. Gondal, I.A. Hydrogen integration in power-to-gas networks. *Int. J. Hydrogen Energy* **2019**, *44*, 1803–1815. [CrossRef]
40. Timmerberg, S.; Kaltschmitt, M. Hydrogen from renewables: Supply from North Africa to Central Europe as blend in existing pipelines–Potentials and costs. *Appl. Energy* **2019**, *237*, 795–809. [CrossRef]
41. Altarazi, Y.S.; Talib, A.R.A.; Gires, E.; Yu, J.; Lucas, J.; Yusaf, T. Performance and exhaust emissions rate of small-scale turbojet engine running on dual biodiesel blends using Gasturb. *Energy* **2021**, *232*, 120971. [CrossRef]
42. Shadidi, B.; Najafi, G.; Yusaf, T. A Review of Hydrogen as a Fuel in Internal Combustion Engines. *Energies* **2021**, *14*, 6209. [CrossRef]
43. Dawood, F.; Anda, M.; Shafiullah, G. Hydrogen production for energy: An overview. *Int. J. Hydrogen Energy* **2020**, *45*, 3847–3869. [CrossRef]
44. Mulder, M.; Perey, P.; Moraga, J.L. Outlook for a Dutch hydrogen market. *Univerity Gron.* **2019**, *16*, 2019. Available online: https://www.rug.nl/ceer/blog/ceer_policypaper_5_web.pdf (accessed on 7 January 2022).
45. Kamarulzaman, M.K.; Hafiz, M.; Abdullah, A.; Chen, A.F.; Awad, O.I. Combustion, performances and emissions characteristics of black soldier fly larvae oil and diesel blends in compression ignition engine. *Renew. Energy* **2019**, *142*, 569–580. [CrossRef]
46. Dicks, A.L.; Rand, D.A. *Fuel Cell Systems Explained*; John Wiley & Sons: Hoboken, NJ, USA, 2018.
47. Chen, Y.-T.; Hsu, C.-W. The key factors affecting the strategy planning of Taiwan's hydrogen economy. *Int. J. Hydrogen Energy* **2019**, *44*, 3290–3305. [CrossRef]
48. Wang, Y.; Huang, R.; Liu, Z. The Relationship between Structural Intensity and Sound Field Characteristics of Cylindrical Shells. *Procedia Eng.* **2017**, *214*, 41–49. [CrossRef]
49. Eftekhari, A.; Fang, B. Electrochemical hydrogen storage: Opportunities for fuel storage, batteries, fuel cells, and supercapacitors. *Int. J. Hydrogen Energy* **2017**, *42*, 25143–25165. [CrossRef]
50. Sadhasivam, T.; Kim, H.-T.; Jung, S.; Roh, S.-H.; Park, J.-H.; Jung, H.-Y. Dimensional effects of nanostructured $Mg/MgH_2$ for hydrogen storage applications: A review. *Renew. Sustain. Energy Rev.* **2017**, *72*, 523–534. [CrossRef]
51. *Hydrogen Energy and Fuel Cells: A Vision of Our Future: Final Report of the High Level Group*; European Commission, Directorate-General for Research: Brussel, Belgium, 2003; Volume 20719, Available online: https://www.fch.europa.eu/sites/default/files/documents/hlg_vision_report_en.pdf (accessed on 27 December 2021).
52. Durbin, D.J.; Malardier-Jugroot, C. Review of hydrogen storage techniques for on board vehicle applications. *Int. J. Hydrogen Energy* **2013**, *38*, 14595–14617. [CrossRef]
53. Parra, D.; Valverde, L.; Pino, F.J.; Patel, M.K. A review on the role, cost and value of hydrogen energy systems for deep decarbonisation. *Renew. Sustain. Energy Rev.* **2019**, *101*, 279–294. [CrossRef]
54. Dias, V.; Pochet, M.; Contino, F.; Jeanmart, H. Energy and economic costs of chemical storage. *Front. Mech. Eng.* **2020**, *6*, 21. [CrossRef]
55. Mah, A.X.Y.; Ho, W.S.; Bong, C.P.C.; Hassim, M.H.; Liew, P.Y.; Asli, U.A.; Kamaruddin, M.J.; Chemmangattuvalappil, N.G. Review of hydrogen economy in Malaysia and its way forward. *Int. J. Hydrogen Energy* **2019**, *44*, 5661–5675. [CrossRef]
56. Robledo, C.B.; Oldenbroek, V.; Abbruzzese, F.; van Wijk, A.J.M. Integrating a hydrogen fuel cell electric vehicle with vehicle-to-grid technology, photovoltaic power and a residential building. *Appl. Energy* **2018**, *215*, 615–629. [CrossRef]
57. Acar, C.; Dincer, I. Investigation of a novel photoelectrochemical hydrogen production system. *Chem. Eng. Sci.* **2019**, *197*, 74–86. [CrossRef]
58. Lyseng, B.; Niet, T.; English, J.; Keller, V.; Palmer-Wilson, K.; Robertson, B.; Rowe, A.; Wild, P. System-level power-to-gas energy storage for high penetrations of variable renewables. *Int. J. Hydrogen Energy* **2018**, *43*, 1966–1979. [CrossRef]
59. Oldenbroek, V.; Verhoef, L.A.; van Wijk, A.J.M. Fuel cell electric vehicle as a power plant: Fully renewable integrated transport and energy system design and analysis for smart city areas. *Int. J. Hydrogen Energy* **2017**, *42*, 8166–8196. [CrossRef]
60. Badea, G.; Naghiu, G.S.; Giurca, I.; Aşchilean, I.; Megyesi, E. Hydrogen Production Using Solar Energy—Technical Analysis. *Energy Procedia* **2017**, *112*, 418–425. [CrossRef]

61. Simonis, B.; Newborough, M. Sizing and operating power-to-gas systems to absorb excess renewable electricity. *Int. J. Hydrogen Energy* **2017**, *42*, 21635–21647. [CrossRef]
62. Olivier, P.; Bourasseau, C.; Bouamama, P.B. Low-temperature electrolysis system modelling: A review. *Renew. Sustain. Energy Rev.* **2017**, *78*, 280–300. [CrossRef]
63. Australia's National Hydrogen Strategy. 2019. Available online: https://www.h2knowledgecentre.com/content/government622 (accessed on 27 December 2021).
64. Szargut, J.; Morris, D.R.; Steward, F.R. *Exergy Analysis of Thermal, Chemical, and Metallurgical Processes*; US Department of Energy: Washington, DC, USA, 1987.
65. Kalinci, Y.; Dincer, I.; Hepbasli, A. Energy and exergy analyses of a hybrid hydrogen energy system: A case study for Bozcaada. *Int. J. Hydrogen Energy* **2017**, *42*, 2492–2503. [CrossRef]
66. Australian Government; Department of Industry, Science, Energy and Resources. Australian Energy Statistics. Available online: https://www.energy.gov.au/government-priorities/energy-data/australian-energy-statistics (accessed on 31 December 2021).
67. MacKay, D. *Sustainable Energy—Without the Hot Air*; UIT Cambridge: London, UK, 2008.
68. Silico. Available online: https://www.silicoai.com/ (accessed on 29 December 2021).
69. Aziz, M.; Wijayanta, A.T.; Nandiyanto, A.B.D. Ammonia as effective hydrogen storage: A review on production, storage and utilization. *Energies* **2020**, *13*, 3062. [CrossRef]
70. Wijayanta, A.T.; Aziz, M. Ammonia production from algae via integrated hydrothermal gasification, chemical looping, $N_2$ production, and $NH_3$ synthesis. *Energy* **2019**, *174*, 331–338. [CrossRef]
71. Guo, H.; Zhou, S.; Zou, J.; Shreka, M. A Numerical Investigation on De-NOx Technology and Abnormal Combustion Control for a Hydrogen Engine with EGR System. *Processes* **2020**, *8*, 1178. [CrossRef]
72. Steyn, J. Hydrogen for Vehicular Transport. In Proceedings of the AIIT 2nd International Congress on Transport Infrastructure and Systems in a changing world (TIS ROMA 2019), Rome, Italy, 23–24 September 2019.
73. Horvath, S.; Fasihi, M.; Breyer, C. Techno-economic analysis of a decarbonized shipping sector: Technology suggestions for a fleet in 2030 and 2040. *Energy Convers. Manag.* **2018**, *164*, 230–241. [CrossRef]

water

MDPI

Article

# Experimental Analysis of a Fish Guidance System for a River Water Intake

**Gabriela Cîrciumaru [1,*], Rareș-Andrei Chihaia [1], Andreea Voina [1], Daniela-Elena Gogoașe Nistoran [2], Ștefan-Mugur Simionescu [2], Lucia-Andreea El-Leathey [1] and Lucian Mândrea [2]**

1 National Institute for R&D in Electrical Engineering ICPE-CA, 030138 Bucharest, Romania; rares.chihaia@icpe-ca.ro (R.-A.C.); andreea.voina@icpe-ca.ro (A.V.); andreea.elleathey@icpe-ca.ro (L.-A.E.-L.)

2 Faculty of Energy Engineering, University Politehnica of Bucharest, 060042 Bucharest, Romania; daniela.nistoran@upb.ro (D.-E.G.N.); s.simionescu@upb.ro (Ș.-M.S.); lucian.mandrea@upb.ro (L.M.)

* Correspondence: gabriela.circiumaru@icpe-ca.ro; Tel.: +40-721-409-547

**Abstract:** This paper describes the experimental study of a behavioural barrier consisting of a bubble curtain placed near an ecological river water intake, aiming to assess the bubble curtain's influence on the velocity field of water flowing in the water intake proximity. The scientific novelty is given by the water intake solution operating in tandem with the dedicated bubble curtain generator used for avoiding fish entrapment. The water intake solution used in the present research is based on a run-of-river intake without a dam in the riverbed, maintaining the mountain rivers longitudinal connectivity. An experimental setup was designed and built to study the behavioural barrier for ecological water intakes in different operating conditions. Thus, tests were performed for two water velocities (0.33 m/s and 0.535 m/s) in the experimental setup, two water depths (92 mm and 119 mm) and two air flow rates (10.5 LPM and 15 LPM). The results were compared with the case when no air is injected through the bubble curtain generator. The assessment of the bubble curtain influence on the water flow inside the channel was made by measuring and analysing the velocity variation in vertical and horizontal direction. The results showed a significant velocity gradient in the bubble curtain proximity, which can contribute to influencing the behaviour of the fish by deterring them from the water intake area.

**Keywords:** bubble curtain; fish guidance system; water intake; fish protection

**Citation:** Cîrciumaru, G.; Chihaia, R.-A.; Voina, A.; Gogoașe Nistoran, D.-E.; Simionescu, Ș.-M.; El-Leathey, L.-A.; Mândrea, L. Experimental Analysis of a Fish Guidance System for a River Water Intake. *Water* **2022**, *14*, 370. https://doi.org/10.3390/w14030370

Academic Editors: Dibyendu Sarkar, Rupali Datta, Prafulla Kumar Sahoo and Mohammad Mahmudur Rahman

Received: 9 December 2021
Accepted: 23 January 2022
Published: 26 January 2022



## 1. Introduction

There are many issues that put the development of society in conflict with the sustainability and conservation of the environment, and one of them is given by the need to use water sources. The two aspects—the use of water resources and the conservation of the environment—must be approached simultaneously by solutions that meet specific situational needs. In recent years, the scientific community has begun to study the behavioural characteristics of native fish species in mountain rivers. Lately, rivers have undergone profound transformations in order to be used for economic and social activities. The general decline of fish populations and aquatic fauna specific to watercourses, or even the extinction of some species, are indicators of environmental degradation. This decline is due to both climate change (especially through changes in the hydrological and thermal regime of the watercourses [1]) and anthropogenic factors; it must be limited by all possible and responsible means. Freshwater species registered a significant decline in the last 50 years. The Living Planet Index, which measures trends in thousands of populations of mammals, birds, reptiles, amphibians and fish across the globe, proposed by the World Wide Fund for Nature [2], shows a decrease of 81% between 1970 and 2012. Among the anthropogenic factors that negatively influence the fishing population of watercourses, the following are worth mentioning: excessive water exploitation, artificialisation of rivers through non-integrated arrangements in the environment, industrialisation, intensive fishing, etc. [3].

These actions significantly impact fish population, the main threats being represented by habitat loss and degradation [2]. Aquatic biodiversity loss is mainly caused by the interruption of longitudinal connectivity due to the presence of hydropower facilities, which block the stream, and to the lack of passageways for migratory species. There are enough known cases where morphological changes of rivers used for economic reasons affected their ecological functions [4]. In the second half of the twentieth century, the hydrotechnical constructions across watercourses have significantly interrupted the longitudinal continuity of many rivers, causing a change in the natural hydrological regime and preventing the migration of fish to areas suitable for wintering, breeding, rearing or feeding. This effect on ecosystems must be taken into consideration, bearing in mind that, according to Grill et al. [5], 48% from the total river volume is subject to flow regulation and/or fragmentation. The longitudinal hydrotechnical constructions (piers) prevented the lateral continuity of the river, as the flooding of the meadows could not take place and the ecological exchange of fertile material and aquatic flora/fauna between the rivers and wetlands in the major riverbeds could not be performed; this changed the hydrological regime [1], the morphology of the riverbed [6–8] and the structure of its bed substrate [8–10].

Another negative impact on fish is created by the operation of water intakes located on rivers. Water intakes contain diversion or bypass installations/constructions for directing and catching river water. The choice of arrangements to ensure the exclusion of fish and their efficiency depends largely on the characteristics of the fish species in the targeted river sections and are related to size, stages of life and reproduction, behaviour and preferred routes. If it is desired to exclude small fish, the slots of the grills must be narrowed and the access speeds reduced to prevent their suction. The screen is the main element of the water intake; it provides protection against the access of solid materials, fish, other impurities and floating debris. From a constructive point of view, the screen consists of metal bars, vertical or inclined at a certain angle, with a rectangular cross section, connected in panels. For metal or reinforced concrete strength beams, a hydrodynamic shape is generally preferred. To avoid fish entrapment, the screen openings must be reduced as much as possible. Often, this is not possible due to the local head losses increase which is proportional to the water velocities; in addition, the risk of clogging also increases. Furthermore, if the screen is not properly designed, the fish can enter through the screen openings into pipelines/channels /penstocks. Thus, for fish protection, other means of deviation should be applied.

The most important provisions on environmental law were imposed by the Water Framework Directive. Under these provisions, rivers must be protected, enhanced and rebuilt in order to achieve both good status of surface water and also a good ecological potential. Barring work of watercourses must be provided with facilities to ensure the required flow downstream, and, where appropriate, special facility for fish migration [11].

Since the 1980s, different solutions for fish screening on water intakes have been adopted. Studies have led to the elaboration of a best practice guide in the UK, in 2005, [12], regarding the use of fish guidance systems. In order to protect fish populations, both from accidentally entrapment in water intakes/diversions and from invasive species, several solutions have been proposed and used. These solutions fall into two broad categories [13]: direct means of excluding fish—physical barriers or positive barriers screens (specially designed screens or sieves, fixed or rotating)—and indirect solutions which influence the behaviour of aquatic population—non-physical barriers or behavioural barriers (designed to keep fish away from the intake area and based on bubble curtains, strobe light, sound, electricity, magnetic fields or by changing direction and velocity of the water flow). The non-physical guidance systems (behavioural barriers) are based on the natural behaviours of fish—of attraction in the fish ladders case or repelling in the water intakes/catchments case.

The water intake solution used in the present research is based on a run-of-river intake without a dam in the riverbed, maintaining longitudinal connectivity of the mountain rivers. In addition, compared to traditional physical systems for fish guidance, the proposed behavioural barrier has the advantage to easily adapt to the flowing conditions of the river. Taking into consideration that each fish species has specific characteristics/features (different size, shape and swimming capabilities, different development stages, habits and life cycle, different features—hearing specialists or generalists respond to certain stimuli, etc.), studies have been undertaken to determine the effectiveness of several types of barriers/configurations in guiding or deterring fish in various operation conditions. This paper approaches the experimental study of a behavioural barrier consisting of a bubble curtain placed near an ecological water intake (perforated plate screen) to evaluate the results obtained and make recommendations for large-scale use. The scientific novelty is given by the water intake solution operating in tandem with a dedicated bubble curtain generator which helps to avoid accidental entrapment of fish fauna.

## 2. Fish Guidance Systems Based on the Use of Non-Physical Barriers

Fish guidance systems have been used over the years to discourage migration and to direct species of interest away from anthropogenic hazards. For example, non-physical guidance/deterrence systems have been used to target commercial or recreational fish around facilities such as hydropower dams, which may threaten their survival, as shown by Taft [14]. There is a critical need for methods and technologies to stop the spread of exotic organisms through natural and human-mediated means, especially in Canada and in the US Great Lakes region (and not limited to these areas), as the area has been invaded by more than 180 invasive species in the last 200 years, according to the US National Oceanic and Atmospheric Administration [15] and to Mills et al. [16]. One method consists in the placement of barriers to intervene in the movement of species. Unlike most terrestrial fauna, fish movement can be discouraged both by physical obstacles placed on waterways and by non-physical methods that change their behaviour. According to Noatch and Suski [17], non-physical deterrence can be defined as "any stimulus or non-solid obstruction that discourages or prevents a selected species from passing through a target region".

Thus, the behavioural barriers represent alternative solutions to the classical barriers to guide fish. Current technologies are not suitable for locations with shallow water depths and rapid change level. As reported by Zielinski 2011 [18], for such locations, bubble curtains may be the right solution. The bubbles generate distinct acoustic and hydrodynamic fields, and by proper handling, could be used to guide juvenile fish.

According to the U.S. Department of Interior [13], the main advantages of behavioural systems are the following: they are easy, safe and cheap to maintain, can also be applied in locations where it would be difficult to install screens; in addition, they are environmentally friendly and do not hurt or kill fish. The disadvantages of behavioural systems include that they do not create an absolute barrier to exclusion and their efficiency varies with the species, fish development stage, environmental conditions (flow and variability), water quality and lighting.

In order to reduce the operating costs, any barrier installed on a watercourse has to not retain the floating debris (to avoid clogging) or to require frequent maintenance. In addition, the remote locations where they are placed restrict the access to the available power supply infrastructure. In addition, the barrier must be flexible to the changes in the water level and flow in order to cope with seasonal floods. There are several guidance systems for fish which are based on stimuli that most fish are able to detect: sound waves, light, chemical stimuli, pressure waves and electricity. As reported by Taylor et al. [19], there are also devices that combine several types of stimuli. Combined barrier systems require more space to place all the equipment and can be difficult to install on shallow watercourses.

Given that behavioural barriers are based on the use of different types of stimuli, this paragraph provides a brief description of the main types of barriers. Electric barriers consist of a series of metal anodes and cathodes placed in water. Electric current passes through the water from an anode to a cathode creating an electric field near the barrier, and their efficiency in guiding fish is due to the behavioural avoidance of electric fields. Electric barriers have two major limitations: they can be disabled by power failure, maintenance (or lack thereof) and human error, and they are not effective for small fish, which are not affected by electric fields. Thus, additional guidance systems are recommended. In the case of visual barriers (strobe lights), ambient light levels influence fish behaviour, including orientation, food location, communication between species, circadian movements and predator avoidance. Strobe lights introduce unnatural light levels relative to the environment, with the ability to induce an avoidance response in fish. For increased efficiency, they should be used as part of an integrated deterrent system (e.g., bubble curtains). Acoustic fish guidance systems are applied when in situ conditions exclude the use of visual stimuli to induce fish avoidance behaviours; thus, sound and pressure waves may be a viable alternative. Sound generated in the water consists of pressure waves that can be detected by fish. The effectiveness of acoustic systems in fish guiding can be influenced by bottom morphology, hydrology and sound waves angle. Low-frequency sound waves do not propagate well in shallow water and on hard substrates. Velocity barriers consist of changing flow regimes within a watercourse or channel so that the water velocity exceeds the swimming capacity of the target species and can only be implemented on a small scale. When dealing with running water, all fish have inherent physiological limitations; thus, high water velocity acts as a barrier to fish movement. For hypoxia and hypercapnia, the method consists of using toxic chemicals to exclude aquatic organisms from specific locations by creating hypoxic/hypercapnic barriers. To ensure efficiency, tolerance thresholds for target and non-target fish must be first quantified. Dissolved gas barriers have the potential to be effective in all sizes of fish once a certain threshold is reached (e.g., DO level of approximately 1.5 mg/L [20]). Pheromones are generally defined as secreted chemical odours that elicit a specific behavioural response in living things of the same species. As a potential chemical barrier for fish, pheromones could be collected or synthesised and released into the water column to drive fish away from a particular location, or to gather fish away from a danger source or an enclosed passageway. The effectiveness of the utilisation of electromagnetic fields barriers is based on the fact that many aquatic organisms, including cartilaginous fish and neopterygian bone fish, have electroreceptive organs capable to detect the electric fields in water. Thus, strong magnetic fields can overstimulate these receptors, causing a deterrent or repellent effect, which can be useful in directing organisms away from their target locations. Behavioural barriers based on the use of bubble curtains are a relatively low cost and maintenance option that do not interfere with the watercourse morphology. They combine two stimuli—velocity and sound—and could be exploited to discourage the migration of fish or their entrapment in river water intakes. The bubble curtain is emitted by air diffusers, perforated pipes, or porous hoses placed along the bottom of the watercourse, perpendicular to the channel or river, respectively. It acts like a continuous screen in the water body and makes an unusual visual effect that fish generally avoid. As bubble curtains do not emit their own light, they should operate in water with low turbidity; in addition, in order to increase their efficiency and to be easily seen from distance, they should operate along with a light source.

First study on non-migratory freshwater fish entrapment was performed by Hadderingh in 1982 [21], reporting intake rates of up to 25 million juveniles per day at a Dutch power plant (Bergum) in Friesland. The fish were entering through the screens of the cooling water inlets especially in no-light conditions. Thus, in order to help fish to orientate and to diminish their entrapment in the intake area, it was artificially illuminated. This led to a significant decrease in the intake rate, especially for ruffe (*Gymnocephalus cernua*) and perch (*Perca fluviatilis*). The interest in using artificial illumination or other means for diminishing the fish entrapment rates has grown steadily worldwide, leading the experts to publish a series of papers or research reports [12,13,21–37] which deal with the effectiveness of different fish guidance systems, either by performing laboratory studies or by evaluation in situ. Besides the intensive fishing, the implementation of water diversions led to the emergence of numerous endangered fish species. The increased intensity of the modern mobility of freight and people, that through the ballast water carried by vessels constantly introduced and spread invasive fish species in different environments, also contributed to this situation. In these conditions, the importance of fish protection at water intakes, small dams, and water diversion structures became significant. This led to the elaboration of guides for fish exclusion facilities. For example, the U.S. Department of the Interior [13] approached a series of fish protection-related subjects such as legislation, designing/choosing fish exclusion facilities depending on the target fish characteristics (species, size, life stage, behaviour, swimming ability), analysis of barriers type (classification, constructive characteristics, velocity/flow-related characteristics, maintenance, advantages and disadvantages, etc.).

Blake et al. [22] performed research concerning the use of a sound–bubble–strobe light (SBSL) barrier to limit the invasion of Asian carp (bighead carp, *Hypophthalmichthys nobilis*, and silver carp, *Hypophthalmichthys molitrix*) in the Mississippi River Basin. These carps influenced the native fishes, especially in Illinois River. A 16 m wide SBSL has been deployed perpendicular to the flow of Quiver Creek. The water flow velocities ranged from 0.4 m/s to 0.8 m/s, while the maximum water depth was 1 m. A hose was used to generate air bubbles, and underwater speakers were used to emit sound frequencies between 500 and 2000 Hz and L.E.D. lights to flash intermittently or permanently. The tests were performed in different conditions/configurations of the barrier, namely, using sound, bubbles, and strobe lights or only sound and bubbles. The effectiveness of the barrier ranged from 94.1% to almost 100%, depending on the fish species (e.g., Cyprinidae 96.9%, Percidae 96%, Centrarchidae–Sunfish 97.5%, Ictaluridae 98.9%, Moronidae 100%), suggesting that the SBSL barriers could be used as a deterrent system to repel Asian carp, to prevent its expansion in Mississippi River Basin. Zielinski and Sorensen [24] explored the theory and application of a bubble curtain deflection screen using a split-passage experimental channel equipped with angled bubble curtains to test two Asian carp species (bighead and silver carp) and the Common carp (*Cyprinus carpio*) to see if they avoid bubble curtains. The bubble curtain was created using a S41 regenerative air blower at 5 kPa to supply 12 L/s of air thorough 1 m of water. It was found that the bubble curtains reduced in 73–80% the passage of all three species through the experimental channel while producing sound between 100 and 1000 Hz at 145 dB. As the bubble curtain deterrence system produced the same effects on the three tested Asian carp species, Zielinski and Sorensen [24] concluded that the Common carp could be used instead of Asian carp when performing field experiments in areas where this invasive species is not yet very numerous, like the case of upper Mississippi River. By following this approach, insight on the efficiency of the bubble curtains in inhibiting the movement of Asian carp was obtained, with minimal impact on other fish living in the studied area. Jesus et al. [26] tested the use of acoustic barriers as an acoustic deterrent for native potamodromous migratory fish species (*Salmo trutta, Pseudochondrostoma duriense*, and *Luciobarbus bocagei*) from Iberia. The tests showed that the endemic cyprinids showed a strong repulse reaction to the frequency sweep-up sound. The 140 Hz stimulus did not significantly alter the behaviour of any of the studied species. These results highlight the potential of acoustic stimuli as fish behavioural barriers and can be used selectively on target species. Leander et al. [27] investigated experimentally in laboratory and in situ the efficiency of bubble barriers in diverting migrating Atlantic salmon (*Salmo salar*). Acoustic telemetry was used to measure the interaction of salmon with the bubble barrier. An efficiency between 90% and 95% in diverting the migrating salmon was found. In the literature, studies and reviews also reported on the use of different behavioural barriers [28,38–40]. Zielinski et al. (2019) [28] presented the effectiveness of current and alternative barrier technologies in invasive sea lamprey (*Petromyzon marinus*) control in the Laurentian Great Lake. The invasion started in the early 1900s and negatively affected the fisheries and local economies. Therefore, since the 1950s an integrated pest management program was initiated that led to a reduction in sea lamprey numbers by 90%. The barriers are critical in the effectiveness of this control program, which is one of the most successful programs. The invasion is controlled by using both lampricide to kill the larvae and physical barriers to block the access of adult sea lamprey to spawning areas. There are almost 100,000 barriers that stops/dimmish the movement of fish between lake and tributary on Great Lakes tributaries. Most of these potential barriers to fish movement have been built for other purposes, while approximately 8000 were built or modified to control the sea lamprey. Jones et al. [39] and Jesus et al. [40] conducted a literature search in different databases to identify the reported research concerning the use of barriers in limiting the invasive species and/or in guiding the freshwater fish. The importance and relevance of this issue is reflected by the significant number of articles identified—87 titles in Jones et al. [39] and 123 in Jesus et al. [40]. The performed research concerns the use, characteristics, efficiency, etc. of different type of barriers operating alone or in different combinations.

The studies reported in literature on the use of bubble curtains to guide fish focus on a wide range of species: Asian carp (*Hypophthalmichthys nobilis, Mylopharyngodon piceus, Ctenopharyngodon Idella, Hypophthalmichthys molitrix*) [19], Atlantic salmon, Eurasian ruffe (*Gymnocephalus cernua*), herring (*Alosa pseudoharengus, Dorosoma cepedianum, Brevoortia tyrannus*) [29]. The use of a bubble curtain in combination with a sound source has been studied by Taylor et al. [19] and by Welton et al. [41]. The bubble curtain was generated by injecting air flow rates of 0.1–1 L/s·m through a PVC pipe with holes drilled in it. Experimental results showed a 95% decrease in Asian carp migration [19], respectively, a 20–40% efficiency in guiding smolt (young *Salmo salar*) during the day and more than 70% at night [40]. These studies did not consider the bubble curtain as the primary guidance solution, nor did they analyse in detail the physical fields generated by it. Studies conducted at the site of a hydroelectric plant in Michigan, USA [13], have shown that choosing a mixed method by coupling bubble curtains with strobe lights can increase the efficiency (fish exclusion rate) of the bubble barrier. For example, in the Richard B. Russell project in the United States [13], the use of high-frequency sounds combined with light systems proved to be very effective in deterring herring from the water intakes pumps in safety areas with lower velocities. Several studies indicate higher rates of deterrence when bubble barriers are associated with an additional light source, such as strobe lights [31] or sound [39]. Welton et al. [39] reported that an experimental barrier consisted of a bubble curtain coupled with deterrent acoustic elements has an efficiency of 95% in limiting the movements of the Asian carp. Therefore, other combinations of guidance systems based on the fish behaviour could also be expected to improve their barrier efficiency.

The data in Table 1 have resulted from analysing different options for non-physical deterrence of fish movements [17,18,29,42–46]. It summarises the different non-physical barriers that could be implemented to deter/guide fish movements, the advantages and disadvantages of different types of barriers, as well as the conditions that lead to their successful implementation.

**Table 1.** Types of behavioural barriers (adapted from the works in [17,18,29,42–46]).

| Barrier Type | Implementation Conditions | Description | Advantages | Disadvantages |
|---|---|---|---|---|
| **Electric** | Site with suitable power supply; Suitable water conductivity | Electric current produced by electrodes located on the channel perimeter | It has proven highly effective against upstream migrating fish | Expensive equipment and maintenance; risk to human and animal health; it is not specific to a species |
| **Visual-strobelights** | Low water turbidity | Strobe light with set water flashing frequency | Relatively small influence on water flow; adaptable to flow; in some species, very effective in combination with bubble curtains | It depends very much on the natural light level; cloudy water reduces its efficiency, and this is not proven in all species |

**Table 1.** *Cont.*

| Barrier Type | Implementation Conditions | Description | Advantages | Disadvantages |
|---|---|---|---|---|
| **Acoustics** | Site with suitable acoustic characteristics | Underwater speaker with specified signal | Flexible to different flow conditions; species specific potential; wide range of sounds | Variable efficiency; the frequencies must be selected according to species; expensive equipment that can clog |
| **Bubble curtains** | Low water turbidity, relatively shallow water | Compressed air emitted as bubbles through diffusers | Flexible in different flow conditions; multiple physical stimuli; relatively simple construction; reduced cost. The efficiency increases in combination with other types of barriers | The bubbles can be "washed" at high flow rates; it cannot work in all conditions |
| **Water velocity** | Target species (a weak swimmer); narrow channel, adequate water flow | Based on local change in water flow velocity | Selectively excludes disturbing species | Major channel change; few sites meet the criteria |
| **Chemical-Hypoxia and hypercapnia** | Relatively shallow water, space needed for bulk gas storage | Based on chemicals that decrease or increase the water oxygen content | It is possible to exclude all fish | Large capital investments and research period |
| **Chemical-Chlorine** | Very limited implementation area | Based on the chlorine entry in water | May selectively exclude all fish | Dangerous to almost all aquatic wildlife; negative public perception |
| **Pheromones** | Small areas and/or short-term applications | Based on the entry of pheromones in water | May selectively exclude certain fish | Time and effort to purchase pheromones in bulk quantity |
| **Electromagnetism** | Narrow areas, narrow spots | Based on the use of electromagnetism | Cost-effective, low environmental impact | Might not work on all teleost fish |

Given that the present paper addresses the study of the behavioural barriers consisting of bubble curtains and aims to determine the influence of the curtain on the velocity field, below we present a brief description of the current state of research on the associated hydrodynamic field. Behavioural barriers consisting of bubble curtains are considered as a deterrent to fish movements in some experimental laboratory or in situ studies reported in literature [20,28,47–49]. The role of a bubble curtain barrier is to create a physical field that can be detected by fish; this determines a deterring response, making the fish avoid that area. Thus, the physical fields generated by a bubble curtain must be measured or characterised in detail in order to make a connection with the response of the fish. According to the study of Webb et al. [20], the hydrodynamic and acoustic field generated by a bubble curtain can guarantee a clear avoidance response from fish. Regarding the hydrodynamic field, the velocity fluctuations and the turbulence created by a bubble barrier are of interest; from an acoustic point of view, the sound pressure level (SPL) generated by the bubbles is analysed. Fish detect the velocity and sound fields by the lateral line (a sensory organ that detects the movements and vibrations in the water, and which helps species to avoid collision, to locate prey) and by the auditory system (inner ear). The lateral lines detect the movement of particles and the inner ear the pressure fluctuations. Experimental data reported by Zielinski et al. 2019 [28] have shown that a coarse bubble curtain creates a lower velocity field but a stronger acoustic field than a fine bubble curtain. Thus, in the research presented in our paper, it was decided to use fine bubbles, which create a higher velocity gradient.

Our own research [47] has shown that the bubble curtain generates a modified velocity field in the surrounding water, creating recirculation (Figure 1).

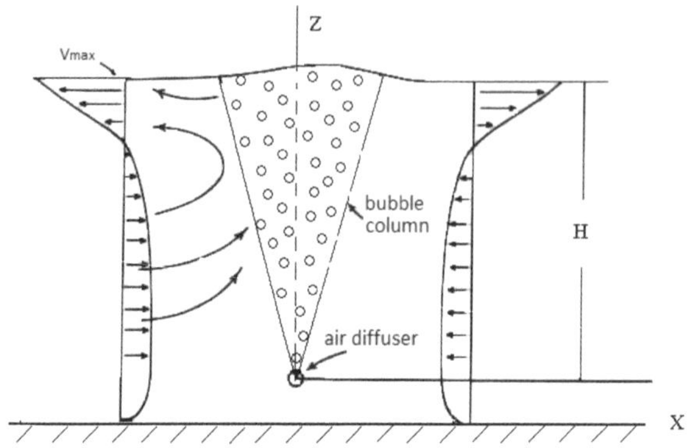


**Figure 1.** The flow induced in water by a bubble column, adapted from Oprina et al. [47].

In addition, the research of Brevik and Kristiansen [48] showed that the velocity field induced by the bubble column can be divided into two subcategories: the near field and the far field. The far field is dominated by a horizontal recirculation cell that extends a distance of about twice the depth, and the near field is dominated by the vertical velocity of the bubble curtain and occurs in the vicinity of the bubbles. The essential difference between the two velocity fields is given by the fact that in the far field, the maximum velocity is in the horizontal plane (along the flow), and in the near field, the maximum velocity is in the vertical plane (perpendicular to the flow). This creates a sharp velocity gradient. Assuming that fish are swimming at some distance from the bubble curtain and moving from the far field to the nearest field, they should be able to detect the velocity gradient and to be discouraged from passing through the bubble curtain.

Regardless of the implemented solutions' multiple advantages (protecting fish populations, maintaining biodiversity in the water stream, etc.), bubble curtains were also used to prevent fish entrapment in port areas, limit access of invasive species in different waters [22], and prevent the access of jellyfish in water supply catchments. In order to improve the efficiency of the guidance system combined solutions, behavioural barriers were used to guide the fish on a particular route. The research on bubble curtains was recently approached under research grants (2012), such as that in [49]. The study aimed at determining the characteristics of hydro-gas dynamics of fine and coarse bubble diffusers and the induced velocity at different operating conditions, emphasising the need of a thorough analysis of bubble curtains and flow spectrum. Recent approaches of this type of technology demonstrates that there is a large interest worldwide on implementing this kind of eco-friendly systems in the benefit of water quality including aquatic fauna.

## 3. Laboratory Testing of a Behavioural Barrier for Ecological Water Intakes

In order to study the behavioural barrier for ecological water intakes, an experimental setup was designed and built envisaging the analysis of a hybrid system–ecological water intake operating in tandem with a bubble curtain.

In the field of water intake arrangements, few solutions stood the test of time, but these also have some shortcomings in terms of captured flow and fish protection (especially when ensuring migration condition by maintaining longitudinal connectivity). The widespread solution for mountain rivers is the Tyrolean intake, which recently started to be upgraded with Coandă screens with better performances. These constructions have complementary fish steps for ensuring migration, which often cannot operate due to several issues like insufficient flow through each step, high slope, clogging with debris, etc. Fish protection in the upstream section of the intake is very difficult because the bottom screen with large openings and high suction velocities acts like a perfect trap for the aquatic fauna present in the stream. The design and sizing of new types of intakes is based on considerations related to operation of existing facilities and aims to ensure both the water needs and the flow conditions of the river.

The ecological water intake is a new concept, that the authors have studied and improved over the last few years. Its main feature consists in keeping the longitudinal connectivity of the river so the fauna can move upstream and downstream without any obstacles. In addition, the intake is easy to be integrated in the surrounding environment, having minimal influence on the watercourse.

The water intake experimental setup consists of an intake chamber with perforated orifices which communicates through an inlet with a lower tank for discharging the flow captured from the main channel of the test stand; the lower tank is placed under the intake chamber, having the role of water storage and ensuring a gravitational flow of the water taken through an upper hole connected to that of the intake chamber. The water taken by the intake is recirculated through a water circuit provided with a pump and a flow meter that enables the measurement of the captured water flow rate. Schematic drawing of the experimental setup of the water intake and behavioural barrier is presented in Figure 2.

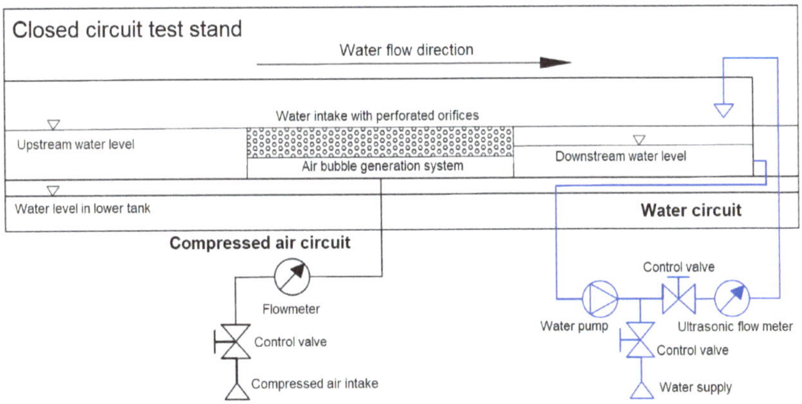


**Figure 2.** Schematic drawing of the experimental setup of the water intake and behavioural barrier.

For performing the tests, a perforated plate with 55 mm width and 400 mm length is used. It has 440 orifices with 4 mm diameter. Figure 3 shows details of the perforated surface of the intake. During the experiments, not all the 440 holes could be used because the tank below and the intake chamber are flooded and switched to suction operating conditions. Therefore, only the first rows at the top (with a number of 80 perforated holes) were used for the preliminary tests. Thus, a constant level could be maintained in the lower tank, ensuring gravitational flow.

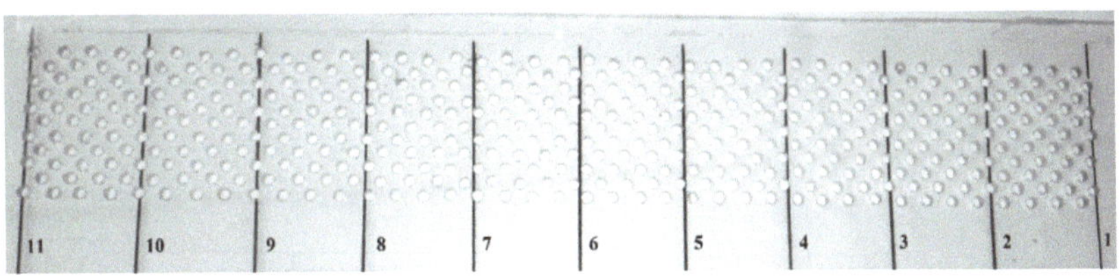


**Figure 3.** Detail on the perforated surface of the intake.

For the fish guidance system, an air bubble generation system was installed at the base of the intake chamber. We first tried ceramic diffusers, then a porous hose specially designed for aerating fishponds. Better results have been obtained with the porous hose that is widely used in water aeration applications (Tanuki Japanese hose). The air bubble generation system is connected to a compressor for supplying compressed air. A precision valve for regulating the flow rate and a flowmeter for measuring its value are installed in the compressed air circuit. The two main components of the hybrid system under study, a behavioural barrier of ecological water intake model, have been integrated in a closed-circuit hydraulic stand (Figure 4) provided with a transparent visualisation area (375 × 300 × 1015 mm) made of Plexiglass (polymethyl methacrylate). The stand is a hydraulic unit made in a watertight and removable modular construction, which circulates clean water in a closed circuit. The main modules of the stand are represented by test section, tank, recirculation pumps, and variable speed motors. The closed-circuit hydraulic stand has a system for measuring and controlling the water flow and velocity in the channel and an acquisition and processing system as well. Thus, it allows water flow velocity adjustment to provide velocities between 0.05 m/s and 1 m/s in the test section.

**Figure 4.** Closed circuit hydraulic stand.

For the experimental model of hybrid ecological intake with behavioural barrier testing, a part of the data acquisition system related to water velocity measurement was used. The water velocity was measured with a Pitot-Prandtl tube positioned in water in parallel direction to the flow. As shown in Figure 5, the tube is fixed by a system that allows sliding horizontally and vertically at certain marked positions detailed in Figure 6.

**Figure 5.** Details on Pitot-Prandtl tube position for measuring water velocity.

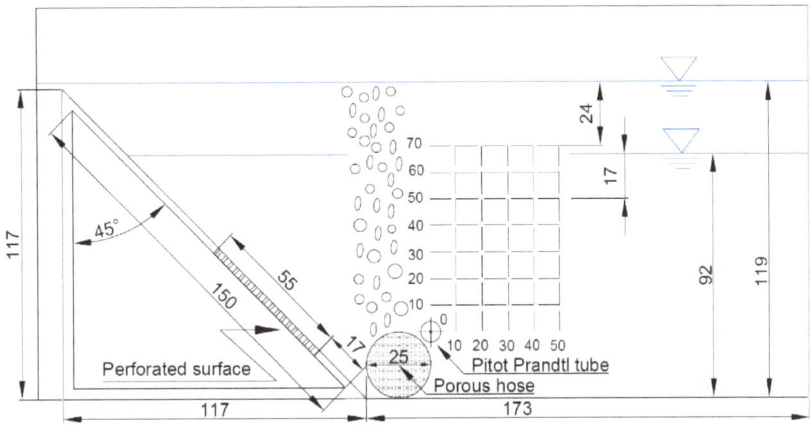


**Figure 6.** Position of the velocity measuring points for two different water depths in the channel.

In order to perform the necessary tests, the pump group converters are started and the required velocity in the experimental stand channel is established before performing the velocity measurements using the Pitot tube. Water levels are measured in all enclosures (intake chamber upstream, downstream and in the drainage area, as well as the lower reservoir level). The water flow rate through the water intake model is measured using an Axioma smart ultrasonic flowmeter, model Qalcosonic W1. The air flow rate through the bubble diffuser is measured using a Cole Parmer flowmeter with a 0 to 20 LPM scale, with an accuracy of ±5% of full scale. The water velocity for each point shown in Figure 6 is measured by the Pitot tube which is connected to a differential pressure transducer model AppliSens, type APRE-2000 with accuracy ±0.1% of calibrated range −5 to 70 mbar. For 92 mm water depth, the velocity was measured in 36 points (from 0 to 50 in horizontal and vertical directions), while for 119 mm water depth the velocity was measured in 42 points (from 0 to 50 in horizontal direction and from 0 to 70 in vertical direction, respectively).

In the Figure 7 the experimental setup of the water intake and behavioural barrier is presented from opposite views. The transparent viewing section allows the phenomena observation and water level measurements. On top of the experimental setup, the Pitot-Prandtl tube slides across the measuring points.

Figure 8 shows that a higher velocity in the main channel determines an increase in the turbulence, especially in the free surface area.

The assessment of the bubble curtain influence on the water flow inside the channel was made by analysing the velocity variation in vertical direction, where certain dependencies between air flow rate, water depth, and velocity have been found.

The bubble curtain influence on the water velocity was investigated for two water depths (H = 92 mm and H = 119 mm), two water velocities inside the channel (0.33 m/s and 0.535 m/s), and two air flow rates (10.5 LPM and 15 LPM). The velocities were compared with the case where no air was injected (0 LPM) in order to determine the bubble curtain influence on the water flow. The water velocity inside the transparent channel of the experimental setup was chosen to vary between 0.33 m/s and 0.535 m/s to cover the typical water velocities of most mountain rivers [50]. A summary of the testing conditions of the hybrid solution model of ecological river water intake with bubble curtain behavioural barrier is presented in Table 2.

a)

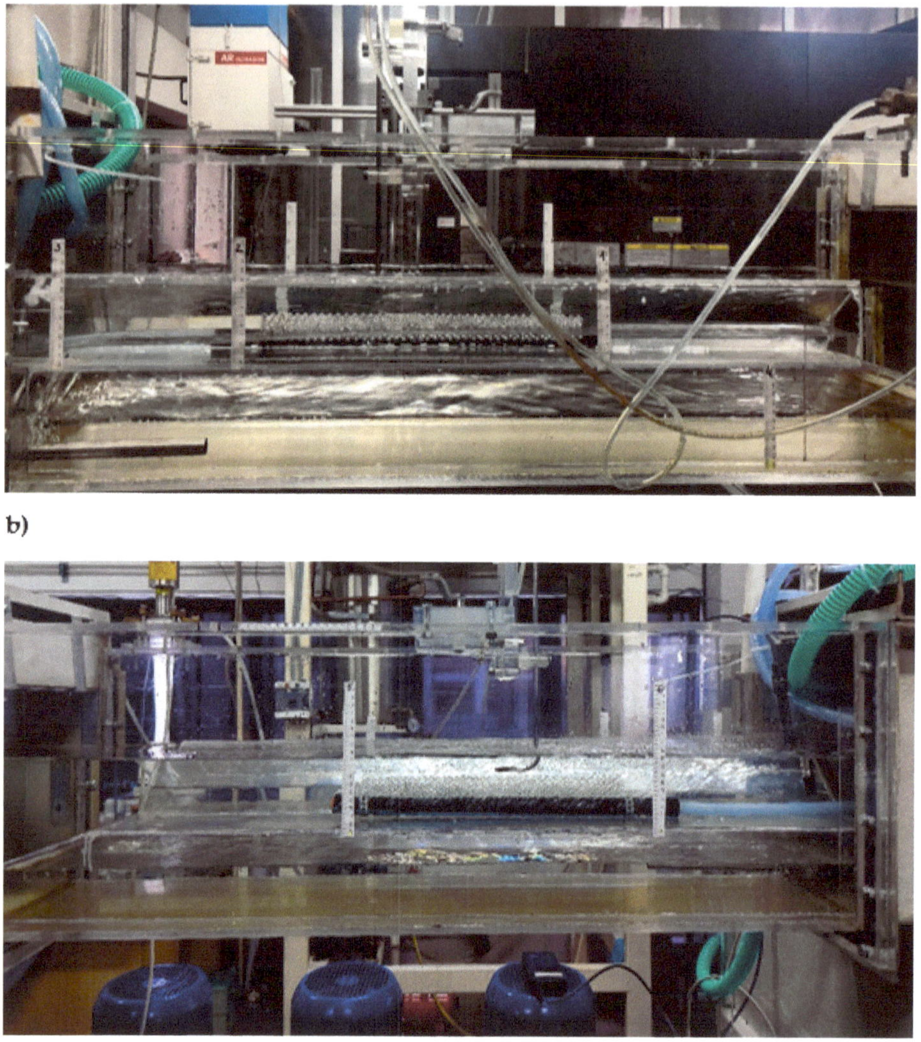

**Figure 7.** Images of the experimental setup of the water intake and behavioural barrier: (**a**) Left side view; (**b**) right side view.

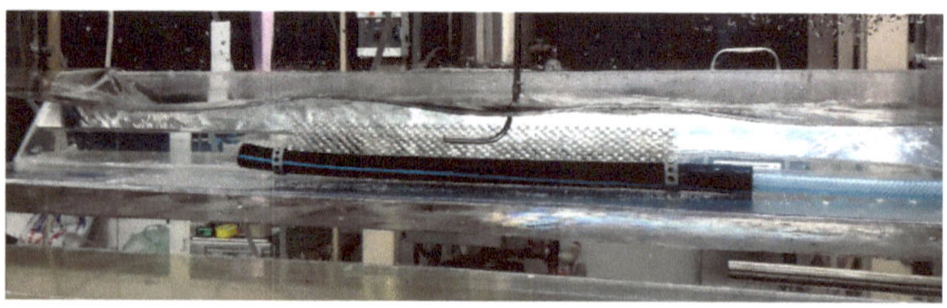

**Figure 8.** Detail on free surface turbulence for v = 0.535 m/s and H = 119 mm.

**Table 2.** Water depths, velocities, and airflow rates used for testing the hybrid solution model.

| $v_{channel}$ [m/s] | H [mm] | Q [LPM] | Figure |
|---|---|---|---|
| 0.33 | 92 | 0 | 9 |
| 0.33 | 92 | 10.5 | 10 |
| 0.33 | 92 | 15 | 15 |
| 0.535 | 92 | 0 | 11 |
| 0.535 | 92 | 10.5 | 12 |
| 0.535 | 92 | 15 | 16 |
| 0.535 | 119 | 0 | 13 |
| 0.535 | 119 | 10.5 | 14 |
| 0.535 | 119 | 15 | 17 |

## 4. Results and Discussion

The velocity variation in the horizontal direction is not significant as the bubbles are generated in vertical plane and they travel to the free surface with reduced lateral movements. This behaviour is beneficial because the bubble curtain is maintained parallel to the water intake area. The experiment showed that the water flow through the channel slightly influences the bubbles trajectories, being carried downstream.

Based on the measurements results, the induced velocity variation curves shown in Figures 9–17 were drafted. The velocity induced in water by the bubble column was denoted by v, the initial velocity in the channel by $v_{channel}$, the air flow rate by Q, and the vertical and horizontal distance from the origin of the chosen coordinate system to the velocity measuring point by Dv and Dh, respectively.

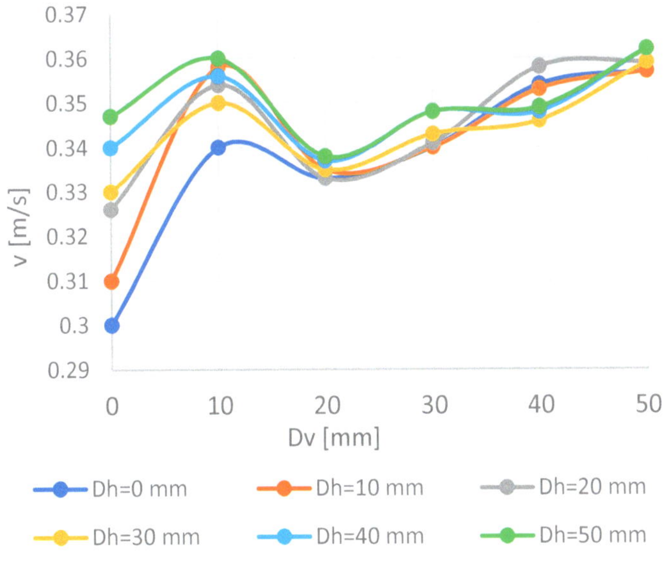


**Figure 9.** Induced velocity variation for $v_{channel}$ = 0.33 m/s, Q = 0 LPM, H = 92 mm.

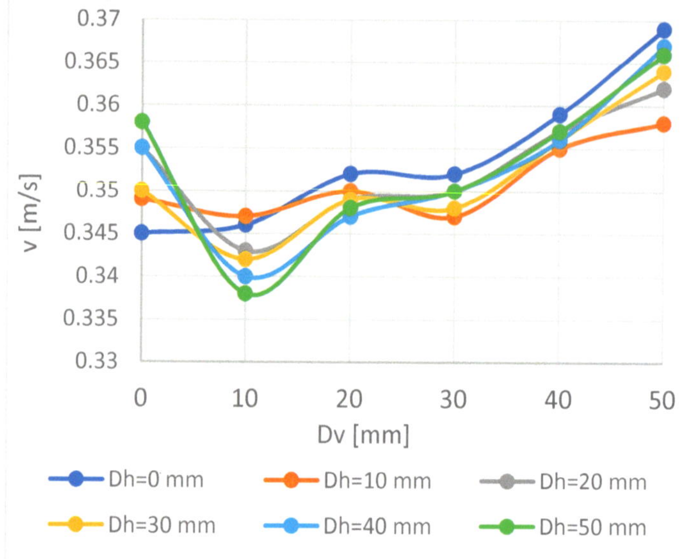


**Figure 10.** Induced velocity variation for $v_{channel}$ = 0.33 m/s, Q= 10.5 LPM, H = 92 mm.

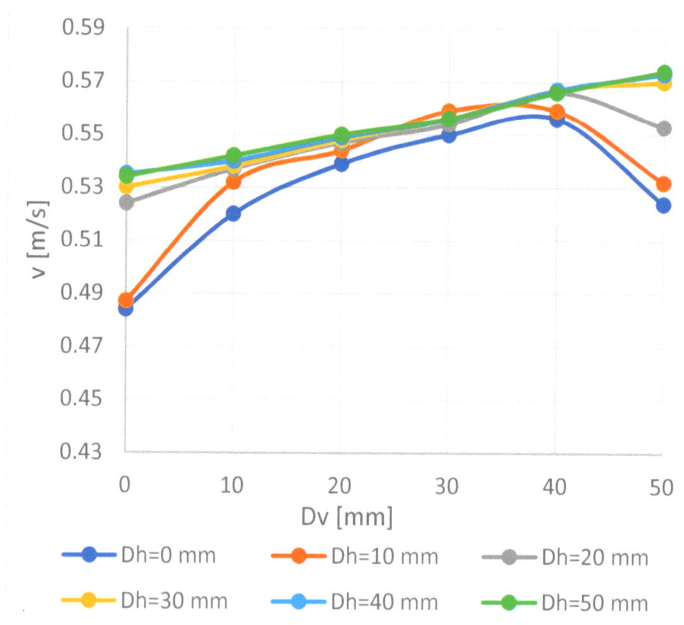


**Figure 11.** Induced velocity variation for $v_{channel}$ = 0.535 m/s, Q = 0 LPM, H = 92 mm.

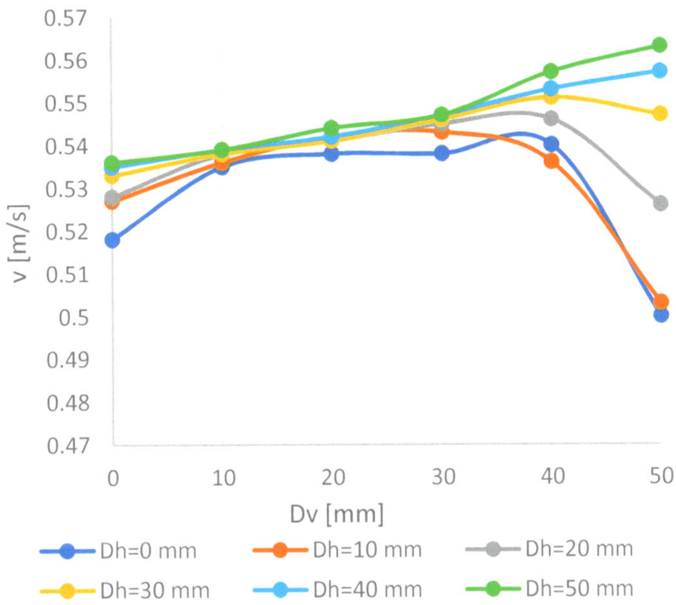


**Figure 12.** Induced velocity variation for $v_{channel}$ = 0.535 m/s, Q = 10.5 LPM, H = 92 mm.

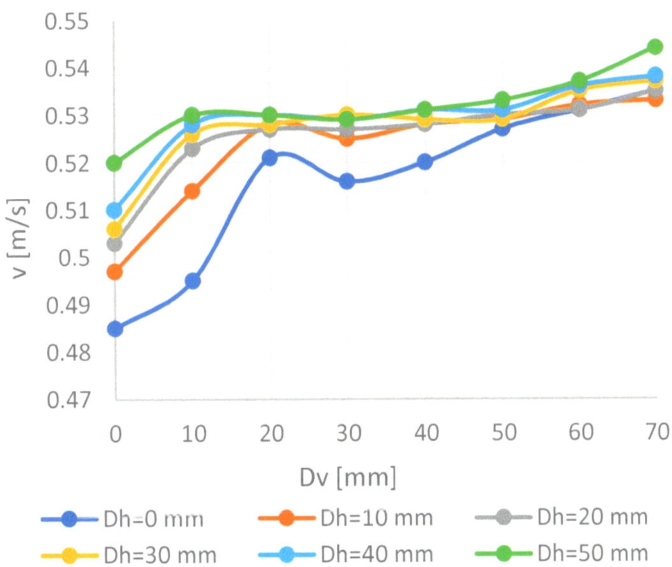


**Figure 13.** Induced velocity variation for $v_{channel}$ = 0.535 m/s, Q = 0 LPM, H = 119 mm.

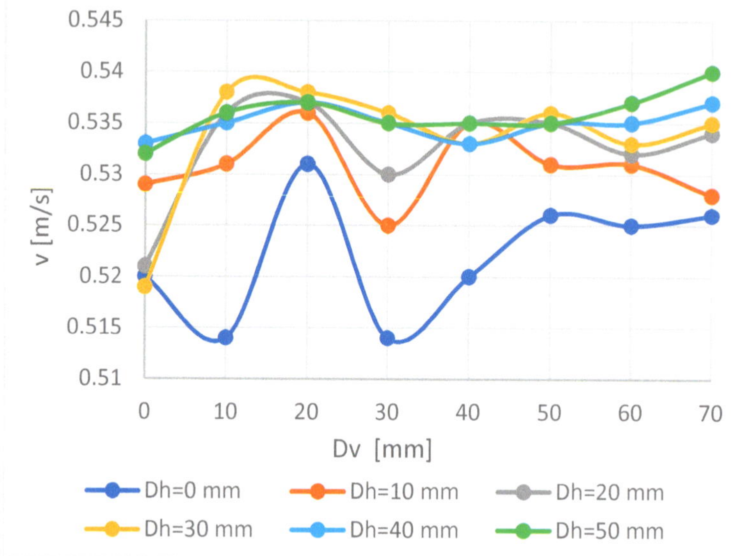


**Figure 14.** Induced velocity variation for $v_{channel}$ = 0.535 m/s, Q = 10.5 LPM, H = 119 mm.

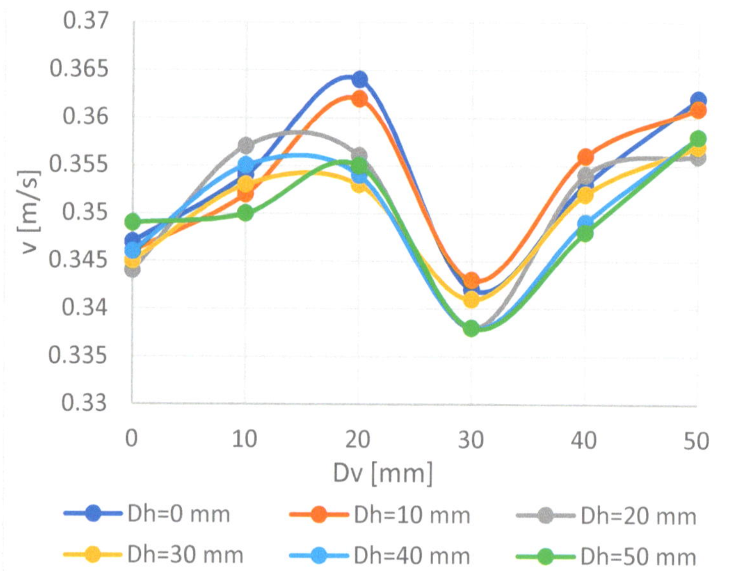


**Figure 15.** Induced velocity variation for $v_{channel}$ = 0.33 m/s, Q = 15 LPM, H = 92 mm.

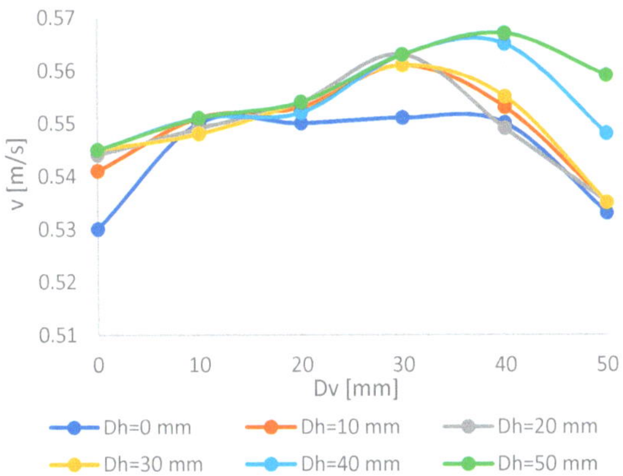


**Figure 16.** Induced velocity variation for $v_{channel}$ = 0.535 m/s, Q = 15 LPM, H = 92 mm.

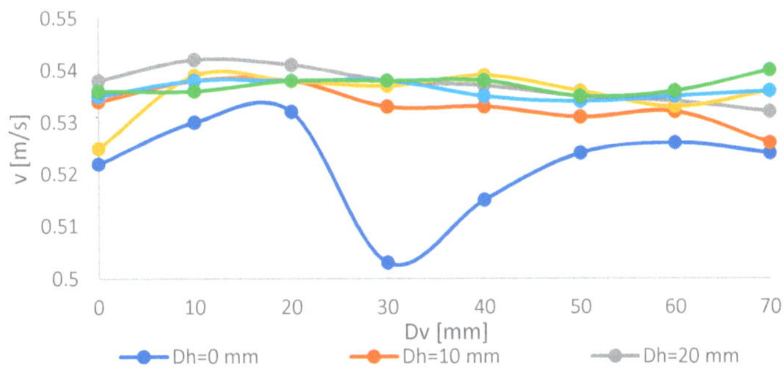


**Figure 17.** Induced velocity variation for $v_{channel}$ = 0.535 m/s, Q = 15 LPM, H = 119 mm.

For the case of 0.33 m/s water velocity in the main channel and 92 mm depth (Figures 9, 10 and 15), the gap between the velocity variation on each vertical is smaller than in the 0.535 m/s velocity case; there is an increasing tendency of the gap as it approaches the free surface. Furthermore, the increase of the air flow favours the increase of the velocity gradient for the 0, 10, 20, and 30 mm verticals.

For 0.535 m/s velocity and 92 mm depth (Figures 11, 12 and 16) case, the higher velocity in the main channel determines a turbulence increase, especially in the free surface area. Thus, in the 40 mm and 50 mm verticals area, there is a decrease in velocity due to the proximity of the Pitot tube to the free surface (17 mm). On the other hand, when the water level in the channel increases to 119 mm, the Pitot tube is placed at a greater depth from the free surface-24 mm respectively (at the point located in vertical direction at a distance of 70 mm from the porous hose), as shown in Figure 6. In this case, the influence of the free turbulence is reduced.

For 0.535 m/s water velocity and 119 mm water depth, with no air flow injected (Figure 13), the sudden increase in the velocity starting with the 10 to 30 mm verticals is due to the disturbed water current in the catchment area of the intake. In vertical plane, the water intake area corresponds to the distance of 10–30 mm.

When injecting air with a flow rate of 10.5 LPM and 15 LPM, respectively (Figures 14, 15 and 17), the velocity shows significant oscillations. Thus, the first three vertical sections located in the proximity of the bubble curtain (0, 10, 20 mm) show the largest oscillations; starting with the vertical of 30 mm, the variation becomes uniform, which is explained by moving away from the bubble curtain and the reduction of its influence on the flow (practically, the induced flow by the bubble curtain is diminished).

In terms of the influence of the water catchment through the intake, we found that near the intake (at a vertical distance of 20–30 mm from the origin of the chosen coordinate system) the longitudinal velocities field (which should have increased towards the free surface) has a decrease due to the increase of the velocity transverse component determined by the lateral catchment of the water. According to Figures 9 and 13, when no air is injected, this influence is noticed both in the low depth (92 mm) and low velocity (0.33 m/s) case, as well as in the high depth (119 mm) and high velocity (0.535 m/s) case. Starting with a distance of 30–40 mm (an average relative depth of 0.33–0.35) upwards, it is found that the longitudinal velocities are no longer influenced by the presence of the intake.

The introduction of air determines in the reduced water depth (92 mm)/reduced velocity (0.33 m/s) case, an increase of the longitudinal velocities field in the vicinity of the intake inlet (15–30 mm) area until a relative depth of 0.33. With the increase of the air flow rate, the movement of the minimum longitudinal velocities towards the free surface is noticed.

We concluded that the generation of a bubble curtain near the intake causes a local modification in the water velocity field. This change also occurs at the introduction of a reduced air flow rate (10.5 LPM), showing that a bubble curtain can be efficient even for reduced air flow rates. Moreover, this effect can be maintained even when the water velocity increases from 0.33 m/s to 0.535 m/s (an increase of approximately 60%). For increased water velocities, the bubble curtain is carried away downstream. Thus, the position of the bubble diffuser should be reconsidered. For increased velocities, the bubble diffuser must be placed with an offset of few centimetres upstream the river water intake. The bubble plume develops more effectively when the water depth is higher. As a general recommendation for real operating conditions, the diffuser should be placed as low as possible in the riverbed in order to benefit of a higher river depth.

In relation to the studies available in the literature, the experiments conducted in the present paper have shown that the bubble curtain generates a modified velocity field in the surrounding water, creating recirculation. The results are encouraging in relation with the findings reported in [48], where it is stated that the fish behaviour can be influenced by this sharp velocity gradient. Unlike the studies in the literature, the present research analysed the changes in the velocity profile under different study conditions. Thus, the water velocity profiles were experimentally determined both at different water depths and at different water flow rates in the testing channel, as well as at different air flow rates injected through the air hose generating the bubble curtain. In addition, the study was conducted on a small-scale hybrid experimental model that incorporates both an ecological river water intake that does not influence river morphology and a fish guidance system based on behavioural barriers; not only the operation of the river water intake model was investigated, but also the tandem operation of the hybrid model components. The water velocity variations induced by the bubble curtain is of interest as, according to the study performed by Mogdans in [50], the lateral-line component of the fish sensory system consists of superficial (hair cell) and canal neuromasts which practically are velocity and pressure gradient detectors respectively, placed on the fish body. They give fish neurological responses to hydrodynamic stimuli of the water due to their own movement, water waves, vortex recirculation, or sounds from air bubble breakup or abiotic sources. As seen from Figures 9–17, longitudinal flow velocity varies with each 10–20 mm cross-channel distance at increasing air flow rate for the same depth. This means that different water velocities may be sensed by the fish either along its body or on its left/right sides. Additionally, the lateral-line and inner ear are sensitive to acoustic underwater sounds up to 100–200 Hz, as reported by Popper and Schilt [51], whereas, according to Braun et al. [52], in fish possessing a swimbladder, hearing is enhanced up to 1000 Hz, which is close to the air bubbles splitting noise analysed by Frizell and Arndt [53]. Both hydrodynamic and sound stimuli give the central neurological system of the fish a signal of dangerous environment and force it to swim away. Given these findings [50–53], we consider that the current paper may contribute to the further understanding of the fish behaviour in relation with bubble curtains. In order to validate the conclusions drawn from the performed studies, the research shall continue with the inclusion of a biotic model. The present research was conducted in the frame of a project which did not consider animal testing for professional ethics reasons, and it focused on other scientific means to demonstrate the system functionality. Biotic model inclusion has not been taken into account yet because the research activity was performed on a state-of-the-art dedicated test stand that can accurately measure velocity fields (including PIV measurements) and other parameters, which can provide reliable results for the preliminary experimental validation of the hybrid solution–ecological intake with behavioural barrier. After this laboratory stage, an installation can be deployed in situ in order to study the interaction of the fish with the bubble curtain in real operating conditions. In this case, special attention will be given when designing the in situ facility in order not to harm the fish.

## 5. Conclusions

The research presented in the paper approached the experimental study of a fish guidance system consisted of a bubble curtain for a river water intake (perforated plate screen) to assess its influence on the water velocity field. In most of the analysed cases, a pronounced velocity gradient is observed in the proximity of the bubble curtain, confirming the hypotheses of previous studies [47,48], according to which the sharp velocity gradient can influence the behaviour of fish.

In addition, compared to traditional physical systems for fish guidance, the proposed behavioural barrier has the advantage to easily adapt to the flowing conditions of the river. Thus, by modifying the air flow rate of the diffusers, it results a bubble curtain characterised by another flow pattern: the bubble plume induces other velocities in the nearby water, the flow regime of the plume changes. Air flow regulation can be done from the river bank, without the need for intervention in the water, on site. If the change of the air flow rate is not enough, then the bubble diffusers can be replaced by other air generating system. To make this fish guidance system energy efficient, it is necessary to identify a minimum required air flow rate that will lead to an increase in local velocity near the intake. Taking into consideration that each fish species has specific characteristics, this solution must be adapted to the envisaged fish features.

During the performed tests, the influence of the water velocity due to the use of the bubble generation system was identified by measuring the water velocities in different points in the cross section of the experimental channel. The presence of the bubble curtain can act as a behavioural barrier for fish due to the modification of the velocity profiles in the considered area and to the sound pressure level. Moreover, by placing a bubble curtain in the admission proximity of the water intake where a suction effect is significant, the streamlines are modified, contributing to diverting this directional flow and deterring the fish from the intake area. Thus, alone or in combination with other behavioural barriers, the bubble curtain can contribute to the safe diversion of the fish into the downstream sector, considerably reducing the risk of accidental entrapment. An in-depth insight of this complex phenomena can be obtained by applying Particle Image Velocimetry measuring technique to determine the direction of the velocity vectors in the proximity of the bubble curtain and water intake.

**Author Contributions:** G.C. and R.-A.C., conceptualisation and methodology, G.C., R.-A.C., A.V. and L.-A.E.-L., writing—review and editing, A.V. and D.-E.G.N., literature review and investigation, Ș.-M.S. and L.M., formal analysis, G.C., R.-A.C. and A.V., investigation and data analysis, G.C., R.-A.C. and D.-E.G.N., results interpretation. All authors have read and agreed to the published version of the manuscript.

**Funding:** This research was funded by Romanian Ministry of Education, Research and Digitalization, UEFISCDI, project number PN-III-P2-2.1-PED-2019-1444–Eco-hybrid Water Intake with Behavioral Barrier to Reduce the Impact on Fish Fauna and River Morphology.

**Institutional Review Board Statement:** Not applicable.

**Informed Consent Statement:** Not applicable.

**Data Availability Statement:** All data generated during the experiments and analysed in the current study are included in this published article.

**Acknowledgments:** This work was supported by the Romanian Ministry of Education, Research and Digitalization, UEFISCDI, project number PN-III-P2-2.1-PED-2019-1444–Eco-hybrid Water Intake with Behavioral Barrier to Reduce the Impact on Fish Fauna and River Morphology and project number 25PFE/30.12.2021 – Increasing R-D-I capacity for electrical engineering-specific materials and equipment with reference to electromobility and "green" technologies within PNCDI III, Programme 1.

**Conflicts of Interest:** The authors declare no conflict of interest.

## References

1. Brasovanu, L.; Gogoașe Nistoran, D.E.; Armaș, I. The hydrological regime of the Prahova River along Carpathian and Subcarpathian stretch. *GtaylorruebuseoPatterns* **2018**, *3*, 32–42. [CrossRef]
2. The world's Biodiversity is Declining at an Alarming Rate. Population Sizes of Vertebrate Species Measured by the Living Planet Index (LPI) Have More than Halved in Little More than 40 Years. Available online: https://wwf.panda.org/discover/knowledge_hub/all_publications/living_planet_index2/ (accessed on 15 November 2021).
3. Äivas, L.; Kesminas, V. Fish distribution and ecological state of the Siesartis River, Lithuania. *Acta Zool. Litu.* **2011**, *21*, 153–162. [CrossRef]
4. Nijland, H.; Menke, U. Communication and Community Involvement: Lessons Learnt from SDF. In Proceedings of the Flood Risk Management and Multifunctional Land Use in River Catchments, Mainz, Germany, 17–19 October 2005.
5. Grill, G.; Lehner, B.; Lumsdon, A.E.; MacDonald, G.K.; Zarfl, C.; Reidy Liermann, C. An index-based framework for assessing patterns and trends in river fragmentation and flow regulation by global dams at multiple scales. *Environ. Res. Lett.* **2015**, *10*, 15001. [CrossRef]
6. Armaş, I.; Gogoaşe Nistoran, D.E.; Osaci-Costache, G.; Braşoveanu, L. Morpho-dynamic evolution patterns of Subcarpathian Prahova River (Romania). *Catena* **2013**, *100*, 83–99. [CrossRef]
7. Nistoran, D.G.; Ionescu, C.; Patru, G.; Armas, I.; Grigorie Omrani, S. One Dimensional Sediment Transport Model to Assess Channel Changes along Olteniţa-Călăraşi Reach of Danube River, Romania. *Energy Procedia* **2017**, *112*, 67–74. [CrossRef]
8. Rodrigues, S.; Breheret, J.-G.; Macaire, J.-J.; Moatar, F.; Nistoran, D.; Juge, P. Flow and sediment dynamics in the vegetated secondary channels of an anabranching river: The Loire River (France). *Sediment. Geol.* **2006**, *186*, 89–109. [CrossRef]
9. Gogoaşe Nistoran, D.E.; Ionescu, C.S.; Brasoveanu, L.; Armas, I.; Opris, I.; Costinas, S. Modeling Hydrodynamic Changes Induced by Run-of-River Hydropower Plants along the Prahova River in Romania. *J. Energy Eng.* **2018**, *144*, 04017078. [CrossRef]
10. Nistoran-Gogoase, D.-E.; Brasoveanu, L.; Ionescu, C.-S.; Armas, I. Downstream coarsening of subsurface sediment grain size along upper Prahova River (Romania). *GeoPatterns* **2019**, *4*, 33–39. [CrossRef]
11. WWF-România. *Analiza Legislaţiei Specifice din Domeniul Planificării şi Emiterii Actelor de Reglementare Aferente Construirii şi Funcţionării Microhidrocentralelor în RO*; WWF: Bucharest, Romania, 2013.
12. Turnpenny, A.W.H.; O'Keeffe, N. *Screening for Intakes and Outfalls: A Best Practice Guide*; Science Report SC030231; Environment Agency: Bristol, UK, 2005; ISBN 1 84432 361 7.
13. U.S. Department of the Interior, Bureau of Reclamation. *Fish Protection at Water Diversions—A Guide for Planning and Designing Fish Exclusion Facilities*; U.S. Department of the Interior, Bureau of Reclamation: Denver, CO, USA, 2006.
14. Taft, E.P. Fish protection technologies: A status report. *Environ. Sci. Policy* **2000**, *3*, 349–359. [CrossRef]
15. Invasive Species. Available online: https://www.regions.noaa.gov/great-lakes/index.php/great_lakes-restoration-initiative/invasive-species/ (accessed on 10 January 2022).
16. Mills, E.L.; Leach, J.H.; Carlton, J.T.; Secor, C.L. Exotic Species and the Integrity of the Great Lakes. *BioScience* **1994**, *44*, 666–676. [CrossRef]
17. Noatch, M.R.; Suski, C.D. Review-Non-physical barriers to deter fish movements. *Environ. Rev.* **2012**, *20*, 71–82. [CrossRef]
18. Zielinski, D. Bubble Barrier Technologies for Common Carp. Master's Thesis, University of Minnesota, Minneapolis, MN, USA, 2011.
19. Taylor, R.M.; Pegg, M.A.; Chick, J.H. Response of bighead carp to a bioacoustic behavioural fish guidance system. *Fish. Manag. Ecol.* **2005**, *12*, 283–286. [CrossRef]
20. Webb, J.F.; Fay, R.R.; Popper, A.N. (Eds.) Fish Bioacoustics. In *Springer Handbook of Auditory Research*; Springer: New York, NY, USA, 2008; Volume 32.

21. Hadderingh, R. Experimental reduction of fish impingement by artificial illumination at Bergum power station. *Int. Rev. Ges. Hydrobiol.* **1982**, *67*, 887–900.
22. Ruebush, B.C.; Sass, G.G.; Chick, J.H.; Stafford, J.D. In-situ tests of sound-bubble strobe light barrier technologies to prevent range expansions of Asian carp. In Proceedings of the 17th International Conference on Aquatic Invasive Species, San Diego, CA, USA, 29 August–2 September 2010; Volume 7, pp. 37–48. [CrossRef]
23. Dawson, H.A.; Reinhardt, U.G.; Savino, J.F. Use of Electric or Bubble Barriers to Limit the Movement of Eurasian Ruffe (*Gymnocephalus cernuus*). *J. Great Lakes Res.* **2006**, *32*, 40–49. [CrossRef]
24. Zielinski, D.P.; Sorensen, P.W. Bubble Curtain Deflection Screen Diverts the Movement of both Asian and Common Carp. *N. Am. J. Fish. Manag.* **2016**, *36*, 267–276. [CrossRef]
25. Zielinski, D.P.; Voller, V.R.; Svendsen, J.C.; Hondzo, M.; Mensinger, A.F.; Sorensen, P. Laboratory experiments demonstrate that bubble curtains can effectively inhibit movement of common carp. *Ecol. Eng.* **2014**, *67*, 95–103. [CrossRef]
26. Jesus, J.; Amorim, M.C.P.; Fonseca, P.J.; Teixeira, A.; Natário, S.; Carrola, J.; Varandas, S.; Pereira, L.T.; Cortes, R.M.V. Acoustic barriers as an acoustic deterrent for native potamodromous migratory fish species. *J. Fish Biol.* **2019**, *95*, 247–255. [CrossRef]
27. Leander, J.; Klaminder, J.; Hellström, G.; Jonsson, M. Bubble barriers to guide downstream migrating Atlantic salmon (*Salmo salar*): An evaluation using acoustic telemetry. *Ecol. Eng.* **2021**, *160*, 106141. [CrossRef]
28. Zielinski, D.P.; McLaughlin, R.; Castro-Santos, T.; Paudel, B.; Hrodey, P.; Muir, A. Alternative Sea lamprey barrier technologies: History as a control tool. *Rev. Fish. Sci. Aquac.* **2019**, *27*, 438–457. [CrossRef]
29. Patrick, P.H.; Christie, A.E.; Sager, D.; Hocutt, C.; Stauffer, J., Jr. Responses of fish to a strobe light/ air-bubble barrier. *Fish. Res.* **1985**, *3*, 157–172. [CrossRef]
30. Bravo, H.R.; Gulliver, J.S.; Hondzo, M. Development of a commercial code-based two-fluid model for bubble plumes. *Environ. Model. Softw.* **2007**, *22*, 536–547. [CrossRef]
31. Turnpenny, A.W.H. *Review of Potential for Coarse Fish Entrainment at Potable Water Intakes on the Lower Freshwater Thames (Hurley-Teddington)*; Client Report to Env. Ag. No. FCR 302/99; Fawley ARL Ltd.: Southampton, UK, 1999.
32. Moursund, R.A.; Dauble, D.D.; Langeslay, M.J. Turbine intake diversion screens: Investigating effects on Pacific lamprey. *Hydro Rev.* **2003**, *22*, 40–46.
33. Turnpenny, A.W.H.; Taylor, C.J.L. An assessment of the effect of the Sizewell power stations on fish populations. *Hydroécologie Appl.* **2000**, *12*, 87–134. [CrossRef]
34. Miranda, L.E.; Hodges, K.B. Role of aquatic vegetation coverage on hypoxia and sunfish abundance in bays of a eutrophic reservoir. *Hydrobiologia* **2000**, *427*, 51–57. [CrossRef]
35. Pegg, M.A.; Chick, J.H. *Aquatic Nuisance Species: An Evaluation of Barriers for Preventing the Spread of Bighead and Silver Carp to the Great Lakes*; Final report for Illinois-Indiana Sea Grant A/SE (ANS)-01–01; Illinois Natural History Survey: Urbana, IL, USA, 2004.
36. Travade, F.; Larinier, M. La migration de devalaison: Problèmes et dispositifs. *Bull. Français Pêche Piscic.* **1992**, *326–327*, 165–176. [CrossRef]
37. Solomon, D.J. *Diversion and Entrapment of Fish at Water Intakes and Outfalls*; R & D Report No. 1; National Rivers Authority: Bristol, UK, 1992; p. 51.
38. Sand, O.; Enger, P.S.; Karlsen, H.E.; Knudsen, F.R. Detection of infrasound in juvenile salmonids and European silver eels: A mini-review. In Proceedings of the American Fisheries Society Symposium, Bethesda, MD, USA, 8 August–13 August 2001; pp. 183–193.
39. Jones, P.E.; Tummers, J.S.; Galib, S.M.; Woodford, D.J.; Hume, J.B.; Silva, L.G.M.; Braga, R.R.; Leaniz, C.G.; Vitule, J.R.S.; Herder, J.E.; et al. The use of barriers to limit the spread of aquatic invasive animal species: A global review. *Front. Ecol. Evol.* **2021**, *9*, 611631. [CrossRef]
40. Jesus, J.; Cortes, R.; Teixeira, A. Acoustic and Light Selective Behavioral Guidance Systems for Freshwater Fish. *Water* **2021**, *13*, 745. [CrossRef]
41. Welton, J.S.; Beaumont, W.R.C.; Clarke, R.T. The efficacy of air, sound and acoustic bubble screens in deflecting Atlantic salmon, *Salmo salar* L., smolts in the River Frome, UK. *Fish. Manag. Ecol.* **2002**, *9*, 11–18. [CrossRef]
42. Katopodis, C.; Koon, E.M.; Hanson, L. *Sea Lamprey Barriers: New Concepts and Research Needs*; Great Lakes Fishery Commission: Ann Arbor, MI, USA, 1994.
43. Sorensen, P.W.; Stacey, N.E. Brief review of fish pheromones and discussion of their possible uses in the control of non-indigenous teleost fishes. *N. Z. J. Mar. Fresh* **2004**, *38*, 399–417. [CrossRef]
44. Gibbs, M.A.; Glenn Northcutt, R. Development of the lateral line system in the shovelnose sturgeon. *Brain Behav. Evol.* **2004**, *64*, 70–84. [CrossRef]
45. Bettoli, P.W.; Maceina, M.J. *Sampling with Toxicants in Fisheries Techniques*, 2nd ed.; Murphy, B.R., Willis, D.W., Eds.; American Fisheries Society: Bethesda, MD, USA, 1996; pp. 303–333.
46. Haymes, G.T.; Patrick, P.H. Exclusion of adult alewife, *Alosa pseudoharengus* using low-frequency sound for application at water intakes. *Can. J. Fish. Aquat. Sci.* **1986**, *43*, 855–862. [CrossRef]
47. Oprina, G.; Pincovschi, I.; Baran, G. *Hidro-Gazo-Dinamica Sistemelor de Aerare Echipate cu Generatoare de Bule*; Politehnica Press: Bucureşti, Romania, 2009; pp. 1–176.
48. Brevik, I.; Kristiansen, Ø. The flow in and around air-bubble plumes. *Int. J. Multiph. Flow* **2002**, *28*, 617–634. [CrossRef]

49. University of Minnesota, Department of Civil Eng. Report regarding Controlling the Movement of Invasive Fish Species Research Contract. 2012. Available online: https://personal.ce.umn.edu/~{}voller/ (accessed on 17 November 2021).
50. Mogdans, J. Sensory ecology of the fush lateral-line system: Morphological and physiological adaptations for the perception of hydrodynamic stimuli. *J. Fish Biol.* **2019**, *95*, 53–72. [CrossRef] [PubMed]
51. Popper, A.N.; Schilt, C.R. Hearing and Acoustic Behavior: Basic and Applied Considerations. In *Fish Bioacoustics*; Webb, F.J., Fay, R.R., Popper, A.N., Eds.; Springer: New York, NY, USA, 2008; pp. 17–48. [CrossRef]
52. Braun, C.B.; Sand, O. Functional Overlap and Nonoverlap between Lateral Line and Auditory Systems. In *The Lateral Line System*; Coombs, S., Bleckmann, H., Fay, R., Popper, A., Eds.; Springer Handbook of Auditory Research; Springer: New York, NY, USA, 2013; Volume 48, pp. 281–312. [CrossRef]
53. Frizell, W.K.; Arndt, R.E.A. *Noise Generation by Air Bubbles in Water: An Experimental Study of Creation and Splitting*; Project Report no. 269; Retrieved from the University of Minnesota Digital Conservancy; St. Anthony Falls Hydraulic Laboratory: Minneapolis, MN, USA, 1987; Available online: https://hdl.handle.net/11299/114029 (accessed on 8 December 2021).

*applied sciences*

MDPI

*Article*

# Evaluating Attraction and Passage Efficiencies of Pool-Weir Type Fishways Based on Hydraulic Analysis

**Kyong-Oh Baek [1], Jeong-Min Lee [2], Eun-Jin Han [3] and Young-Do Kim [4,*]**

[1] Department of Civil and Environmental Engineering, Hankyong National University, 327 Chungang-ro, Anseong-si 17579, Gyeonggi-do, Korea; pko@hknu.ac.kr
[2] Department of Environmental Engineering, Inje University, 197 Inje-ro, Gimhae 50834, Gyengnam-do, Korea; jmlee1515@naver.com
[3] Water Quality Assessment Research Division, National Institute of Environmental Research, Incheon 22689, Gyeonggi-do, Korea; gene0921@korea.kr
[4] Department of Civil and Environmental Engineering, Myongji University, 116 Myongji-ro, Yongin-si 17058, Gyeonggi-do, Korea
* Correspondence: ydkim@mju.ac.kr; Tel.: +82-31-330-6407

**Abstract:** In this study, a hydraulic analysis technique is presented for evaluating the efficiency of a pool-weir type fishway. The River-2D model with the fish physical habitat module was used for the analysis of the attraction efficiency, and the weighted usable area proposed as an index of the efficiency. For the analysis of passage efficiency, the three-dimensional model, FLOW-3D, was used as an evaluating tool to describe the fluid behavior on a hydraulic structure with a free surface. The ice-harbor type fishway at Dalseong weir in the Nakdong River of Korea was selected as a test-site, and the efficiency was estimated using the hydraulic analysis. The results of River-2D analysis showed that the location of the fishway near the right bank now was not appropriate for attracting the target fish to the fishway due to the deep water depth. The 3D analysis results presented that the flow velocity field inside the fishway showed a similar plunging flow in each pool, but various flow patterns appeared locally.

**Keywords:** hydraulic analysis; fishway; attraction efficiency; passage efficiency; multifunctional weir

**Citation:** Baek, K.-O.; Lee, J.-M.; Han, E.-J.; Kim, Y.-D. Evaluating Attraction and Passage Efficiencies of Pool-Weir Type Fishways Based on Hydraulic Analysis. *Appl. Sci.* **2022**, *12*, 1880. https://doi.org/10.3390/app12041880

Academic Editors: Dibyendu Sarkar, Rupali Datta, Prafulla Kumar Sahoo and Mohammad Mahmudur Rahman

Received: 14 January 2022
Accepted: 8 February 2022
Published: 11 February 2022



## 1. Introduction

Various types of devices have been used to facilitate the movement of fish past obstacles in waterways. This type of fish facility is commonly known as a fishway, fish ladder, or fish pass [1]. They are essentially water passages around or through obstacles designed to dissipate the energy of the water in a way that allows fish to ascend without undue stress [2]. Fishways must be designed and positioned based on the hydrodynamic characteristics and behavioral patterns of migrating species. When properly designed and placed, fish moving upstream through fishing structures can bypass river barriers and reach portions of the river suitable for growth and reproduction [2,3].

The efficiency of a fishway must be evaluated separately for attraction and passage efficiencies. The purpose of attraction is to lure fish to the fishway entrance, whilst the passage allows the fish that entered the fishway to pass [4]. Specifically, attraction efficiency can be defined as the ratio of the number of fishes that have been attracted to the fishway entrance among the ascending fish. Passage efficiency can be defined as the ratio of the number of fishes that have entirely escaped the fishway exit among the fish that have entered the fishway entrance. The reason for the separate evaluations of fishways between attraction and passage efficiencies is that these two efficiencies show different values according to the fishway types, as shown in Figure 1. This figure shows the statistical classification of fishway efficiency data that have been collected broadly in papers and reports by Bunt et al. [5]. As shown in Figure 1a, the pool-weir type fishway is relatively advantageous for fish attraction, whereas the nature-like fishway has relatively low attraction efficiency. However, the nature-like fishway shows the highest passage efficiency, and the pool-weir type fishway shows the lowest passage efficiency (see Figure 1b). The nature-like fishway refers to a by-pass fishway that bypasses the river barriers. It is not easy to find the fishway entrance for this type of fishway since it is connected to a riverbank. Furthermore, the attraction efficiency is reduced due to the mild slope of the fishway and the low flow rate at the fishway entrance. By contrast, if there is a low velocity inside a fishway, the passage efficiency becomes high because fishes that have entered the fishway can pass regardless of their swimming capacity. The pool-weir type fishways (including ice-harbor type) have a steeper slope than that of the nature-like fishways, and a strong flow occurs inside them. However, they are more advantageous over the nature-like fishway for attracting fish because a relatively strong flow is formed at the entrance.

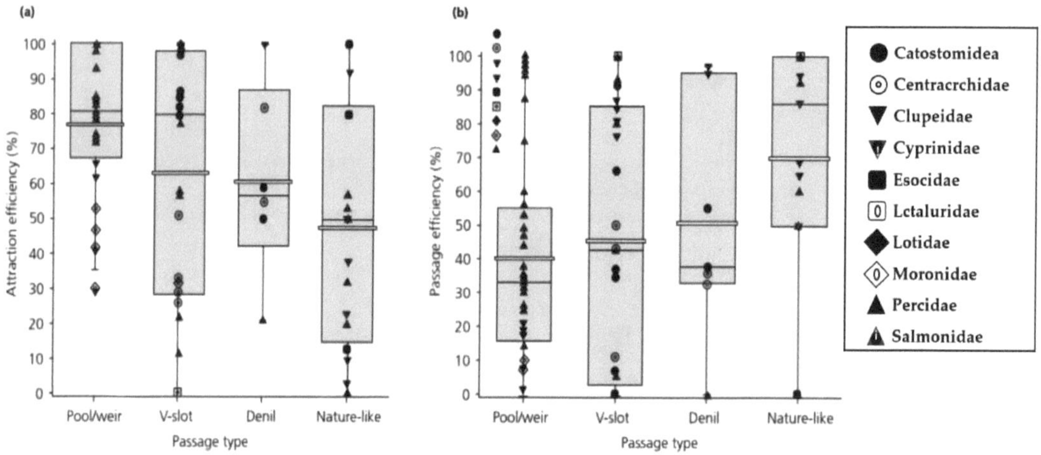


**Figure 1.** Attraction (**a**) and passage efficiency (**b**) according to fishway type [5].

Most fishways installed in small and medium sized rivers in South Korea are pool-weir type. Therefore, the evaluation of fishways should focus on passage efficiency [6]. However, due to hydrological conditions with frequent rainfalls, the riverbed change is considerably higher than that of other countries [7]. This necessitates continuous inspections about the degradation of attraction efficiency. Special care should be taken for maintenance of the vertical slot type with large deviations for both attraction and passage efficiencies. Further, the Denil type shows higher than average attraction and passage efficiencies as compared to other types, as shown in Figure 1.

By evaluating these two efficiencies, the fishway installation and management plans must be established to improve each efficiency. The overall efficiency of existing fishways can be evaluated more directly through fish monitoring. However, when planning for fishways, or when fish monitoring is difficult due to the specific situation, the fishway efficiency can be evaluated indirectly through hydraulic analysis. For attraction efficiency, it is necessary to identify the hydraulic characteristics around the fishway macroscopically. It is essential to analyze the hydraulic characteristics inside the fishway in more detail for passage efficiency. Therefore, two-dimensional (2D) analysis is sufficient for attraction efficiency, whereas the passage efficiency requires three-dimensional (3D) analysis.

Hydraulics in fish passage structures have been investigated for more than 30 years. Most numerical simulations have focused on the flow field inside a fishway according to its type. However, it is surprising to find that there is lesser published research on numerical simulation of flow fields around fishways than those inside [1]. Hereby, the following is a brief look at the previous studies that analyzed the flow velocity structure inside the fishways in three dimensions. Heimerl et al. [8] conducted a three-dimensional numerical simulation on a small-scale fishway (1.13 m in width) installed in the laboratory to identify the flow field in the vertical slot, and drew results that were relatively consistent with the observed flow field. Alvarez-Vazquez et al. [9] used the three-dimensional flow simulation result in the design of appropriate spacing between the slot and the weir of the vertical slot type fishway. Marriner et al. [10] represented the flow pattern in a fishway with a turning pool using a three-dimensional numerical model. If the obstacle's height to be overcome is high, the length of the fishway becomes longer. In the case where the fishway is rotated in the form of a 180-degree curve is called a tunning pool. Studies on this have continued until recently, and Maniecki [11] analyzed the characteristics of the flow field by 3D simulation while changing even the positions of orifices and notch in a fishway with a tunning pool.

In this study, we propose a hydraulic analysis method that can evaluate attraction and passage efficiencies separately. For attraction efficiency analysis, the River-2D model with a fish physical habitat module was used, and the weighted usable area is suggested as an indicator of efficiency. For passage efficiency analysis, the FLOW-3D model, which has a free surface and can analyze the fluid behaviors in hydraulic structures, was used. For the target fishway for evaluation, we selected the ice-harbor type fishway at Dalseong weir in the Nakdong River of Korea.

## 2. Efficiency Evaluation Tools

### 2.1. Attraction Efficiency Evaluation Tool: River 2D

One of the most critical factors in attraction efficiency is the fishway installation location in the weir. In particular, fishways installed in small and medium sized weirs with small steps are more sensitive to location. The optimal method is to install two or more fishways in one barrier because even if one or two fishways do not work well, the other fishways can be used for fish ascending. However, if only one fishway must be installed in one barrier due to limited costs, the following four criteria should be considered [1].

① Close to upstream side: If the weir is not installed perpendicular to the flow direction, the ascending fish try to move upstream along a weir. Thus, the fishway should be installed in riverside close to the upstream sided, so that fish can find the fishway entrance [12].
② Near the thalweg line: During the low water and dry seasons, the flow occurs mainly along the line of maximum depth due to the low water flow. Thus, the fishway should be installed near the deepest line to secure flow in the fishway.
③ Near the line of maximum flow velocity: According to the swimming characteristics of fish, it is known that fish ascend along the line of maximum flow velocity. Therefore, the fishway should be installed near the line of maximum flow velocity [2].
④ The location where scouring and deposition are less inactive: If the area around the fishway entrance is scoured, the structure is separated from the riverbed and prevents the fish from ascending. If deposition occurs in the fishway exit, the fishway loses its function because water does not flow into the fishway.

Therefore, the fishway should be installed in a location where scouring and deposition are less active (Figure 2).

Additionally, Cowx and Welcomme [12], Clay [2], and Lundqvist et al. [13] have proposed various criteria for selecting the location of fishways. The problem is planning and installing fishways in actual weirs in rivers. It is often difficult to follow the criteria due to the complex situation of the river. Moreover, the hydraulic situation that is advantageous for attraction can be different depending on the characteristics of the target fish species. Consequently, it is necessary to develop indicators that can numerically evaluate the attraction efficiency for the target fish species instead of these criteria that state general principles. In this study, we propose the weighted usable area (WUA) as an indicator of attraction efficiency according to the results of Baek et al. [14].

The WUA (Weighted Usable Area) is a primary indicator of the PHABSIM (Physical Habitat Simulation), which is a one-dimensional physical habitat model and based on the Habitat Suitability Index (HSI). The HSI quantitatively represents the relationship between the physical properties of habitat such as flow velocity, water depth, and channel bed. In general, the HSI of fish is indicated as a relative ratio of the number of fish species that appeared in a specific survey location or section to the maximum number of species set as value of 1.0 as shown in Figure 3 [15].

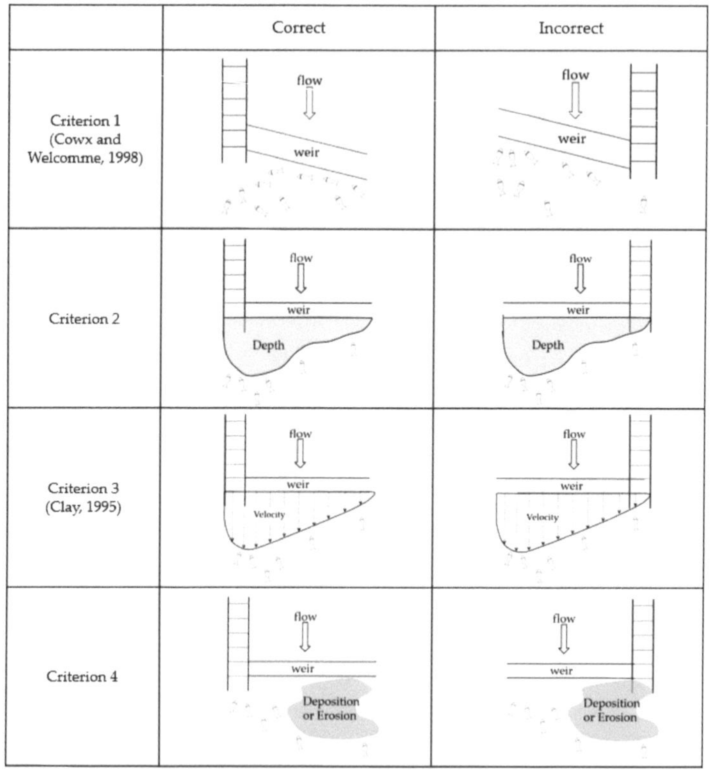


**Figure 2.** Schematic diagram of criteria for installation of fishway on low-head weir [1].

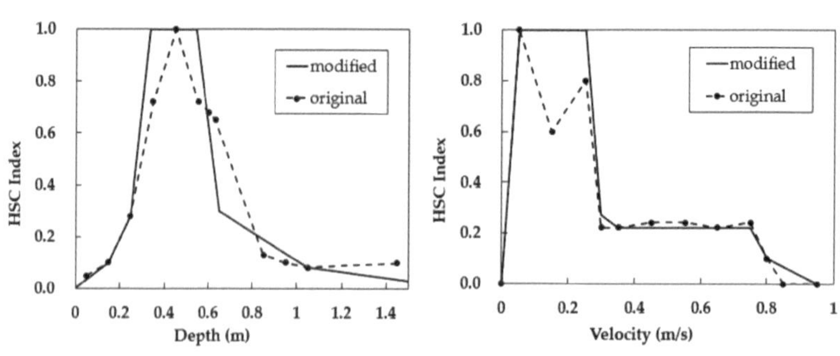


**Figure 3.** Pseudogobio esocinus (Goby minnow) HSI of target fishes in streams [15].

In this figure, the horizontal axis indicates the water depth and flow velocity, namely, physical conditions. The vertical axis indicates the HSI. A value closer to 1.0 means that the water depth and flow velocity are more appropriate for the habitat conditions of the target species. The PHABSIM calculates the Combined Suitability Index (CSI) by calculating the evaluation components, namely flow velocity, water depth, and channel indices. The methods for calculating the CSI include standard computation, geometric mean, lowest limiting factor, and weighting. In this study, we selected the most widely used standard computation method.

The PHABSIM model has two modules: hydraulic simulation and habitat simulation. The primary hydraulic quantities are obtained through hydraulic simulation of the study area. Hence, the average water depth and flow velocity conditions of the channel can be used, or partial simulation can be performed by dividing the channel section into rectangular cells with small areas that can represent a more precise habitat distribution. The habitat simulation calculates the spatial area of the physical habitats of the target fish species by using the water depth, flow velocity, and channel indices obtained from the. hydraulic module. The output is WUA and is calculated by the following equation:

$$\mathrm{WUA} = \sum_{i=1}^{n} A_i \times C_i \quad (1)$$

where $A_i$ is the area obtained by multiplying the distance and width of i cell separated from the channel section. $C_i$ is the CSI of the i cell separated from the channel section as illustrated in Figure 4 [16].

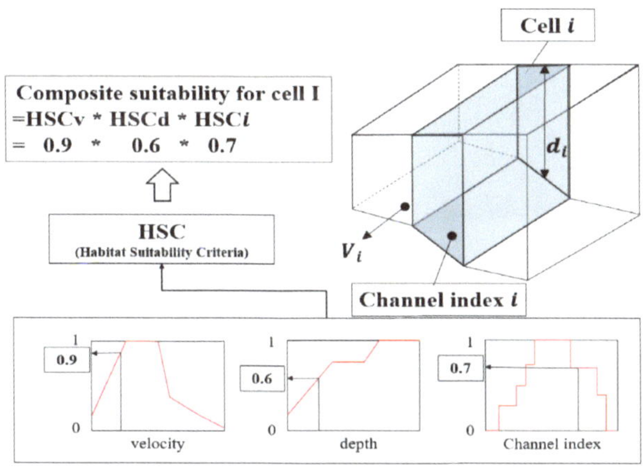


**Figure 4.** Simulation process of WUA [16].

One problem is that it is challenging to select the fishway location where requires two-dimensional consideration since the PHABSIM is a one-dimensional model. Therefore, we used River-2D, which includes PHABSIM as one module. River-2D model was validated through several comparisons with experimental and field results by previous studies [17,18]. River-2D simulates the two-dimensional flow field through the following depth-averaged continuity equation and momentum equation:

$$\begin{gathered}\frac{\partial H}{\partial t} + \frac{\partial q_x}{\partial x} + \frac{\partial q_y}{\partial y} = 0 \\ \frac{\partial q_x}{\partial t} + \frac{\partial}{\partial x}(Uq_x) + \frac{\partial}{\partial y}(Vq_x) + \frac{g}{2}\frac{\partial}{\partial x}H^2 = gH(S_{0x} - S_{fx}) + \frac{1}{\rho}\left(\frac{\partial}{\partial x}(H\tau_{xx})\right) + \frac{1}{\rho}\left(\frac{\partial}{\partial y}(H\tau_{xy})\right) \\ \frac{\partial q_y}{\partial t} + \frac{\partial}{\partial x}\left(Uq_y\right) + \frac{\partial}{\partial y}\left(Vq_y\right) + \frac{g}{2}\frac{\partial}{\partial y}H^2 = gH\left(S_{0y} - S_{fy}\right) + \frac{1}{\rho}\left(\frac{\partial}{\partial x}(H\tau_{yx})\right) + \frac{1}{\rho}\left(\frac{\partial}{\partial y}(H\tau_{yy})\right)\end{gathered} \quad (2)$$

where $q_x = HU$; $q_y = HV$; H is the average depth of the flow; U and V are the depth-averaged velocities of x and y directions, respectively; g is the gravity acceleration; ρ is the water density; $S_{0x}$ and $S_{0y}$ are the riverbed slopes of the x and y directions, respectively; $S_{fx}$ and $S_{fy}$ are the friction slopes of the x and y directions, respectively; and $\tau_{xx}$, $\tau_{xy}$, $\tau_{yx}$ and $\tau_{yy}$ denote the horizontal stress factors [19]. The parameters of River 2D model are the Manning's roughness coefficient. This is a finite element model that can simulate the various flow conditions of a natural river, such as subcritical and supercritical flow.

## 2.2. Passage Efficiency Evaluation Tool: Flow-3D

The most important factor that should be considered concerning passage efficiency is the flow inside the fishway. The flow velocity inside a fishway is not easy to predict because it is highly dependent on external variables such as flow rate and slope. Furthermore, the flow patterns are diverse according to the shape, size, and intervals of structures (e.g., notch, orifice, and weir) installed inside the fishway. Nevertheless, the river design standard [20] in Korea only provides the average flow velocity range (0.5–1.0 m/s) inside the fishway as a design criterion.

The flow structure inside the pool-weir type fishway is generally known to be divided into the streaming flow and the plunging flow [21]. This is briefly illustrated in Figure 5.

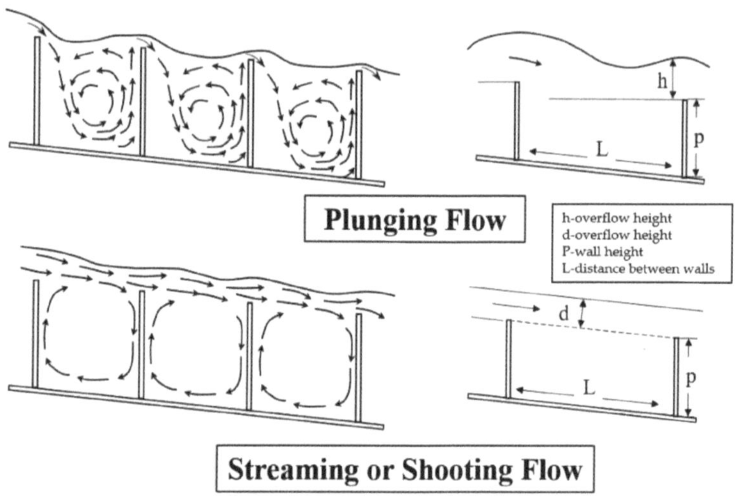


**Figure 5.** Flow pattern in pool/weir fishway [21].

As shown, reverse rotational flow occurs along with streaming flow near the fishway bottom; thus, fish swimming near the bottom are highly likely to lose the upstream direction and fail to ascend. However, fish can easily ascend with plunging flow because of the appropriate flow characteristics. In contrast, if the fish are swimming near the surface, the streaming flow is advantageous for ascending. Therefore, the specific internal flow structure (streaming flow/plunging flow) of the fishway will be an indicator of passage efficiency. A three-dimensional hydraulic analysis is required to discriminate this. Among the various 3D hydraulic models, we used the FLOW-3D model which can quickly analyze the fluid behaviors of hydraulic structures using the free surface. The FLOW-3D model applies the VOF (volume of fluid) theory and is employed in the field of hydraulics where the free surface motion is important. In addition, the FAVOR technique for complex shape expression is added to the FVM (finite volume method) using a rectangular grid for three-dimensional grid division, so that accurate shapes can be expressed quickly [22].

The FLOW-3D model uses the Reynolds equation. The equation has time-averaged the Navier–Stokes equation and the continuity equation as governing equations. Moreover, it provides various options including the k-ε and LES models as the turbulence model.

$$R_{DIF} = \frac{\partial}{\partial x}\left(u_\rho A_x \frac{\partial \rho}{\partial x}\right) + \frac{\partial}{\partial y}\left(v_\rho A_y \frac{\partial \rho}{\partial y}\right) + \frac{\partial}{\partial x}\left(w_\rho A_z \frac{\partial \rho}{\partial z}\right) \tag{3}$$

where $V_F$ is the volume in contact with the fluid; ρ is the density of the fluid; $R_{DIF}$ is the turbulence diffusion term; u, v, and w are the flow velocities in the x, y, and z directions;

$A_x$, $A_y$, and $A_z$ are the contact areas of the fluid in the x, y, and z directions of the cell, respectively [23].

$$\begin{aligned}
\frac{\partial u}{\partial t} + \frac{1}{V_F}\left\{uA_x\frac{\partial u}{\partial x} + vA_y\frac{\partial u}{\partial y} + wA_z\frac{\partial u}{\partial z}\right\} &= -\frac{1}{\rho}\frac{\partial p}{\partial x} + G_x + f_x - b_x - \frac{R_{SOR}}{\rho V_F}u \\
\frac{\partial v}{\partial t} + \frac{1}{V_F}\left\{uA_x\frac{\partial v}{\partial x} + vA_y\frac{\partial v}{\partial y} + wA_z\frac{\partial v}{\partial z}\right\} &= -\frac{1}{\rho}\frac{\partial p}{\partial y} + G_y + f_y - b_y - \frac{R_{SOR}}{\rho V_F}v \\
\frac{\partial w}{\partial t} + \frac{1}{V_F}\left\{uA_x\frac{\partial w}{\partial x} + vA_y\frac{\partial w}{\partial y} + wA_z\frac{\partial w}{\partial z}\right\} &= -\frac{1}{\rho}\frac{\partial p}{\partial z} + G_z + f_z - b_z - \frac{R_{SOR}}{\rho V_F}w
\end{aligned} \quad (4)$$

where $G_x, G_y$, and $G_z$ are body accelerations; $R_{SOR}$ is the mass production term; $f_x$, $f_y$, and $f_z$ are viscous accelerations; and $b_x$, $b_y$, and $b_z$ are the loss generated during passage through the structure.

## 3. Applying Tools to the Fishway and Evaluating Efficiency

The Dalseong Weir, selected as the study site, is located on the Nakdong River in South Korea. As seen in Figure 6, it has two fishways, namely an ice-harbor fishway and a by-pass fishway on the right and left banks, respectively. The Dalseong Weir has a hybrid structure comprising a fixed weir (277 m) and a movable weir (120 m) with a total length of 377 m. Analyze will both the attraction and passage efficiency for the ice-harbor fishway. The ice-harbor fishway is of a set-back type, 104 m long, and 7 m wide. The fishway slope is 1:20. The by-pass fishway is out of scope in this study.

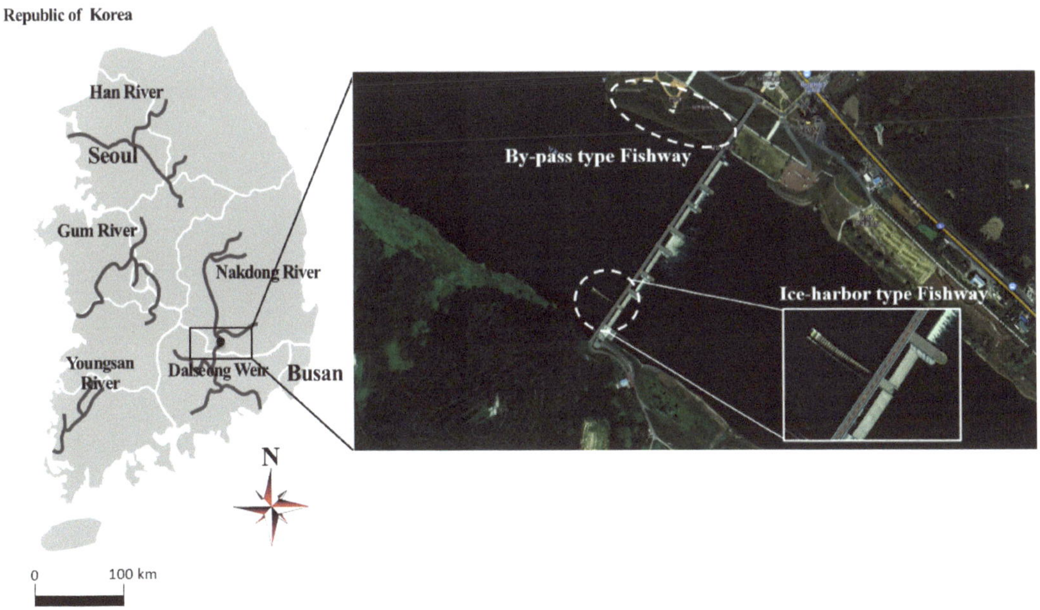


**Figure 6.** Location of ice-harbor type fishway at Dalseong Multifunctional weir.

### 3.1. Evaluation of Attraction Efficiency

To evaluate the attraction efficiency of the ice-harbor type fishway at Dalseong weir, a research area of approximately 1.5 km upstream and downstream of the target fishway was chosen. Furthermore, a finite element mesh was constructed based on the observed river topography data as shown in Figure 7. Fine meshes were constructed for detailed simulation of the fixed weir and fishway. The total number of nodes of the finite element mesh was 4245 and the total number of elements was 8405.

The low water season was selected because the spawning season of most fish inhabiting Korean rivers is spring (April to June); the usage rate of the fishway increases in this season as the upstream and downstream movements of fish increase and this period usually corresponds to the low water season. For upstream boundary conditions for model simulation, we used the Dalseong weir inflow, which is provided in real-time by Korea Water Resources Corporation. The inflow data for April to June (spawning season) were acquired each year between 2013 and 2017. The total number of inflow data were 455. Due to the large differences between maximum and minimum values of daily inflow, using the mode value among the 455 data, resulting in an inflow of 56 $m^3/s$. Furthermore, the management level at the downstream end of the Dalseong weir was found to be always maintained at approximately 13.5 m assigned as the downstream boundary condition. Manning's roughness coefficient was set as 0.020 by referring to the Nakdong River Basic Planning Report [24] (Table 1). Furthermore, the simulation was performed without operating the movable weir as the weir was not operated when the inflow rate was 56 $m^3/s$. Consequently, the water depth and flow velocity field in the low water season were obtained through the hydraulic module of River-2D, as shown in Figure 8a,b.

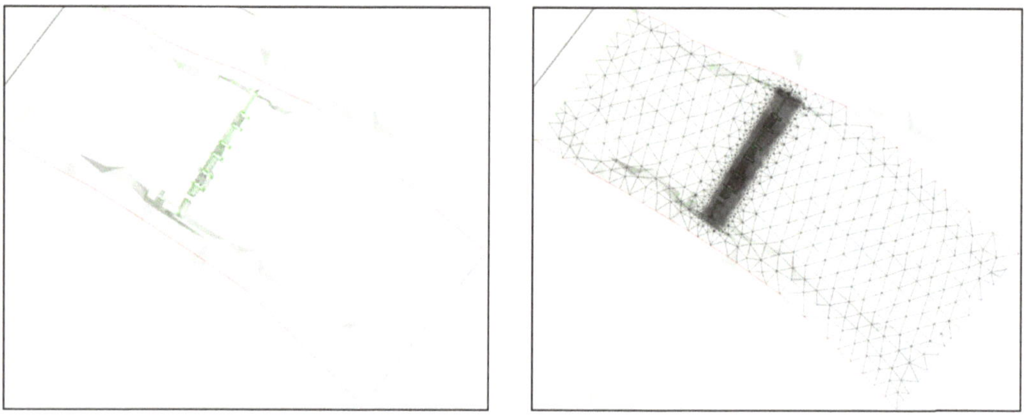

**Figure 7.** Bed contour and finite element mesh at Dalseong weirs.

**Table 1.** Simulation condition of River2D model.

| Section | Upstream Boundary Conditions: Discharge ($m^3/s$) | Downstream Boundary Conditions: Water Surface Elevation (m) | Parameter: Roughness Coefficient |
|---|---|---|---|
| Dalseong weir | 56 | 13.5 | 0.020 |

The HSI information for the target fish species is required to calculate the WUA in the River-2D model. First, the dominant and subdominant fish species were acquired based on the fish monitoring data [25–29] at a point downstream of the Dalseong weir. Goby minnow was selected as the target species because it was the dominant species in general after the installation of the Dalseong weir and subdominant species from 2013 to 2017 (Table 2). Goby minnow is a fish sensitive to water quality, and it is necessary to improve the habitat in the same environment as the fishway [18]. In this study, the research result of Kang et al. [15] was referenced for the HSI of the flow velocity and water depth for Goby minnow, as shown in Figure 2. The simulation result of the WUA for the target species based on the HSI using River-2D is shown in Figure 8c.

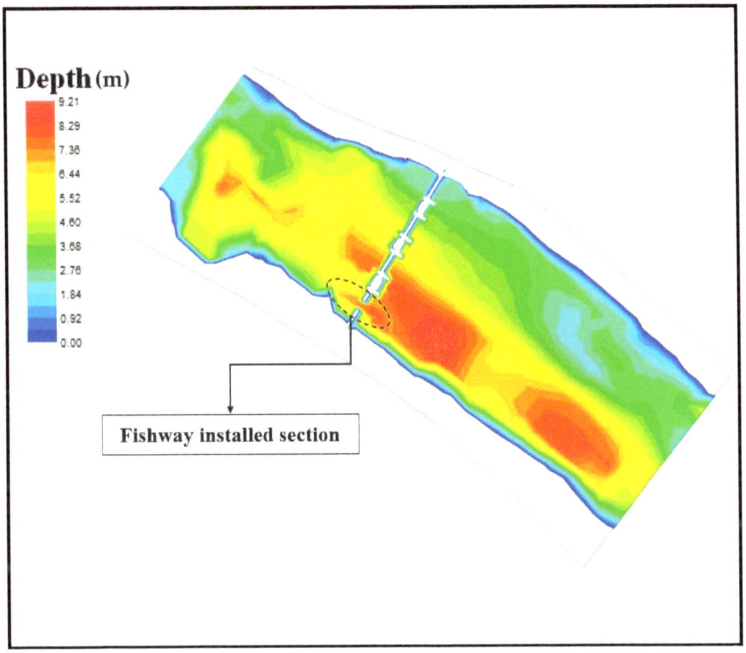


(a)

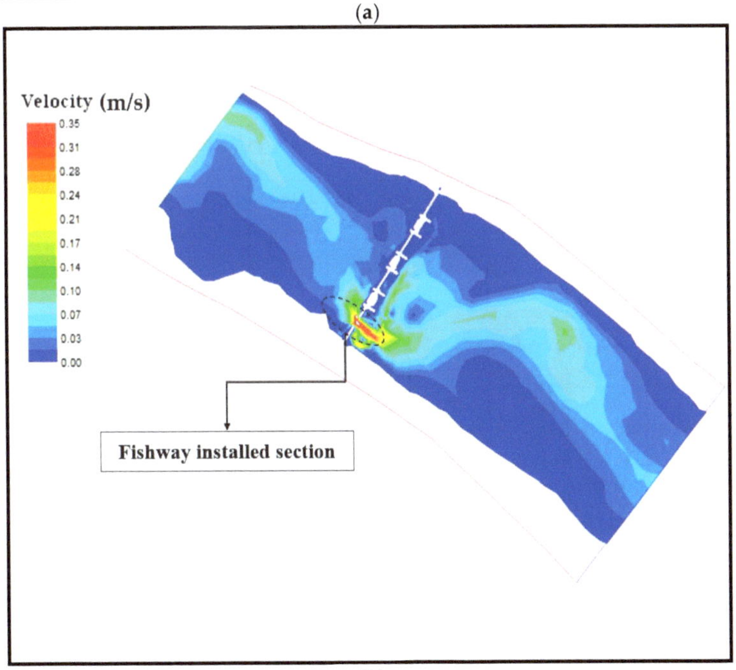


(b)

**Figure 8.** *Cont.*

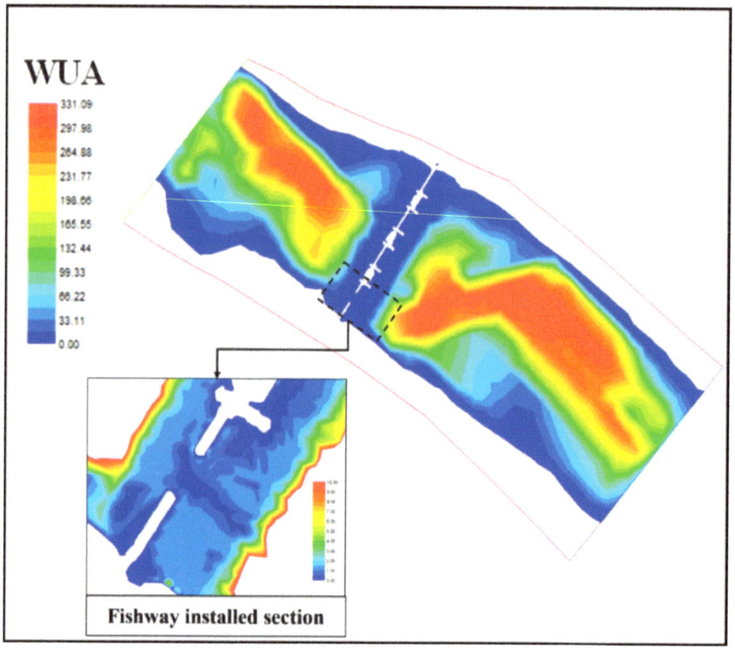


(c)

**Figure 8.** Distribution of bed elevation, depth, velocity and WUA around Dalseong weir. (**a**) Depth distribution; (**b**) Velocity distribution; (**c**) Weighted Usable Area of *Pseudogobio escocinus*.

**Table 2.** Fish monitoring data for Dalseong weir downstream area [25–29].

| Year | Dominant Species | Subdominant Species 1 | Subdominant Species 2 |
|---|---|---|---|
| 2013 | Goby minnow(115) (*Pseudogobio esocinus*) | Korean piscivorous chub (61) (*Opsariichthys uncirostris*) | Korean gudgeon (33) (*Squalidus chankaensis tsuchigae*) |
| 2014 | Goby minnow (25) (*Pseudogobio esocinus*) | Skygager (17) (*Erythroculter erythropterus*) | Light bullhead (14) (*Leiocassis nitidus*) |
| 2015 | Korean sharpbelly (88) (*Hemiculter eigenmanni*) | Bluegill (21) (*Lepomis macrochirus*) | Nile tilapia (15) (*Oreochromis niloticus*) |
| 2016 | Korean gudgeon (39) (*Squalidus chankaensis tsuchigae*) | Bluegill (35) (*Lepomis macrochirus*) | Goby minnow (30) (*Pseudogobio esocinus*) |
| 2017 | Korean sharpbelly (373) (*Hemiculter eigenmanni*) | Korean gudgeon (175) (*Squalidus chankaensis tsuchigae*) | Goby minnow (3) (*Pseudogobio esocinus*) |

As shown in Figure 3, the preferred flow velocity for target fish (Goby minnow) is 0.05–0.25 m/s. In Figure 8b, the flow velocity was 0.30–0.35 m/s inside the fishway and 0.10–0.28 m/s near the fishway entrance. Thus, the preferred flow velocity was generated. Nevertheless, the WUA value, which is an indicator of attraction efficiency, was not large in the fishway entrance (Figure 8c) because of the effect of the water depth. While the preferred water depth of the target fish is 0.3–0.6 m (Figure 3), the water depth of the fishway entrance is more than 6 m (Figure 8a). Consequently, the ice-harbor fishway installed at the Dalseong weir is not effective for attracting the target species. To improve this problem, the location of the fishway entrance can be moved to a location with lower water depth. However, as shown in Figure 8a, the right bank in the downstream of the Dalseong weir at which the fishway is located has a large water depth than the left bank in general. Therefore, a possible method to improve the attraction efficiency of the fishway is to move the fishway from the right bank to the left bank.

### 3.2. Evaluation of Passage Efficiency

The passage efficiency was evaluated based on previous research results [30]. As mentioned earlier, passage efficiency can vary depending on the combination of weir shape, size, spacing, etc. These findings can be found in previous studies [8,31,32]. This study employed the FLOW-3D among commercial CFD models. By applying the RNG turbulent module to the FLOW-3D model, the flow characteristics inside the fishway were analyzed. There are four options for turbulence modules such as LES (Large Eddy Simulation), RNG (Re-Normalization Group), k-ε, and k-ω in FLOW-3D. The flow characteristics are quite sensitive to the turbulence model, so it is necessary to choose an appropriate model in specific situations. In this study, simulation results according to four turbulence models are compared with a measured velocity data acquired from a small-sized fishway which located at the small-sized stream named Gyong-an Cheon. This fishway is same type as the target fishway at Dalseong Weir, but its width and length are short. The inside structure of the small-sized fishway was surveyed as having three rows and 11 columns, whereas the target fishway has seven rows and 37 columns in total. The reason we measured velocity at the small-sized fishway instead of the target fishway is that we could not approach and observe inside of the target fishway due to safety issues. Cross-sectional velocity distribution was measured at the exit-section of the fishway using ADCP (Acoustic Doppler Current Profiler) as shown in Figure 9a. Based on the measured data, the flow rate inflow into the fishway was calculated according to the turbulence model. The flow rates are compared in Table 3. In this table, the inflow flow-rate calculated by the RNG model is the best match with the measured data. In addition, the velocity distribution at the exit-section of measurement by ADCP and that of calculation by the RNG turbulent model are quite similar, as shown in Figure 9. Thus, it can be concluded that the use of the RNG model is reasonable to describe the flow structure inside of ice-harbor type fishways.

**Table 3.** Comparison of measuring and calculating flowrate of inflow into the small-sized fishway.

| | | Flowrate ($m^3/s$) |
|---|---|---|
| **Measurement by ADCP** | | 0.69 |
| **Calculation by** | **RNG** | 0.65 |
| | **LES** | 0.96 |
| | **k-ε** | 0.88 |

The dimensions of the target fishway at Dalsung Weir were approximately 83 m in length and 8 m in width, with a 5% slope. The structure was surveyed as having seven rows and 37 columns in total, with three rows of non-overflow parts and four rows of overflow parts [30]. A movable gate was installed at the fishway exit, which enables flow control in the fishway during the dry season or emergency. The installed fishway is a set-back type, which is different from the existing set-forward type. The set-forward fishway has a protruding entrance in the main body of the weir. Thus, it suffered the disadvantage that fish tend to remain just downstream of the weir, making it difficult for fishes to reach the fishway entrance. The fishway type that has resolved this problem is the set-back type fishway, whose exit is located upstream of the weir and the fishway entrance is located in the main body of the weir [33]. Flowrate into the target fishway was assigned as 0.8 $m^3/s$. The Manning's roughness coefficient was applied as 0.014, which is a commonly used in the concrete structures. Figure 10 shows the snapshot of 3D simulation processes of the target fishway and the number of weirs. The weirs of the fishway were numbered as first from the entrance. There are 35 pools in addition to the movable gate for controlling the flow rate.

(a)

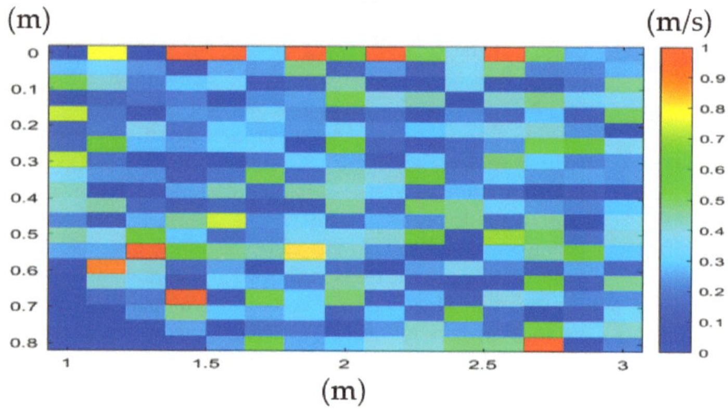


(b)

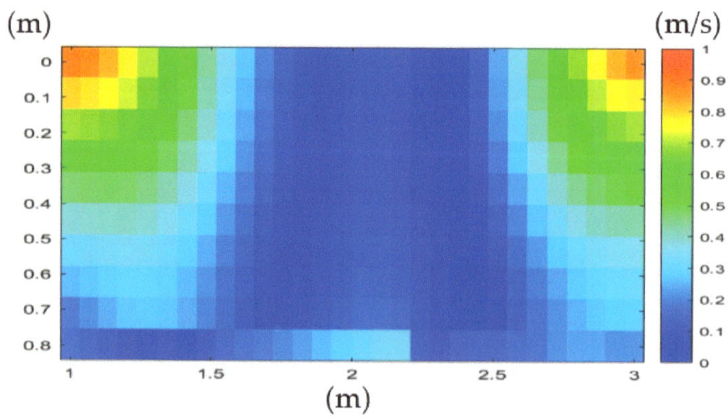


(c)

**Figure 9.** Measuring velocity distribution at small-sized fishway and comparing results by ADCP. (**a**) Velocity measurement by ADCP at exit-section of fishway, (**b**) Velocity distribution at the exit-section of measurement by ADCP, (**c**) Velocity distribution at the exit-section of calculation by the RNG turbulent model.

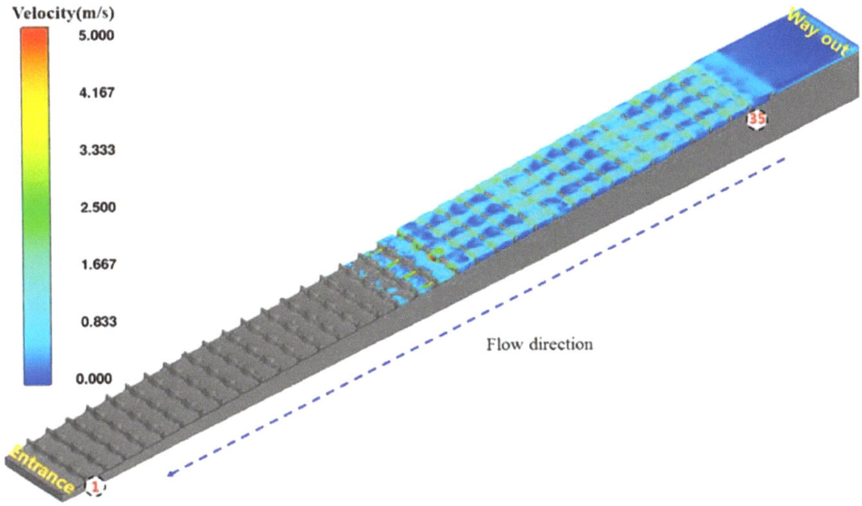


**Figure 10.** Aerial view (snapshot) of flow in the fishway.

Figure 11a shows the results of the second, third, and fourth pools. In general, a plunging flow was formed. There is a rising pattern around the weir, causing complex flow patterns. Figure 11b shows the simulation results side-by-side on the 15th, 16th, and 17th pools, at the mid-point of the fishway. These results showed that the flow pattern is suitable for jumping on fish swimming near the bottom of the pool, such as *Pseudogobio esocinus* (target fish). Figure 11c shows the results for the 32nd, 33rd, and 34th pools. The results are similar to those seen in Figure 11a. A plunging flow is formed in general, but there is a rising flow pattern near the weir, showing complex patterns. The average and maximum flow velocities of the overflow part were 1.10 m/s and 1.24 m/s, respectively, while the average and maximum flow velocities of the undermining were 0.59 m/s and 1.74 m/s, respectively.

In conclusion, the flow velocity field inside the fishway shows that the plunging flow was formed in the fishway middle section, which is advantageous for the ascending of *Pseudogobio esocinus*, the target species. The plunging flow was generally formed at the entrance and exit, but it showed a complex structure as other cells appeared toward the water surface. However, considering that *Pseudogobio esocinus* mainly swim near the flow, the flow velocity field of the surface will not have a significant effect on the passage efficiency.

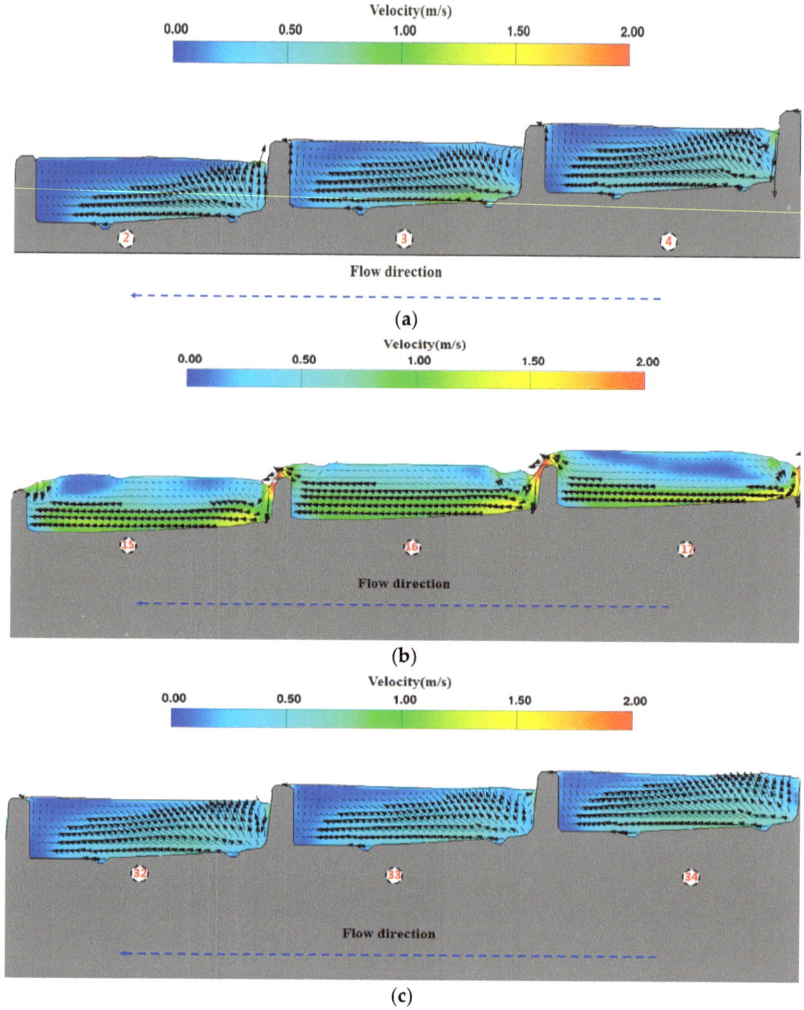


**Figure 11.** Results of ice-harbor type fishway at Dalseong weir: (**a**) Side view of flow structure in pools (2rd, 3th, 4th); (**b**) Side view of flow structure in pools (15th, 16th,17th); (**c**) Side view of flow structure in pools (32th, 33th, 17th).

## 4. Conclusions

A hydraulic analysis method that can indirectly evaluate the attraction efficiency and passage efficiency of fishways was proposed in this study. For attraction efficiency analysis, a River-2D model with a fish physical habitat module was used, and the weighted usable area was suggested as an indicator of efficiency. For passage efficiency analysis, a 3D model with a free surface that can analyze the fluid behaviors of hydraulic structures was used. Moreover, the fishway efficiency was estimated with an ice-harbor fishway installed at Dalseong weir in the Nakdong River. In conclusion, although it was difficult to quantitatively replace the hydraulic analysis results with fishway efficiency, the efficiency was useful for identifying the overall tendency.

The 3D analysis results showed that the flow velocity field inside the fishway showed a similar plunging flow in each pool, but various flow patterns appeared locally. To evaluate and improve the fishway passage efficiency, the average hydraulic values (e.g., average flow velocity, water depth) as well as the results of elaborate modeling should be used. The results of River-2D analysis showed that the fishway located near the right bank was now not appropriate for attracting the target fish to the fishway due to the depth. A possible method to improve the attraction efficiency of the fishway is to move it to the left bank that has a lower depth. The attraction and passage efficiencies certainly require verification through direct fish monitoring in the fishway. However, when planning for a fishway or in difficult fish monitoring circumstances, hydraulic analysis, such as the method used in this study, has the advantage of indirectly evaluating the fishway efficiency.

Although it is possible to roughly confirm the efficiency of fishway through hydraulic analysis, it is difficult to ascertain variations of the fishway efficiency caused by riverbed change. There is a limit to knowing how the riverbed will change. In future work, design and management plans for fishway reflecting riverbed change are needed.

**Author Contributions:** Conceptualizatiation, K.-O.B. and J.-M.L.; methodology, J.-M.L. and E.-J.H.; software, J.-M.L. and E.-J.H.; Validation, J.-M.L. and E.-J.H.; formal analysis, K.-O.B. and Y.-D.K.; investigation, K.-O.B. and J.-M.L. and E.-J.H.; writing-original draft preparation, K.-O.B. and J.-M.L. and E.-J.H.; writing-review and editing, K.-O.B. and Y.-D.K.; visualization, J.-M.L. and E.-J.H.; supervision, K.-O.B. and Y.-D.K.; project administration, K.-O.B. and Y.-D.K. All authors have read and agreed to the published version of the manuscript.

**Funding:** This research was funded by [National Research Foundation of Korea] grant number [NRF-2020R1F1A1076887].

**Institutional Review Board Statement:** Not applicable.

**Informed Consent Statement:** Not applicable.

**Acknowledgments:** This study was carried out with a research grant from the National Research Foundation of Korea (NRF-2020R1F1A1076887).

**Conflicts of Interest:** The authors declare no conflict of interest.

## References

1. Baek, K.O.; Kim, Y.D. A case study for optimal position of fishway at low-head obstructions in tributaries of Han River in Korea. *Ecol. Eng.* **2014**, *64*, 222–230. [CrossRef]
2. Clay, C. *Design of Fishways and Other Fish Facilities*; Lewis Publishers: Boca Raton, FL, USA, 1995.
3. Jungwirth, M.; Schmutz, S.; Weiss, S. *Fish Migration and Fish Bypasses*; Fishing News book: Oxford, UK, 1998.
4. Lucas, M.C.; Baras, E. *Migration of Freshwater Fishes*; Blackwell Science Ltd.: Oxford, UK, 2019. [CrossRef]
5. Bunt, C.M.; Castro-Santos, T.; Haro, A. Performance of fish passage structures at upstream barriers to migration. *River Res. Appl.* **2012**, *28*, 457–478. [CrossRef]
6. Kim, J.; Park, J.C.; Seong, J.; Park, J.H. Classification and Assessment of Fishway in the Tributary of Nakdong River. *Korean J. Ecol. Environ.* **2013**, *46*, 185–191. [CrossRef]
7. Ahn, J.K.; Lee, J.M.; Kim, Y.D.; Kang, B.S. Effect of Climate Change on Long-Term Riverbed Change using GSTARS Model in Nakdong River, Korea. *KSCE J. Civ. Eng.* **2019**, *23*, 1849–1859. [CrossRef]
8. Heimerl, S.; Hangmeyer, M.; Echteler, C. Numerical flow simulation of pool-type fishways: New ways with well-known tools. *Hydrobiologia* **2008**, *609*, 189–196. [CrossRef]
9. Alvarez-Vazquez, L.J.; Martinez, A.; Vazquez-Mendez, M.E.; Vilar, M.A. The importance of design in river fishways. *Procedia Environ. Sci.* **2011**, *9*, 6–10. [CrossRef]
10. Marriner, B.A.; Baki, A.B.M.; Zhu, D.Z.; Thiem, J.D.; Cooke, S.J.; Katopodis, C. Field and numerical assessment of tunning pool hydraulics in a vertical slot fishway. *Ecol. Eng.* **2014**, *63*, 88–101. [CrossRef]
11. Maniecki, L. Numerical modelling of fish passage with turning pools. *Arch. Hydro-Eng. Environ. Mech.* **2018**, *65*, 41–66. [CrossRef]
12. Cowx, I.; Welcomme, R. Rehabilitation of Rivers for Fish. Food and Agriculture Organization of the United Nations, Fishing News Books, Australia, 1998. Available online: https://www.ecrr.org/portals/27/Rehabilitation_of_Rivers_for_fish.pdf (accessed on 13 January 2022).
13. Lundqvist, H.; Rivinoja, P.; Leonardsson, K.; McKinnell, S. Upstream passage problems for wild Atlantic salmon(Salmo salar L.) in a regulated river and its effect on the population. *Hydrobiologia* **2008**, *602*, 111–127. [CrossRef]
14. Baek, K.O.; Ku, Y.H.; Kim, Y.D. Attraction efficiency in natural-like fishways according to weir operation and bed change in Nakdong river, Korea. *Ecol. Eng.* **2015**, *84*, 569–578. [CrossRef]
15. Kang, H.; Im, D.; Hur, H.J.; Kim, K.H. Estimation of Habitat Suitablility Index of Fish Species in the Geum river Watershed. *KSCE J. Civ. Eng. Korean Lit.* **2011**, *31*, 193–203. [CrossRef]

16. US Geological Survey (USGS). *PHABSIM for Windows: User's Manual and Exercises*; Open File Report; Fort Collins Science Center: Fort Collins, CO, USA, 2001; Volume 1, p. 340. [CrossRef]
17. Ghanem, A.; Steffler, P.; Hicks, F. Two-Dimensional hydraulic simulation of physical habitat conditions in flowing 355 streams. *Regul. Rivers Res. Manag.* **1996**, *12*, 185–200. [CrossRef]
18. Baek, K.O.; Lee, J.M.; Ku, T.G.; Kim, Y.D. Evaluation of By-Pass Fishway Operation for Attraction Efficiency Based on GPS Drifter Field Experiments. *Water* **2021**, *13*, 2302. [CrossRef]
19. Lee, S.H.; Oh, K.R.; Cheong, T.S.; Jeong, S.M. An assessment of fish habitat of natural fishway by hydraulic model experiments and numerical analysis. *J. Korea Water Resour. Assoc. KWRA Korean Lit.* **2012**, *45*, 317–329. [CrossRef]
20. Korean Water Resources Association(KWRA). *River Design Standard*; Korean Literature; KWRA: Seoul, Korea, 2009; p. 624.
21. Rajaratnam, N.; Katopodis, C. Plunging and streaming flows in pool and weir fishways. *J. Hydraul. Eng.* **1988**, *114*, 939–944. [CrossRef]
22. Ahn, S.S.; Lee, S.I.; Lee, Z.S. Analysis of Hydraulic Characteristics in Ice-Harbor Fishway. *Korean J. Environ. Sci. Int.* **2012**, *21*, 1395–1406. [CrossRef]
23. Lim, S.Y. An analysis of Hydraulic Characteristics and Fish Swimming Performance Associated with Distance between Baffles in the Ice-Harbor Fishway. Master's Thesis, Myungji University, Yongin-si, Korea, 2010. Korean Literature.
24. Ministry of Land, Infrastructure and Transport (MOLIT). *Nakdong River Basic Plan Report*; Korean Literature; MOLIT: Sejong-si, Korea, 2009; p. 1371.
25. Nakdong River Watershed Management Committee (NRWMC). *Aquatic Ecosystem Monitoring in Weir of the Nakdong River(I)*; Nakdong River Environment Research Center: Sejong-si, Korea, 2013; p. 351.
26. Nakdong River Watershed Management Committee (NRWMC). *Aquatic Ecosystem Monitoring in Weir of the Nakdong River(II)*; Nakdong River Environment Research Center: Sejong-si, Korea, 2014; p. 369.
27. Nakdong River Watershed Management Committee (NRWMC). *Aquatic Ecosystem Monitoring in Weir of the Nakdong River(III)*; Nakdong River Environment Research Center: Sejong-si, Korea, 2015; p. 410.
28. Nakdong River Watershed Management Committee (NRWMC). *Aquatic Ecosystem Monitoring in Weir of the Nakdong River(IV)*; Nakdong River Environment Research Center: Sejong-si, Korea, 2016; p. 371.
29. Nakdong River Watershed Management Committee (NRWMC). *Aquatic Ecosystem Monitoring in Weir of the Nakdong River(V)*; Nakdong River Environment Research Center: Sejong-si, Korea, 2017; p. 399.
30. Jo, J.A.; Han, E.J.; Kim, Y.D.; Baek, K.O. Analysis of Hydraulic Passage Efficiency of Ice-Harbor Type Fishway for Flowrate Change. *J. Korean Soc. Civ. Eng. Korean Lit.* **2013**, *33*, 1841–1850. [CrossRef]
31. Barton, A.; Keller, R. 3D Free Surface Model for a Vertical Slot Fishway. 2003. Available online: https://researchgate.net/publication/253645478_3D_Free_Surfcace_Model_of_a_Vertical_Slot_Fishway (accessed on 13 January 2022).
32. Kim, S.; Yu, K.; Yoon, B.; Lim, Y. A numerical study on hydraulic characteristics in the ice harbor-type fishway. *J. Korean Soc. Civ. Eng. KSCE Korean Lit.* **2012**, *16*, 265–272. [CrossRef]
33. Choi, Y.H. Optimal Fishway Design for Seungchon Small Dam. Master's Thesis, Konkuk University, Seoul, Korea, 2012; Korean Literature, p. 61.

water

MDPI

*Article*

# Changes in Wastewater Treatment Performance and the Microbial Community during the Bioaugmentation of a Denitrifying *Pseudomonas* Strain in the Low Carbon–Nitrogen Ratio Sequencing Batch Reactor

**Tianyuan Chen [1,2], Xiaoyong Yang [1,3], Qian Sun [1,*], Anyi Hu [1], Dan Qin [1], Jiangwei Li [1], Yinhan Wang [4] and Chang-Ping Yu [1,5,*]**

1 CAS Key Laboratory of Urban Pollutant Conversion, Fujian Key Laboratory of Watershed Ecology, Institute of Urban Environment, Chinese Academy of Sciences, Xiamen 361021, China; tychen@iue.ac.cn (T.C.); xyan@env.dtu.dk (X.Y.); ayhu@iue.ac.cn (A.H.); dqin@iue.ac.cn (D.Q.); jwli@iue.ac.cn (J.L.)
2 University of Chinese Academy of Sciences, Beijing 100049, China
3 Department of Environmental Engineering, Technical University of Denmark, DK-2800 Lyngby, Denmark
4 Xiamen Institute of Rare Earth Materials, Haixi Institutes, Chinese Academy of Sciences, Xiamen 361021, China; wangyinhan@fjirsm.ac.cn
5 Graduate Institute of Environmental Engineering, National Taiwan University, Taipei 106, Taiwan
* Correspondence: qsun@iue.ac.cn (Q.S.); cpyu@iue.ac.cn (C.-P.Y.)

**Citation:** Chen, T.; Yang, X.; Sun, Q.; Hu, A.; Qin, D.; Li, J.; Wang, Y.; Yu, C.-P. Changes in Wastewater Treatment Performance and the Microbial Community during the Bioaugmentation of a Denitrifying *Pseudomonas* Strain in the Low Carbon–Nitrogen Ratio Sequencing Batch Reactor. *Water* **2022**, *14*, 540. https://doi.org/10.3390/w14040540

Academic Editors: Dibyendu Sarkar, Rupali Datta, Prafulla Kumar Sahoo and Mohammad Mahmudur Rahman

Received: 29 December 2021
Accepted: 8 February 2022
Published: 11 February 2022



**Abstract:** The low carbon–nitrogen ratio (C/N) of influent wastewater results in the insufficient carbon source for the process of denitrification in urban wastewater treatment plants (WWTPs). A denitrifying bacterial strain, *Pseudomonas* sp. JMSTP, was isolated and demonstrated effective denitrification ability under a low C/N ratio of 1.5–4 (*w*/*w*) in anoxic conditions. Sequencing batch reactor (SBR) studies were conducted to test the bioaugmentation of JMSTP on total nitrogen (TN) removal under the influent COD/N ratio of 3/1. After the second bioaugmentation, the TN of effluent in the bioaugmented SBR was significantly lower than that in the control SBR. Redundancy analysis results showed that there was a positive correlation between *Pseudomonas* sp. abundance and TN removal in the bioaugmented SBR. Microbial community analysis showed that, especially after the second bioaugmentation, the abundance of *Pseudomonas* sp. decreased rapidly, but it was still much higher than that in the control SBR. Correlation network analysis showed that after the addition, *Pseudomonas* sp. had no significant co-occurrence relationship with other native bacteria, owing to the quick increase and decrease. Our results suggest that JMSTP shows the potential to enhance TN removal through bioaugmentation. Since the effect of bioaugmentation gradually diminishes, further research is still needed to investigate its long-lasting applications.

**Keywords:** bioaugmentation; denitrification; network analysis; *Pseudomonas*; low C/N ratio

## 1. Introduction

With the rapid development of the economy and the continuous increase in urbanization, water pollution has been a severe problem in China. Therefore, the environmental standards implemented at existing wastewater treatment plants have become stricter [1]. For example, the effluent standard of total nitrogen (TN) discharged from wastewater treatment plants has been improved from 20 mg/L to 15 mg/L, or even 10 mg/L in certain areas, and therefore a more effective strategy for TN removal will be required. Bioaugmentation, which enhances the removal of pollutants in original systems by adding functional strains or populations [2–5], has shown the potential to improve nitrogen removal in wastewater treatment systems [6,7].

In the full-scale conventional nitrification–denitrification process, removing TN is the conversion of ammonia to nitrate via nitrite with oxygen, and the conversion of nitrate to nitrogen gas or nitrous oxide via nitrite in the anoxic environment [8]. The whole biochemical reaction is catalyzed by several known enzymes, such as ammonia monooxygenase, nitrite oxidoreductase, nitrate reductase, nitrous oxide reductase, and the functional genes, including *amo*A, *nar*G, *nir*, *nor*B, *nos*Z, were characterized [9,10]. In general, heterotrophic denitrifying bacteria are involved in the conventional nitrification–denitrification process. As the carbon-to-nitrogen ratio in China's wastewater influents can be as low as about 3 or less (*w*/*w*), insufficient carbon limits denitrification in the entire nitrogen removal process [2,11]. Therefore, strains with denitrification functions at low C/N ratios have attracted increasing attention. The isolation and characterization of highly functional strains with high denitrification efficiencies at low C/N ratios are important. To date, *Paracoccus*, *Pseudomonas*, *Bacillus*, *Alcaligenes*, and *Rhodococcus* sps. have been reported to have a denitrification efficiency at low C/N [12–15]. However, limited studies have successfully inoculated selected bacteria into activated sludge systems by bioaugmentation to improve nitrogen removal efficiency. Therefore, further bioaugmentation studies are needed to evaluate denitrification performance and understand the mechanism based on the structure and function of microbial communities.

In natural environments, microorganisms do not exist individually, but they directly or indirectly form complex microbial interactions, such as mutualism, symbiosis, parasitism, predation, favoritism, and competition. Interaction has positive, negative, and neutral effects on microbial populations. Denitrification was boosted via inter-genus cooperation, competitive and symbiotic relationships between key taxa (*Azoarcus*, *Paracoccus*, *Thauera*, *Stappia*, and *Pseudomonas* sps.), and other heterotrophic bacteria in aeration reactors to construct the optimal niche contributing to the high TN removal efficiency [16]. Chen et al. used a mixture of heterotrophic nitrification–aerobic denitrification bacteria, including *Agrobacterium tumefaciens* sp. LAD9, *Comamonas testosteroni* sp. GAD3, and *Achromobacter xylosoxidans* sp. GAD4, to enhance the denitrification of a pilot-scale SBR [17]. The results show that when the COD/N ratio is 8, the effluent could meet the first class requirement of the National Municipal Wastewater Discharge Standards of China (TN $\leq$ 15 mg/L). Based on the community analysis results, they considered that although the inoculated mixed bacteria were present in low abundance, they might have a synergistic effect with other indigenous microorganisms in the reactor. Therefore, whether bioaugmented functional microorganisms can have benign interactions with the original microorganisms in the system is of particular concern. High-throughput sequencing data contains large amounts of population information and has the great advantage of exploring microbial interrelationships. The combination of network analysis and high-throughput sequencing provides a valuable research tool for discovering the microbial relationships needed for the construction and stable maintenance of microbial communities [18].

This study aims to improve nitrogen removal efficiency by bioaugmentation in the laboratory-scale bioreactor and understand the underlying mechanism through the changes of microbial communities. First, a denitrifying bacterial strain was isolated from activated sludge, and the denitrification characteristics of the strain were studied. The performance of nitrogen removal ability of the isolated strain was investigated in a laboratory-scale sequencing batch reactor (SBR) by bioaugmentation. The denitrification performance of the inoculated reactor is discussed with reference to the characteristics of the microbial community and the relative abundance of the bioaugmented strain. Furthermore, the shift of microbial communities is linked and discussed.

## 2. Materials and Methods

### 2.1. Media and Isolation Strategy

Three media for isolating facultative anaerobic denitrifying bacteria are as follows:

Enrichment medium ($CH_3COONa$ = 2.84 g; $KNO_3$ = 2 g; $KH_2PO_4$ = 0.5 g; $MgSO_4 \cdot 7H_2O$ = 0.41 g; NaCl = 3.28 g; pH = 7.2–7.5; distilled water = 1000 mL; C/N = 3/1; COD/N = 8/1).

Isolation medium ($CH_3COONa$ = 2.84 g; $KNO_3$ = 2 g; $KH_2PO_4$ = 1 g; $K_2HPO_4$ = 1 g; $FeCl_2{\cdot}6H_2O$ = 0.5 g; $CaCl_2{\cdot}7H_2O$ = 0.2 g; $MgSO_4{\cdot}7H_2O$ = 1 g; NaCl = 10 g; Agar = 20 g; distilled water = 1000 mL; bromothymol blue (BTB, 1% in ethanol) = 1 mL/L; pH = 7.2–7.5).

Fermentation medium (peptone = 10 g/L; yeast extract = 5 g/L; NaCl = 10 g/L; pH = 7.2–7.5).

Denitrifying bacteria were isolated from activated sludge collected from a sewage treatment plant in Xiamen, China. The sludge was slightly treated with a tissue homogenizer. Nitrogen gas was blown into the enrichment medium for more than 15 min, sealed, and sterilized. Then, sludge was added for static culture. The medium was transferred when nitrate was consumed. During the transfer process, the C/N ratio and initial concentrations of $NO_3^-$-N gradually decreased, eventually reaching by as low as 1.5 ($w/w$) and 200 mg/L, respectively. Bacteria in the logarithmic phase in the enrichment medium were diluted with gradients of $10^{-1}$, $10^{-2}$, $10^{-3}$, $10^{-4}$, and $10^{-5}$ and evenly spread on isolation medium plates. The light blue area indicated by bromothymol blue (BTB) was selected for further picking and streaking until pure bacteria were obtained [19].

### 2.2. Batch Experiment

The growth and denitrification capability of isolated strain JMSTP under different dissolved oxygen conditions were examined. The DO of anoxic environment was set to lower than 0.1 mg/L, and that of aerobic environment was set to higher than 8 mg/L. $CH_3COONa$ and $KNO_3$ were used as the only carbon and nitrogen sources in batch experiments with different C/N ratios or different incubation temperatures, and the initial nitrogen concentration of the medium was 200 mg/L. The inoculation volume ratio was 5%. C/N of the batch experiment was set to 1, 1.5, 2, 3, and 4, respectively. The incubation temperature was set as 15 °C, 20 °C, 25 °C, and 30 °C, respectively.

The selected strain JMSTP was applied for the nitrate removal from actual wastewater in shake flasks. When the microorganisms were in the logarithmic growth phase in the fermentation medium at 30 °C, they were collected by centrifugation (8000 rpm and 10 min). A total of 0.39 g of the dry cell weight was added to 500 mL of the actual wastewater from the inlet of the sewage plant spiked with 30 mg/L $NO_3$–N. Control flasks were not inoculated with functional microorganism. They were placed in a shaker at 30 °C and 100 rpm, and duplicate flasks were conducted for each experimental setup. Samples were collected every 24 h for a total of 4 days, and pretreated by centrifugation (8000 rpm and 5 min) before further analysis.

### 2.3. Bioreactor Operation Strategy

Two 6 L SBRs were set up using artificial wastewater as the influent. One was for experimentation (JMSTP SBR), and the other was for control (control SBR). The mixed liquid suspended solid (MLSS) was set to 3000 mg/L using activated sludge collected from the sewage treatment plant in Xiamen, China. The SBR cycle time was 6 h, and the hydraulic retention time (HRT) was 12 h. The operational strategy was as follows: Fill (20 min) → Idle stage (60 min) → Aeration stage (120 min) → Anoxic/Anaerobic stage (90 min) → Settle stage (60 min) → Draw (10 min). The first inoculation of strain JMSTP was performed after 20 days when the two sets of reactors were stable for effluent TN in 18–22 mg/L. The inoculation dosage was 5% MLSS ($w/w$). This means adding strain JMSTP to achieve 150 mg/L in the JMSTP SBR, but no additional bacteria was added to the control SBR. Ten days later, the same inoculation dosage was conducted again for the second bioaugmentation. The effluent was collected once daily on the 1st–20th and 37th–50th days, and effluents from three different sequencing cycles per day were collected on 21–36 days after the inoculation. The composition of the artificial wastewater was as follows: $C_6H_{12}O_6$ (Glucose anhydrate) = 71.52 mg; $(NH_4)_2SO_4$ = 120 mg; $NaHCO_3$ = 220 mg; $KH_2PO_4$ = 24 mg; $MnCl_2$ = 0.19 mg; $CuCl_2$ = 0.022 mg; $MgSO_4$ = 5.6 mg; $FeCl_3{\cdot}6H_2O$ = 0.88 mg; $CaCl_2$ = 1.3 mg; $ZnCl_2$ = 0.0018 mg; distilled water to 1000 mL. The artificial wastewater contained COD of 76.53 mg/L and TN of 25.45 mg/L with COD/N = 3:1 (C/N = 9:8, $w/w$).

### 2.4. Analytical Methods

The cell optical density was measured at 600 nm (OD600) using a spectrophotometer (Metash, UV-5200). COD was detected by a COD analysis kit (5B-3C (V8), Lianhua Technology). $NH_4^+$–N, $NO_3^-$–N, and TP were detected using a flat-bottom 96-well plate method [20]. The MLSS was tested according to standard method procedures.

Polymerase chain reaction (PCR)-agarose gel electrophoresis was used to detect functional genes in the strain, including *nir*S, *nir*K, *cnor*B, *qnor*B, and *nos*Z, according to previous studies [21–23], and the details of primers are shown in Supplementary Materials Table S1. The DNA of strain JMSTP was extracted using FastDNA SPIN Kit for soils (Qbiogene-MP Biomedicals, Irvine, CA, U.S.A). A total of 10–50 ng of DNA was added to 25 μL of reaction solution containing 12.5 μL of TaqMan® Gene Expression Master Mix (2×) (Applied Biosystems, CA, U.S.A.), 1 μM of each primer, and sterile $ddH_2O$.

### 2.5. Microbial Community Analysis

DNA was extracted using FastDNA SPIN Kit for soils (Qbiogene-MP Biomedicals, Irvine, CA, U.S.A.). The V4-V5 region of 16S rRNA genes was amplified using a universal primer pair 515F (5′-GTG YCA GCM GCC GCG GTAA-3′) and 907R (5′-CCG YCA ATT YMT TTR AGT TT-3′). The samples were sequenced on an Illumina Miseq platform (Illumina Inc., San Diego, CA, USA) with the paired-end approach (2 × 300 bp) [24]. Amplicon sequencing data was processed using the LotuS pipeline, as described in the previous study [25]. The quality control settings were adjusted as follows: (1) average sequence quality > 27; (2) sequence length > 170 bp; (3) no ambiguous bases; (4) no barcode and primer mismatches; (5) homopolymer < 8 bp. The obtained qualified reads were chimera-checked and clustered into operational taxonomic units (OTUs) with a 97% identity cutoff using the UPARSE method [26]. Subsequently, the taxonomic assignment of each OTU was made using the RDP classifier in SILVA v132 as the reference database [27], with a bootstrap cutoff of 80% using the QIIME pipeline [28]. All samples were randomly subsampled to the smallest library size (105,000 reads) to standardize the uneven sequencing efforts. Data were uploaded to NCBI with numbers PRJNA706543. Twelve samples of each reactor from the 20th–23rd and 30th–37th days were analyzed (24 samples in total), with labels of B1 (1 day before the first bioaugmentation), D1_1~3 (1st–3rd days after the first bioaugmentation), and B2 (1 day before the second bioaugmentation), and D2_1~7 (1st–7th days after the second bioaugmentation), respectively.

Statistical analysis was performed using R × 64 4.0.2 (http://www.R-project.org/ access on 28 December 2021). The community diversity was calculated using R package vegan [29] and package picante [30]. Network analysis was conducted to evaluate the relationships between the species in the reactors using the R package igraph [31] and Hmisc [32]. The OTUs, with the occurrence of at least 75% in the samples, were used to build the network. The network was visualized using Gephi 0.9.2 (https://gephi.github.io/ access on 28 December 2021). Redundancy analysis (RDA) was analyzed using R package vegan [29] and drawn by Hellinger distance.

## 3. Results

### 3.1. Isolation of Nitrogen Removal Bacteria

After several months of enrichment with $KNO_3$ as the only nitrogen source, one of thirteen isolated strains was selected due to its better nitrogen removal efficiency. This strain belongs to the genus *Pseudomonas*, with high similarity to *Pseudomonas songnenensis* sp. NEAU-ST5-5(T) (99.51%, Figure 1a). Therefore, it is named *Pseudomonas* sp. JMSTP. Based on the scanning electron microscope imaging technique, strain JMSTP has a rod-like shape (Figure 1b). Its growth and denitrification ability was tested under both aerobic and anaerobic conditions (Figure 2). The strain showed exponential growth and reached a stable stage after 12 h of inoculation under aerobic conditions. The denitrification capacity achieved a $NO_3^-$–N removal efficiency of 20.4%. Under anaerobic conditions, the strain grew relatively slowly and reached a stable stage within 30 h, while its denitrification capacity reached 100% within 24 h, with a brief accumulation of nitrite. Denitrification using nitrite was verified using $CH_3COONa$ and $NaNO_2$ as the sources of carbon and nitrogen at the initial concentrations of 300 and 100 mg/L, respectively. The results show that JMSTP could remove more than 97% of nitrite within 12 h (Table S2). As shown in Figure S1, the strain is associated with positive PCR reactions for genes *nir*S, *cnor*B, and *nos*Z. The sequencing of the PCR products suggested that the strain harbors the genes of interest and were deposited in NCBI with accession numbers OL442066, OL442065 and OL456216, respectively.

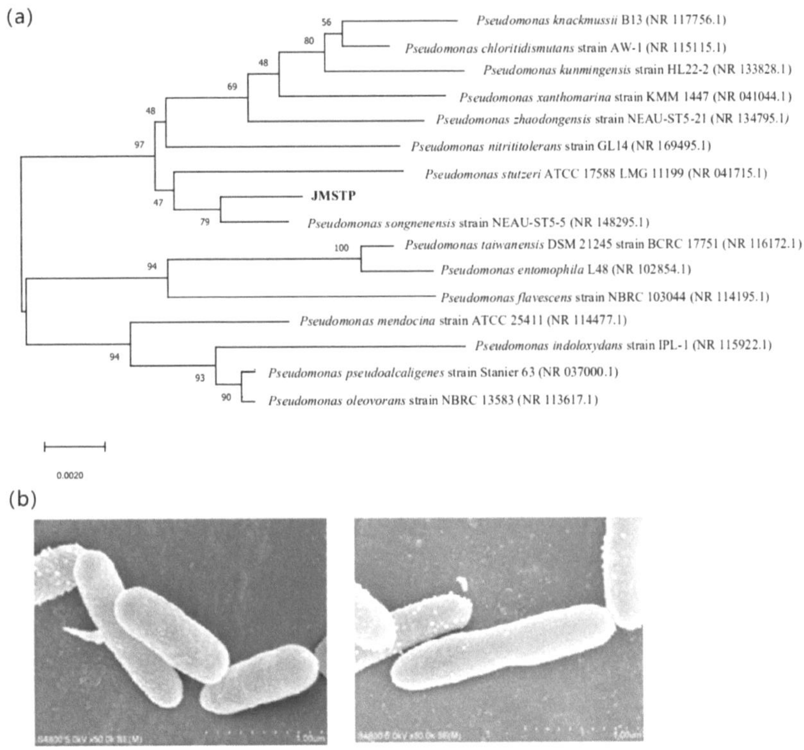


**Figure 1.** Phylogenetic tree derived from 16S rRNA gene sequence of strain JMSTP (**a**) and scanning electron microscope of strain JMSTP (**b**).

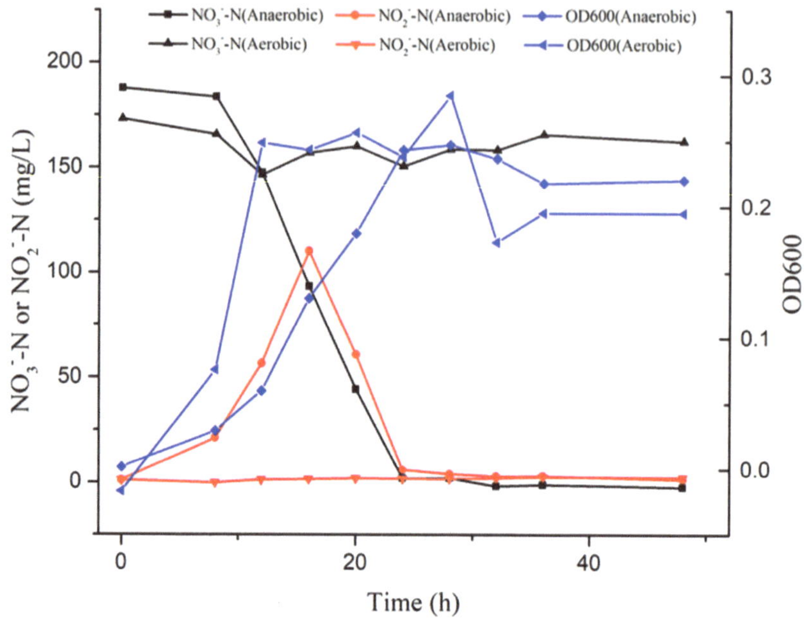


**Figure 2.** Growth and denitrification of strain JMSTP under different dissolved oxygen conditions (25 °C, C/N = 1.5, the DO of anaerobic < 0.1 mg/L, aerobic > 8 mg/L).

*3.2. Denitrification in the Batch Study*

Since the low C/N ratio limits denitrification in wastewater treatment systems, denitrification efficiencies were tested by strain JMSTP at the C/N ratios of 1, 1.5, 2, 3, and 4. As shown in Figure 3a, strain JMSTP has a rapid denitrification ability in the C/N range of 1–4. Removal efficiencies were 100% within 38 h for C/N ratios 3 and 4, but 77.62, 96.45, and 90.43% for C/N ratios 1, 1.5, and 2, respectively. In addition, the denitrification ability of strain JMSTP was tested at 15, 20, 25, and 30 °C with a C/N ratio of 3. As shown in Figure 3b, complete denitrification is achieved in the temperature range of 20–30 °C.

Attempts were made to see if strain JMSTP can be adapted to real urban sewage. As shown in Figure 4, fast denitrification is achieved in the JMSTP group with $NO_3^-$–N concentrations reduced to 1 mg/L within 24 h, but to 12 mg/L in the control group. Nitrate concentrations increased slightly from 24 to 96 h, and this may be due to aerobic nitrification or the release of nitrate due to the death and decomposition of sludge microorganisms [33,34]. The removal efficiency of $NO_3^-$–N was 87.20% at 96 h with strain JMSTP, which was higher than 71.15% in the control group.

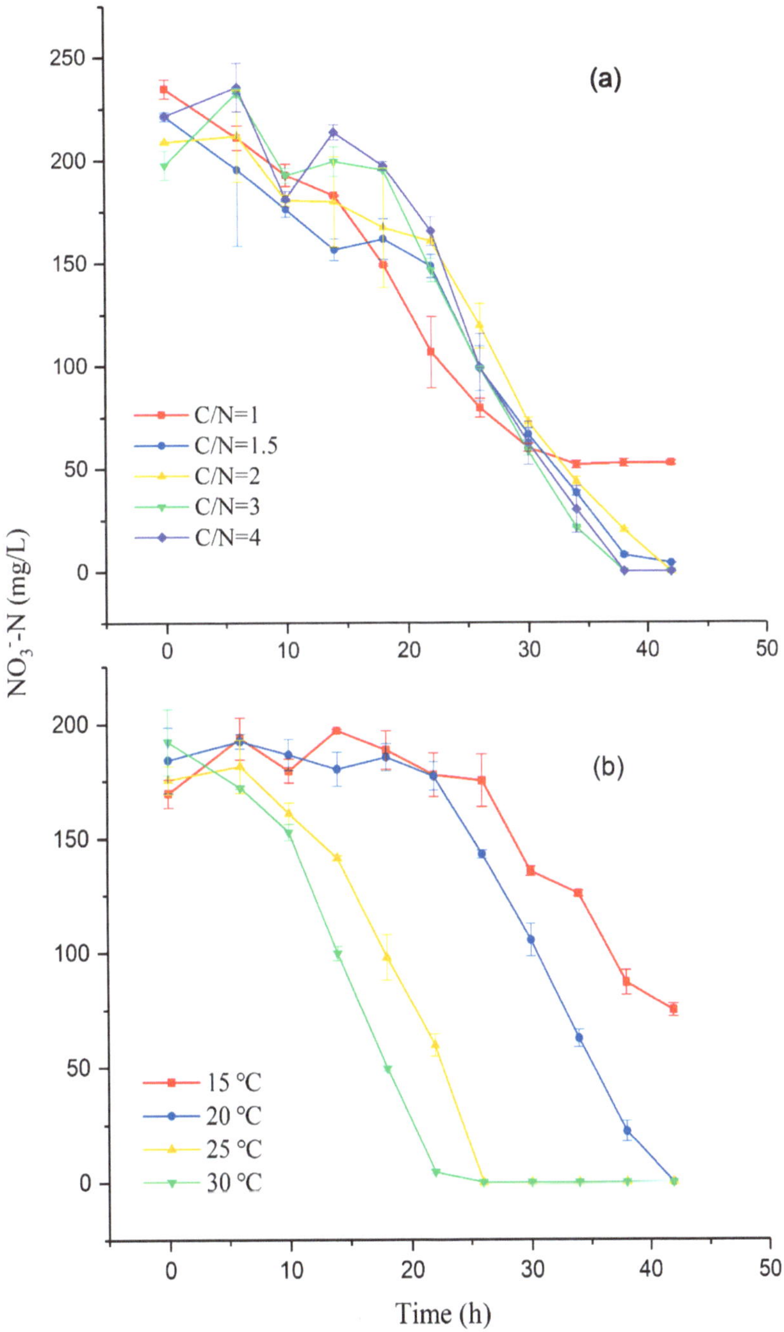


**Figure 3.** Removal of nitrate concentrations by strain JMSTP under different C/N ratios (**a**) and temperatures (**b**).

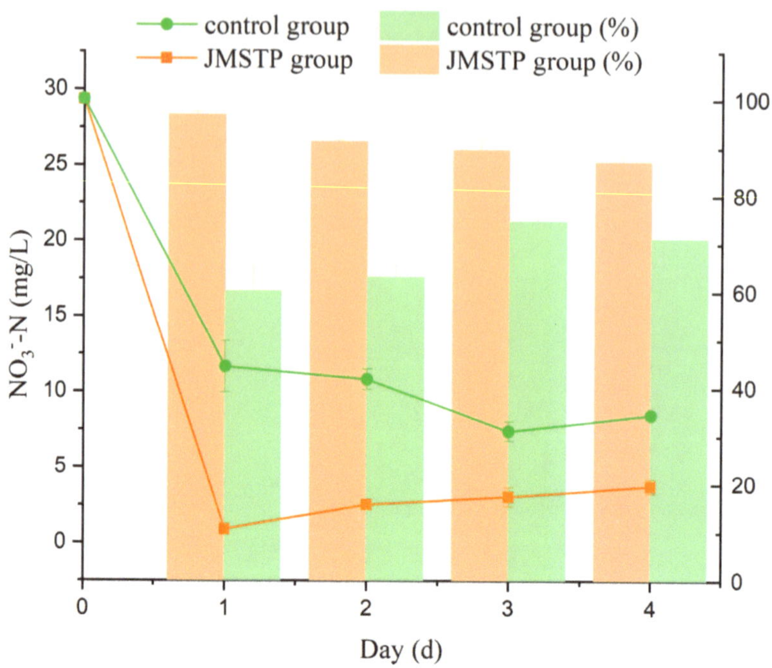


**Figure 4.** Removal of nitrate concentrations by strain JMSTP in real wastewater.

### *3.3. Denitrification in SBR*

Laboratory-scale SBRs were performed to test nitrogen removal using activated sludge with the bioaugmentation of strain JMSTP. Particular emphasis was placed on denitrification efficiencies under harsh conditions with a carbon to nitrogen ratio of 9:8 ($w/w$). Figure 5 shows the results of TN in effluent. A total of 30 cycles were analyzed within the 10 days of the first inoculation, of which 21 cycles of JMSTP SBR had lower TN in effluent than control SBR. However, there was no significant difference in TN of effluent between the two reactors ($t$-test, $p > 0.05$). The effluent of the two cycles after our inoculation was visually turbid. The COD of the effluent in JMSTP SBR can be observed to increase sharply in the cycle after inoculation (Figure 5) and reaches 110 mg/L. However, when the effluent sample was filtered through a 0.45 μm syringe filter, the soluble COD dropped to about 40 mg/L. After the second inoculation, although COD still appeared as a sharp rise after inoculation, the TN concentration in JMSTP SBR was consistently lower than that in the control SBR ($t$-test, $p < 0.05$). This indicates the improvement of nitrogen removal by strain JMSTP bioaugmentation. TN decreased to 10.53–15.54 mg/L from days 4 to 7 after inoculation in the JMSTP SBR, while it was 18.08–22.07 mg/L in the control SBR, under the influent TN of 25.45 mg/L. The TN residue in the effluent, which was lower than 16 mgN/L, and met the criteria of the Grade I standard A (TN $\leq$ 20 mgN/L) in China's wastewater discharge regulation.

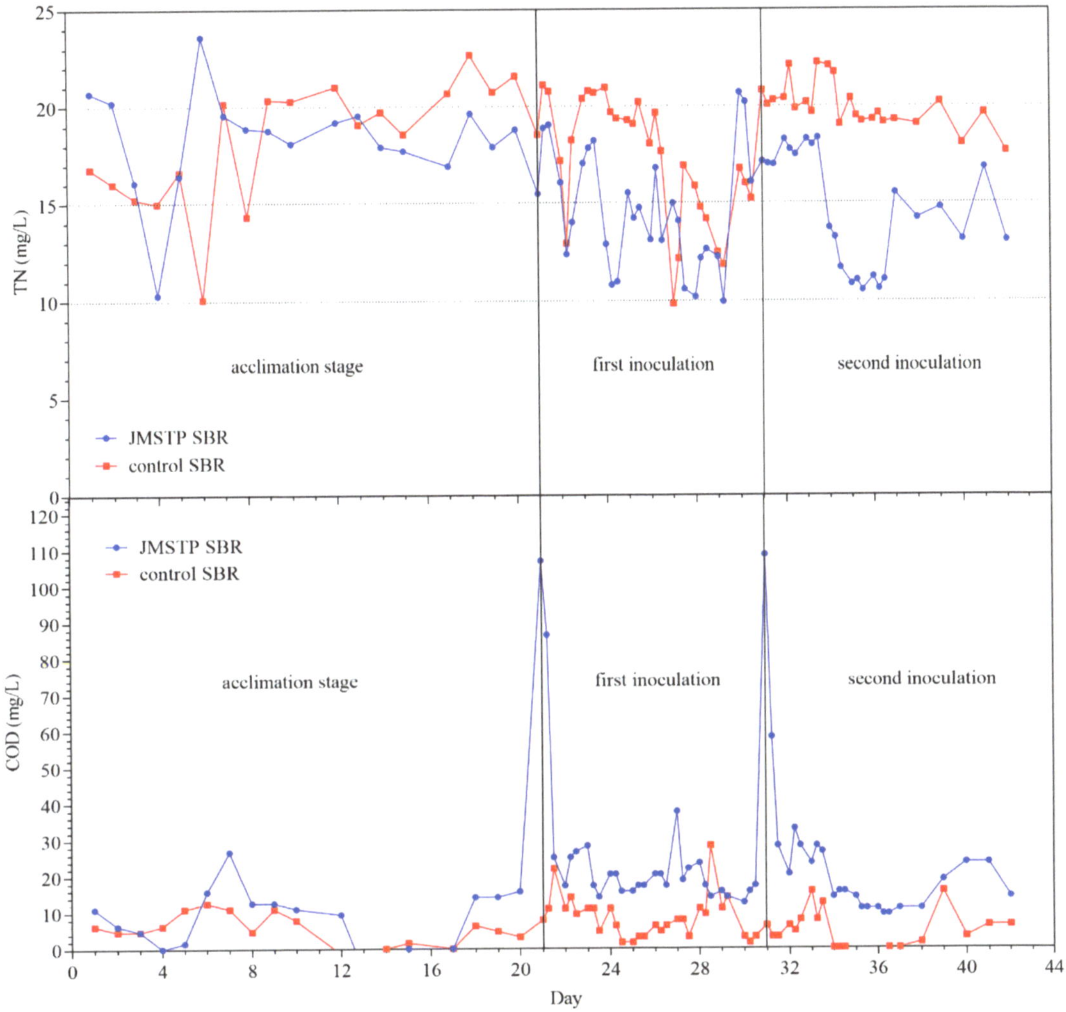


**Figure 5.** Variations of TN and COD concentrations in the effluent of the JMSTP SBR and control SBR using artificial wastewater.

### 3.4. Microbial Community Analysis in Lab-Scale SBR

The diversity of the microbial community and the correlation of different microorganisms in the experimental and control SBRs were analyzed using high-throughput sequencing techniques. Alpha diversity is a comprehensive indicator of richness (e.g., Chao1; Ace), evenness (e.g., Pielou), and the results of microbial diversity, e.g., the Shannon index and Simpson index, are shown in Table S3. The Goods coverage was over 99%, indicating that the sequencing depth basically covered the majority of the species in the samples. These indices of diversity generally showed similar trends between the two reactors. However, the indices after the day of each bioaugmentation in the JMSTP SBR decreased abruptly, but they recovered quickly from the second day. This means that the addition of strain JMSTP could only temporarily affect the diversity of the microbial community within the reactor. Chen et al. operated a bioaugmentation reactor and a control reactor, and they found that the Shannon–Weaver indices of two reactors were similar under long-term operation [17], which is in agreement with our findings. The proportion of OTUs shared between the two SBRs fluctuate between 41.1% and 48.5% within 12 sampling days (shown in Table S4). However, the microbial communities of the two reactors had a similar percentage of shared OTUs before the bioaugmentation of strain JMSTP, implying bioaugmentation did not largely change the OTUs in the reactor [35].

At the phylum level, the dominant phyla in the control reactor are *Bacteroidetes*, *Actinobacteria*, *Proteobacteria*, *Chloroflexi*, *Nitrospirae*, and *Planctomycetes* sps. The JMSTP SBR has one additional dominant phylum, which is *Synergistetes* sp. (relative abundance > 0.01). The types of core microorganisms at the phylum level are generally similar between the two reactors. A bar graph of relative abundance at the genus level is shown in Figure 6. *Pseudomonas* sp. JMSTP showed a significant decrease on the second day after each inoculation, which was consistent with the COD results. However, the relative abundance of *Pseudomonas* sp. JMSTP was much higher than that of the control SBR at the range of 20.98–150.73% from days 4 to 7, after the second inoculation, resulting in the apparent decrease in effluent TN in JMSTP SBR. RDA was used to explore the relationship between environmental parameters and microbial community distribution in the JMSTP SBR. As shown in Figure 7, there is a negative correlation between *Pseudomonas* sp. JMSTP and TN of effluent; in other words, with the increase in JMSTP abundance in the reactor, the TN of effluent decreases.

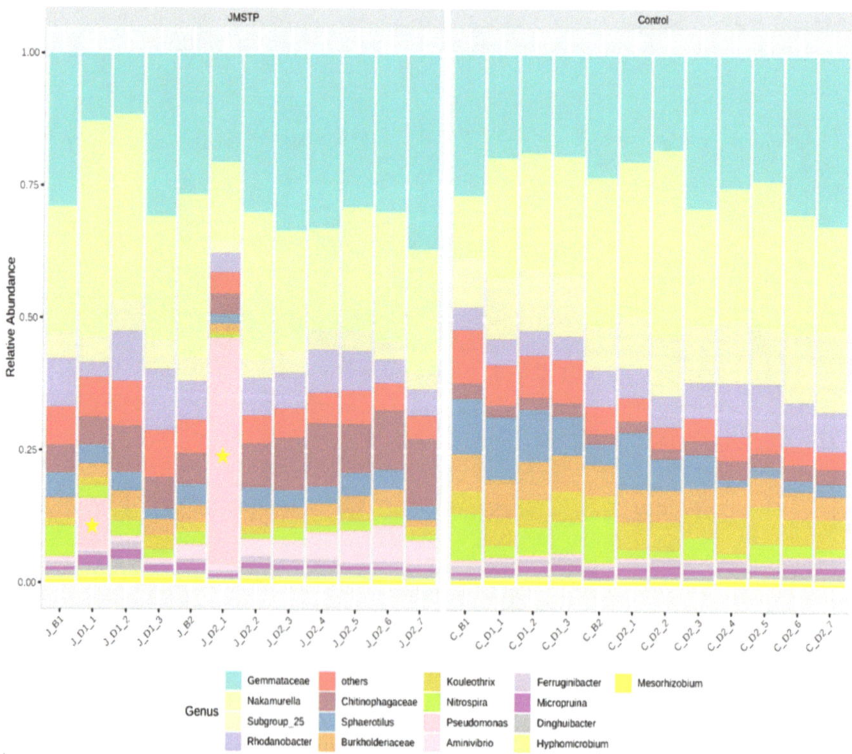


**Figure 6.** Relative abundance of the bacterial genera of the microbial community from reactors. C stands for control SBR, and J stands for JMSTP SBR.

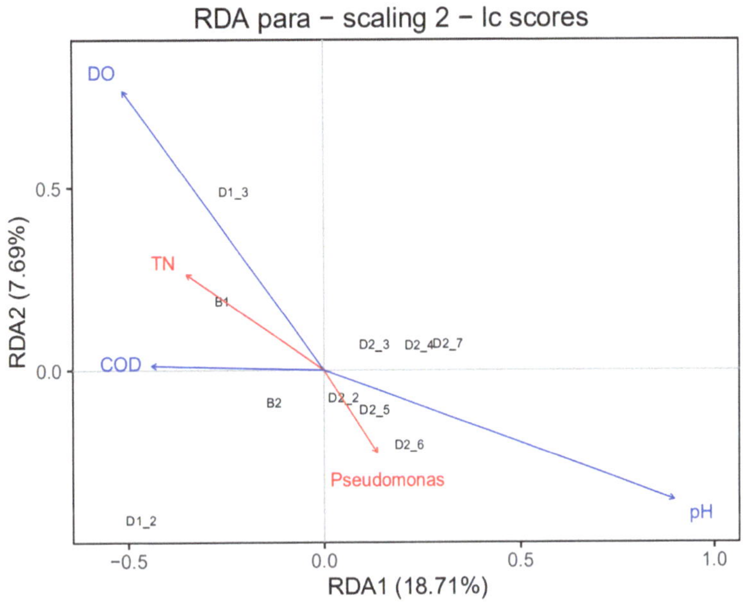


**Figure 7.** RDA between environmental parameters and microorganisms.

Correlation network analysis is used to explore the possible relationships between the microorganisms in two reactors. The R package was used to calculate the Spearman's correlation coefficient. The significance (*p*-value) and correlation (R-value) are controlled at 0.05 and 0.6, respectively. The genus-level visualization results are shown in Figure 8, and each node represents a genus category. The color of the node represents a different phylum. As shown in Figure 8a, the phylum categories that appear on the nodes are *Proteobacteria* (41.18%), *Planctomycetes* (19.61%), *Chloroflexi* (13.73%), *Acidobacteria* (9.8%), *Actinobacteria* (5.88%), and *Bacteroidetes* sps. (3.92%). The intergeneric correlation shown in Figure 8a is mainly positive, with two negative correlations between *Terrabacter* and *Aminivibrio* sps., and *Pseudoxanthomonas* and *Hydrogenophaga* sps. Figure 8b shows the results of network analysis for the control SBR. It is worth noting that *Pseudomonas* sp. JMSTP demonstrated a co-occurrence pattern with other native bacteria in the reactor. In addition, the control SBR showed more genus nodes and relationships than JMSTP SBR.

(a)

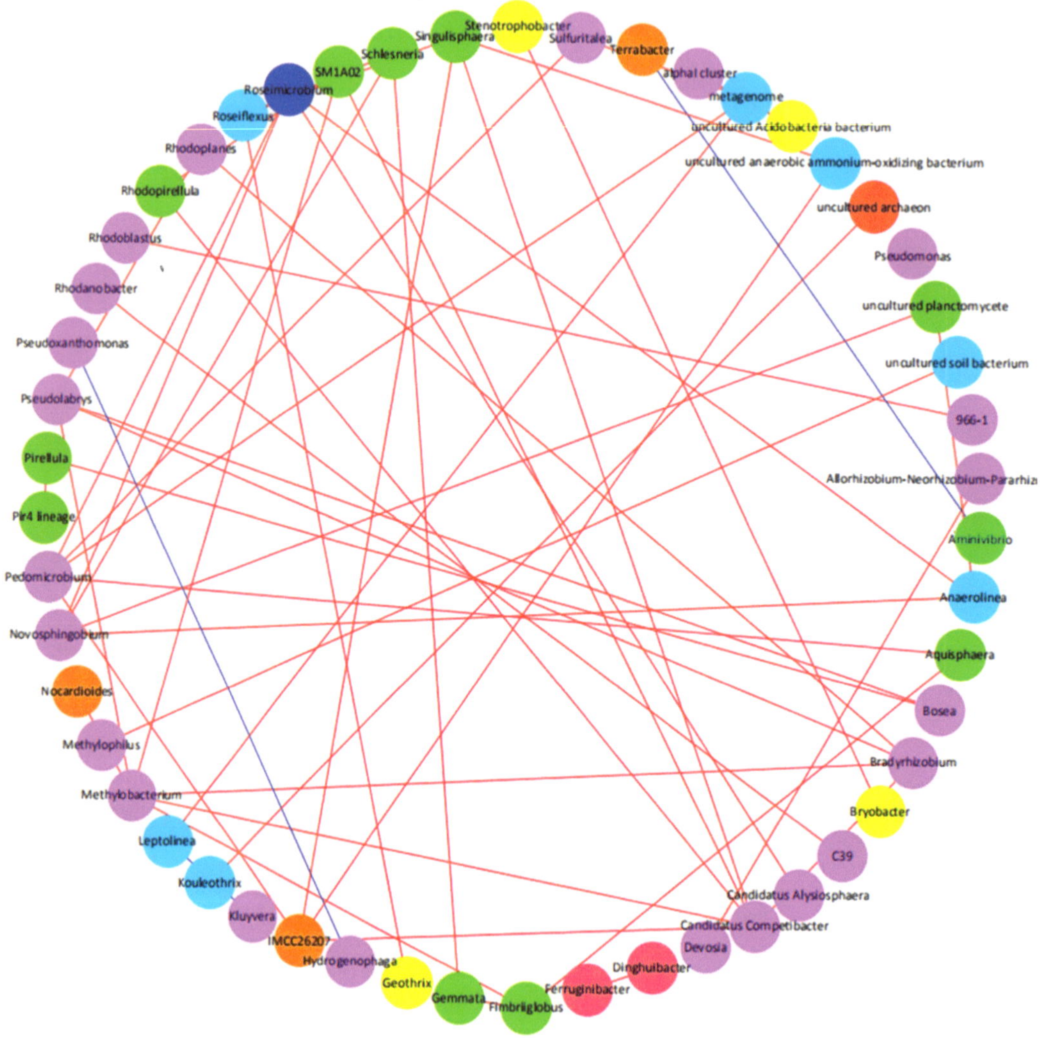


**Figure 8.** *Cont.*

(**b**)

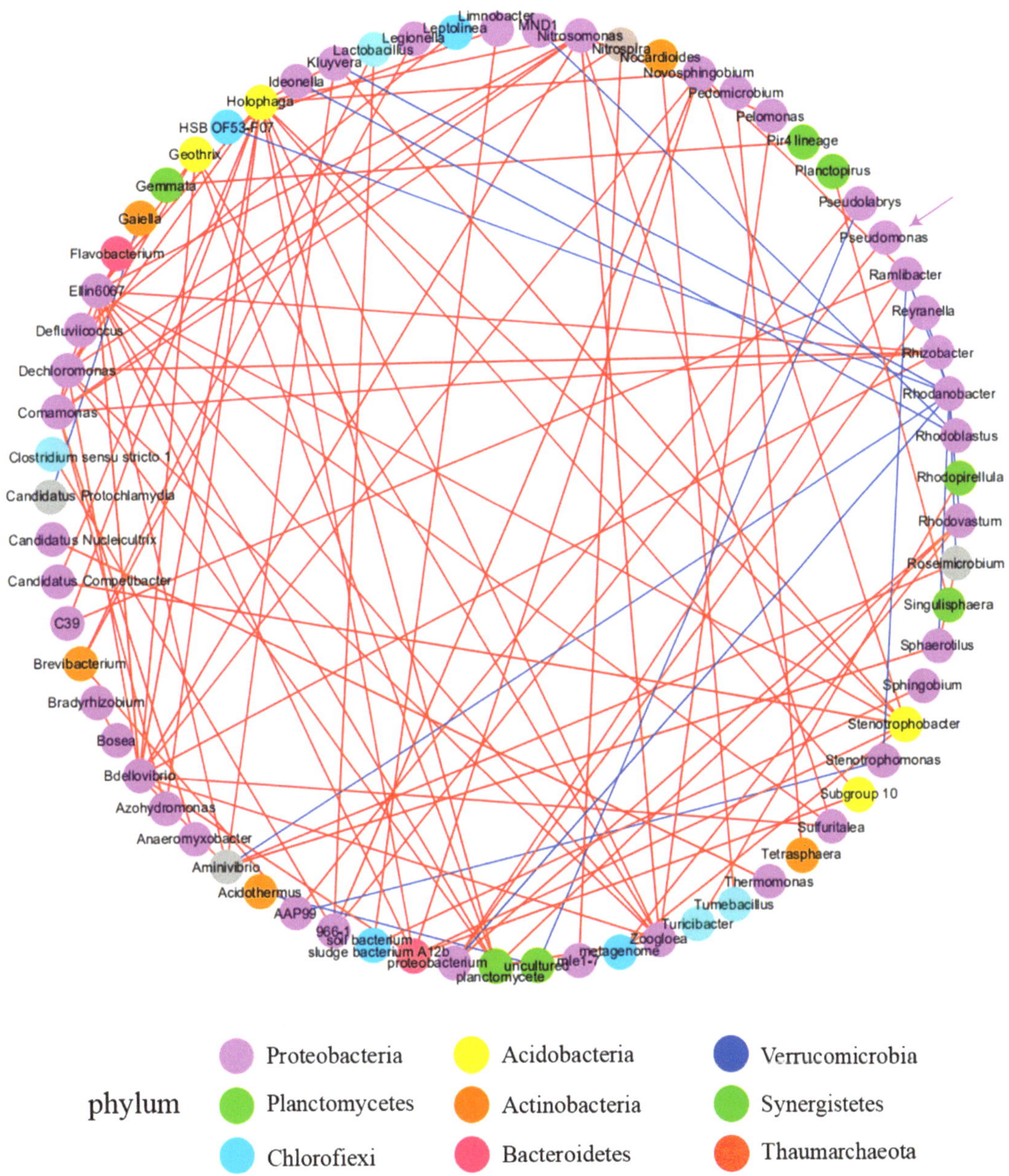


**Figure 8.** The bacterial genus network of microbial communities in JMSTP (**a**) and control (**b**) reactors. Each node represents a genus. The colors of the nodes indicate different major phylum. Red and blue edges indicate the positive and negative correlations, respectively.

## 4. Discussion

In this research, a strain, JMSTP, which can highly denitrify at low C/N, was obtained through laboratory screening. PCR agarose gel electrophoresis and gene sequencing confirm that the strain harbors the functional genes. The *nir*S-encoded cytochrome cd1 enzyme can catalyze the reduction of $NO_2^-$ to NO, which is essential to denitrification [22]. Nitric oxide reductase encoded by *cnor*B is the next functional gene that converts NO to $N_2O$ [21]. The reduction in $N_2O$ is the final step in denitrification and is catalyzed by nitrous oxide reductase encoded by the *nos*Z gene [23]. Gene sequencing results show that strain JMSTP can perform nitrate removal via conventional denitrification pathways through nitrite to nitric oxide, and then via nitric oxide to nitrous oxide and further nitrogen gas. *Pseudomonas stutzeri* GF2 was also found to carry three functional genes and to work in low C/N conditions [36]. High denitrification efficiency at relatively low C/N ratios and sufficient denitrification efficiency under 20–30 °C suggest that strain JMSTP has the potential to improve the denitrification of urban sewage treatment through bioaugmentation. In experiments with actual wastewater, although the concentration of nitrate fluctuated due to the death of microorganisms [33,34], the removal efficiency of the experimental group was still relatively good ($t$-test, $p < 0.01$). These results indicate that strain JMSTP can improve nitrate removal in actual sewage with a complex organic composition [37,38].

In laboratory-scale SBRs with a C/N of 9:8 ($w/w$), which is a harsh environment for microorganisms, JMSTP did not perform very well at the beginning. The result was similar to the first inoculation of two bioaugmentation experiments by Bouchez et al. [37,39]. The insignificant results may be due to the lack of adaptation of the inoculated microorganisms, insufficient substrate, competition between the introduced species and native biomass, and the grazing by protozoa [40]. Bouchez et al. considered that their results were more likely due to the grazing by protozoa. However, through filtration and COD detection, the reason may be that JMSTP did not successfully form flocs with the activated sludge within 6 h and was rapidly washed out from the reactor. After the second inoculation, there was a decrease in JMSTP SBR. The TN residue in the effluent, which was lower than 16 mgN/L, and met the criteria of the Grade I standard A (TN $\leq$ 20 mgN/L) in China's wastewater discharge regulation. The present study suggested that the bioaugmentation of strain JMSTP in the activated sludge could enhance nitrogen removal, even at low carbon sources. The TN in the effluent increased gradually from the 7th day after the second inoculation in the JMSTP SBR, even though it was still lower than that of the control SBR, implying that the effects of bioaugmentation on denitrification gradually reduced. Therefore, it was suggested that the continuous bioaugmentation by cultivating strain JMSTP could further enhance the denitrification effect [38,39]. We have to investigate how functional microorganisms can better use their advantages in the real environmental conditions, and which techniques can facilitate the performance.

The result of the diversity of the microbial community indicates that the sudden introduction in a single microorganism into the activated sludge system does not have a huge impact [40]. Although the relative abundance of *Pseudomonas* sp. decreased significantly after being added, but the decrease in the total nitrogen, which was observed in the reactor, was also related to the addition of JMSTP. In the network analysis diagram, there is no statistical correlation between *Pseudomonas* sp. and other genera in the bioaugmented reactor. This may be due to the large dosage of inoculation and a sharp decrease in relative abundance after inoculation. Namely, no other genus was found to have such a substantial increase and decrease as *Pseudomonas* sp. Previous studies have suggested that although the added bacteria were not detected, the bacteria they added were still believed to have synergistic effects with the native microorganisms [41,42]. Therefore, it could be speculated that strain JMSTP was likely overdosed and it could synergize with native microorganisms after long-term operation. The control SBR showed more genus nodes and relationships than JMSTP SBR. For an ecological network, more nodes and edges are related to the stability of the network structure. The result suggests that the large dosage of strain JMSTP might cause some interference to the stability of the network of the microbial community in JMSTP SBR. The microbial ecology needs a period of stabilization, which might correspond to the variation of effluent TN in the first few days after the bioaugmentation.

## 5. Conclusions

A denitrifying bacterium, *Pseudomonas* sp. JMSTP, was isolated and showed a denitrification ability under low carbon–nitrogen ratios. Under the influent COD/N = 3, the TN of effluent in the 6 L SBR could be less than 15 mgN/L after the bioaugmentation of *Pseudomonas* sp. JMSTP, which was significantly lower than that in the control SBR without bioaugmentation. RDA shows a negative correlation between *Pseudomonas* sp. abundance and TN concentrations in the bioaugmented SBR. The addition of *Pseudomonas* sp. JMSTP in activated sludge temporarily affected the diversity of the microbial community. The relative abundance of *Pseudomonas* sp. could achieve an amount 150% higher than that in the control SBR, after the second bioaugmentation of *Pseudomonas* sp. JMSTP. Correlation network analysis of the bioaugmented SBR found no clear co-occurrence pattern between *Pseudomonas* sp. and other native bacteria, due to the dramatic increase and decrease in *Pseudomonas* sp. abundance. This study suggests that bioaugmentation can be an effective approach for improving TN removal in urban sewage treatment plants. However, since the one-time bioaugmentation of the functional strain can only last for a few days, specific bioaugmentation strategies are needed to be considered further.

**Supplementary Materials:** The following supporting information can be downloaded at: https://www.mdpi.com/article/10.3390/w14040540/s1. Table S1. Primers for functional genes. Table S2. Nitrite removal capacity of JMSTP. Figure S1. The result of PCR sequencing analysis of *nir*S, *nor*B, and *nos*Z. Table S3. Community richness and diversity indices of two reactors. Table S4. OTUs of two reactors at each sampling day during experiment. Text S1. SEM operation process.

**Author Contributions:** Conceptualization, C.-P.Y. and Q.S.; methodology, X.Y.; software, T.C.; validation, C.-P.Y., Q.S. and A.H.; formal analysis, Q.S.; investigation, Y.W.; data curation, T.C.; writing—original draft preparation, T.C.; writing—review and editing, C.-P.Y.; visualization, T.C.; supervision, D.Q.; project administration, J.L.; funding acquisition, C.-P.Y. All authors have read and agreed to the published version of the manuscript.

**Funding:** This research was funded by National Nature Science Foundation of China (41807411), the FJIRSM&IUE Joint Research Fund (RHZX-2019-005), and the STS Project of Fujian-CAS (2016T3004, 2021T3059, 2021T3014).

**Conflicts of Interest:** The authors declare no competing financial interest.

## References

1. Jin, Y.; Ding, D.; Feng, C.; Tong, S.; Suemura, T.; Zhang, F. Performance of sequencing batch biofilm reactors with different control systems in treating synthetic municipal wastewater. *Bioresour. Technol.* **2012**, *104*, 12–18. [CrossRef] [PubMed]
2. Chen, Y.; Li, B.; Ye, L. The combined effects of COD/N ratio and nitrate recycling ratio on nitrogen and phosphorus removal in anaerobic/anoxic/aerobic (A2/O)-biological aerated filter (BAF) systems. *Biochem. Eng. J.* **2015**, *93*, 235–242. [CrossRef]
3. Loperena, L.; Ferrari, M.D.; Saravia, V.; Murro, D.; Lima, C.; Ferrando, L.; Fernandez, A.; Lareo, C. Performance of a commercial inoculum for the aerobic biodegradation of a high fat content dairy wastewater. *Bioresour. Technol.* **2007**, *98*, 1045–1051. [CrossRef] [PubMed]
4. Bella, E.; Smith, P.G. The bioaugmentation of sequencing batch reactor sludges for biological phosphorus removal. *Water Sci. Technol.* **1997**, *35*, 19–26. [CrossRef]
5. Oerther, D.B.; James, D.; Ebru, D.; Eric, L.; Freedman, D.L.; Lutgarde, R. Bioaugmentation of sequencing batch reactors for biological phosphorus removal: Comparative rrna sequence analysis and hybridization with oligonucleotide probes. *Water Sci. Technol.* **1998**, *37*, 469–473. [CrossRef]
6. Singer, A.C.; van der Gast, C.J.; Thompson, I.P. Perspectives and vision for strain selection in bioaugmentation. *Trends Biotechnol.* **2005**, *23*, 74–77. [CrossRef]
7. Boon, N.; Top, E.M.; Verstraete, W.; Siciliano, S.D. Bioaugmentation as a tool to protect the structure and function of an activated-sludge microbial community against a 3-chloroaniline shock load. *Appl. Environ. Microbiol.* **2003**, *69*, 1511–1520. [CrossRef]

8. Wang, F.; Liu, Y.; Wang, J.; Zhang, Y.; Yang, H. Influence of growth manner on nitrifying bacterial communities and nitrification kinetics in three lab-scale bioreactors. *J. Ind. Microbiol. Biotechnol.* **2012**, *39*, 595–604. [CrossRef]
9. Zumft, W.G. Microbiology and Molecular Biology Reviews. *Microbiol. Mol. Biol. Rev.* **1997**, *61*, 533–616.
10. Song, Z.; Zhang, X.; Sun, F.; Ngo, H.H.; Guo, W.; Wen, H.; Li, C.; Zhang, Z. Specific microbial diversity and functional gene (AOB amoA) analysis of a sponge-based aerobic nitrifying moving bed biofilm reactor exposed to typical pharmaceuticals. *Sci. Total Environ.* **2020**, *742*, 140660. [CrossRef]
11. Pelaz, L.; Gomez, A.; Letona, A.; Garralon, G.; Fdz-Polanco, M. Nitrogen removal in domestic wastewater. Effect of nitrate recycling and COD/N ratio. *Chemosphere* **2018**, *212*, 8–14. [CrossRef] [PubMed]
12. Fujiwara, T.; Akiyama, T.; Fukumori, Y. Structure and function of NO reductase with oxygen-reducing activity. In *Oxygen Homeostasis and Its Dynamics. Keio University Symposia for Life Science and Medicine*; Ishimura, Y., Shimada, H., Suematsu, M., Eds.; Springer: Tokyo, Japan, 1998; Volume 1.
13. Zhang, Q.L.; Liu, Y.; Ai, G.M.; Miao, L.L.; Zheng, H.Y.; Liu, Z.P. The characteristics of a novel heterotrophic nitrification-aerobic denitrification bacterium, Bacillus methylotrophicus strain L7. *Bioresour. Technol.* **2012**, *108*, 35–44. [CrossRef] [PubMed]
14. Joo, H.S.; Hirai, M.; Shoda, M. Characteristics of ammonium removal by heterotrophic nitrification-aerobic denitrification by *Alcaligenes faecalis* No. 4. *J. Biosci. Bioeng.* **2005**, *100*, 184–191. [CrossRef]
15. Rijn, J.V.; Tal, Y.; Schreier, H.J. Denitrification in recirculating systems: Theory and applications. *Aquac. Eng.* **2006**, *34*, 364–376. [CrossRef]
16. Yuan, H.; Huang, S.; Yuan, J.; You, Y.; Zhang, Y. Characteristics of microbial denitrification under different aeration intensities: Performance, mechanism, and co-occurrence network. *Sci. Total Environ.* **2020**, *754*, 141965. [CrossRef] [PubMed]
17. Chen, Q.; Ni, J.; Ma, T.; Liu, T.; Zheng, M. Bioaugmentation treatment of municipal wastewater with heterotrophic-aerobic nitrogen removal bacteria in a pilot-scale SBR. *Bioresour. Technol.* **2015**, *183*, 25–32. [CrossRef]
18. Layeghifard, M.; DM Hwang, D.M.; Guttman, D.S. Disentangling interactions in the microbiome: A network perspective. *Trends Microbiol.* **2017**, *25*, 217–228. [CrossRef]
19. Takaya, N.; Catalan-Sakairi, M.A.B.; Sakaguchi, Y.; Kato, I.; Zhou, Z.; Shoun, H. Aerobic denitrifying bacteria that produce low levels of nitrous oxide. *Appl. Environ.* **2003**, *69*, 3152–3157. [CrossRef]
20. Miranda, K.M.; Espey, M.G.; Wink, D.A. A rapid, simple spectrophotometric method for simultaneous detection of nitrate and nitrite. *Nitric Oxide* **2001**, *5*, 62–71. [CrossRef]
21. Braker, G.; Tiedje, J.M. Nitric oxide reductase (norB) genes from pure cultures and environmental samples. *Appl. Environ.* **2003**, *69*, 3476–3483. [CrossRef]
22. Throback, I.N.; Enwall, K.; Jarvis, A.; Hallin, S. Reassessing PCR primers targeting nirS, nirK and nosZ genes for community surveys of denitrifying bacteria with DGGE. *FEMS Microbiol. Ecol.* **2004**, *49*, 401–417. [CrossRef] [PubMed]
23. Henry, S.; Bru, D.; Stres, B.; Hallet, S.; Philippot, L. Quantitative detection of the nosZ gene, encoding nitrous oxide reductase, and comparison of the abundances of 16S rRNA, narG, nirK, and nosZ genes in soils. *Appl. Environ.* **2006**, *72*, 5181–5189. [CrossRef] [PubMed]
24. Hu, A.; Li, S.; Zhang, L.; Wang, H.; Yang, J.; Luo, Z.; Rashid, A.; Chen, S.; Huang, W.; Yu, C.P. Prokaryotic footprints in urban water ecosystems: A case study of urban landscape ponds in a coastal city, China. *Environ. Pollut.* **2018**, *242 Pt B*, 1729–1739. [CrossRef]
25. Hildebrand, F.; T, R.; Voigt, A.Y.; Bork, P.; Raes, J. LotuS: An efficient and user-friendly OTU processing pipeline. *Microbiome* **2014**, *2*, 30. [CrossRef] [PubMed]
26. Edgar, R.C. UPARSE: Highly accurate OTU sequences from microbial amplicon reads. *Nat. Methods* **2013**, *10*, 996. [CrossRef] [PubMed]
27. Quast, C.; Pruesse, E.; Yilmaz, P.; Gerken, J.; Schweer, T.; Yarza, P.; Peplies, J.; Glockner, F.O. The SILVA ribosomal RNA gene database project: Improved data processing and web-based tools. *Nucleic Acids Res.* **2013**, *41*, D590–D596. [CrossRef]
28. Caporaso, J.G.; Kuczynski, J.; Stombaugh, J.; Bittinger, K.; Bushman, F.D.; Costello, E.K.; Fierer, N.; Peña, A.G.; Goodrich, J.K.; Gordon, J.I. QIIME allows analysis of high-throughput community sequencing data. *Nat. Methods* **2010**, *7*, 335–336. [CrossRef]
29. Oksanen, J.; Blanchet, F.G.; Kindt, R.; Legendre, P.; Minchin, P.R.; O'Hara, R.B.; Simpson, G.L.; Solymos, P.; Stevens, H.H.H.; Wagner, H. Vegan: Community Ecology Package. R Package Version 2.5-6. 2019. Available online: http://cran.rproject.org/package=vegan (accessed on 28 December 2021).
30. Kembel, S.W.; Cowan, P.D.; Helmus, M.R.; Cornwell, W.K.; Morlon, H.; Ackerly, D.D.; Blomberg, S.P.; Webb, C.O. Picante: R tools for integrating phylogenies and ecology. *Bioinformatics* **2010**, *26*, 1463–1464. [CrossRef]
31. Nepusz, G.C.T. The igraph software package for complex network research. *Int. J. Complex Syst.* **2006**, *1695*, 1–9.
32. Harrell, F.E., Jr. Hmisc: Harrell Miscellaneous, R Package Version 4.4–0. 2020. Available online: https://cran.r-project.org/package=Hmisc (accessed on 28 December 2021).
33. He, T.; Li, Z.; Sun, Q.; Xu, Y.; Ye, Q. Heterotrophic nitrification and aerobic denitrification by Pseudomonas tolaasii Y-11 without nitrite accumulation during nitrogen conversion. *Bioresour. Technol.* **2016**, *200*, 493–499. [CrossRef]
34. Van Niel, E.W.J.; Braber, K.J.; Robertson, L.A.; Kuenen, J.G. Heterotrophic nitrification and aerobic denitrification in *Alcaligenes faecalis* strain TUD. *Anton Leeuw. Int. J. Gen.* **1992**, *62*, 231–237. [CrossRef]

35. Li, Y.; Zhang, Q.; Li, M.; Sang, W.; Wang, Y.; Wu, L.; Yang, Y. Bioaugmentation of sequencing batch reactor for aniline treatment during start-up period: Investigation of microbial community structure of activated sludge. *Chemosphere* **2020**, *243*, 125426. [CrossRef]
36. Chen, C.; Ali, A.; Su, J.; Wang, Y.; Huang, T.; Gao, J. Pseudomonas stutzeri GF2 augmented the denitrification of low carbon to nitrogen ratio: Possibility for sewage wastewater treatment. *Bioresour. Technol.* **2021**, *333*, 125–169. [CrossRef]
37. Bouchez, T.; Patureau, D.; Delgenes, J.P.; Moletta, R. Successful bacterial incorporation into activated sludge flocs using alginate. *Bioresour. Technol.* **2009**, *100*, 1031–1032. [CrossRef]
38. Boon, N.; De Gelder, L.; Lievens, H.; Siciliano, S.D.; Top, E.M.; Verstraete, W. Bioaugmenting bioreactors for the continuous removal of 3-chloroaniline by a slow release approach. *Environ. Sci. Technol.* **2002**, *36*, 4698–4704. [CrossRef]
39. Bouchez, T.; Patureau, D.; Dabert, P.; Juretschko, S.; Doré, J.; Delgenès, P.; Moletta, R.; Wagner, M. Ecological study of a bioaugmentation failure. *Environ. Microbiol.* **2000**, *2*, 179–190. [CrossRef]
40. Goldstein, R.M.; Mallory, L.M.; Alexander, M. Reasons for possible failure of inoculation to enhance biodegradation. *Appl. Environ. Microbiol.* **1985**, *50*, 977–983. [CrossRef]
41. Bai, Y.; Sun, Q.; Sun, R.; Wen, D.; Tang, X. Bioaugmentation and adsorption treatment of coking wastewater containing pyridine and quinoline using zeolite-biological aerated filters. *Environ. Sci. Technol.* **2011**, *45*, 1940–1948. [CrossRef]
42. Munakata-Marr, J.; Matheson, V.G.; Forney, L.J.; Tiedje, J.M.; McCarty, P.L. Long-term biodegradation of trichloroethylene influenced by bioaugmentation and dissolved oxygen in aquifer microcosms. *Environ. Sci. Technol.* **1997**, *31*, 786–791. [CrossRef]

water

Article

# Adsorption and Its Mechanism of Arsenate in Aqueous Solutions by Red Soil

**Min Guo, Lili Shi, Wen Gu and Wenzhu Wu ***

Key Laboratory of Pesticide Environmental Assessment and Pollution Control, Nanjing Institute of Environmental Sciences, Ministry of Ecology and Environment of the People's Republic of China, Nanjing 210042, China; guomin@nies.org (M.G.); sll@nies.org (L.S.); guwen@nies.org (W.G.)
* Correspondence: wwz@nies.org; Tel.: +86-(25)-8528-7202

**Abstract:** The removal, and its mechanism, of arsenate from aqueous solutions was investigated using Yunnan red soil. A series of adsorption experiments was designed to disclose the effect of key factors (soil types, soil/solution rates, initial arsenate concentrations, and shaking speeds) on the adsorption capacity of Yunnan red soil for arsenate. The soil/solution ratio was optimized as 0.05 g/100 mL to balance the adsorption capacity and removal efficiency. The optimal shaking speed (225 rpm) not only ensured enough contact frequency between the Yunnan red soil and the arsenate, but also reduced the mass transfer resistance. The results from applying an orthogonal array method showed that the most significant factor affecting arsenate removal efficiency was soil type, followed by the soil/solution ratio, contact time, and shaking speed. The IR spectra of the precipitates further confirmed that the metal arsenide was settled by the Yunnan red soil, indicating that the arsenate ion existed on the red soil surface in the form of protonated bidentate surface complexation of $-FeO_2As(O)(OH)^-$ and $FeO_2As(O)^{2-}$. These results indicate that Yunnan red soil is promising for the removal of arsenate from aqueous solutions; it may thus be suitable as a new adsorbent for arsenate removal during water treatment.

**Keywords:** soil; As(V); adsorption; water treatment; $L_9(3^4)$; mechanism

**Citation:** Guo, M.; Shi, L.; Gu, W.; Wu, W. Adsorption and Its Mechanism of Arsenate in Aqueous Solutions by Red Soil. *Water* **2022**, *14*, 579. https://doi.org/10.3390/w14040579

Academic Editors: Dibyendu Sarkar, Rupali Datta, Prafulla Kumar Sahoo and Mohammad Mahmudur Rahman

Received: 20 December 2021
Accepted: 3 February 2022
Published: 14 February 2022



## 1. Introduction

Arsenic is a metalloid that is considered to be a dangerous water pollutant throughout the world. It occurs in natural waters in both inorganic and organic forms, with the inorganic form being more toxic than the organic form. Arsenic usually occurs in two valence states: arsenite (As (III)) and arsenate (As (V)) [1]. Arsenic has been listed as a Class A carcinogen by the US Environmental Protection Agency (US EPA) due to its high toxic effects [2], and the US EPA in 2001 adopted a new standard for arsenic in drinking water at 10 μg/L, replacing the old standard of 50 μg/L. Considering its health and toxicological effects, the World Health Organization has revised the maximum contaminant level (MCL) for arsenic in drinking water from 50 to10 μg/L [3]. Japan and the European Union have complied with the standard of 10 μg/L for arsenic instead of 50 μg/L, which has been used for years [4–6]. China has also set the limit of arsenic concentration in drinking water at 10 μg/L via the China Drinking Water Sanitary Standard (GB5749-2006). However, the current water contamination caused by arsenic is severe, with at least 50 million people in Asia drinking underground water with arsenic concentrations exceeding 50 μg/L [7]. China also faces water contamination problems, from both industrial and natural sources, of enormous proportions. Yunnan Province in southwest China is abundant with mineral resources. The development of the mining industry in recent years has resulted in pollution caused by heavy metals and metalloids becoming increasingly serious. In 2008, the average value of arsenic concentrations in Yangzonghai Lake, Yunnan Province, reached 0.121 mg/L. The water quality of Yangzonghai Lake was identified as inferior grade V, which seriously affected the production and lives of the local population. Later, Datunhai Lake in Honghe

Prefecture, Yunnan Province, was found to have several times the pollution of Yangzonghai Lake, with an arsenic concentration exceeding 96 times the national standards. There are substantial data showing that more than 50% of the lakes in Honghe Prefecture are seriously contaminated by arsenic, and thus require urgent contamination control.

Various treatment methods have been used for the removal of arsenic from water and wastewater, such as chemical oxidation, membrane processes, adsorption, biological methods, ion exchange, and electrocoagulation [8,9]. Adsorption has been recognized as an effective treatment method. It has been observed that arsenate is well adsorbed onto the surface of iron-containing compounds, such as ferric hydroxide, aluminum hydroxide, Fe–Al hydroxide, and manganese dioxide [10–18]. Owing to the complexity of soil composition and structure, the current research on the mechanisms of arsenic sorption in environmental media is still mainly focused on the sorption behaviors of pure minerals or iron-containing compounds; hence, more research is needed on arsenic sorption, especially in soil.

The red soil in Yunnan Province has a high clay content and is rich in Fe and Al oxides; these properties are beneficial for the adsorption of arsenic. Hence, an attempt was made in the present investigations to test the red soil for arsenate adsorption from aqueous solution. In the current study, adsorption equilibrium experiments were conducted under different conditions, which also correspond to the current pollution actuality of Yunnan Province, thus exploring fast, efficient, natural, and harmless arsenic removal technology. The red soil chosen as the sorbent is cheaper than other options and more easily obtained. The red-soil-adsorbed arsenic will sink to the bottom of the lake and thus will not produce large amounts of chemical sludge, which needs further treatment. This technology will have important application value and practical significance for surface water polluted by arsenic, especially for the treatment of arsenic-contaminated lakes in China's southwestern regions.

## 2. Materials and Methods

### 2.1. Materials

All reagents used in this study were of analytical grade. Stock solutions of test reagents were prepared in milli Q water. All glassware and sample bottles were soaked in 10% $HNO_3$ and then rinsed with milli Q water. The resistivity and TOC of milli Q water are 18.2 Ω and <5, respectively. The following reagents were used:

(1) Sodium arsenate heptahydrate ($Na_2HAsO_4 \cdot 7H_2O$) with a purity of 98% was provided by US Alfa Aesar.
(2) A 1% potassium borohydride ($KBH_4$) solution was prepared by dissolving 0.2 g of potassium hydroxide in 100 mL milli Q water, and then 1 g of potassium borohydride was added into the solution and mixed fully.
(3) A 2% hydrochloric acid solution with 10 mL of concentrated hydrochloric acid was diluted to 500 mL with milli Q water. All solutions were prepared using the 2% HCl solution.
(4) Soils and their characterization.

Six typical types of red soil in the southwest of China were chosen as objects for the experiment: four types of Yunnan red soil, one Jiangxi red soil, and one Guangdong latosol. The essential physiochemical properties are shown in Table 1. These soils were air-dried and passed through a mesh of size 200 for further investigation.

**Table 1.** The Essential Physiochemical Properties of Six Typical Soils.

| Soils | pH | Organic Matter (g/kg) | Soil Physical Sand Content (1~0.01 mm) % | Soil Physical Clay Content (<0.01 mm) % | $Fe_2O_3$/% | $Al_2O_3$/% |
|---|---|---|---|---|---|---|
| Yunnan red soil 1 | 6.59 | 9.3 | 70.7 | 29.3 | 14.04 | 28.86 |
| Yunnan red soil 2 | 6.47 | 9.2 | 57.3 | 42.7 | 10.05 | 24.48 |
| Yunnan red soil 3 | 5.93 | 10.2 | 60.8 | 39.2 | 12.59 | 26.70 |
| Yunnan red soil 4 | 5.88 | 11.9 | 73.3 | 26.7 | 12.86 | 26.51 |
| Jiangxi red soil | 6.24 | 17.0 | 87.9 | 12.1 | 8.91 | 18.28 |
| Guangdong latosol | 6.55 | 15.1 | 59.2 | 40.8 | 8.29 | 18.56 |

Note: All information was determined by our laboratory.

### 2.2. Equipment

The equipment used consisted of a PF6 non-dispersive atomic fluorescence spectrometer (Purkinje General, China), a VECTOR-22 infrared spectrometer (Bruker Optics Inc., Ettlingen, Germany), and an E24A-thermostat digital oscillator (New Brunswick Scientific, Edison, NJ, USA).

### 2.3. Batch Adsorption Studies

Equilibrium experiments of arsenate sorption onto the soils were carried out by batch adsorption with aqueous solutions of As(V). The adsorption studies were launched with six different soils, and then the two best soils were chosen for the following experiment according to the adsorption effect. The reaction vessels were equilibrated in a temperature-controlled shaker under different conditions, including different shaking bath speeds ranging from 50 to 300 rpm, different initial concentrations of arsenate ranging from 0.05 to 2.0 mg/L, and different addition ratios of soils ranging from 0.01 to 0.5 g soils/100 mL to study the adsorption behavior of arsenate in soil.

Yangzonghai Lake is one of the largest plateau lakes in Yunnan Province, China. It is the main water source for agriculture, fisheries, and drinking water for its surrounding areas. The arsenic concentrations in Yangzonghai Lake were found to be three times as high as the maximum allowable concentration (MAC) (0.05 mg/L) of grade III of China's Surface Water Environmental Quality Standards, and approximately 1.5-fold higher than that of grade V according to the field research carried out by our workgroup, which was consistent with a previous investigation [19]. Moreover, the arsenic present in the water was mainly in pentavalent form. Therefore, 0.2 mg/L arsenate aqueous solution was chosen to be the suitable concentration in Sections 3.2 and 3.4.

The residual concentrations of arsenate in the filtrate were determined using a PF6 atomic fluorescence spectrometer (Purkinje General), once filtered.

### 2.4. Adsorption Mechanism

Infrared spectra of the red soil before and after arsenate adsorption were recorded using a VECTOR-22 infrared spectrometer.

For IR identification, a 150 g sample consisting of red soil and KBr in a ratio of 1/200 was uniformly mixed and then pressed into one piece under a pressure of 10 t/cm$^2$ for 1 min. A PT-IR spectrometer was then used to measure the sample spectra values within the range of 4000–400 cm$^{-1}$.

The analyses by scanning electron microscopy (SEM) image were completed using a Hitachi S-4800 SEM.

### 2.5. Analytical Methods

Sample pretreatment was carried out with acidified samples of 10% HCl thoroughly mixed with 1% thiourea and 1% ascorbic acid; these were then measured after 30 min.

For the measurement conditions, the negative high voltage on the photomultiplier tube was set to 265 V, the atomization temperature was set to 180 °C, and the lamp current of the As hollow-cathode lamp was set to 40 mA; 300 and 600 mL/min flow rates of carrier gas and shield gas were set, respectively.

The instrument was calibrated from 1.0 to 20.0 μg/L. Other range samples were diluted until results within the calibration range were obtained.

### 2.6. Orthogonal Analysis

The orthogonal array method was used to determine the optimum levels of factors that affect the removal rates. Selected factors (soil type, soil/solution ratio, contact time, and shaking speed) were studied at three levels. According to an $L_9$ ($3^4$) orthogonal array, nine trials with three replications were conducted.

## 3. Results and Discussions

### 3.1. Effect of Soil Types on the Adsorption of Arsenate

In any adsorption process, the adsorption material has a significant effect on the treatment efficiency. Therefore, appropriate selection of the red soil is important. Hence, six different red soils were chosen as adsorbents; these soils are non-toxic and comprise high clay, iron oxide, and aluminum oxide.

Six types of red soil (0.5 g) were added to a reaction vessel containing 100 mL of arsenate solution with a concentration of 10 mg/L. The reaction vessels were then equilibrated in a shaking bath at 225 rpm for a period of 10 h. Figure 1 shows the variation in the arsenate removal rate with different red soils. It can be observed from Figure 1 that the removal of all six adsorption reactions increased over time until a steady state was attained after approximately 2 h, and there was no significant change in the pseudo equilibrium concentration after this time until 10 h. The removal of arsenate when using Yunnan red soil (1, 3, and 4) in the aqueous solution increased significantly compared with the other three soils, with removal rates of up to 49.5%, 44.8%, and 44.0%, and with adsorption capacities of 989.2 μg/g, 895.1 μg/g, and 879.3 μg/g, respectively. Studies on arsenic have indicated that hydrous metal oxides, such as ferric hydroxides, ferrihydrite, and goethite, strongly adsorb arsenic. The adsorption rates of arsenate on Yunnan red soil 2, Jiangxi red soil, and Guangdong latosol were low, which may be related to the relatively low content of iron oxide and aluminum oxide in these three soils. The curves in Figure 1 also show rapid arsenate adsorption in the beginning, which may be attributed to the relatively large number of available vacant sites on the soils compared to those of the later hours.

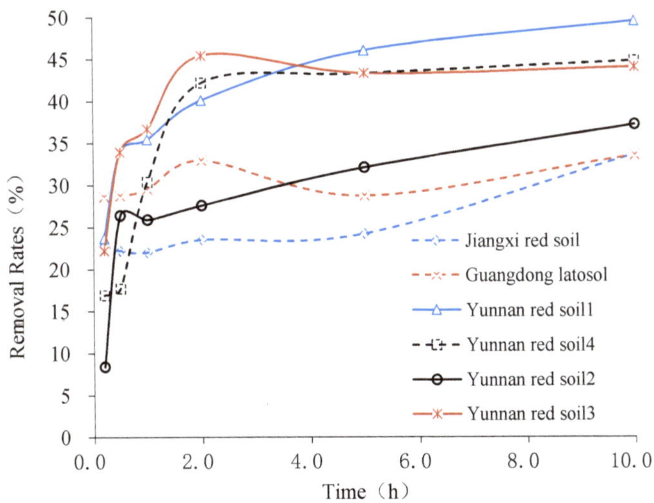


**Figure 1.** Removal Percentage of As(V) on Different Red Soils.

### 3.2. *Effect of Soil/Solution Ratios*

Two soils (Yunnan red soil 1 and 4) with relatively high adsorption capacity were selected for the following experiments.

The sensitivity of As(V) adsorption efficiency to the adsorbent dosage was determined by adding 0.01, 0.02, 0.03, 0.05, 0.1, 0.2, 0.3, and 0.5 g red soil (Yunnan red soil 1 and 4) into 100 mL solutions at a constant test solution As(V) concentration of 0.2 mg/L. The reaction vessels were then equilibrated in a shaking bath at 225 rpm for a period of 2 h. As shown in Figure 2, the adsorption capacity of these two soils all increased with increasing soil/solution ratio. The arsenate adsorption capacity depended profoundly on the soil/solution ratio in the range of 0.01 g/100 mL to 0.05 g/100 mL; the removal rates were 51.1%, 69.2%, 82.2%, and 94.8% (Yunnan red soil 1) and 45.4%, 65.5%, 81.0%, and 93.4% (Yunnan red soil 4) at the soil/solution ratios of 0.01 g/100 mL, 0.02 g/100 mL, 0.03 g/100 mL, and 0.05 g/100 mL, with arsenate adsorption capacities of 1023, 692, 548, and 379 (Yunnan red soil 1) and 908, 655, 540, and 374 (Yunnan red soil 4) μg/g. The removal rates increased with increasing soil dosage, but the removal increased at 0.05 g dosage and leveled off after that point. This behavior implies that the sorption depends on the availability of binding sites.

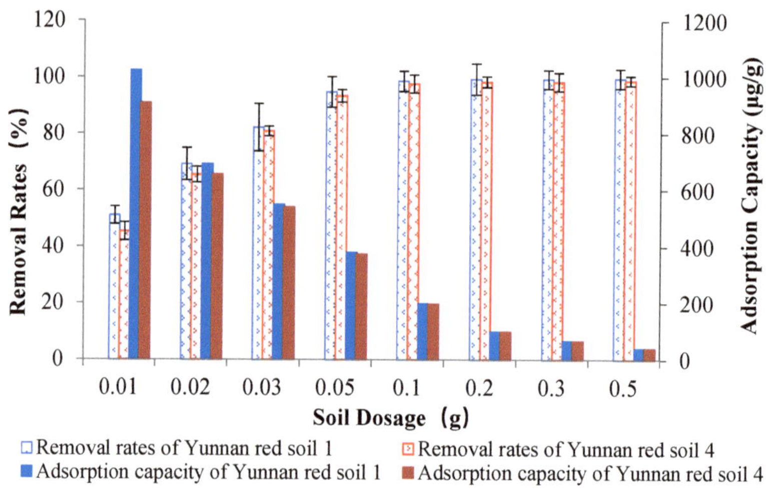


**Figure 2.** Removal Percentage of As(V) at Different Soil/solution Ratios.

### 3.3. *Effect of Initial Arsenate Concentration*

Most surface, ground, and industrial waste waters contain different concentration levels of arsenate; therefore, arsenate concentrations in the range of 0.05 to 2.0 mg/L were chosen for the present findings to match with the natural water system. The adsorption capacity of two different red soils (Yunnan red soil 1 and Yunnan red soil 4) increased with increasing initial arsenate concentration (Table 2). This may be an outcome of the increase in driving force due to the concentration gradient building up between the surface of the red soil and the arsenate solution [20]. Table 2 also reveals that the removal rates of arsenate decreased with increasing initial arsenate concentration. This may be attributed to the fixed quantity of red soils being used during the course of the adsorption process. At the same time, the arsenate adsorption capacity of Yunnan red soil 4 was slightly larger than that of Yunnan red soil 1 at each initial concentration, which indicated that the pH and organic matter content of the soil are also important factors in the adsorption progress.

**Table 2.** Effect of Initial Arsenate Concentration on the Sorption of Arsenate.

| Initial Concentration (mg/L) | Yunnan Red Soil 1 | | | Yunnan Red Soil 4 | | |
|---|---|---|---|---|---|---|
| | Removal Rate (%) | RSD (%) | Adsorption Capacity (μg/g) | Removal Rate (%) | RSD (%) | Adsorption Capacity (μg/g) |
| 0.05 | 95.1 | 4.0 | 158.5 | 95.5 | 5.7 | 159.1 |
| 0.1 | 92.2 | 5.6 | 307.3 | 93.2 | 8.1 | 310.7 |
| 0.2 | 79.5 | 5.4 | 529.8 | 82.5 | 5.2 | 550.3 |
| 0.5 | 58.8 | 3.6 | 979.2 | 59.3 | 0.7 | 987.7 |
| 1.0 | 41.6 | 0.9 | 1387.3 | 42.8 | 1.8 | 1427.7 |
| 2.0 | 20.8 | 3.7 | 1389.2 | 24.6 | 0.1 | 1637.1 |

### 3.4. Effect of Shaking Speed on the Removal of Arsenate

The adsorption was influenced by the mixing conditions of the reaction system. Because the batch adsorption reactions took place in a shaker, the shaking speeds were investigated as an influencing factor. From Figure 3, the adsorption capacities of Yunnan red soil 1 and Yunnan red soil 4 all fell in the range of 50 rpm to 100 rpm. An optimal shaking speed ensured sufficient contact frequency between the sorbent and the adsorbate, which was helpful to improve the adsorption capacity of the sorbent. However, a higher shaking speed might decrease the adsorption capacity by producing a strong shearing effect [21]. Once the shaking speed was over 100 rpm, the removal rates of arsenate increased with increasing shaking speed and reached a peak at 225 rpm, then the adsorption capacity of these two red soils increased slightly with ascending shaking speed. Therefore, a shaking speed of 225 rpm was fixed in the relevant experiments.

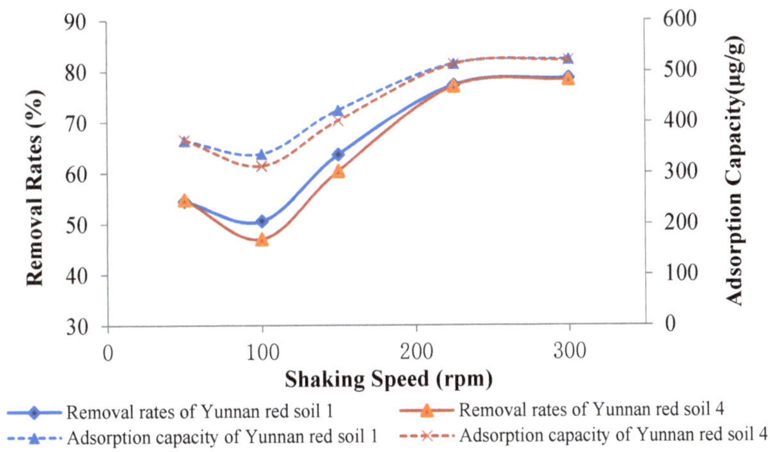


**Figure 3.** Effect of the Shaking Speed on the Adsorption Process.

### 3.5. Optimization of the Factors Affecting Arsenate Removal by Orthogonal Array

Experiments were performed with the soil types of Yunnan red soil 1, 2, and 4; soil/solution ratios of 0.025 g/100 mL, 0.030 g/100 mL, and 0.035 g/100 mL; contact times of 15, 30, and 45 min; and shaking speeds of 200, 225, and 250 rpm. The influences of these factors on the removal rate were determined using the maximum difference method (Table 3).

**Table 3.** The $L_9$ ($3^4$) Orthogonal Array Experimental Design and Arsenate Removal Efficiency As Response.

| No. | A *. Soil Type | B *. Soil/Solution Ratio (g/mL) | C *. Contact Time (min) | D *. Shaking Speed (rpm) | Removal Rate (%) |
|---|---|---|---|---|---|
| 1 | 1 (Yunnan red soil 1) | 1 (0.025/100) | 1 (15) | 1 (200) | 64.1 |
| 2 | 1 | 2 (0.030/100) | 2 (30) | 2 (225) | 76.5 |
| 3 | 1 | 3 (0.035/100) | 3 (45) | 3 (250) | 84.8 |
| 4 | 2 (Yunnan red soil 2) | 1 | 2 | 3 | 47.0 |
| 5 | 2 | 2 | 3 | 1 | 55.8 |
| 6 | 2 | 3 | 1 | 2 | 54.2 |
| 7 | 3 (Yunnan red soil 4) | 1 | 3 | 2 | 70.7 |
| 8 | 3 | 2 | 1 | 3 | 69.6 |
| 9 | 3 | 3 | 2 | 1 | 78.8 |
| *k*1 | 225.4 | 181.8 | 187.9 | 198.7 | |
| *k*2 | 157.0 | 201.9 | 202.3 | 201.4 | |
| *k*3 | 219.1 | 217.8 | 211.3 | 201.4 | |
| *K*1 | 75.1 | 60.6 | 62.6 | 66.2 | |
| *K*2 | 52.3 | 67.3 | 67.4 | 67.1 | |
| *K*3 | 73.0 | 72.6 | 70.4 | 67.1 | |
| R (max-min) | 22.8 | 12.0 | 7.8 | 0.9 | |

* the key factors.

The final removal efficiencies of arsenate from aqueous solution by red soil are presented in Table 3. The range ®, calculated from the results of the orthogonal experiment, indicates that soil type has the largest effect on removal rate, followed by soil/solution ratio, contact time, and shaking speed.

### 3.6. Adsorption Mechanism of the Red Soils

Many studies have shown that iron and aluminum in soil play an important role in arsenic adsorption [22]. Arsenate can form insoluble arsenide with iron and aluminum cations and produce co-precipitation with amorphous iron and aluminum hydroxide. The more amorphous Fe and Al oxides contained in the soil, the stronger the arsenic adsorption capacity [23,24]. Arsenic can be fixed on the surface of metal oxide through specific adsorption and non-specific adsorption. Specific adsorption refers to the entrance of anions into the metal atom coordination shell on the oxide surface to re-coordinate with the ligand hydroxyl or hydrated to combine on the solid surface directly through covalent bonding or coordinate bonding [25,26]. Spectroscopic studies have confirmed that both As(III) and As(V) may form inner-sphere complexes on the surfaces of (hydr)oxides and clay minerals through ligand exchange with OH and $OH^{2+}$ surface functional groups [27]. In situ, Raman and Fourier transform infrared (FTIR) spectroscopic methods, combined with sorption techniques, electrophoretic mobility measurements, and surface complexation modeling, were used by Goldberg [28] to study the interaction of As(III) and As(V) with amorphous oxide surfaces, and the results showed that arsenate formed inner-sphere surface complexes on both amorphous Al and Fe oxides. Arsenite formed both inner- and outer-sphere surface complexes on amorphous Fe oxide and outer-sphere surface complexes on amorphous Al oxides.

In this study, the infrared spectrum was used to examine arsenate adsorption by Yunnan red soil 4 (see Figure 4). It can be seen in Figure 4 that the band intensities of infrared spectra before and after arsenate adsorption by Yunnan red soil 4 were significantly different, indicating that red soil samples before and after adsorption have differences in composition. In the research of iron hydroxide arsenic adsorption and precipitation mechanisms, Liu [29] found that the infrared spectra of hydroxide solid samples before and after arsenic adsorption had –OH bending vibration absorption bands at 1625~1641 $cm^{-1}$, indicating that the arsenate ion might exist on the red soil surface in the form of protonated bidentate surface complexation of -$FeO_2$ As (O) $(OH)^-$ and $-FeO_2As$ (O) $^{2-}$. In this study, the infrared spectra also have a 1641~1686 $cm^{-1}$ characteristic absorption band, indicating that re-coordination occurred between the arsenate and the hydroxyl or hydrated base in the metal oxides, combined on the surface of soil particles via coordination bonds. At the same time, Liu [29] also found that when the initial arsenic concentration was 500 mmol/L, the ferric hydroxide solid after adsorption showed characteristic bands at wavenumber 821.54 $cm^{-1}$, while adsorption appeared at wavenumber 806.11 $cm^{-1}$ for the arsenic initial concentration of 50 mmol/L. Jia's [30] research found that when pH was 3, the arsenic precipitated as ferric arsenate on the surface of ferric hydroxide, and infrared spectrum analysis showed that the As–O stretching vibration band was at approximately 825 $cm^{-1}$. It can be seen from the previous study results that the As–O stretching vibration band was at approximately 825 $cm^{-1}$ and decreased with decreasing arsenic concentration. In this study, the arsenic initial concentration was low (0.1 mmol/L), with complex soil particle composition, and the metal oxide content was much lower than the pure product, so less ferric arsenate was formed. The As–O stretching vibration band reduces to 793.68 $cm^{-1}$ in Figure 4. The results show that arsenate formed a small amount of metal arsenide on the red soil surface, which was consistent with previous findings.

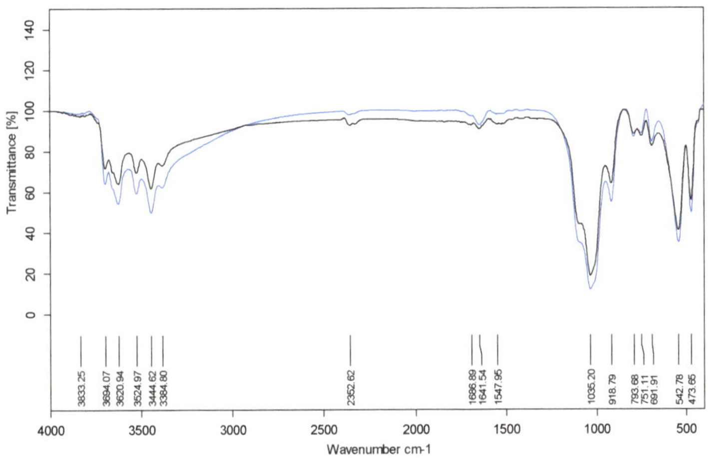


**Figure 4.** IR Spectra of Red Soil 1 Before and After Arsenate Adsorption.

SEM observation was conducted to elucidate the surface morphology of Yunnan red soil 4 after adsorbing arsenate. As can be seen in Figure 5, a precipitate, similar to a well-knitted net was separated from the water after adsorption of the arsenate, thus revealing excellent adsorption performance.

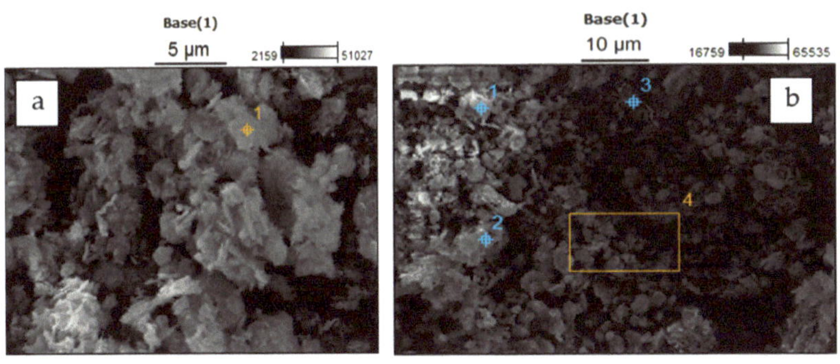


**Figure 5.** SEM Images of Red Soil 1 Before (**a**) and After (**b**) Adsorbing Arsenate.

Our previous studies examined [31] the absorption near-edge structure of arsenic speciation in size-fractions of red soil particles by X-ray. The results indicated that the concentration of arsenic inversely increased with particle size and was positively correlated with the concentrations of Fe and Al oxides and the specific surface area of the particles. The proportion of non-specifically absorbed As was less than 0.6% in all size fractions, while that of specifically-bound forms varied from 25.0 to 38.5%, based on chemical extraction. The proportion of As species associated with hydrous oxides and stable residual arsenic increased inversely with particle size, within the ranges of 31.0–50.8% and 17.8–40.2%, respectively. The XANES spectra of arsenic adsorbed to the particles were linearly fitted with ferrihydrite, goethite, and amorphous iron oxides, and the results showed that arsenate fractions of 1–5 μm and <1 μm were mainly associated with ferrihydrite and goethite. These findings are consistent with the results of our mechanism studies.

## 4. Conclusions

Arsenic is highly toxic to most living organisms and is a known human carcinogen. Therefore, the efficient treatment of arsenate-contaminated water is essential. In this study, Yunnan red soil, which was chosen as a sorbent, was successfully applied to remove arsenate from aqueous solutions. The key influencing factors were investigated. The optimal soil/solution ratio was found to be 0.05 g/100 mL, and an optimal shaking speed for the red soil was found to be 225 rpm. The adsorption capacity of Yunnan red soil increased, and the removal rate of arsenate decreased with increasing initial arsenate concentration. The IR spectra of the precipitates further confirmed that the metal arsenide was settled by the Yunnan red soil, indicating that the arsenate ion existed on the red soil surface in the form of protonated bidentate surface complexation of $-FeO_2As(O)(OH)^-$ and $FeO_2As(O)^{2-}$.

The data obtained indicate that the Yunnan red soil used in this study has excellent potential for use as an unconventional adsorbent that is comparable with commonly used pure chemical adsorbents for As removal, and it may well be suitable as a new adsorbent for As removal during water treatment. More detailed adsorption mechanisms and the release process of As from the sediment should be further examined in future studies.

**Author Contributions:** Conceptualization, methodology, M.G., W.G. and W.W.; data curation, M.G. and L.S.; writing—original draft preparation, M.G.; writing—review and editing, W.W.; supervision, W.W.; project administration, W.W.; funding acquisition, L.S. and W.W. All authors have read and agreed to the published version of the manuscript.

**Funding:** This research was supported by the Major Science and Technology Program for Water Pollution Control and Treatment (No. 2010ZX07212-007).

**Data Availability Statement:** Not applicable.

**Conflicts of Interest:** The authors declare no conflict of interest.

## References

1. Zhang, J.; Ding, T.; Zhang, C. Biosorption and toxicity responses to arsenite (As[III]) in Scenedesmus quadricauda. *Chemosphere* **2013**, *92*, 1077–1084. [CrossRef]
2. Jomova, K.; Jenisova, Z.; Feszterova, M.; Baros, S.; Liska, J.; Hudecova, D.; Rhodes, C.J.; Valko, M. Arsenic: Toxicity, oxidative stress and human disease. *J. Appl. Toxicol.* **2011**, *31*, 95–107. [CrossRef] [PubMed]
3. WHO. *Guidelines for Drinking-Water Quality*, 4th ed.; WHO: Geneva, Switzerland, 2011.
4. Jack, C.; Wang, J.; Shraim, A. A global healthy problem caused by arsenic from natural sources. *Chemosphere* **2003**, *52*, 1353–1359. [CrossRef]
5. Yean, S.; Cong, L.; Yavuz, C.; Mayo, J.; Yu, W.; Kan, A.; Colvin, V.; Tomson, M. Effect of magnetite particle size on adsorption and desorption of arsenite and arsenate. *J. Mater. Res.* **2005**, *20*, 3255–3264. [CrossRef]
6. Gupta, A.; Yunus, M.; Sankararamakrishnan, N. Zerovalent iron encapsulated chitosan nanospheres—A novel adsorbent for the removal of total inorganic Arsenic from aqueous systems. *Chemosphere* **2012**, *86*, 150–155. [CrossRef]
7. Darrell, K.N. Worldwide occurrences of arsenate in groundwater. *Science* **2002**, *296*, 2143–2145. [CrossRef]
8. Hashim, K.S.; AlKhaddar, R.; Shaw, A.; Kot, P.; Al-Jumeily, D.; Alwash, R.; Aljefery, M.H. Electrocoagulation as an Eco-Friendly River Water Treatment Method. *Adv. Water Resour. Eng. Manag.* **2020**, *39*, 219–235. [CrossRef]
9. Usman, M.; Zarebanadkouki, M.; Waseem, M.; Katsoyiannis, I.A.; Ernst, M. Mathematical modeling of arsenic(V) adsorption onto iron oxyhydroxides in an adsorption-submerged membrane hybrid system. *J. Hazard. Mater.* **2020**, *400*, 123221. [CrossRef] [PubMed]

10. Kanel, S.; Greneche, J.; Choi, H. Arsenic(V) Removal from Groundwater Using Nano Scale Zero-Valent Iron as a Colloidal Reactive Barrier Material. *Environ. Sci. Technol.* **2006**, *40*, 2045–2050. [CrossRef] [PubMed]
11. Dinesh, M.; Charles, U.; Pittman, J. Arsenic removal from water/wastewater using adsorbents—A critical review. *J. Hazard. Mater.* **2007**, *142*, 1–53. [CrossRef]
12. Luo, L.; Zhang, S.; Ma, Y. Advance in Research on Arsenic Sorption and Its Affecting Factors in Soils. *Soils* **2008**, *40*, 351–359. [CrossRef]
13. Feng, Q.; Zhang, Z.; Chen, Y.; Liu, L.; Zhang, Z.; Chen, C. Adsorption and Desorption Characteristics of Arsenic on Soils: Kinetics, Equilibrium, and Effect of $Fe(OH)_3$ Colloid, $H_2SiO_3$ Colloid and Phosphate. *Procedia Environ. Sci.* **2013**, *18*, 26–36. [CrossRef]
14. Hashim, K.S.; Ewadh, H.M.; Zubaidi, S.L.; Kot, P.; Muradov, M.; Aljefery, M.; Al-Khaddar, R. Phosphate removal from water using bottom ash: Adsortion performance, coexisting anions and modelling studies. *Water Sci. Technol.* **2021**, *83*, 77–89. [CrossRef] [PubMed]
15. Glass, S.; Mantel, T.; Appold, M.; Sen, S.; Usman, M.; Ernst, M.; Filiz, V. Amine-Terminated PAN Membranes as Anion-Adsorber Materials. *Chem. Ing. Tech.* **2021**, *93*, 1396–1400. [CrossRef]
16. Khan, S.U.; Farooqi, I.H.; Usman, M.; Basheer, F. Energy Efficient Rapid Removal of Arsenic in an Electrocoagulation Reactor with Hybrid Fe/Al Electrodes: Process Optimization Using CCD and Kinetic Modeling. *Water* **2020**, *12*, 2876. [CrossRef]
17. Usman, M.; Katsoyiannis, I.; Mitrakas, M.; Zouboulis, A.; Ernst, M. Performance Evaluation of Small Sized Powdered Ferric Hydroxide as Arsenic Adsorbent. *Water* **2018**, *10*, 957. [CrossRef]
18. Agnieszka, A.; Weronika, S.C.; Grzegorz, J. Arsenate Adsorption on Fly Ash, Chitosan and Their Compasites and Its Relations with Surface, Charge and Pore Properties of the Sorbents. *Materials* **2020**, *13*, 5381. [CrossRef]
19. Wang, Z.; He, B.; Pan, X. Levels, trends and risk assessment of arsenic pollution in Yangzonghai Lake, Yunnan Province, China. *Sci. China Chem.* **2012**, *53*, 1809–1817. [CrossRef]
20. Hamayun, M.; Mahmood, T.; Naeem, A. Equilibrium and kinetics studies of arsenate adsorption by $FePO_4$. *Chemosphere* **2014**, *99*, 207–215. [CrossRef]
21. Zhang, Z.; Wang, P.; Zhang, J.; Xia, S. Removal and mechanism of Cu (II) and Cd (II) from aqueous single-metal solutions by a novel biosorbent from waste-activated sludge. *Environ. Sci. Pollut. Res.* **2014**, *21*, 10823–10829. [CrossRef] [PubMed]
22. Tang, W.; Li, Q.; Gao, S.; Shang, J. Arsenic (III, V) removal from aqueous solution by ultrafine $\alpha$-$Fe_2O_3$ nanoparticles synthesized from solvent thermal method. *J. Hazard. Mater.* **2011**, *192*, 131–138. [CrossRef] [PubMed]
23. Arai, Y.; Sparks, D.; Davis, J. Arsenate adsorption mechanisms at the allophone-water interface. *Environ. Sci. Technol.* **2005**, *39*, 2537–2544. [CrossRef] [PubMed]
24. Vu, T.; Le, G.; Dao, C.; Dang, L.; Nguyen, K.; Nguyen, Q.; Dang, P.; Tran, H.; Duong, Q.; Nguyen, T.; et al. Arsenic removal from aqueous solutions by adsorption using novel MIL-53(Fe) as a highly efficient adsorbent. *RSC Adv.* **2015**, *5*, 5261–5268. [CrossRef]
25. Huang, C. *Soil Science*; China Agriculture Press: Beijing, China, 2000; pp. 169–170.
26. Zhu, J.; Pigna, M.; Cozzolino, V.; Caporale, A.G.; Violante, A. Sorption of arsenite and arsenate on ferrihydrite: Effect of organic and inorganic ligands. *J. Hazard. Mater.* **2011**, *189*, 564–571. [CrossRef]
27. Wang, S.; Mulligan, C. Natural attenuation processes for remediation of arsenic contaminated soils and groundwater. *J. Hazard. Mater.* **2006**, *138*, 459–470. [CrossRef]
28. Goldberg, S.; Johnston, C. Mechanisms of arsenic adsorption on amorphous oxides evaluated using macroscopic measurements, vibrational spectroscopy, and surface complexation modeling. *J. Colloid Interface Sci.* **2001**, *234*, 204–216. [CrossRef]
29. Liu, H.L.; Liang, M.; Zhu, Y.; Cai, F.; Zou, H. The Adsorption of Arsenic by Ferric Hydroxide and its Precipitation Mechanism. *Acta Sci. Circumstantiae* **2009**, *29*, 1011–1020.
30. Jia, Y.; Xu, L.; Fang, Z.; Demopoulos, G. Demopoulos. Observation of surface precipitation of arsenate on ferrihydrite. *Environ. Sci. Technol.* **2006**, *40*, 3248–3253. [CrossRef]
31. Li, S.; Luo, Y.; Zhang, H.; Haung, J.; Li, Z.; Wei, J. Arsenic forms in various particle-size fractions of red soil-Chemical fractionation and speciation using XANES analysis. *Acta Sci. Circumstantiae* **2011**, *31*, 2733–2739. [CrossRef]

*applied sciences*

MDPI

*Article*

# Modeling the Life Cycle Inventory of a Centralized Composting Facility in Greece

**Panagiotis Chazirakis [1,2], Apostolos Giannis [1,*] and Evangelos Gidarakos [1]**

1 School of Chemical and Environmental Engineering, Technical University of Crete, University Campus, 73100 Chania, Greece; pchazirakis@isc.tuc.gr (P.C.); gidarako@mred.tuc.gr (E.G.)
2 Inter-Municipal Enterprise of Solid Waste Management (DEDISA), 50 Grigoriou V, 73100 Chania, Greece
* Correspondence: agiannis@chenveng.tuc.gr; Tel.: +30-2821037819

**Featured Application: The life cycle inventory of a composting facility is expected to predict the fate of individual waste fragments and compost quality.**

**Abstract:** Mechanical composting is a popular treatment method for the mechanically separated organic fraction of municipal solid wastes to stabilize the waste material and reduce its environmental impacts. The model and life cycle inventory database are created based on the existing centralized mechanical composting facility located in Chania (Crete, Greece). This study aims to assess all stages of the composting process, wherein input-output flows are comprehensively analyzed based on specific waste fragments. The transfer coefficients are calculated for each waste fragment throughout the processes. The degradation rate is measured as kg of C and N released per Mg of the treated material. The results show that process degradation rates are independent of the initial fragmental composition. This is the first study that accurately models the fate of distinctive waste fragments in a composting plant, while the developed life cycle inventory (with regard to mass and energy balances) can be applied to estimate the environmental impacts regarding mechanical composting the organic fraction of municipal solid wastes.

**Keywords:** biowaste; composting; input-output flows; energy balance; EASETECH

**Citation:** Chazirakis, P.; Giannis, A.; Gidarakos, E. Modeling the Life Cycle Inventory of a Centralized Composting Facility in Greece. *Appl. Sci.* **2022**, *12*, 2047. https://doi.org/10.3390/app12042047

Academic Editors: Dibyendu Sarkar, Prafulla Kumar Sahoo, Mohammad Mahmudur Rahman and Rupali Datta

Received: 31 December 2021
Accepted: 14 February 2022
Published: 16 February 2022



## 1. Introduction

Biowaste (BW) represents a major fraction of municipal solid waste (MSW), which is consisted of food and kitchen waste from households, HORECA (hotels, restaurants, caterers), and green waste from gardens and parks [1]. It comprises the most significant waste fraction generated by households reaching up to 44% of total household waste globally [2,3]. Recently, biowaste has received considerable attention owing to its environmental, economic, and social impacts. Future trends have shown that municipal solid biowaste (MSBW) will most likely increase due to population growth and rising consumption patterns [4].

According to data from the EU, 86 million Mg of biowaste were generated in 2017, while approximately 40% (up to 100% in some member states) were disposed of in dumpsites and landfills, with a negative impact on the environment [5]. Biowaste recovery and utilization started in the 90s with Directive 1999/31/EC [6]. This directive obliges the member states of the EU to reduce the amount of biodegradable municipal waste and aims for 65% of all MSW produced to be recycled before 2030 (while only 10% should be disposed of in landfills). A feasible approach used for the past 20 years is recovering the organic fraction of municipal solid wastes (OFMSW) from unsorted waste and used in biological processes such as composting and anaerobic treatment. Anaerobic treatment has been studied in several cases since biogas can be produced for energy recovery [7,8]. The research in mechanical closed composting systems has focused on source segregated

OFMSW processes. The mechanical sorting systems vary from simple installation such as shredder, trommel, and magnet, to medium or high complexity OFMSW sorting systems to deliver various quality and purity materials [9].

Diverting OFMSW from landfills and using it as composting material has many environmental benefits (e.g., reducing greenhouse gas emissions [10]), while it can be easily integrated with material recovery facilities (MRFs). It involves the biological aerobic degradation of organic matter under controlled conditions [11], resulting in a nutrient-rich product. The process produces gas emissions composed primarily of $CO_2$, trace amounts of methane, non-methane organic compounds (NMOC), nitrogen, and ammonia [12]. The resulting product when OFMSW originated from unsorted mixed waste is called compost-like output (CLO) [13], and its quality is related to the purity of the initial materials and the pretreatment method [14]. The compost can significantly enhance the fertility of the soil environment by increasing the soil organic carbon (SOC), total N (TN), and soil microbial biomass (SMB). At the same time, it has a positive effect on the activity of enzymes involved in the C, N, and P cycles [15]. However, the impurities and contaminants that are usually released from CLO require increased attention [10]. CLO is considered one of the primary sources of microplastics (MPs) in the agricultural environment, negatively influencing soil microbial processes or plant growth [16]. Therefore, the use of CLO in land applications is limited and regulated to restore quarries, dumping sites, or road slopes [10,13,17].

Societies have started transitioning towards a model based on source segregation of biowaste that can produce higher-quality compost with environmental impacts which are significantly reduced. At the same time, the existing facilities are adapted to accept source segregated biowaste. The evaluation of the composting systems is complicated, with many variables which must be considered. It involves numerous calculations and requires accurate data to model better the variables of each system. The use of life cycle assessment (LCA) is based on the guidelines of ISO 14040 and 14044 [18], and can provide a much-improved viewpoint on waste management by connecting materials, resources, and waste flows with potential environmental impacts. Every LCA study incorporates several available local information and data sets called the life cycle inventory (LCI). In particular, the LCI is a compilation of all mass flows and emissions associated with the activities within the waste management system as well as upstream and downstream activities linked to the management of the waste. It relies on recent, representative, and accurate data such as waste types and their individual material fractions, detailed physico-chemical composition, mass balances for all relevant material fractions, energy balances for all processes and technologies, records of the emissions, and inventories of all relevant upstream and downstream processes [19]. The LCI covers all consumptions and emissions of environmental importance [18]. However, it is difficult to find case-specific data or LCI that include waste composition, energy and resource inputs, and material substitution in an LCA implementation study [20]. Establishing a relevant and high-accurate LCI is often demanding but crucial since it is the technical basis for assessing the waste management system. Existing models and software offer some assistance and databases in setting up the LCI, but it is always important to ensure relevance and consistency in the technical data of the specific study. LCA methodologies and advanced software such as EASETECH (developed by DTU) are based on fragmentation analysis to follow elemental balances throughout the processes.

Inventories on existing facilities managing the OFMSW from unsorted mixed waste are scarce [21]. In most cases, the treatment of such materials is held by private facilities, and the available data concerning full-size treatment are not published. Although several composting technologies have been studied in European countries, a few have developed LCI for composting systems treating segregated biowaste [22]. The available information about the materials of OFMSW and nutrient flows are inconsistent, making it difficult to develop alternative scenarios during urban planning [23]. Therefore, there is an increasing need for predictive models to support environmental policy and decision-making. Few studies have investigated composting of mechanically sorted OFMSW obtained at MBT plants. Thus, this research is urgently needed [24]

This study aims to create a life cycle inventory (LCI) based on the mechanical recycling and composting facility in Chania (Crete, Greece). The objectives are to model the composting unit by mapping the fragmental mass balance between its sub-processes, to monitor the release of C and N as emissions to the environment, and to record the water, electricity, and fuel consumption for the treatment of one Mg of OFMSW introduced in the facility. Two-year waste sampling and data collection are comprehensively analyzed. The outcomes from this study can be used as a tool for the waste management practitioners to foresee the outputs and cost of treating OFMSW and source segregated biowaste.

## 2. Methodology

The LCI model is developed using actual and local data from two-year monitoring study of the composting unit. Material flow analysis (MFA) software STAN is utilized to fill in missing and not-accessible data, while the LCA software EASETECH is used as a tool for the elemental pathway of C and N in the process. The case study facility is in Chania on the island of Crete (Greece). The composting unit is part of the integrated "Mechanical Recycling and Composting Facility—Landfill" of Chania. It serves 156,585 inhabitants [25] and annually treats approximately 91,500 Mg of urban solid waste [26]. The OFMSW is collected from the mechanical recycling facility, and the process is classified as a simplified pretreatment method [27]. Briefly, the comingled waste is fed in the mechanical sorting system, passing throw a bag opener and an automatic rotary sieve (trommel) with a 70 mm diameter mess. A conveyor then drops the undersized material to a magnet for ferrous metals removal, and the remaining is obtained as OFMSW. The oversized material exited from the trommel is driven for recyclables recovery in the facility, while the rejects are disposed of in the nearby landfill.

The system boundary for this study is shown in Figure 1. It includes the composting subsystems (aerobic composting tank, refinery unit, open windrows-maturation), which act as the operational processes after the wastes are delivered to the composting plant. The methodology is based on an in-depth analysis of all of the consisting fragments of OFMSW. The waste fragments are comprehensively characterized throughout the subprocesses until their degradation to greenhouse gas (GHG) emissions released to the environment, rejects disposed of to the landfill, or CLO production. Initially, every subprocess and flow are recognized and recorded, while the monitoring period is two years (2018–2019).

The greenhouse gases (GHG) are also considered and studied as C and N transformations along with the main waste flows. The energy is calculated in the form of electricity, fuel in diesel consumed, and the water consumed in the subprocesses. The green waste (GW) consists of tree branches collected from the municipality bulky collection system. It is shredded in the facility and used as a bulking agent in a ratio of 1:4 by volume. The functional unit is 1 Mg of wet mass OFMSW mixed with green waste entering to the composting unit.

### 2.1. Composting Units

The composting process is divided into three sub-units: (1) the aerobic composting tank (ACT), which is a continuous flow reactor; (2) the refinery unit (RFU); and (3) the open windrows (OPW) for compost maturation. Table 1 provides the main information about the composting conditions. The primary composting process in the ACT sustains aerobic conditions with bottom-up aeration and a leachate draining system. A deodorization system with a biofilter is connected to the air exhaust system of the ACT. The turning, water addition, and movement of the material inside the reactor are controlled by an overhead-suspended bridge system with four screw-shaped turners. A fifth screw turner at the end of the reactor transfers the composted materials to conveyor belts towards the RFU. The RFU comprises flip-flop sieves and gravimetric separators to remove bulky and non-compostable materials. The rejects are diverted for landfilling (landfill cover), and the refined material is sent for secondary composting and maturation at OPW. A hook lift truck, a backhoe loader, a wheel loader, and a compost turner handle the transportation and mixing of rejects and maturation windrows. All above vehicles are considered to use diesel (Euro 5 emission standard engines). The unit employs one (1) senior engineer as operation manager, one (1) heavy machinery operator, and two (2) workers daily, while one (1) truck driver and two maintenance technicians (electrician and mechanic) from the nearby MRF are also involved part-time.

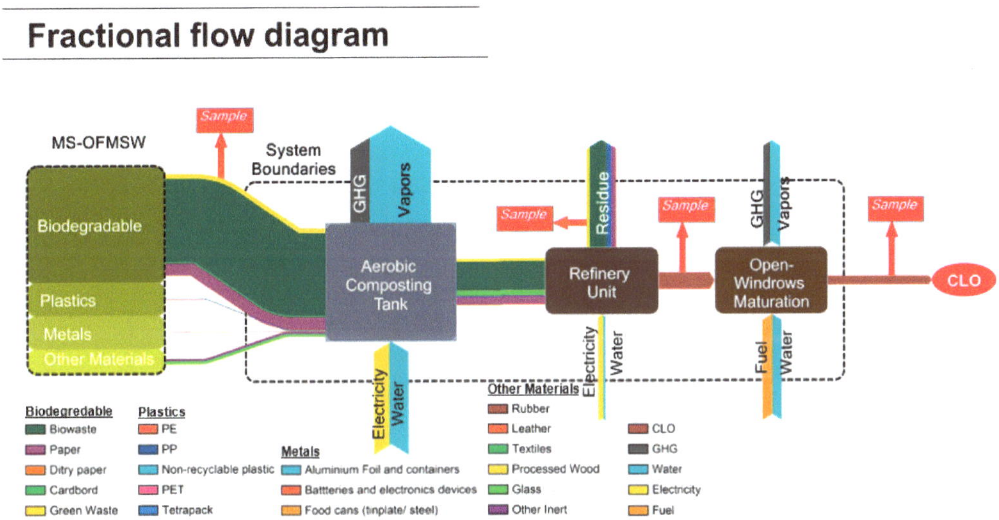


**Figure 1.** Sankey's graphic representation of the system and its boundary includes the composting process and inputs/outputs. The subprocesses (aerobic composting tank, refinery unit, open windrows-maturation) are presented. Colored lines represent the different fractions of MS-OFMSW, while the thickness of the lines is proportional to the mass of each fragment. The resources used (electricity, fuel, and water) are shown with yellow and orange arrows, and the emissions to the atmosphere are shown with dark gray arrows. Water addition and evaporation are shown with light blue arrows. The red arrows indicate the sampling points.

**Table 1.** Composting conditions and involved personnel for CLO production from mechanically separated OFMSW and green waste.

| Composting Conditions | Value—Factor | |
|---|---|---|
| Mixing ratio by volume (OFMSW:GW) | 1:4 | |
| Retention time in the composting tank | 4 weeks | |
| Refinery unit | Yes | |
| Residue after treatment | Used for landfill cover | |
| Maturation time | 6–9 months | |
| **Personel** | **Number of People** | **Involvement** |
| Operation manager (senior engineer) | 1 | Full-time |
| Front line stuff (workers) | 2 | Full-time |
| Heavy machinery operator | 1 | Full-time |
| Mentainence mechanic | 1 | Part-time |
| Mentainence electrician | 1 | Part-time |
| Truck driver | 1 | Part-time |

### 2.2. Sample Collection and Characterization

Sampling was carried out for two consecutive years (2018–2019) and represented the average composition of each material. The samples were collected from the inputs and outputs of each sub-process, and the sampling points are shown in Figure 1. Each sample was then reduced in size using the 'coning and quartering' method at approximately 100 kg and subjected to fragmentation analysis in a nearby area. The samples were sorted into four major categories (biodegradable, plastics, metals, and others-not compostable) and 17 sub-categories based on the research [28] and the recovery potential at the MRF: biowaste, paper, dirty paper, cardboard, garden waste, soft plastic, hard plastic, non-recyclable plastic, aluminum foil and containers, batteries and electronic devices, food cans (tinplate/steel), rubber, leather, textiles, processed wood, glass, and other inert.

Each sample was analyzed for water content and total solids after drying at 105 °C for 24 h (CSN EN 12048), ash content, volatile solids by igniting the dried sample at 550 °C until steady weight (CSN EN 13039), and Kjeldahl Nitrogen in the dried samples using the Kjeldahl method (CSN EN 13654.01).

### 2.3. Life Cycle Inventory (LCI) Modeling

Material flow analysis is used to fill in the missing information. The processed weighting data are fed to the sophisticated software STAN v2.6 (Substance flow Analysis, 2.6), which is developed by the Technical University of Wien [29,30] to complete the missing stream flows and convert them in accordance with the functional unit of the study [31]. The results are fed to the EASETECH model (Environmental Assessment System for Environmental TECHnologies, v2.4.5) to calculate for each fraction the degradation factors and transfer coefficients. EASETECH is a waste-LCA model focusing on managing complex waste streams [19], and it can handle the flow of complex heterogeneous fractions in various bioprocess systems. The framework and calculation structure have been described in detail by [32]. In EASETECH software, the degradation is defined as the reduction of organic dry mass during the composting process. The degradation factor of the fragment '$\alpha$' for the process Df($\alpha$) is defined as the % reduction of the total mass of vs. ascribed to biogenic carbon reduction. In comparison, the transfer coefficient of fragment '$\alpha$' is considered the reduction of the total wet mass due to mechanical separation Tf($\alpha$).

### 2.4. GHG Emissions

GHG emissions related to the processes can be defined to direct emissions, indirect upstream emissions, and indirect downstream emissions. Direct emissions are linked to the composting site and its activities, including waste degradation and emissions from machinery used on the site (fuel consumption). The indirect upstream emissions are related to activities for fuel production, provision of electricity used in the site, and the construction of infrastructure and machinery. Indirect downstream avoided emissions are considered from peat substitution for fertilizer production and the carbon sequestration in the soil when compost is applied to land [12,33]. The indirect emissions related to fuel and electricity production were selected after an extensive literature review to reflect the local fuel and energy mixture. Table 2 shows the emission factors (Efs) used in this study [34,35]. Mass flow analysis was employed to calculate the gases released during the degradation of the materials for the direct and indirect downstream emission, while EASETECH software native database provided machinery emission factors based on the engine euro standard.

**Table 2.** Emission factors (Efs) relevant to GHG during composting.

| Type of Process/Emission | Emission Factor | Reference |
|---|---|---|
| Provision of diesel oil | 0.306 kg $CO_2$-eq/L diesel | [33] |
| Combustion of diesel oil | 2.7 kg $CO_2$-eq/L diesel | [32] |
| Provision of electricity | 0.810 kg $CO_2$-eq/kWh | [33] |

### 2.5. Site-Specific Data

Valuable data from the facility operation are also collected. They concern the primary input and output of each composting unit for the monitoring period, which include weighing data from the daily treated materials, rejects and outputs of the refinery process, daily routes, working hours, annual diesel fuel consumption (L) of every vehicle involved in the composting process, daily electricity consumption from the composting unit (kWh), and daily water consumption (L) in the composting process.

The weighting data are annually averaged for every flow and diverted to the appropriate functional unit and sub-process. The annual electricity consumption is attributed, respectively, to each sub-process and divided by the annual wet mass of the treated material of the specific sub-process. The vehicle diesel consumption is calculated by dividing the annual fuel consumption by attributing working hours and routes for the needs of the composting process. Water is attributed to each subprocess and divided by the mass of the treated materials.

### 2.6. Life Cycle Inventory Boundaries

The LCI boundaries assume a zero-burden approach [36,37] for the received materials at the entrance of the composting facility. Therefore, the facility environmental footprint is not included in the calculations. This excluded component includes emissions from the construction of the facility, equipment, vehicles, and post-processing of the initial material. Also, this study does not consider the environmental impacts associated with the construction of windrow composting facility (equipment and infrastructure).

### 2.7. Sensitivity Analysis

Sensitivity analysis is conducted to examine sensitive inputs and analyze whether the assumptions made in the model influence the results [21]. For this reason, this study uses perturbation analysis, and uncertainty propagation methodology [38,39].

Perturbation analysis identifies the most sensitive parameters of the model. The method calculates each parameter sensitivity ratio (SR) and observes the effect of low but countable changes in the results. Every parameter of the studied system is raised, one at a time by 10% (Δ parameter), the new calculated net result is referred as (Δ result). The SR is the ratio between the relative change of the result and the relative change in the parameter. It is calculated as:

$$\mathrm{SR} = \frac{\frac{\Delta\ \mathrm{result}}{\mathrm{initial\ result}}}{\frac{\Delta\ \mathrm{parameter}}{\mathrm{initial\ parmeter}}}\ \mathrm{NRS}_i = \frac{\mathrm{RS}_i}{\max|\mathrm{RS}_i|}$$

To compare the different SRs in various outputs of the model, the concept of the normalized sensitivity ratio (NSR) has been developed and calculated for each SR. NSR is defined as the ratio of one parameter in one system output divided with the maximum absolute value among all of the SRs in the same output. The concept is a modified adaptation of the methodology of NSRs introduced by [39].

Uncertainty propagation consists in propagating input uncertainties to calculate the result uncertainty. Before propagating them, the practitioner chooses a representation for these input uncertainties. The probability theory was adopted in this case study, and the sampling propagation method of Monte Carlo analysis was selected [40].

## 3. Results

### 3.1. Waste Composition

Table 3 presents the material fraction distribution of the mechanically separated organic fraction of municipal solid wastes (MS-OFMSW) and green waste received for composting. Water constitutes 52.5% of the total wet mass which is higher than MS-OFMSW in other studies [41,42]. The main compostable fragments can be categorized as biowaste (76.5%), paper-like materials (paper, dirty paper, cardboard) (12.9%), and green waste (4.64%). Since mechanical sorting is based on sizing and gravimetric properties, the presence of foreign non-biodegradable materials is justified. According to [43], paper waste and cardboard in various proportions consist of 12–27% of the dry mass of MS-OFMSW treated in similar composting facilities in Spain. Carbon content and its origins, biogenic or fossil, are taken from [44]. These estimates consider that some foreign material may be present in each fragment as suggested by the IPCC 2006 Guidelines [45]. The main greenhouse gases that contribute to global warming are $CH_4$ and $N_2O$, and their release depends on the technology, the waste input, and the management of the process. The above carbon origin is of immense importance in most LCA methodologies since biogenic carbon, when released in the form of $CO_2$ to the environment, is not counted in the impacts, in contrast, when the same portion of the carbon is released in the form of methane in a landfill, for example, it is counted [46,47]. In summary, the initial material chemical composition without the green waste has a TS of 47.16%, vs. 77.03% of TS, ash content of 22.97, biogenic C of 43.82%, and TN of 2.48%. Although the above values vary compared to literature, they are within the same order [10,13]

**Table 3.** Fragmentation composition of OFMSW and green waste, proximate and ultimate analysis for each fraction, and carbon content (divided to biogenic and fossil origin).

| Fraction | OFMWS Composition (%) | OFMSW Water Content (%) | TS (%) | VS (%) | Ash (%) | C Bio (%) | C Fossil (%) | N(%) |
|---|---|---|---|---|---|---|---|---|
| **Biodegradable** | | | | | | | | |
| Biowaste | 76.49 ± 10.55 | 56.04 | 37.40 | 90.00 | 10.00 | 54.60 | 0.60 | 3.72 |
| Paper | 11.40 ± 10.91 | 33.29 | 87.00 | 72.30 | 27.70 | 37.60 | 0.20 | 0.18 |
| Dirty paper | 0.29 ± 0.63 | 53.30 | 75.50 | 91.10 | 8.90 | 44.60 | 0.91 | 0.30 |
| Cardboard | 1.24 ± 17.06 | 39.33 | 89.50 | 84.90 | 15.10 | 41.10 | 0.30 | 0.24 |
| Green waste | 4.64 ± 1.62 | 47.00 | 53.00 | 93.00 | 7.00 | 43.02 | 0.00 | 0.15 |
| **Plastics *** | | | | | | | | |
| Soft plastic (PE) | 0.30 ± 2.90 | 28.25 | 85.89 | 95.60 | 4.40 | 0.41 | 81.60 | 0.20 |
| Hard plastic (PP) | 0.29 ± 2.05 | 22.83 | 96.80 | 97.80 | 2.20 | 0.40 | 79.50 | 5.50 |
| Non-recyclable plastic | 0.30 ± 3.90 | 0.00 | 92.90 | 94.50 | 5.50 | 0.36 | 70.60 | 0.50 |
| **Metals** | | | | | | | | |
| Aluminum foil and containers | 0.14 ± 0.47 | 24.95 | 81.20 | 23.90 | 76.10 | 13.70 | 1.52 | 0.40 |
| Batteries and electronic devices | 0.14 ± 0.93 | 9.72 | 91.10 | 14.20 | 85.80 | 4.35 | 4.35 | 0.10 |
| Food cans (tinplate/steel) | 0.15 ± 1.57 | 7.03 | 86.82 | 0.00 | 100.00 | 0.00 | 0.00 | 0.00 |
| **Other materials** | | | | | | | | |
| Rubber | 0.11 ± 3.58 | 34.42 | 92.30 | 90.30 | 9.70 | 52.30 | 13.10 | 0.60 |
| Leather | 0.11 ± 3.58 | 34.42 | 93.30 | 87.40 | 12.60 | 30.70 | 30.70 | 0.30 |
| Textile | 0.11 ± 3.58 | 34.42 | 94.00 | 96.40 | 3.60 | 39.10 | 13.00 | 3.20 |
| Processed wood | 0.11 ± 3.58 | 34.42 | 84.60 | 96.30 | 3.70 | 49.40 | 0.00 | 0.00 |
| Glass | 3.17 ± 2.05 | 2.23 | 99.70 | 1.20 | 98.80 | 0.00 | 0.00 | 0.10 |
| Other Inert | 1.81 ± 2.87 | 34.71 | 63.40 | 2.30 | 97.70 | 0.65 | 0.65 | 0.00 |

* PET and Tetra pack packaging were monitored but not found.

### 3.2. Material Flow Analysis

The overall process with flow dynamics and mass balance is presented in Figure 2. The estimations of C and N flows are displayed in Figures 3 and 4, respectively, assuming that carbon is 99% oxidized to $CO_2$ while nitrogen is released to the air as $NH_3$ at the ACT process [12]. The modeling of the composting system follows all fractions throughout the processes based on two assumptions: (1) the mass can be transferred between processes, and (2) the carbon of biogenic origin in biodegradable materials is biologically degraded to gases with dominant carbon dioxide. The above transfers and transformations are expressed as degradation factors and transfer coefficients.

The primary process occurs in the continuous-flow aerobic composting tank (ACT), where the materials enter daily, are mechanically mixed, and transferred across the tank. Controlled conditions are provided with aeration, water adjustment, and temperature management. The processed material that exits daily continues to the refinery unit for separation. The calculated retention time of the materials in the composting tank is four to five weeks. Several studies follow a general approach when modeling a composting system and consider the treated materials as a single homogenous mixture appointing one degradation factor [9,48]. That is justified since most studies refer to source-segregated

OFMSW [9,49–51] and only a few to mechanical sorted OFMSW [27]. This research considers the individuality of each of the consisting fragments and its different degradation rates. The degradation factors are calculated utilizing MFA methodology with data of the ash content and mass loss of each fragment in the input and output at the ACT.

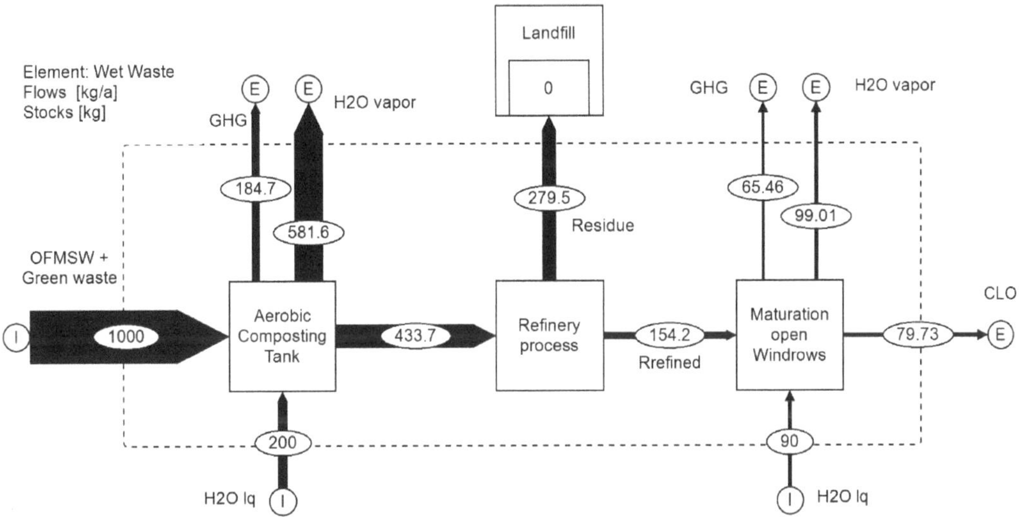


**Figure 2.** Sankey diagram of mass balance for OFMSW + green waste treatment in kg (wet waste) (the lines are proportional to the mass of each flux).

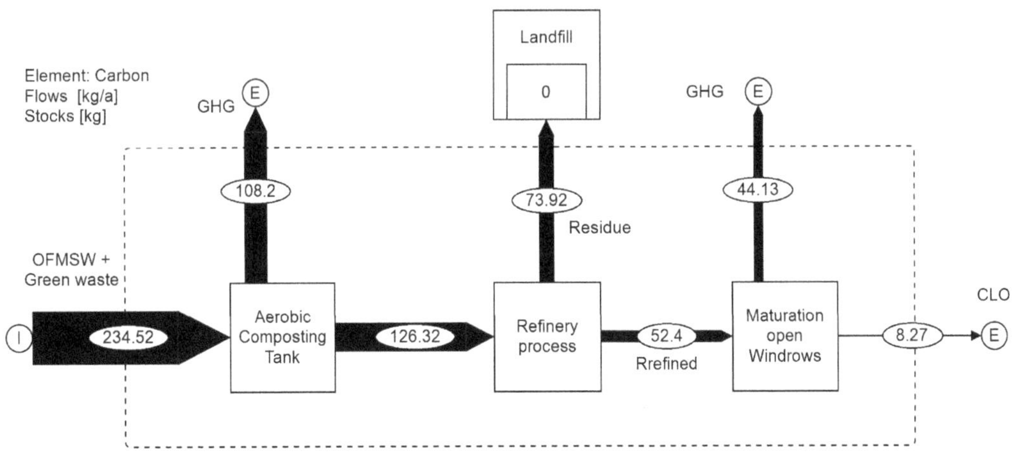


**Figure 3.** Sankey diagram for carbon mass balance in kg for OFMSW + green waste composting processing. The lines are proportional to the mass of each flux.

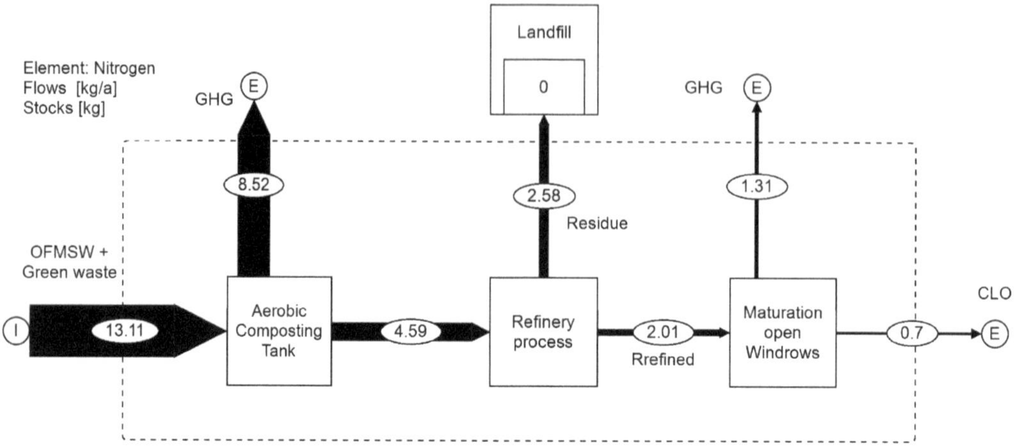


**Figure 4.** Sankey diagram for nitrogen mass balance in kg for OFMSW + green waste composting processing (the lines are proportional to the mass of each flux).

In Table 4, the degradation factors are referred to the biodegradable materials. Paper-like materials present the higher degradation factor with printed paper (61%), cardboard (51%), and dirty paper (43.3%). These materials present high VS, and they consist primarily of cellulose and lignin, organic polymers difficult to decompose. However, the high decompositions are justified by the screening, mixing and the elevated temperatures found during the thermophilic phase of composting process that contribute to the rapid degradation of lignocellulose [52]. Similar high degradation rates have been observed elsewhere [43] concerning various paper-like materials present in OFMSW composting. It is also stated that if the paper-like materials do not exceed 27% of OFMSW, degradations of 36–65% are feasible in controlled conditions with a retention time of 45 days. Concerning biowaste, the degradation factor has an average 48%, while the green waste 15%.

**Table 4.** Degradation factors (%$w/w$) for the volatile solids of waste fractions in the aerobic composting tank process.

| Fraction | Degradation Factor (%) |
|---|---|
| Biowaste | 48 |
| Paper | 61 |
| Dirty paper | 43.3 |
| Cardboard | 51 |
| Green waste | 15 |
| Soft plastic (PE) | 10 |
| Other inert | 5 |
| Other materials | 0 |

The above results suggest that the retention time in the ACT is not sufficient for the complete decomposition of the biodegradable materials of biowaste. Green waste serves as a bulking agent and is shredded to medium-sized particles presenting a low degradation factor [53]. Physical characteristics such as bulk density, particle size, and porosity are important factors for fragment decomposition level. The structure of green waste (containing lignocellulosic materials) appears difficult decomposition and requires specific lignocellulosic microorganisms (and enzymes) to improve it degradation and retention time more than 12 weeks [54,55]. In order to produce mature compost, [12] stated that degradations of 40–83% of the carbon contained in the biowaste are required, while [56] reported 62–66% and 66–77% degradation of carbon for garden waste and food waste, respectively.

Plastic materials present close to zero degradation for this process; only LDPE abandoned in the form of shopping bags can be accounted to have a 10% degradation. Although LDPE films have presented some degradation only in the harsh environment of the composting process [55], and the degraded portion is meager. The mechanical processes employ throughout the composting process, such as turning mixing and screening, can result in polymers being sheared into smaller fragments during the conventional composting process and could explain the above degradation factor, as it is often apparent in household and commercial organic waste [57,58]. An additional issue that must be considered is the rapid increase in biodegradable plastic materials that have started to replace the traditional PE film. Their biodegradability is dependent on the composting conditions and the chemical composition of each material [59].

The second sub-process (refinery unit-RFU) is a mechanical separation stage based on sizing the material using a 'flip-flop' sieve with 10 by 10 mm mesh holes, followed by a gravimetric air separator in line with a gravimetric air cyclone to collect the lightweight material. Table 5 presents the transfer coefficients for each material. Bulky and heavy materials are mainly rejected into residue. Water content is critical in this step since it adds excess weight if not adjusted correctly in the previous process, leading to discarding compostable materials as residue. The fact that the exiting material is collected by its size and gravimetric properties and not its chemical characteristics is advantageous. It provides optimal mechanical characteristics on the collected materials, although it does not prevent the infiltration of unwanted dissolved chemicals such as heavy metals. The collected materials have TS (75.03%), vs. (75.4) of TS, and 19.3% TS carbon of biogenic origin. The rejected material of the process consists of bulky and non-compostable materials. The same principle is followed for the transfer coefficients of the refinery process, and the above assumptions allow experimentation with variations of composition with the same system, providing a handy tool for further research.

**Table 5.** Refinery process transfer coefficients total mass (%$w/w$) for open windrow composting and maturation.

| Fraction | Transfer Coefficients (%) |
|---|---|
| Biowaste | 45 |
| Paper | 60 |
| Cardboard | 100 |
| Green waste | 5 |
| Soft plastic (PE) | 2 |
| Hard plastic (PP) | 1 |
| Non-recyclable plastic | 1 |
| Other materials | 0 |

The material that continues to the final composting/maturation stage has a homogenous texture; the origin fractions are hard to recognize, only some paper-like remains, and some wood fraction with particle size lower than 10 mm are notable. The total vs. is high (69%). The material is accounted as concentrated biodegradable fraction, which justifies the intense composting stage, followed by a prolonged maturation state (composting windrows). It must be stated that open-windrow composting can be challenging owing to variable weather conditions that advance or delay the composting process. Intensive mechanical mixing and constant windrow temperature monitoring which occurs once a week during this process minimize the number of anaerobic pockets in the composting mass. However, it is reasonable to assume that an inevitable release of $CH_4$ occurs. Hence, adopting the lowest emissions values, 0.8–2.5% of degraded C is released as $CH_4$, which seems reasonable [11]. Concerning nitrogen-based GHG, [60] stated that there is no production of $N_2O$ during the thermophilic phase since autotrophic nitrifier activity ceases above 40 °C. Since the maturation phase is considered a continuation of the primary composting process, GHG production is only scarce at the final stages of the process. For this reason, 0.1–0.7% of degrading N is accounted to transform to $N_2O$. Mixing and water addition ensure a partially controlled maturation elongating to six or nine months until the desired physicochemical characteristics are reached.

The quality of MSW compost is dependent on many parameters, including the composting facility design, feedstock source and proportions used, composting procedure, and duration of maturation [61]. The maturation typically requires minor active management. It is a crucial final stage that facilitates the conversion of potentially toxic $NH_4$ to $NO_3$, allows the loss of phytotoxic volatile compounds, and stabilizes the microbial community. At this state, mesophilic fungi and actinomycetes colonize the compost, which is thought to be responsible for the breakdown and transformation of humic substances and lignin. Although, maturation is a vital stage frequently given insufficient time, or is even missed out altogether, to save space and increase the throughput of composting plants. In this case study, the corresponding sub-process can be chronically adjusted depending on the aiming physicochemical characteristics of the final product. The average decomposition rate is calculated to be 75% of the total volatile solids of the initial material. The resulting CLO has 37% water content, vs. of 57%, while the C and N contents are calculated to 56% and 1.9%, respectively. Carabassa et al. (2020) [13] presented CLO with similar physicochemical characteristics ranging from 65 to 70% TS, 44.5 to 64% VS, and 1.4 to 2.17% N, while [1] produced CLO with similar characteristics.

### 3.3. Mass Balance

Material and substance flow analyses are performed based on mass balances. The composting unit is then built graphically and displayed as Sankey plots. Figure 2 presents the mass flows of wet waste throughout the processes and the loss of material and compounds to the atmosphere (in kg). The water content is significant for the proper accounting of the total mass balance. Since it is added during composting and maturation and accounts for 200 L per Mg of treated materials in the ACT and 90 L per Mg of treated materials during maturation, the quantities are not insignificant (while its use is threefold). It provides temperature control by reducing heat due to its evaporation, it acts as a medium for the dilution and exchange of elements. Finally, it regulates the aerobic conditions in the composting mass. During the aerobic process, the evaporated water and mass loss is calculated to 766.3 kg per Mg $w/w$, plus 164.5 kg for the maturation state, while an amount of 279.5 kg is rejected. The resulting CLO material is calculated to be 79.73 kg.

The carbon balance is presented in Figure 3. During the two sub-processes where organic matter degradation occurs, 64.96% of the initial carbon is released into the environment. The primary composting process releases 46.14% of the initial carbon in gaseous form. A significant portion of the initial carbon (31.52%) is diverted to the landfill and contributes to carbon sequestration [62,63]. Finally, 3.52% is included in the CLO destined for land use. Several studies have investigated the degradation of organic matter and C fate during composting. Production of mature compost requires degradation of 40–83% of the carbon contained in the compost [12]. Most of this carbon is emitted as biogenic $CO_2$, and a relatively small portion is emitted as $CH_4$ created in anaerobic pockets in the composting mass.

The total nitrogen loss during the main composting process (ACT) is 64.99% (Figure 4). A portion (19.68%) of the initial nitrogen is landfilled, and 5.3% is bound to the CLO produced mass. The rest is released in gas form during the maturation phase. The controlled conditions in the ACT provide a stable temperature profile of 45–65 °C, favoring the thermophiles phase. The above conditions inhibit the nitrification of produced ammonium to NO2 while the dissociation constant (pKa) of $NH_4^+$ decreases with increasing temperature, meaning that higher temperatures favor evaporation of NH3. Eventually, ammonia is the most emitted form of N [64,65]. However, other by-products have not been investigated (i.e., for ammonia the oxidized forms NO and $N_2O$ are not considered, although aerobic microorganisms may form them). These gases potentially impact the environment. NO may result in ozone depletion in the stratosphere, and $N_2O$ is an effective greenhouse gas [66].

### 3.4. Estimation of Resources Consumed

#### 3.4.1. Electricity

Aeration, deodorization, mixing, transfer, and refining of compost are the main electricity-consuming processes in the ACT and RFU resulting in a 34.56 kWh electricity consumption per Mg of the wet treated material (Table 6). This number is the average electricity consumption for every sub-process for a given volume of the treated material. According to [12], electricity consumption depends mainly on technology use and is higher on closed composting systems, especially reactor technologies, ranging between 9 and 65 kWh/Mg $w/w$ versus 0.023–19.7 kWh/Mg $w/w$ for open technologies. The research by [67] attributes a fourfold electricity consumption to reactor technology than windrow composting, stating that the benefit of reactor composting is covered from N loss by preventing organic contaminants, higher degradation rates, and lower composting periods. Another research in large-scale bioconversion systems based on the aerobic treatment of organic waste implies that the reduction of the produced leachate due to controlled air supply is reduced by 75–99% [68].

**Table 6.** Heavy machinery involved in the composting process (fuel consumption), electricity, and water consumption.

| Row Labels | | Process Attributed | Unit/Mg of Material Treated in the Corresponding Process Material |
|---|---|---|---|
| Backhoe loader (L of diesel) | MIDI wheel loader (L/Mg) | Maturation | 2.201 |
| Wheel loader (L of diesel) | Wheel loader (L/Mg) | Maturation | 1.096 |
| Other tractor (L of diesel) | Hook lift (L/Mg) | Maturation | 0.311 |
| Other drivable machines (L of diesel) | Compost turner (L/Mg) | Maturation | 0.098 |
| Marginal Electricity Consumption (kWh) | Electricity (kWh/Mg) | Aerobic composting Tank and Refinery | 34.56 |
| Water consumption for composting process in aerobic composting tank | L/Mg entering main composting | Aerobic composting Tank and Refinery | 200.0 |
| Water consumption for maturation state in open windrows | L/Mg material in windrows | Maturation | 583.7 |

#### 3.4.2. Fuel

The transportation of the residue to the landfill and the refined material to the maturation area employs a hook-lift truck consuming 0.311 L per Mg of the transferred material. For the management and treatment of the maturing windrows, two wheel-loaders, and one compost turner are involved (Table 6). The fuel consumption for each vehicle is calculated to be 2.201, 1.096, and 0.098 L of diesel consumed, respectively, per Mg of maturing material. The engine technology for all of the vehicles follows the standard of Euro 5 as it has been classified from the European emission standards for heavy-duty diesel engines. The conversion of fuel consumption to the initial wet mass of MS-OFMSW is 0.658 L of diesel per Mg, while the literature review presents a range of 0.4–0.5 L per Mg for similar processes [12].

#### 3.4.3. Water

Water consumption is 200 L per Mg for the ACT and 90 L per Mg of refined material during windrow composting. In many LCA methodologies, water consumption is not included [11]. During the aerobic tank composting, the water addition is constant to substitute the water loses of high composting rates and prevent the compost from overheating, while the only water source is the embedded irrigation system. On the other hand, open composting is exposed to weather conditions and precipitation contributes to the windrows irrigation system.

### 3.5. Sensitivity Analysis

Sensitivity analysis is performed to check the model's robustness and assess overall uncertainty [69]. The parameters tested in the perturbation analysis include the degradation factors for aerobic composting and the transfer coefficients. All parameters are raised, one at a time, by 10% and the resulting change in the three key outputs of the system: the two exits of the refinery process (rejects and refined materials) and the at the end of maturation phase (produced CLO) are recorded as the quantity of wet mass in contrast to the initial quantities. The results as NSRs are presented in Table 7. The resulting NSRs reveal the sensitivity of the model to the degradation rate of biowaste and the maturating CLO.

**Table 7.** Perturbation analysis of NRSs for the main parameters of the model.

| | Refinery Output | Rejects Output | Final CLO |
|---|---|---|---|
| **Degradation Factor** | | | |
| **Aerobic composting tank** | | | |
| Biowaste | 0.8 | 0.9 | 0.2 |
| Paper | 0.2 | 0.1 | 0.1 |
| Dirty paper | 0.0 | 0.0 | 0.0 |
| Cardboard | 0.0 | 0.0 | 0.0 |
| Green waste | 0.0 | 0.0 | 0.0 |
| Soft plastic (PE) | 0.0 | 0.0 | 0.0 |
| Other Inert | 0.0 | 0.0 | 0.0 |
| **Transfer coefficient** | | | |
| **Refinery process** | | | |
| Biowaste | 0.0 | 0.0 | 0.0 |
| Paper | 1.0 | 1.0 | 0.5 |
| Dirty paper | 0.3 | 0.3 | 0.2 |
| Cardboard | 0.0 | 0.0 | 0.0 |
| Green waste | 0.0 | 0.0 | 0.0 |
| Soft plastic (PE) | 0.0 | 0.0 | 0.0 |
| **Degradation Factor** | | | |
| **Open windrows** | | | |
| Degradation Factor | - | - | 1.0 |

The second part of the sensitivity analysis is performed regarding the overall uncertainty propagation for the system outputs. Monde Carlo simulation (MCS) is initialized to generate pseudo-random numbers from the set of the studied parameters. The model degradation parameters and transfer coefficients are attributed with uncertainties of 10% in the form of normal distribution, and the MCS iteration value is set to 10,000 times to obtain the sample distribution of the output parameter [31,69–71]. The results are 6.1% variation for the rejects output, 9.4% for the refinery output, and 15% for the compost output.

## 4. Discussion

This study identifies the dynamics of recovering significant quantities of biogenic carbon and benefitting from the produced compost-like output as a soil conditioner. Although the quality of the produced CLO has several uncertainties due to the origin of the materials, a significant amount of the initial materials is discarded as rejects and usually ends up in landfills. Considering the circular economy perspective, the sustainable treatment of OFMSW requires it to be separated from residual waste at the source to eliminate contaminants remaining in the initial materials [5]. In Spain, samples of produced CLO from 10 MBT plants in Castile and Leon showed heavy metal concentrations below the limits set by the national legislation. However, the percentage of inert impurities, such as plastic or glass, was excessively high, exceeding in some cases the legal limit [72,73]. The same issues were concerned for the CLO produced from the MBT of Attica [1]. The elimination of reject based on the absence of the contaminants mentioned above increases the produced quantities since, in other cases, rejects could be further processed. The restrictions applied to CLO uses do not apply for source segregated biowaste produced compost. The use of the produced material to agriculture, soil improvement, and fertilizer substitution should not be overlooked.

In early 2017, Europe had about 570 active MBT plants with a treatment capacity of 55 million tons [74]. According to the 2020 report from the European Environment Agency (EEA) concerning bio-waste treatment in Europe, the most common treatment methods for biowaste, in line with circular economy principles, were composting and anaerobic digestion. The second was the most preferable in some cases due to benefits from the recovery of material and energy. However, the 22 EU countries average favor composting, with Greece utilizing only composting [5]. In the highest biowaste treatment capacities ranking, Sweden and Croatia present more than 370 kg/capita, followed by Austria, Slovenia, and France near 300 kg/person, while Greece shows the lowest capacities. In the same ranking, comparing source segregation (versus not separately collected biowaste), Greece mainly applies the collection of mixed waste. At the same time, Austria leads the trends with close to 200 kg/capita on separate biowaste collection. Concerning Greece until 2020, six MBT facilities had been constructed and in operation, and ten more are under construction [75].

The LCA study performed by Abeliotis [76] for the MBT of west Attica was based on data provided from the regional administration of Attica, and the native database of the LCA software was used to calculate the produced emissions. In a global LCA review [21] until 2014 (222 case studies), the dominant monitored waste stream was household mixed waste, 70% of the studies concern cases in European countries. Most of the inventory data sources were taken from the literature without addressing the appropriateness of the data used, such as representativeness in time or space of the extracted data compared to the studied system. Composting was the most favorable among the biological treatment methods used in 74 of the above studies. In contrast, anaerobic digestion was used in 53 cases.

In more recent studies, the life cycle inventory analysis is the most time and resource-demanding for the LCA partitioners [72]. The evolution of advanced LCA software with ready-made modules for the composting process may save time and resources. However, it may lead to fault results making mandatory an evaluation step of the primary LCI data.

The goals for a more circular economy in EU by the new revised Waste Framework Directive introduced a new requirement for bio-waste separation. By 31 December 2023, bio-waste must either be separated and recycled at the source or collected separately and not mixed with other types of waste [73]. In addition, as of 2027, compost derived from mixed municipal waste will no longer count towards achieving compliance with the recycling targets for municipal waste. From an LCA perspective, the impacts of a transition from mechanical sorting to source segregated biowaste collection has not yet been studied.

## 5. Conclusions

Most investigations in mechanical composting focus on open composting processes (e.g., windrows or static piles) or laboratory scale enclosed systems due to the difficulty in sampling commercial systems and the heterogeneous nature of MSW. In most cases, the private sector operates the facilities, and scarce data are available. In this paper, a life-cycle inventory (LCI) is created for the first time using primary data collected from an existing mechanical composting facility in Greece. This study uses comprehensive data to systematically model material and substance flows. The results indicate that significant decomposition occurs during the composting and maturation phases with significant conversion of the carbon content into carbon dioxide. It also proves that industrial-scale composting as a set of fragment flows throughout the different sub-processes provides flexibility, robustness, and profound usability to the designed model. The LCI can stand as a platform for environmental assessment. The practitioners can adjust the parameters accordingly to simulate alternative scenarios, providing a holistic view of specific situations and alternative disposal routes.

**Author Contributions:** Conceptualization, P.C. and A.G.; methodology, P.C.; validation, P.C.; investigation, P.C.; writing—original draft preparation, P.C.; writing—review and editing, A.G.; supervision, E.G. All authors have read and agreed to the published version of the manuscript.

**Funding:** This research received no external funding.

**Institutional Review Board Statement:** Not applicable.

**Informed Consent Statement:** Not applicable.

**Data Availability Statement:** Not applicable.

**Acknowledgments:** We would like to acknowledge the Trans-municipal enterprise for solid waste management of Chania (DEDISA S.A. public authorities) for providing valuable data.

**Conflicts of Interest:** The authors declare no conflict of interest.

## References

1. Malamis, D.; Bourka, A.; Stamatopoulou, E.; Moustakas, K.; Skiadi, O.; Loizidou, M. Study and assessment of segregated biowaste composting: The case study of Attica municipalities. *J. Environ. Manag.* **2017**, *203*, 664–669. [CrossRef]
2. Bartocci, P.; Zampilli, M.; Liberti, F.; Pistolesi, V.; Massoli, S.; Bidini, G.; Fantozzi, F. LCA analysis of food waste co-digestion. *Sci. Total Environ.* **2020**, *709*, 136187. [CrossRef]
3. Ardolino, F.; Colaleo, G.; Arena, U. The cleaner option for energy production from a municipal solid biowaste. *J. Clean. Prod.* **2020**, *266*, 121908. [CrossRef]
4. Zeller, V.; Lavigne, C.; D'Ans, P.; Towa, E.; Achten, W.M.J. Assessing the environmental performance for more local and more circular biowaste management options at city-region level. *Sci. Total Environ.* **2020**, *745*, 140690. [CrossRef]
5. EEA. *Bio-Waste in Europe—Turning Challenges into Opportunities*; Europian Environment Agency: Luxemburg, 2020.
6. Council of the European Communities Council Directive 1999/31/EC on the landfill. *Off. J. Eur. Communities* **1999**, L182/1. [CrossRef]
7. Van Fan, Y.; Tin, C.; Perry, S. Anaerobic digestion of municipal solid waste: Energy and carbon emission footprint. *J. Environ. Manag.* **2018**, *223*, 888–897. [CrossRef]
8. Wi, M.; Kulig, A.; Lelici, K. Applied sciences Odour Emissions of Municipal Waste Biogas Plants—Impact of Technological Factors, Air Temperature and Humidity. *Appl. Sci.* **2020**, *10*, 1093.
9. Graça, J.; Murphy, B.; Pentlavalli, P.; Allen, C.C.R.; Bird, E.; Gaffney, M.; Duggan, T.; Kelleher, B. Bacterium consortium drives compost stability and degradation of organic contaminants in in-vessel composting process of the mechanically separated organic fraction of municipal solid waste (MS-OFMSW). *Bioresour. Technol. Rep.* **2021**, *13*, 100621. [CrossRef]
10. Wei, Y.; Li, J.; Shi, D.; Liu, G.; Zhao, Y.; Shimaoka, T. Environmental challenges impeding the composting of biodegradable municipal solid waste: A critical review. *Resour. Conserv. Recycl.* **2017**, *122*, 51–65. [CrossRef]
11. Díaz, M.J.; Ruiz-Montoya, M.; Palma, A.; de-Paz, M.V. Thermogravimetry applicability in compost and composting research: A review. *Appl. Sci.* **2021**, *11*, 1692. [CrossRef]
12. Boldrin, A.; Andersen, J.K.; Møller, J.; Christensen, T.H.; Favoino, E. Composting and compost utilization: Accounting of greenhouse gases and global warming contributions. *Waste Manag. Res. J. Sustain. Circ. Econ.* **2009**, *27*, 800–812. [CrossRef]
13. Carabassa, V.; Domene, X.; Alcañiz, J.M. Soil restoration using compost-like-outputs and digestates from non-source-separated urban waste as organic amendments: Limitations and opportunities. *J. Environ. Manag.* **2020**, *255*, 109909. [CrossRef]
14. Vasileiadou, A.; Zoras, S.; Iordanidis, A. Biofuel potential of compost-like output from municipal solid waste: Multiple analyses of its seasonal variation and blends with lignite. *Energy* **2021**, *236*, 121457. [CrossRef]
15. Bhattacharyya, P.; Chakrabarti, K.; Chakraborty, A. Effect of MSW Compost on Microbiological and Biochemical Soil Quality Indicators. *Compost Sci. Util.* **2003**, *11*, 220–227. [CrossRef]
16. Baiano, S.; Fabiani, A.; Fornasier, F.; Ferrarini, A.; Innangi, M.; Mocali, S.; Morra, L. Biowaste compost amendment modifies soil biogeochemical cycles and microbial community according to aggregate classes. *Appl. Soil Ecol.* **2021**, *168*, 104132. [CrossRef]
17. Palansooriya, K.N.; Shaheen, S.M.; Chen, S.S.; Tsang, D.C.W.; Hashimoto, Y.; Hou, D.; Bolan, N.S.; Rinklebe, J.; Ok, Y.S. Soil amendments for immobilization of potentially toxic elements in contaminated soils: A critical review. *Environ. Int.* **2020**, *134*, 105046. [CrossRef]
18. *BSI 14040*; Environmental Management–Life Cycle Assessment—Principles and Framework. International Standard Organization: London, UK, 2006.
19. Christensen, T.H.H.; Damgaard, A.; Levis, J.; Zhao, Y.; Björklund, A.; Arena, U.; Barlaz, M.A.A.; Starostina, V.; Boldrin, A.; Astrup, T.F.F.; et al. Application of LCA modelling in integrated waste management. *Waste Manag.* **2020**, *118*, 313–322. [CrossRef]
20. Ripa, M.; Fiorentino, G.; Vacca, V.; Ulgiati, S. The relevance of site-specific data in Life Cycle Assessment (LCA). The case of the municipal solid waste management in the metropolitan city of Naples (Italy). *J. Clean. Prod.* **2017**, *142*, 445–460. [CrossRef]
21. Laurent, A.; Bakas, I.; Clavreul, J.; Bernstad, A.; Niero, M.; Gentil, E.; Hauschild, M.Z.; Christensen, T.H. Review of LCA studies of solid waste management systems—Part I: Lessons learned and perspectives. *Waste Manag.* **2014**, *34*, 573–588. [CrossRef]
22. Pini, M.; Neri, P.; Ferrari, A.M. Environmental Performance of Waste Management in an Italian Region: How LCI Modelling Framework could Influence the Results. *Procedia CIRP* **2018**, *69*, 956–961. [CrossRef]

23. Guo, H.; Zhao, Y.; Damgaard, A.; Wang, Q.; Lu, W.; Wang, H.; Christensen, T.H. Material flow analysis of alternative biorefinery systems for managing Chinese food waste. *Resour. Conserv. Recycl.* **2019**, *149*, 197–209. [CrossRef]
24. Mironov, V.; Vanteeva, A.; Sokolova, D.; Merkel, A.; Nikolaev, Y. Microbiota dynamics of mechanically separated organic fraction of municipal solid waste during composting. *Microorganisms* **2021**, *9*, 1877. [CrossRef]
25. EL.STAT. *Revision of the Results of the 2011 Population and Housing Census for the Resident, De Jure (Registered) and De Facto Population of Greece*; Hellenic Statistical Authority: Pireas, Greece, 2014.
26. Prefecture of Crete. *Integration of the Local Waste Plans of the Municipalities of the Region. of Crete in the Regional Waste Management Planning (PESDAK) (Report in Greek Language)*; Prefecture of Crete Authority: Crete, Greece, 2016.
27. Cecchi, F.; Traverso, P.; Pavan, P.; Bolzonella, D.; Innocenti, L. Characteristics of the OFMSW and Behavior of the Anaerobic Digestion Process. *ChemInform* **2003**, *34*, 141. [CrossRef]
28. Edjabou, M.E.; Jensen, M.B.; Götze, R.; Pivnenko, K.; Petersen, C.; Scheutz, C.; Astrup, T.F. Municipal solid waste composition: Sampling methodology, statistical analyses, and case study evaluation. *Waste Manag.* **2015**, *36*, 12–23. [CrossRef]
29. Cencic, O.; Rechberger, H. Material Flow Analysis with Software STAN. *EnviroInfo 2008—Environ. Inform. Ind. Ecol.* **2008**, *2008*, 440–447.
30. Sevigné-Itoiz, E.; Gasol, C.M.; Rieradevall, J.; Gabarrell, X. Methodology of supporting decision-making of waste management with material flow analysis (MFA) and consequential life cycle assessment (CLCA): Case study of waste paper recycling. *J. Clean. Prod.* **2015**, *105*, 253–262. [CrossRef]
31. Henriksen, T.; Levis, J.W.; Barlaz, M.A.; Damgaard, A. Approaches to fill data gaps and evaluate process completeness in LCA—perspectives from solid waste management systems. *Int. J. Life Cycle Assess.* **2019**, *24*, 1587–1601. [CrossRef]
32. Clavreul, J.; Baumeister, H.; Christensen, T.H.; Damgaard, A. An environmental assessment system for environmental technologies. *Environ. Model. Softw.* **2014**, *60*, 18–30. [CrossRef]
33. Favoino, E.; Hogg, D. The potential role of compost in reducing greenhouse gases. *Waste Manag. Res.* **2008**, *26*, 61–69. [CrossRef]
34. Fruergaard, T.; Astrup, T.; Ekvall, T. Energy use and recovery in waste management and implications for accounting of greenhouse gases and global warming contributions. *Waste Manag. Res. J. Sustain. Circ. Econ.* **2009**, *27*, 724–737. [CrossRef]
35. Koffi, B.; Cerutti, A.; Duerr, M.; Iancu, A.; Kona, A.; Janssens-Maenhout, G. *CoM Default Emission Factors for the Member States of the European Union Dataset-Version 2017*; European Commission—Joint Research Centre (JRC): Ispra, Italy, 2017.
36. Djuric Ilic, D.; Eriksson, O.; Ödlund, L.; Åberg, M. No zero burden assumption in a circular economy. *J. Clean. Prod.* **2018**, *182*, 352–362. [CrossRef]
37. Nakatani, J. Life cycle inventory analysis of recycling: Mathematical and graphical frameworks. *Sustainability* **2014**, *6*, 6158–6169. [CrossRef]
38. Clavreul, J.; Guyonnet, D.; Christensen, T.H. Quantifying uncertainty in LCA-modelling of waste management systems. *Waste Manag.* **2012**, *32*, 2482–2495. [CrossRef] [PubMed]
39. Andreasi Bassi, S.; Christensen, T.H.; Damgaard, A. Environmental performance of household waste management in Europe—An example of 7 countries. *Waste Manag.* **2017**, *69*, 545–557. [CrossRef] [PubMed]
40. Hung, M.L.; Ma, H.W. Quantifying system uncertainty of life cycle assessment based on Monte Carlo simulation. *Int. J. Life Cycle Assess.* **2009**, *14*, 19–27. [CrossRef]
41. Burnley, S.J. A review of municipal solid waste composition in the United Kingdom. *Waste Manag.* **2007**, *27*, 1274–1285. [CrossRef]
42. Riber, C.; Rodushkin, I.; Spliid, H.; Christensen, T.H. Method for fractional solid-waste sampling and chemical analysis. *Int. J. Environ. Anal. Chem.* **2007**, *87*, 321–335. [CrossRef]
43. Alvarez, J.V.V.L.L.; Larrucea, M.A.; Bermúdez, P.A.; Chicote, B.L. Biodegradation of paper waste under controlled composting conditions. *Waste Manag.* **2009**, *29*, 1514–1519. [CrossRef]
44. Riber, C.; Petersen, C.; Christensen, T.H. Chemical composition of material fractions in Danish household waste. *Waste Manag.* **2009**, *29*, 1251–1257. [CrossRef]
45. Eggleston, S.; Buendia, L.; Miwa, K.; Ngara, T.; Tanabe, K. *2006 IPCC Guidelines for National Greenhouse Gas. Inventories, Prepared by the National Greenhouse Gas Inventories Programme*; Institute for Global Environmental Strategies: Kanagawa, Japan, 2006.
46. Christensen, T.H.; Gentil, E.; Boldrin, A.; Larsen, A.W.; Weidema, B.P.; Hauschild, M. C balance, carbon dioxide emissions and global warming potentials in LCA-modelling of waste management systems. *Waste Manag. Res. J. Sustain. Circ. Econ.* **2009**, *27*, 707–715. [CrossRef]
47. Saer, A.; Lansing, S.; Davitt, N.H.; Graves, R.E. Life cycle assessment of a food waste composting system: Environmental impact hotspots. *J. Clean. Prod.* **2013**, *52*, 234–244. [CrossRef]
48. López, M.; Soliva, M.; Martínez-Farré, F.X.; Fernández, M.; Huerta-Pujol, O. Evaluation of MSW organic fraction for composting: Separate collection or mechanical sorting. *Resour. Conserv. Recycl.* **2010**, *54*, 222–228. [CrossRef]
49. Campuzano, R.; González-Martínez, S. Characteristics of the organic fraction of municipal solid waste and methane production: A review. *Waste Manag.* **2016**, *54*, 3–12. [CrossRef]
50. Gutiérrez, M.C.; Martín, M.A.; Serrano, A.; Chica, A.F. Monitoring of pile composting process of OFMSW at full scale and evaluation of odour emission impact. *J. Environ. Manag.* **2015**, *151*, 531–539. [CrossRef] [PubMed]
51. Sailer, G.; Eichermüller, J.; Poetsch, J.; Paczkowski, S.; Pelz, S.; Oechsner, H.; Müller, J. Characterization of the separately collected organic fraction of municipal solid waste (OFMSW) from rural and urban districts for a one-year period in Germany. *Waste Manag.* **2021**, *131*, 471–482. [CrossRef] [PubMed]
52. Tuomela, M.; Vikman, M.; Hatakka, A.; Itävaara, M.; It, M. Biodegradation of lignin in a compost environment: A review. *Bioresour. Technol.* **2000**, *72*, 169–183. [CrossRef]

53. Yuan, J.; Zhang, D.; Li, Y.; Chadwick, D.; Li, G.; Li, Y.; Du, L. Effects of adding bulking agents on biostabilization and drying of municipal solid waste. *Waste Manag.* **2017**, *62*, 52–60. [CrossRef]
54. Cerda, A.; Artola, A.; Font, X.; Barrena, R.; Gea, T.; Sánchez, A. Composting of food wastes: Status and challenges. *Bioresour. Technol.* **2018**, *248*, 57–67. [CrossRef]
55. Davis, G.U. Open windrow composting of polymers: An investigation into the operational issues of composting polyethylene (PE). *Waste Manag.* **2005**, *25*, 401–407. [CrossRef]
56. Komilis, D.P.; Ham, R.K. Carbon dioxide and ammonia emissions during composting of mixed paper, yard waste and food waste. *Waste Manag.* **2006**, *26*, 62–70. [CrossRef]
57. Gui, J.; Sun, Y.; Wang, J.; Chen, X.; Zhang, S.; Wu, D. Microplastics in composting of rural domestic waste: Abundance, characteristics, and release from the surface of macroplastics. *Environ. Pollut.* **2021**, *274*, 116553. [CrossRef] [PubMed]
58. Vithanage, M.; Ramanayaka, S.; Hasinthara, S.; Navaratne, A. Compost as a carrier for microplastics and plastic-bound toxic metals into agroecosystems. *Curr. Opin. Environ. Sci. Health* **2021**, *24*, 100297. [CrossRef]
59. Narancic, T.; Verstichel, S.; Reddy Chaganti, S.; Morales-Gamez, L.; Kenny, S.T.; De Wilde, B.; Babu Padamati, R.; O'Connor, K.E. Biodegradable Plastic Blends Create New Possibilities for End-of-Life Management of Plastics but They Are Not a Panacea for Plastic Pollution. *Environ. Sci. Technol.* **2018**, *52*, 10441–10452. [CrossRef] [PubMed]
60. Hellmann, B.; Zelles, L.; Palojärvi, A.; Bai, Q. Emission of climate-relevant trace gases and succession of microbial communities during open-windrow composting. *Appl. Environ. Microbiol.* **1997**, *63*, 1011–1018. [CrossRef] [PubMed]
61. Hargreaves, J.C.; Adl, M.S.; Warman, P.R. A review of the use of composted municipal solid waste in agriculture. *Agric. Ecosyst. Environ.* **2008**, *123*, 1–14. [CrossRef]
62. Kumar, A.; Sharma, M.P. GHG emission and carbon sequestration potential from MSW of Indian metro cities. *Urban. Clim.* **2014**, *8*, 30–41. [CrossRef]
63. Staley, B.F.; Barlaz, M.A. Composition of Municipal Solid Waste in the United States and Implications for Carbon Sequestration and Methane Yield. *J. Environ. Eng.* **2009**, *135*, 901–909. [CrossRef]
64. Andersen, J.K.; Boldrin, A.; Christensen, T.H.; Scheutz, C. Mass balances and life cycle inventory of home composting of organic waste. *Waste Manag.* **2011**, *31*, 1934–1942. [CrossRef]
65. Amlinger, F.; Peyr, S.; Cuhls, C. Green house gas emissions from composting and mechanical biological treatment. *Waste Manag. Res. J. Sustain. Circ. Econ.* **2008**, *26*, 47–60. [CrossRef]
66. Clemens, J.; Cuhls, C. Greenhouse gas emissions from mechanical and biological waste treatment of municipal waste. *Environ. Technol.* **2003**, *24*, 745–754. [CrossRef]
67. Liu, Z.; Wang, X.; Wang, F.; Bai, Z.; Chadwick, D.; Misselbrook, T.; Ma, L. The progress of composting technologies from static heap to intelligent reactor: Benefits and limitations. *J. Clean. Prod.* **2020**, *270*, 122328. [CrossRef]
68. Themelis, N.J.; Kim, Y.H. Material and energy balances in a large-scale aerobic bioconversion cell. *Waste Manag. Res. J. Sustain. Circ. Econ.* **2002**, *20*, 234–242. [CrossRef] [PubMed]
69. Groen, E.A.; Heijungs, R.; Bokkers, E.A.M.; de Boer, I.J.M. Methods for uncertainty propagation in life cycle assessment. *Environ. Model. Softw.* **2014**, *62*, 316–325. [CrossRef]
70. Helton, J.C.; Johnson, J.D.; Sallaberry, C.J.; Storlie, C.B. Survey of sampling-based methods for uncertainty and sensitivity analysis. *Reliab. Eng. Syst. Saf.* **2006**, *91*, 1175–1209. [CrossRef]
71. Groen, E.A.; Bokkers, E.A.M.; Heijungs, R.; de Boer, I.J.M. Methods for global sensitivity analysis in life cycle assessment. *Int. J. Life Cycle Assess.* **2017**, *22*, 1125–1137. [CrossRef]
72. Laurent, A.; Clavreul, J.; Bernstad, A.; Bakas, I.; Niero, M.; Gentil, E.; Christensen, T.H.; Hauschild, M.Z. Review of LCA studies of solid waste management systems—Part II: Methodological guidance for a better practice. *Waste Manag.* **2014**, *34*, 589–606. [CrossRef]
73. *EU 2018/851*; Directive 2018/851 Amending Directive 2008/98/EC on Waste Framework. Official Journal of the European Union: Luxembourg, 2018; L-150/109.
74. Doing, M. *The Market. for Mechanical Biological Waste Treatment in Europe—Locations, Plants, Backgrounds and Market Estimations*; Ecoprog GmbH: Köln, Germany, 2017.
75. Environmental Protection Agency. *National Plan for Waste Management—Including Hazardous Waste (Legislation in Greek Language)*; Environmental Protection Agency: Athens, Greece, 2020; p. 729.
76. Abeliotis, K.; Kalogeropoulos, A.; Lasaridi, K. Life Cycle Assessment of the MBT plant in Ano Liossia, Athens, Greece. *Waste Manag.* **2012**, *32*, 213–219. [CrossRef]

water

*Article*

# Application of Electrocoagulation in Street Food Wastewater

**Wilawan Khanitchaidecha [1,2], Khakhanang Ratananikom [3], Bunyaphon Yangklang [1], Shotita Intanoo [1], Kanokphol Sing-Aed [1] and Auppatham Nakaruk [2,4,*]**

1 Department of Civil Engineering, Faculty of Engineering, Naresuan University, Phitsanulok 65000, Thailand; wilawankl@gmail.com (W.K.); bunyaphony61@nu.ac.th (B.Y.); shootitai61@nu.ac.th (S.I.); kanokphols61@nu.ac.th (K.S.-A.)
2 Centre of Excellence for Innovation and Technology for Water Treatment, Faculty of Engineering, Naresuan University, Phitsanulok 65000, Thailand
3 Department of Science and Mathematics, Faculty of Science and Health Technology, Kalasin University, Kalasin 46000, Thailand; khakhanang_r@yahoo.com
4 Department of Industrial Engineering, Faculty of Engineering, Naresuan University, Phitsanulok 65000, Thailand
* Correspondence: auppathamn@nu.ac.th

**Abstract:** Street food is commonly known as ready-to-eat and go foods and beverages, which is very famous in Thailand and other Asian countries. The street food daily generates high organic content and oily wastewater from washing and rinsing plates. The discharge of street food wastewater to public drains leads to a clogged drain line and unpleasant smell. In this work, an electrocoagulation (EC) system with monopolar aluminum (Al) electrodes was developed to treat two well-known street foods; Hainanese chicken rice (HC) and noodles and dumplings (ND). The results revealed that excellent chemical oxygen demand (COD) and fat, oil, and grease (FOG) removals were achieved under a specific operating condition (i.e., an electric current of 20 mA/cm$^2$ and electrolytic time of 10 min). The initial COD of HC wastewater decreased from 40.6 g/L to 1.9 g/L, approximately 95%, whereas the FOG decreased from 310 mg/L to 50 mg/L, approximately 84%. The lower initial COD and FOG concentrations of ND wastewater obtained approximately 98% for COD removal and 86% for FOG removal; the effluent contained 0.5 g/L of COD and 25 mg/L of FOG. In addition, a relatively low Al concentration of 0.02–0.08 mg/L was observed in the effluents. The appropriate design factors together with ease of use and fast pollutants removal were significant advantages of this study; the EC system has potential to apply to on-site street food treatment.

**Keywords:** street food wastewater; wastewater treatment technology; electrocoagulation

**Citation:** Khanitchaidecha, W.; Ratananikom, K.; Yangklang, B.; Intanoo, S.; Sing-Aed, K.; Nakaruk, A. Application of Electrocoagulation in Street Food Wastewater. *Water* **2022**, *14*, 655. https://doi.org/10.3390/w14040655

Academic Editors: Mohammad Mahmudur Rahman, Dibyendu Sarkar, Rupali Datta and Prafulla Kumar Sahoo

Received: 27 January 2022
Accepted: 18 February 2022
Published: 20 February 2022



## 1. Introduction

Street food is known as ready-to-eat foods and beverages sold on a street or other public space including a mobile vendor selling from a cart and a fixed vendor selling from a stall. The buying and selling of street food is a common economic activity in Thailand and Asian countries. The signature of street food is a variety of food, authentic taste, quick assess, and inexpensive service. Due to food preparation and cleaning processes, a large volume of wastewater is daily generated and discharged to public sanitary sewers. The characteristics of food-service wastewater were high organic (represented in chemical oxygen demand; COD), total suspended solid (TSS), and total fat oil and grease (FOG); these were 1523, 664, and 197 mg/L, respectively [1]. The above concentrations exceeded the regulations for a municipal sewer. In addition, the relatively high organic and FOG levels as well as high flow cause an unpleasant smell and sewer line clogging. In Thailand, the local government spends a large amount of the annual budget for sewer line cleaning to avoid flooding and improve residents' quality of life.

A large production of oily wastewater comes from various sectors; oil and gas, petrochemical, pharmaceutical, and food [2]. Recently, more attention has been focused on enhancing technology for oily wastewater treatment such as flotation, coagulation, biological treatment, membrane separation technology and electrocoagulation [3]. Among the abovementioned options, electrocoagulation (EC) is a potential technology for developing a mobile treatment system for street food. The significant benefits of the EC process include that high COD and FOG removals can be reached in a short electrolytic time due to multilevel treatment processes, such as electrochemical oxidation, coagulation, and flotation. In addition, the EC system required a smaller scale and lower operating cost rather than other oily treatment systems. Furthermore, the EC process has found an efficient pathogens removal [4] as well as potential toxic metals removal such as zinc, copper, and chromium. Large restaurants and households require an oil and grease trapping unit to separate the oil and grease content before discharging wastewater due to Thailand's regulations. However, this oil and grease trapping unit is unsuitable for street food wastewater, because it is a fixed system and requires several hours for gravity FOG separation.

A laboratory-scale EC system with the combination of aluminum (Al, at cathode) and iron (Fe, at anode) was fabricated for domestic wastewater [5]. It achieved 92% COD removal and 97% turbidity removal with a low energy consumption of 0.007 KWh/L. Similarly, in the previous literature [6] applied the EC process using Al and stainless steel as electrodes to treat oily automobile service effluent. A high COD of 94% was eliminated, whereas the FOG of 51% was removed by flocs precipitation and a large Al content of 62% was detected in the sludge. The EC technology was also applied to actual tannery wastewater [7]; the electrolysis time was in the range of 30–45 min to achieve the optimal organic and inorganic pollutants removals including COD, solids, and metals. However, most EC application to actual wastewater in previous studies has been limited to a small capacity. The enlarged capacity of an EC system possibly affects the pollutant removal efficiency and process stability due to homogenous flow and natural variations in wastewater composition. Therefore, the objectives of this study were to develop an enlarged EC system to reduce organic and FOG levels from street food wastewater with the aim of achieving sufficient COD and FOG removals, short treatment time, and simple operation.

## 2. Materials and Methods

### 2.1. Wastewater Sampling Methodology

The wastewater grab samples were collected from the most significant street food in Thailand, Hainanese chicken rice (HC) and noodles and dumplings (ND) in Phitsanulok Province. HC is known as an oily street food and contains very high COD and FOG (as shown in Table 1). NC is also a very common street food and generates a larger volume of wastewater rather than other street food. The sampling collection occurred after the peak hours of vendor operation to incorporate washing and rinsing plates and silverware. Both wastewaters were very turbid, smelly, and oily; their composition varied due to the source of street food and number of customers, as summarized in Table 1.

**Table 1.** Characteristics of street food wastewater *.

| Street Food | pH | COD (g/L) | FOG (mg/L) | Turbidity (NTU) | Conductivity (mS/cm) |
|---|---|---|---|---|---|
| Hainanese chicken rice | 5.6 ± 7% | 40.6 ± 15% | 310 ± 12% | 501 ± 11% | 0.68 ± 4% |
| Noodles and dumplings | 7.2 ± 10% | 27.3 ± 10% | 178 ± 8% | 706 ± 12% | 0.95 ± 5% |

* Data in the table are the average numbers from 10 times of sampling during 2 weeks of observation.

### 2.2. System Design and Operation

The 30 L capacity of a rectangular plastic tank with the dimensions of 31 cm (width) × 42 cm (length) × 23 cm (height) was modified to be the EC system connecting parallel monopolar aluminum electrodes with a size of 21 cm (width) × 20 cm (height) and DC power source (12 V/45 A). The seven aluminum electrodes were expanded along the system length with a 4 cm inter-electrode distance. A space between the electrodes and the side of the system allowed the homogenous mixture to be under no agitation [8]. The EC system configuration is presented in Figure 1. The supplied current density was controlled at approximately 20 $mA/cm^2$, which was the optimal value in terms of pollutants removal and electric consumption for oily wastewater treatment [9]. Due to a low conductivity of the street food wastewater, around 5 g/L of NaCl was added to the actual street food wastewater to enhance the electrochemical reactions and system performance [9]. The actual street food wastewater was prepared by removing large solids such as vegetables and meat. Since the optimal pH for pollutants removal by EC was found to be in the range of 5–7 [5], the street food wastewater without pH adjustment was used in the experiments.

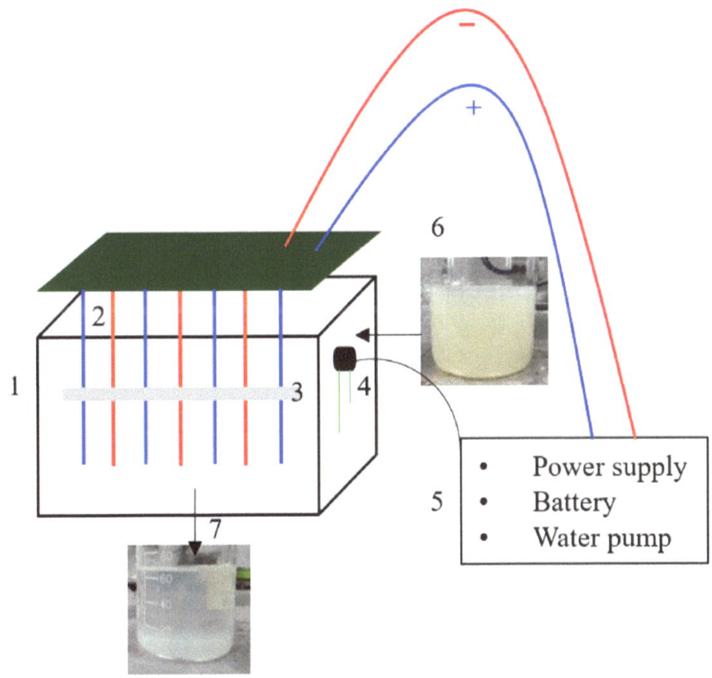


**Figure 1.** Configuration of EC system; (1) rectangle reactor, (2) aluminum electrodes, (3) flocs collector, (4) water level sensor, (5) control box, (6) actual street food wastewater, and (7) effluent.

### 2.3. Analytical Methods

The COD measurement method corresponded to DIN ISO 15705 and was analogous to EPA 410.4, APHA 5220 D, and ASTM D1252-06 B. The COD solutions A + B were purchased from Sigma-Aldrich Canada Co., Ltd. Solution A and solution B were added into the water samples. The water samples were oxidized with a hot sulfuric solution of potassium dichromate, with silver sulfate as the catalyst. The concentration of unconsumed yellow $Cr_2O_7^{2-}$ was then determined photometrically. The FOG was analyzed by the soxhlet extraction method (APHA 5520D). The water samples were acidified by hydrochloric acid, and then n-hexane was added. The acidified samples were transferred to a separatory funnel. After extraction, the extracted FOG was weighted and calculated in FOG concentration. For aqueous Al concentrations, the water samples were digested by concentrated nitric acid. The digestate was filtered and then analyzed by flame atomic absorption spectrometry (FLAA), The pH, conductivity, and turbidity were determined by a pH meter (Proline B210), a conductivity meter (Proline B250) and a turbidity meter (HACH 2100Q).

## 3. Results and Discussion

Two different types of actual street food wastewater (Hainanese chicken rice (HC) and noodles and dumplings (ND)) were selected as the case studies. The analytical data revealed that HC wastewater contained very high COD and FOG concentrations of 40.6 g/L and 310 mg/L, respectively. On the other hand, the ND wastewater contained lower concentrations of 27.3 g/L and 178 mg/L for COD and FOG, respectively. However, it has to be noted that these concentrations were extremely high compared to the domestic effluent standard in Thailand [10]. The COD and FOG from the street food wastewater were in a similar range to industrial waste and wastewater such as waste coolant [11] and high-concentration oil and gas fields [12]. Thus, the direct discharge of street food wastewater can severely impact the environment and water resources.

During EC treatment, flocs were observed at the top of surface and subsequently separated from the effluent. The COD of the HC wastewater immediately decreased from 40.6 g/L to 12.1 g/L after 2 min, approximately 70% (shown in Figure 2a). The COD removal efficiency continuously increased to 85% at 4 min and 94% at 6 min, respectively. A longer electrolytic time of greater than 6 min did not significantly improve the COD removal; the best efficiency was approximately 95%, and an effluent COD of 1.9 g/L remained. The ND wastewater containing lower COD achieved a maximal COD removal efficiency of 93% in 4 min with a low effluent COD of 1.8 g/L. The main removal mechanisms are explained in that the aluminum cations were dissolved from the anode (Equation (1)), negative charged hydroxides were dispersed in the solution, and, consequently, they had a stronger attraction to coagulate the colloidal pollutants (Equations (2) and (3)). Under the electric field of the EC process, the aggregation of colloidal pollutants was higher than the use of coagulant chemicals, leading to the enhancement of coagulation rate and pollutant removal efficiency. In addition, other electrochemical advanced oxidation processes possibly occurred to degrade the pollutants into smaller compounds via hydroxyl radicals (Equations (4) and (5)) [13]:

$$Al_{(s)} \rightarrow Al^{3+}{}_{(aq)} + 3e^{-} \quad (1)$$

$$2H_2O + 2e^{-} \rightarrow H_2 + 2OH^{-} \quad (2)$$

$$Al^{3+}{}_{(aq)} + nOH^{-} \rightarrow Al(OH)_{n\,(s)} \quad (3)$$

$$2H_2O \rightarrow 2OH\bullet + 2H^{+} + 2e^{-} \quad (4)$$

$$2OH\bullet + \text{Organic compounds} \rightarrow \text{intermediate (smaller compounds)} \rightarrow \text{harmless} \quad (5)$$

In the literature, the same current density of 20 mA/cm$^2$ was supplied to treat potato chip manufacturing wastewater; it required a longer electrolytic time of 30 min to achieve 60% COD removal with a second-order kinetic model [14]. Since the potato chip manufacturing wastewater contained several organic compounds including carbohydrates, starches, and protein from process lines, its EC performance for wastewater treatment was lower than this study. Similarly, the continuous increase in COD removal of wastewater containing 0.3–1.0 g/L of COD and 180–200 mg/L of FOG was observed by increasing electrolytic times; the maximal efficiency of 95% was reached in 40 min [15]. Several operating factors including electrode material, inter-electrode distance, pH, and aeration were mentioned to affect the EC performance and extended treatment time [15]. Due to this study, the wastewater characteristics such as pollutants compounds and conductivity were other significant factors for excellent actual wastewater treatment in a short time by the EC process. It should be note that the street food wastewater in this study contained fewer complex compounds and high conductivity from external salts addition.

From Figures 3 and 4, the turbidity and FOG were also effectively removed from the street food wastewater. The turbidity of the HC and ND wastewater was mostly removed in 4 min of electrolytic time, >99% removal. The higher FOG wastewater of HC reached a maximal removal efficiency of 84%, and a concentration of 50 mg/L remained in the

effluent. For the ND wastewater, the FOG removal was slightly higher at 86%, and the effluent FOG was around 25 mg/L. In comparison to the conventional FOG removal of gravity separation (using difference of density between oil and water) and dissolved air flotation (using air bubbles to bring oil droplets to the top), this EC system overcomes the disadvantages of the above two systems including the limited separation capacity, large area requirement, and the high operating cost from constant ultrafine bubbles generation [2]. Using the gravity separation system, the oil content was removed in the range of 65–75% in 20 min of treatment time. However, the performance of the gravity separation system was directly related to the oil types and their different specific gravities compared to water [16]. Furthermore, the FOG as well as COD removal efficiencies decreased with increasing influent flow rates [16].

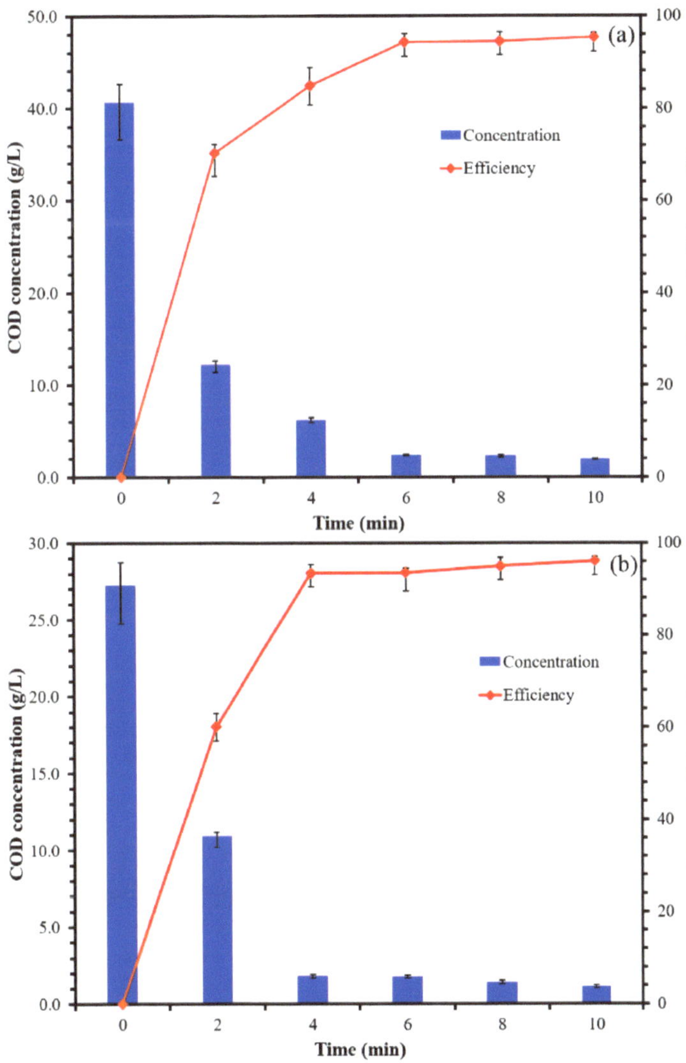


**Figure 2.** COD concentration and removal efficiency; (**a**) Hainanese chicken rice and (**b**) Noodles and dumplings.

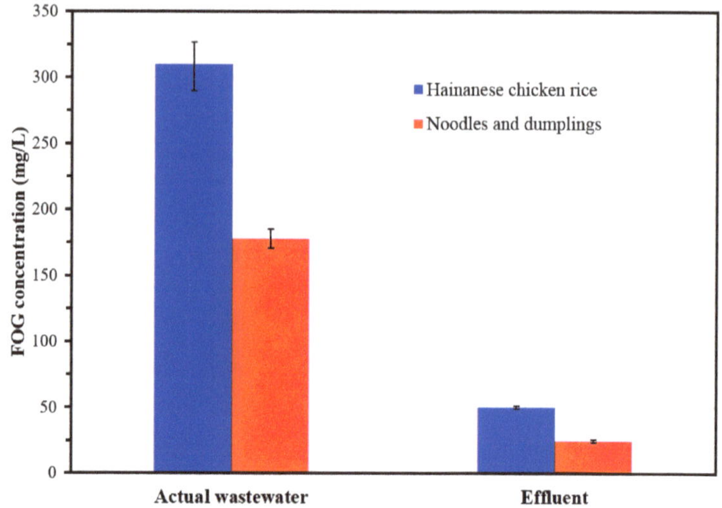


**Figure 3.** Turbidity concentration and removal efficiency.

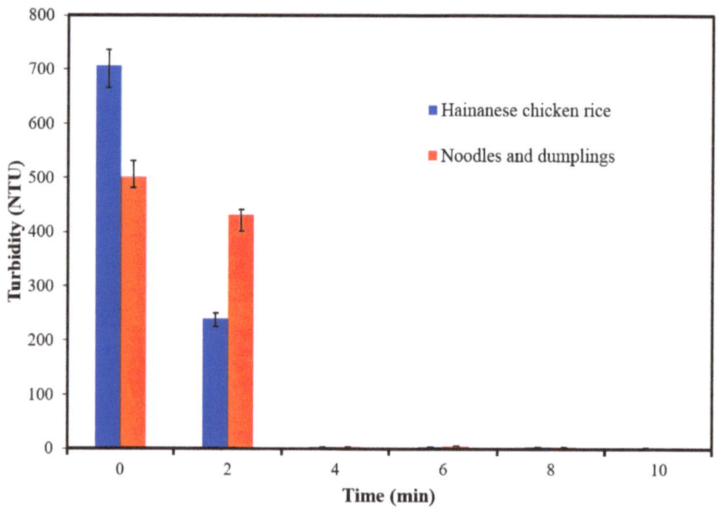


**Figure 4.** FOG concentration and removal efficiency after 10 min of electrolytic time.

In terms of Al ions in the effluent from the anodic dissolution, the effluent HC and ND wastewater were further measured by FLAA. The results revealed that a relatively low Al concentration of 0.08 and 0.02 mg/L was observed in the effluent HC and ND. Although the EC system operated under a high current density, the short treatment time of <10 min caused less aluminum to remain in the effluent. Furthermore, the continuous increase in pH by electrolytic times was observed; the pH increased to 8.5 for HC wastewater and 9.0 for ND wastewater after 8 min (shown in Figure 5). Later, the pH immediately dropped to 5.9 and 7.7 for HC and ND wastewater, respectively, at 10 min. The phenomenon was associated with the increasing electrochemical reactions in the EC system, leading to the change in hydroxide and hydrogen ions.

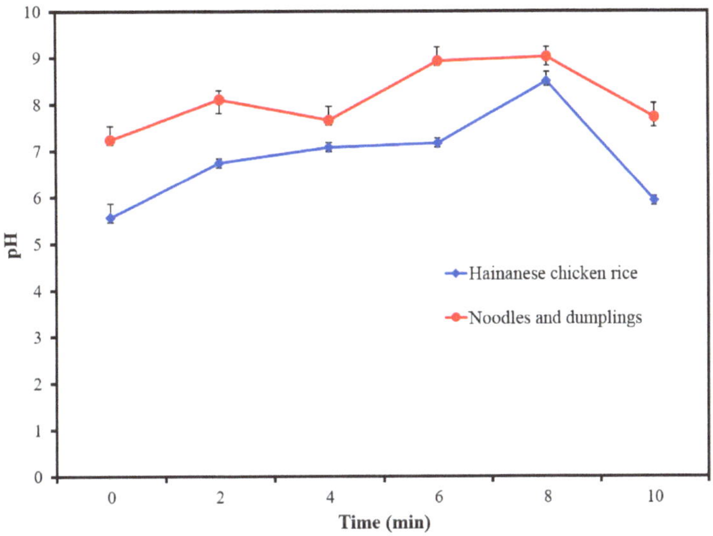


**Figure 5.** Change of pH in the effluent at different electrolytic times.

Overall, the EC system can remove organics, turbidity, and oil and grease from the street food wastewater with high efficiency and minimal treatment time. The effluent HC and ND wastewaters were fairly clear; the average COD, turbidity, and FOG levels were 1.9 g/L, 3 NTU, and 50 mg/L for HC wastewater, respectively, and 0.3 g/L, 1 NTU, and 25 mg/L for ND wastewater, respectively. In comparison, the EC system operating in this study achieved a better pollutant removal (i.e., FOG) in a shorter electrolytic time than previous EC systems in the literature [6,7]. The advantage of this study was that the appropriate design factors encouraged homogeneous flow and performance stability with no mechanic agitation and various wastewater compositions. In addition, its compact system, ease of use, fast and efficient pollutant removal were also significant advantages to further develop the EC system as a mobile unit for on-site street food treatment in Thailand and other Asian countries.

## 4. Conclusions

Oily street food wastewater containing high organic concentrations was efficiently treated by the EC system. High COD and FOG removals of 95% and 84% were obtained for the Hainanese chicken rice wastewater. Slightly higher efficiencies of 98% and 86%, respectively, were obtained for the noodles and dumplings wastewater. Due to the appropriate design and fast pollutants removal, the EC system was able to be operated in a short treatment time of <10 min with high removal efficiency. The effluents were fairly clear and contained <1.9 g/L of COD, <50 mg/L for FOG, <3 NTU for turbidity, and <0.1 mg/L for Al ions. The EC system in this study has the potential to be applied to on-site street food treatment.

**Author Contributions:** Conceptualization, W.K.; Methodology, A.N.; Validation, W.K. and K.R.; Formal analysis, W.K. and K.R.; Investigation, B.Y., S.I. and K.S.-A.; Resources, A.N.; Visualization, A.N.; Writing—original draft, W.K.; Writing—review & editing, K.R. and A.N.; Supervision, W.K. and A.N.; Project administration, W.K.; Funding acquisition, W.K. All authors have read and agreed to the published version of the manuscript.

**Funding:** This research was funded by Thailand Science Research and Innovation (TSRI) through project R2565B038. The APC was partially funded Naresuan University and TSRI (Project No. R2565B038).

**Institutional Review Board Statement:** Not applicable.

**Informed Consent Statement:** Not applicable.

**Data Availability Statement:** Not applicable.

**Acknowledgments:** This work was supported by the Thailand Science Research and Innovation (TSRI) through project R2565B038.

**Conflicts of Interest:** The authors declare no conflict of interest.

## References

1. Lesikar, B.J.; Garza, O.A.; Persyn, R.A.; Kenimer, A.L.; Anderson, M.T. Food-Service Establishment Wastewater Characterization. *Water Environ. Res.* **2006**, *78*, 805–809. [CrossRef] [PubMed]
2. Abuhasel, K.; Kchaou, M.; Alquraish, M.; Munusamy, Y.; Jeng, Y. Oily Wastewater Treatment: Overview of Conventional and Modern Methods, Challenges, and Future Opportunities. *Water* **2021**, *13*, 980. [CrossRef]
3. Yu, L.; Han, M.; He, F. A review of treating oily wastewater. *Arab. J. Chem.* **2017**, *10*, S1913–S1922. [CrossRef]
4. Ghernaout, D.; Elboughdiri, N. Electrocoagulation Process in the Context of Disinfection Mechanism. *OALib* **2020**, *7*, 1–14. [CrossRef]
5. Sharma, L.; Prabhakar, S.; Tiwari, V.; Dhar, A.; Halder, A. Optimization of EC parameters using Fe and Al electrodes for hydrogen production and wastewater treatment. *Environ. Adv.* **2021**, *3*, 100029. [CrossRef]
6. Ibrahim, D.; Sakthipriya, N.; Balasubramanian, N. Electro-coagulation treatment of oily wastewater with sludge analysis. *Water Sci. Technol.* **2012**, *66*, 2533–2538. [CrossRef] [PubMed]
7. Espinoza-Quiñones, F.R.; Fornari, M.M.T.; Módenes, A.N.; Palácio, S.M.; Trigueros, D.E.G.; Borba, F.; Kroumov, A. Electrocoagulation efficiency of the tannery effluent treatment using aluminium electrodes. *Water Sci. Technol.* **2009**, *60*, 2173–2185. [CrossRef] [PubMed]
8. Trinh, D.T.T.; Binh, Q.A.; Van Ty, T.; Channei, D.; Nakaruk, A.; Khanitchaidecha, W. Evaluation of Magnetic Stirring and Aeration on Electrocoagulation Performance in Actual Industrial Treatment. *Front. Environ. Sci.* **2021**, *9*. [CrossRef]
9. Pantorlawn, W.; Threrujirapapong, T.; Khanitchaidecha, W.; Channei, D.; Nakaruk, A. Electrocoagulation for spent coolant from machinery industry. *J. Water Reuse Desalination* **2018**, *8*, 497–506. [CrossRef]
10. Ministry of Natural Resources and Environment. Infection Waste Management Data. 2021. Available online: https://www.pcd.go.th/ (accessed on 15 January 2021).
11. Hilal, N.; Busca, G.; Talens-Alesson, F.; Atkin, B.P. Treatment of waste coolants by coagulation and membrane filtration. *Chem. Eng. Process. Process Intensif.* **2004**, *43*, 811–821. [CrossRef]
12. Wei, Y.; Jin, Y.; Zhang, W. Treatment of High-Concentration Wastewater from an Oil and Gas Field via a Paired Sequencing Batch and Ceramic Membrane Reactor. *Int. J. Environ. Res. Public Health* **2020**, *17*, 1953. [CrossRef] [PubMed]
13. Ghernaout, D. Advanced oxidation phenomena in electrocoagulation process: A myth or a reality? *Desalination Water Treat.* **2013**, *51*, 7536–7554. [CrossRef]
14. Kobya, M.; Hiz, H.; Senturk, E.; Aydiner, C.; Demirbas, E. Treatment of potato chips manufacturing wastewater by electrocoagulation. *Desalination* **2006**, *190*, 201–211. [CrossRef]
15. Priya, M.; Jeyanthi, J. Removal of COD, oil and grease from automobile wash water effluent using electrocoagulation technique. *Microchem. J.* **2019**, *150*, 104070. [CrossRef]
16. Hanafy, M.; Nabih, H.I. Treatment of Oily Wastewater Using Dissolved Air Flotation Technique. *Energy Sources Part A Recover. Util. Environ. Eff.* **2007**, *29*, 143–159. [CrossRef]

applied sciences

MDPI

Article

# Laboratory Test Campaign Aimed at the Analysis of an Uncommon Wear Phenomenon in a Marble Quarry

Alfio Di Giovanni [1], Carmine Todaro [2], Marilena Cardu [1,2,*], Stefano Bianchini [3] and Brunello Forfori [4]

1 DIATI Politecnico di Torino, 10129 Torino, Italy; alfio.digiovanni@polito.it
2 IGG-CNR, 10129 Torino, Italy; carmine.todaro@polito.it
3 Marmotest Srl, 54033 Carrara, Italy; s.bianchini@marmotest.com
4 Luana Marmi Srl, 55030 Vagli Sotto, Italy; studio.forfori@gmail.com
* Correspondence: marilena.cardu@polito.it; Tel.: +39-0110907655

**Featured Application: This paper presents a laboratory test campaign focused on the reduction of the wear phenomenon related to a material extracted from a marble quarry. The aim of the work is to propose a potential solution for the wear reduction pertaining to chainsaw cutting machine tools.**

**Abstract:** The use of ornamental stones has a historical value that makes them strategically precious in Italy; marble can offer high performance in architectural applications, even though the variability of the rock mass requires detailed studies to optimize the exploitation techniques and reduce waste. Italy is world famous for its marble, which is extracted mainly through chainsaw cutting machines, which are currently used intensively due to their high-safety working conditions compared to alternative techniques and for their great versatility, especially in underground applications. Although this cutting technique is well-rooted, an uncommon problem of tool wear was found in the quarry under study, which strongly affected productivity. A series of laboratory test were carried out to estimate the wear potential of the rock and the suitability of the tools. The Cerchar abrasivity test highlighted a mean wear potential for the marble of 2.77, while microhardness outcomes pointed out the presence of quartz veins in the tested material (values over 10,000 MPa). Finally, additives typically used in the conditioning process of EPB machines in tunneling were tested with the purpose of reducing the extent of wear. A reduction of about 50% in the wear (in terms of weight lost) was obtained for a moisture content of 9%.

**Keywords:** chainsaw cutting machine; dimension stone exploitation; tool wear; rock–tool interaction; Cerchar abrasivity test; microhardness; wear reduction

**Citation:** Di Giovanni, A.; Todaro, C.; Cardu, M.; Bianchini, S.; Forfori, B. Laboratory Test Campaign Aimed at the Analysis of an Uncommon Wear Phenomenon in a Marble Quarry. *Appl. Sci.* **2022**, *12*, 2264. https://doi.org/10.3390/app12042264

Academic Editor: Dibyendu Sarkar

Received: 18 January 2022
Accepted: 19 February 2022
Published: 21 February 2022



## 1. Introduction

The exploitation of ornamental stone and marble requires the extraction of intact blocks that can be sawed into slabs and tiles. Therefore, the need to extract intact blocks leads to requirements different from those for, e.g., the extraction of aggregates, and has led to the development of specific exploration and extraction techniques [1–3]. The world of dimension stones, and especially the marble market, is expected to grow, reaching a production value of USD 64 billion in 2023. Italy, mostly the Apuan Alps region, is in the top ten quality-marble-producing countries [4]. The marble quarry under study is located in the geographical context of the Apuan Alps, in the Garfagnana area [5]. The quarry area develops, throughout the project, between the altitudes of 920 and 800 m a.s.l. and has an overall development along the N–S axis, widening towards the west in the northern part. The structural conditions are good, the production area being characterized by a limited number of discontinuities with an average spacing above 1 m, a condition that allows for the obtainment of blocks of good quality [6]. In more detail, according to the surveys in the area being exploited, a geomechanical classification of the site was carried out according to the authors of [7] (RMR) and Hoek and Brown [8] (GSI), obtaining an RMR equal to 62 (Class. N. 2—Good) and a GSI close to 60.

At 869 m a.s.l., where the development was planned for the expansion of the underground exploitation, the presence of grey-white marble with flint nodules was noticed. In this area, during the works started in 2018, a "sheath fold" emerged, with a siliceous marble core in entrochi, characterized by a trend of about 350°, immersion at NW of about 10°, and boudinage. Initially, this structure seemed to close quickly, while in a short space corresponding to the execution of about three development advances in the tunnel, for about 9 m in total, the core opened into a boudinage, and the structure progressed towards the inside of the cluster. The core of this structure proved impossible to cut, due to the fast wear of the tools. The machine used in marble cutting operations is the chain cutting saw (Figure 1), which has the great advantage of not requiring preliminary cutting operations. The arm operates easily, even in the presence of only one free surface. In all cases where it is necessary to perform blind cuts, the chainsaw, alone or in cooperation with the diamond wire saw, is frequently used. At the quarry, the excavation works are carried out with a chainsaw on which a 3.5 m long cutting blade is installed, equipped with guides, where a single-mesh metal chain runs (74 mm pitch, 42 mm cutting width), equipped with polycrystalline diamond plate tools. Seven sets of tool-holders are installed on the cutting chain; each series includes 12 tool holders with different cutting configurations. The tool has a diameter of 8.0 mm, a diamond layer thickness of 1.5 mm, and a tungsten carbide (WC) layer thickness of 6.6 mm, for a total height of 8.1 mm. The WC area is welded directly on the tool holder and does not contribute to the cut (made expressly from polycrystalline diamond) but represents an intermediate support between the diamond and the tool holder. Figure 2 shows a detail of the tools adopted.

**Figure 1.** The chain cutting machine adopted in the quarry. The machine is engaged in a series of horizontal cuts to widen the underground opening (link https://www.fantinispa.it/en/gu70rxc-tunnel-chain-saw-machine/, accessed on 2 October 2021).

In this field, many studies have been conducted to improve the performance of the tools: the complexity of the cutting unit, which includes dozens of devices mounted at different angles, means that the inspection and replacement times are hours instead of minutes, and the environment in which research necessarily develops is a site much less equipped and comfortable than a mechanical workshop.

(a) (b)

**Figure 2.** Detail of the tools and their tool holder: (**a**) front view of the arrangement of the tools in the execution of the cut with a thickness of 40.82 mm, (**b**) cutting chain front view.

As far as the design of marble cutting tools is concerned, it is assumed to be enough to evaluate the medium hardness value (HK, Knoop hardness; or HV, Vickers hardness) for marble characterization. Tool materials should be tested separately for both hardness and toughness [1].

A PCD tool is represented in Figure 3: a new tool is shown above, where the diamond layer and WC are clearly recognizable, and a tool worn out after 0.24 $h/m^2$ is depicted below. The almost total lack of the polycrystalline diamond plate tool can be noticed; in other words, the tool is irreparably damaged, and is characterized by severe wear which has completely consumed the diamond layer.

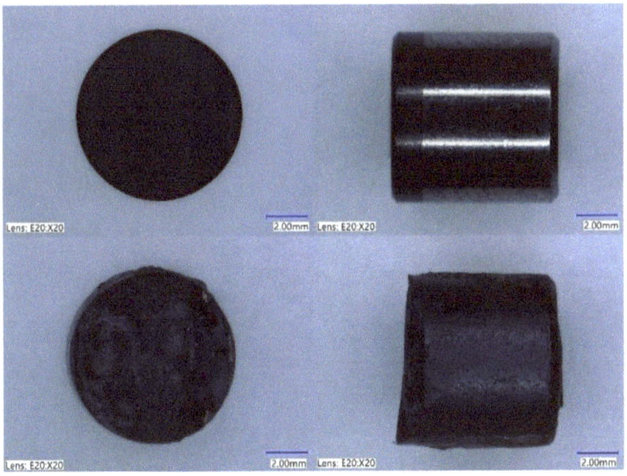


**Figure 3.** Typical tool used in the chainsaw cutting machine. The front and side views are shown on the left and right of the image, respectively. A new tool is shown above: the color difference between polycrystalline diamond (shorter area) and tungsten carbide (longer area) is clearly recognizable from the top right image. Below is shown a tool after 0.24 $h/m^2$ of work: the almost complete absence of the polycrystalline diamond layer is evident.

Currently, the information available in scientific literature on chainsaw cutting machines in carbonate rock applications pertains mainly to studies focused on performance prediction models [9–15], but the potential wear and the quartz content have not been studied in depth. In the context of heavy wear (quite rare for a marble quarry), this research was planned to better understand the phenomenon and try to find a solution capable of reducing the wear. A series of laboratory tests were carried out to characterize and study the characteristics of the marble. By comparing the microhardness results obtained for the quarried material with the typical values of the marbles, two different checks were made. On the one hand, it was verified whether minerals harder than calcite were present in the material tested. On the other hand, by comparing the microhardness of the marble measured with that of the tools, the choice of the quarry managers to use certain tools rather than others was verified. Subsequently, by means of XRD (X-ray diffraction), the presence of these impurities was confirmed and their univocal recognition was carried out. Finally, the Cerchar abrasivity index (CAI) provided an order of magnitude of the wear potential, while the wear test (performed using the soil abrasion test apparatus—SATA) allowed us to find the dependence of the wear phenomenon on the content of humidity, as well as to test the contribution of several types of additives.

## 2. Laboratory Tests and Results

The tests performed are described below and the results obtained are presented. It should be noted that for each test campaign (apart from the microhardness tests), the marble samples were carefully chosen, selecting the material in order to obtain a representative sample of the rock studied. The tests carried out can be sorted into two groups: if the analysis with XRD, microhardness and CAI were programmed to understand the reason for the strong wear of the cutting tools, the wear test was carried out with the aim of proposing a solution able to reduce the wear phenomenon. In addition, it should be noted that unlike the other tests, the wear test performed by SATA is not standardized, although several test campaigns and publications are now available in the literature.

### 2.1. XRD—X-ray Diffraction Analysis

XRD is a versatile, nondestructive analytical technique that provides information on the crystallographic structure, chemical composition, and physical properties of materials. This technique is based on the observation of the scattered intensity of an X-ray beam, which strikes a sample as a function of the incident and scattered angle, of the polarization, and of the wavelength or energy [16]. The analysis was carried out in the laboratory of the Politecnico di Torino, with an XRD—Rigaku SmartLab SE. The selected marble sample was previously ground to a fine powder in order to obtain an almost infinite number of fine crystallites in random orientations. The incidence angles ranged from 4° (start angle) to 84° (stop angle), with a scan speed of 0.1°/s thanks to the continuous scan mode. The output of the analysis was an intensity vs. incidence angle chart, from which the minerals and their abundance could be identified, as each mineral has a specific angle of incidence. The intensity of the peaks, evaluated with the appropriate Rietveld model [17,18], reflects the abundance of each element. The results are shown in Figure 4.

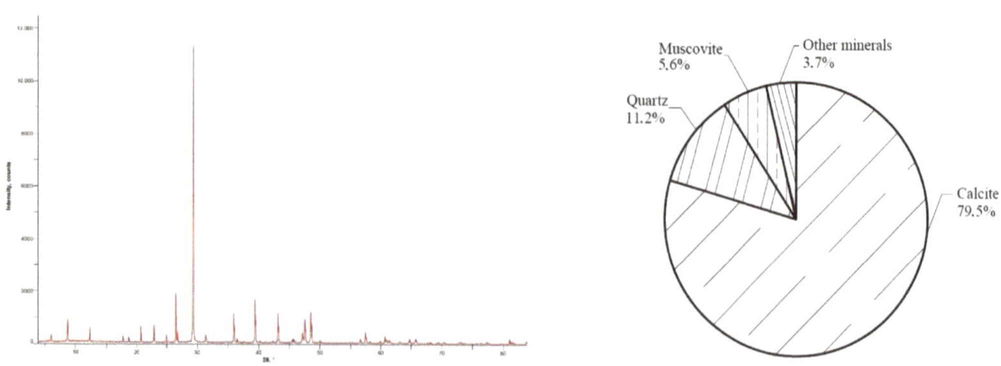


(a) (b)

**Figure 4.** Intensity vs. incidence angle chart (**a**) and mineralogical abundance (**b**).

The associated minerals identified were 79.5% Calcite, 11.2% Quartz, 5.6% Muscovite, and 3.7% other minerals.

### 2.2. Vickers Microhardness Test

Hardness is the measure of the strength to resist localized plastic deformation induced by the mechanical impression due to a given applied load [19] and is obtained by measuring the permanent depth of the impression. The Vickers hardness test method is used for small objects, as well as polished thin sections. Due to the small indentation in a Vickers test, it can test small areas, the surface of a part, or individual microstructures [20]. It can also determine the depth of hardening by dissecting it and creating a series of indentations to find the change in hardness [21]. This method can evaluate almost any type of material, such as ceramics, metals, and rocks. The Vickers microdurometer uses devices developed by the authors of [22], i.e., a pyramid shape is created by a diamond indenter tool (opening of 136°, as shown in Figure 5). By measuring the diagonals of the footprint (d in Figure 5, performed optically as described by the authors of [23]), its surface can be obtained and, consequently, the hardness can be computed as the ratio between the applied load and this area. When the applied load is less than 1 kg, it is common to refer to this test as microhardness. The Vickers test is easy to perform, as the required calculations are independent of the indenter size [24] and the indenter can be used for all types of materials, regardless of hardness. For these reasons, Vickers microhardness tests were chosen for the analysis. The tests were performed according to the authors of [25]. Unlike the standard, the dwell time was slightly longer, close to 40–45 s.

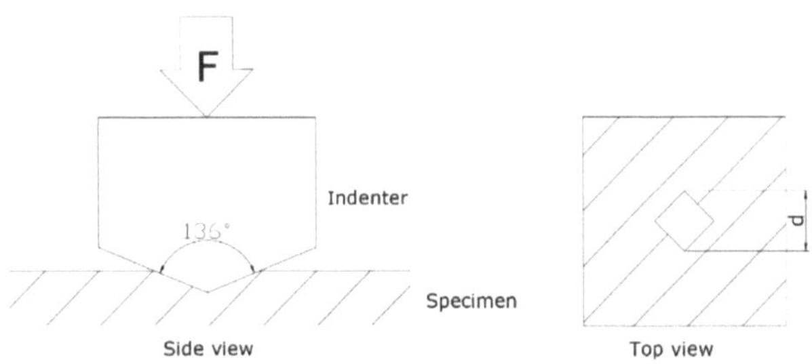


**Figure 5.** Indentation scheme [23]. F is the applied force, while d is the diagonal of the left footprint.

The marble specimens, 50 × 30 × 15 mm in size, had a perfectly polished upper surface. The dimensions were chosen in order to maximize the available surface, compatible with the constraints of the instrument. Four different samples were prepared, taking care to select material with different mineralogy. Considering the grey veins embedded in the calcite matrix (white color of the background of the samples), the patterns were selected by choosing marble pieces with different presences of impurities (grey veins), as shown in Figure 6. For each sample, 80 microhardness measurements were performed by applying a load of 200 g (1.96 N). On average, the distance between two successive footprints was 100 µm. Regarding the outcome analysis, the microhardness values for each sample were sorted and, subsequently, the values in terms of cumulative frequency were displayed, giving a characteristic curve. Figure 7 shows the microhardness characteristic curves obtained.

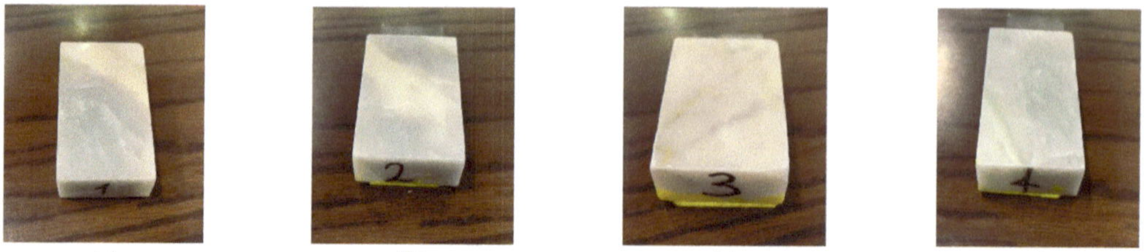

(**a**) (**b**) (**c**) (**d**)

**Figure 6.** Samples analyzed. The grey veins are marginally present in sample 3 (**c**), while in samples 1 and 2 (**a**,**b**), the presence of impurities is greater. Sample 4 (**d**) is almost entirely composed of grey material.

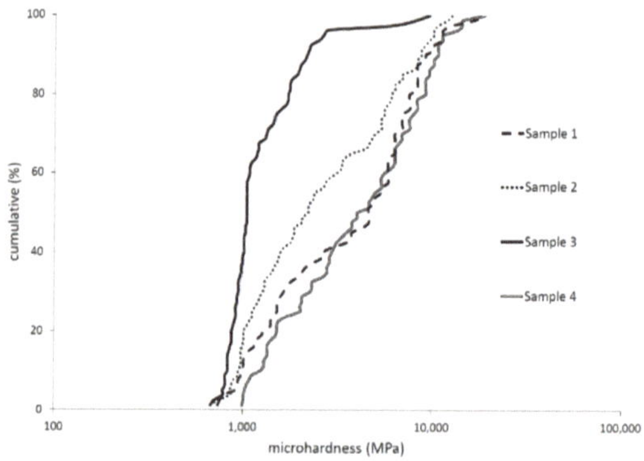


**Figure 7.** Cumulative frequency curves of microhardness (microhardness characteristic curves).

### 2.3. CAI—Cerchar Abrasivity Index

The Cerchar abrasivity test is one of the most common procedures for laboratory measurement of the abrasiveness of rocks [26–32]. An order of magnitude of the wear potential of the rock can be obtained by evaluating this index, which is a tested material's aptitude for wearing steel tools [33,34]. As described in [35], a stylus (composed of HRC 55 steel) loaded with 70 N is scratched for 10 mm on the rock sample at a rate of 1 mm/s. The CAI is evaluated by measuring the wear surfaces experienced by the stylus [36–38]. In this work, Cerchar tests were performed according to [35]. The selected sample had a size of $50 \times 30 \times 15$ mm, Figure 8a, and 10 different tests were performed. In accordance with the reference, five scratches were carried out horizontally and another five vertically (tests 1–5 and tests 6–10, respectively, Figure 8b).

The highlighted CAI values ranged from 2.08 to 3.30, with a mean value of 2.77 (standard deviation = 0.50). The mean CAI value corresponds to the average wear potential.

### 2.4. SATA—Soil Abrasion Testing Apparatus

Wear tests on conditioned material are commonly performed to verify the ability of conditioning agents to prevent or reduce tool wear in tunnelling applications [39,40]. These types of test have been developed for cohesionless soils or crashed rocks but, to the authors' knowledge, have never been applied for quarry applications. In any case, we planned to use the soil abrasion testing apparatus (SATA, an equipment designed and developed at the Politecnico di Torino) on the marble, in order to understand firstly the wear potential of the studied material and secondly the desirable effect potentially obtained by using conditioning agents (water and other additives). By providing more details, the wear test we performed provided insight into the wear potential by evaluating the weight loss. The evaluation of the wear potential by estimating the weight lost is common in scientific literature [40–43]. An aluminum disk rotates around its axis at a constant speed inside a cylindrical tank (Figure 9), embedded with granular material [44]. The rotational speed is kept constant, equal to 160 rpm, for a time span of 10 min. The difference in disk weight before and after the test is the weight loss. As for the aluminum disk, it is characterized by Vickers hardness values of 116 MPa [45]. A normal confining pressure of about 2 kPa is applied to ensure continuous contact between the disk and the soil.

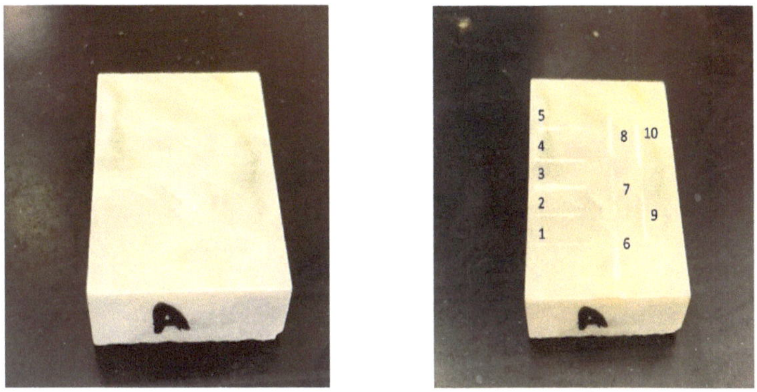


(**a**) (**b**)

**Figure 8.** Sample before the test (**a**) and after the test (**b**).

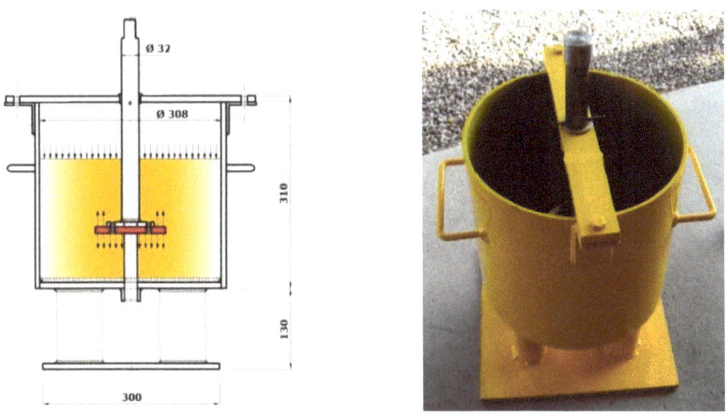


**Figure 9.** Soil abrasion testing apparatus. Image modified from [45].

In this work, wear tests were performed according to [45,46]. Marble was previously crushed, obtaining powdered rock and chips with a maximum size of 20 mm. The particle size distribution is shown in Figure 10.

The first step of wear testing focused on studying the wear bell, while the second phase focused on testing the conditioning additives. The wear bell is a chart with a typical bell shape that shows in the x and y axes the moisture content and weight loss of the tested material [41,42]. This type of curve is very important, since each material exhibits a different curve (depending on the particle size distribution, mineralogy, and grain shape) and by taking a typical wear bell of a highly wearable material as a reference for comparison, the wear potential of the material studied can be deduced. The results for the tests performed with conditioning agents are also shown on the same chart. Once the conditioning amounts were chosen and the weight loss was obtained thanks to the wear test, the moisture content was calculated by considering the water used for producing the foam or activating polymers; then, the representative test point could be drawn on the chart.

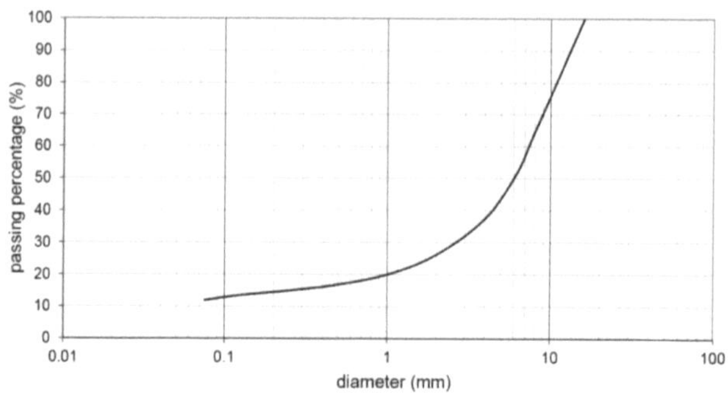


**Figure 10.** Particle size distribution of the crushed marble used in the wear test performed with the SATA.

### 2.4.1. The Wear Bell

After a preliminary analysis, we planned to draw the marble wear bell by testing the moisture contents between 1% and 9%. Three tests were performed for each moisture content and the average was calculated [47–49]. The outcomes of the testing campaign are shown in Figure 11, where, for comparison, the wear bell related to a pure quartzite (98% quartz) is also shown [45].

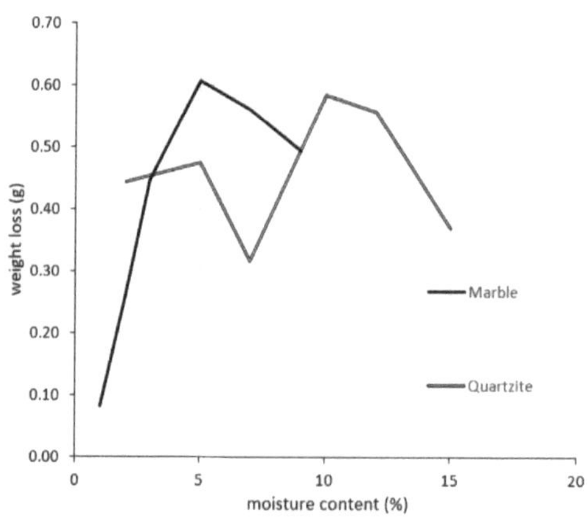


**Figure 11.** Wear bells related to marble and quartzite.

### 2.4.2. Wear Test with Conditioning Agents

In the second phase of the study, the moisture content corresponding to 4% (the average value of the material studied) and 9% (the highest water content analyzed) was considered for testing the additives. Two different products were used: a high-performance liquid foaming agent (Product A) and a natural polymer (Product B) that was commercially available as a dry powder. Due to the lack of representative marble material available, the tests with the highest moisture content were performed with Product B only, the latter having a lower environmental impact and a shorter biodegradation time. For Product A, a water solution (4% by volume of the foaming agent) was prepared, while regarding Product B, a slurry (1% by weight) was produced by activating the polymer through a mixing process in water for 5 min (adding the polymer in the first minute of the mixing process), performed with a concrete mixer. Once both water solutions had been prepared, the appropriate amount of conditioning able to achieve the selected moisture content (4% or 9%) was collected and added to the marble sample. Wear tests were performed following the same procedure used previously. The results are shown in Figure 12.

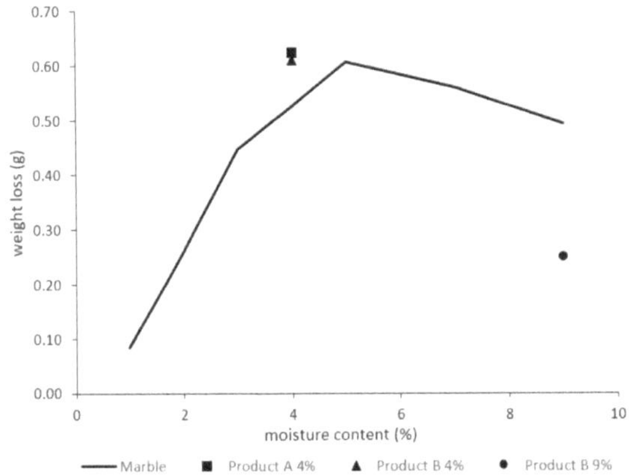


**Figure 12.** Marble wear bell and wear conditioning tests (4% and 9% moisture content).

## 3. Discussion

Taking into account the results obtained from XRD, calcite was the predominant component, even though more than 20% of the marble was made up of other elements. The presence of quartz cannot be neglected, with a percentage of 11.2%. The presence of impurities also strongly influenced the campaign of the microhardness tests. In fact, the results achieved confirmed relatively high microhardness values when compared with the range of marble, typically from 1500 MPa to 3000 MPa [50]. Considering the upper limit of the range, the hardness was below 3000 MPa for 40%, 50%, and 40% of the cumulative curves for samples one, two, and four, respectively. In contrast, in sample three, only 8% of the determinations gave a hardness greater than 3000 MPa. This result was expected: in fact, sample three was characterized by only a tiny vein of grey intrusion. However, the microhardness results justify the company's choice of tool material. The choice to discard tungsten carbide tools in the chainsaw was correct, since the microhardness values of this compound ranged between 10,000 MPa and 14,000 MPa, values that were often exceeded during testing. Consequently, the company's choice to use polycrystalline diamond tools can be considered correct, since this material is characterized by a Vickers hardness of about 55,000 MPa [51].

As for the CAI test campaign, the studied marble was recognized as an average abrasive material, the average CAI value being equal to 2.77. This result highlighted the abrasive potential of the impurities present in the marble and previously recognized by XRD; in fact, the CAI was higher than that of pure marble (between 0.60 and 1.00, as suggested by the Colorado School of Mines). In particular, the sample which had areas where the CAI was higher than 3 could be due to the presence of flinty marble (signifying a highly abrasive material).

Regarding the wear, it should be remarked that tool wear is determined by complex laws [52]. Pertaining to the test performed using the SATA, the tested sample was characterized by a wear bell with a peak at 5% water content. From the results, it was confirmed that the water content deeply affected the weight loss of the test disk, as discussed by the authors of [41,42,46]. Starting from a soil moisture percentage equal to the natural condition (1%), increasing the water content did not provide a beneficial effect until a percentage of 5% was reached. After this value, a further increase in water content resulted in a decrease in weight loss. However, comparing the wear bell of marble with that of quartzite, it was evident that the wear potential was similar, the peaks being of the same order of magnitude (close to 0.6 g). The wear tests with conditioning agents revealed interesting aspects, namely that the additives need a certain dosage to work properly. Tests carried out with Products A and B in solution, dosed and added to the marble sample to achieve 4% moister content, showed no reduction in weight loss, while tests performed with a conditioning agent capable of achieving 9% moisture content provided a reduction in weight loss of 50%. It can be stated that a reduction in the wear was observed when the additive used surpassed a certain threshold. The mechanism of wear reduction is certainly a function of the amount of the additive (9% moisture content provides the lowest wear). Anyway, it should be noticed that additives, in order to work properly, need to be conveyed homogeneously in the conditioned soils. In conclusion, it can be hypothesized that there exists a critical moisture content, recognizable in this case as between 4% and 9%, which, once reached, allows the added product to properly exert its anti-wear action.

## 4. Concluding Remarks

This work was conceived with the aim of deepening the study into the phenomenon of the tool wear of the chain cutting machines used for the exploitation of marble in the quarry studied. The research was planned to optimize the economic and productive sources of the construction site. The tests showed the presence of impurities in the marble that give the material unusual values of microhardness and a high potential for wear. Although the quarry managers adopt polycrystalline diamond and tungsten carbide tools, the study carried out shows that the wear phenomenon cannot be addressed without the addition of conditioning agents during cutting operation. According to the preliminary results obtained on a laboratory scale, conditioning agents appear to be ineffective unless a certain moisture content of the tested material is guaranteed. On the other hand, the polymer added in the form of slurry (1% by weight concentration in water) seems to be strongly effective against wear when used in amounts of 9% of the moisture content of the marble. However, these tests were performed on a laboratory scale, and only by carrying out a test campaign at the real scale can the results can be confirmed. In addition, it should be noted that the cutting operation performed in the quarry is characterized by a different mode of execution and that the tool material is different to that used in the laboratory. However, further tests and an onsite experimental campaign are planned in order to verify the efficiency of the conditioning agents, considering that, to date, quarry activity at the site under study has been interrupted due to the ineffectiveness of the chain cutting machine.

**Author Contributions:** Conceptualization, A.D.G., C.T. and M.C.; methodology, A.D.G., C.T. and M.C.; investigation, A.D.G., C.T. and M.C.; resources, M.C., S.B. and B.F.; data curation, A.D.G.; writing—original draft preparation, A.D.G. and C.T.; writing—review and editing, A.D.G., C.T. and M.C.; supervision, M.C. All authors have read and agreed to the published version of the manuscript.

**Funding:** This research received no external funding.

**Institutional Review Board Statement:** Not applicable.

**Informed Consent Statement:** Not applicable.

**Data Availability Statement:** Not applicable.

**Acknowledgments:** The authors are grateful to A. Boscaro and E. Barbero (Mapei UTT Company) for providing the additives and for their technical and logistical support in carrying out this research.

**Conflicts of Interest:** The authors declare no conflict of interest.

## References

1. Vagnon, F.; Dino, G.A.; Umili, G.; Cardu, M.; Ferrero, A.M. New Developments for the Sustainable Exploitation of Ornamental Stone in Carrara Basin. *Sustainability* **2020**, *12*, 9374. [CrossRef]
2. Careddu, N.; Marras, G. Marble processing for future uses of $CaCO_3$-microfine dust: A study on wearing out of tools and consumable materials in stone working factories. *Miner. Processing Extr. Metall. Rev.* **2015**, *36*, 183–191. [CrossRef]
3. Capitano, C.; Peri, G.; Rizzo, G.; Ferrante, P. Toward a holistic environmental impact assessment of marble quarrying and processing: Proposal of a novel easy-to-use IPAT-based method. *Environ. Monit. Assess.* **2017**, *189*, 1–16. [CrossRef] [PubMed]

4. Mordor Intelligence. *Marble Market-Growth, Trends, COVID-19 Impact, and Forecasts (2021–2026)*; Mordor Intelligence Inc.: Hyderābād, India, 2021.
5. Angotzi, G.; Bramanti, L.; Tavarini, D.; Gragnani, M.; Cassiodoro, L.; Moriconi, L.; Saccardi, P.; Pinto, I.; Stacchini, N.; Bovenzi, M. World at work: Marble quarrying in Tuscany. *Occup. Environ. Med.* **2005**, *62*, 417–421. [CrossRef] [PubMed]
6. Oggeri, C.; Oreste, P. Underground Quarrying for Marble: Stability assessment through modelling and monitoring. *Int. J. Min. Sci.* **2015**, *1*, 35–42.
7. Bieniawski, Z.T. *Engineering Rock Mass Classifications: A Complete Manual for Engineers and Geologists in Mining, Civil, and Petroleum Engineering*; Wiley: New York, NY, USA, 1989.
8. Hoek, E.; Brown, E.T. Practical estimates of rock mass strength. *Int. J. Rock Mech. Min. Sci.* **1997**, *34*, 1165–1186. [CrossRef]
9. Copur, H.; Balci, C.; Bilgin, N.; Tumac, D.; Düzyol, İ. Full-scale linear cutting tests towards performance prediction of chain saw machines. In *Proceedings of the 20th International Mining Congress of Turkey*; The Chamber of Mining Engineers of Turkey: Ankara, Turkey, 2007; pp. 161–169.
10. Sariisik, A.; Sariisik, G. Efficiency analysis of armed-chained cutting machines in block production in travertine quarries. *J. South. Afr. Inst. Min. Metall.* **2010**, *110*, 473–480.
11. Sariisik, A.; Sariisik, G. Investigation of the cutting performance of the natural stone block production in quarries with armed-chain cutting machine. *Proc. Inst. Mech. Eng. Part C J. Mech. Eng. Sci.* **2013**, *227*, 1291–1301. [CrossRef]
12. Copur, H.; Balci, C.; Tumac, D.; Bilgin, N. Field and laboratory studies on natural stones leading to empirical performance prediction of chain saw machines. *Int. J. Rock Mech. Min. Sci.* **2011**, *48*, 269–282. [CrossRef]
13. Copur, H. Linear stone cutting tests with chisel tools for identification of cutting principles and predicting performance of chain saw machines. *Int. J. Rock Mech. Min. Sci.* **2010**, *47*, 104–120. [CrossRef]
14. Hekimoglu, O.Z. Studies on increasing the performance of chain saw machines for mechanical excavation of marbles and natural stones. *Int. J. Rock Mech. Min. Sci.* **2014**, *72*, 230–241. [CrossRef]
15. Tumac, D.; Avunduk, E.; Copur, H.; Bilgin, N.; Balci, C. Estimation of the performance of chain saw machines from shore hardness and the other mechanical properties. *ISRM SINOROCK* **2013**, *1*, 261–265.
16. Milinovic, J.; Dias, Á.A.; Janeiro, A.I.; Pereira, M.F.C.; Martins, S.; Petersen, S.; Barriga, F.J.A.S. XRD Identification of Ore Minerals during Cruises: Refinement of Extraction Procedure with Sodium Acetate Buffer. *Minerals* **2020**, *10*, 160. [CrossRef]
17. Young, R.A. *The Rietveld Method*; Oxford University Press: New York, NY, USA, 1993. [CrossRef]
18. Gualtieri, A. Accuracy of XRPD QPA using the combined Rietveld-RIR method. *J. Appl. Cryst.* **2000**, *33*, 267–278. [CrossRef]
19. Ashby, N.A. The factor of hardness in metals. *N. Z. Eng.* **1951**, *6*, 33–34.
20. Gerberich, W.W.; Tymiac, N.I.; Grunlan, J.C.; Horstemeyer, M.F.; Baskes, M.I. Interpretation of indentation size effect. *J. Appl. Mech.* **2002**, *69*, 433–442. [CrossRef]
21. Yihang, L.; Daxi, G.; Zhenyu, S.; Zehua, Z.; Xinggang, J.; Deyuan, Z. A study on strengthening and machining integrated ultrasonic peening drilling of Ti-6Al-4V. *Mater. Des.* **2021**, *212*, 110238. [CrossRef]
22. Smith, R.L.; Sandland, G.E. An Accurate Method of Determining the Hardness of Metals, with Particular Reference to Those of a High Degree of Hardness. *Proc. Inst. Mech. Eng.* **1922**, *102*, 623–641. [CrossRef]
23. Germak, A.; Herrmann, K.; Dai, G.; Li, Z. Development of calibration methods for hardness indenters. *VDI-Ber* **2006**, *1948*, 13–26.
24. Menčík, J. Determination of mechanical properties by instrumented indentation. *Meccanica* **2007**, *42*, 19–29. [CrossRef]
25. *ASTM E384:2017*; Standard Test Method for Micro-Indentation Hardness of Material. ASTM International: West Conshohocken, PA, USA, 2017.
26. Suana, M.; Peters, T. The Cerchar abrasivity index and its relation to rock mineralogy and petrography. *Rock Mech. Rock Eng.* **1982**, *15*, 1–7. [CrossRef]
27. Al-Ameen, S.I.; Waller, M.D. The influence of rock strength and abrasive mineral content on the Cerchar abrasive index. *Eng. Geol.* **1994**, *36*, 293–301. [CrossRef]
28. Hamzaban, M.T.; Memarian, H.; Rostami, J. Continuous monitoring of pin tip wear and penetration into rock surface using a new Cerchar abrasivity testing device. *Rock Mech. Rock Eng.* **2014**, *47*, 689–701. [CrossRef]
29. Hamzaban, M.T.; Memarian, H.; Rostami, J. Determination of scratching energy index for Cerchar abrasion test. *J. Min. Environ.* **2018**, *9*, 73–89. [CrossRef]
30. He, J.; Li, S.; Li, X.; Wang, X.; Guo, J. Study on the correlations between abrasiveness and mechanical properties of rocks combining with the microstructure characteristic. *Rock Mech. Rock Eng.* **2016**, *49*, 2945–2951. [CrossRef]
31. Majeed, Y.; Abu Bakar, M.Z. Statistical evaluation of CERCHAR abrasivity index (CAI) measurement methods and dependence on petrographic and mechanical properties of selected rocks of Pakistan. *Bull Eng. Geol. Environ.* **2016**, *75*, 1341–1360. [CrossRef]
32. Bakar, M.A.; Majeed, Y.; Rostami, J. Effects of rock water content on CERCHAR Abrasivity Index. *Wear* **2016**, *368*, 132–145. [CrossRef]
33. Rostami, J.; Ozdemir, L.; Bruland, A.; Dahl, F. Review of Issues Related to Cerchar Abrasivity Testing and Their Implications on Geotechnical Investigations and Cutter Cost Estimates. In Proceedings of the Rapid Excavation and Tunnelling Conference, Seattle, WA, USA, 27–29 June 2005; pp. 738–751.
34. Rostami, J.; Ghasemi, A.; Gharahbagh, E.A.; Dogruoz, C.; Dahl, F. Study of Dominant Factors Affecting Cerchar Abrasivity Index. *Rock Mech. Rock Eng.* **2013**, *47*, 1905–1919. [CrossRef]

35. Alber, M.; Yaralı, O.; Dahl, F.; Bruland, A.; Käsling, H.; Michalakopoulos, T.N.; Cardu, M.; Hagan, P.; Aydın, H.; Özarslan, A. ISRM suggested method for determining the abrasivity of rock by the CERCHAR abrasivity test. *Rock Mech. Rock Eng.* **2014**, *47*, 261–266. [CrossRef]
36. Yarali, O.; Yasar, E.; Bacak, G.; Ranjith, P.G. A study of rock abrasivity and tool wear in coal measures rocks. *Int. J. Coal Geol.* **2008**, *74*, 53–66. [CrossRef]
37. Plinninger, R.J.; Kasling, H.; Thuro, K. Wear prediction in hard rock excavation using the Cerchar abrasiveness index (CAI). In Proceedings of the EUROCK 2004 & 53rd Geomechanics Colloquium, Salzburg, Austria, 7–9 October 2004; pp. 599–604.
38. Plinninger, R.J.; Kasling, H.; Thuro, K.; Spaun, G. Testing conditions and geomechanical properties influencing the cerchar abrasivity index (CAI) value. *Int. J. Rock Mech. Min. Sci.* **2003**, *40*, 259–263. [CrossRef]
39. Peila, D.; Martinelli, D.; Todaro, C.; Luciani, A. Soil conditioning in EPB shield tunnelling–An overview of laboratory tests. *Geomech. Tunn.* **2019**, *12*, 491–498. [CrossRef]
40. Salazar, C.G.O.; Todaro, C.; Bosio, F.; Bassini, E.; Ugues, D.; Peila, D. A new test device for the study of metal wear in conditioned granular soil used in EPB shield tunnelling. *Tunn. Undergr. Space Technol.* **2018**, *73*, 212–221. [CrossRef]
41. Rostami, J.; Gharahbagh, E.A.; Palomino, A.M.; Mosleh, M. Development of soil abrasivity testing for soft ground tunneling using shield machines. *Tunn. Undergr. Space Technol.* **2012**, *28*, 245–256. [CrossRef]
42. Jakobsen, P.D.; Langmaack, L.; Dahl, F.; Breivik, T. Development of the Soft Ground Abrasion Tester (SGAT) to predict TBM tool wear, torque and thrust. *Tunn. Undergr. Space Technol.* **2013**, *38*, 398–408. [CrossRef]
43. Zhao, D.H.; Yan, G.Y.; Wu, Q. Research on wear characteristics of CVD composite coated tool in natural marble cutting. *Adv. Mater. Res.* **2014**, *941*, 1644–1649. [CrossRef]
44. Barbero, M.; Peila, D.; Picchio, A.; Chieregato, A.; Bozza, F.; Mignelli, C. Procedura sperimentale per la valutazione dell'effetto del condizionamento del terreno sull'abrasione degli utensili nello scavo con EPB. *Geoing Ambient. Min.* **2012**, *135*, 13–19.
45. Salazar, C.G.O.; Martinelli, D.; Todaro, C.; Peila, D.; Boscaro, A. Study of wear in conditioned granular soil by using a new test device. In Proceedings of the ITA-AITES World Tunnel Congress 2016, San Francisco, CA, USA, 22–28 April 2016; Volume 3, pp. 2445–2454.
46. Salazar, C.G.O.; Martinelli, D.; Todaro, C.; Luciani, A.; Boscaro, A.; Peila, D. Preliminary study of wear induced by granular soil on metallic parts of EPB tunnelling machines. *Geoing Ambient. Min.* **2016**, *148*, 67–70.
47. Alavi Gharahbagh, E.; Rostami, J.; Palomino, A.M. New soil abrasion testing method for soft ground tunneling applications. *Tunn. Undergr. Sp. Tech.* **2011**, *26*, 604–613. [CrossRef]
48. Alavi Gharahbagh, E.; Qiu, T.; Rostami, J. Evaluation of granular soil abrasivity for wear on cutting tools in excavation and tunneling equipment. *J. Geotech. Geoenviron.* **2013**, *139*, 1718–1726. [CrossRef]
49. Alavi Gharahbagh, E.; Rostami, J.; Talebi, K. Experimental study of the effect of conditioning on abrasive wear and torque requirement of full face tunneling machines. *Tunn. Undergr. Sp. Tech.* **2014**, *41*, 127–136. [CrossRef]
50. Zhang, B.; Reza, H.; Gu, S.; Gupta, N. Investigation of physical and chemical characteristics of Masonry Stones and Bricks during Building Cleaning: Part 1. Physical testing. *J. Phys. Sci. Appl.* **2014**, *4*, 207–222.
51. Akaishi, M.; Oshawa, T.; Yamaoka, S. Synthesis of Fine-Grained Polycrystalline Diamond Compact and Its Microstructure. *J. Am. Ceram. Soc.* **1991**, *74*, 5–10. [CrossRef]
52. Antsev, A.V.; Pasko, N.I.; Antseva, N.V. Assessment of wear dependence parameters in complex model of cutting tool wear. *IOP Conf. Ser. Mater. Sci. Eng.* **2018**, *327*, 042005. [CrossRef]

applied sciences

MDPI

Article

# Influence of Wood Properties and Building Construction on Energy Demand, Thermal Comfort and Start-Up Lag Time of Radiant Floor Heating Systems

Álvaro Ruiz-Pardo [1], Enrique Ángel Rodríguez Jara [1], Marta Conde García [2] and José Antonio Tenorio Ríos [3,*]

1 Departamento de Máquinas y Motores Térmicos, University of Cadiz, Av. Universidad de Cádiz, 10, 11519 Cádiz, Spain; alvaro.ruiz@uca.es (Á.R.-P.); enrique.rodriguez@uca.es (E.Á.R.J.)
2 Departamento de Ingeniería Forestal, University of Córdoba, Ctra. Madrid, Km 396, 14071 Córdoba, Spain; marta.conde@uco.es
3 Eduardo Torroja Institute for Construction Sciences, Spanish National Research Council (CSIC), de Serrano Galvache, 4, 28033 Madrid, Spain
* Correspondence: tenorio@ietcc.csic.es

**Abstract:** Radiant floor heating is becoming increasingly popular in cold climates because it delivers higher comfort levels more efficiently than conventional systems. Wood is one of the surface coverings most frequently used in radiant flooring, despite the widely held belief that in terms of thermal performance it is no match for higher conductivity materials if a high energy performance is intended. Given that the highest admissible thermal resistance for flooring finishes or coverings is generally accepted to be 0.15 $m^2K/W$, wood would appear to be a scantly appropriate choice. Nonetheless, the evaluation of the thermal performance of wooden radiant floor heating systems in conjunction with the building in terms of energy demand, thermal comfort, and start-up period, has been insufficiently explored in research. This has led to the present knowledge gap around its potential to deliver lower energy consumption and higher thermal comfort than high-thermal-conductivity materials, depending on building characteristics. This article studies the thermal performance of wood radiant floors in terms of three parameters: energy demand, thermal comfort, and start-up lag time, analysing the effect of wood properties in conjunction with building construction on each. An experimentally validated radiant floor model was coupled to a simplified building thermal model to simulate the performance of 60 wood coverings and one reference granite covering in 216 urban dwellings differing in construction features. The average energy demand was observed to be lower in the wood than in the granite coverings in 25% of the dwellings simulated. Similarly, on average, wood lagged behind granite in thermal comfort by less than 1 h/day in 50% of the dwellings. The energy demand was minimised in a significant 18% and thermal comfort maximised in 14% of the simulations at the lowest thermal conductivity value. The vast majority of the wooden floors lengthened the start-up lag time relative to granite in only 30 min or less in all the dwellings. Wood flooring with the highest thermal resistance (even over the 0.15 $m^2K/W$ cited in standard EN 1264-2) did not significantly affect the energy demand or thermal comfort. On average, wood flooring lowered energy demand by 6.4% and daily hours of thermal comfort by a mere 1.6% relative to granite coverings. The findings showed that wood-finished flooring may deliver comparable or, in some cases, higher thermal performance than high-conductivity material coverings, even when their thermal resistance is over 0.15 $m^2K/W$. The suggestion is that the aforementioned value, presently deemed the maximum admissible thermal resistance, may need to be revised.

**Keywords:** wooden radiant floor heating; radiant floor thermal modelling; energy efficiency of buildings; thermal comfort; natural stone vs. wood radiant floors

**Citation:** Ruiz-Pardo, Á.; Rodríguez Jara, E.Á.; Conde García, M.; Ríos, J.A.T. Influence of Wood Properties and Building Construction on Energy Demand, Thermal Comfort and Start-Up Lag Time of Radiant Floor Heating Systems. *Appl. Sci.* **2022**, *12*, 2335. https://doi.org/10.3390/app12052335

Academic Editors: Dibyendu Sarkar, Rupali Datta, Prafulla Kumar Sahoo and Mohammad Mahmudur Rahman

Received: 8 February 2022
Accepted: 20 February 2022
Published: 23 February 2022

## 1. Introduction

The underfloor heating system is one of the oldest technologies for providing heating in buildings. The Ondol system was probably the first developed (Korea, 1000 BC) [1], and the Kang (China, 500 BC) [2] and the hypocaust (Greece/Rome 300 BC) [3] are other ancient technologies developed for this purpose. After the fall of the Roman Empire, underfloor heating disappeared from Europe during the Middle Ages, re-emerging in France and Britain in the 18th and 19th centuries [4]. In North America, the first radiant heat floor systems date from the end of the 19th century, but the popularization of these systems worldwide probably only began after the Second World War [5]. Radiant floor heating is currently one of the fastest growing heating systems, thanks in part to the introduction of plastic piping [6], systems based on heat pumps [7], and renewable energies.

Radiant floors have a number of advantages over other heating systems, including: high comfort levels [8], resulting from their ideal vertical temperature gradient [8,9]; high energy efficiency, due to the lower air temperature required to meet comfort conditions [9]; silent operation [8]; and their use of low water temperatures that are compatible with renewable energy [10–13]. For these reasons, radiant floor heating is gaining in popularity compared with more conventional systems [14,15].

Numerous investigations have been carried out on the thermal performance of radiant floors from different perspectives, such as control [13,16–18], thermal inertia [19,20], the use of PCMs [21–27], the arrangement of tubes [23,28], and the most suitable radiant surfaces [6], but fewer publications has been found on the influence of floor coverings properties, although, as Sattari and Farhanieh [29] showed in a parametric study, the thickness and thermal properties of the finish or covering have a major bearing on the performance of radiant floor heating.

The coverings most commonly applied in radiant flooring are porcelain or ceramic tiles, natural stone, or wood. Low-thermal-resistance natural stone and ceramic materials may initially be deemed to feature higher thermal performance than wood [30], which is nonetheless used as a covering in radiant floors for subjective reasons: its aesthetics tend to be more highly esteemed and it is deemed more pleasing to the touch. Another advantage, according to Zhao et al. [31], is that with wood 'the surface temperature is more moderate and uniform'. In an empirical study of four types of laminate wood flooring with conductivities ranging from 0.091 W/(m·K) to 0.12 W/(m·K), Seo et al. [32] found that when secured to the system with adhesives, wood delivered better results than when not secured. They contended that although more energy is needed for first-use start-up, such floors maintain a higher temperature for longer after the heating is turned off. That beneficial effect has also been studied in connection with renewable energies. Athienitis and Chen [33], for instance, studied the effect of the solar heat accumulating in wood floors.

European standard EN 1264-2, 'Water based surface embedded heating and cooling systems—Part 2: Floor heating: Prove methods for the determination of the thermal output using calculation and test methods', specifies that thermal resistance values in floor coverings ($R_{\lambda,B}$) exceeding 0.15 m$^2$K/W lie outside its scope, a contention reinforced in Part 3 (on dimensioning), according to which values of $R_{\lambda,B} > 0.15$ m$^2$K/W should be avoided. This has been interpreted to constitute a hard-and-fast maximum and to mean that coverings with thermal resistance higher than this value are unsuitable for radiant floors. Design engineers have consequently challenged the suitability of wood, with its fairly high thermal resistance, as the covering of choice for radiant flooring, particularly in the coldest climates.

In light of the above, although radiant floors with wood covering are widely used for applications other than energy, their use is ruled out when the primary aim is high energy performance. Nonetheless, the evaluation of the thermal performance of wooden radiant floor heating systems in conjunction with the building in terms of energy demand, thermal comfort, and start-up period, has been insufficiently explored in research. This has led to the present knowledge gap around its potential to deliver lower energy consumption and higher thermal comfort than high-thermal-conductivity materials, depending on building characteristics.

This study assesses the impact of different wood coverings on the thermal performance of radiant floors compared to a high conductivity material in terms of three parameters: energy demand, thermal comfort and start-up lag time. The material chosen for comparison in this study was granite, insofar as it is characterised by one of the highest thermal conductivity values of all natural stones. The effect of covering thermal properties in conjunction with dwelling constructional characteristics on these parameters was explored.

Part 1 of the aforementioned European standard (EN 1264-1 [34]) on water-based, surface-embedded heating and cooling systems defines four radiant flooring layouts, which vary in terms of pipe position (Figure 1): embedded (types A and C), underfloor (type B), and heating floor with plane section system (type D). In type A, the pipes are embedded in the thermal diffusion layer, which also affords the system inertial and thermal diffusion. In type B, similar to A, the pipes are laid under the covering, where they rest on diffusor bands. Type C pipes are embedded in an underlayer crowned with a band made of a different material. While wood is explicitly listed as one of the possible covering materials for type B, its use with the other three types is not ruled out. The radiant flooring chosen for this study was standard-compliant for types A and C, configurations widely used in radiant flooring.

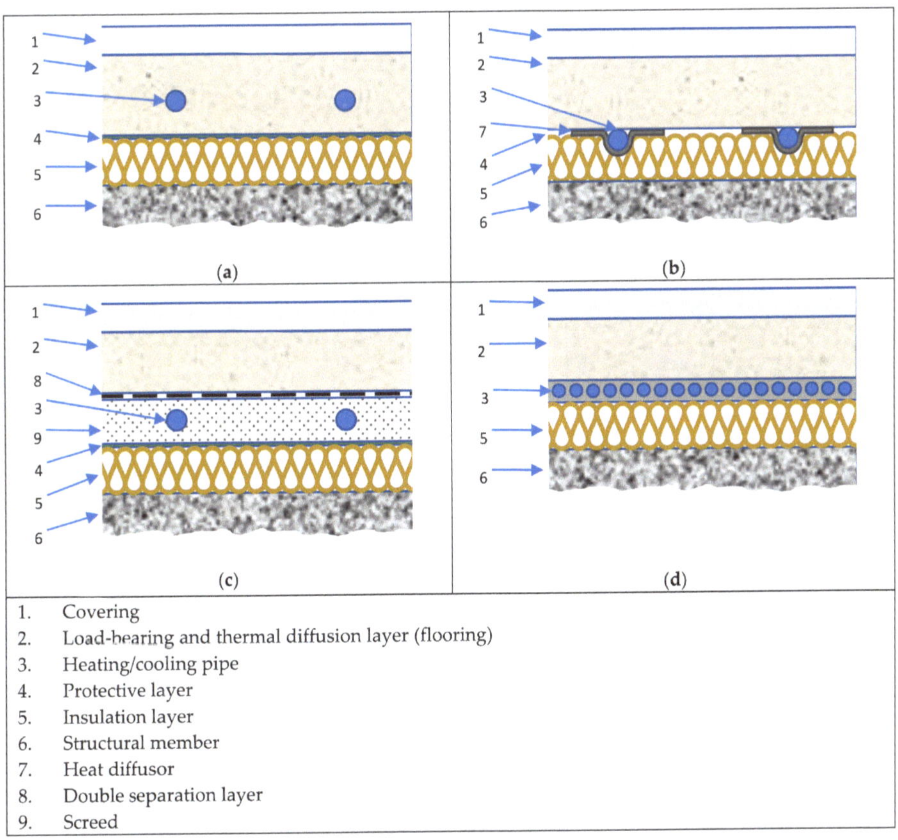


1. Covering
2. Load-bearing and thermal diffusion layer (flooring)
3. Heating/cooling pipe
4. Protective layer
5. Insulation layer
6. Structural member
7. Heat diffusor
8. Double separation layer
9. Screed

**Figure 1.** Types of radiant flooring by pipe layout (standard EN 1264-1). (**a**) Type A: Flooring with embedded pipes. (**b**) Type B: Flooring with underfloor pipes. (**c**) Type C: Flooring with embedded pipes. (**d**) Type D: Plane section systems.

In pursuit of the objectives of the study, a simplified building model was developed and coupled to a detailed, experimentally validated response-factor-method-based radiant floor model. A total of 60 simulations, one for each wood covering, was run with each of the 216 urban dwellings defined, and the results were compared to the performance simulated for a reference granite-finish flooring. Two types of simulation were performed: one to evaluate the energy demand and comfort, and the other to evaluate the start-up lag time. This combination of coverings, dwellings, and simulation types yielded a total of 26,352 simulations. The simulation models and the experimental validation of the radiant floor model are described in Section 2, while the simulation variables are discussed in Section 3, along with the radiant floor coverings, dwelling constructional features, and operating conditions. The findings are discussed in Section 4.

To broaden the applicability of the present findings, the case study was conducted for Madrid, a city comparable in terms of heating-degree days to the climate of a number of the world's major cities (Table 1).

**Table 1.** Cities with a number of heating-degree days at base T = 20 °C similar to Madrid's *.

| | Jan | Feb | Dec |
|---|---|---|---|
| Chengdu (Ch) | 451 | 359 | 400 |
| Istanbul (TK) | 444 | 402 | 363 |
| London (UK) | 467 | 422 | 429 |
| Madrid (SP) | 452 | 364 | 434 |
| Tokyo (JP) | 423 | 367 | 352 |

* Source: author calculations using data drawn from Meteonorm [35], a commercial database.

## 2. Simulation Models

The models described in this section were developed to accurately calculate the thermal performance of radiant floor heating in a large number of cases in a reasonable time. The procedure followed was inspired by an idea put forward by Xu and Wang [36], in which a detailed model is applied for the element at issue (radiant flooring in this case) and a simplified model for the building housing it. The outcome was the development of two low computational-cost models, described in the paragraphs below. The detailed radiant floor model was based on the response factors method and the simplified building model on the lumped parameters or resistance capacity (RC network model).

### 2.1. Detailed Radiant Floor Thermal Modelling

Transient, two-dimensional modelling is required to simulate the thermal behaviour of embedded pipe structures [37]. The well-known response factor method [38] is applicable [39,40] for these purposes, since it delivers high calculation accuracy while ensuring processing speed by minimising the number of computational operations required.

The need to couple the heating to the building model and simulate control strategies entailed the use of a small time step in the simulations. A 15 min time step was defined for the response factor method used here to simulate the transient thermal response in radiant floorings. That the resulting radiant flooring model could be coupled to the simplified building model described in Section 2.2 with no need for iteration or matrix inversion also contributed to lowering computational costs.

The response factor method as initially proposed by Stephenson and Mitalas [41,42] is described in detail in [43]. In essence, it consists of obtaining the heat fluxes across three surfaces of a radiant floor (top, bottom, and pipe) over time in response to a unitary triangular temperature pulse on each surface, generating as output three sets of responses per surface excited. Such heat fluxes are known as response factors and must be determined analytically or numerically. Here, the widely used international ANSYS [44] software was applied to that end, performing the calculations with the finite element method, which is a detailed numerical method for solving the partial differential equations that govern transient two-dimensional heat conduction across radiant floors. ANSYS APDL derives the surface distribution of heat flux on each surface of the radiant floor. It then internally calculates the averaged heat fluxes on each surface weighting the node values by the amount of element surface area associated with each node. These averaged heat fluxes are the response factors.

The response factors calculated were then used to calculate the actual heat fluxes in each surface of the radiant floor based on the superposition principle. Although a more detailed explanation is given in Appendix A, Equation (1), the general expression applied to calculate the heat flux across the three surfaces of the radiant floor is shown below:

$$\dot{q}_{s,k,floor}(n) = f_{1,k}(0)\ T_{s,1}(n) + f_{2,k}(0)\ T_{s,2}(n) + f_{3,k}(0)\ T_{s,3}(n) + h_k \qquad (1)$$

where k denotes the top (1), bottom (2) or pipe (3) surface; n the simulation time step at issue; $f_{j,k}$ the response factor for surface k when a unitary triangular pulse is applied to surface j; and $h_k$ a term grouping the information on the preceding time steps for surface k.

### 2.2. Coupling the Building and Radiant Floor Thermal Models

The building simulations were conducted with a lumped parameter or resistance-capacitance (RC-network) model. 'RC models have proven to be advantageous and are widely used in modeling of thermal dynamics' [45] due to their low computational cost, their capacity to accommodate the main physical information on buildings, and better-than-acceptable accuracy, providing they are properly characterised [46]. This work aims to plausibly simulate the thermal performance of dwellings while using a radiant floor heating system. Vivian J. et al. [47] found that the deviation in the heating peak load of a first-order model (a lumped parameter model with one thermal capacitance) with respect to a detailed model falls between +8% and −6%, and the heating needs approximately +/−5%. They concluded that both first- and second-order models 'appear to reliably calculate the overall energy needs of buildings in both heating and cooling seasons'. On these grounds, the RC-network model presented here appears to be suitable for the purposes of this research. Any number of examples of the use of RC models for thermal simulations in buildings can be found in previous research (see [48–56]), as well as in studies that used such models in building simulations involving radiant flooring. Joe and Karava [57], for instance, developed a thermal resistance and capacitance model to optimise the control of radiant floor heating and Weber et al. [58] combined a detailed model for simulating a thermally activated building component system (TABS) with an RC model for the building to address an issue closely related to that discussed here. Thanks to the RC model applied to the building, which was readily coupled to the radiant floor model described in the preceding sub-section, the output for the 26,352 (Section 3.3) simulations conducted was obtained in a reasonable amount of time.

The simplified thermal model for the dwelling simulation coupled to the radiant floor model is illustrated in Figure 2. The bottom surface of the radiant flooring was assumed to be bounded by an indoor space forming part of the environment-controlled dwelling on the storey below. In the figure, $T_{s2}$ is the temperature of the bottom surface of the flooring and $h_{cr2}$ the convective–radiative heat transfer coefficient between that surface and the indoor space below at temperature $T_i$. The top surface, in turn, was assumed to be bounded by the apartment to be heated, consisting in outer walls and indoor partitions separating it from other dwellings. $T_{cr,I}$ represents the convective–radiative temperature of the space to be heated and hcr1 the convective–radiative heat transfer coefficient between the capacitance node of the space (node C) and the top surface of the flooring at temperature $T_{s1}$. $U_i$ represents the U-value of the indoor partitions between the apartment and the adjacent spaces at temperature $T_i$. $U_e$ represents the U-value of the outer walls, including windows, with the outdoor environment at linear air temperature, $T_{sa}$, which is defined as 'the outside air temperature which, in the absence of solar radiation, would give the same temperature distribution and rate of heat transfer through a wall (or roof) as exists due to the combined effects of the actual outdoor temperature distribution plus the incident solar radiation' [59]. The internal loads and ventilation are represented as $\dot{Q}_{cr}$ at the capacitance node, which also included the thermal loads attributable to the solar radiation entering the space through the windows.

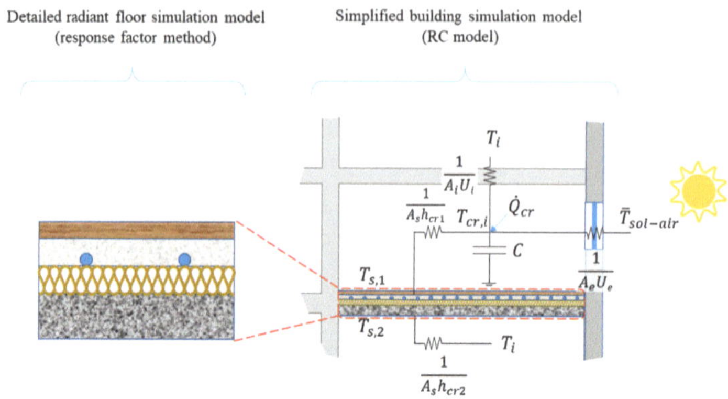


**Figure 2.** Radiant flooring model coupled to the building RC model.

The coupling between the radiant flooring and building model is described below. The boundary conditions in the flooring model were the top ($T_{s,1}$), bottom ($T_{s,2}$), and piping ($T_{s,3}$) surface temperatures and the responses to the heat fluxes across those surfaces ($q_{s,1}$, $q_{s,2}$ and $q_{s,3}$, respectively). The boundary conditions in the simplified building model were the indoor temperatures in the adjacent dwellings ($T_i$), the outdoor sol-air temperature ($T_{sol\text{-}air}$), the top and bottom flooring temperatures ($T_{s,1}$ and $T_{s,2}$, respectively), and the heat flux at the capacitance node ($\dot{Q}_{cr}$). The responses in this model were also the heat fluxes at the top and bottom surfaces of the flooring ($q_{s,1}$ and $q_{s,2}$, respectively), as well as the indoor temperature in the space to be heated, i.e., the temperature of the capacitance node ($T_{cr,i}$). As the figure shows, the two models were coupled via the heat fluxes at the top and bottom surfaces on the radiant flooring. The results of this coupling, i.e., the indoor environmental temperature $T_{cr,I}$ and the radiant floor surface temperatures $T_{s,1}$, $T_{s,2}$, and $T_{s,3}$ were found with no need for iteration or other numerical procedures. The values adopted for each pre-defined building parameter are given in Section 3.

The heat flux at the top surface of the flooring, $q_{s,1,build}$, was calculated with Equation (2):

$$\dot{q}_{s,1,\ build} = h_{cr,1}(T_{cr,i} - T_{s,1}) \quad (2)$$

for which the capacitance node temperature $T_{cr,I}$ needed to be found. The procedure to determine the temperature for the boundary conditions listed earlier is described in [60] and in Appendix A. The excitation was taken to vary linearly at every simulation time step, a reasonable premise inasmuch as excitation data are normally given as discrete values spaced at time-step intervals. Applying continuity across the top surface of the flooring, the conduction heat flux obtained with the flooring model (Equation (1)) is equal to the convective–radiative heat flux found with the RC building model (Equation (3)):

$$\dot{q}_{s,1,floor} = \dot{q}_{s,1,\ build} \quad (3)$$

$$f_{1,1}(0)\ T_{s,1} + f_{2,1}(0)\ T_{s,2} + f_{3,1}(0)\ T_{s,3} + h_1 = h_{cr,1}(T_{cr,i} - T_{s,1}) \quad (4)$$

Substituting the expression for temperature $T_{cr,i}$ obtained with the RC model for the building (Appendix A) and re-grouping terms in Equation (2) yielded the following simple algebraic equation:

$$a_1\ T_{s,1} + b_1\ T_{s,2} + c_1\ T_{s,3} + d_1 = 0 \quad (5)$$

where: $b_1 = f_{2,1}(0)$, $c_1 = f_{3,1}(0)$, $a_1 = f_{1,1}(0) + h_{cr,1}\,(1 - A_1)$ and $d_1 = h_1 - h_{cr,1}\,B_1$; A1 and B1 represent the known values at the current time step that depend on boundary conditions and the model parameters that define the building, i.e., heat capacity, C, and the thermal resistance between the space and the flooring surface. Explicit equations for A1 and B1, given in Appendix A, are omitted here for the sake of simplicity.

Similarly, applying continuity across the bottom and pipe surfaces yields:

$$f_{1,2}(0)\,T_{s,1} + f_{2,2}(0)\,T_{s,2} + f_{3,2}(0)\,T_{s,3} + h_2 = h_{cr,2}(T_i - T_{s,2}) \quad (6)$$

$$f_{1,3}(0)\,T_{s,1} + f_{2,3}(0)\,T_{s,2} + f_{3,3}(0)\,T_{s,3} + h_3 = h_{c,3}(T_w - T_{s,3}) \quad (7)$$

These equations can be re-grouped as shown below:

$$a_2\,T_{s,1} + b_2\,T_{s,2} + c_2\,T_{s,3} + d_2 = 0 \quad (8)$$

$$a_3\,T_{s,1} + b_3\,T_{s,2} + c_3\,T_{s,3} + d_3 = 0 \quad (9)$$

where $a_2 = f_{1,2}(0)$, $c_2 = f_{3,2}(0)$, $b_2 = f_{2,2}(0) + h_{cr,1}$ and $d_2 = h_2 - h_{cr,2}\,T_i$ in Equation (8) and $a_3 = f_{1,3}(0)$, $b_3 = f_{2,3}(0)$, $c_2 = f_{3,3}(0) + h_{c,3}$, and $d_3 = h_2 - h_{c,3}\,T_w$ in Equation (9).

Equations (4), (7) and (8) form a system of equations where the unknowns, $T_{s,1}$, $T_{s,2}$, and $T_{s,3}$, can be represented in a matrix as follows:

$$\begin{bmatrix} a_1 & b_1 & c_1 \\ a_2 & b_2 & c_2 \\ a_3 & b_3 & c_3 \end{bmatrix} \begin{bmatrix} T_{s,1} \\ T_{s,2} \\ T_{s,3} \end{bmatrix} = \begin{bmatrix} -d_1 \\ -d_2 \\ -d_3 \end{bmatrix} \quad (10)$$

The explicit expressions for temperatures $T_{s,1}$, $T_{s,2}$, and $T_{s,3}$, which depend on the response factors for the three flooring surfaces and the building's boundary conditions and parameters, can be readily found with Equation (10) by inverting the matrix. These flooring surface temperatures can then be used to calculate the temperature of indoor space $T_{cr,I}$ by Equation (A12).

As previously stated, the outputs of the model are the surface temperatures of the radiant floor $T_{s,1}$, $T_{s,2}$, and $T_{s,3}$, as well as the temperature of the indoor space $T_{cr,i}$. Another output of interest for the present work is the operative temperature Top. The operative temperature in a room is defined by standards CIBSE [61] and ISO 7730:2005 [62] as follows: 'the operative temperature in a real room is equal to the air temperature in an hypothetical room such that an occupant would experience the same net energy exchange with the surroundings'. These standards also mention that Top is used as an index temperature for comfort where air velocities are low, so it has been used in Section 4 for assessing the indoor thermal comfort. Top can be calculated according to ISO 7726:1998 [63] as:

$$T_{op} = \frac{\overline{T}_r \cdot h_r + T_{air} \cdot h_c}{h_r + h_c} \quad (11)$$

where $\overline{T}_r$ is the mean radiant temperature calculated as:

$$\overline{T}_r = T_{s,1} \cdot F_{s,1} + T_{walls} \cdot F_{walls} \quad (12)$$

In Equation (12) 'walls' refers to all the enclosures composing the space except the top surface of the floor. $F_{s,1}$ and $F_{walls}$ are the view factors of the occupant with floor top surface and space walls, respectively. As an approximation, $F_{s,1} = A_{s,1}/A_{total}$ and $F_{walls} = 1 - F_{s,1}$, where $A_{total}$ refers to the total area of the space enclosures, including the top surface of the floor. In the simplified building model indoor space is assumed to be at uniform temperature and $T_{walls} = T_{air} = T_{cr,i}$, so $T_{op}$ can be determined as:

$$T_{op} = \frac{T_{s,1} \cdot \left(\frac{A_{s,1}}{A_{total}}\right) \cdot h_r + T_{cr,i}\left[\left(1 - \frac{A_{s,1}}{A_{total}}\right) \cdot h_r + h_c\right]}{h_r + h_c} \quad (13)$$

## 2.3. Experimental Validation of the Radiant Floor Model

The radiant floor model developed was experimentally validated by using data from a laboratory trial conducted at Applus laboratories, in Barcelona, further to a request by Uponor Hispania, S.A.U., a specialist in radiant-floor-based HVAC systems. The trial, the details of which were furnished by Uponor, aimed, among other objectives, to experimentally measure the time taken to reach a comfortable temperature in a wood-covered radiant floor heating system. The temperatures recorded in the trials were used to validate the radiant floor model applied here.

The 90-millimetre-thick radiant floor heating system (Figure 3) comprised 19-millimetre-thick varnished oakwood planks secured with adhesives, a 30 mm anhydrite-based mortar slab with a 16 mm diameter embedded cross-linked polyethylene piping, and a smooth 25 mm enhancing-agent-modified EPS self-secured panel. The pipes were spaced at 150 mm in a 2 m × 2 m sample floor.

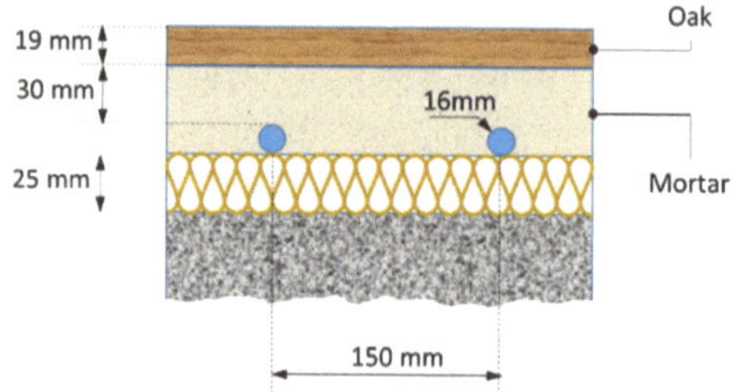


**Figure 3.** Cross-section of radiant flooring used in the model validation trials.

The trial was conducted in a 4.0 m × 4.0 m × 3.0 m chamber built to the specifications laid down in European standard EN 442-2 [64] for radiators and convectors, i.e., with five water-cooled sides and one insulated side fitted with the emitter.

Beginning at a temperature of 1 °C for both flooring and chamber, the trials consisted of pumping a given water flow, at a specified temperature, through the piping and determining the surface temperatures on the flooring and in the chamber until the latter reached 20 °C, measured at a reference point in the centre of the chamber 0.75 m above the flooring. Eight thermocouples were positioned on the wood covering to record the surface temperatures. The three stages of flooring construction depicted in the photographs in Figure 4 show the positions of the piping in the mortar and of the thermocouples on the wood covering. The trials were conducted at inflow water temperatures of 35 °C, 40 °C, and 45 °C, and a flow rate of 200 L/h, in line with European standard EN 1264 [64].

(a) (b) (c)

**Figure 4.** Trial flooring set-up. (**a**) Piping layout; (**b**) mortar layer; (**c**) floor covering and positions of surface-temperature sensors.

The accuracy of the transient thermal model for radiant flooring developed for this study was validated at inflow water temperatures of 35 °C, 40 °C, and 45 °C. The boundary conditions for each were defined by the parameters of inflowing water temperature, water flow rate, and the experimentally measured air and wall temperatures of the chamber ($T_{cr}^{exp}$). The model output consisted of the flooring surface temperatures. For the purposes of validation, the flooring was thermally coupled to the chamber environment and the water. The water-side convective heat transfer coefficient was found with the Gnielinski correlation [65] for inner flow in smooth pipes. The convective–radiative heat transfer coefficient at the covering surface ($h_{cr}$) was found from the mean experimental covering temperature ($T_s^{exp}$) and the chamber temperature, along with the surface heat flux, ($\dot{q}_s$), calculated as laid down in the aforementioned standard EN 1264 [64]. For any given time, then, the convective–radiative coefficient was calculated as $h_{cr} = \dot{q}_s / \left(T_s^{exp} - T_{cr}^{exp}\right)$. For each inflow water temperature, the model-predicted surface temperature was compared to the mean surface temperature on the flooring recorded during the trials (Figure 5). The root mean square (RMS) error was then calculated between the model and experimental values for each water inflow temperature. The use of a single average temperature could have been a limitation of the model, since the tube arrangement in the experiment was serpentine, which could have led to the presentation of a less uniform temperature distribution on the surface compared to a spiral arrangement. However, during the experiments, the difference in water temperature between the inlet and the outlet was small (as an example, when the inlet temperature was 40 °C, the outlet was 39.4 °C, with small variations, during the test), so the temperature on the entire surface was very homogeneous. Therefore, the temperature distribution problem, for this case, was minimal. On these grounds, an average experimental soil temperature was obtained to compare with that of the model. Hence, the a priori limitation of the model, which only uses one temperature for the entire radiant floor surface, is irrelevant to its validation.

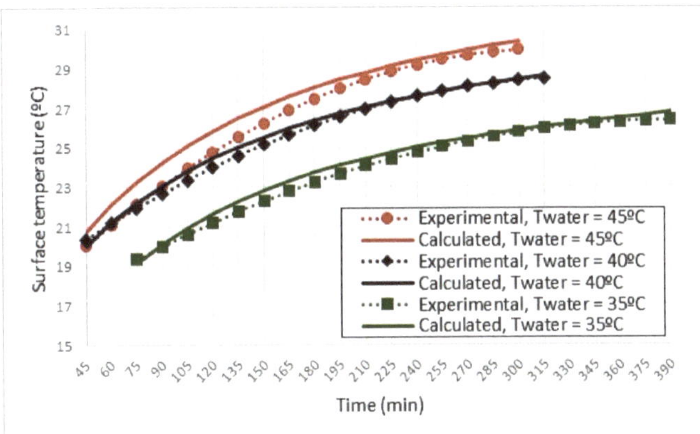


**Figure 5.** Experimental validation of the numerical model for radiant flooring.

The flooring surface temperatures found with the numerical model for water inflow temperatures of 35 °C and 40 °C fitted the experimental measurements very closely, with RMS errors of only 0.33 °C and 0.27 °C, respectively. The numerical and experimental values agreed less closely at a water inflow temperature of 45 °C, particularly in the early time steps, although the RMS error was only 0.75 °C. The less narrow fit for the inflow water temperature at 45 °C may be attributable to the use of an adiabatic boundary condition for the bottom side of the flooring. The downward losses in the flooring, which were greater for the inflow water at 45 °C than for the other values, were not be accounted for by the model. In light of the aforementioned findings, the accuracy of the numerical model was deemed to be valid for the present purposes.

## 3. Simulation Description

This section describes the radiant floor layout, the material properties and building construction designs envisaged in the simulations, and the two types of simulation conducted. The total number of simulations was the result of applying 61 (60 wood and 1 granite) coverings to 216 dwellings under two types of simulation or operating regime ($61 \times 216 \times 2 = 26{,}352$). As the mean time per simulation was around 5 s, the total calculating time amounted to about 37 computer hours (Hewlett-Packard, Intel® Xeon® processor; CPU E5-1620 V3 @ 3.50 GHz; 16.0 GB RAM).

### 3.1. Radiant Floor Layout and Material Properties

The flooring design applied in the simulations, depicted in Figure 6, was patterned on a typical radiant heating system that could be likened to layouts A and C in European standard 1264-1 [34], as noted earlier. The thermal properties of all except the covering (outer-most layer), which is described below, are listed in Table 2. The pipe wall thermal resistance was excluded from the simulations as negligible compared to the resistance of the surrounding mortar and insulation.

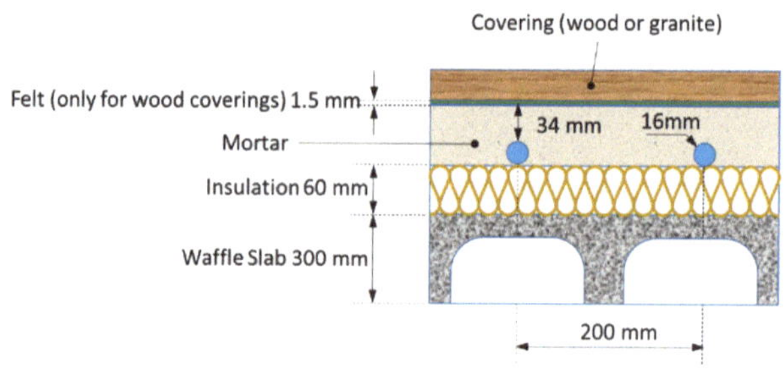


**Figure 6.** Radiant flooring layout.

**Table 2.** Thermal properties of materials comprising the radiant floor.

| | Thermal Conductivity W/(m·K) | Density kg/m$^3$ | Specific Heat J/(kgK) |
|---|---|---|---|
| Felt | 0.033 | 90 | 1000 |
| Mortar | 1.8 | 2100 | 2000 |
| Insulation | 0.033 | 30 | 1200 |
| Waffle slab | 1.22 | 1090 | 1000 |

The values for the thermal properties listed in Table 3 were drawn from an analysis of a number of commercial wood types used in flooring, specifically the types recommended for radiant floors by manufacturers such as Finsa, Pergo, Haro, Meister, and Junckers. Some of those manufacturers recommend a thermal resistance maximum of 0.15 m$^2$K/W for coverings, whereas Spanish standard UNE 56810 on specifications for wood floors [66] sets the limit at 0.17 m$^2$K/W.

The thermal resistance was consistently under 0.15 m$^2$W/K in all the commercial floors analysed, while the thermal conductivity ranged from 0.10 W/m·K to 0.15 W/m·K for the thicknesses listed (Table 3).

Inasmuch as wood is a natural product, its characteristics vary, depending on its species and origin, the individual tree, or even the position of the wood in the trunk of a given tree. Thermal conductivity depends on several parameters, including microstructure, moisture and temperature, along with wood density [67,68], which is also related to most other physical and mechanical properties [69,70], including thermal conductivity.

Table 3. Commercial wood flooring thickness and thermal resistance values.

| Trade Name | Product | Thickness (mm) | Thermal Resistance ($m^2K/W$) |
|---|---|---|---|
| Finsa. Finfloor | Laminate floor coverings | 8–10 | 0.06–0.154 |
| Pergo Lofoten–Senja–Langeland–Svalbard | Multi-layer parquet | 14 | 0.140 |
| Pero Laminate | Laminate floor coverings | 7–9.5 | 0.051–0.07 |
| Haro. Parquet 3000/3500/4000 | Multi-layer parquet | 11–13.5 | 0.063–0.110 |
| Haro. Tritty | Laminate floor coverings | 8 | 0.065 |
| Meister | Multi-layer parquet | 11–14 | 0.084–0.143 |
| Meister | Laminate floor coverings | 9 | 0.09 |
| Junckers | Solid hardwood planks | 15–20.5 | 0.09–0.12 |

The behaviour of possible wood coverings was analysed after compiling information from previous studies on density and thermal conductivity in temperate hardwood (oak, beech, ash, maple, walnut, cherry), softwood (pine, spruce, larch), and tropical wood (iroko, teak, jatoba, Merbau, awong). Denser natural species were not investigated.

On the grounds of the aforementioned inter-relationship between thermal conductivity and density, five mean densities were adopted based on the values found in previous studies (400 kg/m$^3$, 500 kg/m$^3$, 600 kg/m$^3$, 700 kg/m$^3$ and 850 kg/m$^3$) with a view to covering a wide spectrum of wood properties. Three possible thermal conductivity values (consistent with earlier findings [14]) were assigned to each density, calculated from the mean conductivity values and standard deviations reported in previous studies [32,71–80], as well as from the commercial high thermal conductivity materials listed above. The findings are plotted in Figure 7, where the red dots denote the values adopted in this study.

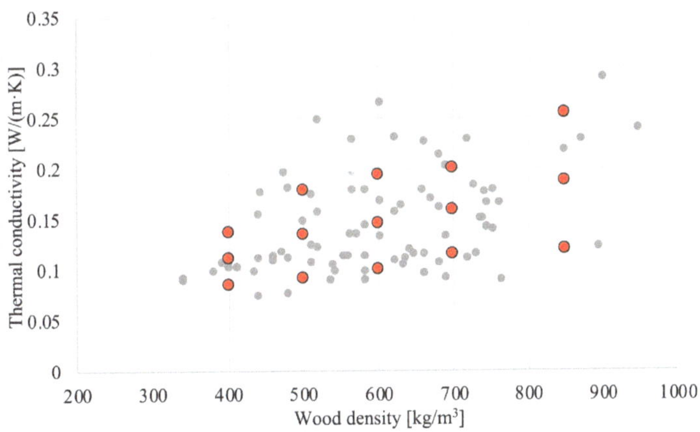


Figure 7. Wood properties: thermal conductivity and density (grey dots: values found in previous studies; red dots: values assumed here).

The fifteen thermal properties of the wood plotted in Figure 7 were each simulated for four thicknesses (10 mm, 15 mm, 19 mm, and 22 mm), yielding a total of 60 wood coverings.

A further simulation was run for a high-conductivity reference, a granite floor characterised by a thermal conductivity of 3.50 W/(m·K), a density of 2850 kg/m$^3$, and a specific heat of 1.0 kJ/(kg·K), for comparison. This inclusion brought the total number of radiant floors simulated to 61.

*3.2. Building Characteristics*

The characteristics of the dwellings analysed are described in this section. The median area of urban dwellings in Madrid is 90 $m^2$ [81].

As a square enclosure measuring 9.5 m on each side, the dwelling simulated here had an area of 90.25 $m^2$, very close to the aforementioned median value. The storey height applied was 3 m. The dwellings were assumed to be located in one of two positions: at mid-building, with just one outer enclosure (hereafter 'interior dwelling'); or on the top storey, in a corner location (hereafter 'corner dwelling') (Figure 8).

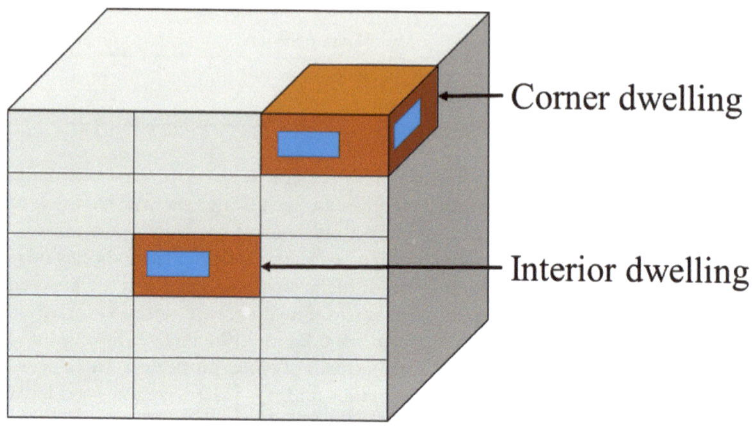


**Figure 8.** Dwelling locations used in simulations.

IIn pursuit of a wide variety of constructional characteristics, four variables were defined: glazed area, envelope insulation, overall heat transfer coefficient (U-value), and heat capacity and orientation, summarised below and substantiated in greater detail in Appendix B.

- Glazed area: 15%, 30%, or 80%;
- Envelope insulation (U-value): window, outer wall, and roof insulation were rated for the simulations as low, medium, or high depending on the respective U-values (Appendix B, Table A2), with high meaning better U-values than required by current legislation; medium meaning compliance-level; and low meaning non-conformity, i.e., as a rule, buildings 20 years old or over;
- Heat capacity: the three levels of heat capacity applied, low, medium, and high, were defined as per standard ISO 52016-1:2017 [82]);
- Orientation: the orientations adopted for interior dwellings were south, east, and west, and for corner dwellings, southeast, southwest, northeast, and northwest.

Consequently, the number of dwellings analysed was the result of combining the above parameters, namely two in-building locations, three percentages of glazing, three insulation levels, three levels of heat capacity, and four orientations, for a total of 216. The dwellings were coded as summarised in Figure 9.

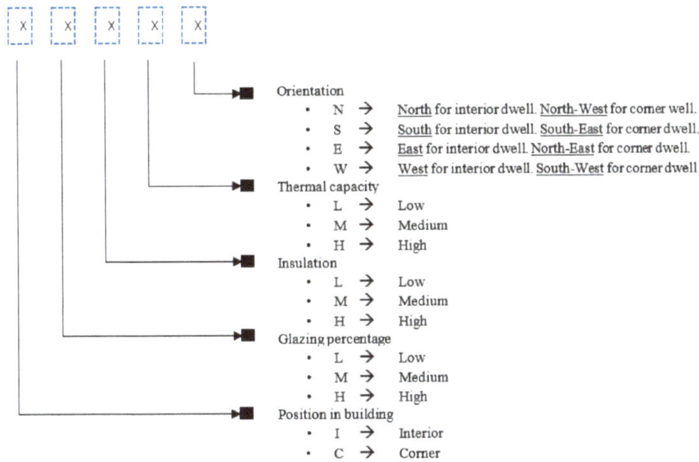


**Figure 9.** Dwelling labelling nomenclature.

### 3.3. Simulation Types

Two types of simulations were conducted, respectively designated 'normal regime' and 'start-up lag time'.

In normal-regime simulations, the radiant flooring was assumed to be operating in the conditions prevailing in January, the coldest month of the year. Under such conditions, the radiant flooring would be on from 08:00 to 23:00 to ensure the convective–radiative temperature, $T_{cr,I}$, required to maintain a set point temperature of 20 °C. The highest top surface temperature allowed by the control strategy was 29 °C, as specified in the standard ISO 1264-2:2009 [83].

In this simulation type, the model output primarily included the heat transferred by the water circulating in the pipes, used to assess the heating demand, and the indoor air and top-floor surface temperatures, to assess thermal comfort.

In the start-up lag time simulations, the radiant floor behaviour was simulated from the time it was switched on under given initial conditions until the 20 °C set point temperature was reached. The starting temperature for this simulation was determined by applying the conditions prevailing on the coldest day in January to a hypothetical three-day (72 h) period to envisage potential long lag times. The simulation ended after 72 h or when the set point temperature was reached, if earlier. To establish the initial conditions, January temperatures were assumed to vary freely and the dwelling to be subject to no occupancy loads. The aim of this simulation type was to calculate the time required to raise the room temperature to the set point from the initial free-float conditions.

## 4. Simulation Results and Discussion

The following discussion of the results is structured around the analysis of the three basic parameters used to assess performance: energy demand, thermal comfort, and start-up lag time.

Energy demand was determined as the amount of energy transferred by the water across the piping for the month of January. As thermal comfort is subjective [84] and cannot be unequivocally measured by any one parameter, a criterion had to be established on which to base the quantification. In this study, the criterion defined for this purpose was a minimum operative temperature of 20 °C, drawn from international standard ISO 7730 [62]. Based on the standard, spaces meeting this minimum wintertime requirement lie under comfort class B, with a PPD (predicted percentage dissatisfied) under 10%. Consequently, any operative temperature greater than or equal to 20 °C was assumed to be comfortable. The start-up lag time, in turn, was measured as the number of hours needed to raise the indoor temperature to 20 °C from certain initial conditions previously determined by simulating dwellings in free-floating conditions for three consecutive days. The conditions determined from these free-floating simulations were then applied as the initial conditions for the start-up lag time simulations.

### *4.1. General Trends in Performance*

The general trends observed in the three basic performance parameters are discussed in this section and plotted in Figure 10. The average, maximum, and minimum energy demand (Figure 10a), mean daily comfort hours (Figure 10b) and start-up lag time (Figure 10c) values were plotted for each dwelling by simulating all 60 woods, along with the value for the granite floor covering. In Figure 10a the dwellings are ranked along the $x$-axis from lowest to highest demand, in Figure 10b from highest to lowest number of comfort hours, and in Figure 10c from shortest to longest start-up lag time. Therefore, the dwellings are arranged in analogous but not identical order on the three figures, although, generally, the apartments with the lowest energy demand were observed to have a higher number of comfort hours and a shorter start-up lag time.

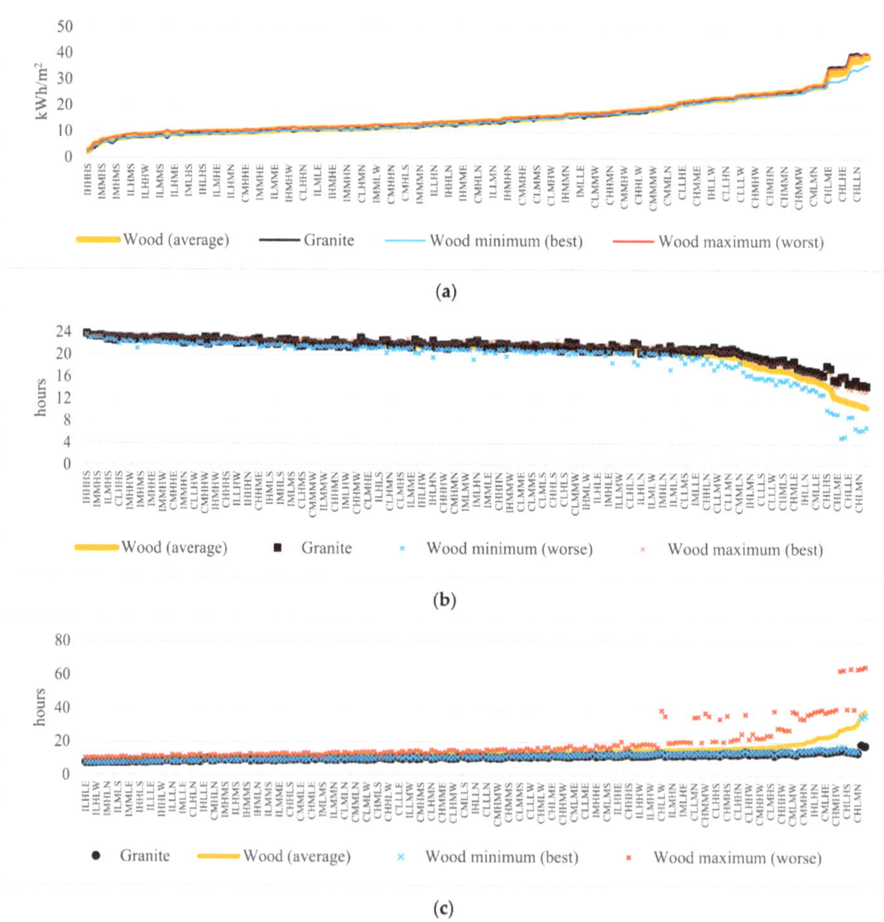


**Figure 10.** General trends of the basic performance parameters. (**a**) Energy demand (January); (**b**) thermal comfort: mean daily hours with $T_{op} \geq 20$ °C (January); (**c**) start-up lag time to reach $T_{op} = 20$ °C.

As expected, the dwellings with the best values for the three parameters (the dwellings placed furthest to the left in the figures) were predominantly those that were insulated the best and located at the interior, whilst the poorest performance (furthest to the right in the figures) was observed for the dwellings with the least insulation and located in corners. This pattern was observed in all the dwellings, whether with wood or granite floor coverings.

As the three figures show, by and large, the difference between using granite or wood covering was not significant, except in the most poorly insulated and corner-located dwellings. In many dwellings, at least one of the wood coverings exhibited lower demand or a greater number of comfort hours than the granite covering, whereas granite behaved consistently better in terms of start-up lag time. All the wood-floored dwellings with lower demand and more comfort hours than their granite-bearing counterparts were the best-insulated and located in the interior. On the whole, in corner-located and poorly insulated dwellings, granite out-performed wood in all three parameters, although in most cases, the values of at least one wood covering were observed to come close to those of the granite.

Specifically, the average energy demand was observed to be lower in the wood than in the granite coverings in 25% of the dwellings simulated. Similarly, on average, wood lagged behind granite in thermal comfort by less than 1 h in 50% of the dwellings. The average difference in start-up lag time between the wood and granite coverings was less than 3 h for 75% of the dwellings.

We found that, on average, with granite floors, the start-up lag time was just over 11 h, while with wooden floors it was 14 h, i.e., 27% longer. The better insulated the building, the lower these times and the difference between them were; in the best case, the time was 7.5 h for granite and 7.75 for wood (3.3% difference). The times were higher for buildings worse insulated, in the worst case 18 h for granite and 36 h for wood (112% difference). These data highlight the importance of the combined analysis of buildings and radiant floor heating. It is difficult to set a limit beyond which the start-up lag time is unacceptable, both in absolute terms, namely, the number of hours it takes to achieve the comfort condition, and in relative terms, namely, the difference between a high-conductivity floor such as granite and a wood floor. This is because the start-up time depends greatly on the specific use, e.g., if it is an occasional-use dwelling, such as a tourist flat, a start-up time of more than a couple of hours is probably unacceptable, but if it is a continuous-use flat, where the heating is usually on for most of the heating season, almost any start-up time can be acceptable.

### 4.2. Effect of Building Construction

The effect of building characteristics on the water-side energy demand (Figure 11) and thermal comfort delivered by the radiant floors (Figure 12) is addressed in this section. The construction parameters assessed here were dwelling heat capacity, insulation level, and percentage of glazed area, classified further according to the values listed in Appendix B. The impact of each parameter was assessed for dwelling orientation and location (corner or interior).

As shown in Figure 11, demand was observed to be greater in the corner-located dwellings. In the interior-located dwellings, whether wood- or granite-floored, demand was consistently highest among the dwellings facing north and lowest for those facing south. By contrast, demand varied very little between east- and west-facing interior-located dwellings. Demand declined and the difference between the average demands of wood and granite covering narrowed with rising building heat capacity, with the average demand exhibited by wood coverings in south-oriented dwellings varying by 37% from lowest to highest building heat capacity.

Demand also declined in all the interior-located dwellings with rising insulation levels. For wood coverings in south-oriented apartments, the average demand was 46% lower in the dwellings with the highest than in those with the lowest insulation; similar values were recorded for the other three orientations.

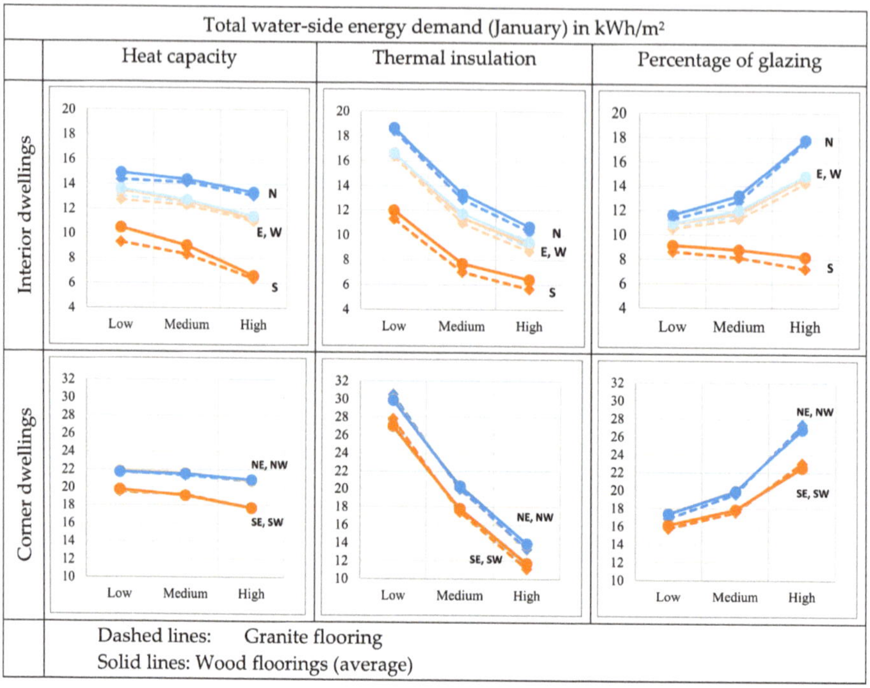


**Figure 11.** Effect of building construction on energy demand.

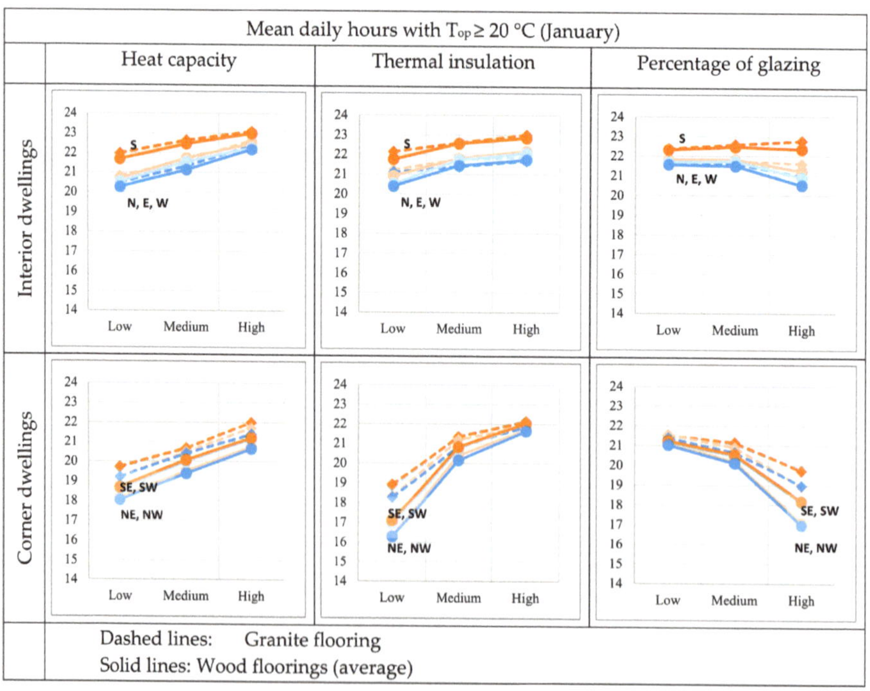


**Figure 12.** Effect of building construction on mean daily hours with Top $\geq$ 20 °C.

The effect of the percentage of glazing on energy demand exhibited a different pattern for south-facing interior dwellings than for the other three orientations. In the former, the solar gains afforded by the higher percentage of glazing offset the greater heat loss across the outer façade wall. In north-facing dwellings with wood coverings and the highest percentage of glazing, the average demand was 53% greater than with the lowest percentage of glazing, whereas in the most-glazed dwellings facing south, the average demand was 10% lower than in the buildings with the least glazing.

Corner-located dwellings facing northeast or northwest, whether wood- or granite-floored, exhibited higher demand than those facing southeast or southwest. Demand declined slightly with building heat capacity, with the average demand for wood coverings in southeast- and southwest-oriented dwellings varying by 10%, from lowest to highest capacity. Similarly, greater insulation levels lowered demand significantly more abruptly than heat capacity, with a 57% decline in average demand found for wood coverings from the lowest to the highest insulation level in southeast- and southwest-oriented dwellings.

The opposite pattern was observed for the percentage of glazing in all orientations: the higher the percentage, the higher the demand, even in south-oriented dwellings, where solar gains failed to offset the outward heat loss across façade walls. In northeast- and northwest-facing corner dwellings with wooden floor coverings, the average demand rose by 54% from the lowest to the highest percentage of glazing.

Based on the graphs in Figure 12, the interior-located dwellings exhibited more mean daily comfort hours than the corner-located dwellings. In the former, whether the flooring was covered with wood or granite, the comfort hours were higher in the south-facing dwellings than in those oriented in any other direction. The higher the heat capacity, the greater the number of comfort hours: in wood-floored, north-oriented dwellings, space conditions were comfortable for around 2 more hours when heat capacity was highest than when it was lowest.

The number of hours also rose consistently with rising insulation level. In north-oriented dwellings with wood floor coverings, space conditions were comfortable approximately 2 more hours in the most than in the least insulated.

In general, the mean number of daily comfort hours was impacted less significantly by the percentage of glazing than by the heat capacity or insulation level. In south-facing dwellings, this number was the same, irrespective of the percentage of glazing, although in wood-covered dwellings facing north, the difference between the lowest and highest percentage was 1 h.

The mean number of daily comfort hours was greatest in corner-located dwellings facing southeast and southwest for wood and granite coverings alike. In such dwellings, a higher heat capacity was also associated with more comfort hours. High-capacity corner dwellings with wood finish floorings had comfortable space conditions for 2.5 h longer than low-capacity dwellings, irrespective of orientation.

The number of comfort hours also rose consistently with rising insulation levels. In all the wood-floored corner dwellings, irrespective of orientation, space conditions were comfortable for 5 h longer when insulation was strongest than when it was weakest.

A higher percentage of glazing in corner locations resulted in a smaller number of hours of comfort in all orientations. For wood coverings in dwellings facing northeast and northwest, the number of comfort hours declined by 4 h between the lowest and highest percentages of glazing.

### 4.3. Effect of Wood Thermal Conductivity

The effects of the wood thermal properties on energy demand, thermal comfort, and start-up lag time are addressed in this section. The conductivity values that performed best on these three parameters are shown.

The thermal inertia (density, specific heat, and conductivity) of all the materials was envisaged in the simulations. Nonetheless, this section addresses the effect of the covering thermal conductivity only in light of the relationship between wood thermal conductivity and density (Section 3.1) and the fact that the same specific heat was considered in all the different wood types. Consequently, variations in thermal conductivity encompass the effect of thermal inertia.

Unfortunately, no simple answers were found to the question of the thermal conductivity values that minimised energy demand or maximised the mean daily number of comfort hours. This lack of cut-and-dried findings, attributable to the difference in the roles played by the dwelling construction characteristics and covering thickness and properties depending on the case, ruled out any straightforward establishment of the conductivities consistently delivering minimum energy demand or maximum comfort.

Figure 13 plots the percentage of cases in which a certain range of thermal conductivities provided the lowest energy demand or the greatest thermal comfort. The bars in Figure 13a denote the percentage of cases in which the lowest energy demand co-existed with operative temperatures of more than or equal to 20 °C for at least 14 h per day. The aim was to find out the conductivity values that reduced energy demand in dwellings where comfort conditions were maintained. Figure 13b, in turn, plots the percentage of cases reaching the highest mean daily number of hours with an operative temperature of higher than or equal to 20 °C. According to the findings in Figure 13a, energy demand was minimised primarily when thermal conductivity was higher than 0.20 W/(m·K), although, in a significant 18% of cases, demand was minimised at conductivities of under 0.1 W/(m·K). The data in Figure 13b show that when wood conductivity was highest (0.20 W/(m·K) to 0.26 W/(m·K)), comfort was maximised in a greater percentage of cases than when it was lowest, although in this case, the highest comfort levels were found in 14% of cases with conductivities under 0.10 W/(m·K) and 16% of cases with values of 0.10 to 0.12 W/(m·K). Therefore, as wood with low thermal conductivity does not necessarily entail poor thermal performance, its use does not need to be ruled out. Rather, depending on the case, it could even be one of the highest performers.

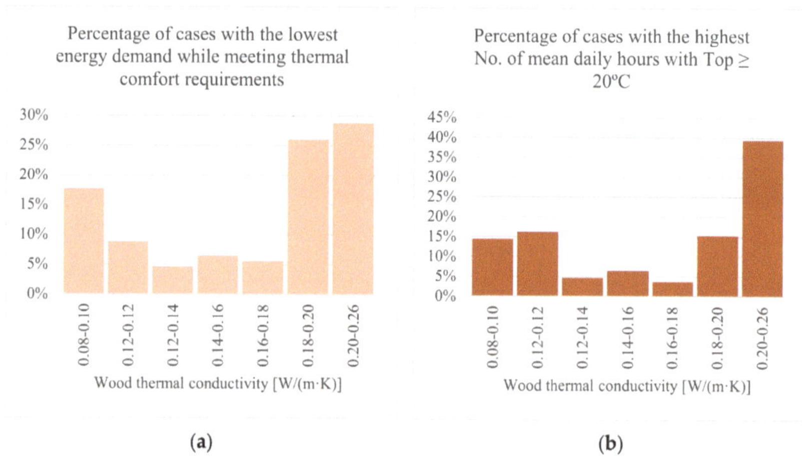


**Figure 13.** Percentage of cases in which a certain range of thermal conductivities performed best in terms of: (**a**) energy demand and (**b**) mean daily comfort hours.

The behaviour observed in Figure 13a is associated with water-side thermal power and the amount of time that the radiant floor heating must be switched on. In some cases, the demand was lower in low-conductivity coverings because they operated at lower power for longer times than high-thermal-conductivity finishes, which operated at higher power for fewer hours. The suggestion is that the use of high-conductivity wood is not always necessary to comply with thermal comfort criteria at low levels of energy demand.

In many cases, low-thermal-conductivity wood meets these criteria while minimising energy demand.

The curves in Figure 14 plot the effect of wood thermal conductivity on start-up lag time for all the dwellings analysed. Line 1 in the figure depicts the conductivity values minimising the start-up lag time; line 2 those prompting a delay of no more than 15 min over the line 1 time (15 min criterion); and line 3 those retarded by no more than 30 min relative to the line 1 time (30 min criterion). The covering consistently observed to minimise start-up lag time (line 1) to comfort conditions in all dwellings was the one with the highest thermal conductivity (0.26 (W/(m·K)). The explanation is simple: raising the temperature in a space in a short period of time depends on the power transferred from the water pipes across the floor covering. Nonetheless, in some dwellings, woods with lower conductivities delivered comfort at lag times that were not substantially longer (lines 2 and 3) and could therefore be adopted as the material of choice. Conductivities of 0.18 W/(m·K) often sufficed to meet the 15-min criterion in well-insulated, low-energy-demand dwellings, and in nearly all cases, such a conductivity value lay within the 30-min criterion. In a few well-insulated, interior-located dwellings, a conductivity value of 0.14 W/(m·K) sufficed to comply with the 30-min criterion.

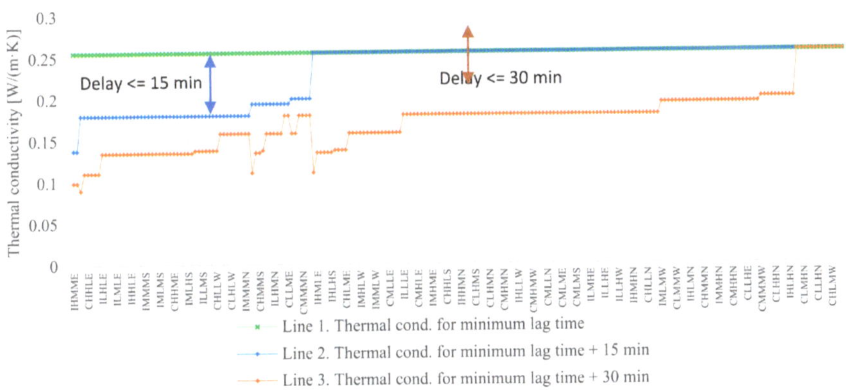


**Figure 14.** Thermal conductivities minimising start-up lag times.

As a rule, conductivity values under 0.26 W/(m·K) complied with the 15-min criterion in well-insulated, interior-located low-thermal-inertia dwellings (the furthest left in the figure), and with the 30-min criterion in nearly all the dwellings.

### 4.4. *Effect of Wood Thermal Resistance*

Given that the thermal resistance in wood coverings of radiant floors is a value routinely used to determine whether these coverings meet the minimum requirements that are thought to ensure correct system operation, the effect of that value on radiant floor performance was studied here.

Floor covering thermal resistance is plotted against energy demand in Figure 15a and thermal comfort in Figure 15b. The points on the three curves in Figure 15a represent, in series 1, the one dwelling of all those simulated that, at the appropriate thermal resistance, exhibited the highest demand; in series 2, these points represent the lowest demand. The series 3 curve denotes the average demand for all the dwellings. Figure 15b plots the mean daily hours during which the operative temperature was higher than or equal to 20 °C for the dwellings depicted in Figure 15a, i.e., each point in a given series in Figure 15b represents the same dwelling in the analogous series in Figure 15a. The hollow blue dots in series 1 denote non-compliance, i.e., cases where the operative temperature was greater than or equal to 20 °C for fewer than 14 h. The lowest thermal resistance value was recorded for the granite floor coverings, while all the other dots plot the findings for the dwellings with wood coverings.

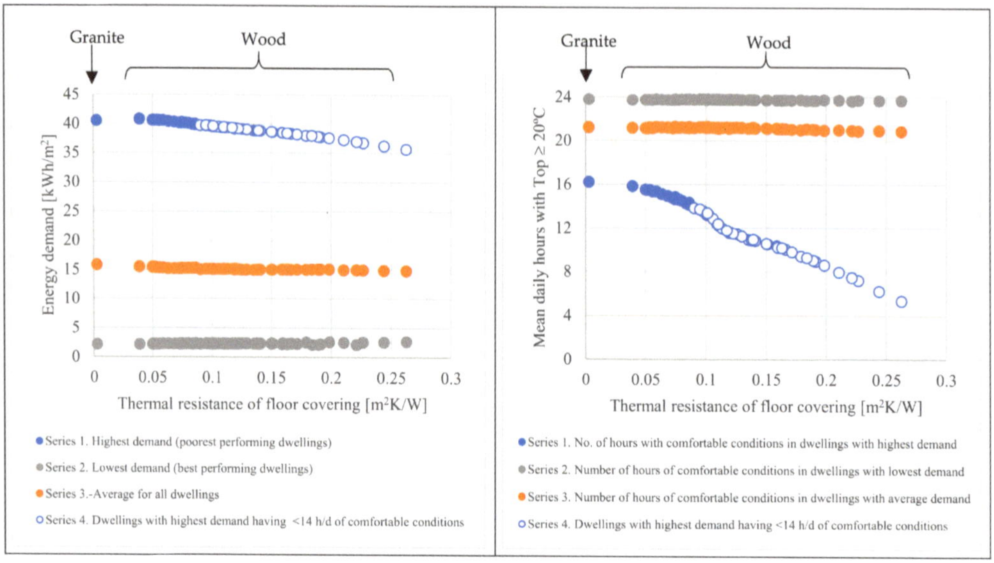


(**a**) (**b**)

**Figure 15.** Effect of covering thermal resistance on (**a**) energy demand and (**b**) mean daily comfort hours in the dwellings plotted in (**a**).

According to Figure 15a, the thermal resistance in the covering had no significant effect on energy demand in the series 2, low-energy-demand (often well-insulated and interior-located) dwellings, or on average energy demand (series 3). By contrast, demand declined with rising thermal resistance in the series 1, high-energy-demand dwellings. The explanation is to be found in the lower amount of energy transferred from the water pipes to the indoor space at higher thermal resistance levels. This decline in heat transfer also translated into a smaller number of comfort hours (series 1, Figure 15b). In these high-energy-demand dwellings, the number of mean daily comfort hours declined with rising resistance and dipped below the minimum 14-h requirement at values of over 0.09 $m^2K/W$. Conversely, in low-demand dwellings (series 2), comfortable conditions prevailed for nearly 24 h daily, irrespective of the thermal resistance of the wood covering, while the average value (series 3) observed was 21 h, also regardless of the thermal resistance values. On average, wood flooring lowered energy demand by 6.4% and daily hours of thermal comfort by a mere 1.6% relative to granite coverings.

As a rule, demand and comfort are inter-related: in buildings with generally low demand, comfort levels are high, and vice-versa. Depending on the dwelling, wood-covering thermal resistance was observed to make practically no difference to the energy demand or mean daily hours of thermal comfort, exhibiting the same, or very nearly the same numbers as granite floors. The conclusion drawn is that a thermal resistance of more than 0.15 $m^2K/W$ in the covering (established as a maximum standard value) is compatible with efficient radiant floor heating.

The effect of wood-covering thermal resistance on start-up lag time is plotted in Figure 16. In this figure, the points on the series 1 curve represent a given dwelling that, at the appropriate thermal resistance, takes the longest, in number of hours, to raise the operative temperature to 20°; the points on the series 2 curve represent the dwellings taking the least time at each thermal resistance. Series 3 plots the average number of hours required to raise all the dwellings to the operative temperature at a given resistance. Again, the granite flooring displayed the lowest thermal resistance value. Thermal resistance was observed to have a non-negligible effect in the dwellings with the shortest start-up lag times (series 2), rising by 37% from the 7.5 h recorded for the granite coverings to 10.25 h for the wood coverings with the highest thermal resistance. The effect on the average start-up lag time (series 3) was also significant, due to the substantial 64% longer lag time in the highest thermal resistance wood cover than in the granite cover. Thermal resistance also induced a steep rise in this parameter in the series 1 dwellings, those with longest start-up lag times, a pattern explained in Appendix C. At resistance values of over 0.14 $m^2K/W$, comfortable conditions were not reached in these dwellings within the 72 h limit established in this study. Wood thermal resistance had a much more significant impact on start-up lag time than it had on energy demand and comfort hours.

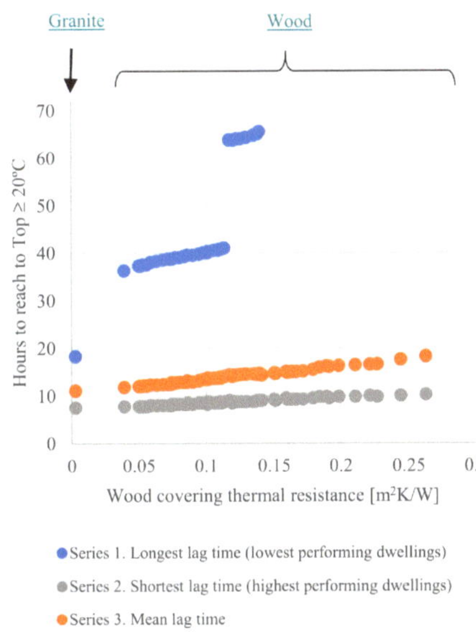


**Figure 16.** Effect of covering thermal resistance on start-up lag time.

## 5. Conclusions

Earlier studies on the most prominent variables involved in radiant flooring design established covering properties and thickness as the primary determinants of thermal performance.

The coverings most commonly laid over radiant flooring are porcelain or ceramic tiles, natural stone, and wood. Given their high thermal conductivity, natural stone and ceramic materials may initially be expected to be more thermally favourable than wood. Design engineers have consequently challenged the suitability of wood, with its fairly high thermal resistance, as the covering of choice in radiant floors, particularly in the coldest climates.

This study assessed the impact of different wood coverings on the thermal performance of radiant floors compared to high-conductivity natural stone (granite) in terms of three parameters: energy demand, thermal comfort, and start-up lag time. The effect of the coverings' thermal properties, in conjunction with the building construction characteristics, on these parameters was explored.

A simplified building model was developed and coupled to an experimentally validated radiant flooring model and used to run a total of 26,352 simulations. The total simulation computing time came to 37 h.

The main conclusions that may be drawn from the results are as follows:

- For most dwellings, the thermal properties of the wood affected energy demand and thermal comfort only scantly. Wood coverings delivered mostly similar and, in some cases, better results than granite coverings in those two respects. The impact of wood properties on demand and comfort was only significant for corner-located, poorly insulated dwellings. In such cases, granite flooring exhibited consistently higher thermal performance, although when appropriate wood properties were chosen, they proved to be a very close competitor to granite. These findings were not always associated with high thermal conductivity only. The average energy demand was observed to be lower in the wood than in the granite coverings in 25% of the dwellings simulated. Similarly, on average, wood lagged behind granite in thermal comfort by less than 1 h in 50% of the dwellings.
- Wood properties played a more substantial role in start-up lag times than in demand or thermal comfort, although the general pattern was much the same: for most dwellings, none of the wood radiant floors simulated lengthened the lag time substantially. As a rule, the dwellings where energy loss was greatest (corner dwellings, those with medium or high percentages of glazing, those that were minimally or only moderately insulated and oriented toward the north, east, or west) required a suitable choice of wood properties to prevent lag times from rising inordinately. The average difference in start-up lag time between the wood and granite coverings was less than 3 h for 75% of the dwellings. It is not possible to set a general time limit beyond which the start-up lag time is unacceptable, as this depends on the use of the dwelling, e.g., whether it is for tourism or for continuous use.
- Despite the scant impact of wood properties in most cases, the pursuit of simple rules to determine which properties would be the most suitable under given circumstances proved to be futile because the combination of wood properties, thickness, and dwelling construction characteristics followed no consistent pattern. The conclusion drawn, therefore, was that cover properties should be studied case-by-case to determine those expected to deliver the best thermal performance.
- In most cases, the highest thermal conductivity values were found to minimise energy demand, maximise comfort, and shorten start-up lag times. Energy demand was minimised primarily when thermal conductivity was higher than 0.20 W/(m·K), although in a significant 18% of cases, demand was minimised at conductivities of under 0.1 W/(m·K). When wood conductivity was highest (0.20 W/(m·K) to 0.26 W/(m·K)), comfort was maximised in a greater percentage of cases than when it was lowest, although in this case, the highest comfort levels were found in 14% of cases with conductivities under 0.10 W/(m·K) and 16% of cases with values of 0.10 to 0.12 W/(m·K). Conductivities of 0.18 W/(m·K) often increased the start-up lag time by only 15 min in well-insulated, low-energy-demand dwellings and in nearly all such dwellings, the conductivity value increases it by 30 min. On these grounds, wood with high thermal conductivity cannot be said to always be necessary for the design of radiant floors.
- One of the conclusions of this study that may be of most immediate interest is that the lowest thermal conductivity and thickest floor covering, i.e., wood flooring with the highest thermal resistance (even more of 0.15 $m^2K/W$ value) does not significantly affect the energy demand or thermal comfort. On average, wood flooring lowered energy demand by 6.4% and daily hours of thermal comfort by a mere 1.6% relative to granite coverings.
- The findings on the thermal resistance of wood coverings provided no justification for establishing an upper limit that must not be exceeded in the selection of woods for radiant floors. Although European standard EN 1264-2 [85] makes no provision for coverings with thermal resistance values of over 0.15 $m^2K/W$, they are not explicitly prohibited. In fact, thermal resistance values higher than 0.15 $m^2K/W$ did not raise energy demand significantly, nor did they lower the number of comfort hours in the vast majority of the conditions simulated. This study consequently suggests that the

standard should be revised and the reference to that value deleted, since manufacturers have misconstrued it to be a limit not to be exceeded in the design of wood-covered radiant flooring.
- It was shown that the thermal behavior of radiant floor heating systems is closely linked to building conditions and, therefore, it is necessary to carry out a technical study for each particular case.

Our approach is based on the following three variables: (1) energy demand, (2) thermal comfort, and (3) start-up lag period. From our point of view, with these three variables, the overall performance of radiant floor heating can be clearly evaluated when its behaviour in transient mode coupled to the building needs to be assessed. The objective, as with any heating system, is to achieve thermal comfort for as long as possible with the lowest energy consumption. This can be analysed by using these two variables: the time in which comfort conditions are achieved and the energy demand of the radiant floor. On the other hand, the time required to achieve thermal comfort from an initial condition that is far from comfortable can be relevant for certain applications. Other approaches, such as the analysis of the power of the systems, or the surface temperature distribution of the floor (for hot and cold spots) may be of interest for other specific studies.

Further work along these lines should be conducted, since the present case study yielded results applicable only to the climate of Madrid and similar locations. The research should be extended to address other locations to determine whether the conclusions are climate-dependent.

**Author Contributions:** Á.R.-P.: methodology, software, investigation, data curation, formal analysis, writing—original draft. E.Á.R.J.: methodology, software and validation, investigation, data curation, formal analysis, writing—review and editing. M.C.G.: investigation, validation, writing—review and editing. J.A.T.R.: conceptualization, methodology, validation, formal analysis, supervision, writing—review and editing. All authors have read and agreed to the published version of the manuscript.

**Funding:** This research was funded by the European Agricultural Fund for Rural Development (EAFRD) and Spain's Ministry of Agriculture, Fishing and Food under the project 'Grupo Operativo Madera Construcción Sostenible', ref. 20180020012335.

**Institutional Review Board Statement:** Not applicable.

**Informed Consent Statement:** Not applicable.

**Acknowledgments:** The information for validating the model furnished by radiant flooring industry leader Uponor is gratefully acknowledged.

**Conflicts of Interest:** On behalf of all the authors, the corresponding author states that there is no conflict of interest.

## Nomenclature

| | | |
|---|---|---|
| $A$ | Area | $m^2$ |
| $ACH$ | Air changes per hour | $h^{-1}$ |
| $C$ | Thermal capacitance | J/K |
| $C_P$ | Specific heat | J/(kgK) |
| $f$ | Response factors of the radiant floor | |
| F | View factor | |
| $GR$ | Solar incident global irradiation | $W/m^2$ |
| $h$ | Convective heat transfer coefficient | $W/(m^2K)$ |
| $\dot{q}$ | Heat flux density | $W/m^2$ |
| $\dot{Q}$ | Heat transfer rate | W |
| $R$ | Thermal resistance | $(m^2K)/W$ |
| $SHGC$ | Solar heat gain coefficient of windows, including frames | |
| $t$ | Time | s |
| $T$ | Temperature | °C |

| | | |
|---|---|---|
| $U$ | Overall heat transfer coefficient | $W/(m^2K)$ |
| $\dot{V}$ | Volumetric air flow rate | $m^3/s$ |
| **Greek symbols** | | |
| $\rho$ | Density | $kg/m^3$ |
| **Subscripts** | | |
| air | Air | |
| build | Building | |
| c | Convective | |
| cr | Convective and radiant | |
| eq | Equivalent | |
| e | Exterior enclosures | |
| floor | Floor | |
| i | Indoor, Interior partitions | |
| r | radiant | |
| s | Surface | |
| sa | Sol-air | |
| w | Window | |
| walls | walls | |
| 1 | Top surface of radiant floor | |
| 2 | Bottom surface of radiant floor | |
| 3 | Pipe surface of radiant floor | |

## Appendix A. Development of the Simulation Models

This appendix describes in more detail the simulation models used in this study.

### *Appendix A.1. Detailed Radiant Floor Thermal Model*

As described in Section 2.1, radiant floor heating was modelled using the response factor method to calculate the heat fluxes across the three surfaces. In this method, the actual excitation of an element is represented as the superposition of triangular pulses spaced at the time-step interval used in the simulation, adopting as the size the real surface temperature at the respective time step. Given that the problem is linear, the superposition principle can also be applied and the real responses on each surface can be determined with simple algebraic equations (Equation (A1)). The outcome is the delivery of the same precision as obtained in detailed methods or high-order models with less computing time.

The heat flux on each surface i can be calculated in a given time-step n from the following equation:

$$\dot{q}_{s,i}(n) = \sum_{j=0}^{\infty} f_{1,i}(j)\ T_{s,1}(n-j) + \sum_{j=0}^{\infty} f_{2,i}(j)\ T_{s,2}(n-j) + \sum_{j=0}^{\infty} f_{3,i}(j)\ T_{s,3}(n-j) \quad \text{(A1)}$$

In Equation (A1) $f_{1,i}$, $f_{2,i}$, and $f_{3,i}$ are the response factors at surface i when the unit triangular pulse excitation is applied at surfaces 1 (top), 2 (bottom), and 3 (pipe), respectively. These response factors represent the averaged heat fluxes for surface i calculated from the surface distribution of the heat flux derived by the transient 2D finite element-based model of ANSYS APDL when unit triangular temperature excitations are applied. To return these surface average values, the node heat fluxes are weighted by the amount of element surface area associated with each node. This operation is internally performed by ANSYS APDL.

As may be inferred from Equation (A1), calculating the heat flux in a given time step calls for a large number of response factors, since the summation must be carried back in time for a period long enough to ensure accurate results. Nonetheless, after a certain number k of response factors, the ratio between the response factor on the surface i when excitation is applied on surface × in a given time step, Ci(k) = fx,i(k)/fx,i(k − 1), and the immediately preceding factoris practically equal: Ci(k) ≈ Ci(k − 1). C is known as the common ratio. It therefore suffices to determine a certain number of response factors to calculate the rest from the common ratio. The summations in Equation (A1) can consequently be simplified by using the common ratio, as exemplified by Equation (A2) for the top surface (1), where the summation is limited to k terms.

$$\begin{aligned}
&\dot{q}_{s,1}(n) = \\
&f_{11}(0)\ T_{s,1}(n) + \sum_{j=1}^{k} T_{s,1}(n-j)[f_{11}(j) - C_1 \cdot f_{11}(j-1)\ ] + C_1 \cdot \dot{q}_{1,1}(n-1) + \\
&f_{21}(0)\ T_{s,2}(n) + \sum_{j=1}^{k} T_{s,2}(n-j)[f_{21}(j) - C_1 \cdot f_{21}(j-1)\ ] + C_1 \cdot \dot{q}_{2,1}(n-1) + \\
&f_{31}(0)\ T_{s,3}(n) + \sum_{j=1}^{k} T_{s,3}(n-j)[f_{31}(j) - C_1 \cdot f_{31}(j-1)\ ] + C_1 \cdot \dot{q}_{3,1}(n-1)
\end{aligned} \tag{A2}$$

To simplify further, the terms referring to time steps prior to the present step can be grouped, yielding:

$$\dot{q}_{s,1}(n) = f_{11}(0)\ T_{s,1}(n) + f_{21}(0)\ T_{s,2}(n) + f_{31}(0)\ T_{s,3}(n) + h_1 \tag{A3}$$

where $h_1$ groups all the terms in Equation (A2) referring to time step n − 1.

Analogously, the heat fluxes across the bottom (2) and pipe (3) surfaces can be calculated as follows:

$$\dot{q}_{s,2}(n) = f_{12}(0)\ T_{s,1}(n) + f_{22}(0)\ T_{s,2}(n) + f_{32}(0)\ T_{s,3}(n) + h_2 \tag{A4}$$

$$\dot{q}_{s,3}(n) = f_{13}(0)\ T_{s,1}(n) + f_{23}(0)\ T_{s,2}(n) + f_{33}(0)\ T_{s,3}(n) + h_3 \tag{A5}$$

*Appendix A.2. Building the Thermal Model*

The radiant floor model described in the preceding section had to be coupled to a building transient thermal behaviour model to analyse the performance of the floors studied when operating in different dwellings. As noted in Section 2.2, with the model used to describe the building, the equivalent temperature (Teq) and resistance (Req) can be defined as:

$$T_{eq}(t) = R_{eq} \cdot \left( \frac{T_{s,1}(t)}{R_{cr,1}} + \frac{T_{eq,i,sa}(t)}{R_{eq,i,sa}} + \dot{q}_{cr,i}(t) \right) \tag{A6}$$

$$R_{eq} = \left( \frac{R_{cr,1} \cdot R_{eq,i,sa}}{R_{cr,1} + R_{eq,i,sa}} \right) \tag{A7}$$

where:

$$T_{eq,i,sa}(t) = R_{eq,i,sa} \cdot \left( \frac{T_i(t)}{R_i} + \frac{T_{sa}(t)}{R_e} \right) \tag{A8}$$

$$R_{eq,i,sa} = \left( \frac{R_i \cdot R_e}{R_i + R_e} \right) \tag{A9}$$

Applying the energy balance at the capacitance node yields:

$$C \cdot \frac{dT_{cr,i}(t)}{dt} = \frac{T_{eq}(t) - T_{cr,i}(t)}{R_{eq}} \tag{A10}$$

The mathematical procedure for analytically finding the capacitance node temperature in the above differential equation, described by one of the present authors in [60], is expressed in Equation (A11). In this procedure, the excitation is assumed to vary linearly at every simulation time step; this is a reasonable premise inasmuch as excitation data are normally given as discrete values spaced at time-step intervals.

$$\begin{aligned}
T_{cr,i}(t) = {} & T_{eq}(t) + \tfrac{\tau}{\Delta t} \cdot \left( T_{eq}(t-\Delta t) - T_{eq}(t) \right) + \\
& \left[ T_{cr,i}(t-\Delta t) - \left[ T_{eq}(t-\Delta t) + \tfrac{\tau}{\Delta t} \cdot \left( T_{eq}(t-\Delta t) - T_{eq}(t) \right) \right] \right] \cdot e^{-\frac{\Delta t}{\tau}}
\end{aligned} \tag{A11}$$

In Equation (A11), τ = C·Req. If Equation (A6) is substituted into Equation (A11) and the terms are regrouped, Tcr,i can be expressed as a function of surface temperature Ts,1 only as:

$$T_{cr,i} = A_1\, T_{s,1} + B_1 \quad \text{(A12)}$$

where:

$$A_1 = \frac{R_{eq}}{R_{cr,1}}\left[\left(1-\frac{\tau}{\Delta t}\right)+\frac{\tau}{\Delta t}e^{-\frac{\Delta t}{\tau}}\right]$$

$$B_1 = \frac{\tau}{\Delta t}(t-\Delta t)T_{eq} + \left(\frac{R_{eq}}{R_{eq,i,sa}}T_{eq,i,sa}(t) + R_{eq}\cdot\dot{q}_{cr,i}(t)\right)\left[\left(1-\frac{\tau}{\Delta t}\right)+\frac{\tau}{\Delta t}e^{-\frac{\Delta t}{\tau}}\right] + \left[T_{cr,i}(t-\Delta t) - T_{eq}(t-\Delta t)\left(1+\frac{\tau}{\Delta t}\right)\right]e^{-\frac{\Delta t}{\tau}}$$

## Appendix B. Values of Building Parameters Chosen for Simulations

### *Appendix B.1. Heat Capacity (C)*

A building's heat capacity is the result of the combination of density, specific heat, building element thickness, and furnishings. In Table A1 in the International and European standard ISO EN 52016-1 [82], five default values for classifying heat capacity from 'very light' to 'very heavy' are established. Only three heat capacity levels were adopted from the standard for this study (Table A1 heat capacity classification from [82]).

**Table A1.** Heat capacity classification from (UNE-EN ISO 52016-1:2017 2017 [82]).

| Class | Effective Heat Capacity [kJ/(m$^2$K)] |
|---|---|
| Very light | 80 |
| Medium | 165 |
| Very heavy | 370 |

### *Appendix B.2. Overall Heat Transfer Coefficient (U-Value)*

In the analysis of a broad spectrum of dwellings, three insulation levels, low, medium, and high, were applied to all the envelope elements (Table A2). The aim was to explore a wide range of overall indoor–outdoor heat transfer coefficients (U-value) from very demanding to medium and even low, the third to cover the conditions prevailing in most existing buildings, where insulation values are lower than presently required by building energy codes. The reference values were drawn from the guidelines set out in Spain's Technical Building Code [86].

**Table A2.** U-values and solar heat gain coefficients used for the simulations.

| | | U-Value [W/(m$^2$K)] | | |
|---|---|---|---|---|
| Insulation Level | Wall | Roof | Windows | Windows SHGC |
| Low | 0.79 | 0.47 | 5.70 | 0.72 |
| Medium | 0.53 | 0.31 | 2.80 | 0.63 |
| High | 0.30 | 0.16 | 1.60 | 0.49 |

### *Appendix B.3. Overall Heat Transfer Coefficient between Indoor Spaces*

The values adopted for the overall heat transfer coefficient for heat transfer between indoor spaces were U = 1.56 W/(m$^2$K) for vertical partitions, characterised by fired clay brick enclosures, and U = 0.71 W/(m$^2$K) for horizontal partitions, comprising floating wood flooring, a layer of mortar, impact insulation, ceramic inter-joist filling, an air chamber, and plasterboard. Both values were drawn from Spanish legislation [80].

*Appendix B.4. Area of Interior and Exterior Enclosures*

The areas of each building element were defined as in Table A3.

**Table A3.** Area of building elements by dwelling location, in $m^2$.

| | Corner Dwelling | Interior Dwelling |
|---|---|---|
| Floor | 90.25 | 90.25 |
| Interior vertical partitions to adyacent spaces | 57 | 85.5 |
| Interior horizontal partitions to upper adyacent spaces | 0 | 90.25 |
| Façade wall | 57 | 28.5 |
| Window area * | 8.55/17.1/34.2 | 4.275/8.55/17.1 |
| Roof | 90.25 | 0 |

* Three values are shown corresponding to 15%, 30% and 80% of glazed area.

*Appendix B.5. Radiant Floor Surface Convection-Radiation Heat Transfer Coefficients*

According to Table 3.8 in CIBSE [61] and Table E.1 in [87], the surface thermal resistance in indoor flooring with upward heat flux reached 0.10 $m^2K/W$, deemed here to be the standard value for radiant floors. The same sources cite surface thermal resistance for an indoor ceiling with downward heat flux as 0.17 $m^2K/W$. These were the conditions assumed here to prevail, for the ceiling at issue covered the apartment immediately below the one studied.

*Appendix B.6. Building Model Boundary Conditions*

Appendix B.6.1. Sol-Air Temperature

The sol-air temperatures were calculated on the grounds of hours of solar irradiance and the outdoor air and sky temperatures drawn from the climate data sheet for Madrid available in [88], subsequently interpolated for conversion to a 15-min time step.

The mean radiant temperature was calculated assuming a view factor for the façade walls relative to both sky and ground of 0.5, and the ground and air temperatures were assumed to be equal. The view factor adopted for the roof relative to the sky was 1 and, relative to the ground, it was 0.

The opaque surface absorptivity was assumed to be 0.7.

Appendix B.6.2. Adjacent Indoor Space Temperatures

The spaces adjacent to the dwelling were assumed to be heated, with the timing the same as in radiant floor system operation, i.e., during the day, from 08:00 to 23:00, the temperature was 20 °C, declining at night to 17 °C, in keeping with Spanish legislation.

Appendix B.6.3. Internal Gain and Ventilation Values

In the model, the convection–radiation heat transfer rate, $\dot{Q}_{cr}$, was calculated as the sum of the indoor source (occupancy and lighting) internal gain values, $\dot{Q}_{int}$ (Table A4), the solar gains due to solar irradiance received through the windows, $\dot{Q}_{sol,int}$, and the ventilation, $\dot{Q}_{vent}$, due to heat gain or loss as a result of indoor–outdoor air changes.

**Table A4.** Indoor sources [W/m$^2$].

| Time of Day | 1 | 2 | 3 | 4 | 5 | 6 | 7 | 8 | 9 | 10 | 11 | 12 | 13 | 14 | 15 | 16 | 17 | 18 | 19 | 20 | 21 | 22 | 23 | 24 |
|---|---|---|---|---|---|---|---|---|---|---|---|---|---|---|---|---|---|---|---|---|---|---|---|---|
| Occupancy | 2.15 | 2.15 | 2.15 | 2.15 | 2.15 | 2.15 | 2.15 | 0.54 | 0.54 | 0.54 | 0.54 | 0.54 | 0.54 | 0.54 | 0.54 | 1.08 | 1.08 | 1.08 | 1.08 | 1.08 | 1.08 | 1.08 | 1.08 | 2.15 |
| Illumination | 0.44 | 0.44 | 0.44 | 0.44 | 0.44 | 0.44 | 0.44 | 1.32 | 1.32 | 1.32 | 1.32 | 1.32 | 1.32 | 1.32 | 1.32 | 1.32 | 1.32 | 1.32 | 2.2 | 4.4 | 4.4 | 4.4 | 4.4 | 2.2 |
| Appliances | 0.44 | 0.44 | 0.44 | 0.44 | 0.44 | 0.44 | 0.44 | 1.32 | 1.32 | 1.32 | 1.32 | 1.32 | 1.32 | 1.32 | 1.32 | 1.32 | 1.32 | 1.32 | 2.2 | 4.4 | 4.4 | 4.4 | 4.4 | 2.2 |
| Total | 3.03 | 3.03 | 3.03 | 3.03 | 3.03 | 3.03 | 3.03 | 3.18 | 3.18 | 3.18 | 3.18 | 3.18 | 3.18 | 3.18 | 3.18 | 3.72 | 3.72 | 3.72 | 5.48 | 9.88 | 9.88 | 9.88 | 9.88 | 6.55 |

The solar gains through the windows, $\dot{Q}_{sol,int}$, were calculated as per Equation (A13):

$$\dot{Q}_{sol,int} = A_w \cdot GR \cdot SHGC \tag{A13}$$

The ventilation-mediated outdoor–indoor heat change, $\dot{Q}_{vent}$, was calculated based on the expression for ventilation gains or losses as:

$$\dot{Q}_{vent} = \dot{V} \cdot \rho_{air} \cdot Cp_{air} \cdot (T_{air,e} - T_{cr,i}) \tag{A14}$$

The air flow rate was estimated to ensure a mean $CO_2$ concentration of under 900 ppm, a requirement that can be met with a flow rate of approximately 12.5 L/(s·person) [61]. The total air flow rate for the dwelling given a mean occupancy of three occupants, a value consistent with the indoor sources attributable to occupancy assumed, came to 3.5 L/s or 0.5 ACH.

*Appendix B.7. Radiant Floor Operation. Timing, Set Points, and Water Temperature*

Heating systems may shut down and started up several times during the day, either after reaching a set point temperature or in line with scheduled operating times. The set point temperature established for this study was 20 °C, measured at the node representing the indoor space in the dwelling (Tcr,i).

The operating times were defined, based on Spanish legislation [89], as 08:00 to 23:00; outside of this window, the heating was assumed to be off.

During radiant floor operation, the mean water temperature in the pipes was assumed to be 40 °C, whereas, when it was off, the water temperature varied freely.

## Appendix C. Explanation of Start-Up Lag Time Discontinuity

The discontinuities in the start-up lag times with rising thermal resistance in the poorest-performing dwellings represented on the curves in Figure 16 are explained graphically in Figure A1. In this figure, the operative temperature is plotted against the lag time for two coverings in one of the poorest-performing dwellings at the point at which one of the discontinuities was observed. The sole (and small) difference between the two cases was the thermal resistance of the covers, 0.114 m$^2$K/W in one and 0.117 m$^2$K/W in the other. Although the operative temperatures were practically identical in the two, enlarging the area of the discontinuity revealed that when the flooring resistance was 0.114 m$^2$K/W, the operative temperature was slightly higher and reached 20 °C within the 41 h limit, whereas, when the resistance was 0.117 m$^2$K/W, the operative temperature reached after 141 h was very close, but not equal, to 20 °C. This took a further 24 h. This behaviour was observed in the poorest-performing dwellings only, where the indoor operative temperatures declined sharply overnight due to outward heat loss.

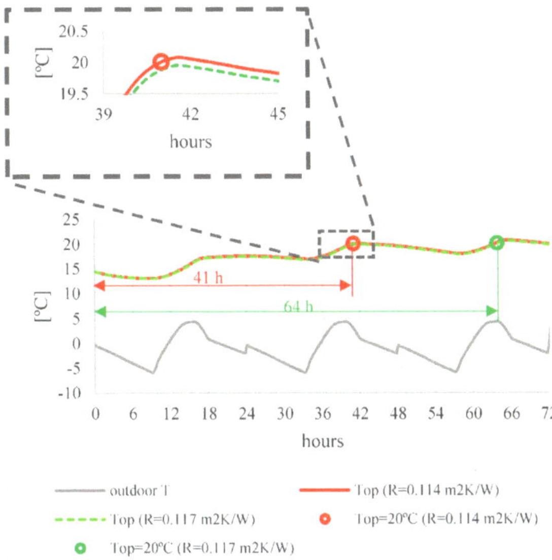


**Figure A1.** Explanation of start-up lag time discontinuity (variation in operative temperature in one of the poorest-performing dwellings with two coverings, differing in thermal resistance: 0.114 $m^2W/K$ and 0.117 $m^2W/K$.

## References


1. Chu, W.-S.; Kim, M.-S.; Lee, K.-T.; Bhandari, B.; Lee, G.-Y.; Yoon, H.-S.; Kim, H.-S.; Park, J.-I.; Bilegt, E.; Lee, J.-Y.; et al. Design and performance evaluation of Korean traditional heating system—Ondol: Case study of Nepal. *Energy Build.* **2017**, *138*, 406–414. [CrossRef]
2. Zhuang, Z.; Li, Y.; Chen, B.; Guo, J. Chinese kang as a domestic heating system in rural northern China—A review. *Energy Build.* **2009**, *41*, 111–119. [CrossRef]
3. Rook, T. The development and operation of Roman hypocausted baths. *J. Archaeol. Sci.* **1978**, *5*, 269–282. [CrossRef]
4. Bean, R.; Olesen, B.W.; Kim, K.W. History of Radiant Heating and Cooling Systems (Part 1). *Ashrae J.* **2010**, *52*, 40–45. Available online: http://search.ebscohost.com/login.aspx?direct=true&db=egs&AN=47756416&site=eds-live (accessed on 1 September 2021).
5. Bean, R.; Olesen, B.W.; Kim, K.W. History of Radiant Heating and Cooling Systems (Part 2). *Ashrae J.* **2010**, *52*, 50–55.
6. Sarbu, I.; Mirza, M.; Crasmareanu, E. Performance of Radiant Heating Systems of Low-Energy Buildings. *IOP Conf. Ser. Mater. Sci. Eng.* **2017**, *245*, 32088. [CrossRef]
7. History of Underfloor Heating. Underfloor Heating Past to Present, (n.d.). Available online: https://www.easyflow.org.uk/history-of-underfloor-heating/ (accessed on 8 July 2021).
8. Rhee, K.-N.; Olesen, B.W.; Kim, K.W. Ten questions about radiant heating and cooling systems. *Build. Environ.* **2017**, *112*, 367–381. [CrossRef]
9. Lin, B.; Wang, Z.; Sun, H.; Zhu, Y.; Ouyang, Q. Evaluation and comparison of thermal comfort of convective and radiant heating terminals in office buildings. *Build. Environ.* **2016**, *106*, 91–102. [CrossRef]
10. Izquierdo, M.; Camacho, P.D.A. Solar heating by radiant floor: Experimental results and emission reduction obtained with a micro photovoltaic–heat pump system. *Appl. Energy* **2015**, *147*, 297–307. [CrossRef]
11. Yang, F.; Liu, J.; Sun, Q.; Cheng, L.; Wennersten, R. Simulation analysis of household solar assistant radiant floor heating system in cold area. *Energy Procedia* **2019**, *158*, 631–636. [CrossRef]
12. Sebarchievici, C.; Dan, D.; Sarbu, I. Performance Assessment of a Ground-coupled Heat Pump for an Office Room Heating using Radiator or Radiant Floor Heating Systems. *Procedia Eng.* **2015**, *118*, 88–100. [CrossRef]
13. Zhang, L.; Huang, X.; Liang, L.; Liu, J. Experimental study on heating characteristics and control strategies of ground source heat pump and radiant floor heating system in an office building. *Procedia Eng.* **2017**, *205*, 4060–4066. [CrossRef]
14. Pajchrowski, G.; Noskowiak, A. Thermal conductivity of wooden floors in the context of underfloor heating system applications. *Drewno* **2018**, *61*, 145–152.
15. Rhee, K.-N.; Kim, K.W. A 50 year review of basic and applied research in radiant heating and cooling systems for the built environment. *Build. Environ.* **2015**, *91*, 166–190. [CrossRef]

16. Zhang, D.; Cai, N.; Cui, X.; Xia, X.; Shi, J.; Huang, X. Experimental investigation on model predictive control of radiant floor cooling combined with underfloor ventilation system. *Energy* **2019**, *176*, 23–33. [CrossRef]
17. Shin, M.-S.; Rhee, K.-N.; Jung, G.-J. Optimal heating start and stop control based on the inferred occupancy schedule in a household with radiant floor heating system. *Energy Build.* **2020**, *209*, 109737. [CrossRef]
18. Chen, Q.; Li, N.; Feng, W. Model predictive control optimization for rapid response and energy efficiency based on the state-space model of a radiant floor heating system. *Energy Build.* **2021**, *238*, 110832. [CrossRef]
19. Tahersima, M.; Tikalsky, P. Experimental and numerical study on heating performance of the mass and thin concrete radiant floors with ground source systems. *Constr. Build. Mater.* **2018**, *178*, 360–371. [CrossRef]
20. Krajčík, M.; Šikula, O. Heat storage efficiency and effective thermal output: Indicators of thermal response and output of radiant heating and cooling systems. *Energy Build.* **2020**, *229*, 110524. [CrossRef]
21. Jeon, J.; Jeong, S.-G.; Lee, J.-H.; Seo, J.; Kim, S. High thermal performance composite PCMs loading xGnP for application to building using radiant floor heating system. *Sol. Energy Mater. Sol. Cells* **2012**, *101*, 51–56. [CrossRef]
22. Zhang, D.; Cai, N.; Wang, Z. Experimental and numerical analysis of lightweight radiant floor heating system. *Energy Build.* **2013**, *61*, 260–266. [CrossRef]
23. Zhou, G.; He, J. Thermal performance of a radiant floor heating system with different heat storage materials and heating pipes. *Appl. Energy* **2015**, *138*, 648–660. [CrossRef]
24. Fu, W.; Lu, Y.; Zhang, R.; Liu, J.; Zhang, T. Developing NaAc·3H2O-based composite phase change material using glycine as temperature regulator and expanded graphite as supporting material for use in floor radiant heating. *J. Mol. Liq.* **2020**, *317*, 113932. [CrossRef]
25. Mohammadzadeh, A.; Kavgic, M. Multivariable optimization of PCM-enhanced radiant floor of a highly glazed study room in cold climates. *Build. Simul.* **2019**, *13*, 559–574. [CrossRef]
26. González, B.; Prieto, M. Radiant heating floors with PCM bands for thermal energy storage: A numerical analysis. *Int. J. Therm. Sci.* **2021**, *162*, 106803. [CrossRef]
27. Larwa, B.; Cesari, S.; Bottarelli, M. Study on thermal performance of a PCM enhanced hydronic radiant floor heating system. *Energy* **2021**, *225*, 120245. [CrossRef]
28. Cho, J.; Park, B.; Lim, T. Experimental and numerical study on the application of low-temperature radiant floor heating system with capillary tube: Thermal performance analysis. *Appl. Therm. Eng.* **2019**, *163*, 114360. [CrossRef]
29. Sattari, S.; Farhanieh, B. A parametric study on radiant floor heating system performance. *Renew. Energy* **2006**, *31*, 1617–1626. [CrossRef]
30. Rambaldi, E.; Prete, F.; Timellini, G. Thermal and Acoustic Performances of Porcelain Stoneware Tiles, n.d. Available online: www.qualicer.org (accessed on 13 March 2020).
31. Zhao, H.Q.; Wang, Z.H.; Zhang, L.S. Influence of Covering Layer on Surface Temperature of Floor Radiant Heating System. *Appl. Mech. Mater.* **2012**, *204–208*, 4260–4263. [CrossRef]
32. Seo, J.; Jeon, J.; Lee, J.-H.; Kim, S. Thermal performance analysis according to wood flooring structure for energy conservation in radiant floor heating systems. *Energy Build.* **2011**, *43*, 2039–2042. [CrossRef]
33. Athienitis, A.; Chen, Y. The effect of solar radiation on dynamic thermal performance of floor heating systems. *Sol. Energy* **2000**, *69*, 229–237. [CrossRef]
34. UNE. *Water Based Surface Embedded Heating and Cooling Systems—Part 1: Definitions and Symbols*; UNE-EN 1264-1:2012; AENOR; UNE: Madrid, Spain, 2012.
35. Meteonorm. 2005. Available online: http://meteonorm.com/ (accessed on 1 September 2012).
36. Xu, X.; Wang, S. A simplified dynamic model for existing buildings using CTF and thermal network models. *Int. J. Therm. Sci.* **2008**, *47*, 1249–1262. [CrossRef]
37. Building Environment Design—Design, Dimensioning, Installation and Control of Embedded Radiant Heating and Cooling Systems—Part 4: Dimensioning and Calculation of the Dynamic Heating and Cooling Capacity of Thermo Active Build. ISO—ISO 11855-4:2012. 2012. Available online: https://www.iso.org/standard/52410.html (accessed on 3 September 2020).
38. Yu, G.; Chen, H.; Xiong, L.; Du, C. A simplified dynamic model based on response factor method for thermal performance analysis of capillary radiant floors. *Sci. Technol. Built Environ.* **2019**, *25*, 450–463. [CrossRef]
39. Strand, R.K. *Heat Source Transfer Functions and Their Application to Low Temperature Radiant Heating Systems*; University of Illinois at Urbana Champaign, Department of Mechanical and Industrial Engineering: Champaign, IL, USA, 1995.
40. Strand, R.K.; Pedersen, C.O. Implementation of a radiant heating and cooling model into an integrated building energy analysis program. *Ashrae Trans.* **1997**, *103*, 949–958. Available online: https://experts.illinois.edu/en/publications/implementation-of-a-radiant-heating-and-cooling-model-into-an-int (accessed on 9 March 2021).
41. MItalas, G.P.; Stephenson, D.G. Room Thermal Response Factors. *Ashrae Trans.* **1967**, *73 Pt 1*, 1.1–1.7.
42. Mitalas, G.P. Calculation of transient heat flow through walls and roofs. *Ashrae Trans.* **1968**, *74*, 182–188.
43. Underwood, C.P.; Yik, F.W.H. *Modelling Methods for Energy in Buildings*; Blackwell Publishing Ltd.: Oxford, UK, 2004.
44. Engineering Simulation & 3D Design Software | Ansys, (n.d.). Available online: http://www.ansys.com (accessed on 3 September 2018).
45. Li, Y.; Castiglione, J.; Astroza, R.; Chen, Y. Real-time thermal dynamic analysis of a house using RC models and joint state-parameter estimation. *Build. Environ.* **2021**, *188*, 107184. [CrossRef]

46. Wang, Z.; Chen, Y.; Li, Y. Development of RC model for thermal dynamic analysis of buildings through model structure simplification. *Energy Build.* **2019**, *195*, 51–67. [CrossRef]
47. Vivian, J.; Zarrella, A.; Emmi, G.; De Carli, M. An evaluation of the suitability of lumped-capacitance models in calculating energy needs and thermal behaviour of buildings. *Energy Build.* **2017**, *150*, 447–465. [CrossRef]
48. Zarrella, A.; Prataviera, E.; Romano, P.; Carnieletto, L.; Vivian, J. Analysis and application of a lumped-capacitance model for urban building energy modelling. *Sustain. Cities Soc.* **2020**, *63*, 102450. [CrossRef]
49. Wang, S.; Xu, X. Simplified building model for transient thermal performance estimation using GA-based parameter identification. *Int. J. Therm. Sci.* **2006**, *45*, 419–432. [CrossRef]
50. Wang, J.; Chen, H.; Yuan, Y.; Huang, Y. A novel efficient optimization algorithm for parameter estimation of building thermal dynamic models. *Build. Environ.* **2019**, *153*, 233–240. [CrossRef]
51. Bueno, B.; Norford, L.; Pigeon, G.; Britter, R. A resistance-capacitance network model for the analysis of the interactions between the energy performance of buildings and the urban climate. *Build. Environ.* **2012**, *54*, 116–125. [CrossRef]
52. Shen, P.; Braham, W.; Yi, Y. Development of a lightweight building simulation tool using simplified zone thermal coupling for fast parametric study. *Appl. Energy* **2018**, *223*, 188–214. [CrossRef]
53. Ogunsola, O.; Song, L. Application of a simplified thermal network model for real-time thermal load estimation. *Energy Build.* **2015**, *96*, 309–318. [CrossRef]
54. Juricic, S.; Goffart, J.; Rouchier, S.; Foucquier, A.; Cellier, N.; Fraisse, G. Influence of natural weather variability on the thermal characterisation of a building envelope. *Appl. Energy* **2021**, *288*, 116582. [CrossRef]
55. Brastein, O.; Perera, D.; Pfeifer, C.; Skeie, N.-O. Parameter estimation for grey-box models of building thermal behaviour. *Energy Build.* **2018**, *169*, 58–68. [CrossRef]
56. Cui, B.; Fan, C.; Munk, J.; Mao, N.; Xiao, F.; Dong, J.; Kuruganti, T. A hybrid building thermal modeling approach for predicting temperatures in typical, detached, two-story houses. *Appl. Energy* **2019**, *236*, 101–116. [CrossRef]
57. Joe, J.; Karava, P. A model predictive control strategy to optimize the performance of radiant floor heating and cooling systems in office buildings. *Appl. Energy* **2019**, *245*, 65–77. [CrossRef]
58. Weber, T.; Jóhannesson, G.; Koschenz, M.; Lehmann, B.; Baumgartner, T. Validation of a FEM-program (frequency-domain) and a simplified RC-model (time-domain) for thermally activated building component systems (TABS) using measurement data. *Energy Build.* **2005**, *37*, 707–724. [CrossRef]
59. O'Callaghan, P.; Probert, S. Sol-air temperature. *Appl. Energy* **1977**, *3*, 307–311. [CrossRef]
60. Rodríguez, E.A.; Sánchez, F.J.; Catalán, A.; Rincón, A.; Fernández, J.M.R.; Galán, C.; Fernández-Nieto, E.D.; Narbona-Reina, G. Building and surroundings. Thermal and aeraulic coupling within the Special Session Applied Mathematics in Architecture. In Proceedings of the XXIV Congress on Differential Equations and Applications/XIV Congress on Applied Mathematics, Cádiz, Spain, 8–12 June 2015; pp. 157–162.
61. CIBSE, Environmental Design: Guide A, 2006. Available online: https://www.cibse.org/knowledge/knowledge-items/detail?id=a0q20000008I79JAAS (accessed on 8 July 2021).
62. UNE. *Ergonomics of the Thermal Environment—Analytical Determination and Interpretation of Thermal Comfort Using Calculation of the PMV and PPD Indices and Local Thermal Comfort Criteria*; EN ISO 7730; AENOR; UNE: Madrid, Spain, 2006.
63. UNE. *Ergonomics of the Thermal Environment—Instruments for Measuring Physical Quantities*; ISO 7726:1998; AENOR; UNE: Madrid, Spain, 2002; Available online: https://www.une.org/encuentra-tu-norma/busca-tu-norma/norma?c=N0037517 (accessed on 14 July 2021).
64. UNE. *Radiadores y Convectores. Parte 2: Métodos de Ensayo y de Evaluación*; UNE-EN 442-2:2015; AENOR; UNE: Madrid, Spain, 2015; Available online: https://www.une.org/encuentra-tu-norma/busca-tu-norma/norma?c=N0055707 (accessed on 24 August 2020).
65. Gnielinski, V. Neue Gleichungen für den Wärme- und den Stoffübergang in turbulent durchströmten Rohren und Kanälen. *Forsch. Ing.* **1975**, *41*, 8–16. [CrossRef]
66. UNE. *Suelos de Madera. Colocación. Especificaciones*; UNE 56810:2013; AENOR; UNE: Madrid, Spain, 2013; Available online: https://www.une.org/encuentra-tu-norma/busca-tu-norma/norma?c=N0051820 (accessed on 11 September 2020).
67. Suleiman, B.M.; Larfeldt, J.; Leckner, B.; Gustavsson, M. Thermal conductivity and diffusivity of wood. *Wood Sci. Technol.* **1999**, *33*, 465–473. [CrossRef]
68. Vololonirina, O.; Coutand, M.; Perrin, B. Characterization of hygrothermal properties of wood-based products—Impact of moisture content and temperature. *Constr. Build. Mater.* **2014**, *63*, 223–233. [CrossRef]
69. Oliva, A.G.; Pulgar, F.P. Physical and mechanical characteristics of the woods of Spain. *Minist. Agric. Inst. For. Investig. Exp.* **1967**. Available online: https://agris.fao.org/agris-search/search.do?recordID=US201300482508 (accessed on 23 July 2020).
70. Tenwolde, A.; McNatt, J.D.; Krahn, L. *Thermal Properties of Wood and Wood Panel Products for Use in Buildings*; DOE/OR/21697-1; 6059532; Forest Service: Madison, WI, USA, 1988. [CrossRef]
71. Leon, G.; Cruz-De-Leon, J.; Villasenor, L. Thermal characterization of pine wood by photoacoustic and photothermal techniques. *Holz Roh Werkst.* **2000**, *58*, 241–246. [CrossRef]
72. Wood Density Chart | Workshop, (n.d.). Available online: https://cedarstripkayak.wordpress.com/lumber-selection/162-2/ (accessed on 28 July 2020).
73. Carman, A.P.; Nelson, R.A. Thermal conductivity and diffusivity of concrete. *J. Frankl. Inst.* **1922**, *193*, 120. [CrossRef]

74. Vay, O.; De Borst, K.; Hansmann, C.; Teischinger, A.; Müller, U. Thermal conductivity of wood at angles to the principal anatomical directions. *Wood Sci. Technol.* **2015**, *49*, 577–589. [CrossRef]
75. Sonderegger, W.; Hering, S.; Niemz, P. Thermal behaviour of Norway spruce and European beech in and between the principal anatomical directions. *Holzforschung* **2011**, *65*, 369–375. [CrossRef]
76. Kol, H.Ş.; Sefil, Y. The thermal conductivity of fir and beech wood heat treated at 170, 180, 190, 200, and 212 °C. *J. Appl. Polym. Sci.* **2011**, *121*, 2473–2480. [CrossRef]
77. European Beech | The Wood Database—Lumber Identification (Hardwood), (n.d.). Available online: https://www.wood-database.com/european-beech/ (accessed on 28 July 2020).
78. Las Propiedades de la Madera de Fresno Europeo, (n.d.). Available online: https://www.castor.es/fresno.html (accessed on 28 July 2020).
79. Ficha Técnica de la Madera de Fresno Europeo, (n.d.). Available online: https://www.maderasmedina.com/fichas-propiedades/maderas-frondosas/fresno-europeo.html# (accessed on 28 July 2020).
80. A. Instituto Eduardo Torroja de Ciencias de la Construcción, CEPCO, Catálogo de Elementos Constructivos, Código Técnico La Edif. CTE. 3 (2010) 141. Available online: http://www.codigotecnico.org/web/recursos/aplicaciones/contenido/texto_0012.html (accessed on 29 July 2020).
81. Alonso, F.; María, L.; Riocerezo, G. Distribución de la Superficie de la Vivienda En España, (n.d.). Available online: https://www.mitma.gob.es/recursos_mfom/pdf/CD7DD6BD-2F2A-4737-B8C3-2113291E69A3/99250/dsv1.pdf (accessed on 15 July 2020).
82. UNE. *Energy Performance of Buildings—Energy Needs for Heating and Cooling, Internal Temperatures and Sensible ans Latent Heat Loads—Part 1: Calculation Procedures (ISO 52016-1:2017)*; UNE-EN ISO 52016-1:2017; AENOR; UNE: Madrid, Spain, 2017.
83. UNE. *Sistemas de Calefacción y Refrigeración de Circulación de agua Integrados en Superficies. Parte 2: Suelo Radiante: Métodos para la Determinación de la Emisión Térmica de los suelos Radiantes por Cálculo y Ensayo*; UNE-EN 1264-2:2009+A1:2013; AENOR; UNE: Madrid, Spain, 2013; Available online: https://www.une.org/encuentra-tu-norma/busca-tu-norma/norma?c=N0051332 (accessed on 24 August 2020).
84. Zhao, Q.; Lian, Z.; Lai, D. Thermal comfort models and their developments: A review. *Energy Built Environ.* **2021**, *2*, 21–33. [CrossRef]
85. UNE. *Water Based Surface Embedded Heating and Cooling Systems—Part 2: Floor heating: Prove Methods for the Determination of the Thermal Output Using Calculation and Test Methods*; UNE-EN 1264-2:2009+A1:2013; AENOR; UNE: Madrid, Spain, 2013.
86. De España, G. Documento Básico HE Ahorro de energía Con comentarios del Ministerio de Fomento HE0 Limitación del consumo energético HE1 Limitación de la demanda energética HE2 Rendimiento de las instalaciones térmicas HE3 Eficiencia energética de las instalaciones de i. 2017. Available online: http://www.codigotecnico.org (accessed on 25 July 2020).
87. Infraestructuras. *Ministerio de Fomento. Secretaría de Estado de Infraestructuras Transporte y Vivienda, Documento Básico HE*; Ahorro de Energía: Madrid, Spain, 2013; pp. 1–129.
88. Documentos CTE, (n.d.). Available online: https://www.codigotecnico.org/index.php/menu-documentoscte/133-ct-documentos-cte/ahorro-de-energia.html (accessed on 10 September 2020).
89. Ministerio de Industria Turismo y Comercio. *Documento de Condiciones de Aceptación de Programas Informáticos Alternativos*; IDAE, Ministerio de Vivienda, Gobierno de España: Madrid, Spain, 2007.

*energies*

MDPI

*Article*

# Investigation of Mechanochemically Treated Municipal Solid Waste Incineration Fly Ash as Replacement for Cement

**Shuping Pan [1], Jiamin Ding [2], Yaqi Peng [3,*], Shengyong Lu [3] and Xiaodong Li [3]**

1 Zhejiang Ecological Environment Monitoring Center, Hangzhou 310007, China; panshuping@zjemc.org.cn
2 Research Institute of Zhejiang University-Taizhou, Taizhou 318000, China; dingjiamin@zju.edu.cn
3 State Key Laboratory of Clean Energy Utilization, Institute for Thermal Power Engineering, Zhejiang University, Hangzhou 310027, China; lushy@zju.edu.cn (S.L.); lixd@zju.edu.cn (X.L.)
* Correspondence: pengyaqi@zju.edu.cn

**Abstract:** Municipal solid waste incineration (MSWI) fly ash has been classified as hazardous waste in China because of the leachable toxic heavy metals and high concentrations of chlorides and polychlorinated dibenzo-p-dioxins and polychlorinated dibenzofurans (PCDD/Fs). Currently, the main treatment method is still landfilling after chemical treatment or cement solidification, and an effective approach to realize fly ash utilization is still lacking. In the present work, the fly ash was firstly water-washed to remove the soluble chlorine salts, which can improve the performance of the produced cement mortar in later work. Mechanochemical pre-treatment was adopted to destroy the PCDD/Fs and improve the heavy metals' stabilization. The results show that 75% of PCDD/Fs can be degraded and that most of the heavy metals are stabilized. After the mechanochemical pre-treatment, the average particle size of the fly ash decreases to 2–5 μm, which is beneficial for promoting the activation energy and accelerating the hydration process in cement mortar production. The compressive and flexural strengths of the fly ash cement mortar improve to 6.2 MPa and 32.4 MPa, respectively, when 35% of the OPC is replaced by treated fly ash. The similarity in the 3-day and 28-day strength with or without the addition of the treated ash shows the light influence of the fly ash addition. Thus, the mechanochemical process can stabilize the heavy metals and activate the fly ash, allowing it to partly substitute ordinary Portland cement in building materials, such as cement raw materials and concrete.

**Keywords:** mechanochemical treatment; fly ash; PCDD/Fs; heavy metals

**Citation:** Pan, S.; Ding, J.; Peng, Y.; Lu, S.; Li, X. Investigation of Mechanochemically Treated Municipal Solid Waste Incineration Fly Ash as Replacement for Cement. *Energies* **2022**, *15*, 2013. https://doi.org/10.3390/en15062013

Academic Editors: Dibyendu Sarkar, Rupali Datta, Prafulla Kumar Sahoo and Mohammad Mahmudur Rahman

Received: 3 February 2022
Accepted: 7 March 2022
Published: 10 March 2022



## 1. Introduction

Incineration has been the major approach to reduce the amount of municipal solid waste in China. A total of 136 million tons of solid waste were incinerated, and more than 500 incineration plants were run in 2021 in China [1]. However, large amounts of fly ash (FA) are produced from municipal solid waste incineration (MSWI), with a production amount of more than 8 million tons per year, which is still increasing due to the increasing amount of solid waste [2]. MSWI FA is intercepted by the baghouse filter in air pollution control devices. MSWI FA is classified as hazardous solid waste because of the leachable toxic heavy metals, such as Cu, Cd, Zn, and Pb, and chloride salts that are leachable; thus, FA can be a serious environmental risk [3–5]. Moreover, polychlorinated dibenzo-p-dioxins (PCDDs) and polychlorinated dibenzofurans (PCDFs) are formed and adsorbed in fly ash in a concentration range of 76–2652 ng TEQ/kg [6,7], which has caused widespread concern due to the harmful effects on human health and the environment [8]. Currently, the main treatment method is still landfilling after chemical treatment or cement solidification [9]. However, the land in East China is currently seriously scant, and landfilling is not a sustainable solution for MSWI FA treatment [10].

Due to the high concentrations of CaO, $SiO_2$, $Al_2O_3$, $Fe_2O_3$, MgO, etc., in fly ash, several researchers have studied the utilization and recovery of MSWI fly ash to prevent adverse environmental risks, including its utilization as a raw material for tile manufacturing and as a substitute for ordinary sand in concrete [11–14]. In previous studies, coal-fired fly ash and MWSI bottom ash have been studied for their potential use in concrete, road pavement, ceramics, and so on [15–20]. When MSWI fly ash is added to cement-based products, to meet the technical and environmental requirements, such as enough strength, durability performance, and leaching limits of heavy metals from the products, pre-treatment is necessary to reduce the leaching of hazardous components [21,22]. Thermal pyrolysis was studied to treat the MSWI fly ash and the PCDD/Fs concentrations can be lower than 20 ng TEQ/kg after the pyrolysis process at 350 °C [8]. However, thermal pyrolysis at a low temperature cannot reduce heavy metal leaching concentrations [23]. Fly ash co-processing in cement kilns has been adopted to destroy hazardous waste in China; however, the restricted addition amount because of the chlorine concentration has limited the utilization of fly ash [24].

Phosphorus-based inhibitors can efficiently prevent dioxin formation in the post-combustion zone but have a poor effect on the heavy metals' solidification in fly ash [25]. To achieve complete detoxification of fly ash and make it recyclable, further detoxification is needed. The mechanochemical (MC) process is simple and can simultaneously solidify heavy metals and degrade PCDD/Fs in fly ash [26]. Birke et al., performed mechanochemical treatment by adding magnesium, alcohols, and ethers to achieve complete dehalogenation of POPs in several minutes [27]. Lu et al., used a mechanochemical method with $CaO/SiO_2$ addition to destroy 2,4,6-trichlorophenol, and a 99% degradation efficiency was achieved [28]. Yan et al., studied the performance of mechanochemical treatment to destruct PCDD/Fs with CaO addition, and 90% of the PCDD/Fs were destroyed [29]. By analyzing the dichlorination and fingerprints of PCDD/Fs, destruction and dichlorination were proposed as major mechanisms for PCDD/Fs degradation [29]. Selena et al., used hydroxyapatite as the agent to treat soil containing heavy metals such as Cd, Zn, and Pb by a mechanochemical method, in which immobilization rates of 99.89%, 97.44%, and 99.7% were obtained, respectively [30]. Chen et al., found that $Ca_3(PO_4)_2$ addition facilitated the mechanochemical stabilization of toxic heavy metals in fly ash, and the Nemerow pollution index reduced from 9.35 to 0.71 [26]. Yuan et al., used mechanochemical treatment to improve heavy metal stabilization in coal-fired fly ash, and the water-soluble and acid-soluble components of Cu, Pb, and Cr decreased by more than 25% [31]. Geng et al., found that the mechanochemical process was beneficial for stabilizing Pb in fly ash because it crushed FA particles and destroyed the crystal structure, further developing its surface area and generating more surface active sites, which was beneficial to increasing the diffusivity and adsorption ability of heavy metals in FA [32].

Moreover, particle refinement contributes to improving the mechanical performance of the produced cement mortar by promoting the pozzolanic and hydration activity [33]. Decreasing the particle size will promote the activation energy and accelerate the hydration process [33,34]. Cordeiro et al., (2008) studied the addition of sugar cane bagasse incineration ash to cement mortar and found that the fine particles boosted the pozzolanic reactions [35].

In this study, a mechanochemical method was adopted to pre-treat municipal solid waste incineration fly ash and promote its characteristics. The evolution of PCDD/Fs and the leaching characteristics of heavy metals in the fly ash were comprehensively studied. In addition, raw and treated fly ash were used as partial substitutes for ordinary Portland cement, and the strength performance and heavy metal leaching of the fly ash cement mortar were investigated.

## 2. Materials and Methods

### 2.1. Fly Ash Samples

The fly ash sample was collected from the baghouse filter in a solid waste incineration plant, which has a total treatment capacity of 2250 t/d with three 750 t/d incinerators. The incinerator studied in this work was equipped with a circulating fluidized bed furnace, and the furnace temperature was kept above 850 °C to prevent dioxin formation. The solid waste was stored and fermented for more than one week before burning to remove the water and increase the calorific value. Typically, the air pollution control devices (APCDs) for this incinerator include selective noncatalytic reduction (SNCR), a semidry scrubber, and activated carbon injection coupled with a baghouse filter. To ensure that the dioxins emissions met the national standard, the APCDs were ungraded with the addition of an inhibitor injection system at the flue pipe, where the flue gas temperature was approximately 500 °C. A novel S–N–P-based inhibitor was chosen with a flow of 20 kg/h. Attributed to the advanced combustion technology and APCDs, the initial dioxin concentration in the fly ash was only 255 ng TEQ/kg, much lower than that reported in reference [7]. Before use, the fly ash was water-washed twice to remove the chlorine with a solid:liquid ratio of 1:6. After water washing, the soluble chlorine concentration was 0.86%, which is lower than the national standard of 1%.

### 2.2. Mechanochemical Treatment of Fly Ash

To study the effect of mechanochemical treatment on the heavy metals and dioxins, planetary ball milling equipment (QXQM-2-DW, Changsha Tianchuang, China) was adopted in this study. A total of 15 g fly ash was put in the zirconia ball milling tank (250 mL) with zirconia spheres (total mass of 375 g, diameters of 10 mm and 5 mm, and mass ratio of 1:1). The rotation speed of the planetary disk was 500 rpm, and the durations were 6 h and 10 h.

CaO was adopted as the addition in ratios of 15%, 30%, and 45%, which was beneficial for destroying the PCDD/Fs and solidifying heavy metals. The sample abbreviations under different operating conditions were water-washed fly ash (WFA) and mechanochemically treated fly ash (MFA-x-y, where x is the amount of CaO addition and y is the milling time).

### 2.3. Fly Ash Cement Mortar Preparation

The effects of WFA or MFA addition on the mechanical properties and the heavy metal leaching concentrations of the fly ash cement mortar were examined. The ordinary Portland cement (OPC) was used as the raw material, which has a density of 3.15 g/cm$^3$ with a surface area of 3460 cm$^2$/g. The mortar was mixed with a mass ratio of 2:6:1 of binder, fine aggregate, and water. In the experiments, WFA, MFA-0%-10 h, MFA-15%-10 h, MFA-30%-10 h, and MFA-45%-10 h were used to replace OPC. To evaluate the effect of the fly ash addition on the mortar performance, fly ash replaced 35%, 50%, and 60% OPC, as shown in Table 1.

**Table 1.** Ratio of fly ash used to replace OPC (wt %).

| | WFA | MFA-0% | MFA-15% | MFA-30% | MFA-45% | OPC |
|---|---|---|---|---|---|---|
| SN0 | | | | | | 100 |
| SN1 | 35 | | | | | 65 |
| SN2 | | 35 | | | | 65 |
| SN3 | | | 35 | | | 65 |
| SN4 | | | 50 | | | 50 |
| SN5 | | | 65 | | | 35 |
| SN6 | | | | 35 | | 65 |
| SN7 | | | | 50 | | 50 |
| SN8 | | | | 65 | | 35 |
| SN9 | | | | | 35 | 65 |
| SN10 | | | | | 50 | 50 |
| SN11 | | | | | 65 | 35 |

In accordance with the Chinese standard (GB/T 17671), after mixing, the mortar was poured into steel molds. Then, the molds were cured at 20 °C with 95% relative humidity for 24 h. After the mortar had cured, it was taken out from the mold and placed in water with a constant temperature of 20 °C. Compressive strength tests of mortar samples were performed after 3 and 28 days according to GB/T 17671. The pretreatment and experiment process is shown in Figure 1. The heavy metal leaching concentrations were also tested by ICP.

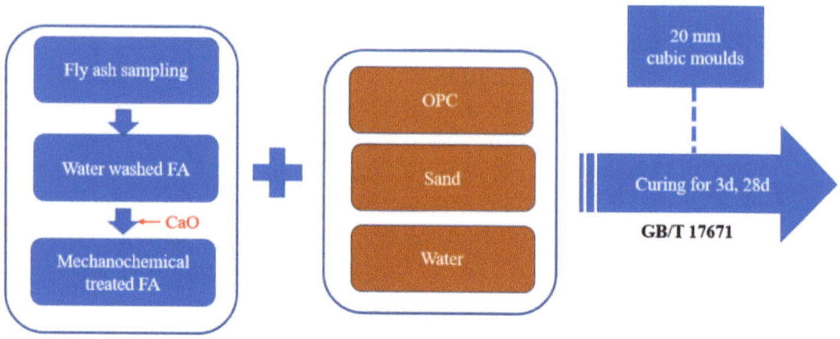


**Figure 1.** The schematic of the experiment procedure.

### 2.4. Characterization and Analysis

The chemical composition of the fly ash samples was determined by X-ray fluorescence spectroscopy (XRF, SMS-Omega XRF, Thermo Fisher, Waltham, MA, USA). The chlorine content was tested by ion chromatography (Integrition, Thermo Fisher).

The fly ash was pre-treated according to US EPA 23, and the PCDD/Fs concentrations were measured by high-resolution gas chromatography/high-resolution mass spectrometry (HRGC/HRMS, JMS-800D, JEOL, Tokyo, Japan). First, the fly ash samples were soaked for 4 h in HCl solution to remove the metallic impurity. Then, the concentrated solutions were transferred to acid silica columns and activated carbon mini-columns to remove metals and other organic pollutants. In the pre-treatment process, samples were spiked with $^{13}C$-labeled standard compounds of PCDD/Fs. Finally, the treated samples were analyzed with HRGC/HRMS with DB-5 MS column (60 m × 0.25 mm × 0.25 mm). HRMS was operated in electron ionization and selected ion monitoring modes with a resolution of >10,000. The detailed pre-treatment and analysis processes were introduced in our previous work [36].

The heavy metals were extracted from the samples following Chinese national standard HJ/T 300-2007 ("Solid Waste-Extraction Procedure for Leaching Toxicity—Acetic Acid Buffer Solution Method"). The samples were detected with an iCAP 6300 instrument (Thermo Fisher).

## 3. Results and Discussion

### 3.1. Characterization of the Raw and Treated Fly Ash

The compositions of the raw fly ash were determined by XRF, and the results show that Ca, Si, and Al dominate fly ash. However, the chlorine content in the present work is only 6 wt %, which is much lower than contents from incinerators without an inhibition system [37]. The low content of chlorine is consistent with the low PCDD/Fs concentration, which is only 255 ng TEQ/kg. The S–N–P inhibitor helps to reduce the PCDD/Fs formation in two ways. On the one hand, the S from the inhibitor is oxidized to $SO_2$, which can inhibit $Cl_2$ conversion from HCl and decrease dioxin formation. On the other hand, the conversion to $CuSO_4$ from $CuCl_2$ with $SO_2$ is an important factor to reduce organic chlorine formation [38].

Figure 2 and Table 2 shows the particle size distribution of the OPC, WFA, and MFA. The average particle size of the WFA is 56 μm, which is larger than that of OPC (8 μm). With the treatment of the mechanochemical method, the average particle size decreases to 2–5 μm, showing three peaks of the treated fly ash. When the average particle size is lower than 10 μm, the force between particles increases significantly, and the particle size does not decrease but remains in a reasonable range [39].

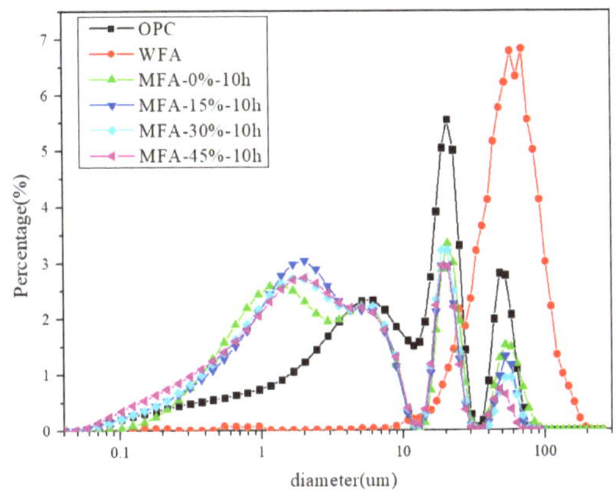


**Figure 2.** Particle size distribution of the samples under different operating conditions.

**Table 2.** D50 and D90 of the samples under different operating conditions (μm).

| Samples | D50 | D90 |
| --- | --- | --- |
| OPC | 16.1 | 45.2 |
| WFA | 49.6 | 96.5 |
| MFA-0%-10 h | 4.2 | 46.8 |
| MFA-15%-10 h | 3.3 | 23.7 |
| MFA-30%-10 h | 3.2 | 23.1 |
| MFA-45%-10 h | 2.7 | 21.4 |

### 3.2. PCDD/Fs and Heavy Metal Concentrations before and after MC Treatment

After being washed with water, the water-soluble chlorides, such as $CaCl_2$, KCl, and NaCl, easily washed away. The total PCDD/Fs concentrations increased from 5035 in the raw fly ash to 5671 ng/kg after water washing. The TEQ concentration of water-washed fly ash (WFA) was 230 ng TEQ/kg. After water washing, the water-soluble chlorides concentration was 0.86%, which can reduce the PCDD/Fs formation in the mechanochemical process.

Figure 3 shows the changes in the PCDD/Fs TEQ of fly ash under different working conditions. Without the addition of CaO to fly ash, the degradation efficiency of PCDD/Fs is quite low. After 10 h of mechanochemical treatment, only 10.9% of the PCDD/Fs was destroyed. When 15% CaO was added to the fly ash, the destruction efficiency reached 22.3% after 6 h of treatment and increased to 35.4% with 10 h of milling. In particular, the PCDD/Fs concentration was only 55.82 ng TEQ/kg with a 45% addition, indicating that a certain amount of CaO can better promote the PCDD/Fs degradation in fly ash by mechanochemical methods. The degradation rate increases with increasing addition ratio and milling time. Compared to the results in a previous work [40], the destruction efficiencies in the present work are much lower. Wei et al., also found a similar phenomenon in their study [41]. In this work, the initial PCDD/Fs concentration is much lower due to the inhibitor addition in the flue gas. The destruction efficiency is much lower when the PCDD/Fs concentration in fly ash is low due to less contact between balls and dioxin pollutants.

By analyzing the PCDD/Fs fingerprints in the raw and treated fly ash, changes in the contents of 17 kinds of 2,3,7,8-polychlorinated toxic PCDD/Fs were observed. According to Figures 4 and 5, after 6 h of ball milling without CaO addition, the OCDD content decreased by 29.6%, while after 10 h of ball milling, the OCDD content continued to decrease to 1.295 ng TEQ/kg, resulting in a degradation rate of 41.2%. In addition, MC treatment has an obvious effect on the degradation of 1,2,3,4,6,7,8-HpCDD, 1,2,3,4,6,7,8-HpCDF, and OCDF. This shows that without additives in the MFA-0%, the PCDD/Fs can still be degraded, and when CaO is added, the CaO in the raw fly ash can help to dechlorinate the PCDD/Fs [40].

Compared to MFA-0%, the degradation efficiency of the samples with the CaO addition obviously improved, and the degradation efficiency increased with increasing addition ratio and ball milling time. When the CaO addition amount was 45%, the degradation rate reached 88.1% after 10 h of ball milling, indicating that CaO has a better promoting effect on PCDD/Fs degradation. Moreover, it is worth noting that when the addition ratio increased from 30% to 45%, the degradation efficiency improvement rate was quite low [28]. In addition, it can be observed from Figures 4 and 5 that the mechanochemical method has a suitable degradation effect on high-chlorinated PCDD/Fs but a mediocre degradation effect on low-chlorinated PCDD/Fs.

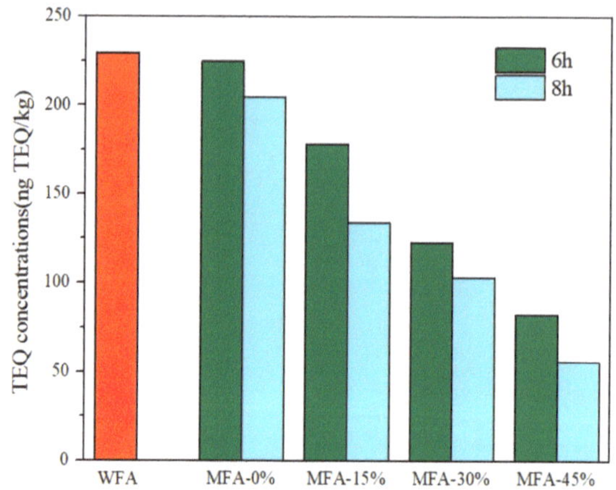


**Figure 3.** TEQ concentrations of the MC-treated fly ash.

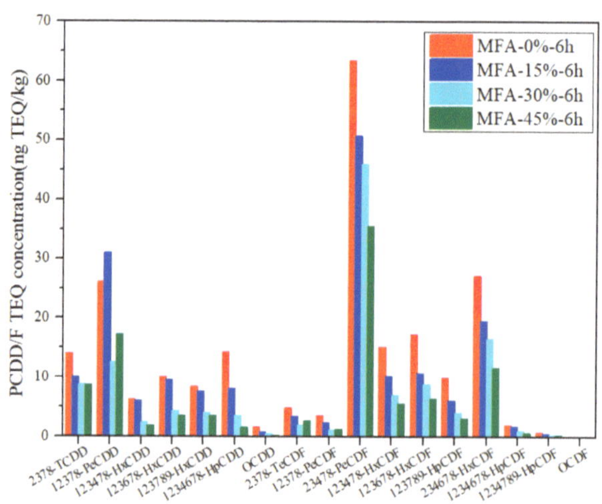


**Figure 4.** Concentrations of 17 2,3,7,8-substitued PCDD/Fs after MC treatment for 6 h.

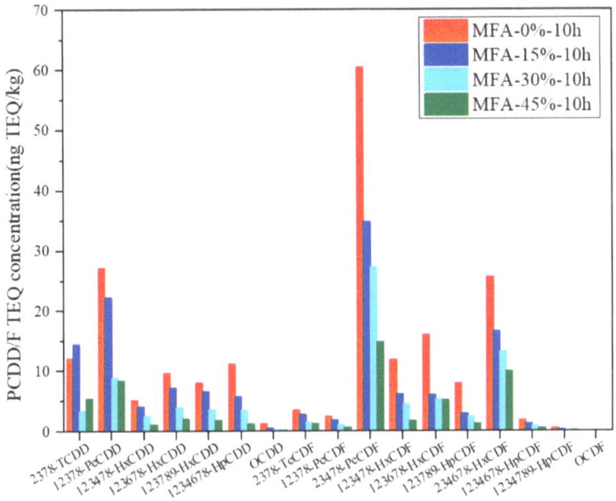


**Figure 5.** Concentrations of 17 2,3,7,8-substitued PCDD/Fs after MC treatment for 10 h.

Figures 4 and 5 show the changes in the toxicity equivalent of 17 toxic PCDD/Fs under four different working conditions. It can be seen from the figure that mechanochemistry treatment has a significant degradation effect on 17 toxic PCDD/Fs toxic equivalents, and the largest contribution of the toxicity equivalent is still PCDFs, especially 2,3,4,7,8-PeCDF, which is approximately 30%. It is worth mentioning that mechanochemistry has a light effect on degrading low-chlorinated PCDD/Fs, especially 2,3,7,8-TCDD, 1,2,3,7,8-PeCDD, and 2,3,4,7,8-PeCDF. The reason may be that on the one hand, high-chlorinated PCDD/Fs are easily converted into low-chlorinated PCDD/Fs, which exhibit higher toxicity [29]. On the other hand, the chlorine at position 2,3,7,8 is not easy to remove, and its toxicity factor is relatively high, resulting in a high TEQ concentration [42].

Table 3 shows the heavy metal leaching concentrations under different working conditions. It shows that after the mechanochemical treatment, the heavy metal leaching concentration in the fly ash decreases, indicating that mechanochemical solidification can block the heavy metals. Li et al., (2010) showed that the change in the crystal structure of fly ash by mechanical force is the main reason for its ability to solidify heavy metals [43]. When CaO is not added, most of the heavy metals are stabilized after mechanochemical treatment, but the Ni content increases, which may be caused by the wear of the grinding balls or tank walls. In addition, the leaching concentration of Cd still exceeds landfill standards. When 15% CaO is added, the leaching concentration of each heavy metal is far lower than the landfilling standard. This shows that CaO has a favorable synergistic effect on the heavy metals in the solidified fly ash. Studies have shown that CaO reduces the leaching concentration of heavy metals in fly ash by increasing the pH value of the leaching solution and leading to part of the heavy metals in the solution being adsorbed [33,43].

**Table 3.** Leaching concentrations of heavy metals in the different fly ash (mg/L).

| Heavy Metal | Cd | Cr | Ni | Cu | Pb | Zn | PH |
|---|---|---|---|---|---|---|---|
| Standard | 0.15 | 1.5 | 0.5 | 40 | 0.25 | 100 | 2–12.5 |
| WFA | 0.41 | 0.66 | 0.40 | 9.45 | 0.68 | 35.73 | 7.2 |
| MFA-0%-6 h | 0.32 | 0.06 | 0.50 | 1.25 | 0.30 | 22.67 | 7.8 |
| MFA-15%-6 h | 0.006 | ND | ND | 0.2 | 0.005 | 0.02 | 8.2 |
| MFA-30%-6 h | 0.002 | ND | ND | 0.06 | 0.002 | 0.004 | 8.9 |
| MFA-45%-6 h | ND | ND | ND | 0.02 | 0.002 | 0.002 | 9.6 |

ND: not detected.

### 3.3. Compressive Strength of the Fly Ash Cement Mortar

Figures 6 and 7 show the compressive and flexural strengths of cement mortar prepared by OPC and partly substituted by fly ash according to GB/T 17671, respectively. The ratios of OPC substituted by raw or treated fly ash are shown in Table 1. For all cement mortar blocks, the flexural and compressive strengths at 28 days are greater than those at 3 days, and their mechanical strength properties decrease with increasing fly ash content. Compared with the cement mortar without fly ash, the flexural and compressive strength of the cement mortar greatly decrease when the amount of added WFA is 35% (decrease from 9.3 and 64.8 MPa to 5.6 and 22.3 MPa, respectively). The water-washed fly ash addition has a notable negative effect on the strength of the cement mortar. The flexural and compressive strengths of the fly ash mortar improve to 6.2 MPa and 32.4 MPa, respectively, when 35% of the OPC is replaced by MFA-0%. When the same proportion of MFA-15%-10 h is added, the strength of the cement mortar improves significantly. Its 28-day flexural and compressive strengths are 7.9 MPa and 45.7 MPa, respectively. When 35% MFA-30%-10 h is added, its 28-day flexural and compressive strengths are 8.2 MPa and 51.6 MPa, respectively. The flexural strength reaches 88.2% of the pure cement mortar, and the compressive strength reaches 79.6% of the pure cement mortar. Even when the addition ratio is increased to 65%, the flexural and compressive strengths at 28 days can reach 7.4 MPa and 43.5 MPa, indicating that the performance of MFA-30%-10 h is greatly improved after mechanochemical treatment with the CaO addition. When 35% MFA-45%-10 h is added, the strength is lower than the MFA-15%-10 h and MFA-30%-10 h cement mortars, and its 28-day flexural and compressive strengths are only 6.1 MPa and 32.7 MPa, respectively. The results indicate that when the amount of free calcium is too high, it will affect the stability of cement and reduce the strength [44]. In conclusion, the performance of fly ash treated by the mechanochemical method is significantly improved. The fly ash after water washing and mechanochemical treatment can be used as cement.

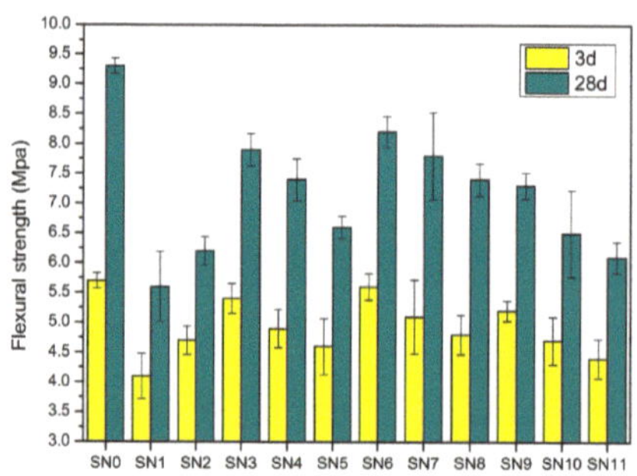


**Figure 6.** Flexural strength of the cement mortar.

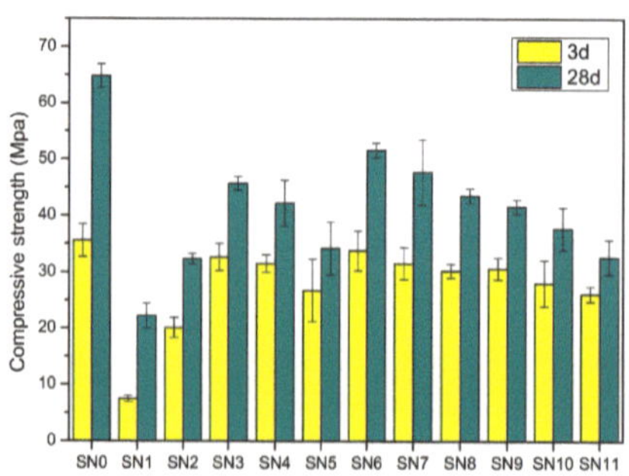


**Figure 7.** Compressive strength of the cement mortar.

### 3.4. Leaching Concentration of Heavy Metals in the Fly Ash Cement Mortar

Table 4 shows the leaching concentrations of heavy metals in the cement mortar under different working conditions. The results show that the leaching concentrations of heavy metals under all working conditions are quite low. Comparing SN1 to SN0, the 35% fly ash addition can increase the leaching concentration of cement mortar. When MC-treated fly ash is added, the leaching concentrations of heavy metals in cement mortar are much lower than those substituted by raw fly ash. The results of SN3–SN5 indicate that the amount of treated fly ash that is added has an important influence on the leaching of heavy metals in the cement mortar, even though all the values meet the standard. With increasing CaO amounts, the performance of the treated fly ash shows a better stabilization effect on heavy metals. The leaching concentrations of heavy metals in SN6–SN11 are much lower than those prepared by pure cement, indicating that MC-treated fly ash can be used in building materials, such as cement raw materials and concrete.

**Table 4.** Leaching concentration of heavy metals in cement mortars (mg/L).

| Heavy Metal | Cd | Cr | Cu | Ni | Pb | Zn |
|---|---|---|---|---|---|---|
| Standard | 0.15 | 1.5 | 0.5 | 40 | 0.25 | 100 |
| 1-7 SN0 | ND | 0.338 | 0.003 | 0.057 | ND | 0.013 |
| SN1 | 0.25 | 2.643 | 0.422 | 0.106 | ND | 0.506 |
| SN2 | 0.005 | 0.228 | 0.040 | ND | ND | 0.312 |
| SN3 | 0.004 | ND | 0.020 | ND | ND | 0.173 |
| SN4 | 0.003 | ND | 0.027 | ND | ND | 0.170 |
| SN5 | 0.048 | ND | 0.048 | ND | ND | 0.723 |
| SN6 | ND | ND | ND | ND | ND | 0.002 |
| SN7 | ND | ND | 0.002 | ND | ND | 0.001 |
| SN8 | ND | ND | 0.007 | ND | ND | 0.013 |
| SN19 | ND | ND | 0.001 | ND | ND | 0.001 |
| SN10 | ND | ND | 0.003 | ND | ND | 0.001 |
| SN11 | ND | ND | 0.002 | ND | ND | 0.002 |

ND: not detected.

## 4. Conclusions

In this work, we used a mechanochemical method with CaO as an additive to destroy PCDD/Fs and stabilize the heavy metals in water-washed fly ash. The results show that the PCDD/Fs concentration decreased to 55.8 ng TEQ/kg after mechanochemical treatment, with a degradation efficiency of 75.6%. After MC treatment with 15% CaO addition, the leaching concentrations of heavy metals met the landfilling standard in China. In particular, the leaching concentration of Pb decreased from 0.68 to 0.005 mg/L. Furthermore, raw fly ash and MC-treated fly ash were used as substitutes for cement in cement mortar production. The performance of fly ash obviously improved after MC treatment. When 35% cement was replaced by MC-treated fly ash, the 28-day flexural and compressive strengths were 8.2 MPa and 51.6 MPa, respectively, reaching 88.2% and 79.6% of those of pure cement mortar, showing a satisfactory mechanical performance. The heavy metal leaching concentrations of the fly ash cement mortar were quite low, close to those of the pure cement mortar. The results indicate that MC-treated municipal solid waste fly ash showed a performance similar to that of cement and can be used in building materials, such as cement raw materials and concrete.

**Author Contributions:** S.P.: investigation, methodology. J.D.: data curation, visualization, writing—original draft. Y.P.: conceptualization, funding acquisition, writing—review and editing. S.L. and X.L.: conceptualization, supervision. All authors have read and agreed to the published version of the manuscript.

**Funding:** This research and the APC was funded by Key R&D Program of Zhejiang (2022C03056).

**Institutional Review Board Statement:** Not applicable.

**Informed Consent Statement:** Not applicable.

**Data Availability Statement:** Not applicable.

**Acknowledgments:** This work was supported by the Key R&D Program of Zhejiang (2022C03056).

**Conflicts of Interest:** The authors declare that they have no known competing financial interests or personal relationships that could have appeared to influence the work reported in this paper.

## References


1. NBSC. *China Statistical Yearbook;* China Statistics Press: Beijing, China, 2021.
2. Zhang, Y.; Wang, L.; Chen, L.; Ma, B.; Zhang, Y.; Ni, W.; Tsang, D.C.W. Treatment of municipal solid waste incineration fly ash: State-of-the-art technologies and future perspectives. *J. Hazard. Mater.* **2021**, *411*, 125132. [CrossRef] [PubMed]
3. Lin, K.; Chang, C. Leaching characteristics of slag from the melting treatment of municipal solid waste incinerator ash. *J. Hazard. Mater.* **2006**, *135*, 296–302. [CrossRef] [PubMed]
4. Ma, W.; Fang, Y.; Chen, D.; Chen, G.; Xu, Y.; Sheng, H.; Zhou, Z. Volatilization and leaching behavior of heavy metals in MSW incineration fly ash in a DC arc plasma furnace. *Fuel* **2017**, *210*, 145–153. [CrossRef]
5. Saqib, N.; Bäckström, M. Chemical association and mobility of trace elements in 13 different fuel incineration bottom ashes. *Fuel* **2016**, *172*, 105–117. [CrossRef]
6. Chen, T.; Yan, J.; Lu, S.; Li, X.; Gu, Y.; Dai, H.; Ni, M.; Cen, K. Characteristic of polychlorinated dibenzo-p-dioxins and dibenzofurans in fly ash from incinerators in China. *J. Hazard. Mater.* **2008**, *150*, 510–514. [CrossRef]
7. Chang, Y.-M.; Fan, W.-P.; Dai, W.-C.; Hsi, H.-C.; Wu, C.-H.; Chen, C.-H. Characteristics of PCDD/F content in fly ash discharged from municipal solid waste incinerators. *J. Hazard. Mater.* **2011**, *192*, 521–529. [CrossRef]
8. Trinh, M.M.; Chang, M.B. Catalytic pyrolysis: New approach for destruction of POPs in MWIs fly ash. *Chem. Eng. J.* **2021**, *405*, 126718. [CrossRef]
9. Zhang, Y.; Cetin, B.; Likos, W.; Edil, T. Impacts of pH on leaching potential of elements from MSW incineration fly ash. *Fuel* **2016**, *184*, 815–825. [CrossRef]
10. Xin-gang, Z.; Gui-wu, J.; Ang, L.; Yun, L. Technology, cost, a performance of waste-to-energy incineration industry in China. *Renew. Sustain. Energy Rev.* **2016**, *55*, 115–130. [CrossRef]
11. Lin, D.-F.; Luo, H.-L.; Zhang, S.-W. Effects of Nano-SiO 2 on tiles manufactured with clay and incinerated sewage sludge ash. *J. Mater. Civ. Eng.* **2007**, *19*, 801–808. [CrossRef]
12. Liu, J.; Yan, K.; Liu, J. Rheological characteristics of polyphosphoric acid–modified asphalt mastic. *J. Mater. Civ. Eng.* **2018**, *30*, 06018021. [CrossRef]
13. Yang, X.; You, Z.; Perram, D.; Hand, D.; Ahmed, Z.; Wei, W.; Luo, S. Emission analysis of recycled tire rubber modified asphalt in hot and warm mix conditions. *J. Hazard. Mater.* **2019**, *365*, 942–951. [CrossRef] [PubMed]
14. Bie, R.; Chen, P.; Song, X.; Ji, X. Characteristics of municipal solid waste incineration fly ash with cement solidification treatment. *J. Energy Inst.* **2016**, *89*, 704–712. [CrossRef]
15. Kalinowska-Wichrowska, K.; Pawluczuk, E.; Bołtryk, M.; Jimenez, J.R.; Fernandez-Rodriguez, J.M.; Suescum Morales, D. The Performance of Concrete Made with Secondary Products—Recycled Coarse Aggregates, Recycled Cement Mortar, and Fly Ash–Slag Mix. *Materials* **2022**, *15*, 1438. [CrossRef] [PubMed]
16. Nowoświat, A.; Gołaszewski, J. Influence of the variability of calcareous fly ash properties on rheological properties of fresh mortar with its addition. *Materials* **2019**, *12*, 1942. [CrossRef]
17. Nguyen, M.; Sokolář, R. Corrosion Resistance of Novel Fly Ash-Based Forsterite-Spinel Refractory Ceramics. *Materials* **2022**, *15*, 1363. [CrossRef]
18. Vanitha, N.; Revathi, T.; Gopalakrishnan, R.; Jeyalakshmi, R. Effect of $TiO_2$, $Al_2O_3$ and $CaCO_3$ nano-additives in singular, binary and ternary forms on the mechanical, thermal and microstructural propertites of fly ash supplemented cement matrix. *Mater. Today Proc.* **2021**, *47*, 871–879. [CrossRef]
19. Goswami, A.P. Determining physico-chemical parameters for high strength ambient cured fly ash-based alkali-activated cements. *Ceram. Int.* **2021**, *47*, 29109–29119. [CrossRef]
20. Jani, P.; Imqam, A. Class C fly ash-based alkali activated cement as a potential alternative cement for $CO_2$ storage applications. *J. Pet. Sci. Eng.* **2021**, *201*, 108408. [CrossRef]
21. Saikia, N.; Mertens, G.; Van Balen, K.; Elsen, J.; Van Gerven, T.; Vandecasteele, C. Pre-treatment of municipal solid waste incineration (MSWI) bottom ash for utilisation in cement mortar. *Constr. Build. Mater.* **2015**, *96*, 76–85. [CrossRef]
22. Gao, X.; Wang, W.; Ye, T.; Wang, F.; Lan, Y. Utilization of washed MSWI fly ash as partial cement substitute with the addition of dithiocarbamic chelate. *J. Environ. Manag.* **2008**, *88*, 293–299. [CrossRef] [PubMed]
23. Trinh, M.M.; Chang, M.B. Transformation of mono- to octa- chlorinated dibenzo-p-dioxins and dibenzofurans in MWI fly ash during catalytic pyrolysis process. *Chem. Eng. J.* **2022**, *427*, 130907. [CrossRef]
24. Liu, G.; Zhan, J.; Zheng, M.; Li, L.; Li, C.; Jiang, X.; Wang, M.; Zhao, Y.; Jin, R. Field pilot study on emissions, formations and distributions of PCDD/Fs from cement kiln co-processing fly ash from municipal solid waste incinerations. *J. Hazard. Mater.* **2015**, *299*, 471–478. [CrossRef] [PubMed]

25. Lu, S.; Xiang, Y.; Chen, Z.; Chen, T.; Lin, X.; Zhang, W.; Li, X.; Yan, J. Development of phosphorus-based inhibitors for PCDD/Fs suppression. *Waste Manag.* **2021**, *119*, 82–90. [CrossRef]
26. Chen, Z.; Lu, S.; Tang, M.; Lin, X.; Qiu, Q.; He, H.; Yan, J. Mechanochemical stabilization of heavy metals in fly ash with additives. *Sci. Total Environ.* **2019**, *694*, 133813. [CrossRef]
27. Birke, V.; Mattik, J.; Runne, D.; Benning, H.; Zlatovic, D. Dechlorination of recalcitrant polychlorinated contaminants using ball milling. In *Ecological Risks Associated with the Destruction of Chemical Weapons*; Springer: Berlin/Heidelberg, Germany, 2006; pp. 111–127.
28. Lu, S.; Huang, J.; Peng, Z.; Li, X.; Yan, J. Ball milling 2, 4, 6-trichlorophenol with calcium oxide: Dechlorination experiment and mechanism considerations. *Chem. Eng. J.* **2012**, *195*, 62–68. [CrossRef]
29. Yan, J.; Peng, Z.; Lu, S.; Li, X.; Ni, M.; Cen, K.; Dai, H. Degradation of PCDD/Fs by mechanochemical treatment of fly ash from medical waste incineration. *J. Hazard. Mater.* **2007**, *147*, 652–657. [CrossRef]
30. Montinaro, S.; Concas, A.; Pisu, M.; Cao, G. Immobilization of heavy metals in contaminated soils through ball milling with and without additives. *Chem. Eng. J.* **2008**, *142*, 271–284. [CrossRef]
31. Yuan, Q.; Zhang, Y.; Wang, T.; Wang, J.; Romero, C.E. Mechanochemical stabilization of heavy metals in fly ash from coal-fired power plants via dry milling and wet milling. *Waste Manag.* **2021**, *135*, 428–436. [CrossRef]
32. Geng, X.; Zhao, W.; Zhou, Q.; Duan, Y.; Huang, T.; Liu, X. Effect of a Mechanochemical Process on the Stability of Mercury in Simulated Fly Ash. Part 1. Ball Milling. *Ind. Eng. Chem. Res.* **2021**, *60*, 14737–14746. [CrossRef]
33. Chen, C.-G.; Sun, C.-J.; Gau, S.-H.; Wu, C.-W.; Chen, Y.-L. The effects of the mechanical–chemical stabilization process for municipal solid waste incinerator fly ash on the chemical reactions in cement paste. *Waste Manag.* **2013**, *33*, 858–865. [CrossRef] [PubMed]
34. Erdoğdu, K.; Türker, P. Effects of fly ash particle size on strength of Portland cement fly ash mortars. *Cem. Concr. Res.* **1998**, *28*, 1217–1222. [CrossRef]
35. Cordeiro, G.; Toledo Filho, R.; Tavares, L.; Fairbairn, E. Pozzolanic activity and filler effect of sugar cane bagasse ash in Portland cement and lime mortars. *Cem. Concr. Compos.* **2008**, *30*, 410–418. [CrossRef]
36. He, H.; Lu, S.; Peng, Y.; Tang, M.; Zhan, M.; Lu, S.; Xu, L.; Zhong, W.; Xu, L. Emission characteristics of dioxins during iron ore Co-sintering with municipal solid waste incinerator fly ash in a sintering pot. *Chemosphere* **2022**, *287*, 131884. [CrossRef]
37. Wang, W.; Gao, X.; Li, T.; Cheng, S.; Yang, H.; Qiao, Y. Stabilization of heavy metals in fly ashes from municipal solid waste incineration via wet milling. *Fuel* **2018**, *216*, 153–159. [CrossRef]
38. Ke, S.; Jianhua, Y.; Xiaodong, L.; Shengyong, L.; Yinglei, W.; Muxing, F. Inhibition of de novo synthesis of PCDD/Fs by SO2 in a model system. *Chemosphere* **2010**, *78*, 1230–1235. [CrossRef]
39. Palaniandy, S.; Azizli, K.A.M.; Hussin, H.; Hashim, S.F.S. Study on mechanochemical effect of silica for short grinding period. *Int. J. Miner. Processing* **2007**, *82*, 195–202. [CrossRef]
40. Zhiliang, C.; Minghui, T.; Shengyong, L.; Buekens, A.; Jiamin, D.; Qili, Q.; Jianhua, Y. Mechanochemical degradation of PCDD/Fs in fly ash within different milling systems. *Chemosphere* **2019**, *223*, 188–195. [CrossRef]
41. Wei, Y.-L.; Yan, J.-H.; Liu, Q.-Y.; Li, X.-D. Decomposition of PCDD/Fs by mechanochemical means with calcium-based additives. *J. ZheJiang Univ.* **2012**, *44*, 991–997.
42. Zhiliang, C.; Minghui, T.; Shengyong, L.; Jiamin, D.; Qili, Q.; Yuting, W.; Jianhua, Y. Evolution of PCDD/F-signatures during mechanochemical degradation in municipal solid waste incineration filter ash. *Chemosphere* **2018**, *208*, 176–184. [CrossRef]
43. Li, M.-G.; Sun, C.-J.; Gau, S.-H.; Chuang, C.-J. Effects of wet ball milling on lead stabilization and particle size variation in municipal solid waste incinerator fly ash. *J. Hazard. Mater.* **2010**, *174*, 586–591. [CrossRef] [PubMed]
44. Polat, R.; Demirboğa, R.; Karagöl, F. Mechanical and physical behavior of cement paste and mortar incorporating nano-CaO. *Struct. Concr.* **2019**, *20*, 361–370. [CrossRef]

water

Article

# The Role of Aquifers in Sustaining the Sponge City Concept in Chinese High-Density Housing

**Mengxiao Jin [1,2,3,†], Michele Lancia [2,3,*,†], Yong Tian [2,3], Stefano Viaroli [4], Charles Andrews [5], Junguo Liu [2,3] and Chunmiao Zheng [2,3,*]**

1 School of Environment, Harbin Institute of Technology, Harbin 150001, China; 11649014@mail.sustech.edu.cn
2 Guangdong Provincial Key Laboratory of Soil and Groundwater Pollution Control, School of Environmental Science and Engineering, Southern University of Science and Technology, Shenzhen 518055, China; tiany@sustech.edu.cn (Y.T.); liujg@sustech.edu.cn (J.L.)
3 State Environmental Protection Key Laboratory of Integrated Surface Water-Groundwater Pollution Control, School of Environmental Science and Engineering, Southern University of Science and Technology, Shenzhen 518055, China
4 Department of Sciences, University of Roma Tre, 00146 Rome, Italy; stefano.viaroli@uniroma3.it
5 S.S. Papadopulos and Associates, Inc., Rockville, MD 20852, USA; candrews@sspa.com
* Correspondence: lancia@sustech.edu.cn (M.L.); zhengcm@sustech.edu.cn (C.Z.)
† These authors contributed equally to this work.

**Abstract:** Chinese cities and megacities occupy large areas with high population densities. These urban sectors have extensive impervious surfaces and import large amounts of water from pristine sources, putting pressure on the urban drainage systems during storms. To reduce flood risk and tackle water scarcity, the Chinese government promulgated the Sponge City concept: an ensemble of absorbing features, including permeable pavements and green roofs, to reduce urban runoff and increase water storage. We evaluate the benefits of merging the urban aquifer with conventional Sponge City features by simulating stormwater storage in aquifers with a groundwater model (MODFLOW) and a stormwater management model (SWMM). The urban aquifers of an ideal district, characterized by high-density housing and a flat morphology, collect and dissipate extreme rains (200 mm in 6 h) when the hydraulic conductivity of the subsoil is higher than $1 \cdot 10^{-5}$ m/s, roughly corresponding to a medium productive aquifer. Economic and ecological benefits at the catchment scale are also discussed. Suitable shallow aquifers are individuated and discussed throughout China via a hydrogeological and geomorphological approach. The proposed method increases the percentage of water that can be controlled above that obtained through conventional Sponge City features.

**Keywords:** groundwater; MODFLOW; numerical analyses; high-density housing; SWMM; waterlogging

**Citation:** Jin, M.; Lancia, M.; Tian, Y.; Viaroli, S.; Andrews, C.; Liu, J.; Zheng, C. The Role of Aquifers in Sustaining the Sponge City Concept in Chinese High-Density Housing. *Water* **2022**, *14*, 929. https://doi.org/10.3390/w14060929

Academic Editors: Dibyendu Sarkar, Rupali Datta, Prafulla Kumar Sahoo and Mohammad Mahmudur Rahman

Received: 14 January 2022
Accepted: 14 March 2022
Published: 16 March 2022



## 1. Introduction

Asian megacities are susceptible to urban flooding due to rapid urban expansion and the increasing concentration of population and property [1]. Existing infrastructure collects a remarkable amount of water due to the sum of urban runoff and imported water for industrial and civil use [2]. Runoff from large storms, however, frequently overwhelms existing sewers and waste-water treatment plants (WWTP). Waterlogging and flooding have become more frequent as cities expand and climate change intensifies extreme weather events, more than anything previously experienced [3–6]. Inspired by strategies and practices from the Western world [7], China introduced the Sponge City concept to mitigate waterlogging and flooding risks [8]. The proposed concept consists of installations on the ground surface that collect and reuse rainfall with additional architectural, ecological, and socio-economic benefits [9]. To retrofit urban areas, installations are connected to pipes and tanks that increase the percentage of collected rainfall. Captured water flow has the dual goal of mitigating waterlogging and flooding risks while simultaneously addressing water scarcity [10].

The goals of implementing the Sponge City concept reflect the duality of the water problem [11] between northern and southern China (Figure 1a). Northern China has a dry climate, and has historically relied on diminishing groundwater resources. Lately, expensive water infrastructures such as the South-to-North Water Diversion Project have alleviated the water scarcity, diverting surface water from southern China catchments [12]. Unfortunately, in northern China the rainfall is concentrated in the summer season, making the urban areas susceptible to waterlogging and flooding (Figure 1a). In southern China, the water demand is satisfied by the numerous reservoirs that gather the abundant sub-tropical rainfalls. During the prolonged wet season, waterlogging and flooding risks are high as the drainage network gets overloaded (Figure 1a). Unfortunately, two thirds of China's land is at risk of flooding, and 62% of Chinese cities have experienced urban rainstorm waterlogging disasters in recent years [13], causing direct economic loss averaging 330 billion RMB per year between 2011 and 2016 [14]. Thus, urban waterlogging is a crucial problem that limits the sustainable and healthy development of Chinese cities and megacities [15].

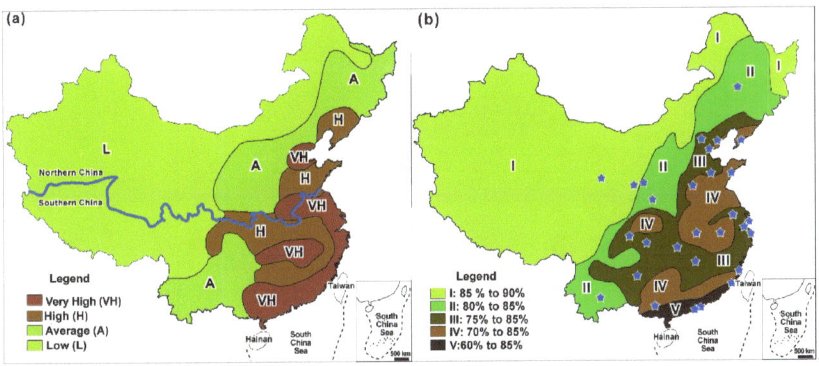


**Figure 1.** Generalized waterlogging risk in China (**a**). Key to the legend: very high (VH), high (H), average (A), low (L). Percentage of the urban runoff aimed to be controlled by the Sponge City plan (**b**). Key to the legend: 85% to 90% (I), 80% to 85% (II); 75% to 85% (III); 70% to 85% (IV); 60% to 85% (V). Blue stars indicate the pilot sites of the Sponge City Project. Redrawn from MOHURD 2014.

In China, residential areas are arranged with high-density housing. This compact urban policy is convenient and feasible [16]. Among the setbacks, high-density housing produces a concentration of clean and black water, which is then managed by pipe networks and waste-water treatment plants (WWTPs). In case of heavy rainfall and storms, pipe network and urban streams come under pressure; the situation worsens during weather warnings as the population is encouraged to stay home, increasing domestic consumption. Despite the government's recommendation to introduce 30% green areas in each compound [17], the management of urban groundwater and waterlogging risk mitigation in these areas is challenging.

The Sponge City concept has the ambitious goal of controlling 65% to 100% of the runoff from urban areas; the goal varies throughout China (Figure 1b). Implementation of the Sponge City initiative began with several pilot sites, and is planned to be extended to many other areas within the next decades [18]. Considering that precipitation is not evenly distributed across the year, numerous and capacious infrastructures are necessary to collect and treat the summer runoff in both northern and southern China. Sponge City installations require pipes and tanks to collect rainfall (Figure 2a). The concept requires large storage volumes that are expensive, and retrofitting existing building and infrastructure is challenging. In addition, retrofitting practices are not adequately compensated through the reuse of collected water [19]. As an alternative, Sponge City installations can be designed to divert rainwater into urban aquifers, restoring natural recharge processes and the groundwater flow regime (Figure 2b). Urban aquifers are underutilized because they largely contain low quality water caused by pollution. Numerical analyses illustrate the effectiveness of urban soil for storing large volumes of water, but quantifying the potential for storage at the catchment scale is challenging [20,21]. The proposed method is inexpensive and can be applied to all minor cities characterized by smaller budgets.

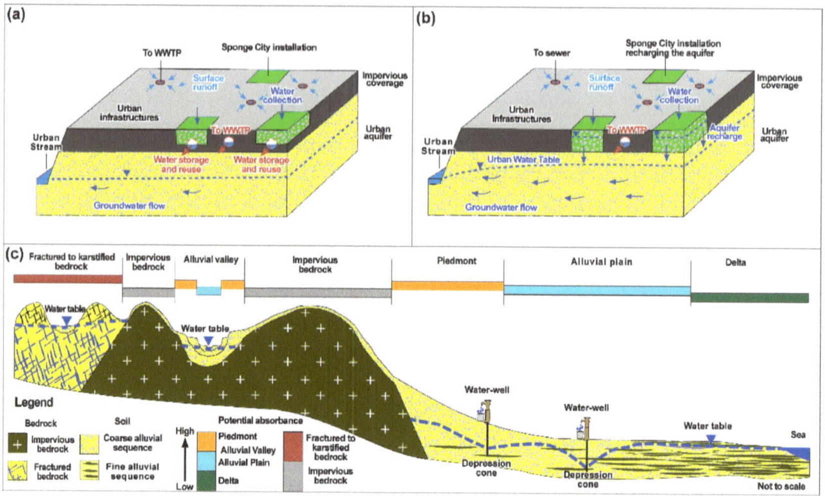


**Figure 2.** Sponge City installations collect and reuse the urban runoff (**a**); Sponge City installations recharge the aquifer, increasing the water table elevation favoring seepage toward urban streams (**b**); conceptual models showing the relationship between the geomorphological zones and potential absorbance according to regional hydrogeological characteristics (**c**).

The major obstacle to the use of urban aquifers as a component of Sponge City implementation is the characterization of the urban hydrogeology. In cities and megacities, groundwater recharge changes in space and time [22]. Impervious surfaces such as building and roads impede rainfall recharge, favoring surface runoff. At the same time, sewers, septic tanks, irrigation excess of parks and flowerbeds, and water supply networks recharge aquifers through infrastructure leakage. Water table maps and groundwater budgets are difficult to compile. In addition, for water storage in aquifers to be feasible (Figure 2b), the soils need to have a sufficiently high hydraulic conductivity and the water table should be located at sufficient depth to not rise to the surface during intense storms or during the prolonged wet season [23]. Ref. [24] suggests the potential absorbance of many Chinese megacities based on groundwater data and geomorphological evidence. However, regional scale analysis is necessary as the application of these practices is of wide interest, especially in smaller cities characterized by less funding.

Because of the duality of Chinese water issues, the intrinsic problem of urban groundwater, and the complexities of city infrastructures, the use of urban aquifers for storage of runoff is not trivial. The feasibility of utilizing aquifers as natural storage tanks in the Sponge City concept is assessed by numerical simulations. Aquifer dynamics underlying a flat residential area of high-density living compounds surrounded by streams are analyzed. The simulated setting represents a typical case in the urbanized alluvial plains and valleys of China. Soils with different hydraulic characteristics are implemented in a MODFLOW groundwater model [25], and results are compared with an ensemble of Sponge City installations, as suggested by [8], using the Storm Water Management Model (SWMM) [26,27]. Via the following analysis, the proposed method (1) quantifies the advantages and limits of the urban aquifers; and (2) compiles an absorbance map for the urban areas of China to assist in the improvement of the Sponge City performance.

## 2. Materials and Methods

### *2.1. The Aquifer Potential Absorbance*

The potential for recharge absorbance at the city scale is difficult to characterize because of heterogeneities in the aquifers and the water infrastructure distribution systems. Variability in the geologic media and the leakage rate from the water infrastructure makes it difficult to characterize hydraulic conductivity and water table oscillations. As the urban expansion of the last decades encompassed large formerly rural areas, urbanization incorporated zones with anthropic modifications such as river diversions, dried up lakes and rivers, rice paddies or other cultivated areas where the natural absorbance capacity of the aquifers has been modified by adding or removing sediments.

Despite the intrinsic uncertainties and difficulties in characterizing the urban subsoil media, a general characterization can be developed by studying the shallow layers at the regional scale. In particular, the connection between the geomorphological and hydrogeological settings remains the driving factor in determining the regional absorbance potential. In accordance with the conceptual model proposed for the hilly terrains of Shenzhen via direct geomorphological and hydrogeological data [23] and major Chinese megacities [24], a conceptual model is proposed to characterize the urban areas. From the mountain sectors to the coastline, each geomorphological sector has unique characteristics of its own:

The bedrock sector is characterized by mountain slopes with a heterogeneous absorbance potential depending on hydraulic conductivity. Metamorphic to igneous rocks are mostly impervious, and infiltrated rain circulates through the shallow weathered blankets (Figure 2c). These blankets represent thin but continuous shallow aquifers where infiltration of precipitation depends on the hydraulic characteristics of the soil. Wet and warm climate areas tend to have thicker soils with a richer component of clays due to increased biological activity; a cold to dry climate, on the other hand, creates thinner but clastic soils [28]. Fractured to karstified sequences can help infiltrate a significant amount of rain via preferential paths (e.g., dolines, polje) or leakage from the upper soil. In this case, the regional water table can be located as deep as hundreds of meters below the ground surface.

The piedmont sector connects the mountain bedrock with alluvial plains and valleys. This sector consists of coarse sequences identifiable as detrital bands and alluvial fans. The sector extent depends on the basin or valley size, and the geological history that caused the depression. The piedmont at the base of the small hills has a width of just tens to hundreds of meters but reaches a width of tens of kilometers in the largest basins. The deposits generally have high conductivity and a flat water table connected to the multilayer aquifers of the alluvial plains and valleys. Moving upstream along the sector, the water table becomes deeper and potential absorbance increases (Figure 2c).

The alluvial valleys consist of coarse fluvial deposits with high hydraulic conductivity alternating with finer lacustrine deposits (natural or anthropic) that can be several orders of magnitude less conductive. Groundwater to surface water exchanges constrain the water table based on the river stage. Thus, the water table is close to the ground near the river and becomes deeper laterally on the fluvial terraces, morphologically higher in elevation (Figure 2c). Thus, potential absorbance generally increases as the distance from the river increases.

The alluvial plain sector is heterogeneous along both vertical and horizontal directions. This sequence is characterized by the alternation of conductive paleochannels, laterally passing to less conductive floodplain deposits. Because of the flat morphology, the water table in connections with rivers and streams is almost coincident with the ground surface. This feature is the main factor limiting the absorbance potential of the plains.

The delta sector is located at the mouth of a river with numerous streams and channels that deposit a huge quantity of fine sediments. Low hydraulic conductivity, flat morphology, and abundant water channel density make the water table almost coincident with the ground level. The potential absorbance capacity is almost null.

The conceptual model of absorbance potential is synthetized in Figure 2c, with a conceptual section from the mountain bedrock sector to the delta. Cities and megacities can occupy many sectors, encompassing more than one catchment. Hilly terrains have diffused small piedmont sectors; major plains and valleys usually have a large piedmont at their borders. The morphology is a direct consequence of the present and former climate, lithology and tectonics characteristics that shaped the landscape in geological times. In addition, the hydrogeological characteristics of the shallow soils in areas of urban expansion play a key role in potential absorbance. Because these parameters may also vary at the district scale, potential absorbance can be further quantified with additional local scale investigations.

### 2.2. Absorbance Zonation of China

A map of the absorbance potential throughout China was constructed by merging hydrogeological and geomorphological information. The hydrogeological map from [29] provides the basic hydrogeological setting of the area. Permafrost areas, karstified terrains, lacustrine to delta sequences, and the piedmont bands of the large plains have been imported from this map. A river database of China, hydrogeological maps at the province scale from the CGS database, and geomorphological and hydrogeological attributes were also used [30]. Unit limits were refined via manual surveys 1:1,000,000, querying satellite multispectral images from LANDSAT-8 and a Digital Elevation Model (DEM) with a resolution of 250 m. Areas with similar color, tone, and pattern were isolated and arranged with contiguous polygons, in accordance with photogeological techniques. ARGIS 10.4 was utilized for the development of the absorbance potential map.

### 2.3. Model Setup

The hydraulic performance of an ideal flat district during and after an intense storm was evaluated using a groundwater model based on the MODFLOW code [25] to quantify the benefit of aquifer storage in the Sponge City design. The model has a square geometry with a total areal extent of 5000 m by 5000 m (25 $km^2$), arranged with 81 real estate compounds of 520 m by 520 m (Figure 3). Each compound simulates high-density living buildings, regularly distributed across the domain and alternated with green spaces equipped with Sponge City installations. High-density living is the most popular construction style in China, and the hydrogeological characterization of these residential areas via numerical analysis is relevant at the national scale. The uniform characteristics of the buildings favor numerical discretization. On the contrary, city centers have unique characteristics due to historical and geographical reasons, and must be characterized at the local scale. In the model, compounds are equipped with absorbing installations covering 30% of the surface. This percentage is the average green space area suggested by the government for each compound [17]. The aquifer characteristics of the district were varied over a reasonable range, and the response of the aquifers to a storm was assessed for different scenarios. Calculated groundwater flows were quantified and compared among scenarios to assess the performance of the Sponge City installations (MOHURD 2014) via SWMM (Figure 3). In the model, MODFLOW and SWMM are loosely coupled as the interaction between groundwater and pipe is neglected.

#### 2.3.1. MODFLOW Setup

The MODFLOW model simulates the groundwater head distribution throughout an ideal district constituted by many high-density living compounds. The model calculates the water flow discharging to the streams located at the domain edges (Figure 3a). It is characterized by 50,000 cells, 10 m width, distributed in 2 layers. Layer 1, specified as one meter thick, consists of absorbing installations and urban impervious infrastructures. The absorbing installations cover 30% of the real estate compounds. Cells are modelled as boxes of sand and gravel recharging the underlying urban aquifer (Figure 3a), arranged with clusters of at least 10 cells and randomly distributed throughout the domain. The remaining cells of Layer 1 are inactive; the urban runoff from the impervious surfaces flows to the sewer, and is not considered in the model. Layer 2 has a thickness of 45 m and represents the urban aquifer, hydraulically connected to the streams. During the MODFLOW simulations, the hydraulic conductivity and specific yield of Layer 2 are varied considering five different materials (Table 1). The UZF (Unsaturated Zone Flow) package in MODFLOW is used where the groundwater table is at land surface.

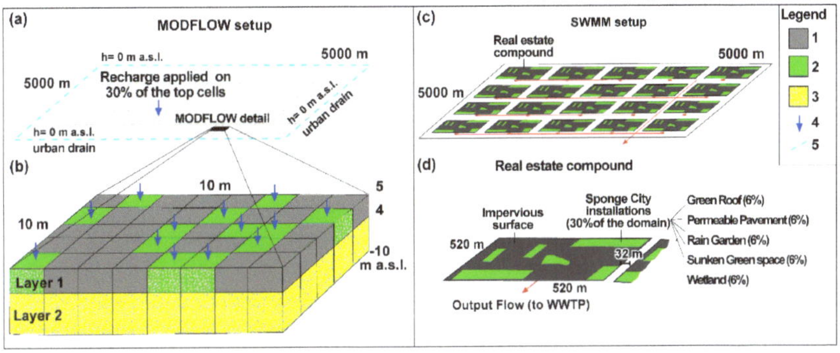


**Figure 3.** MODFLOW model setup of the domain (**a**) and detail (**b**). Key to the legend: impervious urban infrastructures (1); Sponge City installations recharging the aquifer (2); urban aquifer (3); flow recharging the aquifers, 30% of the top cells (4); urban drain, h = 0 m a.s.l. (5). SWMM setup throughout the entire domain (**c**) and detail of the real estate compound (**d**). It is assumed that each compound is equally covered by green roofs, permeable pavements, rain gardens, sunken green spaces, and wetlands, covering 30% of the domain.

**Table 1.** Hydraulic conductivity and specific yield of the materials used for MODFLOW models. Simulations are repeated five times, varying the soil properties of the underlying urban aquifer.

| Material | Type | Layer | Hydraulic Conductivity [m/s] | Specific Yield [%] |
|---|---|---|---|---|
| Sand and gravel box | Sponge City features | 1 | $1{\cdot}10^{-3}$ | 0.33 |
| Soil 1 | Urban Aquifer | 2 | $5{\cdot}10^{-4}$ | 0.31 |
| Soil 2 | Urban Aquifer | 2 | $1{\cdot}10^{-4}$ | 0.27 |
| Soil 3 | Urban Aquifer | 2 | $5{\cdot}10^{-5}$ | 0.23 |
| Soil 4 | Urban Aquifer | 2 | $1{\cdot}10^{-5}$ | 0.21 |
| Soil 5 | Urban Aquifer | 2 | $5{\cdot}10^{-6}$ | 0.19 |

An initial simulation (first time step) of steady-state conditions was conducted with a recharge of 700 mm/a. This recharge rate is equivalent to a sub-tropical wet climate, with permeable terrains characterized by scarce runoff and large infiltration to the subsoil. The recharge rate includes average conditions for rainfall occurring in the area throughout the year, possible urban recharge from pipe leakage, and evapotranspiration. After a manual trial-and-error procedure, selected boundary conditions stressed the Sponge City system, increasing the chance of waterlogging via soil saturation. In fact, the abundant recharge maintains the water table close to ground level throughout the year. Subsequently, a transient simulation was conducted with 12 time steps of 30 min each with a recharge of 200 mm during this six-hour period to simulate an intense storm. Subsequent time steps simulated the 60 days after the storm, with a steady recharge of 700 mm/a. The analysis was repeated with five different materials having decreasing hydraulic conductivity and specific yield (Table 1), simulating different urban aquifers. Along the boundary of the domain, a specified head of 0 m a.s.l. represented the final outflow of the groundwater system (Figure 3a).

2.3.2. SWMM Setup

The Storm Water Management Model (SWMM) is a distributed hydrological model developed by the U.S. Environmental Protection Agency [26]. The model simulates the surface runoff water quantity of a single rainfall event or long-term events in the urban environment. Many studies have demonstrated the utility and usefulness of SWMM for planning and designing Sponge City-type facilities [31].

The domain model coincides with MODFLOW simulations consisting of 81 square catchments of 520 m by 520 m (Figure 3b). Each catchment corresponds to a real estate compound, connected with junctions and conduits, draining the urban runoff. A storm with a steady distribution of 200 mm in 6 h is considered in conjunction with the MODFLOW model. Manning's roughness of impervious N-Imperv and perviousN-Perv; depression storage of impervious Dstore-Imperv and pervious Dstore-Perv; and Horton's infiltration formula, were specified according to the SWMM manual [26]. These parameters affect hydrological processes in the pervious and impervious portions of each sub-catchment, including surface runoff and soil infiltration. Three scenarios were evaluated. In Scenario 1, the sewer drainage system directly collects the rainfall without Sponge City facilities. In Scenario 2, Sponge City installations retain the water attenuating the flow peak of the drainage system, which then flows to the WWTP. A standard module in SWMM simulates the retardation, evaporation, infiltration, and drainage processes of different facilities under rainstorm scenarios. In real estate compounds, five types of Sponge City facilities (Table 2) cover 30% of the top surface. In Scenario 3, Sponge City installations divert the water into the aquifers disconnected from the sewer drainage system.

**Table 2.** SMWW parameters used for the simulations of Scenario 2. Data from [32] [a], [33] [b], [34] [c], [35] [d].

| Layer | Parameter | Green Roof [a] | Permeable Pavement [a] | Rain Garden [b] | Sunken Green Space [c] | Wetland [d] |
|---|---|---|---|---|---|---|
| Surface | Berm Height (mm) | 75 | 0 | 300 | 200 | 400 |
| | Vegetation Volume Fraction | 0.2 | 0 | 0.2 | 0.2 | 0.3 |
| | Surface Roughness | 0.24 | 0.014 | 0.1 | 0.15 | 0.24 |
| | Surface Slope (%) | 1 | 1 | 2 | 1 | 1 |
| Pavement | Thickness (mm) | - | 100 | - | - | - |
| | Void Ratio | - | 0.2 | - | - | - |
| | Permeability (mm/h) | - | 360 | - | - | - |
| Soil | Thickness (mm) | 150 | - | 500 | 500 | 500 |
| | Porosity | 0.45 | - | 0.4 | 0.4 | 0.58 |
| | Field Capacity | 0.2 | - | 0.2 | 0.2 | 0.34 |
| | Wilting Point | 0.1 | - | 0.1 | 0.1 | 0.17 |
| | Conductivity (mm/h) | 120 | - | 72 | 3.6 | 7.62 |
| | Conductivity Slope | 5 | - | 10 | 10 | 8 |
| | Suction Head | 50 | - | 50 | 90 | 90 |
| Storage | Thickness (mm) | 75 | 150 | - | 200 | 250 |
| | Void Ratio | 0.5 | 0.5 | - | 0.5 | 0.5 |

## 3. Results

### 3.1. Drainage Capacity

The MODFLOW model simulates the infiltration and draining capacity of urban aquifers through the year, and in the period following an intense rainfall event. Simulations are repeated for five different soils. In all the simulations, the aquifers discharge water into streams at the model boundaries. In steady-state conditions (1st time step), the hydraulic head has a sub-radial geometry with the maximum located in the domain center, the remotest point from the streams (Figure 4a). The maximum hydraulic head simulated in steady state conditions is 0.5 m a.s.l. for Soil 1, 2.2 m a.s.l. for Soil 2, and 4.2 m a.s.l. for Soil 3, with average gradients of 0.2‰, 0.9‰, and 1.7‰ (Figure 4b), respectively. In steady-state conditions, these soils collect and discharge to urban streams a continuous flow of 1.7 $m^3/s$. Soil 4 and Soil 5 do not discharge all the collected water to the streams as 31% and 42% of the infiltrated water leaves the model as surface runoff throughout the model top, respectively. In fact, soils get saturated when the water table reaches the land surface. These soils are not suited to infiltrate a large amount of water during an exceptional storm. Thus, results of the transient groundwater model are only discussed for Soils 1, 2, and 3.

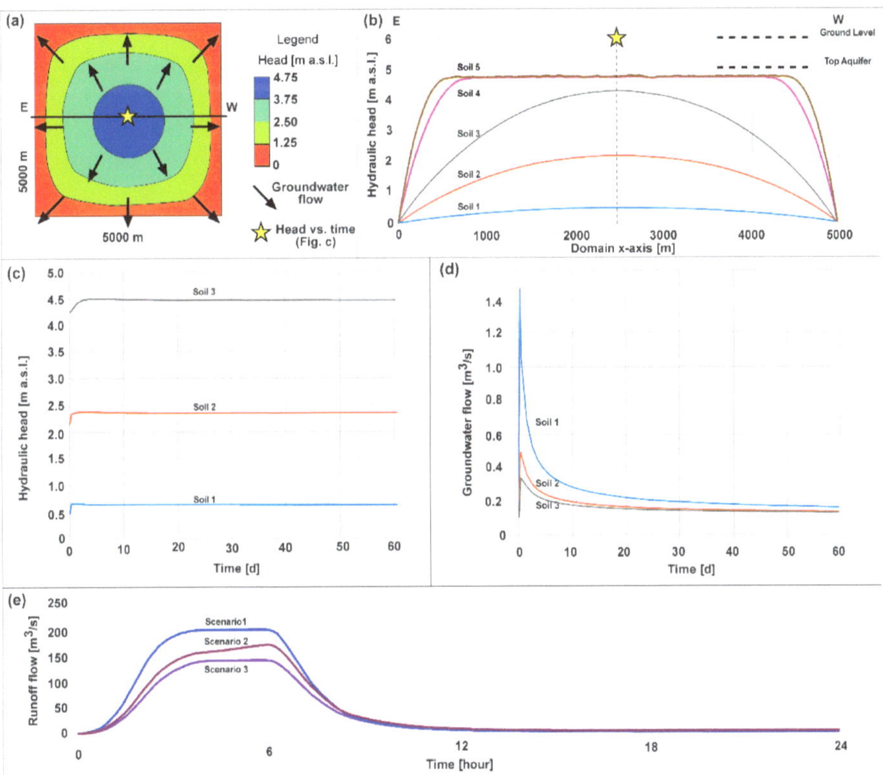


**Figure 4.** MODFLOW groundwater model solution for Soil 3 (**a**); hydraulic head has a radial geometry flowing from the domain center to the model edges; hydraulic head distribution across the domain along the east–west direction for the five considered soils in steady-state conditions (**b**); hydraulic head measured at the domain center (yellow star) during and after the storm (**c**); groundwater flow discharging to the domain edge during and after the storm (**d**); SWMM solutions with total runoff for impervious surface (Scenario 1), Sponge City installations flowing to pipes and tanks, (Scenario 2) and installations recharging the aquifer (Scenario 3) (**e**).

The hydraulic head in the aquifer at the center of the model domain is represented in Figure 4c. The hydraulic head reaches a maximum increment of 35 mm in the first days after the storm. In accordance with soil properties, the incremented hydraulic head slowly declines in the following weeks. Flow is further sustained by continuous recharge after the storm, considering the chance of fresh rainfall after the simulated exceptional storm. Higher increments are observed just underneath the gravel and sand boxes of Layer 1. The shape and number of contiguous active cells of Layer 1 slightly affect the increment variations with time. Peak discharges to the streams of 1.51 $m^3/s$, 0.54 $m^3/s$, and 0.38 $m^3/s$ are simulated for Soil 1, Soil 2, and Soil 3, respectively, in the hours after the storm. After 60 days, the flow is still 17% to 34% higher than the groundwater flow in steady-state conditions (Figure 4d).

The results from the storm water management model provide a broad framework of the performance of the drainage system following a storm. Drainage networks, with no Sponge City features, have an immediate response to the storm (Figure 4e). If a total impervious surface is considered, a total flow peak of 196 $m^3/s$ is reached (Scenario 1). When 30% of the real estate compound surface is improved with Sponge City installations draining to pipes and tanks, a total peak flow of 172 $m^3/s$ is simulated (Scenario 2). The total peak flow of the drainage network is just 12% less than Scenario 1, although the surface recharging the aquifer is about 30% of the real estate compounds. The total flow peak of Scenario 3 is 141 $m^3/s$, a decrement of exactly 30% from Scenario 1. Decremented flow is obtained thanks to the water diverted to the subsoil via the installations.

### 3.2. The Absorbance Potential throughout China

Numerical results point out that in flat urban areas, a significant amount of storm water can infiltrate to the subsoil, reducing the peak flow of the drainage network. A progressive reduction of the absorbance capacity was observed between Soil 3 ($k = 5 \cdot 10^{-5}$ m/s), Soil 4 ($k = 1 \cdot 10^{-5}$ m/s), and Soil 5 ($k = 5 \cdot 10^{-6}$ m/s), and that in the hydrogeological practice corresponded to the passage between an average to a low productive aquifer. Because soils and bedrocks have hydraulic conductivities that span several orders of magnitude, the detected threshold helps assess the absorbance potential. By combining the hydrogeological knowledge of the area with the proposed conceptual model and numerical results, an absorbance map of China is presented.

Piedmont bands, alluvial fans, and terraced alluvial deposits (LS1, in Figure 5) can absorb a significant amount of water. These units strategically extend along the border of the Sanjiang, Songnen, Huai, and North China Plains. The western bounds of Huai and North China Plain are highly urbanized, with many cities and megacities. Due to the alluvial fans, a hilly morphology with fractured to karst bedrock, the Sichuan basin has high potential absorbance. Terraced alluvial valleys of southern China are also mapped with a high potential absorbance capacity. The north-western endorheic basins have coarse sequences that can drain a significant amount of water towards the most depressed points. However, dry climate and lack of significant urbanization nullify the urban waterlogging risk.

The central portion of the alluvial basins and the bottom of river valleys (LS2) extends throughout eastern China (Figure 5). Sandy paleochannels have good hydraulic conductivity [36]. In the last decades, prolonged droughts and massive groundwater withdrawals to sustain urbanization lowered the regional water table, with a maximum drawdown of 1 m/a in North China Plain [11]. The groundwater drawdown caused the progressive disappearance of shallow water bodies. This effect incremented the potential absorbance of the aforementioned areas.

The terminal sector (LS3) is represented by the delta sequence located at the outlet of the major rivers. Lacustrine basins, which dried up in recent history (e.g., Wuhan basin), share the same low absorbance properties as delta sequences. Despite the high-density channels that may shorten the groundwater path from the installation to the streams, low conductive soils do not dissipate infiltration induced by the storm. In fact, densely urbanized areas face recurring waterlogging events, especially along the sub-tropical band, because of unfavorable morphological conditions.

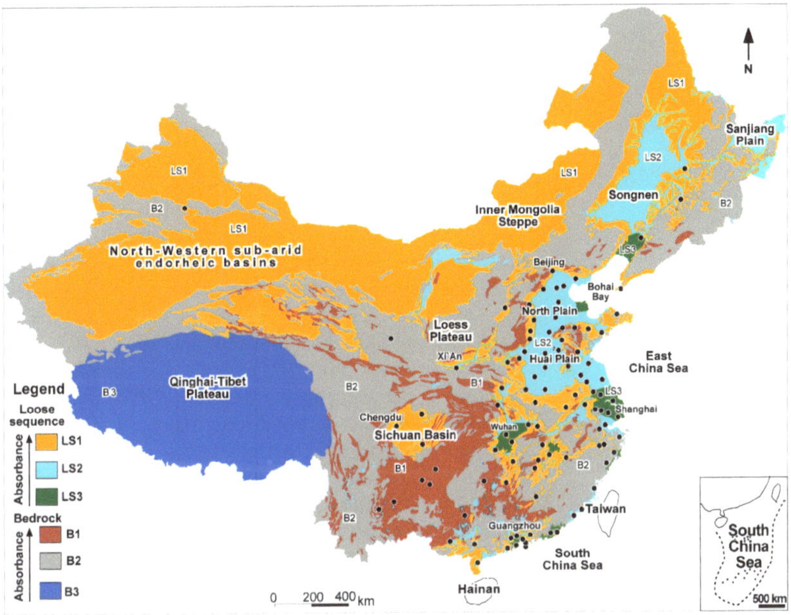


**Figure 5.** Absorbance potential map of China based on hydrogeological and geomorphological criteria. Key to the legend: piedmonts, desertic endorheic basins, alluvial fans, hilly terrains, incised alluvial terraces (LS1); main alluvial plains and valleys, seasonally flooded karst depression (LS2); deltas and lacustrine basins (LS3); karst bedrock (B1); outcrop bedrocks, weathered blankets, and minor alluvial valleys (B2); permafrost and frozen soils (B3). Black dots indicate top 100 most populous cities of China.

Bedrock outcrops are less urbanized and less prone to be affected by severe waterlogging. Karstified to fractured sequences (B1) have a very high absorbing potential, naturally collecting up to 100% of the rainfall. If the karst processes are extremely intense, as in southern China, absorbed storm water is quickly transferred from aquifer to the rivers, increasing flood risks [37]. In fact, the base of the karst basins and valleys can be seasonally flooded as the regional groundwater level rises. Impervious bedrock (B2) has a limited absorbance capacity, driven by the weathering blankets. This unit coincides with the mountain sectors. The cities in the southeast of China, located on the thin coastal sectors, are often urbanized up to the extent of the bedrock (Figure 5). The Qinghai–Tibet Plateau (B3) has limited absorbance due to the permafrost but, at the same time, has scarce urbanization and a dry climate.

## 4. Discussion

Urban expansion changed the original urban pattern of major cities in China, merging small villages and rural areas into contiguous cities and megacities. The original urban cores were typically located near shallow water bodies on alluvial plains and in the delta sectors, and with expansion, have progressively also included upstream sectors characterized by a higher absorbance capacity. This process enormously increased in the last 50 years during China's rapid economic growth. As the most permeable soils are urbanized and made impervious, the total runoff of the catchment increases at the expense of infiltration. In small catchments, the urbanization of the more permeable sectors puts pressure on the drainage network of less permeable sectors, generally located downstream.

Restoration of the natural water cycle in the urban environment is a winning formula to reduce waterlogging and flooding risk. Compared to traditional Sponge City features, installations involving urban aquifers produce better results due to their being disconnected from the drainage system. Numerical analyses have shown that for high-density housing compounds overlying average productive aquifers with hydraulic conductivity higher than $1 \cdot 10^{-5}$ m/s, the area is appropriate for restoration of infiltration potential. Potential absorbance mapping performed at the regional scale (Figure 5) provides a general overview of the areas that are suitable for the proposed approach. In addition, identifying exploited shallow aquifers via water wells, currently or in the past, could represent an indicator of the absorbance potential at the local scale. In fact, a productive aquifer can not only release but also absorb significant volumes of water. However, local aquifer heterogeneities may affect the performance of the installations. Clayey lenses, especially located close to the ground surface, can inhibit seepage toward the discharge points. Before the design and the construction of the installations, preliminary hydrogeological investigations at the local scale can optimize the absorbance performance. In addition, a groundwater conceptual model verifying the preferential flow paths is a value added. Monitoring using a piezometer network can shed solid information about absorbance dynamics.

Taking advantage of the self-treatment capacity of the soil, water collected by the installations may not need treatment, saving money that would otherwise be spent on WWTPs. To guarantee enhanced protection of the aquifers, collected water should come from residential areas; major roads and industrial zones have greater potential for chemical spillage or contaminants. In steady-state conditions and with an aquifer recharge of Sponge City installations at 700 mm/a, the flow discharged to urban streams comes up to 0.17 $m^3$/s. This amount of water bypasses WWTPs, resulting in savings of 6.3 million RMB per year, assuming an average water treatment cost of 1.18 RMB per $m^3$ [38]. From the district to the megacity scale, the economic savings are significant in the long term, especially in wet climates where aquifer recharge via Sponge City installations is remarkable. Reduced initial capital costs and decreased operational and maintenance requirements also make these installations sustainable in minor cities characterized by lower budgets. An increase in the base flow of urban streams provides numerous ecological benefits supporting aquatic ecosystems, especially during the dry seasons. Considering the regional economic inequalities of Chinese cities, the saved money can be reinvested in alternative urban sustainability projects.

Including subsoil in the Sponge City concept can decrease waterlogging and flooding risks. In fact, the total peak flow in the drainage network is reduced proportionally to the extension of the installations, with a gain of more than double if compared to above-ground installations draining to pipes and tanks. However, the right proportion between above-ground installations and those recharging the aquifers will have to be arrived at, based on local conditions. In southern China, where the flood risk is high, the improved design should increase evapotranspiration throughout the installations to lower the water level inside the installation and aquifers, avoiding saturation during wet seasons. In the proposed simulations, an incremented evapotranspiration corresponds to a lower recharge in steady-state conditions. In northern China, characterized by both water scarcity and flood risk, installations diverting water to both pipes and soils may be convenient [39]. Different types of installations can be strategically located along different sectors of the catchment, increasing water availability while decreasing waterlogging and flooding risk. In favorable hydrogeological cases, the limit of the methodology is represented by the number of installations that can be merged in the urban pattern. For the realization of these improvements, local scale absorbance maps based on hydrogeological and geomorphological studies can assist decision-makers, especially during reorganization of urban areas or the retrofitting of existing buildings.

## 5. Conclusions

The inclusion of recharge to urban aquifers in the implementation of the Sponge City concept will mitigate waterlogging and flooding risks more effectively than standard installations. In the simulated domain, numerical analyses suggest that urban aquifers with a hydraulic conductivity higher than $1 \cdot 10^{-5}$ m/s, corresponding to average productive aquifers, can reduce the stormwater flow peak by 18% relative to conventional Sponge City features. Numerous Chinese cities and megacities could benefit from this improved practice. The Sichuan basin, the karst terrains of Southern China, and the major alluvial valleys and hilly sectors have high potential absorbance due to the relatively high hydraulic conductivity of the shallow soils. Similarly, the intensively urbanized alluvial plains of northeastern China have high potential absorbance due to the overexploitation of shallow and deep aquifers.

Pilot projects at the catchment scale, with a combination of enhanced aquifer recharge and above-ground installations, are needed in areas where enhanced aquifer recharge is feasible (e.g., LS1 zone) to demonstrate the efficacy and cost-effectiveness of this approach. Such an approach should be very attractive for minor cities of China with small budgets, as implementation of this approach will be less costly than conventional methods. The costs are lower due to less infrastructure and reduced treatment costs. The proposed procedures can help fill the gap between the rich metropolises of the coast and the poor minor cities of the internal sectors. Direct savings of about 6.3 million RMB/a thanks to the reduced costs of WWTP has been assessed in the performed analysis for a residential area of 25 km$^2$ with a sub-tropical climate. In accordance with the variegated wealth throughout the nation, the saved budget represents an important asset that can be reinvested in other activities. Exploiting urban aquifers as natural sponges has great potential to improve the Sponge City concept, increasing the volume of controlled urban water with possible obtainment of the ambitious goals of the promulged Sponge City concept. Enhancement of the Sponge City with the suggested practices can improve the sustainability of urban areas from an economic, environmental, and architectural point of view.

**Author Contributions:** M.J.: Software, Investigation, Data Curation; M.L.: Writing—original draft, Conceptualization, Methodology, Visualization; Y.T. Software, Investigation, Methodology; S.V.: Conceptualization, Methodology, C.A.: Writing—review & editing, Validation; J.L.: Writing—review & editing; C.Z.: Resources, Funding Acquisition, Supervision. All authors have read and agreed to the published version of the manuscript.

**Funding:** The study was supported by the National Natural Science Foundation of China (Grant No. 41861124003). Additional support was provided by the National Key R&D Program of China (Grant No.2018YFE0206200), and the High-level Special Funding of the Southern University of Science and Technology (Grant Nos. G02296302, G02296402).

**Institutional Review Board Statement:** Not applicable.

**Informed Consent Statement:** Not applicable.

**Data Availability Statement:** The data presented in this study are available on request from the corresponding author.

**Acknowledgments:** Authors are grateful to three anonymous reviewers for the constructive comments.

**Conflicts of Interest:** The authors declare no conflict of interest.

## References

1. Dewan, A.M. *Floods in Megacities: Geospatial Techniques in Assessing Hazard, Risk and Vulnerability*; Springer: Dordecht, The Netherlands, 2013; p. 199. [CrossRef]
2. Barrett, M. Characteristics of urban groundwater. In *Urban Groundwater Pollution*; Lerner, D., Ed.; CRC Press: Boca Raton, FL, USA, 2003; pp. 29–50.
3. Liu, J.; Zang, C.; Tian, S.; Liu, J.; Yang, H.; Jia, S.; You, L.; Liu, B.; Zhang, M. Water conservancy projects in China: Achievements, challenges and way forward. *Glob. Environ. Chang.* **2013**, *23*, 633–643. [CrossRef]

4. Polade, S.D.; Pierce, D.W.; Cayan, D.R.; Gershunov, A.; Dettinger, M.D. The key role of dry days in changing regional climate and precipitation regimes. *Sci. Rep.* **2014**, *4*, 1–8. [CrossRef] [PubMed]
5. Zheng, C.; Gorelick, S. Global change and the groundwater management challenge. *Water Resour. Res.* **2015**, *51*, 1–21. [CrossRef]
6. Huang, Y.; Tian, Z.; Ke, Q.; Liu, J.; Irannezhad, M.; Fan, D.; Hou, M.; Sun, L. Nature-based solutions for urban pluvial flood risk management. *Wiley Interdiscip. Rev. Water* **2020**, *7*, e1421. [CrossRef]
7. Fletcher, T.; Shuster, W.; Hunt, W.; Ashley, R.; Butler, D.; Arthur, S.; Throwsdale, S.; Barraud, S.; Semadeni-Davies, A.; Bertrand-Krajewski, J.; et al. SUSD, LID, BMPs, WSUD and more—The evolution and application of terminology surrounding urban drainage. *Urban Water J.* **2015**, *12*, 525–542. [CrossRef]
8. MOHURD—Ministry of Housing and Urban Rural Development. *Technical Guidelines of the Sponge City Development—Low Impact Development Systems for Storm Water*; MOHURD: Beijing, China, 2014; 88p. Available online: http://www.mohurd.gov.cn/wjfb/201411/W020141102041225.pdf (accessed on 13 January 2022).
9. Nguyen, T.T.; Ngo, H.H.; Guo, W.; Wang, X.; Ren, N.; Li, G.; Ding, J.; Liang, H. Implementation of a specific urban water management-Sponge City. *Sci. Total Environ.* **2019**, *652*, 147–162. [CrossRef] [PubMed]
10. Liu, J.; Zhao, D. Three-dimensional water scarcity assessment by considering water quantity, water quality, and environmental flow requirements: Review and prospect. *Chin. Sci. Bull.* **2020**, *65*, 4251–4261. [CrossRef]
11. Zheng, C.; Liu, J.; Cao, G.; Kendy, E.; Wang, H.; Jia, Y. Can China Cope with Its Water Crisis?—Perspectives from the North China Plain. *Groundwater* **2020**, *48*, 350–354. [CrossRef]
12. Zhang, C.; Duan, Q.; Yeh, P.J.F.; Pan, Y.; Gong, H.; Moradkhani, H.; Gong, W.; Lei, X.; Liao, W.; Xu, W.; et al. Sub-regional groundwater storage recovery in North China Plain after the South-to-North water diversion project. *J. Hydrol.* **2021**, *597*, 126156. [CrossRef]
13. Kong, F.; Sun, S.; Lei, T. Understanding China's urban rainstorm waterlogging and its potential governance. *Water* **2021**, *13*, 891. [CrossRef]
14. Wu, M.; Wu, Z.; Ge, W.; Wang, H.; Shen, Y.; Jiang, M. Identification of sensitivity indicators of urban rainstorm flood disasters: A case study in China. *J. Hydrol.* **2021**, *599*, 126393. [CrossRef]
15. Zheng, Z.; Gao, J.; Wang, Z.; Yang, X.; Luo, X.; Jacquet, T.; Fu, G. Urban flooding in China: Main causes and policy recommendations. *Hydrol. Processes* **2016**, *30*, 1149–1152. [CrossRef]
16. Chen, H.; Jia, B.; Lau, S.S.Y. Sustainable urban form for Chinese compact cities: Challenges of a rapid urbanized economy. *Habitat Int.* **2008**, *32*, 28–40. [CrossRef]
17. MOHURD—Ministry of Housing and Urban Rural Development. *Urban Residential Area Planning and Design Standards*; MOHURD: Beijing, China, 2018; 43p. Available online: http://www.mohurd.gov.cn/wjfb/201811/t20181130_238590.html (accessed on 13 January 2022).
18. MOHURD—Ministry of Housing and Urban Rural Development. *Assessments Standard for Sponge City Construction Effect*; MOHURD: Beijing, China, 2018; 48p. Available online: http://swj.sz.gov.cn/attachment/0/811/811433/9040752.pdf (accessed on 13 January 2022).
19. Wright, T.J.; Liu, Y.; Carroll, N.J.; Ahiablame, L.M.; Engle, B.A. Retrofitting LID practices into existing neighborhoods: Is it worth it? *Environ. Manag.* **2016**, *57*, 856–867. [CrossRef] [PubMed]
20. Zhang, K.; Chui, T.F.M. Linking hydrological and bioecological benefits of green infrastructures across spatial scales—A literature review. *Sci. Total Environ.* **2020**, *646*, 1219–1231. [CrossRef] [PubMed]
21. Zhang, K.; Chui, T.F.M. Assessing the impact of spatial allocation of bioretention cells on shallow groundwater—An integrated surface-subsurface catchment-scale analysis with SWMM-MODFLOW. *J. Hydrol.* **2020**, *586*, 124910. [CrossRef]
22. Lerner, D.N. Identifying and quantifying urban recharge: A review. *Hydrogeol. J.* **2002**, *10*, 143–152. [CrossRef]
23. Lancia, M.; Zheng, C.; He, X.; Lerner, D.N.; Andrews, C.; Tian, Y. Hydrogeological constraints and opportunities for "Sponge City" developments: Shenzhen, southern China. *J. Hydrol. Reg. Stud.* **2020**, *28*, 100679. [CrossRef]
24. Jin, M.; Lancia, M.; Tian, Y.; Viaroli, S.; Andrews, C.; Liu, J.; Zheng, C. Hydrogeological criteria to improve the Sponge City of China. Front. *Environ. Sci.* **2021**, *9*, 700463. [CrossRef]
25. Niswonger, R.G.; Panday, S.; Ibaraki, M. MODFLOW-NWT, A Newton formulation for MODFLOW-2005. *US Geol. Surv. Tech. Methods* **2011**, *6*, 44.
26. Rossman, L.A.; Huber, W.C. *Storm Water Management Model Reference Manual, Volume III—Water Quality*; EPA/600/R-16/093; U.S. Environmental Protection Agency: Cincinnati, OH, USA, 2016.
27. Boroomandnia, A.; Bozorh-Haddad, O.; Bahrami, M.; Goharian, E.; Singh, P.V.; Loaiciga, H.A. Optimizing urban stormwater control strategies and assessing aquifer recharge through drywells in an urban watershed. *Hydrogeol. J.* **2021**, *29*, 1379–1398. [CrossRef]
28. Melesse, A.M.; Abtew, W. *Landscape Dynamics, Soils and Hydrological Processes in Varied Climates*; Springer: Cham, Switzerland, 2016; 893p. [CrossRef]
29. CGS—Chinese Geological Survey. Hydrogeological Map of People's Republic of China, Scale 1:5,000,000. 1 Sheet 1979. Available online: http://en.cgs.gov.cn/ (accessed on 13 January 2022).
30. CGS—Chinese Geological Survey. Repository of Geological and Hydrogeological Maps at Various Scale. 2021. Available online: https://geocloud.cgs.gov.cn/#/home (accessed on 13 January 2022).

31. Chui, T.F.M.; Liu, X.; Zhan, W. Assessing cost-effectiveness of specific LID practice designs in response to large storm events. *J. Hydrol.* **2016**, *533*, 353–364. [CrossRef]
32. Zheng, Y.; Chen, S.; Qin, H.; Jiao, J. Modeling the Spatial and Seasonal Variations of Groundwater Head in an Urbanized Area under Low Impact Development. *Water* **2018**, *10*, 803. [CrossRef]
33. Randall, M.; Sun, F.; Zhang, Y.Y.; Bergen Jensen, M. Evaluating Sponge City volume capture ratio at the catchment scale using SWMM. *J. Environ. Manag.* **2019**, *246*, 745–757. [CrossRef] [PubMed]
34. Liao, X.; Zheng, J.; Huang, C.; Huang, G. Approach for evaluating LID Measure layout scenarios based on random forest: Case of Guangzhou—China. *Water* **2018**, *10*, 894. [CrossRef]
35. Charbonneau, C.; Bradford, A. Wetland modelling in PCSWMM: Exploring options to define wetland features and incorporate groundwater exchanges. *J. Water Manag. Model.* **2016**, *25*, C411. [CrossRef]
36. Wu, C.; Xu, Q.; Ma, Y.; Zhang, X. Palaeochannels on the North China Plain: Palaeoriver geomorphology. *Geomorphology* **1996**, *18*, 37–45.
37. Daoxian, Y. *Karst of China*; Geological Publishing House: Beijing, China, 1991; 224p.
38. Zhang, R. Study on cost structure and Change trend of sewage treatment—Based on data analysis of Shanghai and Shenzhen A-share sewage treatment listed companies. *Price Theory Pract.* **2021**, *113*, 27–30. [CrossRef]
39. Liu, Q.; Cui, W.; Tian, Z.; Tang, Y.; Tillotson, M.; Liu, J. Stormwater management modeling in "Sponge City" construction: Current state and future directions. *Front. Environ. Sci.* **2021**, *9*, 816093. [CrossRef]

*applied sciences*

MDPI

*Article*

# Temporal Variations of Heavy Metal Sources in Agricultural Soils in Malta

**Christine Costa [1,*] and Frederick Lia [2,*]**

[1] Institute of Engineering and Transport, Malta College of Arts, Science and Technology, PLA 9032 Paola, Malta
[2] Institute of Applied Sciences, Malta College of Arts, Science and Technology, PLA 9032 Paola, Malta
* Correspondence: christine.grima@mcast.edu.mt (C.C.); frederick.lia@mcast.edu.mt (F.L.)

**Abstract:** In the opportunity to understand the benefits of Maltese soil and its importance to our climate, the content of heavy metals—including Co, Cr, Cu, Fe, Mn, Ni, Pb, Sr, and Zn—was studied in two fields in proximity in the south-east region of Malta. Analytical determinations were carried out using atomic absorption spectroscopy following heated aqua regia digestion on 50 collected samples using triple repeatability. The decreasing pattern of the concentrations obtained is Fe > Zn > Mn > Sr > Pb > Cu > Ni > Cr > Co. Correlations between pre-harvesting and post-harvesting concentrations were examined to assess lithogenic and anthropogenic relationships. Multivariate analysis including principal component analysis and factor analysis clarified the origin of heavy metals content reviewed. Some of the heavy metals studied showed a dominant relationship between concentration variation and their possible sources. Potential ecological risk assessment demonstrated that the fields reviewed are not contaminated by any of the heavy metals assessed except for Zn which posed a moderate/strong contamination but presented an overall low potential for ecological risk. Concentrations of heavy metals demonstrated no risk to human health and no carcinogenic risk through ingestion and dermal contact with the soil.

**Keywords:** soil; heavy metals; contamination; pollution; ecological assessment; toxicological assessment

**Citation:** Costa, C.; Lia, F. Temporal Variations of Heavy Metal Sources in Agricultural Soils in Malta. *Appl. Sci.* **2022**, *12*, 3120. https://doi.org/10.3390/app12063120

Academic Editor: Dibyendu Sarkar

Received: 2 February 2022
Accepted: 14 March 2022
Published: 18 March 2022

**Publisher's Note:** MDPI stays neutral with regard to jurisdictional claims in published maps and institutional affiliations.



## 1. Introduction

Maltese soils are a natural resource with agricultural, environmental, and cultural value which have undergone human influence over time and due to intensive land-use are experiencing erosion, loss of organic matter, structural deterioration and chemical contamination from excess nitrates and salinity [1]. Desertification is another issue identified within Malta's ecosystem intensifying environmental problems resulting from changes in agricultural practices and increased urban development. Malta unfortunately ranks above the EU average for artificial land coverage with 23.6%, which correlates with a high percentage in population density [2]. The total cultivated area in Malta is around 10,000 hectares amounting to approximately 18.9% of the total land area [1]. An annual soil loss of 19.3% which equates to 10–25 tons per hectare per year indicates that Maltese north-western and Gozitan areas are at the risk of moderate to severe soil erosion [2]. This erosion is very typical of high inclination in arable land areas where poor management and conservation practices are used. The fragmentation of soil reduces its economic viability costing an average agricultural farmer in Malta € 1164.24 per 0.01 km$^2$ per year to replace eroded soils and artificially maintain the soil quality in eroded zones [2]. Malta has registered 135 zones that are undergoing potentially polluting activities in the issued JRC report in the EU and has remediated one registered site. According to the RUSLE2015 model, Malta has an average soil loss rate of 6.02 tons per hectare per year, which is high compared to the EU average of 2.46 tons per hectare per year [2]. Such data depends on the conditions of each individual country and proves that soil degradation is of high importance because of its role in carbon cycling and climate change. The pH level for 56% of Maltese soil lies between 8 and 8.5; thus, soil in Malta has a slightly alkaline pH [3].

There are three main soil types in Malta: Terra Rossa soil, Xerorendzina soil, and Carbonate Raw soil [3]. These have been subdivided into subgroups according to the area of location. As reported by Lang in 1960 [3], the soil types found in the sampling zone are Terra Rossa, Xerorendzina, and the L-Iklin and Tad-Dawl complexes. Terra Rossa soil is a red clay-like soil resulting from both types of coralline limestone existing on the island. Its reddish color is derived from the high iron oxide content and has a calcium carbonate content of 2–15%. Xerorendzina soil ranges from a whitish to red soil with a calcium carbonate content of 58–80%. It is a constituent of Globigerina limestone which is made up of clay and lime [3]. Calcisol soil is characterized by calcium carbonate concentrations which are found as coatings on the soil structure faces. Such Calcisol horizons are present in the lower topsoil or subsoil horizons, are rich in lime and form in dry areas. If their dryness factor is adequately controlled through irrigation, drainage, and fertilization to prevent salinization, Calcisol soil is highly productive for crops [3].

A heavy metal is a dense metal with high atomic weight which is toxic at low concentration. Heavy metals are regarded as pollutants in soil because of their persistence and toxicity upon high concentration or unwanted presence. Since agricultural and industrial processes have changed over time due to modernization, several geo-chemical cycles, degradation processes, and ecosystem functions have been altered leading to significant concern over their polluting effects [4]. Trace elements accumulated in the soil are mainly due to industrial, mining, agricultural practices, and the disposal of waste from communal sources. Several trace elements have essential metabolic functions in living organisms, but some others influence the biological, chemical, and morphological traits in plants, animals and humans upon physical contact, inhalation, or ingestion [5].

Various ecological functions make soil an extremely complex matrix which provides a suitable environment for the development of living organisms, conditions water flow and biological activities, detoxifies and immobilizes organic and inorganic matter, regulates biotic diversity, and transforms potential harmful compounds [5]. Heavy metals are bound in the soil with their oxides or dissolved with aquo complexes, with inorganic complexes or with organic complexes and chelates [6]. Heavy metal dynamics in the soil and their uptake by plants is directly linked to the soil properties, altering their bioavailability. Nortcliff and Schulte-Bisping stated that the pollutant content of soil depends on the background values which represent its appropriate content [6]. Some of the chemical processes involved in the transformation of trace elements in soil include precipitation–dissolution, adsorption–desorption, and complexation. These processes control bioavailability and mobility of such elements to manage both deficiency and toxicity levels in agroecosystems. This allows the buffering of nutrients and pollutants to occur whereby the input of acids in the soil naturally through the absorption of nutrient ions by plants, and anthropogenically through fertilization and atmospheric absorption are balanced out by the corresponding salts maintaining a constant pH in a particular range [6]. The transformation function of soil occurring abiotically or biotically, provides a self-cleansing process from chemical compounds found in the pedosphere. Soils with good nutrient supply, good aeration, and water balance have an optimal microbial activity, and the degradation process of such elements is considerably high. This is essential for plants and microorganisms building up their cellular material, and thus is vital for plant growth [6,7]. The fractional concentration of pollutants entering the soil as soluble compounds depends on the ion affinity with reaction centers in the soil components and are in their metal oxide or in salt form. Their transformation depends on their solubility rate, metal salts being more mobile than their respective oxides, thus having an increased probability of causing an ecological danger [8]. Ghosh and Devi [9] have studied the effect of using inorganic fertilizer on crop yield and have reported that higher yield was attributed to the compost increasing organic carbon, which reduced the release of carbon dioxide from the soil as well as the water retention. This matched the work by Pomazkina and Semenova [10], which outlined that a change in the microbial activity could be observed with the transformation of carbon containing substances, tying in with concepts of mineral transformation processes discussed by Nortcliff and Schulte-Bisping [6] and Mulder and VanVeen [7]. Zhang et al. [11] reported that increasing nitrogen deposition in the soil, increased the amounts of pollutants following a parabolic diffusion relationship. This contrasts conclusions drawn by Ghosh and Devi [9], who documented that high yield is associated with the blending of inorganic fertilizer with compost lowering the total nitrogen content in the soil, thus reducing the need for nitrogen fertilizers, preventing nitrogen percolation and the release of nitrous oxide into the atmosphere. Several studies concluded that fertilizer addition should be appropriate to the needs of the cultivated plants to avoid excess of nutrients in the soil and their eventual release into the environment [6,9].

During crop cultivation, the concentration requirement of non-essential elements varies with the crop growth stage [12]. Mulder and Van Veen, O'Neill and Micó, et al., reported several chemical elements which are essential for the growth of a living organism, especially plants, hence linking sub-optimal growth to the deficiency of major and trace elements [4,7,12]. Some of the elements mentioned include chromium, manganese, iron, cobalt, nickel, copper, and zinc, all of which are trace elements—occurring in plants with less than 0.01% concentrations, but which are important for organism growth. Chemical processes involved in the transformation of trace elements in soil control bioavailability and mobility of such elements to manage both deficiency and toxicity levels in agroecosystems [13]. If these trace elements are in their insoluble state, or are dispersed in other minerals, their availability for uptake depends on the relative stability and mobility of the compound [12]. Sorption interaction with the solid phase may be masked if a significant release of metals into solution from the soil matrix at favorable conditions occurs [14]. This pH adjustment is achieved by shifting the soil pH to alkaline by adding organic matter to the soil, immobilizing heavy metals in their solid state, and thus reducing the leaching probability of heavy metals into the plants [15], altering also the humus content in the soil [10]. The mobility of heavy metal ions decreases slowly with time until equilibrium is achieved possibly after decades while their availability in the soil is highly linked to the origin of the soil and its history [16].

The aim of this study is to provide scientific insight on the chemical pollutants pertaining to Maltese soil through the characterization of heavy metals in two neighboring fields which are worked using traditional methods. The heavy metals characterized include Co, Cr, Cu, Fe, Mn, Ni, Pb, Sr, and Zn. Another study regarding the content of heavy metals on soils in the Maltese Islands carried out by Briffa (2020) has focused on general concentration values of several heavy metals in different locations around the island and only one location has been chosen close to the study areas chosen for this study; thus, comparison of the results obtained and adequate mapping of the contamination effects on the soil was limited. This research should provide knowledge regarding heavy metals concentrations typical of common agricultural practices on the island. Changes in heavy metals content in the soil is to be further analyzed in conjunction with different stages in crop cultivation. Further research and analysis of the concentrations acquired from 50 soil samples with triple repeatability collected from two different fields is to be undergone to assess whether the fields under review are polluted with heavy metals and are in a heavily degraded stage by using multivariate analysis. A classification of the pollution sources outlined in this study is to be done using principal component analysis on the dominant heavy metals being reviewed.

## 2. Materials and Methods

### *2.1. Location of Study Area*

Two fields, Field A and Field B, located in the south-east region of Malta in proximity (about 500 m away from each other) as illustrated in Figure S1 were studied. Field A has an area equal to approximately 1058 $m^2$ and is located on coordinates (35.827699, 14.508289), and Field B has an area of approximately 1248 $m^2$ and is located on (35.828808, 14.509760).

### 2.2. Sampling and Pre-Treatment

Two groups of samples were collected—at the pre-harvesting stage where the soil has already been fertilized, ploughed, and the crops have just been planted and were possibly germinating in November 2020; and at the post-harvesting stage where the crops were fully grown and had just been picked in April 2021. A total of 9.4 mm of rainfall and a temperature of approximately 16 °C was recorded on the day of the pre-harvesting sample collection, while no precipitation and a temperature of approximately 19 °C was documented for the sample collection at the post-harvesting stage [17]. This gives concentrations of the heavy metals after fertilizer addition and before any possible uptake by the crops during the pre-harvesting sampling stage and proves if any differences occur once crops have been grown. Fertilization was done in August 2020 for both fields to allow ample time for the nutrients to dissolve in the soil before sampling. Field A was fertilized using approximately 1.134 $kg/m^2$ of rabbit manure while Field B was fertilized using 0.100 $kg/m^2$ organic fertilizer in pellet form. The same type of crops was grown in both fields, namely, potatoes, beans, onions, spinach, and cauliflower, and potable water used for irrigation was obtained from the same supplier. Such analysis would demonstrate the benefits or drawbacks of the presence of heavy metals in the soil upon product cultivation. Samples were taken from the top 20–25 cm range since upper soil horizons have a wider interaction with atmospheric environmental conditions including dust particles, fertilizers, and pesticides, using a hand shovel, and stored in sealed plastic bags. Fifty samples were collected from both fields for each sampling stage to account for any spatial variability from different zones in the fields. Stratified random sampling system was used, whereby the entire area was divided into smaller sections in a grid pattern, and random sampling was applied on some of the sections. Samples were not collected close to the roots of large trees, buildings, or piles of manure as the natural heavy metal content distribution in such locations is altered. Soil samples were oven-dried at 105 °C until constant mass was obtained [18].

### 2.3. Physical Analysis

Soil physical parameters were analyzed to describe soil texture, friability, structure, and type, as these influence the chemical composition of the soil effecting nutrients and pollutants mobility, and chemical and biological processes involved in soil functions. Each of the physical characteristics tests was carried out on 5 random samples from each field, which were then worked in triplicate to obtain an average value for each field. Physical parameters studied include bulk density, specific gravity, and moisture content. Bulk density is important for root penetration and soil permeability and varies inversely with pore space. Bulk densities for soil generally vary between 0.8 and 1.7 $g/cm^3$ [18]. Bulk density was calculated upon drying 5 collected soil samples from each field to obtain an average representative value.

Specific gravity is directly related to the bulk density and is used to describe soil porosity indicating the amount of water saturation in the soil. Typical values of soil specific gravity lie in between 2.65 and 2.80 [19]. Specific gravity was calculated for 5 soil samples from each field.

Soil moisture content was determined on dried and sieved samples using a 2 mm mesh on 5 soil samples from each field. Water content analysis is important for highlighting any undesirable effects on plant roots and their growth as well as chemical and microbial activities taking place in the soil levels.

### 2.4. Chemical Analysis

All glassware for use during the chemical analysis process was washed with soap and water before soaking in an acid bath for at least 24 h before use to remove contaminants. The acid wash used was made from diluted aqua regia solution which consisted of 1 part concentrated nitric acid and 3 parts concentrated hydrochloric acid, mixed with 1 part water to reduce the generation of chlorine gas since the solution was to be stored over a short period of time [20]. The acid bath was made using the same grades of acids used for analysis. Some soil from each sample was crushed manually using a pestle and mortar and sieved using a 2 mm mesh size sieve, upon cooling. One gram of this sieved soil was taken; with a repeatability of 3. Thus, a total of 150 samples were then available for processing, each of which was adequately labelled for full traceability. Digestion was carried out using aqua regia solution made from nitric acid with 69.5% concentration RPE ACS-Reag.PH.EUR.-Reag. USP grade and hydrochloric acid with 37% concentration RPE ACS-Reag.PH.EUR.-Reag. USP grade in molar ratio of 1:3. 15 mL of aqua regia were added gradually in steps (1 mL, 1 mL, 1 mL, 2 mL, 10 mL) to control the reaction of the acid solution with the calcareous stones in the soil in order to reduce soil sample loss as much as possible. In between each addition of solution, each sample was slightly swirled to completely wet the sample, speed up the reaction and entirely dissolve the calcareous deposits for easier processing. Digestion was carried out, heating the samples with acid solution in heating blocks for 30 min at 50 °C and 2 h at 120 °C. Once cooled, each was filtered using vacuum filtration setup with a sintered ceramic membrane and a filter paper Grade 393 which retains 1–2 μm. The filtrate was transferred to a 50 mL plastic falcon tube, and each sample was reconstructed using 10 mL nitric acid.

### 2.5. Analysis of Heavy Metals Using Spectroscopy

Upon digestion to extract heavy metals from the soil samples collected, analysis using Atomic Absorption Spectroscopy (AAS) was used to give the concentrations of such metallic elements, based on the quantity of light being absorbed at a specific wavelength corresponding to the known characteristics of the element being tested. AAS gives readings with a high degree of accuracy, results generally fall in the range of 0.5–5% accuracy, and is a highly sensitive method of analysis [21]. The amount of light absorbed by the sample is proportional to the concentration of metal ion in solution. Since AAS is used to measure trace elements, contamination from equipment or storage of the sample can lead to severe sources of error. The sample must be conserved in terms of pH, and constituent elements. Since the material of the storage containers may absorb some of the analyte, perflouroalkoxy (PFA) polymers, or glass were used as storage materials to affect the analyte as least possible. To prevent analyte ions from adhering to the walls of the containers used, solution acidification with nitric acid was carried out. A continuous flame-type Atomic Absorption Spectrophotometer produced by Shimadzu Europa GmbH model AA-7000 was used. The lamp mode was set to BGC-D2 type for all elements tested except for Sr for which a Non-BGC lamp mode was used. Burner height was set at 7 mm for all elements tested except for Cr and Fe for which this parameter was set to 9 mm. For all elemental readings, burner lateral and burner angle were set at 0 pulse and 0 degrees. The Air-Acetylene flame was used, which supply flow rate was in the range of 1.6 L/min to 2.8 L/min, and the support gas was supplied with a flow rate of 15.0 L/min for all elements. Wavelength settings used were: 240.7 nm Co, 324.8 nm Cu, 357.9 nm Cr, 248.3 nm Fe, 279.5 nm Mn, 232.0 nm Ni, 283.3 nm Pb, 460.7 nm Sr, 213.9 nm Zn respectively. Slit widths were set at 0.2 nm for Co, Fe, Mn, and Ni respectively, and at 0.7 nm for Cu, Cr, Pb, Sr, and Zn respectively.

Upon sample analysis using AAS, absorbance values were obtained. To obtain concentration of analytes in the solution, standard calibration curves of absorbance against concentration were created with water. Percentage recovery values computed showed a substantial amount of matrix interference; thus, calibration curves were then set up using sample analyte solution to reduce the matrix effect to a minimum using standard addition calibration method. All analysis was carried out in triplicates, against a method blank.

A calibration check was carried out prior to each experiment and included every 6–10 samples. Apart from that precision was checked by preparing 6 standards of same concentration to check that the Percentage Relative Standard Deviation (% RSD) was always less than 2%.

### 2.6. Statistical Analysis

IBM SPSS Statistics 24 software was used to carry out correlation analysis, regression analysis and normality tests with a 95% confidence interval on the data obtained. JMP Trial 16 software was utilized to perform parametric tests, multivariate analysis, PCA, and FA. PCA was used to identify elemental dominance when variation of their respective concentrations was tested, and to distinguish possible elemental sources. In carrying out factor analysis, a rotating component matrix was used to separate the possible elemental combinations and two factors with accumulative variance contributions were chosen for analyses.

### 2.7. Health Risk Assessment

Exposure to heavy metals and other chemicals present in the environment such as arsenic, cadmium, chromium, cobalt, copper, lead, mercury, molybdenum, selenium, nickel, and zinc [22] may pose risks to human health through inhalation, dermal contact, and ingestion. Impacts of the content of various heavy metals in soil samples are assessed by calculating the average daily doses (ADD) by ingestion and by dermal contact as shown in the below equations.

$$ADD_{ingestion} = \frac{C_s \times IR_{ingest} \times EF \times ED \times CF}{BW \times AT} \quad (1)$$

$$ADD_{dermal} = \frac{C_s \times SA \times AF \times ABS \times EF \times ED \times CF}{BW \times AT} \quad (2)$$

where $C_s$ is the average heavy metal concentration in the sample; $IR_{ingest}$ is the ingestion rate of heavy metals; EF is the frequency of exposure; ED is the duration of exposure; BW is the average body weight for adults; AT is the averaging time for carcinogens and non-carcinogens; CF is a unity conversion factor; SA is the estimated average exposed skin area; AF is the adherence factor; and ABS is the dermal absorption fraction [22]. Typical values used to compute $ADD_{ingestion}$ and $ADD_{dermal}$ are included in Table 1.

**Table 1.** Parameter values used to compute average daily dose by ingestion and dermal contact [22,23].

| Parameter | Symbol | Unit | Adult Value | Children Value |
|---|---|---|---|---|
| Ingestion Rate | $IR_{ingest}$ | mg/day | 100 | 200 |
| Exposure Frequency | EF | days/year | 350 | 350 |
| Exposure Duration | ED | years | 30 | 6 |
| Average Body Weight | BW | kg | 70 | 15 |
| Non-Carcinogens Averaging Time | AT = ED × 365 | days | 10,950 | 2190 |
| Carcinogens Averaging Time | AT = 70 years × 365 | days | 25,550 | 25,550 |
| Unity Conversion Factor | CF | kg/mg | $1 \times 10^{-6}$ | $1 \times 10^{-6}$ |
| Exposed Skin Area | SA | $cm^2$ | 5700 | 2800 |
| Adherence Factor | AF | $mg/cm^3$ | 0.07 | 0.07 |
| Dermal Absorption Fraction | ABS | / | 0.001 | 0.001 |

The Hazard Quotient (HQ) shows the tendency of non-carcinogenic toxicity which may come about from a long-time exposure of humans to a non-carcinogenic heavy metal. The HQ and the Hazard Index, HI, for all metallic elements in the soil samples analyzed are calculated using:

$$HQ = \frac{ADD}{RFD} \quad (3)$$

$$HI = \sum_{i=0}^{n} HQ_i \quad (4)$$

where RFD is the reference dose for each heavy metal that an individual is exposed to per day either by ingestion or dermal contact throughout his entire life without any harm;

HI is the overall toxic risk and n is the total number of metals under consideration as stated in the United States Environmental Protection Agency IRIS (USEPA IRIS) (2011), and referenced by Agoro, et al. (2020), Anyanwu and Nwachukwu (2010), and Liang, et al. (2017) [22,24–26] shown in Table 2. If HQ > 1, it represents adverse non-carcinogenic effects and if HQ < 1, it shows an acceptable level of these concentrations without any concerns. If the Hazard Index, HI < 1, the non-carcinogenic adverse effects are assumed to be negligible [24,25].

**Table 2.** RFD and SF values for heavy metals [22,24–26].

| Metal | RFD [1] by Ingestion (Mg/Kg/Day) | RFD [1] by Dermal Contact (Mg/Kg/Day) | Slope Factor (Mg/Kg/Day) |
|---|---|---|---|
| Fe | 0.007 | Not found | Not found |
| Mn | 0.014 | Not found | Not found |
| Zn | 0.300 | 0.060 | Not found |
| Cr | $3.000 \times 10^{-3}$ | $6.000 \times 10^{-5}$ | 0.500 |
| Ni | 0.020 | $5.400 \times 10^{-3}$ | 0.910 |
| Pb | $3.5 \times 10^{-3}$ | $5.25 \times 10^{-4}$ | 0.085 |
| Cu | 0.040 | 0.012 | Not found |

[1] RFD is the reference dose for each heavy metal.

Cancer risk (CR) is the incremental risk of a person having cancer over a lifetime by means of exposure to carcinogenic substance (like heavy metals) because of the exposure of twenty-four hours per day for almost a full year for a lifetime i.e., seventy years [27]. The following equation is used to calculate this incremental probability:

$$\mathrm{CR} = \mathrm{ADD} \times \mathrm{SF} \tag{5}$$

$$\mathrm{RI} = \sum_{i=0}^{n} \mathrm{CR_i} \tag{6}$$

where SF is the slope factor (mg/kg/day) defined as an incremental rate of cancer development originating from a lifetime exposure to contaminants and is represented by the upper bound estimate approximating 95% confidence limit of the increased cancer risk [28]. SF values for ingestion are outlined in Table 2, based on USEPA IRIS (2011) standards [25]. RI is the cancer risk index including the summation of individual cancer risk attributing to each metallic element in the soil sample analyzed [27].

## 2.8. Potential Ecological Risk Assessment

To assess the pollution level of an environment, the identification between anthropogenic activities and the natural level of a given substance is important. Soil pollution assessment with heavy metals is possible using pollution indices to determine the provenance of such trace elements accumulated in the soil and calculating the respective ecological risk. The geo-accumulation index gives a direct evaluation of the extent of heavy metal accumulation in the soil. Such values are important to quantify the degree of anthropogenically or lithogenically accumulated contaminants in the studied areas. The Müller relationship is used to calculate the geo-accumulation index [29]. Different background concentration values for different world sediments are outlined in the reference document issued by Turekian and Wedepohl [30].

$$\mathrm{I_{geo}} = \log_2\left(\frac{\mathrm{C_n}}{1.5 \times \mathrm{B_n}}\right) \tag{7}$$

where $C_n$ is the metal concentration in the sample for each metal from this study, and $B_n$ is the background level for individual metals, taken as 0.1 ppm, 11 ppm, 4 ppm, 1100 ppm, 20 ppm, 9 ppm, 20 ppm for Co, Cr, Cu, Mn, Ni, Pb, and Zn, respectively [30]. Seven classes of the geo-accumulation index are listed in Table 3.

**Table 3.** Classification of geo-accumulation index ($I_{geo}$) [29].

| Geo-Accumulation Index Level | $I_{geo}$ Class | $I_{geo}$ Value |
|---|---|---|
| Uncontaminated | 0 | $I_{geo} \leq 0$ |
| Uncontaminated/moderately contaminated | 1 | $0 < I_{geo} < 1$ |
| Moderately contaminated | 2 | $1 < I_{geo} < 2$ |
| Moderately/strongly contaminated | 3 | $2 < I_{geo} < 3$ |
| Strongly contaminated | 4 | $3 < I_{geo} < 4$ |
| Strongly/extremely contaminated | 5 | $4 < I_{geo} < 5$ |
| Extremely contaminated | 6 | $5 < I_{geo}$ |

The Contamination Factor (CF) represents the relationship between the concentration of the metal being tested and its corresponding background level. The Potential Ecological Risk Index is used to assess the degree of metal pollution in sediments according to the toxicity of metals and the response of the environment. It is calculated using the equation:

$$CF = \frac{C_{metal}}{C_{background}} \tag{8}$$

$$E_i = T_i CF_i \tag{9}$$

$$PERI = \sum_{i=0}^{n} E_i \tag{10}$$

where, $E_i$ is the monomial potential ecological risk factor; $T_i$ is the toxic response factor for a given substance; typical values include Co = 5, Cr = 2, Cu = 5, Pb = 5, Mn = 1, Ni = 5, Zn = 1 [29,31,32], and $CF_i$ is the contamination factor for each metal under study.

Table 4 describes the classification of resulting values for the risk index to better understand the overall effect of environmental contamination by the respective heavy metals.

**Table 4.** Classification for the Potential Ecological Risk Factor ($E_i$) [29].

| Risk Factor Level | $E_i$ Value |
|---|---|
| Low potential ecological risk | $E_i < 40$ |
| Moderate potential ecological risk | $40 \leq E_i < 80$ |
| Considerable potential ecological risk | $80 \leq E_i < 160$ |
| High potential ecological risk | $160 \leq E_i < 320$ |
| Very high potential ecological risk | $320 \leq E_i$ |

## 3. Results and Discussion

### 3.1. Physical Analysis

Physical studies of bulk density and specific gravity on the soil tested indicated a high organic content owing to the soil natural nutrients, high porosity, and a crumbly texture, proving it healthy and abundant in microorganisms making it suitable for crop cultivation. This ties in with the alkalinity level found in literature for soil in Malta [3,15]. Results of the water content from both fields differed with changes in the weather conditions, since samples collected at the pre-harvesting stage were taken during the precipitation season and the post-harvesting stage were taken during summer. An average water content difference of 48.5% was recorded; hence, the solubility of soil minerals [7,11] could affect the bioavailability of the heavy metals concentrations reviewed.

### 3.2. Chemical Analysis

#### 3.2.1. Descriptive Statistics

The concentrations of the heavy metals studied listed in decreasing order are Fe > Zn > Mn > Sr > Pb > Cu > Ni > Cr > Co as shown in the descriptive statistics for the pre- and post-harvesting results in Table 5. The concentration results obtained indicate primarily that the fields under study are not polluted by heavy metals since none of the concentrations quantified exceed reference limits set by legislation as documented in Table 6.

**Table 5.** Descriptive statistics of heavy metal contents in soil in μg/g—Pre-harvesting and Post-harvesting stages.

| Variable | Symbol | Unit | Descriptive Statistics | | | |
|---|---|---|---|---|---|---|
| | | | Pre-Harvesting Concentrations | Post-Harvesting Concentrations | Mean | Standard Deviation |
| Cobalt | Co | μg/g | 0.150 | 0.047 | 0.100 | 0.055 |
| Chromium | Cr | μg/g | 0.450 | 0.197 | 0.339 | 0.214 |
| Copper | Cu | μg/g | 1.456 | 0.108 | 0.818 | 1.092 |
| Iron | Fe | μg/g | 3814.549 | 729.829 | 2594.401 | 1290.501 |
| Manganese | Mn | μg/g | 66.338 | 15.118 | 41.898 | 27.021 |
| Nickel | Ni | μg/g | 0.824 | 0.0915 | 0.461 | 0.370 |
| Lead | Pb | μg/g | 4.373 | 1.331 | 3.034 | 4.026 |
| Strontium | Sr | μg/g | 10.050 | 10.197 | 9.969 | 1.236 |
| Zinc | Zn | μg/g | 254.888 | 157.437 | 207.262 | 59.237 |

**Table 6.** Heavy metals contents in Mediterranean soils.

| Element | Mean Concentrations in This Study (mg/g) | Study From Spain by Micó, et al. (2006) (mg/Kg) [4] | Study from Iraq by Salah, et al. (2015) (mg/Kg) [33] | Study from Greece by Serelis, et al. (2010) (mg/Kg) [34] | Study from Portugal-Spain Border by Nunes, et al. (2014) (mg/Kg) [35] | Study from Egypt by Khatita, et al. (2020) (ppm) [36] | Concentration target Values for Soil (ppm) [37] | Target Values for Malta by 86/278/EEC (mg/Kg) [38] |
|---|---|---|---|---|---|---|---|---|
| Co | 0.100 | 7 | 3.430 | 8.300 | - | 31 | 20 | - |
| Cr | 0.339 | 27 | 11.590 | - | 0.850 | 128 | 100 | - |
| Cu | 0.818 | 23 | 2.010 | - | 1.230 | 61.800 | 36 | 100 |
| Fe | 2594.401 | 13,608 | 235.770 | - | - | 62,583 | - | - |
| Mn | 41.898 | 295 | - | - | - | 1099 | 200 | - |
| Ni | 0.461 | 21 | 8.960 | 9.430 | 1.950 | 70.600 | 35 | 70 |
| Pb | 3.034 | 23 | 3.820 | 161.360 | 3.160 | 30.700 | 85 | 100 |
| Sr | 9.969 | - | - | - | - | 308 | - | - |
| Zn | 207.262 | 53 | 5.500 | 3.520 | 0.640 | 143 | 50 | 200 |

With reference to Table 6, micronutrients Fe, Zn, and Mn presented higher concentrations in agricultural soil while Cr and Co had the lowest values, which results match findings from Spain and Iraq reported by Micó, et al., (2006) and Salah, et al., (2015) [4,33] respectively. Upon comparison of means for each element reviewed during the post-harvesting stage, it is observed that all concentrations have decreased considerably, except for Sr content which has slightly increased. This reduction in concentration may be attributed to several physical environmental changes such as the weather, water content variation due to differences in seasons, pH levels in the soil and crop uptake of the heavy metals analyzed which varies according to the macronutrient content and the time of crop harvesting [11,15,39].

The average heavy metals concentrations for all samples collected over both sampling stages do not exceed the reference values for Malta outlined in the 86/278/EEC Directive for agricultural soils except for Zn content which average exceeds the limit values marginally. Another study carried out by Briffa (2020) about agricultural soil in Malta used Dutch and Finland reference values for concentration limits [37]. The concentration results obtained in this study are within these limit values except for Zn content. Compared to other studies carried out in the Mediterranean region shown in Table 6, Zn content was higher but still within the legislation range outlined. The closest documented results (with an exception to Zn) are the ones reported by Nunes, et al., (2014), which focused on soil from the Caia area in Portugal and Spain [35]. Further statistical analysis shall confirm whether the fields in this study are in fact polluted with high Zn content and aid in outlining possible Zn sources to remediate and reduce such contamination. The only study that determined Sr content is by Khatita, et al., (2020) but the values documented by this work are by far higher than found in this study [36]. No limit reference values were found for Sr content in agricultural soil.

The distributions at the pre-harvesting stage of Co and Ni are symmetrical since the skewness coefficient is close to zero. The other distributions are asymmetrical since a discrepancy between the mean and median values could be observed, and high values for the skewness coefficients resulted. From the post-harvesting stage analysis, the distributions of Sr and to a certain extent of Zn for Field A resulted symmetrical proving homogenous distributions. Cr is the only elemental distribution that is symmetrical and thus homogenous over Field B in both sampling stages. Non-point sources could be attributed to the origin of these homogenous distributions, such as air pollution effects from vehicle, aircraft emissions, industrial emissions, and urban dust. Serelis, et al., (2010) suggested that Co and Ni are derived from the parent material since their highest concentrations were recorded at a lower sampling depth (<25 cm) and both metals are usually rich in soil with high organic matter [34]. The other distributions for Fe, Zn, Mn, Sr, Pb, Cu, and Cr could be linked to the dumping of waste or inorganic material, falling hunting shots, and the use of paint and oil on site.

Upon analyzing the pre-harvesting and post-harvesting samples with the Kolmogorov–Smirnov tests, Cr is the only normal distribution in all the data with $p = 0.143$. This result could originate from the fact that Cr is found in an insoluble fraction of soil which makes up structural silicate minerals that are inert [16]. This links to the homogeneity of the Cr distribution across both fields reported in the descriptive statistics and matches similar results documented by Orrono and Lavado (2009) [16]. Sr content is marginally significant with $p = 0.031 < 0.05$.

Various hypotheses were analyzed using the Mann–Whitney U-test to check whether elemental concentrations had the same distributions across both fields at the pre-harvesting stage and the post-harvesting stage, or the same across Field A and Field B distinctively at the pre-harvesting and post-harvesting stages combined.

As shown in Table 7, the null hypothesis stating that the elemental distributions were the same across both fields was retained for Ni and Zn at the pre-harvesting stage (Mean Rank = 5.96, N = 50, U = 245, z = −1.089, $p = 0.276$ for Ni and Mean Rank = 24.97, N = 50, U = 364, z = 1.267, $p = 0.205$ for Zn) and for Fe and Zn at the post harvesting stage (Mean Rank = 25.59, N = 50, U = 289, z = −0.218, $p = 0.828$ for Fe and Mean Rank = 24.685, N = 50, U = 398, z = 1.941, $p = 0.052$ for Zn). It can be observed that Zn content is the same across Field A and Field B both at pre-harvesting and post-harvesting stages; thus, crop growth has not altered the widespread distribution of this element in the existing soil even though its concentration has decreased. Sr content has remained unchanged upon crop growth over Field A but has shown a slight increase in content between the pre-harvesting stage and post-harvesting stage. This could be affected by Sr solubility since this element tends to be highly mobile in the soil profile, depending on the physicochemical and mineralogical characteristics of the soil. Dubchak (2018) documented that an increase in Sr content is directly linked to an increase in exchangeable calcium content, an increase in soil acidity and an increase in organic matter content [40]. Dubchak (2018) stated that only a small fraction is taken up by crops from the soil to supply the physiochemical and physiological needs of the plant for adequate growth, which mostly affects root crops and legumes [40].

#### 3.2.2. Correlation Analysis

A combined correlation analysis of the results obtained for the pre-harvesting and post-harvesting concentrations proved very strong linear relationships with $r > 0.7$ among most elements, hinting towards a common source for these naturally occurring elements as shown in Figure 1. Thus, upon the reduction of concentrations of elements in the soil, correlations among elements increased. This matches studies by Salah, et al. (2015) showing moderate correlations among Cu, Co, Pb, and Zn [33], and those published by Micó, et al. (2006) and Khatita, et al. (2020) [4,36].

**Table 7.** Hypothesis Test Summary for Mann–Whitney U-Tests for Field A and Field B—Pre-harvesting and Post-harvesting stages.

| Elements | Hypothesis 1: The Distribution Is the Same Across Field A & Field B. | | Hypothesis 2a: The Distribution Is the Same Across Field A. | Hypothesis 2b: The Distribution Is the Same Across Field B. |
|---|---|---|---|---|
| | Pre-Harvesting | Post-Harvesting | Pre- & Post-Harvesting | Pre- & Post-Harvesting |
| Co | Rejected | Rejected | Rejected | Rejected |
| Cr | Rejected | Rejected | Rejected | Rejected |
| Cu | Rejected | Rejected | Rejected | Rejected |
| Fe | Rejected | Retained (0.828) | Rejected | Rejected |
| Mn | Rejected | Rejected | Rejected | Rejected |
| Ni | Retained (0.276) | Rejected | Rejected | Rejected |
| Pb | Rejected | Rejected | Rejected | Rejected |
| Sr | Rejected | Rejected | Retained (0.165) | Rejected |
| Zn | Retained (0.205) | Retained (0.052) | Rejected | Rejected |

Test significance level is 0.05.

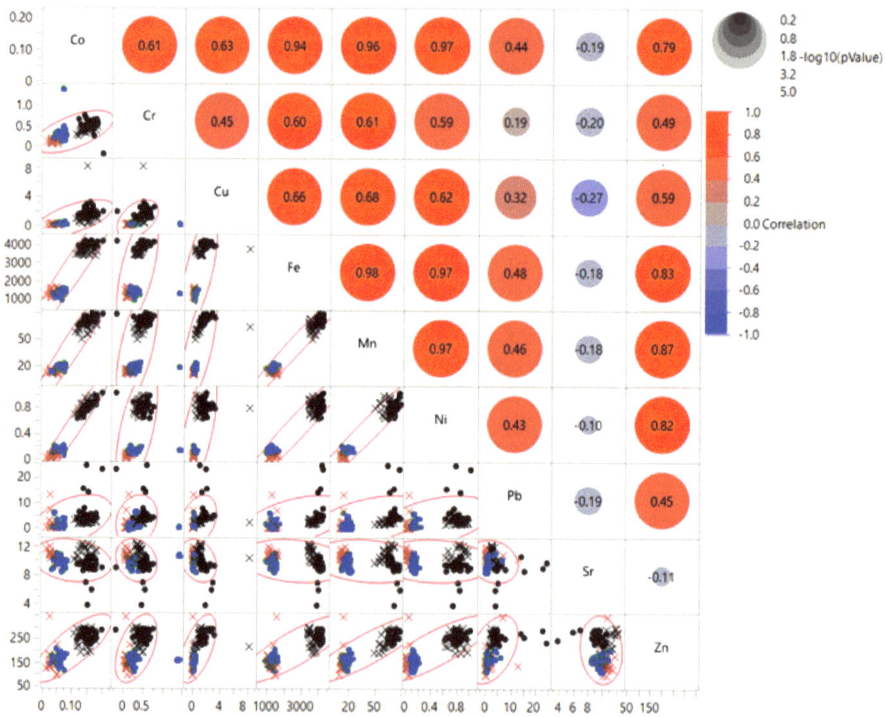

**Figure 1.** Scatter plot on multivariate correlations—Pre-harvesting and Post-harvesting stages. (Key: × represents data from Field A pre-harvesting stage; • represents data from Field B pre-harvesting stage; × represents data from Field A post-harvesting stage; • represents data from Field B post-harvesting stage.)

Weak negative relations between each of the elements with Sr are observed consolidating the outcomes of the individual analysis carried out on the each of the sampling stages whereby, relations with Sr were also negatively represented. This proves that the behavior of Sr is different to that of the other elements, coinciding also with the results obtained from the Mann–Whitney U-test showing that Sr content remained the same or increased slightly across seasons in each of the fields. Weaker but positive relationships with Pb are also observed for all elements, the weakest being between Cr and Pb, with r = 0.19 from the multivariate correlations having a 95% confidence interval. This contrasts the results of the study by Stofejova', et al. (2021) for which the correlation between Cr and Pb was a slightly weak negative relation with r = −0.343 [41]. These interactions have been reported using the multivariate correlations and match the results obtained using the Spearman's Rho correlations analysis.

### 3.2.3. Multivariate Statistical Analysis

Chemometric studies showed an interdependency of the elemental concentrations in the soil matrix and their possible sources. The concentration outcomes were combined for PCA analysis to review the effect of varying concentration on the element dominance in the soil matrix. PCA was then conducted for the sets of results individually to analyze differences in loadings to obtain the elemental sources.

An eigenvalue of 380.277 with 98.322% and a cumulative percentage of 98.322% for PC1, while an eigenvalue of 6.488 with 1.678% and a cumulative percentage of 100.000% for PC2 for the combined analysis of pre-harvesting stage and post-harvesting stage were obtained. A variance of 5.911 and 1.017, a percentage variance of 65.678% and 11.296% and a cumulative percentage of 65.678% and 76.974% for Factor 1 and Factor 2, respectively, for the combined pre-harvesting and post-harvesting analysis of results was obtained as tabulated in Table 8a.

An eigenvalue of 22.707 with 65.859% and a cumulative percentage of 65.859% for PC1, while an eigenvalue of 11.7711 with 34.14% and a cumulative percentage of 100.000% for PC2 at pre-harvesting stage were obtained. A variance of 3.287 and 1.817, a percentage variance of 36.520% and 20.192% and a cumulative percentage of 36.520% and 56.712% for Factor 1 and Factor 2 respectively for the pre-harvesting results analysis shown in Table 8b.

An eigenvalue of $3.27 \times 10^8$ with 100.000% and a cumulative percentage of 100.000% for PC1, while an eigenvalue of 5.518 with 0.000% and a cumulative percentage of 100.000% for PC2 at post-harvesting stage were obtained. A variance of 2.499 and 2.174, a percentage variance of 27.763% and 24.160% and a cumulative percentage of 27.763% and 51.924% for Factor 1 and Factor 2 respectively for the post-harvesting results analysis Table 8c.

**Table 8.** PCA initial loading matrix and rotated factor loadings matrix [2].

| | **(a) Pre- & Post-Harvesting Stages** | | | | **(b) Pre-Harvesting Stage** | | | | **(c) Post-Harvesting Stage** | | | |
|---|---|---|---|---|---|---|---|---|---|---|---|---|
| | **$PC1_a$** | **$PC2_a$** | **$F1_a$** | **$F2_a$** | **$PC1_b$** | **$PC2_b$** | **$F1_b$** | **$F2_b$** | **$PC1_c$** | **$PC2_c$** | **$F1_c$** | **$F2_c$** |
| Co | **0.952** | *0.124* | **0.967** | *0.100* | 0.443 | **0.824** | *0.241* | **0.835** | **0.822** | −0.343 | **0.709** | **0.410** |
| Cr | **0.919** | *0.134* | **0.861** | 0.329 | **0.709** | *0.114* | 0.530 | *0.214* | **0.843** | *0.111* | **0.663** | **0.448** |
| Cu | **0.893** | *−0.125* | **0.807** | *0.271* | **0.854** | *−0.159* | **0.896** | *0.030* | **0.712** | *0.284* | *0.212* | **0.777** |
| Fe | **0.893** | *−0.207* | **0.735** | **0.592** | **0.810** | *−0.117* | **0.669** | *0.037* | *0.229* | **−0.637** | *0.199* | *0.072* |
| Mn | **0.970** | *0.111* | **0.958** | *0.216* | **0.864** | *0.170* | **0.857** | 0.309 | **0.828** | *0.172* | 0.310 | **0.951** |
| Ni | **0.985** | *0.121* | **0.961** | *0.265* | *0.060* | **0.951** | *−0.150* | **0.964** | **0.864** | *−0.158* | **0.818** | 0.359 |
| Pb | **0.954** | *0.219* | **0.995** | *0.053* | 0.529 | −0.373 | 0.468 | *−0.172* | *0.272* | 0.558 | *0.205* | *0.031* |
| Sr | **0.618** | *−0.281* | **0.458** | 0.394 | **−0.716** | *0.297* | **−0.648** | *0.096* | **−0.684** | 0.351 | **−0.812** | *−0.066* |
| Zn | *−0.223* | **0.932** | *−0.022* | **−0.446** | 0.458 | *−0.019* | 0.550 | *0.094* | 0.407 | 0.527 | *0.078* | 0.399 |

[2] PC1a, PC1b, PC1c—Principal Component 1 for the three results configurations; PC2a, PC2b, PC2c—Principal Component 2 for the three results configurations; F1a, F1b, F1c—Factor 1 for the three results configurations; F2a, F2b, F2c—Factor 2 for the three results configurations.

PC1 in Table 8a indicates the dominant element concentrations. Large loadings for Cr, Cu, Mn, and Sr prove that concentration effects are strong. PC1 in Table 8b,c can be interpreted as the indicator of anthropogenic sources for these elements, while PC2 suggests a lithogenic origin of the elements with high loadings.

It can be observed that the concentration of the elements has affected considerably the classification of the elements with their respective possible sources. Fertilizer addition and surrounding environmental pollutants have unbalanced the ecosystem, which was possibly at least partly restored after crop growth. Results from the pre-harvesting stage indicate that Cr, Cu, Fe, Mn, Pb, Sr, and Zn are associated to anthropogenic sources available which were present due to the precipitation of pollutants in the rain from the atmosphere, fertilizer addition to prepare the soil for crop cultivation, urban dust, and vehicle and industrial emissions. Post-harvesting results for the same elements suggest a prevalent lithogenic source or mixed with an anthropogenic one in the case of Cr, Fe, and Mn. It also resulted that Fe is only partially affected by the concentration available in the soil matrix. This is due to the high natural Fe content in the soil which was proven in the high loading assigned to Fe in the post-harvesting PCA and rotated factor loadings. These results tie in with studies from Micó, et al. (2006) [4] and Nunes, et al. (2014) [35], who suggest that Cr, Zn, Fe, and Mn are precipitated from the oxides and hydroxides of sedimentary rocks within the Mediterranean region and are present in areas which are of calcareous nature [4,41]; hence, consolidating a lithogenic nature towards these elements when a low quantity of added nutrients is present in the soil. A similar pattern is observed with Pb which results match perfectly those issued by Khatita, et al. (2020) [36].

Zn is the only element which for which a change in concentration between pre-harvesting and post-harvesting stage showed no particularly dominant link to its possible source. This is due to having high Zn concentration even available after growing crops. Such concentrations are probably affected by the accumulation of Zn content added anthropogenically over the years. Lithogenically, this owes its origin to the high natural Zn content in typical Mediterranean soils, while anthropogenic sources accounting for such accumulations including vehicle exhaust, combustion of fossil fuels, industrial emissions, and the leaching of wastewater [21,36].

Concentration variations for Sr content are dominant and PC2 values for pre-harvesting and post-harvesting stages prove that its sources are of natural form. The negative correlations obtained with the Spearman's Rho and multivariate correlation analyses prove an inverse relation with the other elemental concentrations. Sr was attributed to the anthropogenic component in the work carried out by Khatita, et al. (2020) [36], and to radioactive sources by Dubchak (2018) [40]. As shown in the 3D scatter plots (Figure 2), Sr has a different coordinate loading from all other elements whose coordinates are closely grouped together. Sr is the most mobile element and since it can be present in a soluble or insoluble form, its uptake by plants during cultivation depends on its solubility state and thus its bioavailability [40].

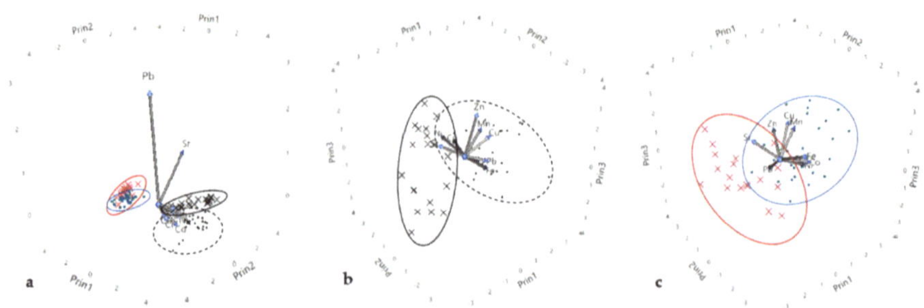

**Figure 2.** 3D Scatterplot showing Principal 1, Principal 2, and Principal 3: (**a**) Pre-harvesting and Post-harvesting stages; (**b**) Pre-harvesting stage; (**c**) Post-harvesting stage. (Key: × represents data from Field A pre-harvesting stage encircled in a black cluster; • represents data from Field B pre-harvesting stage encircled in a black dashed cluster; × represents data from Field A post-harvesting stage encircled in a red cluster; • represents data from Field B post-harvesting stage encircled in a blue cluster.

Co and Ni loadings in both the PCA and factor loading matrix show that their sources differed with fluctuating concentrations. Initial loadings and factor rotations demonstrate a dominant lithogenic origin at the pre-harvesting stage with loadings >0.8, although an anthropogenic origin is present. This confirms results documented by Micó, et al. (2006) and Khatita, et al. (2020) but contrasts with those of Salah, et al. (2015) in the case of Co origins [4,33,36]. Loadings for Co and Ni with the post-harvesting stage analysis associate these elements to the anthropogenic source, although these are at a lower concentration. Such concentration variation in the results could have come about upon the combination with the surrounding activities in the vicinity of the fields, namely, from vehicular exhaust from neighbouring main roads, from the use of agricultural fertilizers which include natural manure for Field A and mineral fertilizer for Field B, from the leaching of wastewater which could possibly occur during precipitation seasons from neighbouring rabbit farms in the case of Field A, and from residues of fossil fuel combustion [18,33,35]. The latter source could be originating from an industrial zone in proximity of the fields, and from low flying aircraft at the airport which is close by. Increased activity during spring/summer months from such industry could possibly bring about such results. The overall analysis proves that Co and Ni content is not placing an anthropogenic negative effect on the soil because their effect was balanced out upon crop uptake.

### 3.3. Health Risk Assessment

HQ values calculated for adults and children (shown in Table 9) are all less than the benchmark value of unity, thus as referenced in literature by Anyanwu and Nwachukwu (2010), none of the heavy metals present in the soil for Field A and Field B present quantifiable adverse non-carcinogenic risks [24]. Since the Hazard Index (HI) is also less than unity, these non-carcinogenic adverse effects which may originate through soil ingestion and dermal contact are assumed to be negligible.

**Table 9.** Results for the Health Risk Assessment and the Potential Ecological Risk Assessment variables [3].

| Element | Mean Concentration (mg/g) | $HQ_{adults}$ | $CR_{adults}$ | $HQ_{children}$ | $CR_{children}$ | $I_{geo}$ | CF | $E_i$ |
|---|---|---|---|---|---|---|---|---|
| Co | 0.100 | N/A * | N/A * | N/A * | N/A * | −0.584 | 1.001 | 5.005 |
| Cr | 0.339 | $7.951 \times 10^{-5}$ | $9.982 \times 10^{-8}$ | $1.298 \times 10^{-4}$ | $1.858 \times 10^{-7}$ | −5.606 | 0.031 | 0.062 |
| Cu | 0.818 | $1.216 \times 10^{-5}$ | N/A * | $2.248 \times 10^{-5}$ | N/A * | −2.875 | 0.204 | 1.022 |
| Fe | 2594.401 | N/A * | N/A * | N/A * | N/A * | −1.136 | 0.683 | N/A ** |
| Mn | 41.898 | N/A * | N/A * | N/A * | N/A * | −5.299 | 0.038 | 0.038 |
| Ni | 0.461 | $1.374 \times 10^{-5}$ | $2.475 \times 10^{-7}$ | $2.537 \times 10^{-5}$ | $4.606 \times 10^{-7}$ | −6.023 | 0.023 | 0.115 |
| Pb | 3.034 | $5.225 \times 10^{-4}$ | $1.520 \times 10^{-8}$ | $9.563 \times 10^{-4}$ | $2.829 \times 10^{-8}$ | −2.154 | 0.337 | 1.686 |
| Sr | 9.969 | N/A * | N/A * | N/A * | N/A * | −6.520 | 0.016 | N/A ** |
| Zn | 207.262 | $4.137 \times 10^{-4}$ | N/A * | $7.608 \times 10^{-4}$ | N/A * | 2.788 | 10.363 | 10.363 |
| $HI_{adults}$ | | | | $1.042 \times 10^{-3}$ | | | | |
| $HI_{children}$ | | | | $1.895 \times 10^{-3}$ | | | | |
| $RI_{adults}$ | | | | $3.625 \times 10^{-7}$ | | | | |
| $RI_{children}$ | | | | $6.746 \times 10^{-7}$ | | | | |
| PERI | | | | 18.291 | | | | |

[3] HQ represents the Hazard Quotient; CR represents the Cancer Risk values; $I_{geo}$ represents the geo-accumulation index for each heavy metal; CF represents the carcinogenic factor; $E_i$ represents the Potential Ecological Risk Index. * Ingestion Rate values required for the computation of HQ and CR respectively were not documented. ** Toxic response factor values for the respective elements were not documented in other researched studies.

Cancer Risk (CR) values in Table 8 demonstrate that concentrations obtained through analysis pose very low cancer risk in both adults and children. The RI is also very low since the results obtained are below $10^{-6}$ [22,26], which suggests that the risk of cancer development is extremely low in any person exposed to the metallic elements reviewed, thus ensuring safety to the landowner and any persons farming the land, together with children who might visit the respective fields.

This health risk assessment study matches the outcome of concentration results for the heavy metals studied. None of the elements was higher than limit values set, and although Zn content was higher than documented in other studies, its content was still within the acceptable range; thus, its level of contamination was minimal.

### 3.4. Potential Ecological Risk Assessment

The calculations for the potential ecological risk assessment were based on the mean concentration values of the metals studied and the respective baseline values for carbonate sedimentary rock which is typical in the Maltese Islands [42]. With reference to the geo-accumulation index calculated, Zn is the only element posing contamination of a moderate/strong level on the fields studied, with a Class 3 level as shown in Table 2. All other elements have given a negative geo-accumulation index, proving not to be contaminating the fields. Zn is also the element with higher Contamination Factor (CF) and Potential Ecological Risk Index ($E_i$) values, thus demonstrating that Zn is the element, which is mostly contaminating the soil, but is overall posing a low potential ecological risk, since its $E_i = 10.363 < 40$, as established in Table 9.

These results tie in with those from the statistical analysis since Zn behaviour was not dependent on concentration variation and is of an anthropogenic accumulated nature. CF results also show that Zn is the most metal contaminating the soil, followed by Co. Statistical results for Co at the post-harvesting stage show that anthropogenic sources for this heavy metal were affecting the soils more than at the pre-harvesting stage. It is important to note that its geo-accumulation index is negative; thus, although its sources have not accumulated over time, these could still be anthropogenic. Its concentration factor and ecological risk factor prove that it possesses a very low potential of ecological risk, even though its value is higher than that of the other heavy metals assessed.

The value of Potential Ecological Risk Index (PERI) calculated for the fields under review demonstrates that the overall effect of the heavy metals in the soil portrays a low ecological risk potential to the environment. This proves that there was not any significant pollution level in the area studied and the pollution indices assessed showed a moderate pollution level of Zn.

## 4. Conclusions

The results obtained in this research increase the knowledge of heavy metals and their possible sources in Maltese agricultural soils. Physical studies have proven that the soil tested has a high organic content proving it healthy and abundant in natural nutrients making it adequate for crop cultivation. Concentrations obtained indicate that the fields under study are not polluted by heavy metals since none of the concentrations quantified exceed reference limits set by legislation. All elemental concentrations have decreased between the pre-harvesting and the post-harvesting sampling stages, proving the uptake of heavy metals by the growing crops from the soil matrix.

The PCA carried out on the nine heavy metals reviewed at the pre-harvesting and post-harvesting stages, identified by two principal components, demonstrate the concentration dependency of all the elements to their possible sources affecting agricultural soils, except for Zn, whose concentration was marginally high even after harvesting the crops. The geo-accumulation index proved that the fields are moderately/strongly contaminated with Zn content.

Results from the pre-harvesting stage indicate that Cr, Cu, Fe, Mn, and Pb are associated with anthropogenic sources available including the precipitation of pollutants in the rain from the atmosphere, fertilizer addition to prepare the soil for crop cultivation, urban dust, and vehicle and industrial emissions. Post-harvesting results for the same elements suggest a prevalent lithogenic source mixed with an anthropogenic one in the case of Cr, Fe, and Mn. Fe is only partially affected by the concentration available in the soil matrix due to the high natural Fe content typical of the soil. Results for Co and Ni reflect that, at the pre-harvesting stage, these elements were linked to a lithogenic origin, while at the post-harvesting stage their content was linked to an anthropogenic origin which could be originating from surrounding activities in the vicinity of the fields by neighboring industry, and other farming practices. Sr content, although concentration dependent, varies inversely to the other elements concentrations. Sr content remained relatively unchanged between the pre-harvesting and post-harvesting sampling stages hinting at natural sources.

A health risk assessment study associated with the heavy metal content observed, showed that the concentrations available in both fields do not pose a carcinogenic risk or a hazard to human health upon ingestion or inhalation of the soil being studied during exposure. An assessment of the potential ecological risks associated with heavy metals content in the soil have demonstrated that the area studied is not contaminated by any of the metals except for Zn which creates a moderate to strong contamination but gives an overall low potential ecological risk.

**Supplementary Materials:** The following supporting information can be downloaded at: https://www.mdpi.com/article/10.3390/app12063120/s1, Figure S1: Geographical information of the study areas and positioning of sampling points.

**Author Contributions:** Conceptualization, C.C. and F.L.; methodology, C.C. and F.L.; formal analysis, C.C. and F.L.; writing—original draft preparation, C.C. and F.L.; writing—review and editing, C.C. and F.L. All authors have read and agreed to the published version of the manuscript.

**Funding:** This research received no external funding.

**Institutional Review Board Statement:** Not applicable.

**Informed Consent Statement:** Not applicable.

**Data Availability Statement:** Not applicable.

**Conflicts of Interest:** The authors declare no conflict of interest.

## References

1. Hallett, S.; Sakrabani, R.; Thompson, A.; Deeks, L.; Knox, J. Improving Soil and Water Management for Agriculture: Insights and Innovation from Malta. *MCAST J. Appl. Res. Pract.* **2017**, *1*, 40–59. [CrossRef]
2. Union, E. *The EU Environmental Implementation Review 2019 Country Report—MALTA*; European Commission: Brussels, Belgium, 2019.
3. Environment and Resources Authority. Maltese Soils. 2020. Available online: https://era.org.mt/topic/maltese-soils/ (accessed on 20 May 2021).
4. Micó, C.; Recatalá, L.; Peris, M.; Sánchez, J. Assessing heavy metal sources in agricultural soils of an European Mediterranean area by multivariate analysis. *Chemosphere* **2006**, *65*, 863–872. [CrossRef] [PubMed]
5. Paul, E. *Soil Microbiology, Ecology and Biochemistry*, 4th ed.; Academic Press: Cambridge, MA, USA, 2015.
6. Nortcliff, B.; Schulte-Bisping, L. Soil Definition, Function, and Utilization. In *Ullmann's Encyclopaedia*; John Wiley & Sons: Hoboken, NJ, USA, 2011; pp. 400–420.
7. Mulder, E.G.; VanVeen, W.L. *Effect of Microorganisms on the Transformation of Mineral Fractions in Soil*; Agricultural University: Wageningen, The Netherlands, 2010.
8. Ladonin, D.V.; Karpukhin, M.M. Fractional Composition of Nickel, Copper, Zinc, and Lead Compounds in Soils Polluted by Oxides and Soluble Metal Salts. *Eurasian Soil Sci.* **2011**, *44*, 874–885. [CrossRef]
9. Ghosh, M.; Devi, A. Assessment of crop growth, soil properties and crop yield in an upland acidic soil with inorganic fertilizer blended with organic amendments in summer rice cropping seasons. *Int. J. Recycl. Org. Waste Agric.* **2019**, *8*, S1–S9. [CrossRef]
10. Pomazkina, L.V.; Semenova, Y.V. Effects of Climate Changes and Pollution with Heavy Metals on the Transformation of Carbon Compounds in Different Soil Types of Agroecosystems in the Forest-Steppe of Baikal Region. *Eurasian Soil Sci.* **2018**, *51*, 576–587. [CrossRef]
11. Zhang, L.; Fu, K.; Yang, F.; Chen, Y.; Fu, C.; Huang, Y.; Guo, Z.; Li, T. Migration and Transformation of Heavy Metals in the Soil of the Water-Level Fluctuation Zone in the Three Gorges Reservoir under Simulated Nitrogen Deposition. *J. Chem.* **2021**, *2021*, 6660661. [CrossRef]
12. O'Neill, P. *Environmental Chemistry*, 3rd ed.; Blackie Academic & Professional: London, UK, 1998.
13. He, Z.L.; Yang, X.E.; Stoffella, P.J. Trace elements in agroecosystems and impacts on the environment. *J. Trace Elem. Med. Biol.* **2005**, *19*, 125–140. [CrossRef]
14. Markiewicz-Patkowska, J.; Hursthouse, A.; Przybyla-Kij, H. The interaction of heavy metals with urban soils: Sorption behaviour of Cd, Cu, Cr, Pb and Zn with a typical mixed brownfield deposit. *Environ. Int.* **2005**, *31*, 513–521. [CrossRef]

15. Zwolak, A.; Sarzyńska, M.; Szpyrka, E.; Stawarczyk, K. Sources of Soil Pollution by Heavy Metals and Their Accumulation in Vegetables: A Review. *Water Air Soil Pollut.* **2019**, *230*, 164. [CrossRef]
16. Orroño, D.I.; Lavado, R.S. Distribution of extractable heavy metals in different soil fractions. *Chem. Speciat. Bioavailab.* **2009**, *21*, 193–198. [CrossRef]
17. CustomWeather, Valletta Weather History. 2021. Available online: https://www.timeanddate.com/weather/malta/valletta/historic?month=11&year=2020 (accessed on 28 July 2021).
18. Radojevic, M.; Bashkin, V. *Practical Environmental Analysis*, 2nd ed.; The Royal Society of Chemistry: Dorset, UK, 2006.
19. Sarkar, D.; Haldar, A. *Physical and Chemical Methods in Soil Analysis*; New Age International (P) Ltd.: New Delhi, India, 2005.
20. Shugar, G.; Ballinger, J. *Chemical Technicians' Ready Reference Handbook*, 4th ed.; McGraw-Hill Inc.: New York, NY, USA, 1996.
21. Sanz-Medel, A.; Pereiro, R. *Atomic Absorption Spectrometry: An Introduction*, 2nd ed.; Momentum Press, LLC: New York, NY, USA, 2014.
22. Agoro, M.A.; Adeniji, A.O.; Adefisoye, M.A.; Okoh, O.O. Heavy Metals in Wastewater and Sewage Sludge from Selected Municipal Treatment Plants in Eastern Cape Province, South Africa. *Water* **2020**, *12*, 2746. [CrossRef]
23. Adamiec, E.; Jarosz-Krzemińska, E. Human Health Risk Assessment associated with contaminants in the finest fraction of sidewalk dust collected in proximity to trafficked roads. *Sci. Rep.* **2019**, *9*, 16364. [CrossRef]
24. Anyanwu, E.D.; Nwachukwu, E.D. Heavy metal content and health risk assessment of a South-eastern Nigeria River. *Appl. Water Sci.* **2020**, *10*, 209–210. [CrossRef]
25. USEPA IRIS (US Environmental Protection Agency)'s Integrated Risk Information System. *Environmental Protection Agency Region I*; United States Environmental Protection Agency: Washington, DC, USA, 2011.
26. Liang, Y.; Yi, X.; Dang, Z.; Wang, Q.; Luo, H.; Tang, J. Heavy Metal Contamination and Health Risk Assessment in the Vicinity of a Tailing Pond in Guangdong, China. *Int. J. Environ. Res. Public Health* **2017**, *14*, 1557. [CrossRef]
27. Mohammadi AA Zarei, A.; Majidi, S.; Ghaderpoury, A.; Hashempour, Y.; Saghi, M.H.; Ghaderpoori, M. Carcinogenic and non-carcinogenic health risk assessment of heavy metals in drinking water of Khorramabad, Iran. *MethodsX* **2019**, *6*, 1642–1651. [CrossRef]
28. Salhotra, A.M. Chapter Ten—Human Health Risk Assessment for Contaminated Properties. *Prog. Mol. Biol. Transl. Sci.* **2012**, *112*, 285–306.
29. Rouhani, A.; Shahivand, R. Potential ecological risk assessment of heavy metals in archaeology on an example of the Tappe Rivi (Iran). *SN Appl. Sci.* **2020**, *2*, 1277. [CrossRef]
30. Turekian, K.; Wedepohl, K. Distribution of the Elements in Some Major Units of the Earth's Crust. *Geol. Soc. Am. Bull.* **1961**, *72*, 175–192. [CrossRef]
31. Negahban, S.; Mokarram, M. Potential Ecological Risk Assessment of Ni, Cu, Zn, Cd, and Pb in Roadside Soils. *Earth Space Sci.* **2021**, *8*, e2020EA001120. [CrossRef]
32. Al-Edresy, M.; Wasel, S.; Al-Hagibi, H. Ecological risk assessment of heavy metals in coastal sediments between Al-Haymah and Al-Mokha, south red sea, Yemen. *Int. J. Hydrol.* **2019**, *3*, 159–173. [CrossRef]
33. Salah, E.; Turki, A.; Mahal, S.N. Chemometric Evaluation of the Heavy Metals in Urban Soil of Fallujah City, Iraq. *J. Environ. Prot.* **2015**, *6*, 1279–1292. [CrossRef]
34. Serelis, K.; Kafkala, I.; Parpodis, K.; Lazaris, S. Anthropogenic and Geogenic Contamination Due to Heavy Metals in the Vast Area of Vari, Attica. *Bull. Geol. Soc. Greece* **2017**, *43*, 2390–2397. [CrossRef]
35. Nunes, J.R.; Ramos-Miras, J.; Lopez-Piñeiro, A.; Loures, L.; Gil, C.; Coelho, J.; Loures, A. Concentrations of Available Heavy Metals in Mediterranean Agricultural Soils and their Relation with Some Soil Selected Properties: A Case Study in Typical Mediterranean Soils. *Sustainability* **2014**, *6*, 9124–9138. [CrossRef]
36. Khatita, A.; Koch, R.; Bamousa, A. Sources identification and contamination assessment of heavy metals in soil of Middle Nile Delta, Egypt. *J. Taibah Univ. Sci.* **2020**, *14*, 750–761. [CrossRef]
37. Briffa, J. Heavy Metals in Maltese Agricultural Soil, Malta. University of Malta Library. 2020. Available online: https://www.um.edu.mt/library/oar/handle/123456789/73391 (accessed on 17 May 2021).
38. European Commision, DG Environment. Final Implementation Report for Directive 86/278/EEC on Sewage Sludge: 2013–2015. 2018. Available online: https://ec.europa.eu/environment/archives/waste/reporting/pdf/Final_Implementation_Report_2013_2015_Sewage_Sludge.pdf (accessed on 11 June 2021).
39. Xie, Y.; Fan, J.; Zhu, W.; Amombo, E.; Lou, Y.; Chen, L.; Fu, J. Effect of Heavy Metals Pollution on Soil Microbial Diversity and Bermudagrass Genetic Variation. *Front. Plant Sci.* **2016**, *7*, 755. [CrossRef] [PubMed]
40. Dubchak, S. Distribution of Strontium in Soil: Interception, Weathering, Speciation, and Translocation to Plants. In *Behaviour of Strontium in Plants and the Environment*; Gupta, D., Walther, C., Eds.; Springer: Cham, Switzerland, 2018. [CrossRef]
41. Štofejová, L.; Fazekaš, J.; Fazekašová, D. Analysis of Heavy Metal Content in Soil and Plants in the Dumping Ground of Magnesite Mining Factory Jelsava-Lubenik (Slovakia). *Sustainability* **2021**, *13*, 4508. [CrossRef]
42. Pedley, H.M.; House, M.R.; Waugh, B. The geology of Malta and Gozo. *Geol. Assoc.* **1976**, *87*, 325–341. [CrossRef]

*energies*

MDPI

*Review*

# Reuse of Water Contaminated by Microplastics, the Effectiveness of Filtration Processes: A Review

**Juan A. Conesa [1,2,*] and Nuria Ortuño [1,2]**

1 Department of Chemical Engineering, University of Alicante, P.O. Box 99, E-03080 Alicante, Spain; nuria.ortuno@ua.es

2 Institute of Chemical Process Engineering, University of Alicante, P.O. Box 99, E-03080 Alicante, Spain

* Correspondence: ja.conesa@ua.es; Tel.: +34-9-6590-3400

**Abstract:** Water treatment generally does not specifically address the removal of microplastics (MPs). Nevertheless, treatment plants process water effectively, and the number of synthetic microparticles in effluents is usually very low. Still, discharge volumes from water-treatment plants are often elevated (reaching around $10^8$ L/day), leading to the daily discharge of a substantial number of MPs and microfibers. Furthermore, MPs accumulate in the primary and secondary sludge, which in the end results in another environmental problem as they are currently used to amend soils, both for cultivation and forestry, leading to their dispersion. Something similar occurs with the treatment of water intended for human consumption, which has a much lower but still significant number of MPs. The amount of these pollutants being released into the environment depends on the processes that the water undergoes. One of the most-used treatment processes is rapid sand filtration, which is reviewed in this article. During the filtration process, MPs can break into smaller pieces, resulting in a greater number of plastic particles which mainly accumulate in sewage sludge. Thermal processes, such as incineration, carried out in facilities with the best available techniques in practice, could guarantee the safe disposal of highly MP-contaminated sewage sludges.

**Keywords:** drinking water; wastewater; sewage sludge; microplastics; microfibers; sand filtration

**Citation:** Conesa, J.A.; Ortuño, N. Reuse of Water Contaminated by Microplastics, the Effectiveness of Filtration Processes: A Review. *Energies* **2022**, *15*, 2432. https://doi.org/10.3390/en15072432

Academic Editors: Dibyendu Sarkar, Rupali Datta, Prafulla Kumar Sahoo and Mohammad Mahmudur Rahman

Received: 28 February 2022
Accepted: 24 March 2022
Published: 25 March 2022



## 1. Introduction

The global demand for water is expected to increase by nearly one-third by 2050. Investing in water recovery and reuse could save up to 90% in energy and 70% in water usage, according to the United Nations (UN) [1]. Several of the Sustainable Development Goals (SDGs) are related to effective wastewater treatment. In objective 6, "Clean water and sanitation", one of its goals is to improve water quality by reducing pollution, eliminating discharge and minimizing the emission of chemical products and hazardous materials, increasing safe reuse worldwide. Likewise, objective 11, "Sustainable cities and communities", advocates for reducing the negative environmental impact per capita of cities, including paying special attention to air quality and the management of various toxic effluents.

Typical water-purification systems consist of pretreatment units (generally screening and clarification) and primary treatment that continues in a biological treatment unit (which constitutes secondary treatment). The primary treatment of wastewater consists of coagulation or flotation (for the removal of fats) followed by sedimentation. Afterwards, the organic matter in suspension is eliminated through a secondary treatment in the activated sludge reactor [2]. Many treatment systems also have tertiary units (often consisting of sand filtration or membrane-based filtration), ensuring a high-quality final effluent, whether for reuse in agriculture or safe disposal. Finally, a disinfection process is carried out (with chlorine, ozone or ultraviolet light) to eliminate pathogenic pollutants before discharging the treated water into the effluents. A simplified schematic representation of the processes that are generally present in a wastewater-treatment system is provided in Figure 1.

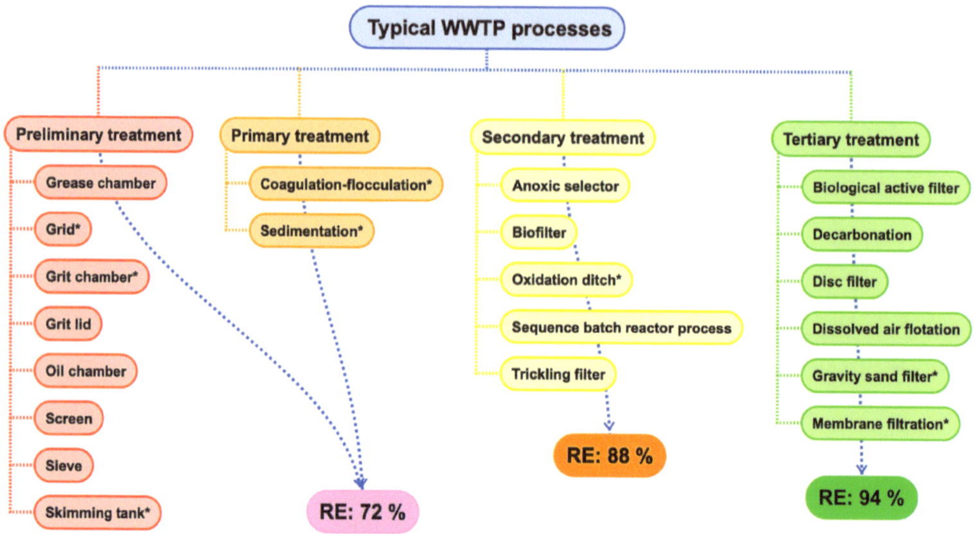


**Figure 1.** Main systems present in a wastewater-treatment plant (WWTP) [3]. Rough data on the removal efficiency (RE) of MPs of the different stages of treatment as a global, from Iyare et al. [4]. Those processes marked with an asterisk are present in the majority of plants.

If the treatment is focused on the purification of riverwater or water from desalination plants for human consumption, water-quality standards are much more stringent due to the expected lower concentration of foreign substances in drinking water. As a result, the preparation of drinking water generally requires the use of advanced water-treatment processes, such as membrane filtration and advanced oxidation. In fact, water and wastewater-treatment processes are similar, except for the biological processes that are required for wastewater due to the high content of organic matter [5].

Wastewater-treatment systems produce sludge, which must be treated before disposal or reuse [6,7]. The sludge needs to be concentrated to avoid the transport of extremely wet materials, among other reasons, and must also go through a process in order to destroy pathogens, and collect by-products such as biogas, in order to dispose of the sludge in an ecologically acceptable way [8].

On the other hand, as is well-known, plastics are present in almost all items related to human life, such as clothing, cosmetics, food packaging, equipment and instruments, etc. The annual global production of plastics has increased considerably in recent years, reaching 367 million tonnes in 2020 (only 1 million tonnes less than in 2019) [9]. It is important to add to these numbers the manufacture of synthetic fibers, since they are used in textiles, cords, fabrics, and other products, representing 61 million tonnes per year.

According to some estimates, around 10 million tonnes of plastic enter the oceans every year [10], accounting for 1.5 kg of plastic for each person on this planet, each year. Note that most plastics end up in sinks such as water collectors, rivers and seas until they drain into the oceans [11].

However, the problem of plastic is no longer that of "macroplastics" but of the plastic wastes being debased and reduced in size until they form what are called "microplastics", i.e., tiny particles. Plastics, although resistant, are exposed to various factors that cause them to break into smaller pieces, producing microparticles that are present globally.

Microplastics (MPs) are generally understood to be plastic waste with sizes smaller than 5 mm [10]. The effect of human exposure to MPs is not yet known, which raises many unresolved questions [12]. For this reason, several studies are currently being carried out on the presence and influence of MPs, especially in the marine food chain, since when MPs reach the sea, they can be consumed by organisms such as zooplankton, thus entering the food chain.

Today, MPs are present everywhere, and come mainly from single-use plastics, fishing gear, clothing and cosmetics, paints, tires and urban dust [13]. These MPs easily reach rivers and seas due to their generally low density.

So far, three possible toxic effects of MPs have been indicated: the plastic particles themselves, their ability to adsorb persistent organic pollutants (POPs) and other substances present in aquatic systems around the world and the additives that these materials contain. Namely, the impact on human health stems from the fact that they can carry potentially toxic chemicals and microorganisms [14–17]. MPs could adsorb (or absorb) these dangerous chemicals, such as polyaromatic hydrocarbons (PAHs), metals and dioxins, from the environment and transport them to food products, which are ultimately consumed by humans [3,4,9,17].

MPs are classified as primary and secondary, as well as according to the polymer that constitutes them: polyethylene (PE), polyvinyl chloride (PVC), etc. Primary MPs are those that are already manufactured as MPs. More restrictions now apply to their manufacturing, but they have generally been used in hygiene products (exfoliating creams, toothpaste, gels, ... usually known as personal care and cosmetic products, or PCCPs). Secondary MPs are those derived from the wear or degradation of large plastics (heat, UV, mechanical or biological degradation) [13,18]. Similarly, nanoplastics, with a size between 1 and 100 nm (i.e., 0.001–0.1 μm), can be produced by degradation of MPs or can be derived directly from industrial or domestic activity.

MPs are also classified according to their shape, and fall into the following categories (see Figure 2): fragments or microflakes (larger pieces, they could almost be called macroplastics); microfibers (with an elongated shape when observed under a microscope); microspheres or microbeads (which are small hard spheres); foams (sponge-like mass) and nurdles or granules (usually used in manufacturing). The terms nurdles and granules are used in many types of MPs when they do not fit in any of the others. The differences between pellets, nurdles, microbeads and microspheres are almost imperceptible. In much research, only the distinction between elongated microfibers and microparticles of a spherical or undefined shape is considered.

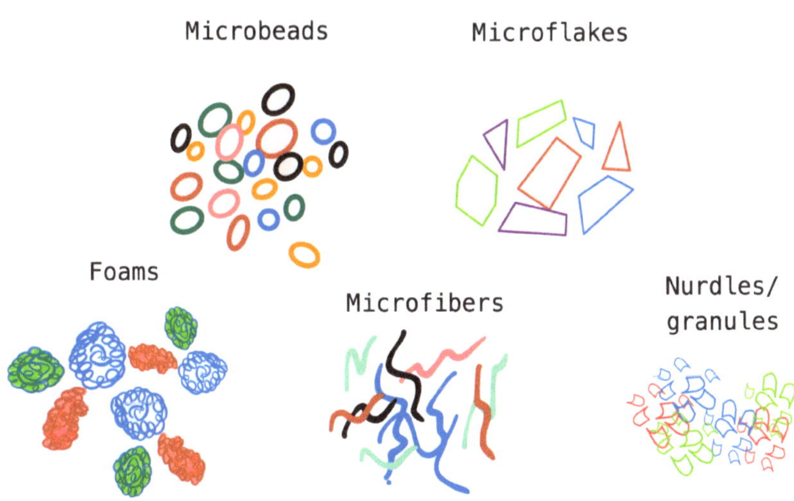


**Figure 2.** Main shapes presented by microplastics.

Each type of microplastic is associated with a main source. For example, microspheres are related mainly to the production of PCCPs, and microfibers are related to actions on synthetic polymers in clothing. These are considered the most common examples of microplastic contamination in the environment [14]. Fragments and fibers are the two dominant shapes of plastic particles in wastewater [15].

Nowadays, one of the sources of MPs entering the ocean are treated wastewaters. According to Bayo et al. [19] the effluent of a treatment plant can vary between 5900 MPs/$m^{-3}$ on wet-weather days to 3000 MPs/$m^{-3}$ on dry ones. As commented above, many synthetic-fiber fabrics (e.g., nylon, polyesters, polyacrylamides, etc.) derive from washing processes [15]. In this way, textile fibers, which account for around 70% of the MPs in WWTPs, are also commonly found in wastewater. Many clothes are made of polyester, which is a plastic very similar to polyethylene terephthalate (PET) (that of the majority of water bottles, for example) and it is already in the form of fibers, demonstrating the huge potential problem they can pose [10,16].

Murphy et al. [20] pointed out the problem represented by WWTP, which can be considered as sources of MPs in the aquatic environment. The study, performed in a plant in Glasgow, showed that the removal efficiency of MPs was higher than 98%, but still the release of 65 million MPs into the receiving water was denounced.

In the present work, a review of the main findings related to the elimination of MPs in wastewaters is carried out, with special interest in elimination by filtration. The removal efficiency of several systems will be discussed, as well as the effect of the shape and size of the MPs to be eliminated and the nature of the polymers.

## 2. Microplastic Fate in Conventional Water/Wastewater-Treatment Processes

The fate of MPs entering WWTPs is currently a very relevant research topic. Many authors have investigated the role of MPs in the different stages of wastewater treatment. It has been observed that the presence of MPs reduces the efficiency of wastewater and sludge-treatment processes and increases sludge volume [6,21], affecting methane production in bioreactors, for instance. The presence of MPs could make it necessary to change the configuration of the treatment plant. For example, the presence of a large number of MPs could oblige to the use of certain reagents. Similarly, a decrease in the nitrogen conversion rate and fouling of the membrane in wastewater-treatment processes have been described [22]. As a result, it is difficult to accurately assess MP-removal efficiency (RE) in WWTPs. At present, there are no policies or legislation that require a minimum for MP removal during wastewater treatment.

Johnson et al. [23] looked for the presence of MPs in potable water. They found extremely low amounts, less than 2 particles/$m^3$, showing that there was an effective removal of this type of particles in the water-treatment plants.

Table 1 shows a summary of the most important data summarized in the present review.

In their review, Liu et al. [24] presented the numbers of MPs in the influent of different WWTPs. They showed that the abundance of MPs in the influent was between 0.28 and 31,400 MPs/L, while that in the effluent was between 0.01 and 297 MPs/L. They found quantities of MPs in the sludge varying within the range $4.40 \cdot 10^3$–$2.40 \cdot 10^5$ MPs/kg.

Rasmussen et al. [25] calculated the masses of plastics of >500 μm in a WWTP. They showed that the plant load was 202.2 kg/day, of which 73% was retained. The remaining plastic was found in the sludge (13.6%) and in the effluent (0.4%).

Tadsuwan et al. [26] pointed out that the RE of a grit trap alone was >33%, followed by 14.2% removal in a secondary treatment. In the sludge, the number of MPs was 8120 particles/kg.

Even considering that conventional WWTPs usually eliminate more than 90% of the incoming MPs, these systems continue to be the main source of introduction of MPs into the environment, as the high volumes of generated effluents represent a very large contribution [27]. Simon et al. [28] noted that Danish WWTPs discharge around 3 t/year of MPs in the size range of 10–500 μm, despite the 98% RE of this size of particles. Bretas-Alvim et al. [27] presented data on the discharge of MPs in different countries, observing important differences (±1000% approx.) among them.

Recently [29] a study investigating the MP removal at different treatment stages showed that the primary treatment process has an effectiveness of about 75%, and that it increases to 91.9% in the secondary stage. Removal efficiency further increased to >98% after tertiary treatment (coagulation, ozonation, disc membrane filters and rapid sand filters).

Iyare et al. [4] studied the fate of MPs in 21 WWTPs around the world. The authors show that the preliminary and primary treatment of wastewater remove 72% of the MPs, and that the removal efficiency increases to 88% (of the number of microplastic particles present in sewage influent) in secondary treatment and 94% in the tertiary. A major part of the MPs eliminated are accumulated in the sludge, which has lately been used for soil enrichment, causing accumulation in terrestrial ecosystems [4]. Ou and Zeng [30] pointed out the minor presence of MPs at night, showing an uneven temporal distribution of MPs.

The largest WWTP in P. R. China was studied by Yang et al. [31]. A concentration of approx. 12 MPs/L was detected in the influent, reducing the amount to 0.6 MPs/L by employing a conventional WWTP that was equipped with ultrafiltration as the final tertiary treatment. This study shows a decrease in the MP concentration throughout the treatment processes, i.e., the RE in the primary treatment was ca. 59%; it was 72% after the secondary stage; and that of the overall process was 95%. This research also investigates the role of the MP size and found an entering average size of ca. 1110 μm, being quite similar to the size in the effluent. The main polymers detected, as in many other studies, were PET, PS (polystyrene) and PP (polypropylene). The fact that they find similar sizes at the inlet and the outlet of the plant after an ultrafiltration process suggests that MPs might not be properly retained in the filter, or that a mix of filtered and unfiltered waters was produced.

Nevertheless, other authors [32] observed a lower removal efficiency in WWTPs located in Thailand. In this work, the overall RE was between 67 and 57% in the dry and rainy sessions, respectively.

Recently, Wu et al. [22] pointed out that secondary treatment was, within the WWTP, the most efficient process to eliminate MPs from wastewater. According to their data, the removal efficiency can be as high as 98%.

**Table 1.** Summary of data found in literature. MBR: membrane bioreactor; RSF: rapid sand filter; UF: ultrafiltration; WWTP: wastewater-treatment plant; DWTP: drinking-water-treatment plant.

| Country | Influent | Treatment | Effluent | Reference |
|---|---|---|---|---|
| Spain | 4.40 MPs/L<br>4.40 MPs/L | MBR<br>RSF | 0.92 MPs/L, average size = 1.39 mm<br>1.08 MPs/L, average size = 1.15 mm | Bayo et al. [18] |
| Spain | | Global | Effluent with 5.9 MPs/L in wet weather and 3.0 MPs/L in dry ones | Bayo et al. [19] |
| Israel | 74% fibrous MPs | Global<br>RSF<br>RSF | RE = 97%<br>1.97 MPs/L<br>91% fibrous MPs | Ben-David et al. [33] |
| Spain | 90% fibers (9% MPs) | | | Bretas-Alvim et al. [27] |
| USA | | Primary and secondary | 0.00144 MPs/L | Carr et al. [34] |
| (several) | 61–5600 μg/L | Global<br>Global | RE > 93%, by weight<br>0.5–170 μg/L | Cheng et al. [3] |
| Italy | | Global | RE > 90% | Cristaldi et al. [15] |
| Spain | | RSF | RE > 78% in DWTP | Dalmau-Soler et al. [35] |
| France | | Primary<br>Primary and secondary | RE = 80% for fibrous materials<br>RE = 95% for fibrous materials | Dris et al. [36] |
| | | Global | Up to 88% of MPs removed go to sludge | Freeman et al. [37] |
| Israel | 32.4% suspected MPs were plastics | Primary and secondary<br>Global | 65.6% fibers, 28.1% fragments, 5.4% pellets<br>RE = 97–99% | Gies et al. [17] |
| Greece | 3160 MPs/L | Global<br>RSF | 125 MPs/L<br>RE = 98.9% | Gatidou et al. [38] |
| South Korea | | Primary<br>Secondary | RE = 79%<br>RE = 92% | Hidayaturrahman et al. [29] |
| (several) | | Preliminary and primary<br>Secondary<br>Tertiary | RE = 72%<br>RE = 88%<br>RE = 94% | Iyare et al. [4] |
| China | 126 MPs/L | Global | 30,6 MPs/L in the effluent<br>>75% of MPs transferred to the sludge<br>Accumulation of $7.74 \cdot 10^{12}$ particles/year in the sludges. | Jiang et al. [39] |

**Table 1.** *Cont.*

| Country | Influent | Treatment | Effluent | Reference |
|---|---|---|---|---|
| Thailand | | Global<br>Global<br>Global<br>Global<br>RSF<br>RSF | RE = 67% (drain season)<br>RE = 57% (dry season)<br>RE = 60% for sizes 6.5–53 µm<br>RE = 80% for size > 500 µm<br>RE = 26.8% for the smallest particles<br>RE = 60.7% for the biggest particles | Kankanige [32] |
| Singapore | | Membrane filter | RE < 7% for nanoparticles (100–500 nm) | Li et al. [40] |
| Denmark | 0.917 MPs/L | Biofilter | 0.179 MPs/L<br>RE = 100% for sizes lower than 0.1 mm | Liu et al. [41] |
| Denmark | 0.28–31,400 MPs/L | Global | 0.01–297 MPs/L | Liu et al. [24] |
| China | | Primary<br>Secondary<br>Tertiary<br>MBR<br>Activated sludge | RE = 75%<br>RE = 95%<br>RE = 98%<br>RE = 99.5%<br>RE = 97% | Lv et al. [42] |
| USA | | Global | Effluent with less than 1 particle per liter<br>$5\cdot10^4$–$1.5\cdot10^7$ particles/day | Mason et al. [43] |
| USA | | Global | 60% fibers in the effluent | Michielssen et al. [44] |
| Germany | | Use of magnetic ionic liquids | RE = 100% (PS) | Misra et al. [45] |
| Canada | | Filtration with 0.22 µm | RE > 99% | Murray et al. [46] |
| Scotland | 15.7 MPs/L | Global | 0.25 MPs/L (RE > 98%) | Murphy et al. [20] |
| China | | (Simulated runs) | Smaller-sized MPs are more mobile in RSF<br>A greater number of wet–dry cycles increases MPs' penetration depth | O'Connor et al. [47] |
| China | | | Grease on the sludge contains many more MPs than grit and sludge cake. | Ou and Zeng [30] |
| Sweden | 202.2 kg MPs/day | Global | RE = 72% | Rasmussen et al. [25] |
| Germany | | Seeds of $Fe_3O_4$ and application of magnetic field | 95% | Rhein et al. [48] |
| Indonesia | | RSF | RE = 50% for plastic flakes, in DWTP<br>RE = 96% for tires, in DWTP | Sembiring et al. [49] |
| India | | RSF | RE = 85%, by weight<br>RE = 90%, by number | Seth et al. [50] |
| Denmark | | Disk filter | RE = 76% | Simon et al. [51] |
| Denmark | | | 3 tonnes MPs/year, size range 10–500 um, 98% RE | Simon et al. [28] |
| Thailand | | Grit trap<br>Secondary treatment | RE > 33%<br>RE > 14.2% (additional) | Tadsuwan et al. [26] |
| (several) | | Primary<br>Primary and secondary<br>Secondary<br>MBR<br>RSF<br>Dissolved air flotation<br>Disc filtration | RE = 98%<br>RE > 99%<br>RE = 88%<br>RE = 99.9%<br>RE = 97%<br>RE = 95%<br>RE = 40–98% | Talvitie et al. [52] |
| Spain | | | SS has a lighter plastic load, much lower than heavy plastic load (18,000 vs 32,070 MPs/kg) | Van der Berg et al. [53] |
| China | | Primary and secondary | RE > 98% | Wu et al. [22] |
| China | 12 MPs/L | UF<br>Primary<br>Primary and secondary<br>Global | 0.6 MPs/L<br>RE = 59%<br>RE = 72%<br>RE = 95% | Yang et al. [31] |
| India | | Hollow-fiber membrane | RE = 99.3% | Yaranal et al. [54] |
| USA | | RSF<br>RSF<br>RSF | RE = 86.9% for small particles<br>RE = 87% for critical size (10–20 µm)<br>RE > 99.9% for particles larger than 100 µm | Zhang et al. [6] |

Talvitie et al. [52] investigated the behavior of advanced treatment technologies in various WWTPs. The highest removal efficiency was achieved by the membrane bioreactor (MBR) with a RE of 99.9%, followed by gravity or rapid sand filter (RSF) (RE = 97%) and air flotation (RE = 95%).

In the study led by Lv et al. [42] two parallel systems of wastewater treatment were studied, namely oxidation ditch treatment and MBR. The MPs observed by these authors at the plant entrance consisted of PET (47%), PS (20%), PE (18%) and PP (15%). Regarding the shape of the microparticles, a greater proportion of fragments was found (65%), followed by fibers (21%, mainly PET), films (12%) and foams (2%). No microbeads were observed. Regarding the size of the particles, the dominant size was >500 µm (40%) and between 62.5–125 µm (29%). These researchers observed that MP concentrations increased in the treatment systems as a function of the ease of the treatment process. In this way, the MPs were eliminated by 99.5% in the MBR system compared to 97% in the oxidation system (by weight). These percentages indicate the tendency to retain the largest fragments.

The study developed in Finland by Talvitie et al. [55] also showed the effectiveness of WWTPs for the elimination of MPs. The results show that conventional primary and secondary treatments can effectively eliminate (>99%) the MPs that reach the WWTP. The majority (98%) of the MPs were eliminated during the primary treatment. The activated sludge (secondary) process further decreased (~88% from process input) the MP concentration. During wastewater treatment, most MPs (>99.5%) were retained in the primary and secondary sludge; however, part (~20%) of the retained MPs was recirculated back to the treatment process as part of the water was rejected.

The authors [55] stated that the removal of MPs can be further improved with the implementation of a specific treatment system. The proposed equipment consists of a membrane bioreactor (additional RE = 99.9%), sand filtration (RE = 97%), dissolved air flotation (RE = 95%), or disc filtration (RE = 40–98.5%). Biologically active filtration had no impact on the concentration of MPs. Nevertheless, Bayo et al. [12] found persistent low-density polyethylene (LDPE), nylon and PVC in the RSF and melamine after MBR treatment.

Gies et al. [17] studied the presence and removal of MPs in a Canadian WWTP. They found that only 32.4% of the suspected MPs were plastic polymers, and that the global removal efficiency of the plant varied between 97.1% and 99.1%. Nevertheless, the number of MPs released to the environment was concentrated in the primary and secondary sludges.

Ben-David et al. [33] studied the elimination of MPs in a WWTP sited in northern Israel, observing a removal efficiency higher than 97%, with most MPs eliminated in the secondary stage and observing an average of 1.97 MPs/L after sand filtration.

Carr et al. [34] showed that existing treatment processes are effective at removing MPs in WWTPs, and found no MP presence after tertiary treatment in seven plants, and only 0.00114 MPs/L for effluent after secondary treatment. In this way, MP particles were found to be removed mainly in the primary treatment zones during the processes of solid skimming and sludge settling. These authors mention that some consumer products, as the toothpaste containing MPs may be contributing much more than others to the amounts of MPs reaching the WWTPs.

Cheng et al. [3] found a global removal efficiency of MPs higher than 93%, by weight, going from 61–5600 μg/L to 0.5–170 μg/L. These researchers showed that preliminary and primary treatments contribute to the removal of MPs, and that the efficacy of secondary and tertiary treatments was, understandably, highly dependent on the applied techniques.

Cristaldi et al. [15] reviewed information on the removal efficiency of different WWTPs around the world, showing a removal efficiency greater than 90%. This seems to be a good situation, but the high amount of water treated in the plants means that millions of MPs continue to be released every day into the aquatic environment.

Biofilters have also been tested for the removal of MPs from treated wastewater. In their study, Liu et al. [41] showed that a raw effluent containing 0.917 MPs/L was easily filtered to 0.179 MPs/L in the first stage of a conventional biofilter. Nevertheless, small sizes, as expected, cannot be completely removed, although no MPs higher than 100 μm in size were found (i.e., RE = 100% for sizes lower than 0.1 mm).

Ngo et al. [56] showed that the existing WWTPs are inefficient to completely remove the MPs and there is a risk that they may be discharged into the ambient water.

In many studies, filtration has been shown to play an important role in removing MPs. However, this method has its drawbacks [21,57] because the mechanical stress generated during treatment can cause plastics to wear away, resulting in smaller particles that are likely to be released into the environment without restriction.

### 2.1. Microplastic Accumulation in Sludges

Zhang et al. [6] insisted that in a WWTP, microplastics are simply transferred from sewage to sludge; this could be a problem during anaerobic digestion, as MPs can contain significant amounts of toxic substances such as persistent organic pollutants. Gatidou et al. [38] showed that MP content ranges up to 3160 MPs/L in raw water, 125 MPs/L in treated wastewater and $170.9 \cdot 10^3$ MPs/kg total dry solid in the sludge. Gies et al. [17] observed that primary sludge had approx. $14.9 \cdot 10^3$ MPs/kg, with fibers being more abundant than other particles. Likewise, secondary sludge had $4.4 \cdot 10^3$ MPs/kg, with fibers also being the most abundant shape. The dominance of fibers in sludge samples has been previously reported [58,59]. On their hand, Liu et al. [24] indicated that the microplastic abundance in the sludge was within the range of $4.40 \cdot 10^3$–$2.40 \cdot 10^5$ particles/kg.

Because MPs accumulate in biological sludge, and because these sludges are mainly used to improve agricultural or forest soil, studies should be carried out on the incorporation of these particles in meat from land animals [27].

Regarding the MP content in sewage sludges (SS), Ou and Zeng [30] indicated that the grease on the SS samples contained a number of MPs that was significantly higher than the grit and sludge-cake samples. PE microbeads from PCCPs were dominant in grease samples.

Habib et al. [60] observed an abundance of synthetic fibers in soil conditioners, fertilizers and similar biosolid materials, denouncing the presence of MPs in sludges.

Jiang et al. [39] studied the fate of MPs in sewage and sludge filter cake. These authors showed that there were 126.0 MP particles/L in the influent and 30.6 in the effluent, transferring about 75% of the MPs to the sludge. In their study, the abundance of MPs in dewatered sludge and filter cake was 36.3 and 46.3 particles/g. These figures are equivalent to the accumulation of $7.74 \cdot 10^{12}$ particles/year in the sludges.

This implies that most of the MPs removed during wastewater treatment will end up in the open environment [33,61]. For this reason, Wolff et al. [62] indicated that WWTPs are a good sink for MPs only when the sludge is thermally treated and not used for agricultural amendment. Incineration and production of chemicals by pyrolysis of sludges from WWTPs are very promising current lines of research [63] that would avoid the accumulation of MPs in the environment. These techniques would, furthermore, reduce the energetic demand on the treatment plants, increasing the potential for chemical-energy recovery from wastewater pollutants [64].

In this way, an alternative route for disposal, incineration of SS with heat recovery, has been pointed out as a good practice [65]. Note that in the European legislation [66] incineration is preferred to landfilling. This is because landfill is much less controllable, producing gas release and in many cases spontaneous combustion of the biogas, in addition to the fact that the necessary technology for the treatment of the gas produced in the incinerator is already available. Through incineration, MPs contained in the SS would definitively be eliminated from the environment, in what emerges as a safe alternative for disposal.

### 2.2. Need for Standardized Methods

There is no standard protocol for MP measurement in WWTPs. Many authors [3,27,67–70] insist that the differences between studies are related to the different analytical techniques employed for measuring MP contamination.

Lusher et al. [71] insisted on the necessity of harmonization in the isolation and extraction of MPs from environmental samples, including the sample-collection techniques and data analysis. On the other hand, Oβmann [72] insisted that the harmonization of methods for the quantification of MPs in drinking water and other food should be developed and defined.

Bayo et al. [19] indicated that the concentrations reported by different authors on the presence of MPs in WWTPs are difficult to compare, because different sampling techniques, volumes and analytical methods are used. Furthermore, the use of MP concentrations on a numerical basis has been criticized for overstating MP abundance because this involves counting the number of suspected particles [3].

Sun [73] pointed out that since MP sampling and detection methods can significantly affect the result of their quantification and identification, harmonization is urgent. Hong et al. [74] proposed a method for estimating the mass of MPs in sewage, based on the quantification of total organic carbon (TOC).

Moreover, differences can be found due to the difficulties present in the sampling of MPs in waters. Lenz et al. [75] designed a water-sampling system specially dedicated to the collection of MPs in surface and groundwater for environmental pollution studies. It is designed specifically for sampling small MPs (size fraction <300 μm) in different source areas. Li et al. [40] mentioned that in many MP studies, these particles are just sampled by nets with a typical mesh size of 330 μm. The sampled MPs are usually purified and digested, and lately analyzed by different methods, but there are no universally accepted sampling and quantification tools, so important differences among studies can be found.

Another interesting topic is the treatment of samples in the laboratory. Elkhatib et al. [69] insisted on the need for quality-assurance methods. This would include the use of a wet filter exposed to air during sample processing to take into account air contamination, the analysis of blank samples using deionized water, the addition of a sample with a known polymer, and the analysis of samples in triplicate. Ou and Zeng [30] showed that the use of $H_2O_2$ to preoxidize the particles can remove surface biogenic materials from the polymer. However, the preoxidation combined with heating induced the melting of some MPs, especially microbeads. Dyachenko et al. [76] also used peroxide oxidation to eliminate interferences in the analysis.

Gatidou et al. [38] also insisted that special care should be taken on the types of clothing used. Gies et al. [17] observed a deposition of 36 fibers on a 1 μm pore-size × 47 mm-diameter filter membrane over an 8 h period.

## 3. Filtration as a Route of Elimination

Liu et al. [24] compared the performance of 38 WWTPs in different countries and showed that filter-based treatment processes attained the highest microplastic-removal efficiency.

In the RSF, the sand bed captures the particles when the secondary wastewater effluent passes on, and the MPs can be removed by backwashing the sand; this process can remove as much as 97% of the MPs from the wastewater [37]. Bayo et al. [12] discussed the role of two different treatment technologies in the abatement of MPs, showing a RE of ca. 80% for MBR and of 76% for RSF.

The study by Hidayaturrahman and Lee [29] used a sand bed for treating wastewater from the coagulation process in one sand filter by gravity. In this study, the MP retention rate was lower compared to ozone and membrane disc-filter technologies, but the efficiency was still 98.9%. It was observed that MPs whose dimensions are smaller than the diameter of the sand are more likely to pass through the filter.

In the work published by Seth and Shriwastav [50], reference was also made to the filtration of waters with a high content of MPs. In this case, a sand filter was used with water that had been artificially contaminated, with removal efficiencies of around 85% by weight, which represents 90% by number. The authors tested nine different commercial polymers. MPs were quantified before and after sand to establish filtration efficiency. For all thicknesses of the sand bed, the removal efficiency (% by weight) was greater than 85% and continued to improve with subsequent cycles, while it remained at 87% after 25 cycles. The RE of particles of different sizes was also estimated, finding RE (% by number) higher than 90% for all sizes, although larger particles (>500 μm) showed a higher filtration efficiency.

In the study by O'Connor et al. [47] the vertical mobility of MPs in sand columns was investigated. The study focused on the penetration of rain-drying cycles in a terrain exposed to climatic variations, but it may be interesting from the point of view of the efficiency of filtration in a sand bed. The study carried out experiments in a column of sandy soil, checking the mobility of five different MPs, which consisted of PE and PP particles of various sizes and densities. The experiments were mounted on a transparent PVC column and several various amounts of liquid were fed to the column, with the objective of examining the effect of the filtration volume on the migration of MPs. The study revealed that the smallest PE were those with the greatest mobility among the five MPs evaluated. It was also concluded that subjecting these MPs to a greater number of wet–dry cycles increased their depth of penetration in the sand columns, with a practically linear relationship between the depth of penetration and the number of the wet–dry cycles; in comparison, almost no variations in the migration of the microparticles were observed when the volume of the filtration liquid or the initial concentration of MPs in the column varied. In this study, the quantities of MPs introduced into the bed were very high compared to the other studies mentioned, which may bias the results.

Michielssen et al. [44] showed that efficiency of a WWTP was high (>95%) in removing small anthropogenic litter, but this litter consisted of many different materials, including cellulosic products manufactured from natural materials, among others. Indeed, the authors insist that tertiary granular-sand filtration exhibited a high RE for the different materials and recommend the installation of this device in the WWTPs, mentioning a possible decrease of more than 50% in the presence of MPs. On the contrary, Carr et al. [34] found that WWTP using RSF as a tertiary treatment showed low removal efficiency of MPs.

Other authors maintained that MPs can be adsorbed by silica grains due to hydrophilic interaction, thereby clogging the initial layers and reducing RSF performance faster than expected. Thus, even though backwashing is performed regularly, MP adsorption can be difficult to reverse due to the presence of hydroxyl groups on the surface of MPs as a result of weathering [34,77]. Different new techniques with high potential for removing MPs have been tested, such as dynamic membranes, enhanced flocculation/coagulation systems and electrocoagulation.

RSF has also been used to remove MPs from drinking water. Sarkar et al. [78] tested a drinking-water facility for the removal of MPs, indicating that the influent has a very little amount of MPs (17.8 MPs/L), and that pulse clarification and sand filtration removed 63 and 85% of the particles, respectively. Dalmau-Soler et al. [35] studied the removal of MPs in a DWTP close to Barcelona (Spain). In this research, the authors showed a predominance of PS and PP in the intake water from a river, with a mean concentration of 0.96 MPs/L, while at the outlet it was $0.06 \pm 0.04$ MPs/L, with an overall removal efficiency of 93%. The authors mention the importance of the sand filtration stage, calculating a mean removal efficiency of 78%. This research points out the effectivity of methods such as ultrafiltration/reverse osmosis, much more effective than ozonation/carbon filtration stages in the MP removal.

Sembiring et al. [49] performed tests on the RSF of drinking water containing MPs of plastic from shopping bags (probably PE) and tires that were crushed using a grinder. The produced particles were 10 to 500 μm in size and the authors found a RE of about 50% for plastic flakes and about 96% for tire particles. For the study, filtration rates between 4 and 10 m/h were used, and the runs were performed in a RSF unit constructed ad hoc. Samples were taken at different times in order to study the possible differences produced by fouling of the filtration media. The authors also studied the effect of the filter media porosity on the RE for both kinds of MPs, comparing the RE obtained for the effective size of the filtration media of 0.39 mm and 0.68 mm. The authors understandably found a decrease in the RE as the size of the filtration media increased, but in all runs the particles bigger than 200 μm were completely removed.

Other filtration techniques have also been used for MP removal. A disk filter was studied by Simon et al. [51], where they captured ca. 76% of the particles contained in the secondary effluent of a WWTP.

Ding et al. [79] showed the great effectivity of membrane filters for the removal of MPs, but discussed the possibility of releasing nano- and microplastics from organic membranes in drinking-water-treatment plants. They insist on this possibility and demand for more research on this topic.

Ahmed et al. [80] reported that membrane technologies such as UF, microfiltration and MBR increase the removal efficiency of MPs and nanoparticles.

On the other hand, Li et al. [81] showed that many nanoparticles are not retained by membrane filters with a pore size as small as 1 μm. In their runs, using spherical PS and poly(methyl methacrylate) (PMMA) plastics as models, the efficiency of membrane filtration was lower than 7% for nanoparticles (size of 100–500 nm). These authors propose cloud-point extraction to concentrate these small particles for analysis.

Finally, an interesting study was published by Rhein et al. [48]. The authors proposed a magnetic seeded filtration to improve the RE of MPs. The study uses PVC and PMMA particles of a very low diameter (mean diameters of 2.06 and 5.98 μm, respectively), together with seeds of $Fe_3O_4$. The application of a magnetic field seems to efficiently improve the removal of these very small particles (RE > 95%), which are usually problematic. Rhein et al. determined the optimal working conditions studying pH, intensity, rotary speed and concentration of particles. Misra et al. [45] also studied the removal of MPs using magnetic techniques, in the case of ionic liquid phases, that furthermore present bactericidal properties. The removal efficiency of their system was 100% for tested PS beads.

## 4. Differences among MPs Being Filtrated

The size and shape of the MPs being eliminated can play an important role, as well as the nature of the polymer that makes up the plastic. It has been indicated in literature that the morphological characteristic is one of the most important factors that contribute to the removal process. This is mainly due to the interaction between plastic waste and other pollutants, as well as microorganisms in wastewater [56]. Murray et al. [46] studied the RE for nanoplastics (<400 nm) in water and sewage-treatment plants, showing that filtration with a 0.22 μm filter removed 99% of the particles, and ballasted flocculation removed 88%.

Fibers and fragments are the two dominant types of plastic particles in wastewater [15]. In many cases, MP fragments are associated to the presence of PE and PP (63% and 25%, respectively [27]), whereas microfibers related to PE probably originated in washing machines. The type of polymer most present was PE in almost all WWTPs investigated [15].

Gies et al. [17] indicated that in the primary and secondary effluent-wastewater samples, the presence of MPs was dominated by fibers (65.6%), followed by fragments (28.1%) and pellets (5.4%), while only a few were classified as foams, granules and sheets. (0.22%, 0.45% and 0.20%, respectively).

Bayo et al. [82] calculated a RE for MPs of 64.22% after the tertiary treatment, one of the lowest figures found in the literature. Fibers were less-efficiently removed (56.16%) than particulate MPs (90.03%)

Different researchers point out that wastewater-treatment plants are capable of removing large amounts of larger MP particles but are ineffective at removing particles with any dimension of less than 100 μm, and tributaries and effluents tend to have similar amounts of these smaller particles [37]. In this sense, Michielssen et al. [44] calculated that the microparticles in the final effluents of two WWTPs in the USA included more than 60% fibers, with estimated concentrations of 3.58 and 5.25 microfibers/L and 1.94 and 0.8 fragments/L. Contrarily, Sun et al. [73] considered that the pretreatment in a WWTP was more effectively removing fibers than other types of MPs, and they found that the relative abundance of fibers decreased after the pretreatment.

Freeman et al. [37] affirmed that there are no methods capable of efficiently removing very small plastic particles and fibers in a way that is technically, environmentally and economically sustainable in wastewater-treatment plants.

Bretas-Alvim et al. [27] investigated the presence of different types of microparticles in a WWTP sited in Valencia (Spain), finding a much higher presence of microfibers than other shapes (90%), but only 9% of these were confirmed to be MPs.

Hu et al. [14] showed that most MPs from PCCPs (usually in the shape of microballs) can be efficiently removed during wastewater treatment, although the efficiency was not so high for MPs in the shape of fibers, mainly coming from the domestic washing of synthetic fibers.

Furthermore, Talvitie et al. [52] reported that the primary sedimentation acts mainly on textile fibers, while the contribution of this step to the elimination of synthetic particles was almost negligible. Similarly, Liu et al. [24] mentioned that primary treatment has superiority over secondary and tertiary treatments for plastic-fiber removal, indicating that fibers were easily trapped during primary treatment due to flocculation and settling. On the other hand, fragments exhibited excellent removal efficiency during the secondary treatment process. Dris et al. [83] observed a RE > 85% for MPs in the form of fibers during primary treatment, while this efficiency increased to almost 95% after biological treatment.

Dalmau-Soler et al. [35] linked the MPs in the shape of fibers to PS and PP, and the shape of fragments to particles of PE. The presence of PS fibers was related to the washing of synthetic clothes. The authors found a slight decrease in fibers of 20–500 μm after sand filtration but with an increase in larger fibers thereafter.

Simon et al. [28] centered their study in the shape and weight of the MPs found in a WWTP. They found no significant differences between the mass of the particles in the wastewater samples before and after the treatment, showing a median particle mass of 13.45 ng in raw wastewater and 3.52 ng in treated wastewater. Furthermore, they insisted that research should focus on both the length and the diameter of the fibers, and mentioned that a single dimension is not representative of the features of the particles. For fibers with small diameters (<50 μm), a single dimension could not adequately define size or provide sufficient information on its contribution to the total number of MPs in a sample.

In a very recent study, Wolff et al. [62] showed that the presence of an RSF removed on average around 99.2% of the total discharge of MPs, in experiments where the majority of the particles had a diameter of <100 μm. The authors showed that after sand filtration, PE was the only polymer found among the MPs. All particles >500 μm were removed by the sand filter and the predominant one in the RSF effluent was 10–50 μm in size.

Ben-David et al. [33] observed a lower efficiency for the removal of fibers in a sand filtration unit, representing fibers in 74% of MPs in the influent, and increasing to 91% after filtration.

Kankanike et al. [32] found a majority of fibers in the treated waters, with PE, PP, PET and PVC commonly identified. These authors showed removal efficiency was a function of the MP size, as expected. RE increased with MP size, going from 60.0% (dry season) for a size between 6.5–53 μm (46.9% in the rainy season), to 80.2% (dry season) for a size >500 μm (67.5% rainy season). With respect to the filtration step, Kankanike also showed an increase in RE with size, but very low values were found in comparison to other studies. In this case, reported RE for filtration was 26.8% for the smallest particles (12.7% in the rainy season), and 60.7% (52.1% rainy) for the biggest. The filtration unit of the plant included two layers: one of anthracite coal as the top layer (500–600 μm porosity) and the bottom layer being graded sand (400–500 μm porosity). Both layers can retain ≥500 μm MPs, thus showing a good retaining efficiency for higher sizes. In contrast to filtration, Kankanike et al. showed that clarification presented a higher removability for smaller-sized particles [32].

With regard to the removal efficiency of granular filters, Zhang et al. [84] found that it was much more effective to filter micro- and nanoplastics (RE = 86.9% for small particles, RE >99.9% for particles larger than 100 μm); conversely, a critical size exists (10–20 μm) where RE was quite poor, namely 87%. This means that the RE is not proportional to MP size, contrarily to that found by many researchers, and a smaller size does not necessarily mean a lower RE.

Finally, Ma et al. [85] observed that coagulation processes are more effective in the elimination of smaller particles of PE compared to bigger ones. In addition, they obtained a complete removal of all PE particles by ultrafiltration, observing a slight membrane fouling due to the large diameter of MPs compared to the ultrafiltration-membrane pores.

## 5. Conclusions

A bibliographic review has been carried out on aspects related to the presence of MPs in water and wastewater-treatment plants, with special attention given to the efficiency of the filtration process in removing MPs.

Although in most current systems the removal of MPs is effective, important differences have been observed in the removal efficiencies of smaller plastics, which are not removed effectively. In addition, it has been shown that the filtration process can break up the particles, creating larger amounts of very small MPs.

The quantities of MPs eliminated from the waters mostly end in the sewage sludge. These are generally used for agricultural applications, which results in their dispersion through the environment in a very worrying way. Only thermal treatment of this sludge in well-designed systems will be a desirable process from the point of view of their disposal.

Even with large removal rates, the small quantities of MPs that remain in the effluents can pose a serious problem, given the large volumes of effluents that are discharged into the aquatic environment.

Some of the most important points can be summarized as follows:

- Although the removal efficiency of MPs in WWTP is relatively high, the quantity of these particles discharged into the environment is still very high, becoming the largest source of the introduction of MPs into the environment.
- Primary and secondary treatment processes effectively remove MPs from wastewater with removal efficiencies ranging from 75% to 91.9%. RE can be increased to >98% after tertiary treatment.
- Most of the MPs eliminated from sewage accumulate in the sewage sludge. Incineration/thermal treatment of SS is the only effective way to destroy these plastic particles.
- During sewage treatment, part of the retained MPs can be recirculated back to the treatment process as part of the water is rejected.
- There is a need for harmonization of the techniques used for MP analysis and counting.
- Larger particles exhibit greater filtration efficiency. Smaller particles have more mobility.
- RSF shows a good performance for MPs in the form of beads or granules. However, it does not guarantee microbeads (nor microfibers) would be absent from the effluent.
- Efficiency of membrane filtration can be as low as 7% for nanoparticles.

**Author Contributions:** Conceptualization, N.O. and J.A.C.; writing—original draft preparation, J.A.C.; writing—review and editing, N.O.; funding acquisition, J.A.C. All authors have read and agreed to the published version of the manuscript.

**Funding:** This research was funded by MINISTRY OF SCIENCE AND INNOVATION (Spain), grant number PID2019-105359RB-I00, and by the UNIVERSITY OF ALICANTE, grant number UAUSTI20-05.

**Institutional Review Board Statement:** Not applicable.

**Informed Consent Statement:** Not applicable.

**Data Availability Statement:** Not applicable.

**Conflicts of Interest:** The authors declare no conflict of interest.

## References

1. UNEP Green Economy | UNEP-UN Environment Programme. Available online: https://www.unep.org/explore-topics/green-economy/what-we-do/economic-and-fiscal-policy/fiscal-policy/policy-analysis-4 (accessed on 16 February 2022).
2. Pal, P. Selection of Water-Treatment Technology. In *Industrial Water Treatment Process Technology*; Elsevier: Amsterdam, The Netherlands, 2017; pp. 537–544.
3. Cheng, Y.L.; Kim, J.-G.G.; Kim, H.-B.B.; Choi, J.H.; Fai Tsang, Y.; Baek, K. Occurrence and removal of microplastics in wastewater treatment plants and drinking water purification facilities: A review. *Chem. Eng. J.* **2021**, *410*, 128381. [CrossRef]
4. Iyare, P.U.; Ouki, S.K.; Bond, T. Microplastics removal in wastewater treatment plants: A critical review. *Environ. Sci. Water Res. Technol.* **2020**, *6*, 2664–2675. [CrossRef]
5. Spellman, F.R. *Handbook of Water and Wastewater Treatment Plant Operations*; CRC Press: Boca Raton, FL, USA, 2003.

6. Zhang, X.; Chen, J.; Li, J. The removal of microplastics in the wastewater treatment process and their potential impact on anaerobic digestion due to pollutants association. *Chemosphere* **2020**, *251*, 126360. [CrossRef]
7. Cydzik-Kwiatkowska, A. Advanced wastewater treatment and biomass energy. *Energies* **2022**, *15*, 173. [CrossRef]
8. Zhang, Q.; Hu, J.; Lee, D.J.; Chang, Y.; Lee, Y.J. Sludge treatment: Current research trends. *Bioresour. Technol.* **2017**, *243*, 1159–1172. [CrossRef] [PubMed]
9. PlasticEurope-Association of Plastics Manufactures. *Plastics—The Facts*; European Association of Plastics Recycling and Recovery Organisations: Wemmel, Belgium, 2021.
10. Iñiguez, M.E.; Conesa, J.A.; Fullana, A. Marine debris occurrence and treatment: A review. *Renew. Sustain. Energy Rev.* **2016**, *64*, 394–402. [CrossRef]
11. Conesa, J.A.; Iñiguez, M.E. Analysis of Microplastics in Food Samples. In *Handbook of Microplastics in the Environment*; Rocha-Santos, T., Costa, M., Mouneyrac, C., Eds.; Springer: Berlin/Heidelberg, Germany, 2020. [CrossRef]
12. Bayo, J.; López-Castellanos, J.; Olmos, S. Membrane bioreactor and rapid sand filtration for the removal of microplastics in an urban wastewater treatment plant. *Mar. Pollut. Bull.* **2020**, *156*, 111211. [CrossRef] [PubMed]
13. Prajapati, S.; Beal, M.; Maley, J.; Brinkmann, M. Qualitative and quantitative analysis of microplastics and microfiber contamination in effluents of the City of Saskatoon wastewater treatment plant. *Environ. Sci. Pollut. Res.* **2021**, *28*, 32545–32553. [CrossRef]
14. Hu, Y.; Gong, M.; Wang, J.; Bassi, A. Current research trends on microplastic pollution from wastewater systems: A critical review. *Rev. Environ. Sci. Bio Technol.* **2019**, *18*, 207–230. [CrossRef]
15. Cristaldi, A.; Fiore, M.; Zuccarello, P.; Conti, G.O.; Grasso, A.; Nicolosi, I.; Copat, C.; Ferrante, M.; Oliveri Conti, G.; Grasso, A.; et al. Efficiency of wastewater treatment plants (Wwtps) for microplastic removal: A systematic review. *Int. J. Environ. Res. Public Health* **2020**, *17*, 8014. [CrossRef]
16. De Falco, F.; Di Pace, E.; Cocca, M.; Avella, M. The contribution of washing processes of synthetic clothes to microplastic pollution. *Sci. Rep.* **2019**, *9*, 6633. [CrossRef] [PubMed]
17. Gies, E.A.; LeNoble, J.L.; Noël, M.; Etemadifar, A.; Bishay, F.; Hall, E.R.; Ross, P.S. Retention of microplastics in a major secondary wastewater treatment plant in Vancouver, Canada. *Mar. Pollut. Bull.* **2018**, *133*, 553–561. [CrossRef]
18. Bayo, J.; López-Castellanos, J.; Olmos, S. Abatement of Microplastics from Municipal Effluents by Two Different Wastewater Treatment Technologies. *Water Pollut. XV* **2020**, *1*, 15–26. [CrossRef]
19. Bayo, J.; Olmos, S.; López-Castellanos, J. Removal of Microplastics from Wastewater. In *Handbook of Microplastics in the Environment*; Rocha-Santos, T., Costa, M., Mouneyrac, C., Eds.; Springer: Berlin/Heidelberg, Germany, 2020. [CrossRef]
20. Murphy, F.; Ewins, C.; Carbonnier, F.; Quinn, B. Wastewater Treatment Works (WwTW) as a Source of Microplastics in the Aquatic Environment. *Environ. Sci. Technol.* **2016**, *50*, 5800–5808. [CrossRef]
21. Zhang, Z.; Chen, Y. Effects of microplastics on wastewater and sewage sludge treatment and their removal: A review. *Chem. Eng. J.* **2020**, *382*, 122955. [CrossRef]
22. Wu, M.; Tang, W.; Wu, S.; Liu, H.; Yang, C. Fate and effects of microplastics in wastewater treatment processes. *Sci. Total Environ.* **2021**, *757*, 143902. [CrossRef]
23. Johnson, A.C.; Ball, H.; Cross, R.; Horton, A.A.; Jürgens, M.D.; Read, D.S.; Vollertsen, J.; Svendsen, C. Identification and Quantification of Microplastics in Potable Water and Their Sources within Water Treatment Works in England and Wales. *Environ. Sci. Technol.* **2020**, *54*, 12326–12334. [CrossRef]
24. Liu, W.Y.; Zhang, J.L.; Liu, H.; Guo, X.N.; Zhang, X.Y.; Yao, X.L.; Cao, Z.G.; Zhang, T.T. A review of the removal of microplastics in global wastewater treatment plants: Characteristics and mechanisms. *Environ. Int.* **2021**, *146*, 106277. [CrossRef]
25. Rasmussen, L.A.; Iordachescu, L.; Tumlin, S.; Vollertsen, J. A complete mass balance for plastics in a wastewater treatment plant-Macroplastics contributes more than microplastics. *Water Res.* **2021**, *201*, 117307. [CrossRef]
26. Tadsuwan, K.; Babel, S. Microplastic contamination in a conventional wastewater treatment plant in Thailand. *Waste Manag. Res.* **2021**, *39*, 754–761. [CrossRef]
27. Bretas Alvim, C.; Bes-Piá, M.A.; Mendoza-Roca, J.A. Separation and identification of microplastics from primary and secondary effluents and activated sludge from wastewater treatment plants. *Chem. Eng. J.* **2020**, *402*, 126293. [CrossRef]
28. Simon, M.; van Alst, N.; Vollertsen, J. Quantification of microplastic mass and removal rates at wastewater treatment plants applying Focal Plane Array (FPA)-based Fourier Transform Infrared (FT-IR) imaging. *Water Res.* **2018**, *142*, 1–9. [CrossRef] [PubMed]
29. Hidayaturrahman, H.; Lee, T.-G.G. A study on characteristics of microplastic in wastewater of South Korea: Identification, quantification, and fate of microplastics during treatment process. *Mar. Pollut. Bull.* **2019**, *146*, 696–702. [CrossRef] [PubMed]
30. Ou, H.; Zeng, E.Y. *Occurrence and Fate of Microplastics in Wastewater Treatment Plants*; Elsevier Inc.: Amsterdam, The Netherlands, 2018; ISBN 9780128137475.
31. Yang, L.; Li, K.; Cui, S.; Kang, Y.; An, L.; Lei, K. Removal of microplastics in municipal sewage from China's largest water reclamation plant. *Water Res.* **2019**, *155*, 175–181. [CrossRef] [PubMed]
32. Kankanige, D.; Babel, S. Contamination by ≥6.5 μm-sized microplastics and their removability in a conventional water treatment plant (WTP) in Thailand. *J. Water Process Eng.* **2021**, *40*, 101765. [CrossRef]

33. Ben-David, E.A.; Habibi, M.; Haddad, E.; Hasanin, M.; Angel, D.L.; Booth, A.M.; Sabbah, I. Microplastic distributions in a domestic wastewater treatment plant: Removal efficiency, seasonal variation and influence of sampling technique. *Sci. Total Environ.* **2021**, *752*, 141880. [CrossRef] [PubMed]
34. Carr, S.A.; Liu, J.; Tesoro, A.G. Transport and fate of microplastic particles in wastewater treatment plants. *Water Res.* **2016**, *91*, 174–182. [CrossRef]
35. Dalmau-Soler, J.; Ballesteros-Cano, R.; Boleda, M.R.; Paraira, M.; Ferrer, N.; Lacorte, S. Microplastics from headwaters to tap water: Occurrence and removal in a drinking water treatment plant in Barcelona Metropolitan area (Catalonia, NE Spain). *Environ. Sci. Pollut. Res.* **2021**, *28*, 59462–59472. [CrossRef]
36. Dris, R.; Gasperi, J.; Mirande, C.; Mandin, C.; Guerrouache, M.; Langlois, V.; Tassin, B. A first overview of textile fibers, including microplastics, in indoor and outdoor environments. *Environ. Pollut.* **2017**, *221*, 453–458. [CrossRef]
37. Freeman, S.; Booth, A.M.; Sabbah, I.; Tiller, R.; Dierking, J.; Klun, K.; Rotter, A.; Ben-David, E.; Javidpour, J.; Angel, D.L. Between source and sea: The role of wastewater treatment in reducing marine microplastics. *J. Environ. Manag.* **2020**, *266*, 110642. [CrossRef]
38. Gatidou, G.; Arvaniti, O.S.; Stasinakis, A.S. Review on the occurrence and fate of microplastics in Sewage Treatment Plants. *J. Hazard. Mater.* **2019**, *367*, 504–512. [CrossRef] [PubMed]
39. Jiang, J.; Wang, X.; Ren, H.; Cao, G.; Xie, G.; Xing, D.; Liu, B. Investigation and fate of microplastics in wastewater and sludge filter cake from a wastewater treatment plant in China. *Sci. Total Environ.* **2020**, *746*, 141378. [CrossRef] [PubMed]
40. Li, J.; Liu, H.; Paul Chen, J. Microplastics in freshwater systems: A review on occurrence, environmental effects, and methods for microplastics detection. *Water Res.* **2018**, *137*, 362–374. [CrossRef] [PubMed]
41. Liu, F.; Nord, N.B.; Bester, K.; Vollertsen, J. Microplastics Removal from Treated Wastewater by a Biofilter. *Water* **2020**, *12*, 1085. [CrossRef]
42. Lv, X.M.; Dong, Q.; Zuo, Z.Q.; Liu, Y.C.; Huang, X.; Wu, W.M. Microplastics in a municipal wastewater treatment plant: Fate, dynamic distribution, removal efficiencies, and control strategies. *J. Clean. Prod.* **2019**, *225*, 579–586. [CrossRef]
43. Mason, S.A.; Garneau, D.; Sutton, R.; Chu, Y.; Ehmann, K.; Barnes, J.; Fink, P.; Papazissimos, D.; Rogers, D.L. Microplastic pollution is widely detected in US municipal wastewater treatment plant effluent. *Environ. Pollut.* **2016**, *218*, 1045–1054. [CrossRef]
44. Michielssen, M.R.; Michielssen, E.R.; Ni, J.; Duhaime, M.B. Fate of microplastics and other small anthropogenic litter (SAL) in wastewater treatment plants depends on unit processes employed. *Environ. Sci. Water Res. Technol.* **2016**, *2*, 1064–1073. [CrossRef]
45. Misra, A.; Zambrzycki, C.; Kloker, G.; Kotyrba, A.; Anjass, M.H.; Franco Castillo, I.; Mitchell, S.G.; Güttel, R.; Streb, C. Water Purification and Microplastics Removal Using Magnetic Polyoxometalate-Supported Ionic Liquid Phases (magPOM-SILPs). *Angew. Chem. Int. Ed.* **2020**, *59*, 1601–1605. [CrossRef]
46. Murray, A.; Örmeci, B. Water with Separation Processes Used for Water and Wastewater Treatment. *Water* **2020**, *12*, 635. [CrossRef]
47. O'Connor, D.; Pan, S.; Shen, Z.; Song, Y.; Jin, Y.; Wu, W.M.; Hou, D. Microplastics undergo accelerated vertical migration in sand soil due to small size and wet-dry cycles. *Environ. Pollut.* **2019**, *249*, 527–534. [CrossRef]
48. Rhein, F.; Scholl, F.; Nirschl, H. Magnetic seeded filtration for the separation of fine polymer particles from dilute suspensions: Microplastics. *Chem. Eng. Sci.* **2019**, *207*, 1278–1287. [CrossRef]
49. Sembiring, E.; Fajar, M.; Handajani, M. Performance of rapid sand filter—single media to remove microplastics. *Water Supply* **2021**, *21*, 2273–2284. [CrossRef]
50. Seth, C.K.; Shriwastav, A. Contamination of Indian sea salts with microplastics and a potential prevention strategy. *Environ. Sci. Pollut. Res.* **2018**, *25*, 30122–30131. [CrossRef] [PubMed]
51. Simon, M.; Vianello, A.; Vollertsen, J. Removal of >10 μm Microplastic Particles from Treated Wastewater by a Disc Filter. *Water* **2019**, *11*, 1935. [CrossRef]
52. Talvitie, J.; Mikola, A.; Koistinen, A.; Setälä, O. Solutions to microplastic pollution—Removal of microplastics from wastewater effluent with advanced wastewater treatment technologies. *Water Res.* **2017**, *123*, 401–407. [CrossRef]
53. van den Berg, P.; Huerta-Lwanga, E.; Corradini, F.; Geissen, V. Sewage sludge application as a vehicle for microplastics in eastern Spanish agricultural soils. *Environ. Pollut.* **2020**, *261*, 114198. [CrossRef] [PubMed]
54. Yaranal, N.A.; Subbiah, S.; Mohanty, K. Identification, extraction of microplastics from edible salts and its removal from contaminated seawater. *Environ. Technol. Innov.* **2021**, *21*, 101253. [CrossRef]
55. Talvitie, J.; Heinonen, M.; Pääkkönen, J.P.; Vahtera, E.; Mikola, A.; Setälä, O.; Vahala, R.; Paakkonen, J.P.; Vahtera, E.; Mikola, A.; et al. Do wastewater treatment plants act as a potential point source of microplastics? Preliminary study in the coastal Gulf of Finland, Baltic Sea. *WATER Sci. Technol.* **2015**, *72*, 1495–1504. [CrossRef]
56. Ngo, P.L.; Pramanik, B.K.; Shah, K.; Roychand, R. Pathway, classification and removal efficiency of microplastics in wastewater treatment plants. *Environ. Pollut.* **2019**, *255*, 113326. [CrossRef]
57. Enfrin, M.; Dumée, L.F.; Lee, J. Nano/microplastics in water and wastewater treatment processes—Origin, impact and potential solutions. *Water Res.* **2019**, *161*, 621–638. [CrossRef] [PubMed]
58. Lares, M.; Ncibi, M.C.; Sillanpää, M.; Sillanpää, M. Occurrence, identification and removal of microplastic particles and fibers in conventional activated sludge process and advanced MBR technology. *Water Res.* **2018**, *133*, 236–246. [CrossRef] [PubMed]
59. Mahon, A.M.; O'Connell, B.; Healy, M.G.; O'Connor, I.; Officer, R.; Nash, R.; Morrison, L. Microplastics in sewage sludge: Effects of treatment. *Environ. Sci. Technol.* **2017**, *51*, 810–818. [CrossRef] [PubMed]

60. Habib, D.; Locke, D.C.; Cannone, L.J. Synthetic fibers as indicators of municipal sewage sludge, sludge products, and sewage treatment plant effluents. *Water. Air. Soil Pollut.* **1998**, *103*, 1–8. [CrossRef]
61. Milojevic, N.; Cydzik-kwiatkowska, A. Agricultural use of sewage sludge as a threat of microplastic (Mp) spread in the environment and the role of governance. *Energies* **2021**, *14*, 6293. [CrossRef]
62. Wolff, S.; Weber, F.; Kerpen, J.; Winklhofer, M.; Engelhart, M.; Barkmann, L. Elimination of microplastics by downstream sand filters in wastewater treatment. *Water* **2021**, *13*, 33. [CrossRef]
63. Jellali, S.; Khiari, B.; Usman, M.; Hamdi, H.; Charabi, Y.; Jeguirim, M. Sludge-derived biochars: A review on the influence of synthesis conditions on pollutants removal efficiency from wastewaters. *Renew. Sustain. Energy Rev.* **2021**, *144*, 111068. [CrossRef]
64. Yan, P.; Shi, H.X.; Chen, Y.P.; Gao, X.; Fang, F.; Guo, J.S. Optimization of recovery and utilization pathway of chemical energy from wastewater pollutants by a net-zero energy wastewater treatment model. *Renew. Sustain. Energy Rev.* **2020**, *133*, 110160. [CrossRef]
65. Durdević, D.; Blecich, P.; Jurić, Ž. Energy Recovery from Sewage Sludge: The Case. *Energies* **2019**, *12*, 1927. [CrossRef]
66. The European Parliament; The Council of the European Union. European Parliament and of the Council Directive 2008/98/EC. *Off. J. Eur. Union* **2008**, *L 312*, 3–30.
67. Bannick, C.G.; Szewzyk, R.; Ricking, M.; Schniegler, S.; Obermaier, N.; Barthel, A.K.; Altmann, K.; Eisentraut, P.; Braun, U. Development and testing of a fractionated filtration for sampling of microplastics in water. *Water Res.* **2019**, *149*, 650–658. [CrossRef]
68. Bretas Alvim, C.; Mendoza-Roca, J.A.A.; Bes-Piá, A. Wastewater treatment plant as microplastics release source—Quantification and identification techniques. *J. Environ. Manage.* **2020**, *255*, 109739. [CrossRef] [PubMed]
69. Elkhatib, D.; Oyanedel-Craver, V. A Critical Review of Extraction and Identification Methods of Microplastics in Wastewater and Drinking Water. *Environ. Sci. Technol.* **2020**, *54*, 7037–7049. [CrossRef] [PubMed]
70. Ziajahromi, S.; Neale, P.A.; Rintoul, L.; Leusch, F.D.L. Wastewater treatment plants as a pathway for microplastics: Development of a new approach to sample wastewater-based microplastics. *Water Res.* **2017**, *112*, 93–99. [CrossRef] [PubMed]
71. Lusher, A.L.; Munno, K.; Hermabessiere, L.; Carr, S. Isolation and Extraction of Microplastics from Environmental Samples: An Evaluation of Practical Approaches and Recommendations for Further Harmonization. *Appl. Spectrosc.* **2020**, *74*, 1049–1065. [CrossRef]
72. Oßmann, B.E. Microplastics in drinking water? Present state of knowledge and open questions. *Curr. Opin. Food Sci.* **2021**, *41*, 44–51. [CrossRef]
73. Sun, J.; Dai, X.H.; Wang, Q.L.; van Loosdrecht, M.C.M.M.; Ni, B.J. Microplastics in wastewater treatment plants: Detection, occurrence and removal. *Water Res.* **2019**, *152*, 21–37. [CrossRef]
74. Hong, Y.; Oh, J.; Lee, I.; Fan, C.; Pan, S.Y.; Jang, M.; Park, Y.K.; Kim, H. Total-organic-carbon-based quantitative estimation of microplastics in sewage. *Chem. Eng. J.* **2021**, *423*, 130182. [CrossRef]
75. Lenz, R.; Labrenz, M. Small Microplastic Sampling in Water: Development of an Encapsulated Filtration Device. *Water* **2018**, *10*, 1055. [CrossRef]
76. Dyachenko, A.; Mitchell, J.; Arsem, N. Extraction and identification of microplastic particles from secondary wastewater treatment plant (WWTP) effluent. *Anal. Methods* **2017**, *9*, 1412–1418. [CrossRef]
77. Cooper, D.A.; Corcoran, P.L. Effects of mechanical and chemical processes on the degradation of plastic beach debris on the island of Kauai, Hawaii. *Mar. Pollut. Bull.* **2010**, *60*, 650–654. [CrossRef]
78. Sarkar, D.J.; Das Sarkar, S.; Das, B.K.; Praharaj, J.K.; Mahajan, D.K.; Purokait, B.; Mohanty, T.R.; Mohanty, D.; Gogoi, P.; Kumar, V.S.; et al. Microplastics removal efficiency of drinking water treatment plant with pulse clarifier. *J. Hazard. Mater.* **2021**, *413*, 125347. [CrossRef] [PubMed]
79. Ding, H.; Zhang, J.; He, H.; Zhu, Y.; Dionysiou, D.D.; Liu, Z.; Zhao, C. Do membrane filtration systems in drinking water treatment plants release nano/microplastics? *Sci. Total Environ.* **2021**, *755*, 142658. [CrossRef] [PubMed]
80. Ahmed, M.B.; Rahman, M.S.; Alom, J.; Hasan, M.S.; Johir, M.A.H.; Mondal, M.I.H.; Lee, D.Y.; Park, J.; Zhou, J.L.; Yoon, M.H. Microplastic particles in the aquatic environment: A systematic review. *Sci. Total Environ.* **2021**, *775*, 145793. [CrossRef] [PubMed]
81. Li, Q.; Lai, Y.; Yu, S.; Li, P.; Zhou, X.; Dong, L.; Liu, X.; Yao, Z.; Liu, J. Sequential Isolation of Microplastics and Nanoplastics in Environmental Waters by Membrane Filtration, Followed by Cloud-Point Extraction. *Anal. Chem.* **2021**, *93*, 4559–4566. [CrossRef]
82. Bayo, J.; Olmos, S.; López-Castellanos, J. Assessment of microplastics in a municipal wastewater treatment plant with tertiary treatment: Removal efficiencies and loading per day into the environment. *Water* **2021**, *13*, 1339. [CrossRef]
83. Dris, R.; Gasperi, J.; Rocher, V.; Saad, M.; Renault, N.; Tassin, B. Microplastic contamination in an urban area: A case study in Greater Paris. *Environ. Chem.* **2015**, *12*, 592–599. [CrossRef]
84. Zhang, Y.; Diehl, A.; Lewandowski, A.; Gopalakrishnan, K.; Baker, T. Removal efficiency of micro- and nanoplastics (180 nm–125 μm) during drinking water treatment. *Sci. Total Environ.* **2020**, *720*, 137383. [CrossRef] [PubMed]
85. Ma, B.; Xue, W.; Ding, Y.; Hu, C.; Liu, H.; Qu, J. Removal characteristics of microplastics by Fe-based coagulants during drinking water treatment. *J. Environ. Sci.* **2019**, *78*, 267–275. [CrossRef]

*energies*

MDPI

*Article*

# Nexus between Technological Innovation, Renewable Energy, and Human Capital on the Environmental Sustainability in Emerging Asian Economies: A Panel Quantile Regression Approach

**Miguel Angel Esquivias [1,*], Lilik Sugiharti [1], Hilda Rohmawati [1,2], Omar Rojas [1,3] and Narayan Sethi [4]**

1 Faculty of Economics and Business, Airlangga University, Surabaya 60286, Indonesia; sugiharti.lilik@feb.unair.ac.id (L.S.); hilda.rohmawati-2016@feb.unair.ac.id or hilda.rohmawati@support.bappenas.go.id (H.R.); orojas@up.edu.mx (O.R.)
2 Ministry of National Development Planning, Jakarta 10310, Indonesia
3 Facultad de Ciencias Económicas y Empresariales, Universidad Panamericana, Álvaro del Portillo 49, Zapopan 45010, Mexico
4 Department of Humanities and Social Sciences, National Institute of Technology Rourkela, Rourkela 769008, India; sethin@nitrkl.ac.in
* Correspondence: miguel@feb.unair.ac.id

**Citation:** Esquivias, M.A.; Sugiharti, L.; Rohmawati, H.; Rojas, O.; Sethi, N. Nexus between Technological Innovation, Renewable Energy, and Human Capital on the Environmental Sustainability in Emerging Asian Economies: A Panel Quantile Regression Approach. *Energies* **2022**, *15*, 2451. https://doi.org/10.3390/en15072451

Academic Editor: Alberto-Jesus Perea-Moreno

Received: 22 February 2022
Accepted: 25 March 2022
Published: 26 March 2022

**Publisher's Note:** MDPI stays neutral with regard to jurisdictional claims in published maps and institutional affiliations.



**Abstract:** The goal of this study was to examine the interlinkage of renewable energy, technology innovation, human capital, and governance on environment quality by using a panel quantile regression in Asian emerging economies over the period of 1990–2019. The results indicated that higher economic growth, population density, technological innovation in renewable energy, and exploitation of natural resources have significantly raised $CO_2$ emissions in emerging Asia. Furthermore, larger capital, more use of renewable energy, green technology, and human capital development can improve environmental sustainability in Asia. As for governances, proxied by corruption rates, no evidence indicated that it has resulted in more damage, unlike earlier studies have suggested. The findings indicated that the three channels exposed in the Kuznets hypothesis can serve as a reference for proposals for environmental policies (scale of consumption, energy composition, and choice of technologies). There are opportunities to reduce $CO_2$ emissions through investments in human development, investing in new technologies to increase efficiency in energy (generation and consumption), increasing working capital (GCF), and migrating to more environmentally friendly energy. The negative link between carbon dioxide emissions and economic growth, increases in population density, and exploitation of natural resources can compromise the achievement of sustainable environmental goals.

**Keywords:** technological innovation; panel quantile regression; environmental sustainability; $CO_2$ emissions; renewable energy; emerging Asian economies

## 1. Introduction

The increase in carbon dioxide ($CO_2$) emissions has prompted extensive research, especially due to its impact on the environment. The general belief is that excessive emissions of $CO_2$ deteriorate life quality and environmental sustainability. Studies have also shown how economic growth prompts more $CO_2$ emissions and contributes to environmental degradation [1,2]. Other factors identified in previous research include energy consumption [3], population growth [4], capital accumulation [5], and corruption [6]. On the other side of the equation, improvements in human resources [7], technological innovation [3], and renewable energies [8] can reduce $CO_2$ and support environmental sustainability.

In the last three decades, rapid economic growth in emerging economies has led to increased emissions in Asia. Evidence suggests a negative effect of economic growth on environmental sustainability in Brazil, Russia, India, China, and South Africa (BRICS) countries [3,5]. The surge in emissions in China [9], South Asia [10], and other large emerging economies was linked to high energy consumption as noted in Adebayo et al. [11]. Other factors affecting environmental quality include population growth, industrial activity, and poor governance [12].

The use of innovative technologies and renewable energies, the rate of corruption, and the improvement of human capital, are emerging issues in Asian economics, but research has yet to uncover their impacts on environmental sustainability. Research in this area is fundamental in growing economies such as China, South Korea, India, Pakistan, Indonesia, Malaysia, and Thailand, as they shape the development model in the region as noted by Danish et al. [13]. Besides, they have substantial natural resource reserves that will significantly impact the environment if not appropriately managed as pointed out for cases on emerging countries [5,14].

Carbon dioxide emissions in emerging Asian countries increased by nearly 400% from 1990 to 2019, likely driven by the growing economic activities, population increase, and higher energy consumption. According to the International Energy Agency (IEA, Paris, France), the total renewable energy consumption in Asia (excluding China) increased from 85 GWh in 1990 to 63,119 GWh in 2019. Altogether there has been a shift towards renewable sources, although many countries, including China, Indonesia, Pakistan, Thailand, Malaysia, and India, continue relying on non-renewable sources. These countries are also lagging in technological innovation, alternative energies, good governance, and human resource development, as pointed out by Balsalobre-Lorente et al. [15], even when these factors are critical in supporting environmental sustainability (e.g., Brazil in [11]).

This study investigated the nexus between $CO_2$ emissions and technological innovation, renewable energy use, and human capital development in Asian emerging countries—China, South Korea, India, Pakistan, Indonesia, Thailand, Malaysia, and the Philippines—from 1990 to 2019. The findings were expected to extend the literature focusing on the link between the environment and economic growth, population, corruption, and use of natural resources. We used a panel quantile regression approach to identify the variables' heterogeneous structure and capture the market structure's differences of the sample data. As demonstrated by Allard et al. [16], the quantile approach is a powerful tool to evaluate the environmental Kuznets hypothesis in a panel context. Unlike systems based on mean values (e.g., pooled OLS, FEM, or REM approaches), the quantile method estimates different points on the conditional distribution of the chosen dependent variable (see [17,18]).

The current study offered the following contributions. First, we applied a panel quantile regression that allows for an estimation of the conditional distribution, as opposed to conditional mean in OLS-based approaches. This allowed us to examine the link across quantiles, and hence, the $CO_2$ emissions' rates across countries. Second, we investigated the countries in Asia with a great demand for energy and growing innovation, renewable energy use, human development, and good governance as pointed out by Zafar el al. [19]. Third, we employed variables that drive emissions (GDP, energy consumption, population, and corruption) and those that reduce emissions (innovation, renewable energies, human development, and capital growth). Fourth, we tested Kuznets hypothesis, which states that economic expansion propels emissions in three channels (Halliru et al. [7]): by scale, i.e., consumption of energy associated with larger growth; by composition, i.e., choice of energies; and by technique, i.e., technology choices. The proposed model aimed to capture all three channels, which has not been attempted in past research. Additionally, the region has been engaging in sustainable practices in the last decades, so an assessment of whether the development and policy have been supportive is needed (see [15,18,19]).

## 2. Literature Review

Previous research examined the drivers of carbon emissions from an economic perspective, mainly under the Kuznets hypothesis, which identifies how economic activities impact the environment in three channels. The first is through the scale of economic growth, which is synonymous with greater energy demand. The second is the composition of renewable and non-renewable energy adoptions. There has been plenty of research in this area of study in advanced and emerging countries, e.g., Adebayo et al. [11] for Brazil, Halliru et al. [7] for West African states, and Yao et al. [8] for major developing countries. The general assumption of the Kuznets hypothesis is that income growth drives energy demand, hence increasing $CO_2$ emissions at a rapid rate until it reaches a certain threshold. The threshold that leads to recovery varies depending on different factors. Azam and Khan's [6] study in Southeast Asia confirmed the hypothesis applying ordinary least squares (OLS). Other studies using augmented mean group (AMG) [3] and bootstrap ARDL [20] in BRICS, and panel quantile regression in Africa [7], revealed that capital inflows (domestic or foreign) have driven environmental degradation forward.

In emerging Asian countries, $CO_2$ emissions have increased due to the rising demand for energy and the intensive use of non-sustainable energy resources as noted by Hanif et al. [21] who applied an ARDL approach. The adoption of greener technologies has lowered the emissions rate in emerging countries as confirmed by Adams and Nsiah [22] using a GMM model, although not as low as in advanced countries. A shift towards more renewable sources could also decrease $CO_2$ emissions as supported in cases for emerging countries [10,11], G7 nations [23], and large economies including India and China [24,25]. However, this was not the case in other Asian countries such as India, Pakistan, and Bangladesh [26]. This begs the question as to whether adopting renewable technologies at a larger scale can offer a positive impact on environmental sustainability in emerging Asian countries.

Meanwhile, innovative technology has proven to be effective in improving the environment in many countries and economic regions, including Brazil [11], APEC [19], the G7 [23], and BRICS [3]. This is because innovative technologies can make power generation and energy use more efficient, which is good for the environment as found by Yao et al. [5] for major developing countries using a GMM model. Innovative technologies can also regulate the exploration of natural resources and monitor governance, leading to better environmental conditions.

Good governance and human resource development also reduce $CO_2$ emissions as supported by Zafar et al. [19] for Asia Pacific countries. Corruption, on the other hand, may hinder sustainability program delivery. In 16 OECD countries, a higher corruption perception index was found to be unfavorable to achieving reduction targets of $CO_2$ emissions [15]. Corruption may also lower the impacts of innovative programs and other policies to improve the environment [8,12]. In Southeast Asia, Azam and Khan [6] found a significant impact of corruption on the effort to reduce $CO_2$ in Malaysia, but not in Thailand and Indonesia. Applying quantile regression, Luqman et al. [27] found that anticorruption policies and human capital improvements supported environmental sustainability in Pakistan. Studies covering China, India, and other emerging countries tended to support that improvements in governance can moderate energy consumption [2,5,10].

In terms of wealth in natural resources, using an ARDL approach, Tufail et al. [28] found that developed countries with rich natural resources but poor governance emit more $CO_2$. They also believed that decentralization could improve the situation because local authorities have more power to address environmental challenges more effectively. As for developing countries, evidence suggested that abundant natural resources and poor governance are likely to harm the environment [12,14,29]. Therefore, controlling corruption is necessary to prevent the exploitation of natural resources and reduce the carbon footprint [5].

Considering the mixed results found in the literature, this study provided new evidence on the three channels described in the Kuznets hypothesis with respect to the means by which economic development could impact the environment. The variables chosen in this study attempted to provide evidence on each of the three channels: scale was approximated by GDP and population density; the composition channel was captured through the consumption of renewable and non-renewable energy; the choice of energies and technique channel was approximated with gross capital formation and patents in renewable and non-renewable energies. Finally, corruption control and natural resources were two variables that could alter the environmental curve, either by adding or diminishing the effects. As described, there is no absolute evidence on the impact of the proposed variables in the context of emerging Asian countries.

## 3. Materials and Methods

This study used cross-sectional data from eight countries: China, India, Indonesia, Korea, Malaysia, Pakistan, the Philippines, and Thailand, from 1990 to 2019. Table 1 describes the variables employed in this study and the corresponding data sources. Panel quantile regression—initially introduced by Koenker and Bassett [30]—was used in the estimations following earlier studies [7,31]. As noted by Zhu et al. [18], by using fixed effects panel quantile regressions, we could estimate how the variables modulated $CO_2$ emissions 'throughout a conditional distribution', including in countries with high and low levels of emissions.

**Table 1.** The variables and data Sources.

| Variables | Description | Source |
|---|---|---|
| EQ | Carbon dioxide ($CO_2$) emissions in Kt units as the environment's quality | EIA |
| COR | Control of corruption | WGI |
| GCF | Gross capital formation (percent of GDP) | WDI |
| HDI | Human development index | UNDP |
| GDP | Economic growth | WDI |
| TI_R | Technology patent application for renewable energy | WDI |
| TI_NR | Technology patent application for non-renewable energy | WDI |
| RE | Total of renewable energy consumption | EIA |
| NRE | Non-renewable energy to primary energy | EIA |
| NR | Total natural resources rent (percent of GDP) | WDI |
| PO | Population density | WDI |

Note: Energy Information Administration (EIA), World Governance Indicator (WGI), World Development Indicator (WDI), United Nations Development Programme (UNDP).

According to De Silva et al. [32] and Salman et al. [33], OLS regression assumptions such as the random disturbance is zero-mean and identically distributed, with error terms usually identically distributed. This may not occur in real life, as economic and social indicators may be distributed differently in a dataset. Other OLS assumptions, namely, heteroscedasticity, unbiased estimation, and minimum variance, are also unrealistic [34]. Therefore, panel quantile regression is a more reasonable approach as it can accommodate distribution differences [33].

Panel quantile regression is based on the regression of an independent variable with the conditional distribution of the dependent variable, estimated to produce a regression model for all quantiles [34]. It is not limited by classical distribution assumptions, which are unlikely to be met [35] and can better capture the tail characteristics of a variable distribution [34].

To analyze the effect of an independent variable on the dependent variable, the following model was adopted:

$$\begin{aligned} &\mathrm{EQ}_{\mathrm{CO2}_{it}}(\pi_k/\alpha_i,\ x_{it}) \\ &= \alpha_i + \beta_1\mathrm{COR}_{it} + \beta_2 Ln\mathrm{GCF}_{it} + \beta_3 Ln\mathrm{HDI}_{it} + \beta_4 Ln\mathrm{GDP}_{it} + \beta_5 Ln\mathrm{TI}_{\mathrm{R}it} + \beta_6 Ln\mathrm{TI}_{\mathrm{NR}it} + \beta_7 Ln\mathrm{RE}_{it} \\ &+ \beta_8 Ln\mathrm{NRE}_{it} + \beta_9\mathrm{NR}_{it} + \beta_{10}\mathrm{PO}_{it} \end{aligned} \quad (1)$$

where $i$ and $t$ are country and year (fixed effects), $\pi_k$ represents the quantile number, $Ln$ is the natural logarithm, and EQ is the dependent variable ($CO_2$). The independent variables are control of corruption (COR) as a proxy of good governance, gross capital formation (GCF) as a proxy of physical capital, human development index (HDI) as a proxy of human capital, gross domestic product (GDP) as a measure of economic growth, technology patent application for renewable energy (TI_R) as a measure of innovation technology, technology patent application for non-renewable energy (TI_NR), total renewable energy consumption (RE), non-renewable energy to primary energy (NRE), total natural resources rent (NR), and population density (PO). See Table 1 for details.

The control of corruption index is compiled by the World Bank under the World Governance Indicators (WGI). Control of corruption is one of the six indices to generate an indicator of governance. The WGI defines control of corruption as "perceptions of the extent to which public power is exercised for private gain, including both petty and grand forms of corruption, as well as "capture" of the state by elites and private interests". The index of control of corruption ranged from a weak point of –2.5 to a strong point of 2.5. A survey of experts on the perception of control of corruption is applied to obtain these indices.

As the statistical distribution of the data often follows a non-equal variation, the location of the conditional distribution of the data is therefore unequal. In a quantile regression, the conditional distribution is spread across different quantiles, where a mean is represented by the 50th quantile [36]. As such, quantile regressions are more suitable to evaluate samples in the presence of outliers than any estimation techniques based on mean conditional distribution. A quantile approach allows capturing the different slopes for the other set of quantiles.

The initial hypothesis of this study was that increasing GDP would increase $CO_2$ emissions as energy demand also rises; by contrast, more use of renewable energy would lead to a decrease in $CO_2$ emissions. The use of technology would also reduce emissions as production would be more efficient. As for good governance (proxied by low corruption level) and human capital improvement (HDI), we hypothesized that both would reduce emissions. Finally, greater non-renewable energy use, population density, and abundance of natural resources were likely to increase emissions. As for gross capital formation, the result could be either positive or negative, depending on whether it was directed for more efficient technology or the other way round.

Thus, the panel quantile regression model was used to examine the link among the variables at different points of the distribution (across quantiles). The standard regression model represents the association between an explanatory variable ($x$) and the dependent variable ($y$) following E ($y \mid x$). By contrast, the quantile regression approach constricts the relationship based on the conditional distribution on the variable $y$ [37]. Considering that not all the variables follow an equal statistical distribution, the relationship among variables may be subject to the conditional distribution function of the dependent variable. The quantile regression proposed by Koenker and Bassett [30], thus, estimates the conditional distribution of output variables [17]. A standard quantile regression can be presented as

$$y_i = x'\beta_q + e_i, \tag{2}$$

where $\beta_q$ indicates the vector(s) of parameters to be estimated pertaining to the percentile quantile $q$th [38]. The conventional least squares regression estimates the conditional expectation of the response variable by minimizing the sum of squares of the unobserved factors, or error, $e_i\left(\sum_i e_i^2\right)$. The quantile regression estimates the conditional median or other quantiles of the response variable by targeting the minimization of the objective function:

$$Q\left(\beta_q\right) = \sum_{i:y_i \geq x_i'\beta}^{N} q\left|y_i - x_i'\beta_q\right| + \sum_{i:y_i < x_j'\beta}^{N} (1-q)\left|y_i - x_i'\beta_q\right|, \tag{3}$$

for $0 < q < 1$. As suggested in the literature [38], the quantile approach of Powell [39] was used to validate results. The panel quantile regression of Powell follows a non-additive regression model at various distribution points (percentiles) of $CO_2$ release. This model was suggested by Powell [39] and proven to offer robust inference when examining long-run paths of $CO_2$ emission [38].

The procedure to compute the parameters in the model proposed in Equation (1) is as follows: the first step taken before estimating the panel quantile regression is to test the cross-sectional dependence on the panel data in order to determine the homogeneity of the data and determine if the univariate time series possess a unit root. The rejection of the null hypothesis in the cross-sectional dependence test allows for the heterogeneity panel estimation condition (Halliru et al. [7]). Furthermore, it is necessary to check the stationarity of the variables to avoid spurious regressions. A cointegration test is then carried out to see if there is a long-term relationship between the variables. The results of the panel integration test using Johansen Fisher depend on the lag order of the VAR system fitted to all variables (Zhu et al. [18]).

After the aforementioned tests were carried out and the proper hypothesis was tested, we used a quantile regression panel model with fixed effects in order to predict the effect of conditional heterogeneous covariance of the dependent variable so as to control for individual heterogeneity that is not observed [37]. The fixed effect panel quantile regression model is as follows:

$$Qy_{it}(\tau_k|\alpha_i, x_{it}) = \alpha_i + x_{it}^T\beta(\tau_k) \tag{4}$$

where $i = 1, \ldots, N$; $t = 1, \ldots, T$. In this formula, $\alpha_i$ has a pure location shift effect on the conditional quantile of the response. The effect of the $x_{it}$ covariate may depend on the $\tau$ quantile. $i$ is individual and $t$ is time. N is the number of observations on individual $i$. T is the number of observations at time $t$.

The unique characteristic of this method is the existence of a penalty term in minimizing computational problems in estimating the mass of parameters specifically (Chen & Lei [40]; Zhu et al. [18]). The parameter estimation is calculated as follows:

$$\min_{(\alpha,\beta)} \sum_{k=1}^{K}\sum_{t=1}^{T}\sum_{i=1}^{N} w_k\rho\tau_k\left(y_{it} - \alpha_i - x_{it}^T\beta(\tau_k)\right) + \lambda\sum_{i}^{N}|\alpha_i| = 1, \ldots, N;\ t = 1, \ldots, T \tag{5}$$

where K is the quantile index, $x$ is the explanatory variable matrix, $\rho\tau_k$ is the quantile loss function, $w_k$ is the relative weight assigned to the quantile ($w_k = 1/K$) which controls the contribution of the $k$th quantile to the fixed effect estimate, and $\lambda$ is a tuning parameter that reduces individual effects to zero to improve estimation performance ($\lambda = 1$).

## 4. Results

### *4.1. Empirical Results*

Table 2 shows the original data's descriptive statistics before transformation using natural logarithms. Prior to the estimations, the Jarque–Bera test was applied as a goodness-of-fit test to validate that the data had the skewness and kurtosis following a normal distribution. As indicated in Table 2, only the HDI variable was not normally distributed, indicated by the insignificant result of the Jarque–Bera test.

**Table 2.** Descriptive statistics.

| Variables | Mean | Median | Maximum | Minimum | Std. Dev. | Skewness | Kurtosis | Jarque–Bera |
|---|---|---|---|---|---|---|---|---|
| EQ | 1,125,592.00 | 254,777.70 | 10,291,927.00 | 41,763.46 | 2,227,455.00 | 2.994 | 11.385 | 1061.747 *** |
| COR | −0.39 | −0.34 | 0.45 | −1.59 | 0.46 | −0.7 | 2.73 | 20.13 *** |
| GCF | 28.79 | 27.62 | 46.66 | 14.12 | 8.38 | 0.26 | 2.06 | 11.60 *** |
| HDI | 0.65 | 0.65 | 0.92 | 0.402 | 0.12 | 0.08 | 2.49 | 2.93 |
| GDP | $1.04 \times 10^{12}$ | $3.97 \times 10^{11}$ | $1.15 \times 10^{13}$ | $7.93 \times 10^{10}$ | $1.84 \times 10^{12}$ | 3.623 | 16.925 | 2464.4 *** |
| TI_R | 54,932.4 | 856 | 1,393,815 | 0 | 193,235.3 | 5.175 | 30.82 | 8817.8 *** |
| TI_NR | 17,188.89 | 4970.5 | 157,093 | 0 | 29,063.53 | 2.81 | 11.03 | 961.2 *** |
| RE | 28.32 | 30.37 | 58.65 | 0.44 | 18.06 | −0.096 | 1.73 | 16.44 *** |
| NRE | 3.69 | 2.24 | 18.49 | 0.39 | 4.61 | 2.05 | 5.83 | 248.1 *** |
| NR | 3.78 | 2.27 | 25.86 | 0 | 4.11 | 1.88 | 7.55 | 348.9 *** |
| PO | $3.78 \times 10^{8}$ | $1.08 \times 10^{8}$ | $1.40 \times 10^{9}$ | $1.80 \times 10^{7}$ | $4.90 \times 10^{8}$ | 1.167 | 2.527 | 56.8 *** |

Note: *** significant at 1% level of significance. Carbon dioxide $CO_2$ (EQ); Control of corruption (COR); Gross capital formation (GCF); Human development index (HDI); Economic growth (GDP); Technology for renewable energy (TI_R); Technology for non-renewable energy (TI_NR); Renewable energy consumption (RE); Non-renewable energy to primary energy (NRE); Total natural resources rent (NR); Population density (PO).

The results of the cross-sectional dependence tests using the Pesaran CD test [41], the bias-correlated scaled LM, the Pesaran-scaled LM, and the Breusch–Pagan LM are shown in Table 3. There was sufficient cross-sectional dependence in the eight countries involved in this study. Using Pesaran and Yamagata [42], we tested the homogeneity conditions. Apergis et al. [31] stated that the cross-dependence test is crucial before applying the unit root test. The null hypothesis was that there was no sufficient cross-sectional dependence of the variables used in this study.

**Table 3.** Cross-dependency test results.

| Variables | Pesaran CD Test | | Bias-Corrected Scaled LM | | Pesaran Scaled LM | | Breusch–Pagan LM | |
|---|---|---|---|---|---|---|---|---|
| | Statistic | *p*-Value | Statistic | *p*-Value | Statistic | *p*-Value | Statistic | *p*-Value |
| *Ln*EQ | 27.45 *** | 0.000 | 96.90 *** | 0.000 | 97.04 *** | 0.000 | 754.2 *** | 0.000 |
| COR | 14.24 *** | 0.000 | 62.91 *** | 0.000 | 63.05 *** | 0.000 | 499.8 *** | 0.000 |
| *Ln*GCF | 3.99 *** | 0.000 | 22.11 *** | 0.000 | 22.24 *** | 0.000 | 194.5 *** | 0.0001 |
| *Ln*HDI | 28.73 *** | 0.000 | 106.4 *** | 0.000 | 106.5 *** | 0.000 | 825.2 *** | 0.000 |
| *Ln*GDP | 28.60 *** | 0.000 | 105.5 *** | 0.000 | 105.6 *** | 0.000 | 818.2 *** | 0.000 |
| *Ln*TI_R | 24.22 *** | 0.000 | 75.98 *** | 0.000 | 76.12 *** | 0.000 | 597.6 *** | 0.000 |
| *Ln*TI_NR | 14.67 *** | 0.000 | 32.37 *** | 0.000 | 32.51 *** | 0.000 | 271.3 *** | 0.000 |
| *Ln*RE | 12.69 *** | 0.000 | 59.18 *** | 0.000 | 59.32 *** | 0.000 | 471.9 *** | 0.000 |
| *Ln*NRE | 9.78 *** | 0.000 | 27.57 *** | 0.000 | 27.70 *** | 0.000 | 235.3 *** | 0.000 |
| NR | 14.82 *** | 0.000 | 32.93 *** | 0.000 | 33.06 *** | 0.000 | 275.4 *** | 0.000 |
| PO | 28.79 *** | 0.000 | 106.9 *** | 0.000 | 107.0 *** | 0.000 | 829.1 *** | 0.000 |

Note: *** significant at 1% level of significance. *Ln* is the natural logarithm. Carbon dioxide $CO_2$ (EQ); Control of corruption (COR); Gross capital formation (GCF); Human development index (HDI); Economic growth (GDP); Technology for renewable energy (TI_R); Technology for non-renewable energy (TI_NR); Renewable energy consumption (RE); Non-renewable energy to primary energy (NRE); Total natural resources rent (NR); Population density (PO).

Data tests against unit root problems are needed to avoid results from spurious regressions. As suggested by earlier studies [7,16], it is possible to test for the presence of unit roots by using the Levin, Lin, and Chu (LLC), the Im, Pesaran, and Shin (IPS), the Augmented Dickey–Fuller (ADF) Fisher, and the PP-Fisher. The panel unit root test (see Table 4) showed that all variables were stationary at the first difference except *Ln*RE using LLC. Using IPS and ADF-Fisher, all variables were stationary at the first difference. Meanwhile, the PP-Fisher suggested that all variables were stationary at the first difference except for PO. In general, it could be said that all variables were free from the unit root problem, so that were eligible for regressions and would not produce spurious regressions.

**Table 4.** Panel unit root test.

| Variables | LLC | | IPS | | ADF-Fisher | | PP-Fisher | |
|---|---|---|---|---|---|---|---|---|
| | I(0) | I(I) | I(0) | I(I) | I(0) | I(I) | I(0) | I(I) |
| *Ln*EQ | −2.31 ** | −5.71 *** | 0.03 | −6.45 *** | 18.29 | 73.78 *** | 36.70 *** | 132.36 *** |
| COR | −1.75 ** | −2.83 *** | −0.46 | −8.79 *** | 14.09 | 100.95 *** | 60.27 *** | 195.99 *** |
| *Ln*GCF | −1.46 * | −4.34 *** | −0.99 | −7.25 *** | 17.69 | 82.01 *** | 14.82 | 142.02 *** |
| *Ln*HDI | −5.29 *** | −1.79 ** | −1.57 * | −4.04 *** | 28.97 ** | 47.75 *** | 54.84 *** | 109.20 *** |
| *Ln*GDP | −1.69 ** | −4.12 *** | 2.53 | −5.59 *** | 12.77 | 61.55 *** | 37.41 *** | 90.13 *** |
| *Ln*TI_R | −0.38 | −5.75 *** | 2.38 | −7.85 *** | 11.32 | 88.49 *** | 42.69 *** | 144.51 *** |
| *Ln*TI_NR | −3.16 *** | −9.48 *** | −1.31 * | −9.31 *** | 22.36 | 106.51 *** | 25.11 * | 174.29 *** |
| *Ln*RE | −2.20 ** | −0.96 | −1.05 | −4.04 *** | 31.62 ** | 44.77 *** | 45.01 *** | 100.07 *** |
| *Ln*NRE | −0.08 | −7.77 *** | 1.52 | −9.64 *** | 5.85 | 111.14 *** | 7.68 | 172.02 *** |
| NR | −0.01 | −4.58 *** | −0.74 | −8.69 *** | 17.89 | 99.87 *** | 41.69 *** | 171.24 *** |
| PO | −5.09 *** | −4.68 *** | −0.74 | −3.99 *** | 31.69 ** | 58.18 *** | 86.48 *** | 9.23 |

Note: *, **, *** significant at 10, 5, and 1% levels of significance. *Ln* is the natural logarithm.

For the cointegration test, we used the Johansen–Fisher Test following Halliru et al. [7], aiming to validate a long-term relationship between variables [33]. The null hypothesis was that there was no long-term cointegration between variables. The Johansen–Fisher Test for cointegration, as shown in Table 5, rejected the null hypothesis, which indicated that all variables in the sample countries were cointegrated, which meant that a long-run relationship existed between the variables during the period of analysis (1990–2019).

**Table 5.** Panel cointegration test results.

| Hypothesized No. of CE(s) | Fisher Stat. (From Trace Test) | Prob. | Fisher Stat. (From Max-Eigen Test) | Prob. |
|---|---|---|---|---|
| None | 9.704 | 0.7835 | 9.704 | 0.7835 |
| At most 1 | 9.704 | 0.7835 | 9.704 | 0.7835 |
| At most 2 | 8.318 | 0.8721 | 26.74 | 0.0208 |
| At most 3 | 1.386 | 1 | 111.9 | 0 |

Table 6 shows some selected indicators to better understand the situation of the sample countries. The EQ column indicates the compound average growth rate (CAGR) of the $CO_2$ emissions. With the exception of Thailand and South Korea, the remaining countries grew a more than 4% annual average in $CO_2$ emissions. India and Malaysia grew at more than 5% CAGR. Regarding the corruption index (COR), South Korea had the best corruption control index with 0.74, while Pakistan (−0.87) registered the lowest level, followed by the Philippines (−0.57) and Indonesia (−0.42). Regarding gross capital formation (GCF), China maintained an average of 41% in GCP, while the Philippines (21%) and Pakistan (17%) registered the lowest percentages of GFC. In terms of human capital (HDI), South Korea registered the maximum index (0.91) while India (0.65) and Pakistan (0.56) reached the lowest HDI index in 2019. Indonesia, the Philippines, and Thailand showed an average of 0.74 in the HDI Index, with a substantial increase from the beginning to the end of the study period. In terms of economic growth (GDP), China registered a 9.2% annual average growth, while India reached 6%. Indonesia, the Philippines, Thailand and Pakistan registered the lowest GDP growth, on average 3.9% to 4.6% of annual growth.

**Table 6.** Selected indicators.

| Variables | EQ CAGR | COR Avrg | GCF Avrg % | HDI Last | GDP CAGR | TI_R Total (000) | TI_NR Total (000) | RE Avrg % | NRE Avrg % | NR Avrg % | PO CAGR |
|---|---|---|---|---|---|---|---|---|---|---|---|
| Philippines | 4.3% | −0.57 | 21% | 0.72 | 4.4% | 6.46 | 82.34 | 30.9% | 3.1% | 1.6% | 1.9% |
| China | 4.8% | −0.30 | 41% | 0.76 | 9.2% | 9927.08 | 2064.26 | 18.8% | 3.6% | 3.8% | 0.7% |
| India | 5.2% | −0.25 | 32% | 0.65 | 6.0% | 187.59 | 544.01 | 44.0% | 2.3% | 3.0% | 1.5% |
| Indonesia | 4.2% | −0.42 | 29% | 0.72 | 4.6% | 14.57 | 129.38 | 40.8% | 0.7% | 7.3% | 1.3% |
| S Korea | 3.2% | 0.74 | 33% | 0.92 | 4.8% | 3005.29 | 993.67 | 1.6% | 15.8% | 0.0% | 0.6% |
| Malaysia | 5.1% | 0.25 | 28% | 0.81 | 5.4% | 18.90 | 146.18 | 5.5% | 1.0% | 11.5% | 1.9% |
| Pakistan | 4.0% | −0.82 | 17% | 0.56 | 4.0% | 2.89 | 25.94 | 48.1% | 3.4% | 1.6% | 2.4% |
| Thailand | 3.8% | −0.29 | 29% | 0.78 | 3.9% | 21.00 | 139.57 | 22.0% | 0.9% | 1.8% | 0.7% |

Note. Compound average growth rate (CAGR); Average (Avrg); Last period (Last); Carbon dioxide $CO_2$ (EQ); Control of corruption (COR); Gross capital formation (GCF); Human development index (HDI); Economic growth (GDP); Technology for renewable energy (TI_R); Technology for non-renewable energy (TI_NR); Renewable energy consumption (RE); Non-renewable energy to primary energy (NRE); Total natural resources rent (NR); Population density (PO).

The total number of patents related to renewable energy (TI_R) and non-renewable energy (TI_NR) are also shown in Table 6. China and South Korea accumulated the highest number of patents in the period (30 years), while Pakistan registered the lowest number in both indicators. Regarding renewable energy (RE) consumption, Pakistan and India consumed on average about 48% to 44% of energy in renewable types. S Korea, Malaysia, and China had the lowest share of RE among the sample countries. In reference to natural resources (NR), Indonesia and Malaysia had the highest percentages of resources to total GDP. Finally, the population density (PO) increased in all the countries in the sample, but to a greater extent in Pakistan, Malaysia, the Philippines, and India.

### *4.2. Determinants of Carbon Dioxide ($CO_2$) Emissions*

Table 7 shows the panel quantile regression results. The quantitative regression estimates showed that higher control of corruption (COR) had a negative effect on $CO_2$ emissions (better environment). However, the results were not significant. A possible explanation for the statistically non-significant result may be because the impact of corruption on $CO_2$ emissions is indirect, observed through specific policies, as indicated for the case of the G7 countries [43]. Similarly, Sinha et al. [2] stressed that the effects of corruption on $CO_2$ emissions are manifested indirectly (mediation) through a reduction in the impact of the use of renewable energy and an increase in the negative effects of dirty energy.

**Table 7.** Panel quantile regression estimates.

| Quantiles | COR | *Ln*GCF | *Ln*HDI | *Ln*GDP | *Ln*TI_R | *Ln*TI_NR | *Ln*RE | *Ln*NRE | NR | PO |
|---|---|---|---|---|---|---|---|---|---|---|
| τ = 0.05 | 0.001 | −0.20 * | −1.58 *** | 0.79 *** | 0.13 *** | 0.016 | −0.21 *** | −0.25 *** | 0.01 * | $6.92 \times 10^{-10}$ *** |
| | (0.03) | (0.12) | (0.25) | (0.05) | (0.03) | (0.05) | (0.02) | (0.04) | (0.01) | $(1.16 \times 10^{10})$ |
| τ = 0.10 | −0.04 | −0.14 | −1.21 *** | 0.77 *** | 0.12 *** | −0.04 | −0.21 *** | −0.23 *** | 0.005 | 8.39 × 10−10 *** |
| | (0.04) | (0.12) | (0.19) | (0.05) | (0.02) | (0.04) | (0.03) | (0.04) | (0.01) | (1.08 × 10−10) |
| τ = 0.20 | −0.02 | −0.1 | −1.27 *** | 0.73 *** | 0.13 *** | −0.01 | −0.19 *** | −0.27 *** | 0.001 | 7.60 × 10−10 *** |
| | (0.03) | (0.06) | (0.16) | (0.03) | (0.02) | (0.03) | (0.03) | (0.02) | (0.003) | (9.98 × 10−11) |
| τ = 0.30 | −0.01 | −0.103 | −1.34 *** | 0.72 *** | 0.14 *** | 0.01 | −0.18 *** | −0.28 *** | 0.002 | 6.89 × 10−10 *** |
| | (0.04) | (0.07) | (0.19) | (0.04) | (0.02) | (0.04) | (0.03) | (0.02) | (0.003) | (1.10 × 10−10) |
| τ = 0.40 | −0.01 | −0.15 ** | −1.57 *** | 0.73 *** | 0.17 *** | −0.002 | −0.14 *** | −0.26 *** | 0.01 | 5.58 × 10−10 *** |
| | (0.04) | (0.07) | (0.18) | (0.04) | (0.03) | (0.05) | (0.03) | (0.02) | (0.004) | (9.71 × 10−11) |
| τ = 0.50 | 0.01 | −0.13 * | −1.76 *** | 0.71 *** | 0.22 *** | −0.06 | −0.10 *** | −0.24 *** | 0.01 ** | 5.29 × 10−10 *** |
| | (0.04) | (0.08) | (0.22) | (0.06) | (0.05) | (0.14) | (0.03) | (0.03) | (0.01) | (1.29 × 10−10) |
| τ = 0.60 | −0.01 | −0.1 | −1.37 *** | 0.72 *** | 0.22 *** | −0.17 *** | −0.11 *** | −0.22 *** | 0.02 ** | 7.51 × 10−10 *** |
| | (0.05) | (0.08) | (0.31) | (0.06) | (0.03) | (0.04) | (0.04) | (0.03) | (0.01) | (1.54 × 10−10) |
| τ = 0.70 | −0.05 | −0.15 ** | −1.16 *** | 0.72 *** | 0.19 *** | −0.16 *** | −0.15 *** | −0.22 *** | 0.01 | 9.20 × 10−10 *** |
| | (0.05) | (0.08) | (0.34) | (0.05) | (0.04) | (0.03) | (0.05) | (0.04) | (0.01) | (1.94 × 10−10) |
| τ = 0.80 | −0.06 | −0.07 | −0.45 | 0.60 *** | 0.15 *** | −0.14 *** | −0.10 * | −0.19 *** | 0.002 | 1.22 × 10−9 *** |
| | (0.04) | (0.12) | (0.55) | (0.09) | (0.04) | (0.04) | (0.06) | (0.06) | (0.01) | (2.25 × 10−10) |
| τ = 0.90 | −0.02 | 0.03 | −0.08 | 0.56 *** | 0.15 *** | −0.13 *** | −0.02 | −0.14 *** | −0.002 | 1.25 × 10−9 *** |
| | (0.03) | (0.08) | (0.39) | (0.08) | (0.04) | (0.03) | (0.04) | (0.05) | (0.01) | (1.99 × 10−10) |
| τ = 0.95 | −0.01 | 0.08 | 0.03 | 0.49 *** | 0.17 *** | −0.14 *** | 0.03 | −0.11 *** | 0.001 | 1.28 × 10−9 *** |
| | (0.03) | (0.07) | (0.33) | (0.05) | (0.03) | (0.03) | (0.03) | (0.04) | (0.01) | (1.64 × 10−10) |

Note: *, **, *** significant at 10, 5, and 1% levels of significance. Numbers in brackets are standard error values. Carbon dioxide CO2 (EQ); Control of corruption (COR); Gross capital formation (GCF); Human development index (HDI); Economic growth (GDP); Technology for renewable energy (TI_R); Technology for non-renewable energy (TI_NR); Renewable energy consumption (RE); Non-renewable energy to primary energy (NRE); Total natural resources rent (NR); Population density (PO).

Regarding gross capital formation (GCF), increases in capital helped lower emissions and improved the environment. The results showed that at the 0.05, 0.4, 0.5, and 0.7 quantiles, Asia's gross capital formation was negatively related to $CO_2$. Higher capital formation relates to increases to capital assets for the country in the form of land improvements, additions of equipment, infrastructure expansion (roads, rails, schools, hospitals, private and public buildings), as well as increases in inventories held by firms. The results suggested that increased capital enables investment in superior technology, undertaking projects, and the upgrading of industrial activities that help emerging Asian countries reduce $CO_2$ emissions.

As for the human development index (HDI), we found a significant and negative correlation between FDI and $CO_2$, suggesting that improvements in human capital will improve the environmental sustainability. The results were significant in quantiles 0.1 to 0.7, which suggested that the impact was greater in low-medium quantiles where most of the sample distribution was found. The HDI captures development in human capabilities related to health, education, and quality of life. A negative correlation between the HDI and EQ suggested that as people in emerging Asian countries gain access to better health systems, the more opportunities they have to be more knowledgeable, and as they improve their standards of living, it lowers environmental degradation ($CO_2$ emissions). These results indicated that investments in human development in Asia had a positive influence on environmental quality.

GDP affected $CO_2$ emissions as indicated by the positive and significant relationship found in all quantiles. The results revealed that growth in domestic product and income is positively associated with higher carbon dioxide emissions. In this sense, the results suggest that emerging Asia has not exceeded the threshold of environmental degradation exposed in the environmental Kuznets curve. The Kuznets hypothesis indicates that in the initial phase of development, countries experience a positive link between economic growth and environmental degradation up to a point where the growth–$CO_2$ link becomes negative.

Technology patent application for renewable energy (TI_R) positively affected emissions, meaning that it does not improve environmental quality. The results were significant for all quantiles (0.05–0.95). The results were counterintuitive, as we expected that R&D on renewable energy could help reduce emissions. It is important to note that new technologies in renewable sources such as biomass can generate larger $CO_2$ emissions than fossil fuels [43,44]. In other words, technologies must be concerned more about making a greener environment and not merely focused on increasing energy production. These results were important considering that the increase in patents is associated with a more extensive use of renewable energies as noted in Tee et al. [45]. However, as the results of this study suggest, further development of renewable technologies does not reduce $CO_2$ emissions.

Technology patent application of non-renewable energy (TI_NR) had a negative impact on $CO_2$ emissions, meaning that R&D projects can positively affect environmental sustainability. The results were significant in high quantiles (0.6–0.95).

Renewable energy consumption (RE) had a negative effect on $CO_2$ emissions as indicated in the statistically significant results in quantiles 0.05 to 0.8. In other words, increasing the consumption of renewable sources helped decrease $CO_2$ emissions, offering opportunities to increase environmental quality. Increased use of renewable energy (i.e., geothermal, solar, hydroelectric, biomass waste, biofuels, or wood) could help Asian countries reduce $CO_2$ emissions. As for the share of non-renewable energy to primary energy (NRE), it had a negative effect on $CO_2$ emissions on all quantiles (0.05 to 0.95). The result was somehow opposite of previous research stating that non-renewable energy use leads to higher emissions in developing countries [2,10]. However, some emerging Asian countries are reducing non-renewable sources and substituting them with renewable sources, signaling that energy policies have helped achieve higher environmental quality. A recent study provided evidence in the case of Malaysia [46], where emissions were reduced through diversification in the production and use of energy towards sources other than oil and coal.

Meanwhile, abundance in natural resources (NR) positively affected environmental degradation. The results were significant in the low (0.05) and middle quantile (0.5 and 0.6) and not significant for the higher quantiles.

Population density (PO) positively affected carbon emissions, which meant higher density led to environmental degradation. The results were significant for all quantiles. Population density refers to the number of inhabitants per square kilometer of area. In China, the population density increased from 120 to 149 inhabitants per sq km (similar to Indonesia). In Pakistan, PO increased from 139 to 280, while in India, the increase was from 293 to 460 inhabitants per sq km. Thus, it could be inferred that increasing levels of population density had supported larger $CO_2$ emissions in Asia.

### 4.3. Discussion

Table 8 summarizes the estimates (a sign of impact) of the quantile results in Table 7.

The results of this study did not provide evidence of a statistically significant relationship between corruption control (COR) and environmental quality (EQ). Previous research also showed mixed results. Using the environmental performance index as an indicator of environmental quality for a large sample of countries, Lisciandra and Migliardo [12] found that increases in levels of corruption reduce environmental quality. Other studies also found a positive relationship between corruption and environmental damage as in Rehman et al. [47] (South Asian countries) and Sinha et al. [2] (BRICS). Some studies found significant results at high quantiles as in Luqman et al. [27] for the case of Pakistan, while others did not find any significant results, as in Azam and Khan [6] for the case of Malaysia, Indonesia, and Thailand, or Haldar and Sethi [10] for a sample of developing countries. One possible explanation is linked to the improvement in the corruption control index that was observed in the Asian countries covered in this study. However, it is also true that the index showed large fluctuations during the period of analysis, without showing a constant trend.

**Table 8.** Summary of quantile regression estimates.

| Variables | Low (τ = 0.05 until 0.30) | Middle (τ = 0.40 until 0.60) | High (τ = 0.70 until 0.95) |
|---|---|---|---|
| COR | / | / | / |
| *Ln*GCF | − | − | − |
| *Ln*HDI | − | − | − |
| *Ln*GDP | + | + | + |
| *Ln*TI_R | + | + | + |
| *Ln*TI_NR | / | − | − |
| *Ln*RE | − | − | − |
| *Ln*NRE | − | − | − |
| NR | + | + | / |
| PO | + | + | + |

Note: *Ln* is the natural logarithm, "/" indicates no significant effect, "−" indicates a negative impact, and "+" indicates a positive effect of the independent variable on the dependent variable. Carbon dioxide $CO_2$ (EQ); Control of corruption (COR); Gross capital formation (GCF); Human development index (HDI); Economic growth (GDP); Technology for renewable energy (TI_R); Technology for non-renewable energy (TI_NR); Renewable energy consumption (RE); Non-renewable energy to primary energy (NRE); Total natural resources rent (NR); Population density (PO).

Regarding gross capital formation (GCF), the results of this study supported earlier evidence, such as the one presented by Shahbaz et al. [43] stating that capital formation helps lower emissions in high-income countries (G7) and APEC countries [19]. Other studies showed that capital formation can drive emissions, while others claimed a bidirectional relation between finance and $CO_2$ (e.g., BRICS in Rafique et al. [3]). In the BRICS countries, foreign capital improved the environmental sustainability in China, but the remaining countries (Brazil, India, Russia, and South Africa) showed mixed results. This study (Tables 7 and 8) provided new evidence on the positive effect of capital on the environmental quality of emerging countries in Asia, previously not found in studies using a similar method (e.g., a quantile approach in Haldar & Sethi [10]). It is possible that earlier studies [10] over specified the model by including variables that are overlapped (FDI, domestic credit, gross capital formation, among others).

Regarding the HDI—$CO_2$ nexus, this study validated policies in support of human development in emerging Asia as they had effects in favor of environmental quality. Earlier studies also considered HDI or sub-components, e.g., education in APEC ([19] or [48]), finding that higher levels of HDI or sub-components help lower emissions. In China, human capital improvement (e.g., education) lowered $CO_2$ emissions [49], but in other studies, human capital improvement resulted in higher emissions as in West African states [7] or Pakistan [50].

The results provided evidence on the positive nexus between economic growth (GDP) and $CO_2$ emissions in the eight Asian countries examined in this study. The findings of this study are in line with earlier evidence, stating that economic growth in Asia continues to be a significant contributor to environmental degradation (see, e.g., [10,19,21,51]). In Pakistan, India, and Bangladesh, Mehmood et al. [26] found similar results to the ones in this study, stating that good governance should accompany economic growth for emissions to decline.

Regarding the link between patents in renewable energies and $CO_2$, the results indicated that $CO_2$ emissions are linked to the increase in patents related to renewable energies. R&D or higher technology in APEC countries result in larger $CO_2$, as noted by Zafar el al. [19]. The results presented in this study also align with Allard et al. [16] who used quantile regression and found that R&D was positively associated with $CO_2$ emissions in low and middle-income countries. Improvements in technology can help decrease emissions in higher income economies in the APEC region [19,52], G7 countries [23,51], or for high-income level countries [16], but not in emerging Asia. It is perhaps necessary to accompany the development of renewable energy technologies with efficiency in generation, distribution, and consumption of energy.

Considering that technology for non-renewable energy (TI_NR) showed a negative relation to $CO_2$ emissions, this study then suggests that energy policies in Asia may shift towards cleaner energy by supporting technological development. Access to more advanced technologies can help reduce emissions, probably associated with more efficient production, distribution, and use of energy. In line with the Kuznets hypothesis, economic expansion in Asia may be associated with an increase in per capita energy consumption due to higher population and income, but also with the inappropriate use of technologies to produce and consume energy. The findings of this study supported previous evidence that showed that technological innovation can reduce $CO_2$ in Brazil [11], the BRICS [3], and G7 countries [43]. Energy policies in Asia may lean towards cleaner energy by supporting technological improvements in power generation, not necessarily green sources. Nevertheless, the results were also counterintuitive to the findings of Jiao et al. [53] in China, Santra [54] for some BRICS countries, and Zafar et al. [19] for APEC, who found that innovation in non-renewable technologies drives $CO_2$ emissions.

Regarding renewable energy consumption (RE), this study found results that align with earlier studies in advanced and BRICS countries (see [10,23,48,51,55]). Employing a similar approach on a large dataset of countries, Allard et al. [16] also found a negative correlation between renewable sources and $CO_2$, maintaining that green energy sources can improve the environment. Within Asia, the findings differed from a recent study for India, Pakistan, and Bangladesh, whereby non-renewable and renewable energy use drives more emissions [26].

Among the eight countries included, the available natural resources are abundant in countries like Indonesia, Malaysia, and China, suggesting the need for environmental protection policies that are directly linked to natural resources. Previous studies in BRICS countries yielded mixed results [13,56]. Evidence from this study supported those of Masron & Subramaniam [29] who pointed out the need for more robust policies to manage natural resources to reduce pollution and greenhouse emissions in developing countries. Furthermore, earlier studies in the BRICS [13] and other regions [57], looking into the biocapacity perspective (productive land and available water for human consumption), noted that exploitation of resources can accelerate environmental degradation as the capacity to absorb pollutants decreases. As such, Asian countries need a more active policy in protecting the ecology in the region as emissions can lower the biological ability to absorb pollutants.

The evidence of this study is in line with previous findings stating that a rise in population and urban development was identified as one of the major determinants of $CO_2$ emissions in the BRICS [2,3], G7 [43], and emerging Asian countries [10].

Regarding the Kuznets hypothesis, which establishes that economic expansion drives emissions in three channels, this study provided evidence on the following aspects. First, in the eight Asian countries examined, energy consumption is associated with higher economic growth and population density (scale effect in Kuznets). Second, the fact that $CO_2$ emissions are negatively associated with renewable energy consumption suggests that better energy choices can help improve sustainable development in Asia. Finally, this study provided evidence that emissions in Asia can be improved through better choice of technologies (production, distribution, and use of energy). Increasing the use of working capital and the use of more advanced energies can reduce the environmental impact. The positive link between GDP growth and $CO_2$ emissions suggests that the eight countries observed in this study appear not to have reached the threshold of the environmental curve.

The three channels exposed in the Kuznets hypothesis can serve as a reference to proposals for environmental policies. There are opportunities to reduce $CO_2$ emissions through the three channels: scale of consumption, energy composition, and choice of technologies. For Asian countries, a way to reach the environmental threshold in the Kuznets hypothesis could be by increasing investments in human development, investing in new technologies to increase efficiency in power generation (i.e., TI_NR), increasing working capital (GCF), and migrating to more environmentally friendly energy consumption (RE).

### 4.4. Policy Recommendations and Implications

In accordance with the findings of this study, the following policy implications were drawn: Firstly, the investment in research and development is not enough. Hence, the government should focus on R&D investment to improve their technology innovation effectiveness to achieve the dual effect of growth with environmental quality. Secondly, the empirical results proved that the patents on renewable energy contribute more to carbon emission. On the other hand, the patents on non-renewable energy improve the quality of the environment by reducing carbon ejection. Although energy technology patents have significantly increased, many energy-efficient technologies have not been widely adopted. Hence, the government should design energy structures and develop zero carbon-emitting energy technologies. Thirdly, low income and middle-income countries should pay particular concern to the benefits of technology innovation and proactively participate with other developed countries to stimulate the utilization of environmentally friendly technologies. Countries should shift from an energy-driven economy to an innovation-driven economy to curb emissions and achieve sustainable development, specifically high carbon-emitting countries.

Fourthly, development in human capital formation is a critical requirement to achieve economic growth and sustainable development. Human capital plays a vital role, by stimulating research and development on energy-efficient technology. The results also support improving the level of the HDI, which causes the quality of human capital to flourish through education. Fifthly, the results also found that both renewable and non-renewable energy negatively affect carbon emissions, which was quite unexpected. Energy consumption, especially non-renewable energy, is the main driver for economic growth for emerging countries, which causes environmental degradation. So, it is essential to maintain the energy sector to achieve sustainability without compromising growth. Therefore, the policymakers should prioritize obtaining an optimum mixture of both renewable and non-renewable energy, aiming to meet the nation's requirements and environmental needs. Lastly, to achieve sustainability, a long-term strategic policy is needed to improve the efficiency of energy by investing in renewable energy sources like solar, wind, geothermal, and biofuels to achieve sustainable development in emerging Asian countries

## 5. Conclusions

We estimated the link between technological innovation, renewable energy use, governance, and human capital and the carbon emissions in China, South Korea, India, Pakistan, Indonesia, Thailand, Malaysia, and the Philippines, with data from 1990 to 2019. We employed a quantile regression approach in the analysis. The findings suggested that improvements in technological innovation in non-renewable energies, human resource development, capital formation, and renewable energy use can help reduce emissions. Meanwhile, patents (proxies for R&D) positively relate to $CO_2$ emissions, which means that not all technological developments are linked to better environmental quality.

On the other hand, economic growth, population density, and energy consumption drive emissions. Therefore, investments in human capital, technological innovation, and renewable energies need to go hand in hand with economic growth to reduce the impact on the environment. The use of greener energy and good governance could also help lower emissions. Finally, we also found that energy consumption was not the only source of emissions. The extraction of natural resources can also cause more carbon footprint. Asian countries, rich in natural resources, need to reconsider policies towards more sustainable management of natural resources as this is closely linked to $CO_2$ emissions (at least in lower and middle quantiles).

There are opportunities to reduce $CO_2$ emissions through the three channels exposed in the Kuznets hypothesis: scale of consumption, energy composition, and choice of technologies. For Asian policy makers, policies should focus on human capital and R&D investment to improve technology innovation effectiveness to achieve the dual effect of growth with environmental quality. More attention paid to increasing investments in human development, investing in higher technologies to increase efficiency in generation, distribution, and use of power can help to lower $CO_2$ emissions. Migrating to more environmentally friendly energy consumption can help to improve environmental quality, although choosing appropriate technologies to generate clean energies is needed.

Regarding the choice of panel quantile regression as a research tool, the approach was selected as it could accommodate differences in the distribution of variables so that better information could be extracted from them. Panel quantile regression provides a regression model at all quantile levels, based on a conditional distribution of the data, suitable for samples with outliers. As a result, a quantile regression captures a different slope for each quantile, a consideration often missed in previous studies.

**Author Contributions:** M.A.E. carried out the conceptualization, writing—review, and funding acquisition; L.S. carried out the supervision, project administration, and validation; H.R. handled the methodology and formal analysis. O.R. carried out the writing—review and editing. N.S. focused on data curation. All authors have read and agreed to the published version of the manuscript.

**Funding:** This research was supported by the Research Grant Universitas Airlangga, Hibah Penelitian Kerja Sama Luar Negeri, 2021.

**Institutional Review Board Statement:** Not applicable.

**Informed Consent Statement:** Not applicable.

**Data Availability Statement:** Not applicable.

**Conflicts of Interest:** The authors declare no conflict of interest.

## References

1. Hassan, A.A.G.; Sethi, N.; Halliru, A.M.; Loganathan, N. FDI inflows, energy consumption and sustainable economic growth for ECOWAS countries: The pollution haven hypothesis approach. *Int. J. Green Econ.* **2020**, *14*, 1. [CrossRef]
2. Sinha, A.; Gupta, M.; Shahbaz, M.; Sengupta, T. Impact of corruption in public sector on environmental quality: Implications for sustainability in BRICS and next 11 countries. *J. Clean. Prod.* **2019**, *232*, 1379–1393. [CrossRef]
3. Rafique, M.Z.; Li, Y.; Larik, A.R.; Monaheng, M.P. The effects of FDI, technological innovation, and financial development on $CO_2$ emissions: Evidence from the BRICS countries. *Environ. Sci. Pollut. Res.* **2020**, *27*, 23899–23913. [CrossRef]
4. Agabo, T.; Abubakar, I.F.; Adedoyin, F.F. The anthropogenic consequences of energy consumption and population expansion in Africa? Do governance factors make any difference? *Environ. Sci. Pollut. Res.* **2021**, *28*, 27109–27118. [CrossRef] [PubMed]
5. Yao, X.; Yasmeen, R.; Hussain, J.; Hassan Shah, W.U. The repercussions of financial development and corruption on energy efficiency and ecological footprint: Evidence from BRICS and next 11 countries. *Energy* **2021**, *223*, 120063. [CrossRef]
6. Azam, M.; Khan, A.Q. Growth-corruption-health triaca and environmental degradation: Empirical evidence from Indonesia, Malaysia, and Thailand. *Environ. Sci. Pollut. Res.* **2017**, *24*, 16407–16417. [CrossRef]
7. Halliru, A.M.; Loganathan, N.; Golam Hassan, A.A.; Mardani, A.; Kamyab, H. Re-examining the environmental kuznets curve hypothesis in the economic community of West African states: A panel quantile regression approach. *J. Clean. Prod.* **2020**, *276*, 124247. [CrossRef]
8. Yao, S.; Zhang, S.; Zhang, X. Renewable energy, carbon emission and economic growth: A revised environmental Kuznets Curve perspective. *J. Clean. Prod.* **2019**, *235*, 1338–1352. [CrossRef]
9. Yin, Y.; Xiong, X.; Ullah, S.; Sohail, S. Examining the asymmetric socioeconomic determinants of $CO_2$ emissions in China: Challenges and policy implications. *Environ. Sci. Pollut. Res.* **2021**, *28*, 57115–57125. [CrossRef] [PubMed]
10. Haldar, A.; Sethi, N. Effect of institutional quality and renewable energy consumption on $CO_2$ emissions−an empirical investigation for developing countries. *Environ. Sci. Pollut. Res.* **2021**, *28*, 15485–15503. [CrossRef]

11. Adebayo, T.S.; Adedoyin, F.F.; Kirikkaleli, D. Toward a sustainable environment: Nexus between consumption-based carbon emissions, economic growth, renewable energy and technological innovation in Brazil. *Environ. Sci. Pollut. Res.* **2021**, *28*, 52272–52282. [CrossRef]
12. Lisciandra, M.; Migliardo, C. An Empirical Study of the Impact of Corruption on Environmental Performance: Evidence from Panel Data. *Environ. Resour. Econ.* **2017**, *68*, 297–318. [CrossRef]
13. Danish; Baloch, M.A.; Wang, B. Analyzing the role of governance in $CO_2$ emissions mitigation: The BRICS experience. *Struct. Chang. Econ. Dyn.* **2019**, *51*, 119–125. [CrossRef]
14. Akalin, G.; Erdogan, S.; Sarkodie, S.A. Do dependence on fossil fuels and corruption spur ecological footprint? *Environ. Impact Assess. Rev.* **2021**, *90*, 106641. [CrossRef]
15. Balsalobre-Lorente, D.; Shahbaz, M.; Chiappetta Jabbour, C.J.; Driha, O.M. The Role of Energy Innovation and Corruption in Carbon Emissions: Evidence Based on the EKC Hypothesis. In *Energy and Environmental Strategies in the Era of Globalization*; Shahbaz, M., Balsalobre, D., Eds.; University of Lincoln: Lincoln, UK, 2019; pp. 271–304.
16. Allard, A.; Takman, J.; Uddin, G.S.; Ahmed, A. The N-shaped environmental Kuznets curve: An empirical evaluation using a panel quantile regression approach. *Environ. Sci. Pollut. Res.* **2018**, *25*, 5848–5861. [CrossRef] [PubMed]
17. Cade, B.S.; Noon, B.R. A gentle introduction to quantile regression for ecologists. *Front. Ecol. Environ.* **2003**, *1*, 412–420. [CrossRef]
18. Zhu, H.; Duan, L.; Guo, Y.; Yu, K. The effects of FDI, economic growth and energy consumption on carbon emissions in ASEAN-5: Evidence from panel quantile regression. *Econ. Model.* **2016**, *58*, 237–248. [CrossRef]
19. Zafar, M.W.; Sinha, A.; Ahmed, Z.; Qin, Q.; Zaidi, S.A.H. Effects of biomass energy consumption on environmental quality: The role of education and technology in Asia-Pacific Economic Cooperation countries. *Renew. Sustain. Energy Rev.* **2021**, *142*, 110868. [CrossRef]
20. Yilanci, V.; Bozoklu, S.; Gorus, M.S. Are BRICS countries pollution havens? Evidence from a bootstrap ARDL bounds testing approach with a Fourier function. *Sustain. Cities Soc.* **2020**, *55*, 102035. [CrossRef]
21. Hanif, I.; Faraz Raza, S.M.; Gago-de-Santos, P.; Abbas, Q. Fossil fuels, foreign direct investment, and economic growth have triggered $CO_2$ emissions in emerging Asian economies: Some empirical evidence. *Energy* **2019**, *171*, 493–501. [CrossRef]
22. Adams, S.; Nsiah, C. Reducing carbon dioxide emissions; Does renewable energy matter? *Sci. Total Environ.* **2019**, *693*, 133288. [CrossRef] [PubMed]
23. Khan, Z.; Ali, S.; Umar, M.; Kirikkaleli, D.; Jiao, Z. Consumption-based carbon emissions and International trade in G7 countries: The role of Environmental innovation and Renewable energy. *Sci. Total Environ.* **2020**, *730*, 138945. [CrossRef] [PubMed]
24. Kirikkaleli, D. Do public-private partnerships in energy and renewable energy consumption matter for consumption-based carbon dioxide emissions in India? *Environ. Sci. Pollut. Res.* **2021**, *28*, 30139–30152. [CrossRef]
25. Yang, J.; Cai, W.; Ma, M.; Li, L.; Liu, C.; Ma, X.; Li, L.; Chen, X. Driving forces of China's $CO_2$ emissions from energy consumption based on Kaya-LMDI methods. *Sci. Total Environ.* **2020**, *711*, 134569. [CrossRef] [PubMed]
26. Mehmood, U.; Tariq, S.; Ul-Haq, Z.; Meo, M.S. Does the modifying role of institutional quality remains homogeneous in GDP-CO2 emission nexus? New evidence from ARDL approach. *Environ. Sci. Pollut. Res.* **2021**, *28*, 10167–10174. [CrossRef] [PubMed]
27. Luqman, M.; Li, Y.; Khan, S.U.-D.; Ahmad, N. Quantile nexus between human development, energy production, and economic growth: The role of corruption in the case of Pakistan. *Environ. Sci. Pollut. Res.* **2021**, *28*, 61460–61476. [CrossRef] [PubMed]
28. Tufail, M.; Song, L.; Adebayo, T.S.; Kirikkaleli, D.; Khan, S. Do fiscal decentralization and natural resources rent curb carbon emissions? Evidence from developed countries. *Environ. Sci. Pollut. Res.* **2021**, *28*, 49179–49190. [CrossRef] [PubMed]
29. Masron, T.A.; Subramaniam, Y. The environmental Kuznets curve in the presence of corruption in developing countries. *Environ. Sci. Pollut. Res.* **2018**, *25*, 12491–12506. [CrossRef]
30. Koenker, R.; Bassett, G. Regression Quantiles. *Econometrica* **1978**, *46*, 33. [CrossRef]
31. Apergis, N.; Christou, C.; Gupta, R. Are there Environmental Kuznets Curves for US state-level $CO_2$ emissions? *Renew. Sustain. Energy Rev.* **2017**, *69*, 551–558. [CrossRef]
32. De Silva, P.N.K.; Simons, S.J.R.; Stevens, P. Economic impact analysis of natural gas development and the policy implications. *Energy Policy* **2016**, *88*, 639–651. [CrossRef]
33. Salman, M.; Long, X.; Dauda, L.; Mensah, C.N.; Muhammad, S. Different impacts of export and import on carbon emissions across 7 ASEAN countries: A panel quantile regression approach. *Sci. Total Environ.* **2019**, *686*, 1019–1029. [CrossRef]
34. Lin, B.; Xu, B. Factors affecting $CO_2$ emissions in China's agriculture sector: A quantile regression. *Renew. Sustain. Energy Rev.* **2018**, *94*, 15–27. [CrossRef]
35. Sherwood, B.; Wang, L. Partially linear additive quantile regression in ultra-high dimension. *Ann. Stat.* **2016**, *44*, 288–317. [CrossRef]
36. Hübler, M. The inequality-emissions nexus in the context of trade and development: A quantile regression approach. *Ecol. Econ.* **2017**, *134*, 174–185. [CrossRef]
37. Giles, D.E. Roger Koenker, Victor Chernozhukov, Huming He and Limin Peng (2017): Handbook of Quantile Regression. *Stat. Pap.* **2018**, *59*, 849–850. [CrossRef]
38. Dash, D.P.; Dash, A.K.; Sethi, N. Designing hydro-energy led economic growth for pollution abatement: Evidence from BRICS. *Environ. Sci. Pollut. Res.* **2022**. [CrossRef]
39. Powell, D. Quantile regression with nonadditive fixed effects. *Quantile Treat. Eff.* **2016**, *1*, 1–33. [CrossRef]

40. Chen, W.; Lei, Y. The impacts of renewable energy and technological innovation on environment-energy-growth nexus: New evidence from a panel quantile regression. *Renew. Energy* **2018**, *123*, 1–14. [CrossRef]
41. Pesaran, M.H. General diagnostic tests for cross-sectional dependence in panels. *Empir. Econ.* **2021**, *60*, 13–50. [CrossRef]
42. Pesaran, M.H.; Yamagata, T. Testing slope homogeneity in large panels. *J. Econom.* **2008**, *142*, 50–93. [CrossRef]
43. Shahbaz, M.; Balsalobre, D.; Shahzad, S.J.H. The Influencing Factors of $CO_2$ Emissions and the Role of Biomass Energy Consumption: Statistical Experience from G-7 Countries. *Environ. Model. Assess.* **2019**, *24*, 143–161. [CrossRef]
44. Solarin, S.A. Does tourism-led growth hypothesis exist in Mauritius? Evidence from disaggregated tourism markets. *Curr. Issues Tour.* **2018**, *21*, 964–969. [CrossRef]
45. Tee, W.-S.; Chin, L.; Abdul-Rahim, A.S. Determinants of Renewable Energy Production: Do Intellectual Property Rights Matter? *Energies* **2021**, *14*, 5707. [CrossRef]
46. Rahman, A.R.A.; Shaari, M.S.; Masnan, F.; Esquivias, M.A. The Impacts of Energy Use, Tourism and Foreign Workers on $CO_2$ Emissions in Malaysia. *Sustainability* **2022**, *14*, 2461. [CrossRef]
47. Rehman, F.-U.-; Ali, A.; Nasir, M. Corruption, Trade Openness, and Environmental Quality: A Panel Data Analysis of Selected South Asian Countries. *Pak. Dev. Rev.* **2007**, *46*, 673–688. [CrossRef]
48. Zafar, M.W.; Qin, Q.; Malik, M.N.; Zaidi, S.A.H. Foreign direct investment and education as determinants of environmental quality: The importance of post Paris Agreement (COP21). *J. Environ. Manag.* **2020**, *270*, 110827. [CrossRef]
49. Hua, Y.; Xie, R.; Su, Y. Fiscal spending and air pollution in Chinese cities: Identifying composition and technique effects. *China Econ. Rev.* **2018**, *47*, 156–169. [CrossRef]
50. Bano, S.; Zhao, Y.; Ahmad, A.; Wang, S.; Liu, Y. Identifying the impacts of human capital on carbon emissions in Pakistan. *J. Clean. Prod.* **2018**, *183*, 1082–1092. [CrossRef]
51. Ding, Q.; Khattak, S.I.; Ahmad, M. Towards sustainable production and consumption: Assessing the impact of energy productivity and eco-innovation on consumption-based carbon dioxide emissions ($CCO_2$) in G-7 nations. *Sustain. Prod. Consum.* **2021**, *27*, 254–268. [CrossRef]
52. Wang, Z.; Rasool, Y.; Zhang, B.; Ahmed, Z.; Wang, B. Dynamic linkage among industrialisation, urbanisation, and $CO_2$ emissions in APEC realms: Evidence based on DSUR estimation. *Struct. Chang. Econ. Dyn.* **2020**, *52*, 382–389. [CrossRef]
53. Jiao, J.; Jiang, G.; Yang, R. Impact of R&D technology spillovers on carbon emissions between China's regions. *Struct. Chang. Econ. Dyn.* **2018**, *47*, 35–45. [CrossRef]
54. Santra, S. The effect of technological innovation on production-based energy and $CO_2$ emission productivity: Evidence from BRICS countries. *Afr. J. Sci. Technol. Innov. Dev.* **2017**, *9*, 503–512. [CrossRef]
55. Zhang, Y.-J.; Wang, W. Do renewable energy consumption and service industry development contribute to $CO_2$ emissions reduction in BRICS countries? *Environ. Sci. Pollut. Res.* **2019**, *26*, 31632–31643. [CrossRef]
56. Wang, Z.; Danish; Zhang, B.; Wang, B. The moderating role of corruption between economic growth and $CO_2$ emissions: Evidence from BRICS economies. *Energy* **2018**, *148*, 506–513. [CrossRef]
57. Ulucak, R.; Bilgili, F. A reinvestigation of EKC model by ecological footprint measurement for high, middle and low income countries. *J. Clean. Prod.* **2018**, *188*, 144–157. [CrossRef]

applied sciences

MDPI

Article

# Influence of Five Additives on No Loading Swelling Potential of Red Clay

**Bing Yue, Ziye Zhao and Zengzhen Qian ***

School of Engineering and Technology, China University of Geosciences (Beijing), Beijing 100083, China; yue@email.cugb.edu.cn (B.Y.); ziye_zhao@cugb.edu.cn (Z.Z.)
* Correspondence: zzqian@cugb.edu.cn

**Abstract:** In this study, cement, zeolite powder and three industrial by-products (blast furnace slag, steel slag, and fly ash) were selected as additives to study their effects on the swelling potential of red clay with different curing ages and dosages. At the same time, the mechanism of additives reducing the swelling potential of red clay was analyzed by scanning electron microscope (SEM) and X-ray fluorescence (XRF) tests. The X-ray diffraction (XRD) test was used to detect the clay mineral content of the red clay specimens before and after the modification to determine the change in the clay mineral content of the specimens. The direct shear test was used to explore the influence of additives on the strength of red clay. The results show that with 9% cement content, the no loading swelling potential of specimens can be reduced by 82.5% under 28 days of curing, and the cohesion of the specimens can be greatly increased by 82%. However, the specimens with cement have an increase in no loading swelling potential under the condition of no curing. In contrast, when steel slag is used as an additive to modify the swelling properties of red clay, the swelling potential can be reduced without curing, but the addition of steel slag will reduce the cohesive strength of specimens. XRD testing shows that the clay mineral composition in cement-modified specimens and steel slag-modified specimens experienced a relative change, the relative content of montmorillonite and illite decreased, and the relative content of kaolinite increased. Combined with SEM and XRF test results, it is concluded that cement's reduction in the swelling potential of red clay depends on pozzolanic reaction products filling the pores in specimens and bonding clay particles, so as to reduce the permeability of red clay and increase the resistance during swelling. The addition of cement can also convert hydrophilic clay minerals into nonhydrophilic clay minerals. Compared with cement, the reduction in swelling potential caused by steel slag mainly depends on the adsorption of ions to reduce the adsorption of water molecules on the surface of clay slices.

**Keywords:** red clay; expansive soil; cement; steel slag; modification; swelling potential

**Citation:** Yue, B.; Zhao, Z.; Qian, Z. Influence of Five Additives on No Loading Swelling Potential of Red Clay. *Appl. Sci.* **2022**, *12*, 3455. https://doi.org/10.3390/app12073455

Academic Editor: Fernando Rocha

Received: 22 February 2022
Accepted: 22 March 2022
Published: 29 March 2022



## 1. Introduction

Red clay is widely distributed in southern China. It has good engineering mechanical properties under the condition of low water content, but the clay will expand and contract under the conditions of water absorption and loss. When red clay is used as a foundation, it will deform due to the dry–wet cycle, which will have an adverse impact on the upper pavement and other structures [1]. With the construction of infrastructure and the expansion of urban areas in recent years, engineering construction increasingly needs to be carried out in areas covered with red clay. Reducing the influence of red clay swelling and shrinkage has become a common problem in engineering construction. At present, the method of replacing foundation soil with other soil is often used to eliminate the influence of expansive soil. However, the replacement method not only has the problem of high cost, but also has the problem of environmental protection; that is, the excavated earthwork is difficult to dispose of. The use of additives to modify the engineering properties of soil can fully take advantage of the role of in situ soil as much as possible, reduce the excavation and disposal of soil, and achieve the dual benefits of reducing expenses and protecting the environment. The use of additives to modify the swelling properties of expansive soil mainly considers two aspects: one is to reduce the hydrophilicity of clay particles in soil by cation exchange, and the other is to bond soil particles by the pozzolanic reaction to increase the resistance of soil expansion. Lime is one of the most commonly used additives to modify expansive soil. Yongzhen Cheng et al. studied the modification effect of lime, pozzolana, and their combination on the expansibility of black cotton soil in Kenya using the liquid plastic limit test, California bearing ratio (CBR) test, and Unconfined Compressive Strength (UCS) test. Combined with the analysis of pH, electrical conductivity (EC), XRD, and transmission electron microscope (TEM) test results, lime was found to significantly modify the physical and mechanical properties of black cotton soil. Pozzolana was also found to modify the physical and mechanical properties of black cotton soil to a certain extent, but it is not as effective as lime. The combination of lime and pozzolan to modify the physical and mechanical properties of black cotton soil can achieve better results than adding one of them alone [2]. Muzahim al-Mukhtar et al. studied the treatment of bentonite by lime under high-temperature curing conditions and found that a certain amount of lime is used to destroy the silicate layer of bentonite and form different hydrates, whereas more than this quantity of lime results in a pozzolanic reaction to produce hydrated calcium aluminate (CAH) and hydrated calcium silicate (CSH), and a high temperature can greatly accelerate the rate of pozzolanic reaction [3]. Blast furnace slag (BFS) is a nonmetallic by-product obtained during the manufacturing process of cast iron. It consists primarily of silicates, aluminosilicates, and calcium aluminosilicates [4]. Different forms of slag products are manufactured depending on the method used to cool the molten slag. These products include air-cooled BFS, expanded or foamed slag, pelletized slag, and granulated blast furnace slag (GBFS). When crushed or ground into very fine particles, blast furnace slag has cementitious properties [5]. Mohammad Hamed Fasihnikoutala studied the unconfined compression strength of the olive–GGBS treated soil, and the experimental results showed that ground granulated blast furnace slag (GGBS) has soluble calcium, which can generate CSH gel to improve soil properties [6].

Cement is widely used in soil modification and solidification because of its large output, low cost, and small side effects on the environment. Cement has been used in improving the water stability of expansive soil, and the UCS tests and swelling potential tests show that the addition of cement can significantly improve the strength of soil and modify the swelling and shrinkage properties [7,8]. Steel slag is a common solid waste produced by steel-making, and its output accounts for 15–20% of the total crude steel production [9]. In 2020, Chinese crude steel production was 1053 million tons, accounting for 56.5% of global production [10]. Massive quantities of steel slag are discarded because it cannot be effectively utilized, which has a negative effect on the environment. The efficient utilization of steel slag has become an urgent problem to be solved. Previous studies have shown that steel slag has a similar chemical and mineral composition to that of Portland cement. In many cases, steel slag can be used as an additive for soil modification, so as to save resources and reduce carbon emission in cement production. Nicholas E. Malasavage conducted field experiments to study the engineering properties of a steel slag mixed soil embankment. The Cone Penetration Test (CPT) results of the studied embankment with different steel slag contents showed that the average tip resistance of the test embankment improved by steel slag was greatly improved, and there was no expansion and heavy metal leaching in the experiment [11]. Poh H. Y. found that the swelling properties and long-term strength of soil treated with basic oxygen steel slag were better than those treated with lime [12]. Akinwumi Isaac proposed that the cation exchange between soil and steel slag is the main factor affecting the modification of soil engineering properties by steel slag. The addition of steel slag reduces the diffusion water layer and agglomerates the clay particles in the soil–slag mixture, to improve the strength and plasticity of the soil [13]. Thermal power generation results in large quantities of coal ash. Fly ash, which is produced from the combustion of pulverized coal, has an important place among industrial by-products.

A large amount of fly ash cannot be used effectively, which results in environmental problems. B. A. Mir and A. Sridharan pointed out that the addition of fly ash to clayey soils significantly reduces their swelling due to the reduction in plastic fines of clay by non-plastic fines of fly ash. The swell potential decreases significantly as the percentage of fly ash increases [14,15]. K. Ramu and R. Dayakarbabu investigated the influence of fly ash on the free swell index of expansive soil. The differential free swell was reduced by 28.3% with 20% fly ash as a replacement in the expansive clay [16]. Zeolite and other pozzolanic materials may catalyze the occurrence of pozzolanic reactions due to their high $SiO_2$ and $Al_2O_3$ content. The addition of natural zeolite to cement kiln dust-stabilized expansive clayey soil has been proven to greatly improve the strength of the stabilized soil [17,18].

In the current study, paper, cement, zeolite powder, and three industrial by-products (blast furnace slag, steel slag, and fly ash) were selected as additives. The swelling potential of mixed specimens with different additive contents at different curing ages was tested through no loading swelling tests. The no loading swelling test measures the ratio of the unidirectional expansion in the height direction of the specimen after immersion in water to the original height value under the condition of no-load and side limit. This ratio is called the expansion rate and is expressed by a dial indicator. It is suitable for measuring the no loading swelling rate of undisturbed soil and compacted soil specimens, which can be used as a reference for evaluating the expansion potential energy of cohesive soil. Specimens with significantly reduced expansion potential were selected. Referring to the previous tests on soil modification, SEM tests and XRD tests were carried out on the red clay before and after modification to determine the changes in soil structure and mineral composition [19–21]. Direct shear tests are used to verify the effect of additives on soil strength [22]. XRD and SEM tests were carried out on these specimens to determine the changes in clay minerals in the specimens and observe the micromorphology of the specimens. Combined with the XRF test results, the modification mechanism of additives on the swelling potential of red clay was comprehensively analyzed. Then, the additives selected according to the above experiments to effectively decrease the swelling potential of red clay were used to make specimens for the direct shear test. The influence of additives on the shear strength of red clay was verified through the changes in cohesive force and internal friction angle. Through the experiments, it was verified that adding cement and steel slag to the red clay can effectively improve red clay's water stability. This paper proposes a means for the reuse of waste and provides a reference for reducing the impact on the environment in engineering construction.

## 2. Materials and Methods

### 2.1. Materials

The red clay used in this test was sourced from Honggutan District, Nanchang City, Jiangxi Province, China. Undisturbed red clay specimens were used for physical and mechanical property tests (the liquid limit and plastic limit were measured using a photo-electrical liquid plastic limit tester, and an electric strain-controlled direct shear apparatus was used for the direct shear test) and the XRF test (XRF-1800, Shimadzu, Kyoto City, Japan). The tests were conducted in accordance with the National Standard of the People's Republic of China [23] and American Society of Testing Materials standards [24,25]. The basic physical and mechanical properties of the red clay are shown in Table 1. The chemical composition was determined by the XRF test, and the results are shown in Table 2.

The five additives were ordered online from building material suppliers, and their XRF test results are shown in Table 2.

### 2.2. Specimen Preparation

The specimen used for the no loading swelling potential test was a cylinder with a diameter of 61.8 mm and a height of 20 mm. Before making the specimens, the red clay was crushed and dried naturally. The specimens were made from the clay passing through the 2 mm standard sieve. When making specimens, distilled water was added to the clay to ensure the clay had the best moisture content. The quantities of additives were calculated as follows: the mass of additives accounted for 3%, 6%, 9%, 12% and 15% of the dry weight of the clay and additives mixture. The dry density of each specimen was controlled at 1.6 $g/cm^3$. Three curing ages of 0, 7 and 28 days were taken for each dosage of different additives. The clay specimen was kept in shape in a stainless steel ring, sealed and wrapped with plastic film, and then cured in a sealed box at room temperature.

**Table 1.** Properties of the tested red clay.

| Physical and Index Properties of the Red Clay | Quantity |
|---|---|
| Natural density ($g/cm^3$) | 1.92 |
| Dry density ($g/cm^3$) | 1.64 |
| Liquid limit (%) | 63.30 |
| Plastic limit (%) | 11.31 |
| Plasticity index (%) | 51.99 |
| Optimum moisture content (%) | 9.55 |
| Unified soil classification | CH |
| Free swelling ratio (%) | 68.00 |
| Total cohesion (kPa) | 21.51 |
| Total angle of internal friction (°) | 30.90 |

**Table 2.** Chemical properties of the red clay, cement, gypsum, steel slag, blast furnace slag, fly ash and zeolite.

| Chemical Composition | Red Clay (%) | Cement (%) | Steel Slag (%) | BFS (%) | Fly Ash (%) | Zeolite (%) |
|---|---|---|---|---|---|---|
| CaO | 0.74 | 47.40 | 41.22 | 59.31 | 3.60 | 2.80 |
| $SiO_2$ | 52.99 | 15.82 | 6.32 | 16.31 | 35.71 | 65.41 |
| $Al_2O_3$ | 18.78 | 10.67 | 2.88 | 10.24 | 37.34 | 11.02 |
| $Fe_2O_3$ | 13.48 | 4.96 | 22.44 | 1.46 | 9.86 | 2.53 |
| MgO | 1.04 | 4.49 | 5.68 | 5.63 | 0.46 | 0.89 |
| $SO_3$ | - | - | 0.59 | - | - | - |
| $Na_2O$ | 0.07 | 0.65 | 0.37 | 0.43 | 0.19 | 1.37 |
| $K_2O$ | 5.06 | 0.96 | 0.33 | 0.57 | 1.66 | 6.85 |
| $P_2O_5$ | 0.08 | 0.06 | - | - | 0.43 | 0.04 |
| MnO | 0.17 | 0.37 | 3.80 | 1.00 | 0.09 | 0.13 |
| $ZrO_2$ | 0.52 | 0.32 | 0.14 | 0.45 | 1.49 | 0.14 |
| Cl | - | 1.24 | 0.06 | 0.06 | - | 0.08 |
| S | 0.02 | 1.78 | - | 1.22 | - | 0.02 |
| SrO | 0.06 | 0.67 | 0.17 | 0.67 | 1.63 | 0.11 |
| Loss of ignition | 4.97 | 8.20 | 12.77 | - | 2.05 | 7.86 |

### 2.3. Test Procedures

The no loading swelling ratio test refers to the National Standard of the People's Republic of China: GB/T 50123-2019 Standard for geotechnical testing method. The red clay specimen and schematic of the test setup used for the no loading swelling ratio test are shown in Figure 1.

Through the no loading swelling test of red clay, the height of the red clay specimen increased by 1.89 mm three hours after adding water. The swelling of red clay specimens tended to be stable six hours after adding water, and the subsequent height increase was less than 0.1 mm. Therefore, the swelling ratio of the specimen was calculated according to the reading at 12 h after adding water. Three parallel tests were carried out for each specimen with a specific dosage and age, and the average value was taken. The swelling ratio was defined as the height change in the specimen after swelling after water addition divided by the original height of the specimen, and the result was taken as a percentage. The size, manufacturing method, and curing method of the specimen for the direct shear test were the same as those for the no loading swelling ratio test. Four specimens were used for each set of direct shear tests. The vertical pressures applied during the test were

100, 200, 300 and 400 kPa. The cohesion and internal friction angle of the specimen were tested by the fast shear test. XRD tests used an X-ray diffractometer D8 Advance (Bruker, Rheinstetten, Germany) to analyze the relative content of clay. SEM tests used a JSM-IT300 (JEOL, Tokyo, Japan) scanning electron microscope working at 15.0 kV to observe the microstructure of the section after the direct shear test.

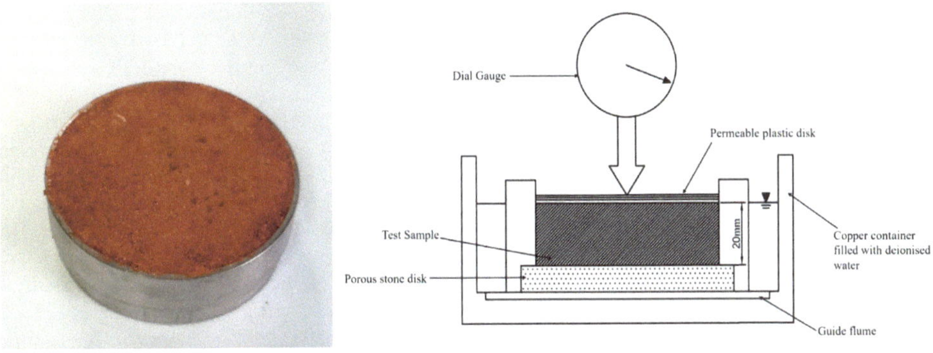


**Figure 1.** The specimen in the stainless steel ring (**left**) and schematic of the test setup used for the no loading swelling ratio test (**right**).

## 3. Results and Discussion

### 3.1. No Loading Swelling Ratio Test

The variation in the swelling potential (the relative change in the no loading swelling ratio of specimens containing different additives compared with red clay specimens, taking a positive value when increasing and a negative value when decreasing) is shown in Figure 2.

Under three different curing conditions, the no loading swelling potential of steel slag-modified clay generally showed a more obvious decrease with the increase in steel slag content. Compared with the red clay without additives, the swelling potential of the fly ash-modified specimens and zeolite powder-modified specimens did not achieve the effect of reducing the swelling potential because of the lack of lime in the clay. The XRF test showed that the main components of zeolite and fly ash are $SiO_2$ and $Al_2O_3$. This indicates that their functions in pozzolanic reactions are similar. When zeolite powder and fly ash are used as additives, due to the lack of calcium ions, the pozzolanic reaction is not sufficient to produce CSH and CAH colloids, so it cannot effectively reduce the swelling potential of red clay. Because of the hygroscopicity of $Al_2O_3$, the soil expansion potential increased slightly after the addition of zeolite powder and fly ash. The no loading swelling potential of the blast furnace slag-modified specimen did not show a significant decline. Combined with the XRF test results, the blast furnace slag used in this test was alkaline slag with high CaO content, and its composition is similar to that of cement, but they have different effects on the swelling potential of red clay. This shows that the pozzolanic activity of blast furnace slag is far lower than that of cement. If blast furnace slag is used to improve the swelling potential of red clay, some methods need to be used to stimulate its reaction potential. It can be observed from the above figure that the cement can significantly inhibit the swelling potential of red clay under the curing times of 7 and 28 days, but cement-modified clay specimens without curing resulted in swelling potential that exceeded that of the no-additive clay specimens. Adding steel slag can greatly reduce the no loading swelling potential of specimens without curing. However, with the increase in curing age, the swelling potential of specimens to which steel slag was added did not decrease further. This shows that the mechanisms of cement and steel slag used as additives to reduce the swelling potential of red clay are different. The cement-modified specimens having a curing age of 28 days showed a further decrease in the swelling potential compared to the specimens with the same cement content with 7 days' curing. In other words, when cement is used as an additive, it needs a certain curing time to show the effect of reducing the swelling potential of red clay, and the longer the curing time, the better the effect. Without curing, or when the curing time is insufficient, a large number of unhydrated cement clinkers will no longer hydrate with the water distribution in the soil after immersion, but will directly absorb the outside water to form CSH and CAH colloids, which will promote the expansion of the clay. After curing the cement-modified specimens for a long enough period, sufficient parts of the clinker can react with the water in the clay. The hydration products are generated to fill the pores of the red clay, bond the red clay particles, reduce the water absorption of the red clay, and increase the resistance of the red clay expansion. When the cement content exceeded 9%, the swelling rate of specimens rose slightly, which indicates that the hydration reaction of cement is limited due to the limited water content in the soil. When the curing time was 28 days, the swelling ratio of specimens with 3% cement content was lower than that of specimens with 15% cement content and a curing age of 7 days, which indicates that prolonging the curing age is a better choice to improve the cement modification effect than increasing the cement content. If the curing time is not long enough to hydrate the cement clinker and water in the red clay, increasing the cement content will result in a waste of cement clinker. In addition, a portion of the cement does not produce CSH and CAH colloids, and this portion of the cement clinker will absorb water from the outside of the clay and continue to react when the clay meets water, which will result in negative effects on the water stability of the red clay. Under 28 days of curing, the change in the soil swelling ratio is no longer obvious when the cement content reaches 6%. Therefore, under the condition of long-term curing, if the water content of the soil is not sufficient, a large amount of cement is not required to modify red clay. The appropriate amount of cement can fully enable the reaction potential of water content in clay and reduce possible cement waste. Compared with the case in which the swelling ratio of specimens with cement increased under the condition of no curing, the swelling ratio of specimens with steel slag was lower than that of red clay specimens without additives under the condition of no curing. In addition, the swelling ratio of specimens with steel slag in all three curing ages showed an overall trend of decreasing with the increase in steel slag content. There was no situation in which the modification effect of the clay swelling potential was no longer obvious after the content of the additive reached 9% when the cement-modified specimens were cured for 7 and 28 days.

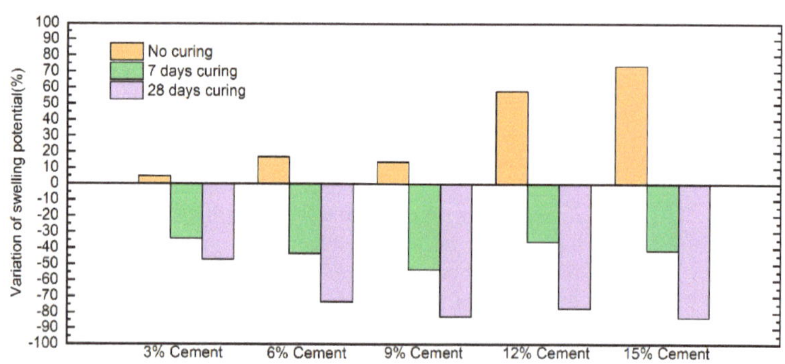


(a)

**Figure 2.** *Cont.*

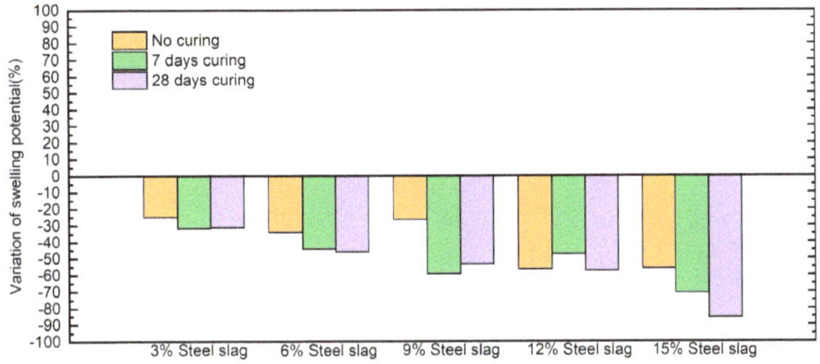


(**b**)

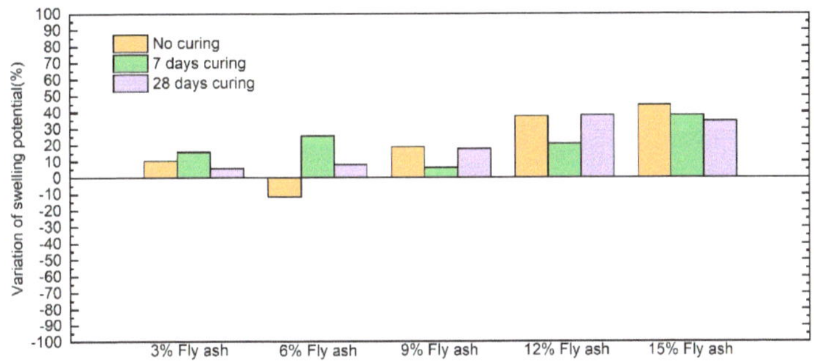


(**c**)

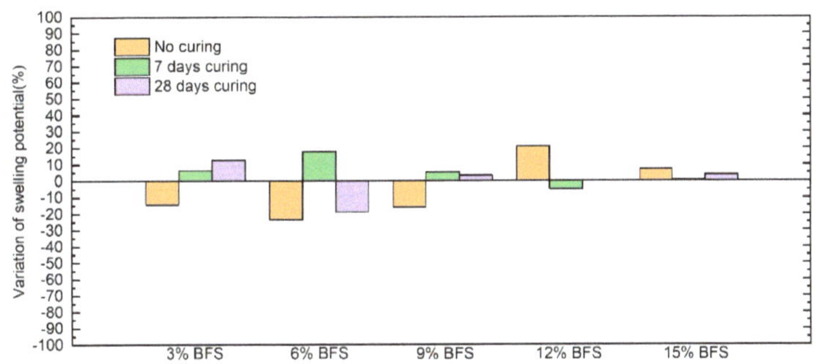


(**d**)

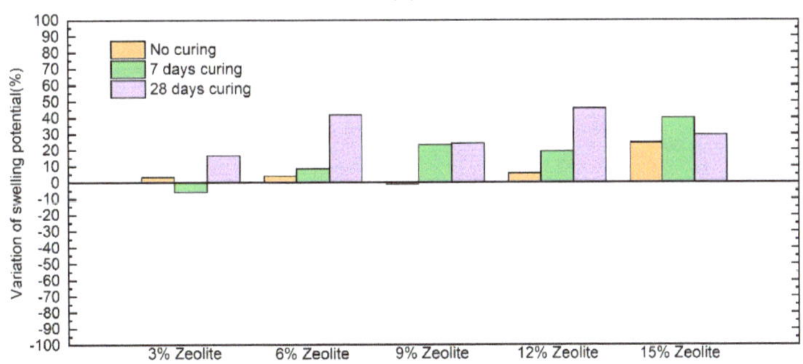


(**e**)

**Figure 2.** The change in the no loading swelling ratio of specimens with different additives relative to the red clay specimen: (**a**)-cement, (**b**)-steel slag, (**c**)-blast furnace slag, (**d**)-fly ash, (**e**)-zeolite.

The value of the change in the no loading swelling ratio of the cement and steel slag-modified specimens compared with the red clay specimen without additives was set as the dependent variable (the swelling ratio decreases to a positive value and increases to a negative value) in the binary linear regression equation between the improvement effect of the two additives, and the curing time and additive contents. The linear regression coefficient was calculated according to the following equations:

$$a_1 = -\frac{l_{1y}l_{22} - l_{2y}l_{12}}{l_{11}l_{22} - l_{12}^2} \tag{1}$$

$$a_2 = -\frac{l_{2y}l_{11} - l_{1y}l_{21}}{l_{11}l_{22} - l_{12}^2} \tag{2}$$

where:

$$l_{ij} = l_{ji} = \sum(x_i x_j) - (\sum x_i \sum x_j)/k$$

$$l_{iy} = l_{ji} = \sum(x_i y_i) - (\sum x_i \sum y_i)/k$$

and $k$ is the number of test groups. After calculation using Equations (1) and (2), the values of $a_1$ and $a_2$ of cement-modified specimens were 3.26 and $-1.05$, respectively. The values of $a_1$ and $a_2$ of steel slag-modified specimens were 0.34 and 3.54, respectively.

Through calculation, it was again verified that, when steel slag is used to modify red clay, the influence of the steel slag content is more significant than that of curing time, and for cement, the influence of curing time is more significant than that of the cement content. From the above results, it can be concluded that, when steel slag modifies the swelling potential of red clay, the main modification mechanism is not a hydration reaction like that in cement. Combined with the XRF test results, both cement and steel slag contain a large amount of the Ca element, and the contents of Si and Al in steel slag are lower than those in cement, but there are sufficient quantities of Si and Al in red clay. This shows that the relative lack of Al and Si in steel slag is not the main reason for its weak pozzolanic reaction. It can be considered that a portion of Ca in steel slag does not exist in the form of calcium oxide or calcium hydroxide, but in the form of calcium carbonate and other forms that are not prone to pozzolanic reaction. This situation should also exist in blast furnace slag. As a result, the ability of steel slag to produce CSH and CAH colloids in hydration reaction is less than that of cement. In addition, the Fe and Mn contents in steel slag are higher than those in cement. Therefore, it can be considered that a part of the ability of steel slag to modify red clay is related to the ions of Fe and Mn. According to the double layer theory of clays [26], shear displacement occurs between clay particles when the volume of clay changes. Because clay molecules contain a large number of oxygen atoms, the surface of clay particles has a fixed negative electric charge related to the crystal lattice. When the red clay encounters fresh water, a large number of water molecules are polarized, and a large number of hydrogen bonds are formed on the surface of the clay slices to adsorb water molecules on the surface of the clay slices. This leads to an increase in the volume of clay due to the filling of water [27–29]. For the steel slag modified clay, after the steel slag is mixed with clay, the Fe and Mn ions in the steel slag are adsorbed on the clay slices. This reduces the number of hydrogen bonds formed on the surface of the clay slices when the clay meets water, thus reducing the swelling of the clay caused by the filling of water molecules. Ion adsorption can be considered to occur instantaneously when red clay and steel slag are mixed, so the curing time does not affect the ion adsorption. With the increase in steel slag content, more cations are adsorbed on the surface of clay slices, so as to neutralize more negative charges on the surface of clay slices. This can further reduce the hydrogen bonds formed between the surface of clay slices and water molecules when clay meets water. Hence, when steel slag improves the swelling of red clay, the greater the steel slag content, the better the modification effect.

### 3.2. X-ray Diffraction Quantitative Analysis of Clay Minerals

After comprehensively considering the effect and economy of additives, a red clay specimen, a specimen having 15% steel slag content and 28 days curing, and a specimen having 9% cement content and 28 days curing, were selected to make air drying-oriented slices (the glass slides were coated with a clay suspension and dried naturally), ethylene glycol treatment-oriented slices (i.e., air drying-oriented slices after saturation treatment in a ethylene glycol steam), and heat treatment-oriented slices (obtained by natural cooling of ethylene glycol tablets after treatments at a high temperature and a constant temperature), respectively, for the XRD analysis of clay minerals. The XRD patterns of clay minerals in specimens are shown in Figure 3.

Comparing the relative intensity of each diffraction peak of the XRD pattern, the relative intensity of the theta angles of 12.27° and 25.00° in the XRD patterns of the red clay, steel slag-modified, and cement-modified specimens shows a gradually increasing trend. Therefore, it can be concluded that the relative content of kaolinite in the red clay specimen, the steel slag-modified specimen, and the cement-modified specimen increases gradually. The specific relative contents of clay minerals of the red clay, steel slag-modified, and cement-modified specimens are shown in Table 3.

The content of montmorillonite in the illite-montmorillonite mixed layer in specimen 1 is 10%. The content of montmorillonite in the illite-montmorillonite mixed layer in specimen 2 is 10%. Comparing the relative content of montmorillonite, that of specimen 1 is 5.9%, that of specimen 2 is 5.7%, and that of specimen 3 decreases to 2%. Comparing the relative content of illite, that of specimen 1 is 72.1%, that of specimen 2 is 67.3%, and that of specimen 3 is 48%. It can be seen that, under the condition of 28 days curing, the addition of cement can greatly reduce the content of montmorillonite and illite resulting in strong hydrophilicity, and increase the content of kaolinite resulting in poor hydrophilicity. The addition of steel slag to red clay can also reduce the relative content of montmorillonite and illite in clay minerals, and increase the relative content of kaolinite, but the effect is less than that of cement. From the analysis of XRD test results, it can be inferred that when cement is used to modify the swelling properties of red clay, it can change the relative content of clay minerals in the specimens. The relative content of hydrophilic clay minerals decreases, whereas the relative content of nonhydrophilic clay minerals increases, which reduces the water absorption of red clay.

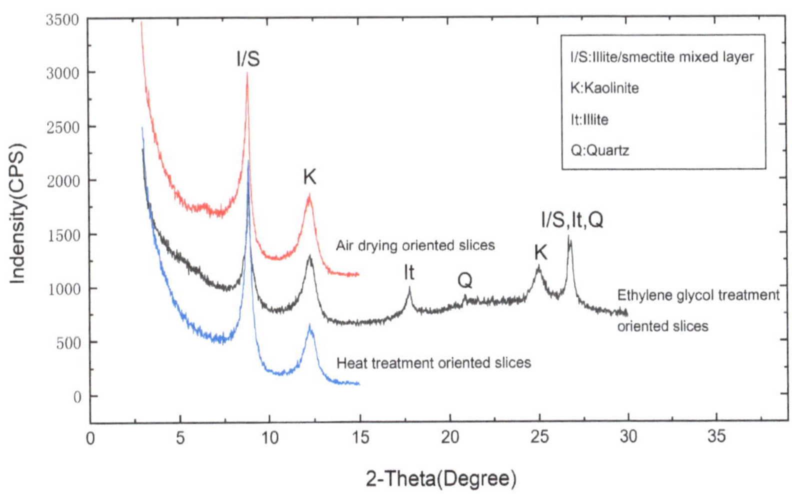


(a)

**Figure 3.** *Cont.*

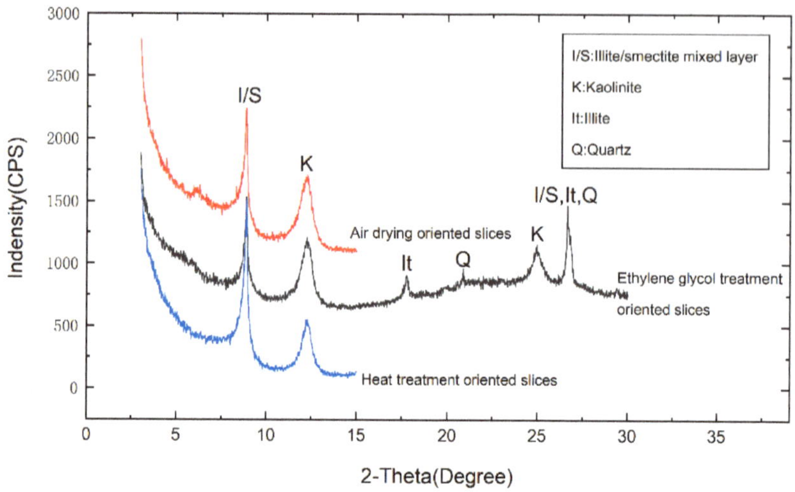


(b)

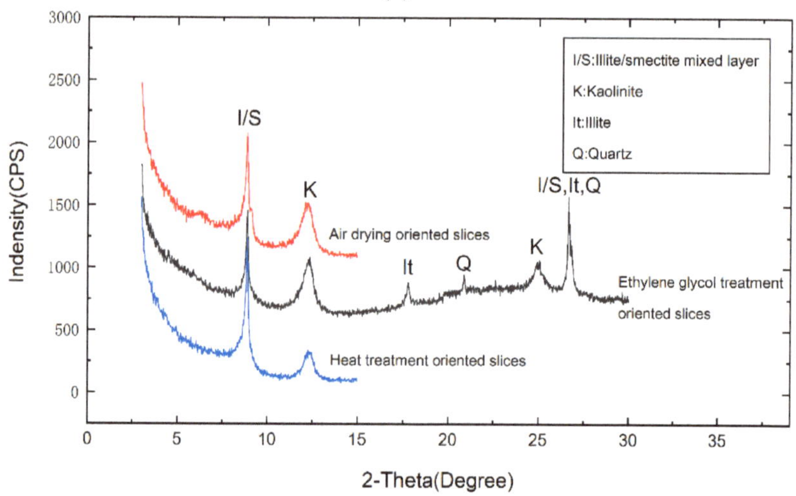


(c)

**Figure 3.** XRD patterns of clay minerals (smectite, kaolinite, illite, and their mixed layer) in specimens at 28 days: (**a**) red clay; (**b**) 15% steel slag; (**c**) 9% cement.

**Table 3.** The relative content of clay minerals in the red clay, steel slag-modified and cement-modified specimens.

| Specimen Number | Additives | The Relative Content of Clay Minerals | | | | | |
|---|---|---|---|---|---|---|---|
| | | S [1] | I/S [1] | It [1] | Kao [1] | C [1] | C/S [1] |
| 1 | - | 1 | 49 | 28 | 22 | - | - |
| 2 | 15% Steel Slag | 1 | 47 | 22 | 30 | - | - |
| 3 | 9% Cement | 2 | - | 48 | 50 | - | - |

[1] S is montmorillonite, It is illite, Kao is kaolinite, C is chlorite, I/S is the illite montmorillonite mixed layer, and C/S is the chlorite montmorillonite mixed layer.

### 3.3. Scanning Electron Microscopy

Figure 4 shows the SEM images of sections of the red clay specimen, the steel slag-modified specimen with 15% steel slag content cured for 7 and 28 days, and the cement-modified specimen with 9% cement content cured for 28 days.

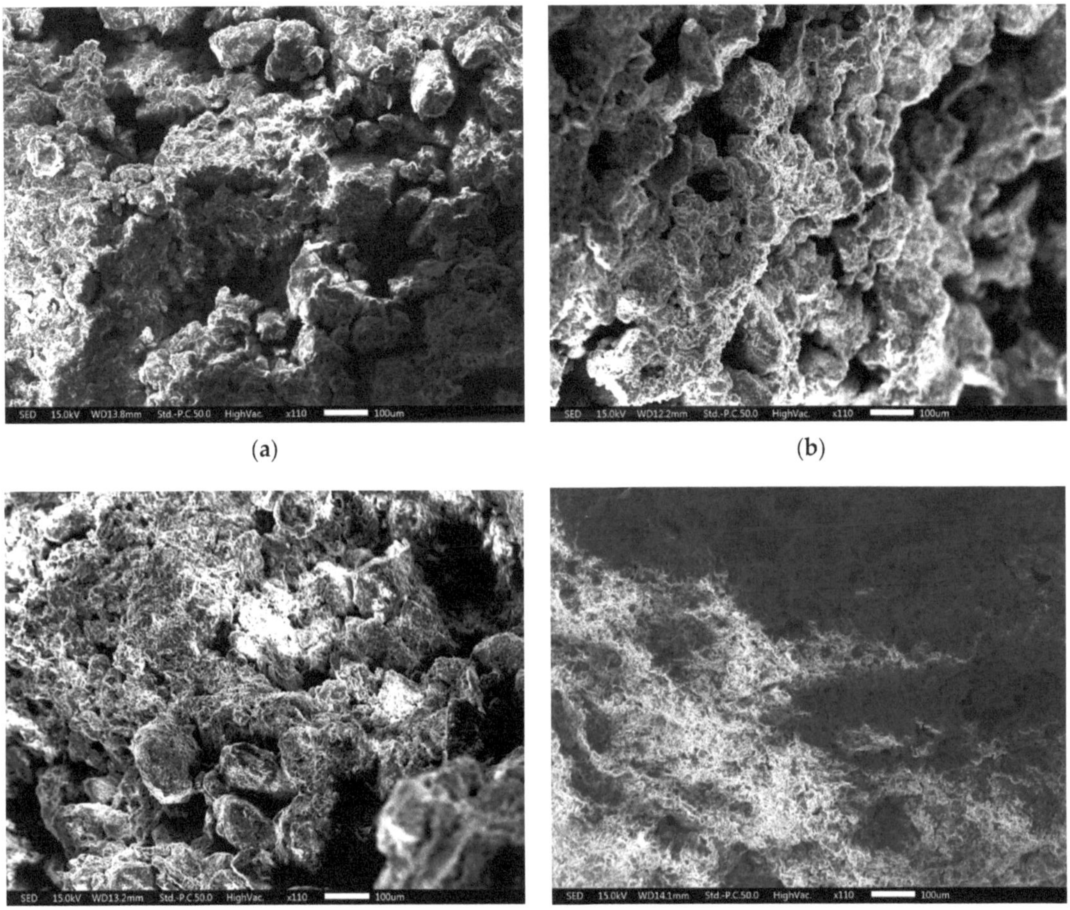


**Figure 4.** SEM micrographs at 100 µm of (**a**)-red clay, (**b**)-15% steel slag-modified specimen at 7 days curing, (**c**)-15% steel slag-modified specimen at 28 days curing, (**d**)-9% cement-modified specimen at 28 days curing.

It can be seen from the image that there are many pores in the red clay. In the clay with steel slag, some smaller pores have been filled, but the pores with larger diameters are still not filled. In the red clay with cement, almost all pores are effectively filled. As a result, it is more difficult for the water to penetrate the interior of the cement-modified clay, and the number of pores in the clay that can hold water is greatly reduced.

### 3.4. Direct Shear Test

Through the above tests, it was found that, when cement or steel slag is used as a single additive, 9% cement content and 15% steel slag content are more suitable for reducing the swelling potential of red clay. Therefore, we selected the specimens with 9% cement content cured for 7 and 28 days, and the specimens with 15% steel slag content cured for 0, 7 and 28 days for the direct shear test. The obtained cohesive strength and internal friction angle were compared with those of the red clay specimen to verify the influence of the above treatment scheme, which can effectively reduce the swelling potential of red clay, on the strength of the specimens. By comparing the values of cohesive strength and internal friction angle obtained from the direct shear test given in Figure 5, it can be seen that the addition of steel slag leads to the loss of cohesion of the clay. When the specimen is not cured, the cohesive strength of the modified clay containing 15% steel slag is reduced to 5.37 kPa, which is 75% lower than that of red clay. After a period of curing, the cohesive strength of the steel slag modified clay is restored. When the curing time is 28 days, the cohesive strength of the steel slag modified clay can be restored to 94% of the cohesive strength of the red clay before adding steel slag. The cohesive strength of the cement-modified specimens can be greatly improved after a certain curing time. After curing for 7 days, the cohesive strength of the modified clay with 9% cement content is 43% higher than that of the red clay. After curing for 28 days, the cohesive strength of the modified clay with 9% cement content is 82% higher than that of the red clay. Regardless of whether the additive is cement or steel slag, and the curing age is long or short, no particularly significant effect on the internal friction angle $\varphi$ can be observed. Based on the XRF test results and SEM images, the ability of steel slag hydration to generate CAH and CSH colloids is far lower than that of cement, so the cementation and filling effect of steel slag on red clay particles is not obvious because the quantity of reactants for the pozzolanic reaction in steel slag is less than that in cement. When steel slag is added to the red clay, most of the calcium oxide in steel slag first reacts with the water in the clay. Then, CSH and CAH colloids are gradually formed, which takes a period of time. The above reaction process leads to a significant reduction in the water content of the clay in a short period after the steel slag is added to the clay. The decrease in water content in red clay will result in two problems. The first is the decrease in water tension between clay particles. The second is that part of the cohesion of clay depends on the hydrogen bond, and the water plays the role of the connecting node between the two clay slices [21]. The decrease in water leads to the loss of the bond strength between the clay particles, which relies on the hydrogen bond. This effect is more obvious in illite and montmorillonite, which are more hydrophilic. Previous experiments show that the shear strength of illite and montmorillonite will decrease with the decrease in water content when the clay's water content is lower than a certain level [6]. At the same time, because there is not enough CSH colloid in the clay to provide cementation between particles, the cohesion of the clay will be lost to a large extent. Then, with the increase in curing time, the pozzolanic reaction in steel slag is carried out gradually, and a part of the newly generated CSH colloid provides the adhesive force between some soil particles, which shows that the cohesion of the clay modified by steel slag measured by the direct shear test is partially restored after a period of curing. The quantity of reactants of the volcanic ash reaction in cement-modified clay is much greater than that in steel slag-modified clay. After a certain period of curing, a large number of generated CAH and CSH colloids directly fill the internal pores of the specimens and cement the clay particles. After 28 days of curing, the soil has formed a compact structure of the combination of soil particles and hydration products. Particles in the clay are connected as a whole. The hardened CSH and CAH colloids that fill the pores play an integral role in bearing the external force with the clay particles. Hence, the cohesion of cement-modified clay obtained from direct shear tests is greatly enhanced.

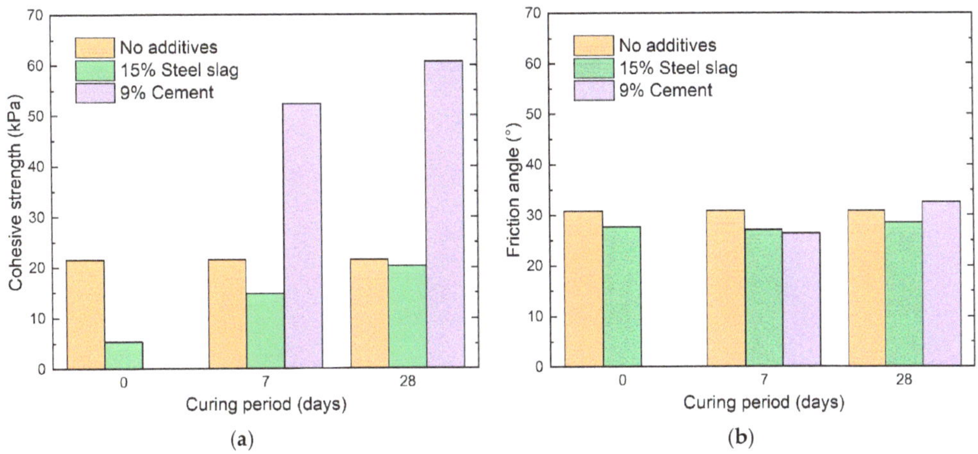


**Figure 5.** Cohesive strength and internal friction angle of red clay and cement/steel slag-modified specimens: (**a**) the cohesive strength of specimens; (**b**) internal friction angle of specimens.

## 4. Conclusions

In this study, the influences of different additives on the swelling potential of red clay were investigated. The following conclusions can be summarized from this paper:

1. For cement, zeolite powder, blast furnace slag, steel slag and fly ash were added separately to red clay. Cement and steel slag can significantly reduce the swelling potential of red clay compared with the other three additives. The modification effect of cement can be reflected after a certain period of curing, whereas the modification effect of steel slag depends more on the increase in steel slag content. When cement is added to red clay, the cement clinker reacts with the water in the red clay to form a large number of CSH and CAH colloids. These colloids fill the pores in the red clay and bond the soil particles as a whole, which greatly improves the compactness of the clay, reduces the water absorption capacity of the clay, and improves the resistance to inhibit the swelling of the clay.
2. Adding steel slag can reduce the swelling potential of red clay without curing. The main reason for this is that Fe and Mn ions are adsorbed on the surface of negatively charged clay slices, so as to reduce the number of water molecules adsorbed by hydrogen bonds formed on the surface of the clay slices. Because the adsorption of ions on the surface of clay slices can be considered to occur at the same time when steel slag is added to red clay, the steel slag can immediately modify the swelling properties of red clay. SEM images show that the pozzolanic reaction in steel slag-modified specimens is similar to that in cement-modified specimens. A certain quantity of CSH and CAH colloids are formed, but this quantity is far less than that of cement.
3. According to the position and intensity of diffraction peaks of clay minerals shown in XRD patterns, the diffraction peak relative intensity of kaolinite in steel slag-modified clay and cement-modified clay is higher than that in red clay. Therefore, the addition of steel slag and cement can change the relative content of clay minerals in red clay, reduce the relative content of hydrophilic clay minerals such as montmorillonite and illite, and increase the relative content of kaolinite. The content of nonhydrophilic clay minerals in red clay increases relatively, and the water stability of red clay is also improved. The above phenomenon is more obvious in cement-modified clay than in steel slag-modified clay.
4. When steel slag is added alone, it causes a decrease in the water content in the specimen, and then causes a large loss of cohesion in the short term. After a certain curing period, the cohesion of the steel slag-modified specimen can be restored due

to the bonding effect of CSH and CAH colloids caused by the pozzolanic reaction in the steel slag modified-specimen. The production of CSH and CAH colloids in the cement-modified specimen is much higher than that in the steel slag-modified specimen, so the cohesion of the cement-modified specimen can be greatly improved after a certain period of curing.

**Author Contributions:** Conceptualization, Z.Q. and B.Y.; methodology, B.Y.; validation, B.Y. and Z.Z.; formal analysis, B.Y.; investigation, B.Y. and Z.Z.; resources, Z.Q.; data curation, B.Y.; writing—original draft preparation, B.Y.; writing—review and editing, B.Y.; visualization, B.Y.; supervision, Z.Q.; project administration, Z.Q.; funding acquisition, Z.Q. All authors have read and agreed to the published version of the manuscript.

**Funding:** This research received no external funding.

**Institutional Review Board Statement:** The study did not require ethical approval.

**Informed Consent Statement:** The study did not involve humans.

**Data Availability Statement:** Not applicable.

**Conflicts of Interest:** The authors declare no conflict of interest.

## References


1. She, J.; Lu, Z.; Yao, H.; Fang, R.; Xian, S. Experimental Study on the Swelling Behavior of Expansive Soil at Different Depths under Unidirectional Seepage. *Appl. Sci.* **2019**, *9*, 1233. [CrossRef]
2. Cheng, Y.; Huang, X. Effect of Mineral Additives on the Behavior of an Expansive Soil for Use in Highway Subgrade Soils. *Appl. Sci.* **2019**, *9*, 30. [CrossRef]
3. Al-Mukhtar, M.; Lasledj, A.; Alcover, J.-F. Behaviour and mineralogy changes in lime-treated expansive soil at 50 °C. *Appl. Clay Sci.* **2010**, *50*, 199–203. [CrossRef]
4. TFHRC. Transportation and Federal Highways Research Council-U.S. Department of Transportation. Blast Furnace Slag, Material Description Guideline. 2004. Available online: http://www.tfhrc.gov/hnr20/recycle/waste/bfs1.html (accessed on 22 December 2021).
5. Cokca, E.; Yazici, V.; Ozaydin, V. Stabilization of Expansive Clays Using Granulated Blast Furnace Slag (GBFS) and GBFS-Cement. *Geotech. Geol. Eng.* **2009**, *27*, 489–499. [CrossRef]
6. Fasihnikoutalab, M.H.; Pourakbar, S.; Ball, R.; Unluer, C.; Cristelo, N. Sustainable soil stabilisation with ground granulated blast-furnace slag activated by olivine and sodium hydroxide. *Acta Geotech.* **2020**, *15*, 1981–1991. [CrossRef]
7. Chen, L.; Chen, X.; Wang, H.; Huang, X.; Song, Y. Mechanical Properties and Microstructure of Lime-Treated Red Clay. *KSCE J. Civ. Eng.* **2021**, *25*, 70–77. [CrossRef]
8. Wang, J.; Wang, S.; Hong, M.; Li, Y.; Wu, G.; Li, L.; Zhang, J. Correlation Analysis between Clay Mineral Composition and Shear Strength. *J. Southwest Jiaotong Univ.* **2018**, *53*, 1033–1038. [CrossRef]
9. Yu, H.M.; Wang, Q. *Steel Slag: Treatment and Resource Utilization*; Metallurgical Industry Press: Beijing, China, 2015.
10. Global Crude Steel Output Decreases by 0.9% in 2020. Available online: https://www.worldsteel.org/media-centre/press-releases/2021/Global-crude-steel-output-decreases-by-0.9--in-2020.html (accessed on 22 December 2021).
11. Malasavage, N.E.; Jagupilla, S.; Grubb, D.G.; Wazne, M.; Coon, W.P. Geotechnical Performance of Dredged Material—Steel Slag Fines Blends: Laboratory and Field Evaluation. *J. Geotech. Geoenviron. Eng.* **2012**, *138*, 981–991. [CrossRef]
12. Poh, H.Y.; Ghataora, G.S.; Ghazireh, N. Soil Stabilization Using Basic Oxygen Steel Slag Fines. *J. Mater. Civ. Eng.* **2006**, *18*, 229–240. [CrossRef]
13. Akinwumi, I. Soil Modification by the Application of Steel Slag. *Period. Polytech. Civ. Eng.* **2014**, *58*, 371–377. [CrossRef]
14. Mir, B.A.; Sridharan, A. Mechanical behaviour of fly ash-treated expansive soil. *Ground Improv. Proc. Inst. Civ. Eng.* **2019**, *172*, 12–24. [CrossRef]
15. Mir, B.A.; Sridharan, A. Volume change behavior of clayey soil–fly ash mixtures. *Int. J. Geotech. Eng.* **2014**, *8*, 72–83. [CrossRef]
16. Ramu, K.; DayakarBabu, R. A Laboratory Study on the Stabilized Expansive Soil with Partial Replacement of Fly Ash and Palm Oil Fuel Ash. *Lect. Notes Civ. Eng.* **2022**, *152*, 69–78. [CrossRef]
17. Sharo, A.A.; Shaqour, F.M.; Ayyad, J.M. Maximizing Strength of CKD—Stabilized Expansive Clayey Soil Using Natural Zeolite. *KSCE J. Civ. Eng.* **2021**, *25*, 1204–1213. [CrossRef]
18. Eyo, E.U.; Ng'Ambi, S.; Abbey, S.J. Performance of clay stabilized by cementitious materials and inclusion of zeolite/alkaline metals-based additive. *Transp. Geotech.* **2020**, *23*, 100330. [CrossRef]
19. Goodarzi, A.R.; Goodarzi, S.; Akbari, H.R. Assessing geo-mechanical and micro-structural performance of modified expansive clayey soil by silica fume as industrial waste. *Iran. J. Sci. Technol.-Trans. Civ. Eng.* **2015**, *39*, 333–350. [CrossRef]
20. Xu, B.; Yi, Y. Soft Clay Stabilization Using Three Industry Byproducts. *J. Mater. Civ. Eng.* **2021**, *33*, 06021002. [CrossRef]
21. Yang, Q.; Du, C.; Zhang, J.; Yang, G. Influence of Silica Fume and Additives on Unconfined Compressive Strength of Cement-Stabilized Marine Soft Clay. *J. Mater. Civ. Eng.* **2020**, *32*, 04019346. [CrossRef]

22. Huang, K.; Wan, J.-W.; Chen, G.; Zeng, Y. Testing study of relationship between water content and shear strength of unsaturated soils. *Rock Soil Mech.* **2012**, *33*, 2600.
23. *GB/T 50123-2019*; National Standard of the People's Republic of China. Standard for Geotechnical Testing Method. CSBTS & Ministry of Construction: Beijing, China, 2019.
24. *ASTM D2487*; Standard Practice for Classification of Soils for Engineering Purposes (Unified Soil Classification System). ASTM International: West Conshohocken, PA, USA, 2017.
25. *ASTM D4318*; Standard Test Methods for Liquid Limit, Plastic Limit, and Plasticity Index of Soils. ASTM International: West Conshohocken, PA, USA, 2017.
26. Sridharan, A.; Jayadeva, M.S. Double layer theory and compressibility of clays. *Geotechnique* **1982**, *32*, 133–144. [CrossRef]
27. Ikari, M.J.; Kopf, A.J. Cohesive strength of clay-rich sediment. *Geophys. Res. Lett.* **2011**, *38*, L16309. [CrossRef]
28. Komine, H. Simplified evaluation for swelling characteristics of bentonites. *Eng. Geol.* **2004**, *71*, 265–279. [CrossRef]
29. Zeng, Z.; Cui, Y.-J.; Conil, N.; Talandier, J. Experimental study on the aeolotropic swelling behaviour of compacted bentonite/claystone mixture with axial/radial technological voids. *Eng. Geol.* **2020**, *278*, 105846. [CrossRef]

water

MDPI

*Article*

# Seasonal Enhancement of Nitrogen Removal on Domestic Wastewater Treatment Performance by Partially Saturated and Saturated Hybrid Constructed Wetland

**José Contreras [1], Daniela López [1,2], Gloria Gómez [1] and Gladys Vidal [1,*]**

1 Engineering and Environmental Biotechnology Group, Environmental Science Faculty & Center EULA-Chile, Universidad de Concepción, Concepción 4070386, Chile; contreraspincheira1@gmail.com (J.C.); spdanielita@gmail.com (D.L.); gloriagomezosorio@gmail.com (G.G.)

2 Facultad de Ingeniería y Ciencias, Universidad Adolfo Ibáñez, Avda. Padre Hurtado 750, Viña del Mar 2562340, Chile

* Correspondence: glvidal@udec.cl; Tel.: +56-41-2204067; Fax: +56-41-2207076

**Abstract:** The aim of this study is to evaluate seasonal enhancement of nitrogen removal on domestic wastewater treatment performance by partially saturated and saturated HBCWs. To achieve this, two HBCWs consisting of a vertical subsurface flow constructed wetland, followed by a horizontal subsurface flow constructed wetland (VSSF-HSSF) were evaluated. Two saturation levels were used: (a) partially saturated HB1:VSSF1 (0.6 m)-HSSF1 (0.15 m), (b) saturated HB2: VSSF2 (0.8 m)-HSSF2 (0.25 m). Each unit was planted with *Schoenoplectus californicus* and was operated for 297 days. The removal efficiencies in HB1 and HB2 were above 70%, 86%, 77% and 55% for chemical oxygen demand (COD), total suspended solids (TSS), nitrogen as ammonium ($NH_4^+$-N), and total nitrogen (TN), respectively. For VSSF, a higher level of saturation (from 0.6 to 0.8 m) meant a decrease of 17% in the TN removal efficiencies, and for HSSF, an increase from 0.15 to 0.25 m of saturation meant a decrease of 11 and 10% in the $NH_4^+$-N and TN removal efficiencies, respectively. Thus, the increase of saturation level in HBCWs reduces the transformation and/or removal of components of the wastewaters to be treated, particularly nitrogen. Through this research, the possibility of optimizing the transformation of nitrogen with partially saturated hybrids can be examined.

**Keywords:** hybrid constructed wetland; saturation level; organic matter; ammonia

**Citation:** Contreras, J.; López, D.; Gómez, G.; Vidal, G. Seasonal Enhancement of Nitrogen Removal on Domestic Wastewater Treatment Performance by Partially Saturated and Saturated Hybrid Constructed Wetland. *Water* **2022**, *14*, 1089. https://doi.org/10.3390/w14071089

Academic Editors: Dibyendu Sarkar, Rupali Datta, Prafulla Kumar Sahoo, Mohammad Mahmudur Rahman and Christos S. Akratos

Received: 27 January 2022
Accepted: 20 March 2022
Published: 30 March 2022



## 1. Introduction

Hybrid constructed wetlands (HBCWs) are based on the sequential utilization of several constructed wetlands (CWs), allowing the combination of each system advantage. Moreover, several configurations have been used, VSSF + VSSF, HSSF + HSSF and VSSF + HSSF, the latter being the most commonly used to treat municipal and industrial wastewater [1–5]. VSSFs are characterized by an oxygen transfer of 28–100 $gO_2/m^2{\cdot}d$, maintaining aerobic conditions and allowing processes such as nitrification with elimination rates of 0.01–4.17 $gNH_4^+/m^2{\cdot}d$ [6–9]. In the case of HSSF, oxygen transfer occurs at 0.3–3.2 $gO_2/m^2{\cdot}d$, maintaining anaerobic conditions and allowing processes such as denitrification if nitrate is present with elimination rates of 0.47–0.83 $gNO_3^-/m^2{\cdot}d$ [6,8–10]. Several studies have reported that HBCWs present removal efficiencies for COD (58–97.6%), TSS (94–96%), TN (62.8–97%) and $NH_4^+$-N (81.1–99.3%) [7,11–14].

Among the design parameters, it has been stated that the wastewater saturation level impacts the different processes that occur in the HBCWs. [15,16]. Various authors have investigated the influence of partial saturation on the component's transformation in HSSF or VSSF separately, evidencing the potentialities to further improve the elimination efficiencies of these components. That is how Sanchez-Ramos et al. [17] determined higher removal efficiencies (94% vs. 83% for COD and 74% vs. 32% for $NH_4^+$-N) in an HSSF under different saturation levels (0.25 and 0.50 m) than a system that presented a lower saturation level.

Moreover, García et al. [15] determined in a partially saturated HSSF (0.27 m) and in a saturated HSSF (0.50 m) that the best removal efficiencies for COD and $NH_4^+$-N were in the partially saturated system with values up to 23.83 and 69.35% higher than those in a saturated HSSF. Huang et al. [18] used a VSSF with different saturation levels (0.05, 0.30, 0.45 and 0.60 m) to improve nitrogen removal. The results indicated that the modification of the saturation level significantly affected the transformation of nitrogen, and the optimal level of saturation was achieved at 0.45 m, where the authors reported that transformation/removal efficiencies of TN and $NH_4^+$-N of 76.65 and 82.91% enabled the generation of other processes, such as anammox. Furthermore, Saeed and Sun [19] employed a VSSF under unsaturated and partially saturated conditions with saturation levels of 0.14, 0.29 and 0.43 m, respectively, and conducted a comparative analysis on contaminant removal. They observed that VSSF with a lower saturation level had higher removal efficiencies of $BOD_5$ (47%), TN (52%) and $NH_4^+$-N (53%), as it presented an aerobic environment (2.2 mg/L), favoring processes such as nitrification with mean removal rates of 1.30 $gNH_4^+/m^2{\cdot}d$.

Specifically, partial saturation in CWs allows the development of an anoxic/anaerobic zone inside the saturated support medium in VSSF and the penetration of atmospheric oxygen toward the saturated zone through overlaid unsaturated bulk media volume without altering commonly practiced overall main mean depth of traditional fully saturated HF wetlands [20]. Despite the advances in research that have been made regarding partial saturation and single-stage CWs, investigations about transformation of nitrogen with partially saturated HBCWs is very limited. Given all that has been mentioned thus far, the main objective of this study is to evaluate the saturation level in the in the treatment of municipal wastewater through hybrid constructed wetlands. Specifically, we will evaluate the saturation of wastewater level in the transformation of nitrogen through hybrid constructed wetlands.

## 2. Materials and Methods

### 2.1. Hybrid Constructed Wetlands

The wastewater influent was collected after primary treatment from the wastewater treatment plant in Hualqui, Concepción Province, Biobío Region (Chile) (36°59′26.93″ from south latitude and 72°56′47.23″ from west latitude), which can serve 20,000 inhabitants [21,22]. The influent was transported and stored in 20 L tanks and refrigerated at 4 °C in the dark [23].

Two hybrid constructed systems (HBCWs) were used to scale at a laboratory at Universidad de Concepción. The systems were implemented in a greenhouse-type laboratory with semicontrolled temperatures with between 1–20 °C corresponding to a temperate maritime climate with Mediterranean influence. These HBCWs consisted of a vertical subsurface flow constructed wetland, followed by a horizontal subsurface flow constructed wetland (VSSF-HSSF). In addition, different saturation levels were used. For the VSSF, a partially saturated model (0.6 m) and a saturated model (0.8 m) were employed. Similarly, for the HSSF, partially saturated (0.15 m) and saturated models (0.25 m) were also employed. Therefore, the experimental units for the HBCWs consisted of (a) HB 1: VSSF 1 (0.6 m)-HSSF 1 (0.15 m) and (b) HB 2: VSSF 2 (0.8 m)-HSSF 2 (0.25 m). Each of the four units was planted with the macrophyte species (*Schoenoplectus californicus*). The VSSF and HSSF systems were inoculated with 4 and 8 seedlings distributed homogeneously, respectively.

Each system was fed discontinuously with 3.6–4.2 L/d for VSSF 1 and VSSF 2 and 1.1–1.7 L/d for HSSF 1 and HSSF 2, respectively. The VSSF had a hydraulic retention time (HTR) of 1 d, and the HSSF had an HRT of 7 d. The effluent was collected using pipeline systems located underneath each wetland. The main design and operational characteristics of the CW are shown in Table 1. Moreover, Figure 1 shows a schematic representation of the HBCWs.

**Table 1.** Main characteristics of a hybrid constructed wetland.

| Characteristics | Unit | VSSF | HSSF |
|---|---|---|---|
| | Design parameters | | |
| Surface area | $m^2$ | 0.025 | 0.17 |
| Average height | m | 0.85 | 0.30 |
| Water table height | m | 0.6–0.8 | 0.15–0.25 |
| Total volume | $m^3$ | 0.02 | 0.05 |
| | Support medium | | |
| Type | - | Sand/Gravel | Gravel |
| Size | mm | 1–4/19–25 | 19–25 |
| Porosity | - | 0.2/0.4 | 0.4 |
| | Operation parameters | | |
| HRT | d | 1 | 7 |
| HLR | L/d | 3.6–4.2 | 1.1–1.7 |

HRT: Hydraulic retention time; HLR: Hydraulic loading rate.

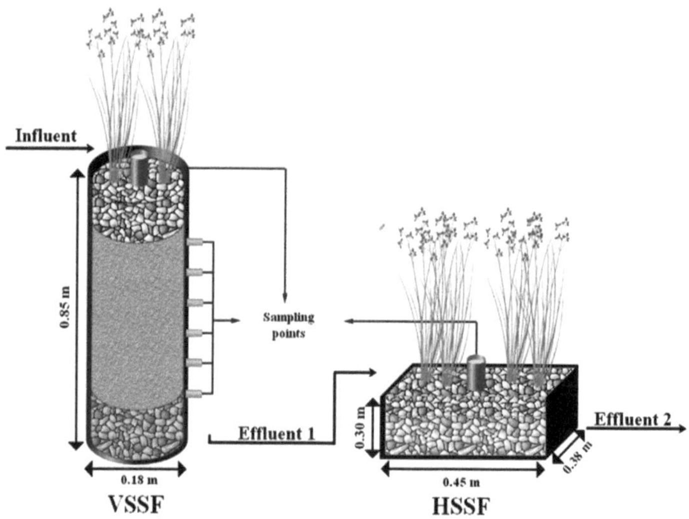


**Figure 1.** Schematic representation of hybrid constructed wetlands.

### 2.2. Sampling Strategy

During the operational period (a total of 297 days), which included summer, autumn, winter, and spring, the in situ parameters, including pH temperature, electrical conductivity, dissolved oxygen (DO) and oxide reduction potential (ORP), were measured weekly using the samplers located at the center of each wetland. Physicochemical parameters, including chemical oxygen demand (COD), total organic carbon (TOC), total suspended solids (TSS), volatile suspended solids (VSS), nitrate-nitrogen ($NO_2^-$-N), the nitrogen of the nitrate ($NO_3^-$-N) and ammonium nitrogen ($NH_4^+$-N), were measured weekly. Total nitrogen (TN) was measured monthly [22,24]. Removal efficiencies were determined according to the Equation (1) used by Tuttolomondo et al. [25].

$$R(\%) = ((Qi \cdot Ci - Qe \cdot Ce)/(Qi \cdot Ci)) \cdot 100 \quad (1)$$

where Qi is the influent flow rate, Qe is the effluent flow rate, Ci and Ce are the influent and effluent pollutant concentration, respectively.

### 2.3. Analitycal Methods

To monitor the HBCWs, samples were taken of the influent and the output of each unit, which were filtered through a Whatman 0.45 mm membrane pore size. All protocols described in the standard methods were followed for the physicochemical parameters [26]. Combustion catalytic oxidation at 680 °C and detection by a nondispersive infrared sensor (NDIR) were used to determine TOC. The COD was determined using the colorimetric method (5210-B). The TSS and VSS were determined by the gravimetric method. The nutrients $NO_3^-$-N were determined by UV, $NO_2^-$-N and $NH_4^+$-N spectrophotometry, which were determined using the colorimetric method, and TN was determined by Spectroquant-Nova 60 (kits of Merck, Kenilworth, NJ, USA). For the in situ parameters, OAKTON portable multiparameter equipment (PC650-480485, Charleston, SC, USA) was used, and for DO, a portable oximeter (HANNA OXI 330i/set HI 9146-04, Villafranca, Italy) was used.

Statistical analysis was performed to compare the removal efficiencies of pollutants of the different CW units. Data were grouped, and the Shapiro–Wilk normality test was performed. A paired t-test was used if the data were normally distributed, and data without normal distribution were analyzed with a Wilcoxon test. The effect of the saturation level of the HBCWs was compared; if they were characterized by a normal distribution, an ANOVA test was performed, but if data were not normally distributed, the Kruskal–Wallis test was used to conduct the analysis. All statistical analyses were performed using the InfoStat statistical program with a significance level of 0.05 [27].

## 3. Results and Discussion

### 3.1. Influent Physicochemical Characterization

Table 2 shows the results of the physicochemical characterization of the influent throughout the duration of this study. The influent had an average COD value of 211.67 mg/L and a range of 127.37–401.17 mg/L, reflected in the variations in COD concentrations when considering municipal wastewater [28]. There were applied organic load rates (OLRs) in the range of 18.23–57.76 $g/m^2 \cdot d$ for HB 1 and 21.27–67.40 $g/m^2 \cdot d$ for HB 2. Similar results were obtained by García-Ávila et al. [29], who registered an influent with an average value of 222.44 mg/L for municipal wastewater. Likewise, Al-Ajalin et al. [30] reported influents with COD concentrations of 234.00 mg/L regarding municipal wastewater. Additionally, the highest concentrations of COD were recorded during autumn with an average of 296.94 mg/L. In contrast, the lowest concentrations of COD were recorded during spring with a mean of 178.51 mg/L. Thus, the highest variation between seasons was recorded during autumn and spring at 44%.

**Table 2.** Physicochemical characterization of the influent.

| | Concentration (mg/L) ± SD | | | | | | | |
|---|---|---|---|---|---|---|---|---|
| | Summer | | Autumn | | Winter | | Spring | |
| | Average ± SD | Range | Average ± SD | Range | Average ± SD | Range | Average ± SD | Range |
| COD | 200.67 ± 66.49 | 139.93–314.73 | 296.94 ± 66.75 | 193.93–401.17 | 183.53 ± 42.77 | 132.35–251.49 | 178.51 ± 69.48 | 127.37–257.61 |
| TOC | 45.72 ± 22.31 | 16.35–85.13 | 87.09 ± 21.41 | 52.45–116.00 | 49.64 ± 11.57 | 35.80–68.02 | 48.28 ± 18.73 | 34.45–69.68 |
| TSS | 98.96 ± 11.23 | 20.94–106.92 | 42.94 ± 12.67 | 25.20–61.00 | 38.25 ± 13.82 | 27.00–59.33 | 32.67 ± 2.52 | 30.00–35.00 |
| VSS | 24.25 ± 9.31 | 18.75–35.00 | 40.26 ± 11.97 | 23.20–58.00 | 33.25 ± 14.18 | 15.50–57.33 | 28.20 ± 1.08 | 27.00–29.10 |
| $NO_2^-$-N | 0.54 ± 0.04 | 0.51–0.57 | 0.42 ± 0.18 | 0.17–0.72 | 0.31 ± 0.05 | 0.25–0.49 | 0.36 ± 0.12 | 0.27–0.49 |
| $NO_3^-$-N | 1.69 ± 0.33 | 1.26–2.15 | 1.98 ± 0.53 | 1.34–2.86 | 1.22 ± 0.26 | 0.58–1.83 | 0.33 ± 0.07 | 0.28–0.38 |
| $NH_4^+$-N | 81.79 ± 12.44 | 66.34–96.38 | 103.66 ± 7.91 | 92.71–123.78 | 95.81 ± 9.90 | 81.73–106.03 | 97.23 ± 15.74 | 80.05–110.98 |
| TN | 105.00 ± 19.00 | 86.00–124.00 | 130.67 ± 5.69 | 126.00–137.00 | 137.67 ± 6.66 | 132.00–145.00 | 130.00 ± 2.83 | 128.00–132.00 |

COD: chemical oxygen demand; TOC: total organic carbon; TSS: total suspended solids; VSS: volatile suspended solids; $NO_2^-$-N: nitrite nitrogen; $NO_3^-$-N: nitrate nitrogen; $NH_4^+$-N: ammonium nitrogen; TN: total nitrogen.

The TSS presented a range of 32.67–98.96 mg/L with a mean of 53.20 mg/L. Higher concentrations were recorded during summer with a mean of 98.96 mg/L. Moreover, the highest observed variation was recorded between summer and spring at 66%. Similar results were reported by Zurita et al. [31] with a TSS average concentration value of 57.50 mg/L for domestic wastewater.

In terms of nitrogen, the TN recorded a concentration range from 86 to 145 mg/L with higher concentrations during winter (137.67 mg/L average) and up to 23.73% higher than during summer (105.00 mg/L average). $NH_4^+$-N had a concentration range from 66.34 to 123.78 mg/L, and a variation of 21.10% was registered in summer (81.79 mg/L average) and autumn (103.66 mg/L average). The applied nitrogen load rates (NLRs) were, on average, 12.80 and 15.01 g $NH_4^+$-N/$m^2$·d for VSSF 1 and VSSF 2, respectively. On the other hand, HSSF 1 and HSSF 2 registered applied nitrogen loads of 0.44 and 0.67 g $NH_4^+$-N/$m^2$·d, respectively. $NO_2^-$-N registered an average value of 0.40 mg/L with higher concentrations during summer (0.54 mg/L average) and lower concentrations during spring (0.25 mg/L on average). Regarding $NO_3^-$-N, a variation of 83.33% during autumn and spring was registered with a concentration range of 0.28 to 2.86 mg/L. Other studies have used influent of the same source and recorded 60.9–79.9 and 76–119 mg/L for $NH_4^+$-N and TN, respectively [22,32]. This could be explained since, as it is rural wastewater, the nitrogen contribution is higher because of the different agricultural activities in the area [33].

### 3.2. Parameters in HBCWs

Figure 2 shows the variation of the in situ parameters for the HBCWs systems. The average temperature was 15.0 °C with a minimum of 13.9 °C in winter and a maximum of 28.2 °C in summer. These data are consistent with Leiva et al. [22] who obtained similar results. They used a laboratory-scale VSSF and reported a temperature range that varied between 8.2 and 22.9 °C. There were no significant differences between the VSSF and HSSF groups ($p > 0.05$) regarding temperature. In the case of the incidence of the temperature in the biodegradation processes of organic components in anaerobic conditions, studies have concluded that there are no significant effects because bacteria are capable of performing elimination processes even at temperatures as low as 5 °C. Additionally, roots and porous media allow temperatures inside the HBCWs to be 2.3 °C above the outside temperature, thus allowing microbial activity to continue to function properly [34–36]. The pH (Figure 2a) presented values that ranged from 5.9 to 8.9, which is considered to be an optimal range for the survival of bacteria in HBCWs (4.0–9.5) [8,37]. Furthermore, higher levels of up to 10% were found in the VSSF (6.5–8.2) than in the HSSF (5.9–7.9). Likewise, the highest values of pH were registered during the autumn-winter period (up to 8.20 and 7.93 for VSSF and HSSF), and a significant difference ($p < 0.05$) was found when comparing these values to those obtained during the spring-summer period (up to 7.66 and 7.15 for VSSF and HSSF). Similar results were obtained by Marzec et al. [11]. In their study, they reported a pH range of 6.68 and 8.70 for HBCWs with a VSSF-HSSF layout that treated domestic wastewater.

Regarding conductivity, ranges of up to 1814 and 1760 µS/cm were found for VSSF 1 and VSSF 2, respectively. Conversely, HSSF 1 and 2 registered up to 1423 µS/cm and 1402 µS/cm, respectively (Figure 2b). There was a reduction of up to 34% between VSSF and HSSF, which would be explained by the absorption of micro- and macroelements and ions by plants and bacteria and their elimination through adsorption by the action of the roots of plants and sedimental suspended particles [38]. Moreover, a similar decrease was observed by Kyambadde et al. [37], who used an HBCW composed of six staggered VSSFs. In their case, a decrease of 20.72% regarding conductivity concentrations was found, going from 796.3 µS/cm to 631.5 µS/cm.

The DO concentrations in the HBCWs registered a range of 0.1–0.6 mg/L. Average values of 0.22 and 0.23 mg/L were registered for VSSF 1 and VSSF 2 (Figure 2c), respectively. No significant differences ($p > 0.05$) were found between seasons. In relation to other investigations, the concentrations in VSSF 1 and VSSF 2 were up to 89.52% lower than those reported by Sgroi et al. [39]. They recorded DO concentrations between 0.5 and2.1 mg/L for a VSSF that treated domestic wastewater with an OLR of 40 g COD/$m^2$·d. Regarding HSSF 1 and HSSF 2 (Figure 2c), there were average concentrations of 0.4 mg/L, values up to 78.57% lower than those obtained by Caselles–Osorio et al. [40], who reported DO concentrations of 2.1 mg/L in an HSSF that treated domestic wastewater with an OLR between 4 and 5 g COD/m2·d. Higher concentrations of DO were registered during winter with values up to 0.4 and 0.5 mg/L for VSSF 1 and VSSF 2, respectively, and 0.6 mg/L for HSSF 1 and HSSF 2. Thus, it can be determined that anaerobic conditions prevail in the HBCWs, registering concentrations < 2 mg/L of DO [41].

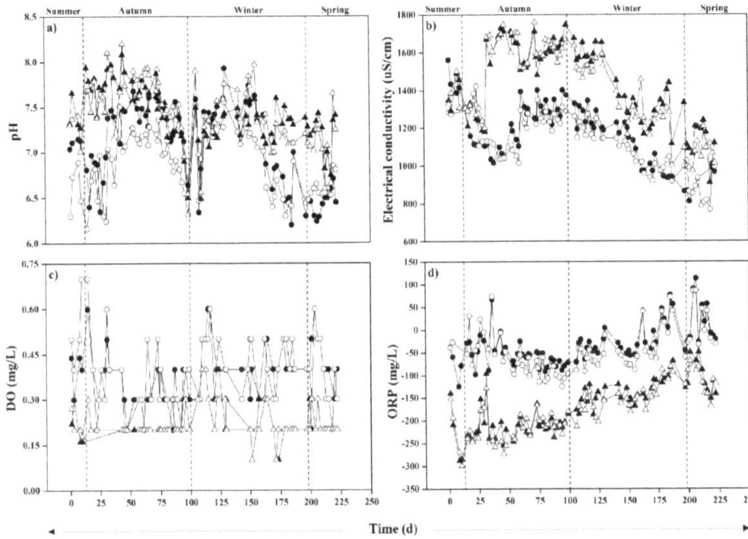


**Figure 2.** In situ parameters variation: (**a**) pH; (**b**) electrical conductivity; (**c**) dissolved oxygen; and (**d**) oxide reduction potential. Where VSSF 1 (△); VSSF 2 (▲); HSSF 1(○); and HSSF 2 (•).

Regarding the measured ORP, the VSSF (Figure 2d) scored a range from 288 to −25 mV with means of −174.9 and −169.1 mV for VSSF 1 and VSSF 2, respectively. The HSSF registered values ranged from −125 to 77 mV with means of 34.7 and −35.5 mV for HSSF 1 and HSSF 2 (Figure 2d), respectively. A value higher than 100 mV is required to consider an aerobic environment, while a value lower than −100 mV indicates anaerobic conditions [42]. Consequently, the HBCWs were characterized by an anaerobic environment in both the VSSF and HSSF. The oxygen diffusion was not high enough to increase the oxygen concentration within the HBCWs and thus modify the oxidation conditions [43].

### 3.3. Concentrations and Removal Efficiencies of Organic Matter and Suspended Solids for the HBCWs System

In the VSSF, the average concentration of COD (Table 3) was kept in ranges between 70.25–92.57 mg/L with an average OLR of 23.84 and 27.81 g COD/$m^2$·d for VSSF 1 and VSSF 2, respectively. Significant differences ($p < 0.05$) were found for VSSF 1 and VSSF 2 between autumn and the rest of the seasons, causing an influent concentration of 296.94 mg/L with concentrations up to 44% higher than those in spring. In the VSSF, the average removal efficiencies for COD (Figure 3a,b) were 56.67 and 57.94% for VSSF 1 and VSSF 2, reaching higher levels of 78.41 and 82.68%, respectively, in autumn. In a recent study, Chang et al. [44] used a laboratory-scale VSSF and reached an average removal efficiency level of 59.9 and 62.8% for an influent with a COD of 288.66 mg/L.

Previous research has reported removal efficiencies that fluctuate between 41 and 92.90% [45–49]. There was only a difference of 2.21% for COD between the partially saturated VSSF system and saturated VSSF with no significant differences ($p > 0.05$). The results are consistent with the findings of Huang et al. [18]. In their study, no significant differences were found regarding the removal efficiencies of COD when using VSSF at different saturation levels (0.05, 0.30, 0.45 and 0.60 m). Therefore, it can be observed that the removal of organic matter was not affected by the alteration of the saturation level because organic matter decomposes under aerobic/anaerobic conditions as well as by sedimentation and filtration of organic particles [50]. However, different results were reported by Bassani et al. [51]. In their study, they used an unsaturated (0.1 m) and a partially saturated VSSF (0.4 m), and significant differences ($p < 0.05$) were reported regarding the removal efficiencies of COD (87.9% and 94.3% for the unsaturated and saturated systems, respectively). These results differ from the present study since in the Bassani et al. [51] study, different HRTs (0.6–0.8 d for unsaturated VSSF and 2–2.5 d for partially saturated VSSF) were used. Therefore, a higher HRT allowed a longer period of microorganism interaction with organic matter, thus enabling better removal efficiencies of COD.

**Table 3.** Physicochemical characterization of the effluent for different CW.

| | | Concentration ± SD | | | |
|---|---|---|---|---|---|
| **Parameter** | **Period** | **VSSF 1** | **VSSF 2** | **HSSF 1** | **HSSF 2** |
| COD | Summer | 81.23 ± 25.31 | 70.25 ± 17.70 | 58.92 ± 18.23 | 57.38 ± 16.16 |
| | Autumn | 81.74 ± 18.38 | 92.57 ± 24.69 | 35.97 ± 8.84 | 41.25 ± 12.12 |
| | Winter | 84.33 ± 17.39 | 76.45 ± 8.36 | 35.25 ± 9.31 | 33.40 ± 6.11 |
| | Spring | 85.63 ± 13.78 | 84.69 ± 21.10 | 35.98 ± 3.79 | 25.07 ± 3.06 |
| TOC | Summer | 30.91 ± 4.54 | 27.29 ± 5.47 | 30.25 ± 13.44 | 13.40 ± 2.31 |
| | Autumn | 18.22 ± 4.10 | 26.96 ± 7.19 | 5.20 ± 2.01 | 5.56 ± 1.83 |
| | Winter | 18.80 ± 3.88 | 22.26 ± 2.44 | 4.57 ± 1.21 | 4.72 ± 0.86 |
| | Spring | 19.09 ± 3.07 | 24.67 ± 6.14 | 4.67 ± 0.49 | 3.54 ± 0.43 |
| TSS | Summer | 7.27 ± 2.07 | 5.41 ± 1.41 | 4.71 ± 0.42 | 3.76 ± 0.71 |
| | Autumn | 7.92 ± 2.68 | 8.33 ± 3.23 | 3.48 ± 0.53 | 3.66 ± 0.61 |
| | Winter | 5.63 ± 1.00 | 5.79 ± 1.50 | 2.93 ± 0.79 | 3.25 ± 0.94 |
| | Spring | 5.60 ± 0.10 | 6.10 ± 0.10 | 3.50 ± 0.20 | 3.27 ± 0.25 |
| VSS | Summer | 6.87 ± 1.55 | 8.19 ± 3.51 | 3.24 ± 0.85 | 3.04 ± 1.12 |
| | Autumn | 7.37 ± 2.63 | 6.03 ± 1.47 | 3.35 ± 1.91 | 2.73 ± 0.53 |
| | Winter | 4.56 ± 1.09 | 5.59 ± 1.27 | 2.38 ± 0.71 | 2.15 ± 0.77 |
| | Spring | 4.53 ± 0.06 | 4.73 ± 0.25 | 2.67 ± 0.15 | 2.73 ± 0.21 |
| | Summer | 63.50 ± 8.26 | 54.55 ± 14.36 | 13.06 ± 5.25 | 16.07 ± 7.62 |
| | Autumn | 83.25 ± 7.92 | 83.91 ± 10.18 | 18.26 ± 6.84 | 24.10 ± 10.40 |
| | Winter | 68.87 ± 15.27 | 78.89 ± 12.33 | 22.30 ± 5.32 | 34.43 ± 6.33 |
| | Spring | 33.35 ± 3.06 | 34.62 ± 7.43 | 8.86 ± 1.35 | 13.87 ± 3.09 |
| $NH_4^+$-N | Summer | 63.50 ± 8.26 | 54.55 ± 14.36 | 13.06 ± 5.25 | 16.07 ± 7.62 |
| | Autumn | 83.25 ± 7.92 | 83.91 ± 10.18 | 18.26 ± 6.84 | 24.10 ± 10.40 |
| | Winter | 68.87 ± 15.27 | 78.89 ± 12.33 | 22.30 ± 5.32 | 34.43 ± 6.33 |
| | Spring | 33.35 ± 3.06 | 34.62 ± 7.43 | 8.86 ± 1.35 | 13.87 ± 3.09 |
| $NO_2^-$-N | Summer | 0.89 ± 0.28 | 1.72 ± 0.53 | 0.06 ± 0.02 | 0.19 ± 0.09 |
| | Autumn | 0.34 ± 0.09 | 0.31 ± 0.08 | 0.21 ± 0.07 | 0.24 ± 0.06 |
| | Winter | 9.24 ± 1.97 | 4.16 ± 1.98 | 0.49 ± 0.10 | 0.32 ± 0.05 |
| | Spring | 22.79 ± 3.55 | 25.21 ± 9.38 | 1.26 ± 0.48 | 0.34 ± 0.06 |
| | Summer | 7.44 ± 2.65 | 7.76 ± 2.17 | 16.48 ± 1.25 | 8.82 ± 0.99 |
| | Autumn | 0.64 ± 0.14 | 0.69 ± 0.25 | 25.94 ± 10.11 | 16.53 ± 5.97 |
| | Winter | 2.65 ± 0.91 | 2.08 ± 0.69 | 39.37 ± 12.93 | 24.36 ± 11.15 |
| | Spring | 5.18 ± 2.75 | 5.49 ± 2.12 | 62.14 ± 1.27 | 35.63 ± 5.35 |
| TN | Summer | 82.00 ± 10.46 | 81.20 ± 17.44 | 30.40 ± 3.13 | 28.60 ± 4.34 |
| | Autumn | 96.67 ± 11.50 | 98.67 ± 11.68 | 57.00 ± 5.29 | 51.33 ± 13.32 |
| | Winter | 114.00 ± 16.37 | 113.67 ± 6.43 | 74.33 ± 7.51 | 73.33 ± 6.66 |
| | Spring | 96.00 ± 5.66 | 103.00 ± 11.31 | 70.00 ± 5.66 | 68.50 ± 9.19 |

VSSF 1: partially saturated vertical constructed wetland; VSSF 2: saturated vertical constructed wetland; HSSF 1: partially saturated horizontal constructed wetland; HSSF 2: saturated horizontal constructed wetland.

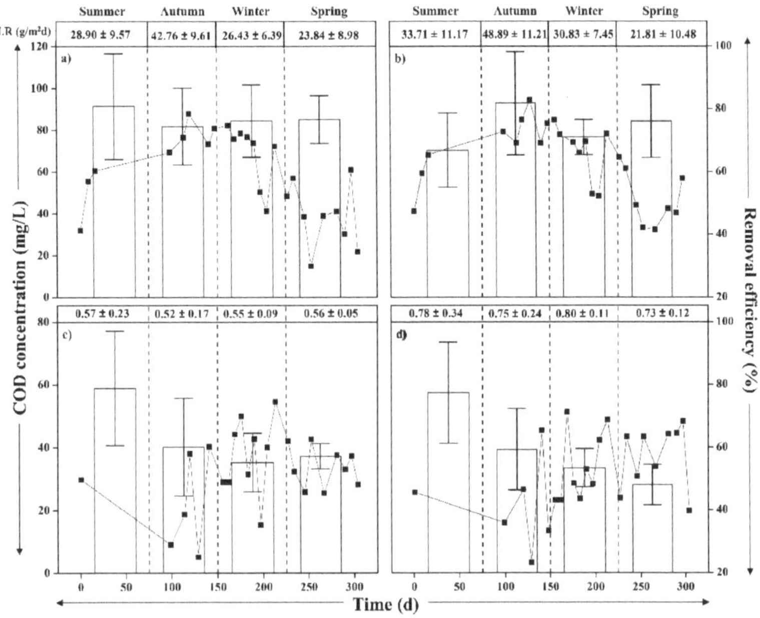

**Figure 3.** Average of effluent COD concentration (bars) and removal efficiencies (points): (**a**) VSSF 1: (**b**) VSSF 2: (**c**) HSSF 1; and (**d**) HSSF 2. OLR: organic loading rate.

HSSF registered average COD concentrations in a range of 25.07–58.92 mg/L (Table 3) with an average OLR of 0.43 and 0.59 g COD/m$^2$·d for HSSF 1 and HSSF 2, respectively. Removal efficiencies (Figure 3c,d) fluctuated between 23.2 and 74.5% with means of 49.53 and 48.56% for HSSF 1 and HSSF 2, registering removal efficiencies of up to 37% higher during winter and spring (up to 74.52 and 71.18%, respectively). These values are lower than those reported by Haddis et al. [52] who used an HSSF that treated municipal wastewater and recorded an average removal efficiency of 65%. Other studies have reported organic matter removal efficiencies in HSSF that range between 33.60 and90.22% [53,54]. A difference of 1.90% for COD was observed between the partially saturated HSSF and the saturated HSSF; therefore, no significant differences ($p > 0.05$) were found. These results differ from those reported by García et al. [15]. In their study, significant differences ($p < 0.05$) were found for COD when using a saturated HSSF (0.50 m) and a partially saturated HSSF (0.27 m). Furthermore, in their study, the partially saturated system registered removal efficiencies of up to 83% versus removal efficiencies of 65% for the saturated system. Similarly, Al-Ajalin et al. [30] reported significant differences ($p < 0.05$) using an HSSF with two saturation levels (0.35 and 0.45 m) with higher removal efficiencies of COD in the system with a lower saturation level (up to 96.94%). On the other hand, Benvenuti et al. [55] indicated that a DO concentration higher than 1.5 mg/L is needed to allow aerobic microbial metabolism. Therefore, it can be expected that in a VSSF and HSSF that included DO < 1.5 mg/L concentrations and a reductive environment (between −288 and 77 mV), both in the partially saturated and saturated systems, organic matter was removed mainly by anaerobic processes [56], thus explaining that there were no significant differences between the saturation levels.

The removal efficiencies of COD in the HBCWs were 79.01 for HB 1 and 79.79% for HB 2. There were no significant differences ($p > 0.05$) between HB 1 and HB 2 regarding removal efficiencies. In a recent study by Kraiem et al. [57], removal efficiencies of COD of 86% were reported when using an HBCW with a VSSF-HSSF layout that treated rural wastewater. In both VSSF and HSSF, higher removal efficiencies of COD were achieved during cold seasons (autumn-winter), thus concluding that temperature did not affect the removal processes of organic matter. Likewise, Rozema et al. [58] evaluated the effect of temperature in a VSSF, concluding that cold weather did not affect the VSSF treatment efficiencies; furthermore, efficiencies reached up to 99% for COD.

These were removed mainly by physical mechanisms such as sedimentation and filtration in relation to the TSS. Filtration occurred in the roots and stems of macrophyte plants and in the support medium [29]. Additionally, VSSF 1 and VSSF 2 presented removal efficiencies of 83.55 and 82.24%, respectively, and no significant differences were found ($p > 0.05$) between seasons and amid the partially saturated VSSF and the saturated VSSF. The values reported are higher than those reported by Abdelhakeem et al. [59]. They reported an average removal efficiency of 75% for a laboratory-scale VSSF that used municipal wastewater. HSSF 1 and HSSF 2 recorded removal efficiencies of 57.24 and 50.60%, respectively. Jácome et al. [60] registered removal efficiencies of 77% in an HSSF that treated municipal wastewater with concentrations of TSS in the influent of 41–48 mg/L (up to 22% lower than the current study). There were no significant differences ($p > 0.05$) between seasons, but there was a significant difference ($p < 0.05$) of 11.60% between the partially saturated HSSF and saturated HSSF systems, mainly because of the increase in the saturation level, which in turn generated a free flow in the upper part of the support medium. As a consequence, the removal capacity of plants decreased when removing TSS through the roots [61]. Additionally, HB 1 and HB 2 reached average removal efficiencies of 89.81 and 90.53%, respectively. Moreover, there were no significant differences ($p > 0.05$) between HB 1 and HB 2 regarding the removal efficiencies of TSS. The values reported are consistent with previous research that registered average removal efficiencies of 94–96% [11,12].

### *3.4. Concentrations and Removal Efficiencies of Nitrogen for the HBCWs System*

VSSF 1 (Figure 4a) registered lower concentrations of $NH_4^+$-N during spring, at 29.97 mg/L. On the other hand, VSSF 2 (Figure 4b) $NH_4^+$-N concentrations fluctuated between 28.53 and96.18 mg/L with lower average concentrations (Table 3) during spring (34.62 mg/L). Additionally, in HSSF 1 (Figure 5c), the minimum concentration of $NH_4^+$-N was recorded during summer (6.92 mg/L). In the case of HSSF 2 (Figure 4d), the $NH_4^+$-N remained between 7.61 and 42.67 mg/L, recording the lowest concentration during spring (13.87 mg/L).

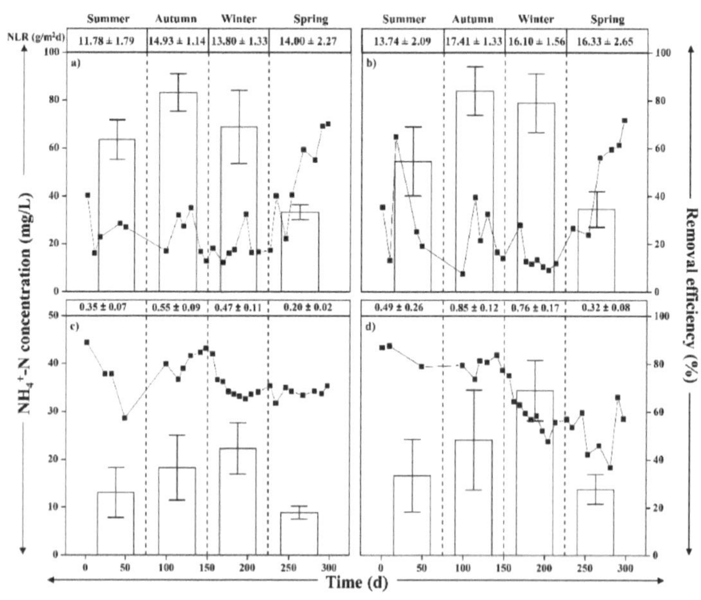


**Figure 4.** Effluent average $NH_4^+$-N concentration (bars) and removal efficiencies (points): (**a**) VSSF 1; (**b**) VSSF 2; (**c**) HSSF 1; and (**d**) HSSF 2.NLR: nitrogen loading rate.

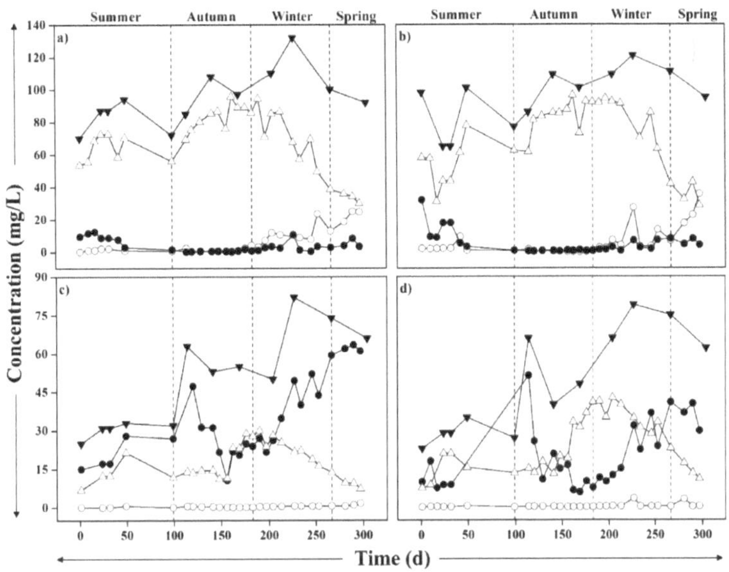


**Figure 5.** Concentrations of different forms of nitrogen (mg/L) $NO_2^-$-N (○), $NO_3^-$-N (•), $NH_4^+$-N (△) and TN (▼): (**a**) VSSF 1; (**b**) VSSF 2; (**c**) HSSF 1; and (**d**) HSSF 2.

For the removal efficiencies of $NH_4^+$-N, VSSF 1 (Figure 4a) registered a range of 12.28–70.22% with an average of 34.47%. For VSSF 2, a range of 7.52–71.66% (Figure 4b) was recorded with higher efficiencies during spring (up to 71.66%) and an average of 33.40%. Furthermore, significant differences ($p < 0.05$) were found between VSSF 1 and VSSF in spring and the rest of the seasons. In a similar study, Abdelhakeem et al. [59] reported removal efficiencies of $NH_4^+$-N of 36% for a laboratory-scale VSSF that registered an effluent with $NH_4^+$-N concentrations of 21 mg/L (37.5% lower than the average concentrations of this study). Moreover, several studies have reported removal efficiencies of $NH_4^+$-N that have reached values up to 88.0–97.8% [62–64]. Additionally, a difference of only 3.01% for $NH_4^+$-N was found between the partially saturated VSSF and the saturated VSSF; thus, no significant differences ($p > 0.05$) were reported, and nitrification was not affected when modifying the saturation level. In contrast, Huang et al. [18] reported significant differences ($p < 0.05$) concerning the removal efficiencies of $NH_4^+$-N when using different saturation levels (0.05, 0.30, 0.45 and 0.60 m). Furthermore, when changing the saturation level from 0.45 to 0.60 m, a decrease of 12.71% in removal efficiencies was reported. This can be explained by the difficulty of diffusing oxygen at 0.45 m of saturation.

HSSF registered average removal efficiencies regarding $NH_4^+$-N of 73.95–65.93% for HSSF 1 and HSSF 2. For HSSF 1 (Figure 4c), removal efficiencies of $NH_4^+$-N higher than 60% were recorded, reaching 88.98% in summer. Similarly, HSSF 2 (Figure 4d) registered removal efficiencies in the range of 36.82–87.49%, reaching 87.49% in summer. Furthermore, significant differences ($p < 0.05$) were found for HSSF 1 and HSSF 2 during the autumn-winter period. The values reported are higher than those obtained by Zurita et al. [31]. In their study, HSSF was used, reaching $NH_4^+$-N removal efficiencies of 45.8–48.6%. Additionally, for the partially saturated and saturated HSSFs, although there was a difference of 10.85% in $NH_4^+$-N, there were no significant differences ($p < 0.05$). Sanchez-Ramos et al. [17] obtained similar results when using HSSF at different saturation levels (partially saturated (0.27 m) and saturated (0.50 m)). They reported significant differences ($p < 0.05$) in $NH_4^+$-N since the partially saturated system registered $NH_4^+$-N removal efficiencies up to 56% higher than those of the saturated system. Moreover, a higher level of saturation generated changes at the redox state level, creating more reducing conditions and changing oxygen availability as a result of a shorter diffusion of oxygen from the surface and the air-water interface, which are fundamental in the transformation processes of nitrogen components [65].

DO concentrations (<2 mg/L), particularly in the VSSF, would be caused by the oxygen demands for the organic matter oxidation of (min 1.5 mg/L DO), generating a competition of the DO, thus decreasing the ammonium-oxidizing bacteria's performance and, as a consequence, nitrification [66,67]. According to Liu et al. [67], DO concentrations higher than 1.5 mg/L are needed for nitrification to occur. The previously mentioned factors would help explain the VSSF's lower performance in removing $NH_4^+$-N (34.47% for VSSF vs. 73.95% for HSSF), owing to lower concentrations of available DO (0.2 mg/L for VSSF vs. 0.4 mg/L for HSSF). In addition, different studies have suggested that temperature affects the nitrification process with the optimal temperature ranging from 16 to 32 °C [54,68,69]. With regard to VSSF and HSSF, the removal efficiencies of $NH_4^+$-N were higher, reaching 70.22 and 37.00% in the warmer seasons (spring-summer), respectively. In their study, Hua et al. [68] reported a decrease in removal efficiencies of $NH_4^+$-N of up to 65% to 20% from summer to winter, caused by the reduction of activity and proliferation of nitrification microorganisms amid a decrease in the temperature [70].

The $NO_2^-$-N registered values ranged from 0.26–23.53 mg/L for VSSF 1 (Figure 5a) and 0.23–35.63 mg/L for VSSF 2 (Figure 5b). In the case of HSSF 1 (Figure 5c), $NO_2^-$-N concentrations lower than 2 mg/L were recorded throughout the monitoring period with values ranging between 0.035 and1.61 mg/L. On the other hand, HSSF 2 (Figure 5d) had an average value for $NO_2^-$-N of 0.019 mg/L with a range of 0.01–3.35 mg/L. The $NO_3^-$-N in VSSF 1 (Figure 5a) remained at values lower than 13 mg/L, in a range of 0.07–12.56 mg/L. In the case of VSSF 2 (Figure 5b), in general, $NO_3^-$-N concentrations remained lower than 5 mg/L, except in summer, where concentrations up to 32 mg/L were recorded. The HSSF 1 (Figure 5c) registered concentrations of $NO_3^-$-N ranged between 10.65 and 63.48 mg/L. On the other hand, HSSF 2 (Figure 5d) recorded $NO_3^-$-N values that reached 51.32 mg/L in autumn. Furthermore, in the VSSF, there were lower concentrations of $NO_3^-$-N than in the HSSF (up to 97% in autumn), which could have been removed through denitrification under anaerobic conditions [66]. Alternatively, the decrease in $NO_3^-$-N could also be explained by simultaneous nitrification/denitrification (SND) processes [71]. This happens as the biofilm is formed in the support medium; thus, an anaerobic microenvironment is created inside, and an aerobic microenvironment is created outside, allowing the coexistence of facultative aerobic and anaerobic microorganisms [72,73].

Conversely, the highest concentrations of $NO_3^-$-N in HSSF (up to 63.48 mg/L) could be caused by the suppression of denitrification, owing to the lack of an organic carbon source, since it significantly affects the efficiency of denitrification, causing an accumulation of $NO_3^-$-N in the medium [8,74]. Ding et al. [74] used different C/N relationships (0, 2, 4, 6 and 9) and established that the optimal removal appeared when increasing the C/N relationship to 9, where there were concentrations of $NO_3^-$-N in the effluent of only 0.03 mg/L for an influent of 19.98 mg/L.

In this study, the lowest concentrations of C/N were found in the HSSF. Furthermore, HSSF 1 registered a C/N relation of 0.8 during the monitoring period, and HSSF 2 registered C/N relations of 1.0, 0.8, 0.7 and 0.7 during summer, autumn, winter, and spring, respectively, thus explaining the accumulation of $NO_3^-$-N in these systems.

In VSSF 1 (Figure 5a), the TN remained in the range of 70–132 mg/L, and in winter, lower concentrations (70 mg/L) were registered. On the other hand, in VSSF 2 (Figure 5b), TN concentrations fluctuated between 65.00–121.00 mg/L, registering a similar concentration in summer (65.00 mg/L). Additionally, in HSSF 1 (Figure 5c) and HSSF 2 (Figure 5d), the lowest concentrations of TN were recorded in summer, at 33 and 23 mg/L, respectively. Regarding removal efficiencies, VSSF recorded average efficiencies of 22.64 and 18.78% for VSSF 1 and VSSF 2, respectively. These values were up to 62.44% lower than those reported by Zurita et al. [31], who used a VSSF that treated domestic wastewater and reported removal efficiencies of 50% for an influent with concentrations of TN of 28.7 mg/L (up to

77% lower than the concentrations of TN in this study). Additionally, significant differences ($p < 0.05$) were found between the partially saturated VSSF and the saturated VSS with a difference of 17.05% for TN. This could be caused by changes in the saturation levels creating changes at the DO level (0.62 to 2.2 mg/L) and ORP (−5.39 to 90.93 mV), causing the different microbial communities that act simultaneously in the medium to be affected, thus altering nitrogen removal processes such as nitrification and denitrification [75–77]. In the case of the HSSF, average removal efficiencies of TN of 37.03 and 39.40% were recorded for HSSF 1 and HSSF 2, respectively. In addition, no significant differences ($p > 0.05$) were found between the systems. In a recent study, Jácome et al. [60] used HSSF and established a TN removal efficiency of 37%. On the other hand, Zhu et al. [78] reported removal efficiencies of TN that reached 89.9%; however, this efficiency was due to the TN concentration in the influent of 40 mg/L, which was up to 68% lower than the average concentration of TN in this study (126 mg/L).

In HBCWs, the removal efficiencies of $NH_4^+$-N reached 83.50 for HB 1 and 77.28% for HB 2. Saeed et al. [14], who used partially saturated HBCWs, obtained similar results. They established removal efficiencies of $NH_4^+$-N of 81.1%. On the other hand, these values are lower than those reported by Ávila et al. [79], who registered values of up to 94%. Regarding TN, average removal efficiencies of 55.12 for HB 1 and 57.40% for HB 2 were reported. Moreover, Nguyen et al. [13] reported lower values than those previously mentioned, where removal efficiencies of 79% were reported using municipal wastewater for HBCWs with a layout of VSSF-HSSF.

### 3.5. In Situ Spatial Variations and Physical-Chemical Parameters

Table 4 shows the in situ spatial variations and physicochemical parameters. In the case of the COD, it can be observed that during 2019 (data with influent concentrations), the greatest decrease in concentrations occurred in the upper zone with reductions of 58.64 and 53.54 mg/L for VSSF 1 and VSSF 2, namely, removal efficiencies of 34 and 31%, respectively. As the zone being measured decreased, removal efficiencies declined since only 16% was recorded in the lower zone. In a study conducted by Yi et al. [80], it was established that COD concentrations were mainly removed at the entrance with a removal efficiency of 64.23% in the upper zone. Previous research has reported that the anaerobic routes of organic matter removal are slower than aerobic routes [19], which would explain the decrease in the removal efficiency in the lower zones when dealing with systems with a low HRT.

**Table 4.** Spatial variations of in situ and physicochemical parameters.

| Year | CW | Location | Concentration (mg/L) | | | | | | |
|---|---|---|---|---|---|---|---|---|---|
| | | | COD (mg/L) | $NH_4^+$-N (mg/L) | Temperature (°C) | pH | ORP (mV) | DO (mg/L) | EC (μS/cm) |
| 2018 | VSSF1 | Upper | - | - | 10.9 ± 1.3 | 7.2 ± 0.2 | −166.9 ± 44.6 | 0.5 ± 0.2 | 1247.3 ± 393.1 |
| | | Middle | - | - | 10.7 ± 1.4 | 7.1 ± 0.1 | −194.1 ± 47.5 | 0.3 ± 0.1 | 1232.2 ± 360.5 |
| | | Bottom | - | - | 10.6 ± 1.2 | 7.1 ± 0.1 | −162.4 ± 127.9 | 0.3 ± 0.0 | 1034.0 ± 79.2 |
| | VSSF2 | Upper | - | - | 10.9 ± 0.5 | 7.4 ± 0.2 | −162.9 ± 99.3 | 0.4 ± 0.2 | 1170.3 ± 167.2 |
| | | Middle | - | - | 10.5 ± 0.6 | 7.2 ± 0.3 | −164.3 ± 127.1 | 0.3 ± 0.1 | 1139.3 ± 111.1 |
| | | Bottom | - | - | 10.0 ± 0.5 | 7.2 ± 0.2 | −193.2 ± 83.2 | 0.2 ± 0.1 | 1031.3 ± 122.7 |
| 2019 | VSSF1 | Upper | 111.9 ± 14.5 | 65.8 ± 12.6 | 11.1 ± 1.1 | 7.4 ± 0.2 | −112.4 ± 24.8 | 0.3 ± 0.1 | 1339.0 ± 95.3 |
| | | Middle | 104.7 ± 19.4 | 68.9 ± 13.5 | 11.0 ± 1.2 | 7.4 ± 0.2 | −135.5 ± 36.6 | 0.3 ± 0.1 | 1341.0 ± 98.7 |
| | | Bottom | 87.6 ± 18.9 | 58.8 ± 13.2 | 10.9 ± 1.2 | 7.3 ± 0.1 | −141.2 ± 33.3 | 0.3 ± 0.1 | 1332.0 ± 111.2 |
| | VSSF2 | Upper | 117.0 ± 17.0 | 73.7 ± 7.1 | 11.2 ± 1.1 | 7.4 ± 0.2 | −133.6 ± 14.9 | 0.3 ± 0.1 | 1361.7 ± 65.6 |
| | | Middle | 92.2 ± 20.2 | 62.9 ± 17.7 | 11.1 ± 1.2 | 7.4 ± 0.2 | −154.7 ± 11.9 | 0.3 ± 0.1 | 1354.0 ± 54.6 |
| | | Bottom | 77.2 ± 20.2 | 55.9 ± 15.3 | 10.9 ± 1.2 | 7.1 ± 0.0 | −153.9 ± 11.9 | 0.3 ± 0.1 | 1182.3 ± 38.8 |

Similarly, for $NH_4^+$-N, the highest concentration decline was recorded in the upper zone with 31.43 mg/L for VSSF 1 and 23.53 mg/L for VSSF 2, namely, removal efficiencies of 32% and 24%, respectively. Sanchez-Ramos et al. [17] found that aerobic bacteria are distributed in the first centimeters of the wetland, thus taking advantage of the atmospheric diffusion of oxygen. Therefore, it can be expected in his study that nitrification bacteria were developed in the first centimeters of the wetland, thus explaining the removal rates for $NH_4^+$-N despite the ORP and DO values obtained, which hinder the nitrification process.

If we consider the in situ parameters, the pH presented a decrease of 0.1 pH units when the measurement zone descended vertically with a range between 7.1–7.4 for VSSF 1 and VSSF 2, and no significant differences ($p > 0.05$) were found. Thus, the pH was not vertically affected. Regarding DO, during 2018, there were higher concentrations in the upper zone with 0.5 mg/L for VSSF 1 and 0.4 mg/L for VSSF 2. Moreover, as the measured zone descended, the DO concentrations decreased, reaching 0.3 mg/L for VSSF 1 and 0.2 mg/L for VSSF 2. However, during 2019, the DO concentration remained consistent as the vertical measurement decreased. This vertical decline could be caused by the removal process of organic matter and nitrogen transformation, which in turn would cause a rapid consumption of DO through aerobic respiration and chemical oxidation [42]. In the case of the ORP, a range of values registered from −112.4 mV in the upper zone to −162.4 mV in the lower zone of VSSF 1. On the other hand, VSSF 2 registered a range from −133.6 mV in the upper zone to −193.2 mV in the lower zone. This decline could be caused by the oxygen diffusion mainly in the upper zone of the gravel beds, decreasing as the depth increases, thus creating more reducing conditions [15].

## 4. Conclusions

In this study, the medium saturation level effect in the VSSF and HSSF does not show significant difference when the COD removal efficiencies are considered. However, when the saturation level of the medium increased, the TSS removal efficiency decreased in 11% in the HSSF. Moreover, this also meant a 17% decrease in the average removal efficiencies of TN in VSSF and a 10% decrease for $NH_4^+$-N at HSSF. Due to this, we conclude that when designing HBCWs, it is essential to consider the medium saturation level because it affects the transformation and/or removal of wastewater's components to be treated. Even more so, the increase of saturation level in HBCWs reduces the transformation of the $NH_4^+$-N. This research shows the possibility of optimizing the transformation of nitrogen with partially saturated hybrids constructed wetlands.

As future prospects for this work, it is important to enhance a mass balance of the different forms of nitrogen ($NH_4^+$-N, $NO_3^-$-N $NO_2^-$-N), considering the microbiological transformations and plant assimilation. In this same line, in future research it would be important to evaluate the gas production and emissions such as $CH_4$, $CO_2$ or $N_2O$ due to the use of partially saturated wetlands and also optimize the system to control said emissions.

**Author Contributions:** Conceptualization, D.L. and G.V.; methodology, D.L. and J.C.; software, G.G.; validation, G.V., D.L. and J.C.; formal analysis, J.C. and D.L.; investigation, J.C. and D.L.; resources, G.V.; data curation, J.C. and G.G.; writing—original draft preparation, J.C. and G.G.; writing—review and editing, G.V., D.L. and G.G.; visualization, G.G.; supervision, G.V.; project administration, G.V. and D.L.; funding acquisition, G.V. All authors have read and agreed to the published version of the manuscript.

**Funding:** This research was funded by ANID/FONDAP/15130015 and Grant No. 3170295 from ANID-FONDECYT (Chile).

**Conflicts of Interest:** The authors declare no conflict of interest.

## References

1. Ye, F.; Li, Y. Enhancement of nitrogen removal in towery hybrid constructed wetland to treat domestic wastewater for small rural communities. *Ecol. Eng.* **2009**, *35*, 1043–1050. [CrossRef]
2. Vera, I.; García, J.; Sáez, K.; Moragas, L.; Vidal, G. Performance evaluation of eight years experience of constructed wetland systems in Catalonia as alternative treatment for small communities. *Ecol. Eng.* **2011**, *37*, 364–371. [CrossRef]

3. Ayaz, Ç.; Aktaş, Ö.; Akça, L.; Fındık, N. Effluent quality and reuse potential of domestic wastewater treated in a pilot-scale hybrid constructed wetland system. *J. Environ. Manag.* **2015**, *156*, 115–120. [CrossRef] [PubMed]
4. Barco, A.; Borin, M. Treatment performance and macrophytes growth in a restored hybrid constructed wetland for municipal wastewater treatment. *Ecol. Eng.* **2017**, *107*, 160–171. [CrossRef]
5. Vymazal, J. The use of hybrid constructed wetlands for wastewater treatment with special attention to nitrogen removal: A review of a recent development. *Water Res.* **2013**, *47*, 4795–4811. [CrossRef]
6. Butterworth, E.; Dotro, G.; Jones, M.; Richards, A.; Onunkwo, P.; Narroway, Y.; Jefferson, B. Effect of artificial aeration on tertiary nitrification in a full-scale subsurface horizontal flow constructed wetland. *Ecol. Eng.* **2013**, *54*, 236–244. [CrossRef]
7. Gajewska, M.; Skrzypiec, K.; Jóźwiakowski, K.; Bugajski, P. Kinetics of pollutants removal in hybrid treatment wetlands-Case study comparison. *Ecol. Eng.* **2018**, *120*, 222–229. [CrossRef]
8. Saeed, T.; Sun, G. A review on nitrogen and organics removal mechanisms in subsurface flow constructed wetlands: Dependency on environmental parameters, operating conditions and supporting media. *J. Environ. Manag.* **2012**, *112*, 429–448. [CrossRef]
9. Vymazal, J.; Kröpfelová, L. Multistage hybrid constructed wetland for enhanced removal of nitrogen. *Ecol. Eng.* **2015**, *84*, 202–208. [CrossRef]
10. Hickey, A.; Arnscheidt, J.; Joyce, E.; O'Toole, J.; Galvin, G.; O'Callaghan, M.; Walsh, K. An assessment of the performance of municipal constructed wetlands in Ireland. *J. Environ. Manag.* **2018**, *210*, 263–272. [CrossRef]
11. Marzec, M.; Gizińska-Górna, M.; Jóźwiakowski, K.; Pytka-Woszczyło, A.; Kowalczyk-Juśko, A.; Gajewska, M. The efficiency and reliability of pollutant removal in a hybrid constructed wetland with giant miscanthus and Jerusalem artichoke in Poland. *Ecol. Eng.* **2019**, *127*, 23–35. [CrossRef]
12. Melián, J.; Rodríguez, A.; Araña, J.; Díaz, O.; Henríquez, J. Hybrid constructed wetlands for wastewater treatment and reuse in the Canary Islands. *Ecol. Eng.* **2010**, *36*, 891–899. [CrossRef]
13. Nguyen, X.; Chang, S.; Nguyen, T.; Ngo, H.; Kumar, G.; Banu, J.; Nguyen, D. A hybrid constructed wetland for organic-material and nutrient removal from sewage: Process performance and multi-kinetic models. *J. Environ. Manag.* **2018**, *222*, 378–384. [CrossRef]
14. Pat-Espadas, A.M.; Rene Loredo Portales, R.; Amabilis-Sosa, L.E.; Gómez, G.; Vidal, G. Review on constructed wetlands for acid mine drainage treatment. *Water* **2018**, *10*, 1685. [CrossRef]
15. García, J.; Aguirre, P.; Barragán, J.; Mujeriego, R.; Matamoros, V.; Bayona, J. Effect of key design parameters on the efficiency of horizontal subsurface flow constructed wetlands. *Ecol. Eng.* **2005**, *25*, 405–418. [CrossRef]
16. Mander, Ü.; Dotro, G.; Ebie, Y.; Towprayoon, S.; Chiemchaisri, C.; Nogueira, S.; Mitsch, W. Greenhouse gas emission in constructed wetlands for wastewater treatment: A review. *Ecol. Eng.* **2014**, *66*, 19–35. [CrossRef]
17. Sanchez-Ramos, D.; Agulló, N.; Samsó, R.; García, J. Effect of key design parameters on bacteria community and effluent pollutant concentrations in constructed wetlands using mathematical models. *Sci. Total Environ.* **2017**, *584*, 374–380. [CrossRef]
18. Huang, M.; Wang, Z.; Qi, R. Enhancement of the complete autotrophic nitrogen removal over nitrite process in a modified single-stage subsurface vertical flow constructed wetland: Effect of saturated zone depth. *Bioresour. Technol.* **2017**, *233*, 191–199. [CrossRef]
19. Saeed, T.; Sun, G. Pollutant removals employing unsaturated and partially saturated vertical flow wetlands: A comparative study. *Chem. Eng. J.* **2017**, *325*, 332–341. [CrossRef]
20. Saeed, T.; Haque, I.; Khan, T. Organic matter and nutrients removal in hybrid constructed wetlands: Influence of saturation. *Chem. Eng. J.* **2019**, *371*, 154–165. [CrossRef]
21. Burgos, V.; Araya, F.; Reyes-Contreras, C.; Vera, I.; Vidal, G. Performance of ornamental plants in mesocosm subsurface constructed wetlands under different organic sewage loading. *Ecol. Eng.* **2017**, *99*, 246–255. [CrossRef]
22. Leiva, A.; Núñez, R.; Gómez, G.; López, D.; Vidal, G. Performance of ornamental plants in monoculture and polyculture horizontal subsurface flow constructed wetlands for treating wastewater. *Ecol. Eng.* **2018**, *120*, 116–125. [CrossRef]
23. López, D.; Fuenzalida, D.; Vera, I.; Rojas, K.; Vidal, G. Relationship between the removal of organic matter and the production of methane in subsurface flow constructed wetlands designed for wastewater treatment. *Ecol. Eng.* **2015**, *83*, 296–304. [CrossRef]
24. López, D.; Sepúlveda-Mardones, M.; Ruiz-Tagle, N.; Sossa, K.; Uggetti, E.; Vidal, G. Potential methane production and molecular characterization of bacterial and archaeal communities in a horizontal subsurface flow constructed wetland under cold and warm seasons. *Sci. Total Environ.* **2019**, *648*, 1042–1051. [CrossRef] [PubMed]
25. Tuttolomondo, T.; Leto, C.; La Bella, S.; Leone, R.; Virga, G.; Licata, M. Water balance and pollutant removal efficiency when considering evapotranspiration in a pilot-scale horizontal subsurface flow constructed wetland in Western Sicily (Italy). *Ecol. Eng.* **2016**, *87*, 295–304. [CrossRef]
26. American Public Health Association (APHA). *Standard Methods for the Examination of Wastewater*, 20th ed.; APHA: Washington, DC, USA, 1998.
27. Di Rienzo, J.; Casanoves, F.; Balzarini, M.; Gonzalez, I.; Tableda, M.; Robledo, C. *Infostat Statistical Software*; FCA National University of Córdoba: Córdoba, Argentina, 2011; p. 336. Available online: https://www.researchgate.net/publication/233842986_InfoStat_Version_2011 (accessed on 19 March 2022).
28. Al-Isawi, R.; Ray, S.; Scholz, M. Comparative study of domestic wastewater treatment by mature vertical-flow constructed wetlands and artificial ponds. *Ecol. Eng.* **2017**, *100*, 8–18. [CrossRef]

29. García-Ávila, F.; Patiño-Chávez, J.; Zhinín-Chimbo, F.; Donoso-Moscoso, S.; Del Pino, L.F.; Avilés-Añazco, A. Performance of Phragmites Australis and Cyperus Papyrus in the treatment of municipal wastewater by vertical flow subsurface constructed wetlands. *Int. Soil Water Conserv. Res.* **2019**, *7*, 286–296. [CrossRef]
30. Al-Ajalin, F.; Idris, M.; Abdullah, S.; Kurniawan, S.; Imron, M. Effect of wastewater depth to the performance of short-term batching-experiments horizontal flow constructed wetland system in treating domestic wastewater. *Environ. Technol. Innov.* **2020**, *20*, 101106. [CrossRef]
31. Zurita, F.; De Anda, J.; Belmont, M. Treatment of domestic wastewater and production of commercial flowers in vertical and horizontal subsurface-flow constructed wetlands. *Ecol. Eng.* **2009**, *35*, 861–869. [CrossRef]
32. Sepúlveda, R.; Leiva, A.M.; Vidal, G. Performance of Cyperus papyrus in constructed wetland mesocosms under different levels of salinity. *Ecol. Eng.* **2020**, *151*, 105820. [CrossRef]
33. Li, W.; Zhai, L.; Lei, Q.; Wollheim, W.M.; Liu, J.; Liu, H.; Liu, S. Influences of agricultural land use composition and distribution on nitrogen export from a subtropical watershed in China. *Sci. Total Environ.* **2018**, *642*, 21–32. [CrossRef] [PubMed]
34. Akratos, C.; Tsihrintzis, V. Effect of temperature, HRT, vegetation and porous media on removal efficiency of pilot-scale horizontal subsurface flow constructed wetlands. *Ecol. Eng.* **2007**, *29*, 173–191. [CrossRef]
35. Steer, D.; Fraser, L.; Boddy, J.; Seibert, B. Efficiency of small constructed wetlands for subsurface treatment of single-family domestic effluent. *Ecol. Eng.* **2002**, *18*, 429–440. [CrossRef]
36. Steinmann, C.R.; Weinhart, S.; Melzer, A. A combined system of lagoon and constructed wetland for an effective wastewater treatment. *Water Res.* **2003**, *37*, 2035–2042. [CrossRef]
37. Kyambadde, J.; Kansiime, F.; Gumaelius, L.; Dalhammar, G. A comparative study of Cyperus papyrus and Miscanthidium violaceum-based constructed wetlands for wastewater treatment in a tropical climate. *Water Res.* **2004**, *38*, 475–485. [CrossRef] [PubMed]
38. Kyambadde, J.; Kansiime, F.; Dalhammar, G. Nitrogen and phosphorus removal in substrate-free pilot constructed wetlands with horizontal surface flow in Uganda. *Water Air Soil Pollut.* **2005**, *165*, 37–59. [CrossRef]
39. Sgroi, M.; Pelissari, C.; Roccaro, P.; Sezerino, P.; García, J.; Vagliasindi, F.; Ávila, C. Removal of organic carbon, nitrogen, emerging contaminants and fluorescing organic matter in different constructed wetland configurations. *Chem. Eng. J.* **2018**, *332*, 619–627. [CrossRef]
40. Caselles-Osorio, A.; Vega, H.; Lancheros, J.; Casierra-Martínez, H.; Mosquera, J. Horizontal subsurface-flow constructed wetland removal efficiency using *Cyperus articulatus* L. *Ecol. Eng.* **2017**, *99*, 479–485. [CrossRef]
41. Vymazal, J.; Kröpfelová, L. Is concentration of dissolved oxygen a good indicator of processes in filtration beds of horizontal flow constructed wetlands? In *Wastewater Treatment, Plant Dynamics and Management*, 1st ed.; Vymazal, J., Ed.; Springer: Dordrecht, The Netherlands, 2008; pp. 311–317.
42. Ding, Y.; Wang, W.; Song, X.; Wang, Y. Spatial distribution characteristics of environmental parameters and nitrogenous compounds in horizontal subsurface flow constructed wetland treating high nitrogen-content wastewater. *Ecol. Eng.* **2014**, *70*, 446–449. [CrossRef]
43. Carballeira, T.; Ruiz, I.; Soto, M. Effect of plants and surface loading rate on the treatment efficiency of shallow subsurface constructed wetlands. *Ecol. Eng.* **2016**, *90*, 203–214. [CrossRef]
44. Chang, J.; Wu, S.; Dai, Y.; Liang, W.; Wu, Z. Treatment performance of integrated vertical-flow constructed wetland plots for domestic wastewater. *Ecol. Eng.* **2012**, *44*, 152–159. [CrossRef]
45. Abou-Elela, S.; Golinielli, G.; Abou-Taleb, E.; Hellal, M. Municipal wastewater treatment in horizontal and vertical flows constructed wetlands. *Ecol. Eng.* **2013**, *61*, 460–468. [CrossRef]
46. Martín, M.; Gargallo, S.; Hernández-Crespo, C.; Oliver, N. Phosphorus and nitrogen removal from tertiary treated urban wastewaters by a vertical flow constructed wetland. *Ecol. Eng.* **2013**, *61*, 34–42. [CrossRef]
47. Sylla, A. Domestic wastewater treatment using vertical flow constructed wetlands planted with Arundo donax, and the intermittent sand filters impact. *Ecohydrol. Hydrobiol.* **2020**, *20*, 48–58. [CrossRef]
48. Zhao, Y.; Liu, B.; Zhang, W.; Ouyang, Y.; An, S. Performance of pilot-scale vertical-flow constructed wetlands in responding to variation in influent C/N ratios of simulated urban sewage. *Bioresour. Technol.* **2010**, *101*, 1693–1700. [CrossRef]
49. Zheng, X.; Zhuang, L.; Zhang, J.; Li, X.; Zhao, Q.; Song, X.; Liao, J. Advanced oxygenation efficiency and purification of wastewater using a constant partially unsaturated scheme in column experiments simulating vertical subsurface flow constructed wetlands. *Sci. Total Environ.* **2020**, *703*, 135480. [CrossRef]
50. Saeed, T.; Paul, B.; Afrin, R.; Al-Muyeed, A.; Sun, G. Floating constructed wetland for the treatment of polluted river water: A pilot scale study on seasonal variation and shock load. *Chem. Eng. J.* **2016**, *287*, 62–73. [CrossRef]
51. Bassani, L.; Pelissari, C.; da Silva, A.; Sezerino, P. Feeding mode influence on treatment performance of unsaturated and partially saturated vertical flow constructed wetland. *Sci. Total Environ.* **2020**, *754*, 142400. [CrossRef]
52. Haddis, A.; Van der Bruggen, B.; Smets, I. Constructed wetlands as nature based solutions in removing organic pollutants from wastewater under irregular flow conditions in a tropical climate. *Ecohydrol. Hydrobiol.* **2020**, *20*, 38–47. [CrossRef]
53. Kaseva, M. Performance of a sub-surface flow constructed wetland in polishing pre-treated wastewater-a tropical case study. *Water Res.* **2004**, *38*, 681–687. [CrossRef]
54. Ma, Y.; Zhai, Y.; Zheng, X.; He, S.; Zhao, M. Rural domestic wastewater treatment in constructed ditch wetlands: Effects of influent flow ratio distribution. *J. Clean. Prod.* **2019**, *225*, 350–358. [CrossRef]

55. Benvenuti, T.; Hamerski, F.; Giacobbo, A.; Bernardes, A.; Zoppas-Ferreira, J.; Rodrigues, M. Constructed floating wetland for the treatment of domestic sewage: A real-scale study. *J. Environ. Chem. Eng.* **2018**, *6*, 5706–5711. [CrossRef]
56. Huang, Y.; Ortiz, L.; Aguirre, P.; García, J.; Mujeriego, R.; Bayona, J. Effect of design parameters in horizontal flow constructed wetland on the behaviour of volatile fatty acids and volatile alkylsulfides. *Chemosphere* **2005**, *59*, 769–777. [CrossRef] [PubMed]
57. Kraiem, K.; Kallali, H.; Wahab, M.; Fra-vazquez, A.; Mosquera-Corral, A.; Jedidi, N. Comparative study on pilots between annamox favored conditions in a partially saturated vertical flow constructed wetland and a hybrid system for rural wastewater treatment. *Sci. Total Environ.* **2019**, *670*, 644–653. [CrossRef]
58. Rozema, E.; Rozema, L.; Zheng, Y. A vertical flow constructed wetland for the treatment of winery process water and domestic sewage in Ontario, Canada: Six years of performance data. *Ecol. Eng.* **2016**, *86*, 262–268. [CrossRef]
59. Abdelhakeem, S.; Aboulroos, S.; Kamel, M. Performance of a vertical subsurface flow constructed wetland under different operational conditions. *J. Adv. Res.* **2016**, *7*, 803–814. [CrossRef]
60. Jácome, J.; Molina, J.; Suárez, J.; Mosqueira, G.; Torres, D. Performance of constructed wetland applied for domestic wastewater treatment: Case study at Boimorto (Galicia, Spain). *Ecol. Eng.* **2016**, *95*, 324–329. [CrossRef]
61. Stecher, M.; Weaver, R. Effects of umbrella palms and wastewater depth on wastewater treatment in a subsurface flow constructed wetland. *Environ. Technol.* **2003**, *24*, 471–478. [CrossRef]
62. Dan, T.; Chiem, N.; Brix, H. Treatment of high-strength wastewater in tropical constructed wetlands planted with Sesbania sesban: Horizontal subsurface flow versus vertical downflow. *Ecol. Eng.* **2011**, *37*, 711–720. [CrossRef]
63. Dąbrowski, W.; Karolinczak, B.; Gajewska, M.; Wojciechowska, E. Application of subsurface vertical flow constructed wetlands to reject water treatment in dairy wastewater treatment plant. *Environ. Technol.* **2017**, *38*, 175–182. [CrossRef]
64. Maina, C.; Mutua, B.; Oduor, S. Evaluating performance of vertical flow constructed wetland under various hydraulic loading rates in effluent polishing. *J. Water Sanit. Hyg. Dev.* **2011**, *1*, 144–151. [CrossRef]
65. Nivala, J.; Wallace, S.; Headley, T.; Kassa, K.; Brix, H.; van Afferden, M.; Müller, R. Oxygen transfer and consumption in subsurface flow treatment wetlands. *Ecol. Eng.* **2013**, *61*, 544–554. [CrossRef]
66. Fan, J.; Zhang, B.; Zhang, J.; Ngo, H.; Guo, W.; Liu, F.; Wu, H. Intermittent aeration strategy to enhance organics and nitrogen removal in subsurface flow constructed wetlands. *Bioresour. Technol.* **2013**, *141*, 117–122. [CrossRef]
67. Liu, L.; Zhao, X.; Zhao, N.; Shen, Z.; Wang, M.; Guo, Y.; Xu, Y. Effect of aeration modes and influent COD/N ratios on the nitrogen removal performance of vertical flow constructed wetland. *Ecol. Eng.* **2013**, *57*, 10–16. [CrossRef]
68. Hua, Y.; Peng, L.; Zhang, S.; Heal, K.V.; Zhao, J.; Zhu, D. Effects of plants and temperature on nitrogen removal and microbiology in pilot-scale horizontal subsurface flow constructed wetlands treating domestic wastewater. *Ecol. Eng.* **2017**, *108*, 70–77. [CrossRef]
69. Katayon, S.; Fiona, Z.; Megat Mohd Noor, M.; Abdul Halim, G.; Ahmad, J. Treatment of mild domestic wastewater using subsurface constructed wetlands in Malaysia. *Int. J. Environ. Stud.* **2008**, *65*, 87–102. [CrossRef]
70. Faulwetter, J.; Gagnon, V.; Sundberg, C.; Chazarenc, F.; Burr, M.; Brisson, J.; Stein, O. Microbial processes influencing performance of treatment wetlands: A review. *Ecol. Eng.* **2009**, *35*, 987–1004. [CrossRef]
71. Hu, Y.; Zhao, Y.; Rymszewicz, A. Robust biological nitrogen removal by creating multiple tides in a single bed tidal flow constructed wetland. *Sci. Total Environ.* **2014**, *470*, 1197–1204. [CrossRef]
72. Jiang, X.; Tian, Y.; Ji, X.; Lu, C.; Zhang, J. Influences of plant species and radial oxygen loss on nitrous oxide fluxes in constructed wetlands. *Ecol. Eng.* **2020**, *142*, 105644. [CrossRef]
73. Zhuang, L.; Yang, T.; Zhang, J.; Li, X. The configuration, purification effect and mechanism of intensified constructed wetland for wastewater treatment from the aspect of nitrogen removal: A review. *Bioresour. Technol.* **2019**, *293*, 122086. [CrossRef]
74. Ding, Y.; Song, X.; Wang, Y.; Yan, D. Effects of dissolved oxygen and influent COD/N ratios on nitrogen removal in horizontal subsurface flow constructed wetland. *Ecol. Eng.* **2012**, *46*, 107–111. [CrossRef]
75. Adrados, B.; Sánchez, O.; Arias, C.; Becares, E.; Garrido, L.; Mas, J.; Morató, J. Microbial communities from different types of natural wastewater treatment systems: Vertical and horizontal flow constructed wetlands and biofilters. *Water Res.* **2014**, *55*, 304–312. [CrossRef]
76. Pelissari, C.; Guivernau, M.; Viñas, M.; García, J.; Velasco-Galilea, M.; Souza, S.; Ávila, C. Effects of partially saturated conditions on the metabolically active microbiome and on nitrogen removal in vertical subsurface flow constructed wetlands. *Water Res.* **2018**, *141*, 185–195. [CrossRef]
77. Sun, G.; Zhu, Y.; Saeed, T.; Zhang, G.; Lu, X. Nitrogen removal and microbial community profiles in six wetland columns receiving high ammonia load. *Chem. Eng. J.* **2012**, *203*, 326–332. [CrossRef]
78. Zhu, H.; Yan, B.; Xu, Y.; Guan, J.; Liu, S. Removal of nitrogen and COD in horizontal subsurface flow constructed wetlands under different influent C/N ratios. *Ecol. Eng.* **2014**, *63*, 58–63. [CrossRef]
79. Ávila, C.; Garfí, M.; García, J. Three-stage hybrid constructed wetland system for wastewater treatment and reuse in warm climate regions. *Ecol. Eng.* **2013**, *61*, 43–49. [CrossRef]
80. Yi, X.H.; Jing, D.D.; Wan, J.; Ma, Y.; Wang, Y. Temporal and spatial variations of contaminant removal, enzyme activities, and microbial community structure in a pilot horizontal subsurface flow constructed wetland purifying industrial runoff. *Environ. Sci. Pollut. Res.* **2016**, *23*, 8565–8576. [CrossRef]

*environments*

MDPI

*Article*

# Stabilisation of Ozone in Water for Microbial Disinfection

**Emmanuel I. Epelle [1,2], Andrew Macfarlane [2], Michael Cusack [2], Anthony Burns [2], Ngozi Amaeze [3], Kenny Richardson [3], William Mackay [3], Mostafa E. Rateb [1] and Mohammed Yaseen [1,*]**

1 School of Computing, Engineering & Physical Sciences, University of the West of Scotland, Paisley PA1 2BE, UK; emmanuel.epelle@uws.ac.uk (E.I.E.); mostafa.rateb@uws.ac.uk (M.E.R.)

2 ACS Clothing, 6 Dovecote Road, Centralpark ML1 4GP, UK; andrewmacfarlane@acsclothing.co.uk (A.M.); michaelcusack@acsclothing.co.uk (M.C.); anthonyburns@acsclothing.co.uk (A.B.)

3 School of Health & Life Sciences, University of the West of Scotland, Paisley PA1 2BE, UK; ngozi.amaeze@uws.ac.uk (N.A.); kenny.richardson@uws.ac.uk (K.R.); w.mackay@uws.ac.uk (W.M.)

* Correspondence: mohammed.yaseen@uws.ac.uk

**Abstract:** In current times of increasing global decontamination concerns, sustainable and environmentally-friendly technologies that possess rapid and effective disinfection capabilities are necessary for public health and safety. In this study, we evaluate the potential of ozone-based technology to reveal its immense potential in disinfection applications. Ozonated water generated by an electrolytic method was utilised to quantify ozone retention as a function of mineralogical composition for microbial decontamination. The impacts of temperature and detergent concentration on ozone concentration are critically analysed, as well as ozone's decomposition and stain removal characteristics. In addition, fabric swatches inoculated with known concentrations of environmental microbes (*Escherichia coli*, *Staphylococcus aureus*, *Candida albicans*, and *Aspergillus fumigatus*) are washed with ozonated water to ascertain the impact of wash duration on bacterial removal efficiency. The results show significant improvement in the stability and retention potential of ozone in mineral water at low temperature and in the presence of a detergent. The experiments demonstrate first-order decomposition kinetics of ozone in aqueous formulations. The disinfection potency of ozone is also substantiated by a significant removal of microbiota on the fabric utilised (up to 7 log reduction for the bacteria analysed), thus making it effective for sterilisation applications. This also reduces the need for toxic chemicals or chemicals with toxic by-products (e.g., chlorine) for large-scale decontamination operations in various industries.

**Keywords:** ozone concentration; microbial disinfection; water mineralogy

**Citation:** Epelle, E.I.; Macfarlane, A.; Cusack, M.; Burns, A.; Amaeze, N.; Richardson, K.; Mackay, W.; Rateb, M.E.; Yaseen, M. Stabilisation of Ozone in Water for Microbial Disinfection. *Environments* **2022**, *9*, 45. https://doi.org/10.3390/environments9040045

Academic Editors: Dibyendu Sarkar, Rupali Datta, Prafulla Kumar Sahoo and Mohammad Mahmudur Rahman

Received: 26 February 2022
Accepted: 29 March 2022
Published: 1 April 2022



## 1. Introduction

Most industrial deployments of ozone are often connected to its antimicrobial properties and its ability to degenerate organic compounds via oxidation. Its rapid reactivity and non-selectivity to different classes of microorganisms are particularly advantageous for a diverse range of applications [1–7]. Compared to ozone, which does not form any long-term toxic byproducts, the use of chlorine (a widely used disinfectant) is associated with the production of carcinogenic byproducts, including trihalomethanes and haloacetic acids [8]. Furthermore, while ozone directly oxidises the constituents (such as proteins and amino acids) in cell walls and membranes of spores, chlorine is thought to initially diffuse into their cell protoplasm, through the cell wall before inactivating the enzymes [9–11], thus implying that a lower concentration and shorter contact time is required for their inactivation with ozone, compared to chlorine. This is mainly desirable in large-scale applications, where rapid processing and worker safety are primary concerns. The use of ozone in garment processing (particularly laundry systems) has received considerable attention over the past decade, given its potential for energy savings compared to conventional thermal laundry systems. Ozonated water production for laundry purposes is mainly generated by injecting ozone gas into water at a controlled pressure, followed by mechanical mixing/atomisation of the gas to ensure homogeneity. This is particularly important because of ozone's low solubility in water—the solubility of ozone in aqueous solutions is 14 mmol/L at 20 °C and it is documented to be more soluble in hydrophobic organic solvents [12,13].

Rice et al. [14] describe four main techniques of applying this method of ozonated water generation in different laundry establishments. Ozonated water may also be produced by direct electrolysis, for instance, via a polymer electrolyte membrane. This process involves the electrolytic decomposition of water at the anode of the cell to produce ozone, which is then mixed with water [13]. Compared to the injection method, which requires separate ozone-generating and mixing equipment, the electrolysis method is more compact, as shown in Figure 1. Several studies which apply a variety of ozone generation methods have examined the inactivation kinetics of bacteria in ozone; dissolved molecular ozone has been found to be primarily responsible for bacterial inactivation [15–17]. The kinetics of decolourisation of reactive-dyed fabrics and the treatment of dyed textile effluents has also been extensively studied [18,19]. In addition, the cytocompatibility and biocidal action of ozonized water (at 4 °C and pH of 5) was recently demonstrated in the work of dos Santos et al. [20].

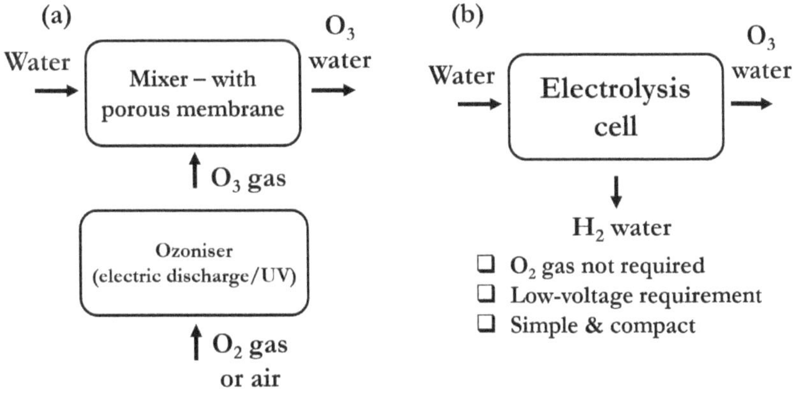


**Figure 1.** Two methods of producing ozonated water using a gas phase ozoniser with water (**a**) and water electrolysis (**b**) (adapted with permission from Okada and Naya [13]).

Despite the many documented successful applications of ozone washing mentioned in Rice et al. [14], Körlü [7], Neral [21], and the above-cited studies, a wholistic textile-related study that collectively explores the influence of water mineralogy on ozone retention, ozone's thermal degradation kinetics, its stain-dependent (synthetic/solvent-based and natural) removal efficiency, and bacterial inactivation properties is scarce. No significant collective analysis of these parameters using fabric substrates has been carried out in the literature. In this paper, we address these gaps by providing insights into these key attributes. To achieve this, we apply the electrolytic method for ozonated water generation and quantify the ozone retention and disinfection characteristics under different processing conditions, necessary for improved large-scale laundry operations. Furthermore, it remains unclear if the use of ozonated water with a detergent in the washing phase of laundry operations provides additional benefit compared to its usage during the rinsing phase alone. The analysis of ozone stabilisation as a function of detergent concentrations presented in this work allows recommendations to be made in this regard. In addition, the disinfection efficiency of an ozonated wash is comparatively analysed with an ordinary wash using water only.

## 2. Materials and Methods

Figure 2 illustrates the setup for the experiments performed herein. An Electrolysis Oxygen Radical Generator (EORG™—Novus Clean Tech Ltd., Inverurie, UK) was employed for the generation of ozone using available oxygen molecules in water. The ozonated aqueous solutions were gently but continuously stirred during the ozone generation and decomposition phases to ensure mixture homogeneity. Rotation and temperature control were achieved using a magnetic stirrer equipped with a temperature probe and a controller; this allowed the specification of the desired temperature setpoint. Given ozone's partial solubility in water, the tests were carried out in a fume cupboard to limit ozone exposure. The Palintest® procedure (subsequently described in Section 2.2) was utilised over others for ozone concentration measurements.

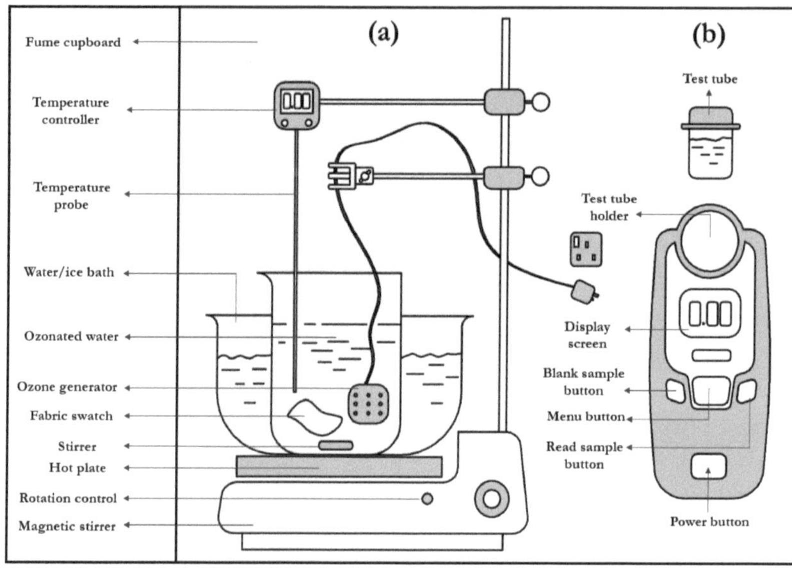


**Figure 2.** Experimental setup for (**a**) lab-scale ozonated washing of fabric substrates, showing (**b**) the Palintest kit employed for ozone concentration measurements.

### 2.1. Ozone Concentration Measurement (Palintest®)

Palintest is a colorimetric ozone measurement method that depends on the intensity of colours produced by the Palintest reagent (*N,N*-diethyl-p-phenylenediamine, DPD No. 4 tablet) when using Palintest photometers. The test procedure quantifies the amount of coloured light (530 nm wavelength) absorbed by a sample in reference to an untreated sample (blank) and displays the corresponding ozone concentration using calibration data programmed into the instrument. The water samples tested were correspondingly used as the blank during measurements. The Palintest instrument applied in this study has a measuring range of 0.01–3 ppm and utilises a wavelength of 530 nm for ozone concentration measurement. Thus, where higher concentrations beyond the measuring limit were expected, 1 in 10 dilutions was performed, and the actual concentration was reconstituted using the dilution factor. When working with the detergent and mineral water mixtures in this study, the dilution was carried out using pure mineral water (i.e., 1 mL of ozonated detergent–mineral water mixture in 9 mL of mineral water). Palintest measurements (within the measuring range of the instrument) performed on samples without dilution, with 1 in 2 and with 1 in 10 dilutions, all gave similar readings within an error range of 3%.

It is worth mentioning that ozone concentration measurement is challenging. The reaction of organic compounds with ozone carbonyl groups may produce carboxylic acids, carbonyl groups, and peroxides; these, together with free chlorine or bromine in water, may interfere with ozone concentration measurements. Most of these effects can be mitigated using glycine tablets as carried out in this work [22]. Foam formation is also particularly challenging when aqueous solutions containing detergents are ozonated. These interferences usually exist, irrespective of the measurement technique employed. Firstly, bubbles formed when detergents are present tend to scatter UV light, and byproducts of ozonation may absorb UV radiation in the same region as ozone [10]. Titrimetric methods show significant sensitivity to pH, buffer concentration, and temperature; moreover, the fast decomposition rate of ozone also makes it difficult for near-real-time measurements to be obtained for kinetic studies.

Although electrometric methods are suitable for continuous data collection using ozone-specific electrodes, they suffer from relatively slow response time, pronounced sensitivity to stirring, and the need for frequent calibration. Furthermore, foam formation may affect the sensitivity of membrane electrodes. Other measurement techniques that utilise thermal conductivity, gas-phase titration, and isothermal pressure changes are not commonly used [7]. The indigo method, which utilises sulfonated indigo compounds, is also a common ozone-measuring technique but is usually adversely affected by the age and purity of the indigo-based compounds as well as the presence of the $Mn^{+2}$ ion, which is capable of destabilising indigo trisulfonate [7,23]. Thus, the Palintest procedure (a widely applied and readily available ozone measurement procedure) sufficed for the measurements performed in this study. However, to further mitigate the effect of possible interferences, three independent runs were carried out for each measurement presented (with standard deviations reported throughout); gentle stirring and low detergent concentrations were also employed to minimise micelle formation. The wide density difference between any foam layer formed and the main liquid phase caused a clear separation, thus allowing for foam-free sample collection in the determination of ozone concentration during the Palintest procedure. All glassware and plasticware were sterilised before use to mitigate interferences in ozone concentration measurements from impurities (via intractable changes in ozone demand).

### 2.2. Water Characterisation

As it was desired to examine the ozone retention capabilities of different water samples, with varying mineralogical and ionic compositions, inductive coupled plasma (ICP) (Perkin Elmer (Waltham, MA, USA) Avio500 ICP-OES) was employed for the determination of metal compositions of the water samples, whereas Dionex ICS-1100 ion chromatography (Thermo Scientific, Waltham, MA, USA) was utilised for anion concentration measurements. The pH and conductivity of the water samples before ozonation were also measured and recorded using meters equipped with probes. These measurements were taken for all water samples utilised, including those containing a known mass of the applied detergent.

A commercial powder-based "*Daz*" detergent formulation (5–15% anionic surfactants, <5% non-ionic surfactants, phosphonates, polycarboxylates, zeolites, enzymes, optical brighteners, perfumes, citronellol, hexyl cinnamal, limonene) was applied in this study. Rather than a pure surfactant, this readily available detergent formulation was of particular interest to the authors, because of its applicability and prevalence in commercial laundry establishments in the UK. By applying a commercial detergent formulation, the adaptability of the obtained results in this work to industrial settings is demonstrated.

### 2.3. Preparation of Microorganisms

The preparation process of the bacteria (*Escherichia coli* NTCC1290, *Staphylococcus aureus* ATCC25923) applied in this study began by transferring a representative colony into 10 mL of Luria broth (Sigma Aldrich, St. Louis, MO, USA), and incubating in a shaker at 37 °C and 150 rpm for 24 h. A total of 1 mL of the bacteria suspension was subsequently transferred into a 1.5 mL microcentrifuge tube and centrifuged for 5 min at 10,000 rpm. This was followed by the washing of the harvested cells with ozone demand-free phosphate buffer saline (PBS) solution and the adjustment of the suspensions' absorbance (at 570 nm) to an optical density (OD) of 0.2 (±0.02).

The preparation of the fungal inoculum (*Candida albicans* and *Aspergillus fumigatus*) involved growing them on ISP2 agar plates (International *Streptomyces* Project-2 Medium) for 48 h. A 1 cm × 1 cm section of the agar piece of each fungal culture was inoculated into 100 mL of ISP2 broth to prepare seed cultures of each species by shaking them for 48 h. This was followed by an adjustment of the optical density of the two-day-old seed cultures to an optical density of 0.2 (±0.02). Required volumes of the 0.2 OD bacterial and fungal suspensions were applied onto the sterilized fabric swatches aseptically, which were subsequently ozonated.

### 2.4. Disinfection

Square-shaped fabric swatches (6.5 cm by 6.5 cm) were first sterilised in 70 % ($v/v$) ethanol, autoclaved, and then air-dried. Using an Eppendorf pipette, the swatches were inoculated at their centres, on a sterile non-absorbent surface with 80 µL of prepared *E. coli* bacterial suspension (at an optical density of 0.2 at 570 nm, which is approximately $10^9$ CFU/mL). A total of 80 µL of the suspension applied was sufficient for full coverage of the swatches' surface area. Using sterile forceps, the inoculated swatches were transferred into 100 mL of ozonated water, and washed for varying durations, using the magnetic stirring system shown in Figure 2. After washing, the swatches were transferred onto a sterile surface, and dipslides (Dip-Slides, Glenrothes, UK) were placed onto the fabric surface with gentle pressure applied for 10 s. The slides were subsequently incubated at 37 °C for 24 and 48 h (bacteria and fungus, respectively). To also evaluate residual bacteria in the water used for washing, the agar regions of the slides were immersed in it for 10 s, after which they were also incubated. The slides contain a red spot dye—2,3,5-triphenyltetrazolium chloride (TTC), which allows the enumeration of microbial colonies in solid structure media [24]. The *Colour Thresholder* and *Image Region Analyser* toolboxes of MATLAB (R2020b) were applied to process the images of the dipslides and for the enumeration of the number of bacteria colonies. Section 3 shows pictures of the slides, tested on contaminated swatches before and after ozonated washing. The number of CFUs per $cm^2$ of the dipslide agar area (5 cm by 2 cm → 10 $cm^2$) was calculated according to Equation (1), where the corrected pick-up rate (CPUR) = 2. This correction factor is applied because approximately 50% of bacteria is picked up by the slide from a surface [25].

$$\text{CFUs/cm}^2 = \frac{\text{Number of colonies} \times \text{CPUR}}{\text{Agar area}} \quad (1)$$

The Miles and Misra method was also employed for the enumeration of bacteria count in the water after washing the fabric swatches. A sequential 1:10 dilution of bacterial suspension in water was made in 6 places. A total of 20 µL of each diluent was dropped in 3 places on sterile nutrient agar plates. The plates were allowed to dry for about 10 min, after which they were incubated at 37 °C overnight. An average of counts obtained from the three droplets was taken and expressed as $\text{Log}_{10}$CFU/mL.

## 3. Results and Discussion

### 3.1. Impact of Water Composition on Ozone Retention

Different water samples were explored for their ozone-retention capabilities in this study. Tap water collected from ACS Clothing, a garment processing company in Motherwell, Scotland, was utilised in this study and is referred to as *ACS water*. Mineral water supplied by *AquAid* water coolers (Glasgow, UK) was also applied for generating ozonated water. Ultra-high-quality/purity (UHQ) and distilled water were also used. Table 1 shows the measured metallic and anionic distributions of ACS and mineral water, respectively. As expected, mineral water shows significantly higher ionic concentrations compared to ACS water, with calcium being the most dominant metal ion present.

**Table 1.** Ionic composition of water samples as obtained via ICP analysis.

| Water Type | Ca (mg/L) | K (mg/L) | Mg (mg/L) | Na (mg/L) | $Cl^-$ (mg/L) | $SO_4^{2-}$ (mg/L) | $NO_3^-$ (mg/L) |
|---|---|---|---|---|---|---|---|
| ACS water | 6.801 | 0.415 | 0.754 | 4.827 | 6.44 | 11.24 | 0.00 |
| Mineral water | 46.945 | 0.969 | 17.141 | 12.466 | 17.23 | 20.66 | 13.95 |

This concentration distribution translates to the pH and conductivity values (Table 2), which yielded a better understanding of the water's ozone retention capability. As observed in Table 2, the addition of salts and detergents inevitably increases the pH and conductivity of the solution. This presents some desirable effects in terms of ozone retention, as shown in Figure 3. Acero and von Gunten [26] and Eriksson [10] have reported bicarbonates to have both inhibition and promoting characteristics on ozone decomposition. According to Acero and von Gunten [26], carbonate radicals formed from the reaction of carbonates and OH radicals serve as chain carriers for increased ozone decomposition, whereas Eriksson [10] pointed out that, in the presence of other ions in solution, carbonate ions may prevent an increase in alkalinity, thus retarding ozone decomposition. Thus, it was of particular interest to investigate the existence of possible concentration thresholds (for our application) beyond which either of these positive or negative effects is observed. According to Table 1, sodium is the second-highest metal ion present in ACS water and was the most desirable metal to focus on as far as the metallic concentration and conductivity improvements are concerned. Carbonates of sodium also represent one of the most widely applied and studied compounds in terms of their interactions with water.

**Table 2.** The pH and conductivity measurements of water samples.

| Water Type/Composition | pH | Conductivity, σ (mS) |
|---|---|---|
| ACS water | 7.16 | 0.10 |
| Mineral water | 8.07 | 0.46 |
| Distilled water | 7.13 | 0.00 |
| UHQ water | 7.30 | 0.00 |
| 0.25 g detergent in 900 mL mineral water | 8.90 | 0.78 |
| 0.50 g detergent in 900 mL mineral water | 9.16 | 1.13 |
| 0.75 g detergent in 900 mL mineral water | 9.37 | 1.39 |
| 0.1 g $Na_2CO_3$ in 900 mL mineral water | 9.56 | 0.59 |
| 2.0 g $Na_2CO_3$ in 900 mL mineral water | 10.89 | 4.14 |
| 0.1 g $Na_2CO_3$ in 900 mL UHQ water | 10.71 | 0.37 |
| 2.0 g $Na_2CO_3$ in 900 mL UHQ water | 11.27 | 3.95 |
| 0.1 g $Na_2CO_3$ in 900 mL ACS water | 10.54 | 0.34 |
| 2.0 g $Na_2CO_3$ in 900 mL ACS water | 11.27 | 4.01 |

The relatively low toxicity of sodium bicarbonate compared to potassium carbonate and solubility relative to carbonates of calcium and magnesium also made it the preferred candidate for further exploration [27,28]. As observed in Figure 3, the application of 0.1 g of sodium bicarbonate improves the ozone concentration in solution by more than 200% (0.10 to 0.32 ppm). However, increasing the amount of $Na_2CO_3$ dissolved did not correspondingly increase the ozone concentration in the solution. It can be observed that the dissolution of 2 g of $Na_2CO_3$ (964.4 mg/L of Na) has an adverse effect (63% reduction) on ACS water, despite the 36% increase observed with 0.1 g $Na_2CO_3$ solution (48.2 mg/L of Na). The enhancement of ozone concentration via the addition of carbonate compounds may be attributed to its scavenging effect (its consumption of $OH^-$ radicals, responsible for ozone decomposition to yield products that do not further react with ozone in solution) [29]. The results also demonstrate the existence of a critical ionic concentration, beyond which a negative impact on ozone retention ensues.

Compared to mineral water (the best performing sample), water samples void of ions (distilled and UHQ) do not facilitate ozone generation and retention. The lowest ozone concentrations were observed in the distilled and UHQ water samples. This is inevitably due to the decreased potential for electrolytic decomposition of water into adsorbed OH* and H* radicals required for ozone production [30]. For significant amounts of ozone to form in pure water, prolonged ozonation periods or a very large electric potential (>1.511 V) is required to initiate the autoionization of water. We also hypothesize that an increased concentration of the sodium carbonate salts (>>0.1 g/900 mL) in solution yields a high concentration of sodium ions, which may cause increased displacement of the ozone molecules in solution; the effect of this displacement increases with the presence of other ions in solution (UHQ vs. ACS water—Figure 3). Despite the ozone concentration increments (0.1 g/900 mL), it is worth mentioning that retention and stability of ozone may be negatively affected by weak alkalinity (pH 8–11) [10,31], which some of the water samples with $Na_2CO_3$ display (Table 2). Greenwood and Earnshaw [32] stated that this effect may be reversed in strongly alkaline solutions, via the formation of ozonide. Nonetheless, the results presented herein demonstrate that an added dimension to this complex relationship exists—that of water conductivity.

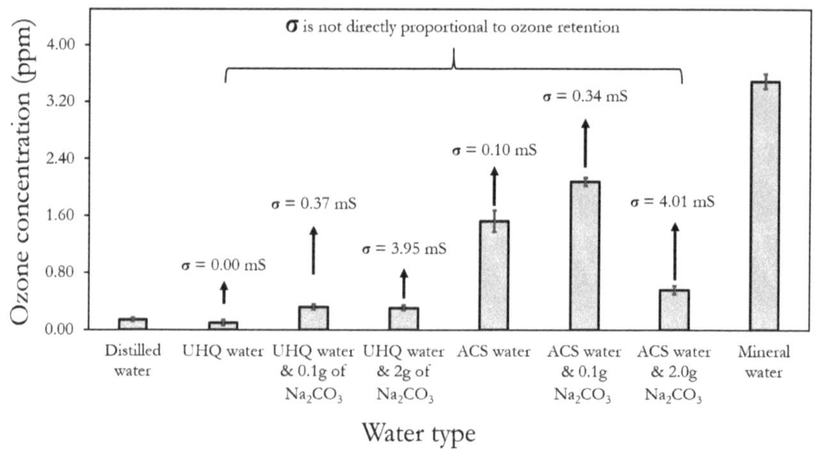


**Figure 3.** Ozone retention of different water samples, showing the influence of $Na_2CO_3$ concentration on ozone concentration. Experimental conditions: 900 mL of water ozonated for 11 min at 18 °C, where $\sigma$ is the conductivity.

### 3.2. *Effect of Detergent Concentration on Ozone Retention*

Figure 4 shows the impact of the applied detergent on the obtained ozone concentration. Compared to pure mineral water, it can be observed that ozone concentration increases significantly with the addition of the detergent (up to three times its original value when 0.75 g of detergent is used). A possible reason for this increased concentration with detergent usage is the formation of micellar enclosures, and nanobubbles that trap ozone molecules (given their co-hydrophobicity) during ozonation. Furthermore, IR spectroscopy results from Bulanin et al. [33,34] and Ward et al. [12] have shown that ozone in aqueous solutions adsorbs onto the surface of zeolites (which are present in the detergent applied herein), thus stabilizing ozone concentration. Eriksson [10] studied the decomposition of ozone in aqueous solutions containing 0.1mM Dodecyl-β-D-Maltoside, DDM (a neutral/non-ionic surfactant) and observed nearly double the ozone concentration in comparison to a pure aqueous ozone solution after 18 min of decomposition. Sodium dodecyl sulphate, SDS (an anionic surfactant), and dodecyl tri-methyl ammonium acetate, DAA (a cationic surfactant), were also tested, with similar results obtained.

To confirm the absence of any colour interference by the DPD No. 4 tablets (used in our measurements) on the detergent solution, the tablets were dissolved in 10 mL of the detergent–mineral water mixtures (without ozone), as shown in Figure 4. No colour change was observed. Despite the presence of enzymes in the detergent, the 11 min ozonation time was sufficient to neutralise these organic compounds while allowing further ozone production and retention. As we demonstrate in subsequent sections of this paper, a duration of 5 min yielded 100% removal of *E. coli* in water.

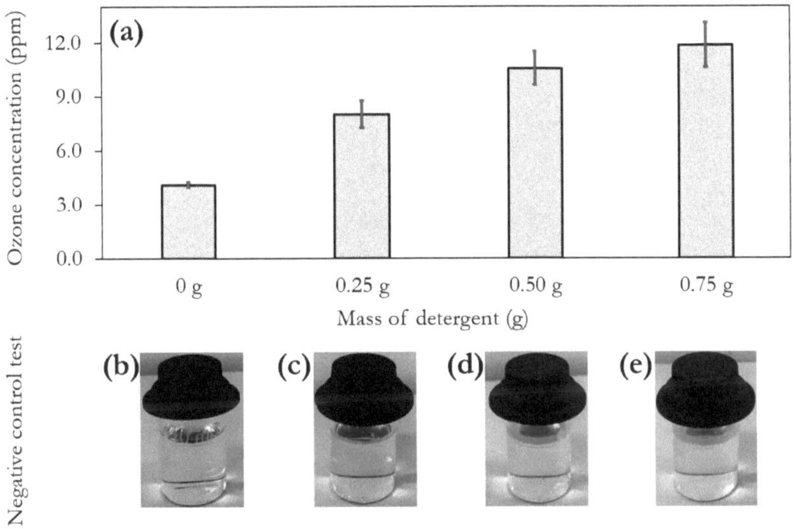


**Figure 4.** (**a**) The effect of detergent concentration on ozone concentration in water. Experimental conditions: 900 mL of mineral water ozonated for 11 min at 18 °C. Panels (**b–e**) show that no colour change is observed on dissolving the Palintest (DPD No. 4 tablets) in the water–detergent mixtures. Error bars represent the standard deviation of three separate runs.

### 3.3. Kinetics of Ozone Generation and Decomposition in Different Water Types

The generation rates of ozone are shown in Figure 5a,c,e for ACS water, mineral water, and the detergent–mineral water mixture, respectively. Compared to ACS water, which showed potential to harbour more ozone with time (a steadily rising profile), the generation profiles of mineral water and the detergent–water mixture seemed to be approaching a saturation point. While it is important to retain produced ozone and increase the contact time, the time taken to attain the desired ozone concentration (in the interest of adequate disinfection) cannot be overlooked.

Thus, conventional 10 min washing or rinsing cycles may be insufficient to yield the desired disinfection levels when ordinary tap water (with low mineral content) is used, especially if the gas is bubbled through water rather than electrolytically produced. This also necessitates the modification of the water's properties, for example, through the addition of salts, alkyl compounds, or acetic acid, which are capable of boosting ozone retention. Detergent concentrations (>>critical micelle concentration—CMC) utilised in commercial laundry processing industries are significantly higher than those explored in this work; this may adversely affect ozone's availability in solution for disinfection via over-entrapment of the produced ozone or significantly increased biological material (enzymes) from the detergent, which reduce the available ozone in solution for disinfection; non-biological detergents may be utilised to mitigate the inactivation effects of ozone on the enzymes. Thus, there is a need to investigate the specific critical concentrations of required additives (e.g., detergents and carbonate salts) for the rapid boosting of ozone's concentration, improving its stability and availability, using the water samples of interest. This is an important consideration when planning large-scale microbial decontamination processes.

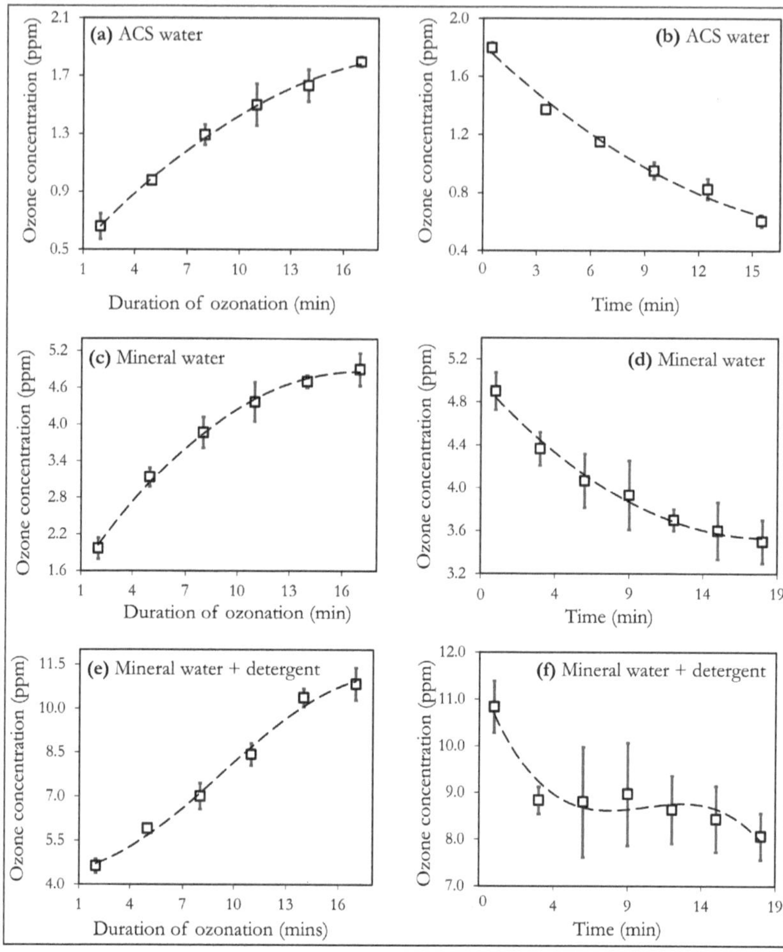


**Figure 5.** Ozone generation and decomposition curves using (**a**,**b**) ACS water; (**c**,**d**) mineral water; and (**e**,**f**) mineral water + detergent. Experimental conditions: 900 mL of water ozonated at 18 °C; where a detergent was applied, 0.25 g was utilised. Error bars represent the standard deviation of three separate runs. The lines through the mean points are drawn to help the reader follow the profile of the data.

Two popular decomposition mechanisms of aqueous ozone have been proposed by Hoigné–Staehelin–Bader (HSB) and Gordon–Tomiyasu–Fukutomi (GTF) [29,31,35]. Despite the differences in their reaction steps, they propose that the presence of OH* radicals initiates the reaction, subsequently forming superoxide ($O_2^-$) and hydroxyl ($HO^-$) ions through a series of steps. These are the main ozone-consumption radicals, with rates of reaction between $1.6 \times 10^9$ and $4.2 \times 10^9$ $M^{-1}$ $s^{-1}$, according to the HSB and GTF mechanisms. Thus, the removal of these ions retards the decomposition rate. In addition, the lower the concentration of the main initiators—the $OH^-$ radical, the slower the decomposition reaction, thus suggesting that alkaline solutions facilitate ozone decomposition. However, as demonstrated in Heidt and Landi [36] and Greenwood and Earnshaw [32], strongly alkaline solutions show the opposite behaviour.

The rates of ozone decomposition shown in Figure 5b,d,f are generally observed to be more rapid in the first 4 min, after which the rate reduces. Based on the HSB and

GTF decomposition mechanisms, the faster decline in the first 4 min may be attributable to the presence of $OH^-$ radicals, given the weak alkalinity of the solution (Table 2). As demonstrated in the work of Eriksson [10], this rapid drop in the first few minutes is not present due to the acidity of the aqueous solutions employed (pH = 5). Figure 5f illustrates that the decomposition profile is followed by an abrupt rise in ozone concentration in the 9th minute. Compounds containing alkyl groups (into which the surfactants in the detergents fall) are known inhibitors of ozone decomposition (radical scavengers). They do so via the consumption of $HO^-$ radicals, which are primarily responsible for rapid ozone decomposition [29]. Some of these compounds are also capable of reforming ozone in solution and, thus, are good ozone stabilisers [10]. This is the likely reason for the profile observed in Figure 5f—the radical scavenging effect of detergent's alkyl compounds. Furthermore, the generated foam (via the action of the detergent) forms a dense thick top layer, leaving a clearer solution beneath. This layer may have also contributed to the entrapment of ozone in solution, thus reducing its removal rate.

The ozone decomposition profiles of ACS and mineral water both strongly follow first-order kinetics, as shown in Figure 6a,b, with mineral water showing a significantly longer half-life compared to ACS water. This indicates the increased ozone retention, brought about by an increase in water's mineral content. With the detergent–mineral water mixture (Figure 6c), the coefficient of correlation (the $R^2$ value) is affected by the presence of surfactants and possible ozone regeneration; the longer half-life can be attributed to the stabilising effects of the detergent on ozone concentration, and the thick foam layer on top of the solution. The presence of nanobubbles in solution also has a stabilising effect.

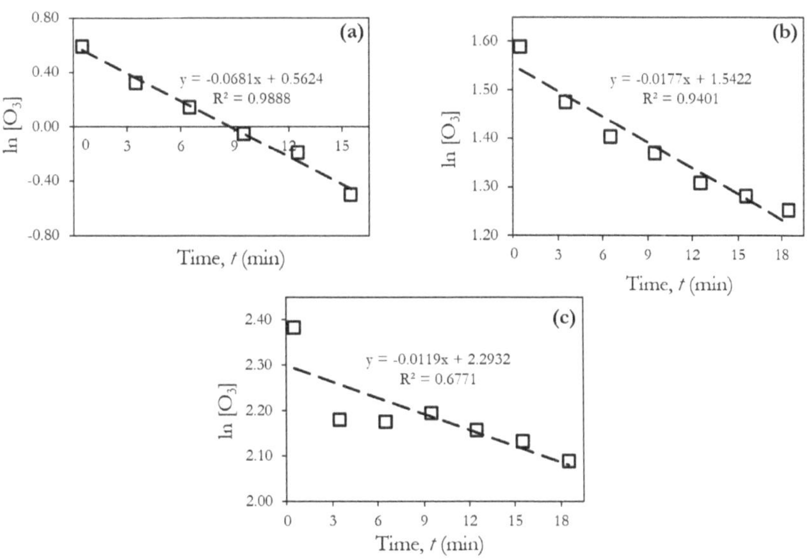


**Figure 6.** Firstorder kinetics of ozone decomposition in (**a**) ACS water with a half-life of 10 min (**b**), mineral water with a halflife of 39 min, and (**c**) mineral water and detergent with a halflife of 58 min. Experimental conditions: 900 mL of water ozonated at 18 °C.

### 3.4. *Effect of Temperature on Ozone Retention*

Since mineral water was most effective in ozone retention, it was used to analyse the effect of temperature. As observed in Figure 7, decreasing the temperature of the solution aids the production of ozone in mineral water. A possible explanation for ozone's decomposition resistance at low temperatures is the resonance stability that the molecule possesses [37]. Furthermore, its reactivity increases significantly at high temperatures, causing a reduction in its stability. Thus, the benefits of ozonated washing (in terms of improved stability) are better experienced at low temperatures. The electrical costs involved with the large-scale production and maintenance of ozone at a desired concentration may increase if the process temperature is unfavourably high. Besides ozone's disinfection capabilities, one of the main incentives for its adoption in the textile processing industry is the reduced energy cost expended in raising the temperature of large volumes of water by approximately 30–40 °C during conventional washing cycles; high-temperature washes are conventionally applied to destroy bacteria in garments. Low-temperature wash cycles can be performed for cleaning purposes only, while ozone may be applied separately in the rinsing phase for garment disinfection. Additionally, ozone may be utilised in the low-temperature wash cycle along with a detergent for improved benefits (higher concentrations and better stability).

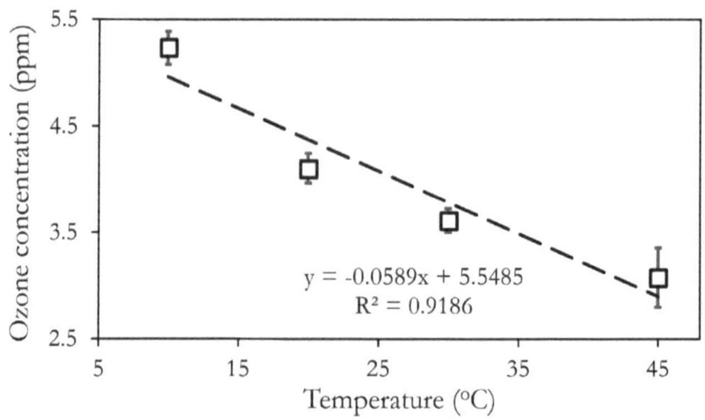


**Figure 7.** Effect of temperature on ozone concentration. Experimental conditions: 900 mL of mineral water ozonated at 18 °C for 11 min. The decomposition rate, on increasing temperature, is approximately 0.06 ppm/°C. Error bars represent the standard deviation of three separate runs.

### *3.5. Stain Removal Efficacy of Ozonated Washing*

To evaluate the stain removal efficiency of ozone, two stain types were investigated; firstly, a mixture of blue food colouring (containing glycerine, water, spirulina concentrate, and citric acid), sunflower oil, and turmeric powder (BCSOT) was applied to a fabric swatch. The second stain utilised was a solvent-based, red-coloured temporary board marker ink (RTBM). The fabric swatches were stained and stored for a duration of 48 h to allow sufficient bonding to the fabric fibres. Figure 8 shows the stained swatch areas before washing was carried out, and the extent of stain removal after washing and drying the fabric.

As observed with the BCSOT stain (Figure 8a), there are three layers; the outer/bigger blue food colouring, followed by the oil stain, and then the central core yellow stain caused by the turmeric powder. Washing with water alone removes most of the food colouring but does not break down the oil stain marked by the dotted line. With the *detergent only*, *ozone only*, and *ozone + detergent* scenarios (Figure 8b–e), the food colouring is completely removed. The oil stain, however, proved more difficult for the *ozone only* wash, compared to the *detergent only* and *ozone + detergent* wash. This is evidently attributable to the action of the surfactants in the detergent. More importantly, the core yellow turmeric stain on the fabric proved difficult for all wash types, except with the *ozone + detergent wash*, where a bleaching effect can be faintly seen qualitatively (Figure 8e) and more evidently (Figure 8k). The RTBM stain on the other hand also proved tough for ozone, and the *detergent only* scenario was the best-performing (83.4% removal), as shown in Figure 8l. The increased colour removal percentages of the BSCOT scenarios in comparison to the RTBM (Figure 8k,l) are attributable to the disappearance of the blue colouring stain in virtually all wash scenarios.

These results demonstrate the combined potential of ozone and a detergent to enhance stain removal on fabric swatches; however, this efficiency is stain-composition-dependent.

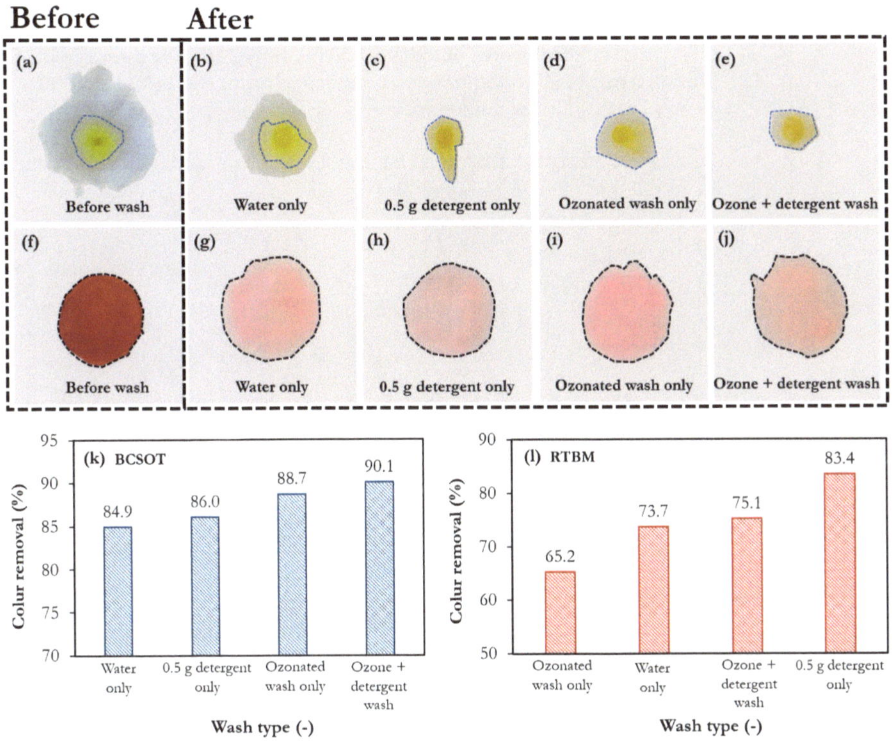


**Figure 8.** Processed images of stained fabric regions pre- and post-ozonation. Experimental conditions: washing with 900 mL of mineral water for 11 min; where ozone was used, ozonation at 18 °C for 11 min was performed. The colour intensity analysis on the tested fabric swatches was carried out using MATLAB's (R2020b) image processing toolbox (the *Colour Thresholder*). BCSOT represents a mixture of blue food colouring, sunflower oil, and turmeric powder; whereas, RTBM represents solvent-based, red-coloured temporary board marker ink.

### *3.6. Analysis of Bacteria Concentration*

To evaluate the attainable disinfection level with ozone, it was important to first evaluate the performance of the dipslides used in this study. The fabric samples to be disinfected were originally inoculated with different volumes of bacterial suspension ($10^9$ CFUs/mL of *E. coli*), as shown in Figure 9, and the dipslides were applied to obtain the bacterial count. The bacteria population recovered from the swatch via the dipslides was found to increase with increasing inoculum volumes of the same concentration. A similar observation was made when the respective volumes of bacterial suspension (10, 25, 40, 65, 80, and 100 µL), were individually dissolved in 100 mL of sterile distilled water. For *E. coli*, it can be observed that the contamination levels observed upon applying 100 mL of the suspension are similar (Figure 9d). This observation was similar for all organisms utilised in this work.

By applying the dipslides into the contaminated microbial solutions, the number of CFUs/mL of solution was also obtained. Besides the application of the dipslides, the Miles and Misra method [38] was also employed to enumerate the residual bacteria in the water used in washing the fabric, and both methods yielded consistent results. To enumerate the fungal removal/reduction efficiency of ozone, the dipslides were also applied; however, rather than counting the number of formed colonies, the area fraction of the contaminated slide was evaluated using MATLAB (R2020b). For the disinfection results presented next, the fabric swatches were inoculated with 100 µL of the respective organisms. This volume was sufficient to cover the entire swatch (6.5 cm by 6.5 cm), which was, in turn, able to accommodate the full area of the dipslide (5 cm by 2 cm), via three separate applications. This high microbial concentration was tested to prove the disinfection efficiency of ozone in very harsh aqueous conditions.

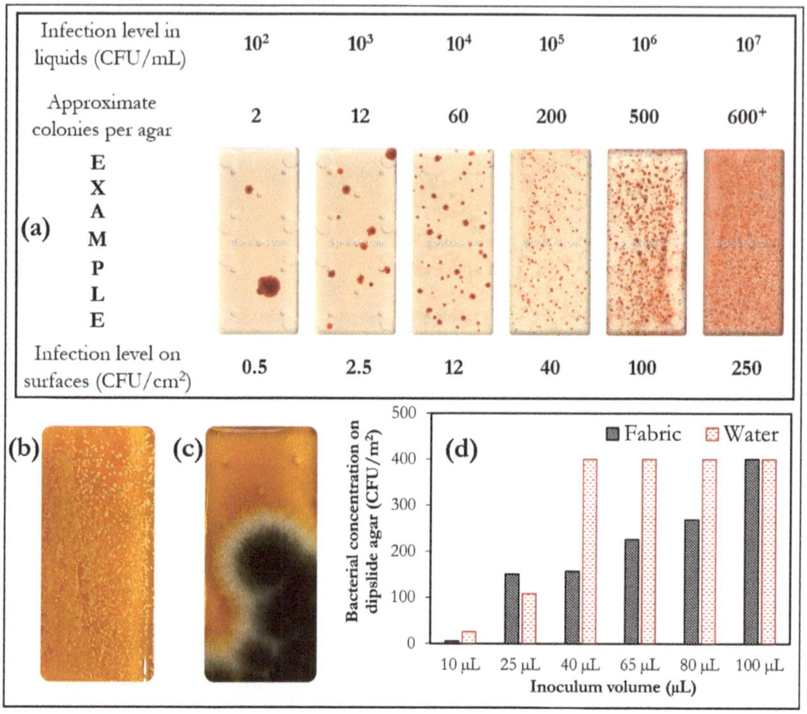


**Figure 9.** (**a**) Enumeration of bacterial concentration on the dipslides on surfaces and in solution (adapted with permission from [25]); number of CFUs on the dipslide—applicable to *E. coli* and *S. aureus*; contaminated regions of the dipslide for (**b**) *C. albicans* and (**c**) *A. fumigatus*, determined via area fraction analysis. (**d**) Obtained concentrations of *E. coli* at different volumes of utilised bacterial suspensions: 10 µL, 25 µL, 40 µL, 65 µL, and 80 µL. Dipslides were immersed into respective mixtures of the known volume of bacteria suspension and 100 mL of sterile distilled water, and also applied directly onto contaminated swatches with the respective suspension volumes.

Before washing with ozonated water, the contaminated fabric swatches (inoculated with 100 mL of microbial suspension) were washed in ordinary mineral water, and the dipslides were used to test for the residual bacteria in the water and on the fabric swatch. Upon testing the fabric swatch, there was no apparent reduction in the contamination level of the swatch after washing with ordinary water, relative to the contamination level before the wash. This was observed for all organisms tested. Upon applying the dipslide in the water (after taking out the swatch), a significant microbial recovery was observed (similar to that of the fabric swatch) for all organisms except *S. aureus*. A 1 log reduction was observed in water relative to the fabric swatch, with *S. aureus*. Hence, more bacterial colonies were recovered from the fabric (the originally contaminated substrate) than the water into which the substrate was placed. This observation may be attributed to the mass transfer peculiarities of *S. aureus* (binding more to the fabric material than migrating into solution).

### 3.7. Disinfection Efficiency of Ozonated Water

To study the disinfection efficiency of ozonated water, two key factors were analysed—the contact duration of the fabric swatch in ozonated water at a fixed ozone concentration, and the ozone concentration utilised at a fixed contact time. Figures 10–13 illustrate the influence of these factors for *E. coli*, *S. aureus*, *C. albicans*, and *A. fumigatus*, respectively. Furthermore, the impact of ozone on the contaminated fabric and water (after the wash) were individually analysed. This analysis provided insights into the transfer and survival of microbes from the fabric substrate into water (in which it was washed).

Based on the results presented in Figure 10a,c, complete inactivation of all *E. coli* cells is obtained by washing with 3 ppm of ozonated water for 5 min, both in water and on the fabric surface. This is a strong indication of the rapid reactivity of ozone, for disinfection and sterilisation of contaminated substrates. As expected, ozonating for 8 min and 11 min produced the same results as that obtained with 5 min—there was no bacterial recovery, as evidenced by a sterile dipslide. Upon analysing the disinfection results using the Miles and Misra method, no colonies were observed with ozonated water. A contact duration of 2 min at 3 ppm (Figure 10a,c) was insufficient to yield complete bacterial removal. As expected, increased contact duration enabled better disinfection of the fabric swatches by ozone (this can be more clearly observed in the log reduction plot of Figure 10c).

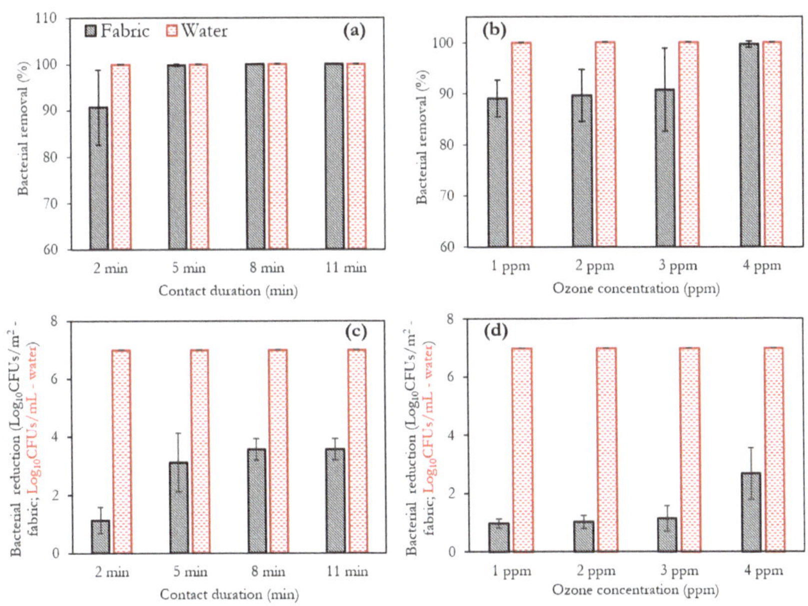


**Figure 10.** Effect of (**a**) contact duration at 3 ppm and (**b**) ozone concentration at 2 min exposure duration on ***E. coli*** removal efficiency after ozonated water treatment. Panels (**a**,**b**) illustrate the percentage removal, whereas (**c**,**d**) illustrate the log reductions for the displayed conditions.

Upon analysing the effect of ozone concentration with a 2 min exposure duration (Figure 10b,d), it can be observed that significant improvements in the disinfection efficiency are observed after 3 ppm, as ozone concentrations of 1, 2, and 3 ppm yield similar disinfection efficiencies. It is expected that higher dissolved ozone concentrations will yield even better performance, particularly if ozone retention is enhanced via the optimal addition of salts and surfactants, as demonstrated in Figures 3 and 5. Contrary to the mentioned survival of microbes transferred from the fabric into solution in Section 3.6, when ordinary water was used, the generation of ozone eliminated this occurrence. Irrespective of the treatment condition (Figure 10a–d), *E. coli* inactivation in water was 100%, yielding approximately a 7 log reduction in solution, in each case.

The observed behaviour of *S. aureus* on interacting with ozone under the same conditions as shown in Figure 10 is shown in Figure 11. Contrary to the 5 min duration required for the total inactivation of *E. coli* (Figure 10a,c), 8 min was required in the case of *S. aureus* (Figure 11a,c); however, a contact duration of 2 min eliminated more *S. aureus* cells than *E. coli* cells (Figures 10c and 11c).

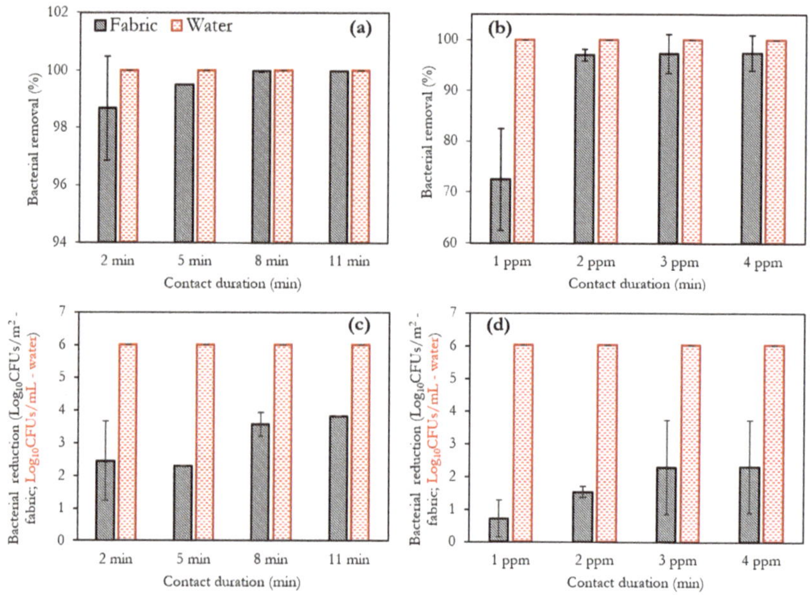


**Figure 11.** Effect of (**a**) contact duration at 3 ppm and (**b**) ozone concentration at 2 min exposure duration on ***S. aureus*** removal efficiency after ozonated water treatment. Panels (**a**,**b**) illustrate the percentage removal, whereas (**c**,**d**) illustrate the log reductions for the displayed conditions.

To enable a better comparison of ozone's performance on the Gram-negative (*E. coli*) and Gram-positive bacteria (*S. aureus*) utilised in this study, the mean log reduction across all test conditions (on the fabric swatch) can be computed, i.e., mean values from Figure 10b,d compared to those from Figure 11b,d. Computing the mean yields 3.9 and 4.3 for *E. coli* and *S. aureus*, respectively, indicating a more significant antimicrobial action of ozone on *S. aureus* than *E. coli*. As observed with *E. coli* (Figure 10), the water samples after the wash were all free of *S. aureus*, irrespective of the treatment condition (Figure 11).

The performance of ozonated water was also tested on the fungal species *C. albicans*. As generally observed in Figure 12, this organism proved significantly more difficult for ozone to inhibit than the bacteria tested in this study. A 100% removal efficiency in water could only be achieved at 11 min of ozone contact at 3 ppm (Figure 12a), compared to the complete removal observed (in water) at all conditions for *E. coli* and *S. aureus* (Figures 10 and 11).

As observed in Figure 13, *A. fumigatus* proved the most difficult for ozonated water at the concentrations and exposure durations tested. The survival and growth of this organism in a wide range of environmental conditions make it a famously challenging pathogen to inactivate. At best, the suppression of its growth was observed, rather than the complete inhibition observed with bacteria. Again, as earlier pointed out, higher concentrations and longer exposure times will have a significant impact; the half-life of ozone in mineral water obtained herein is also an important factor to consider, as far as long exposure durations are required. The 11 min maximum utilised in this study would not have resulted in a significant decline in ozone concentration during washing (half-life in mineral water is 39 min—Figure 6b). Unlike the bacterial species, the log reductions for the fungi utilised herein are not reported because the percentage reductions clearly show the trend in fungal growth suppression as a function of the treatment conditions applied.

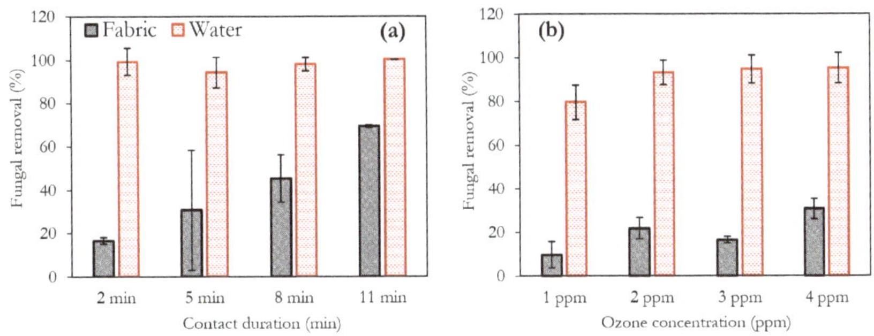


**Figure 12.** Effect of (**a**) contact duration at 3 ppm and (**b**) ozone concentration at 2 min exposure duration on ***C. albicans*** removal efficiency after ozonated water treatment.

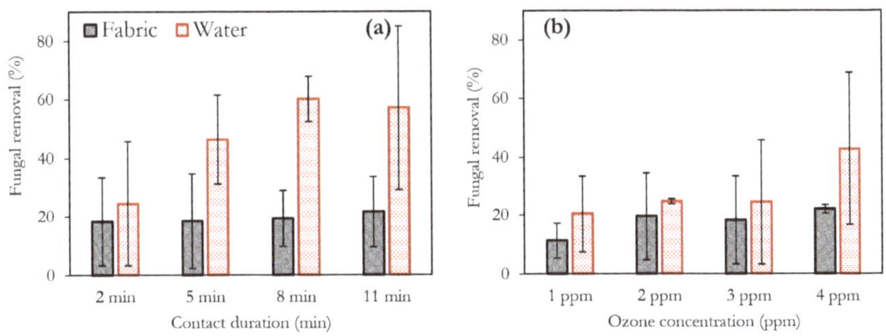


**Figure 13.** Effect of (**a**) contact duration at 3 ppm and (**b**) ozone concentration at 2 min exposure duration on ***A. fumigatus*** removal efficiency after ozonated water treatment. The larger error bars associated with these measurements may be attributed to the non-uniform growth pattern of this fungus in liquid suspensions, which inherently affects the microbial load used for inoculation.

For textiles with heavy fungal contamination, ozonation in air may be utilised for rapid disinfection at higher ozonation concentrations. Compared to water, ozone concentrations as high as 20 ppm (in air) may be easily attained in a couple of minutes via UV radiation, with complete removal of *A. fumigatus* in as little as 4 min readily achieved [39]. Follow-on research from this work will involve a detailed comparison of ozonation efficiency in air versus water. It is also worth mentioning that the bacterial log reductions (in solution) obtained in this study are in line with those of the European Standard BS EN16616 values (≥7 log reduction) and the EPA (≥6 log reduction under 10 min for hard surfaces) [40,41]; however, this work has not considered all bacteria types required by these standard test procedures. Future work should constitute a replication of these findings for more organisms and, importantly, a comparative assessment of the fungicidal effects of ozonated water, a surfactant solution, and a mixture of ozonated water and a surfactant (utilising *A. fumigatus* and *C. albicans*).

## 4. Conclusions

In this study, a parametric assessment of ozone concentration, generation and decomposition kinetics, stain removal, and attainable disinfection level was performed, with demonstrable applications for controlled decontamination in the textile processing industry (as a case study).

The results show that low concentrations of detergent and salts (mono- and divalent) promote ozone stability within water. Ozone decomposition in water follows first-order kinetics with a ~10 min half-life in ordinary tap water (ACS water) and ~39 min in the mineral water used herein. Interestingly, this approximately corresponds to a 3–4 times higher retention and is comparable to the relative ionic concentration of Na ions. Thus, increased mineral content favours ozone retention; however, this occurs until a critical point, beyond which the outcome is reversed. This can be attributed to potential salting-out effects or water structure-breaking influence of ions, hence promoting ozone decomposition as shown, with the addition of sodium bicarbonate. Further analysis is required for a robust determination of this critical point for different salts.

The presented results show that the resistance of the tested organisms to ozonated water is in the order *A. fumigatus* > *C. albicans* > *E. coli* > *S. aureus*. A 7 log reduction in *E. coli* bacteria (100% removal) was observed in water after ozone treatment at 3 ppm for only 5 min. In comparison, *C. albicans* required a higher level of ozone treatment, with complete removal in water at 3 ppm observed after 11 min. Microbial inactivation in water was generally higher than that achieved on the fabric swatch. This demonstrates ozone's efficacy in mitigating the transfer and survival of microbes from a substrate immersed in water.

The results of the presented study demonstrate the potential of aqueous ozone treatment for large-scale industrial applications. Furthermore, the benefits of using ozone for bleaching and textile decontamination purposes also include the reduction in detergent usage, which mitigates negative environmental impact of the resultant wastewater. The number of necessary changes when retrofitting existing industrial textile processing systems to accommodate ozone are minimal, thus creating an incentive for its adaptation and continued application.

**Author Contributions:** Conceptualisation, E.I.E., A.M., M.C., A.B., W.M. and M.Y.; methodology, E.I.E., N.A., K.R., M.E.R., W.M. and M.Y.; software and data processing, E.I.E.; writing original draft, E.I.E.; writing review draft, E.I.E., A.M., M.C., A.B., N.A., K.R., W.M., M.E.R. and M.Y.; project lead and principal investigator, M.Y. All authors have read and agreed to the published version of the manuscript.

**Funding:** This research was funded by Innovate UK and ACS Clothing (KTP 12079).

**Institutional Review Board Statement:** Not applicable.

**Informed Consent Statement:** Not applicable.

**Data Availability Statement:** Data are contained within the article.

**Acknowledgments:** The authors gratefully acknowledge the financial support of Innovate UK (KTP 12079), as well as ACS Clothing and the University of the West of Scotland, for providing the required equipment used in the experiments. The authors also acknowledge the technical and experimental support of Charles McGuiness and Claire Easdon.

**Conflicts of Interest:** The authors declare that they have no known competing financial interests or personal relationships that could have appeared to influence the work reported in this paper.

## Abbreviations

| | |
|---|---|
| CFU | Colony-forming units |
| CPUR | Corrected pick-up rate |
| DAA | Dodecyl tri-methyl ammonium acetate |
| DDM | Dodecyl-β-D-maltoside |
| DPD | *N,N*-diethyl-p-phenylenediamine |
| BCSOT | Blue colouring, sunflower oil, and turmeric powder |
| GTF | Gordon–Tomiyasu–Fukutomi |
| HSB | Hoigné–Staehelin–Bader |
| ICP | Inductive coupled plasma |
| IR | Infrared spectroscopy |
| SDS | Sodium dodecyl sulphate |
| RTBM | Red temporary board marker |
| TTC | 2,3,5-triphenyltetrazolium chloride |
| UHQ | Ultra-high quality |

## References

1. Guzel-Seydim, Z.B.; Greene, A.K.; Seydim, A.C. Use of ozone in the food industry. *LWT* **2004**, *37*, 453–460. [CrossRef]
2. Summerfelt, S.T.; Sharrer, M.J.; Tsukuda, S.M.; Gearheart, M. Process requirements for achieving full-flow disinfection of recirculating water using ozonation and UV irradiation. *Aquac. Eng.* **2009**, *40*, 17–27. [CrossRef]
3. Lee, Y.; von Gunten, U. Advances in predicting organic contaminant abatement during ozonation of municipal wastewater effluent: Reaction kinetics, transformation products, and changes of biological effects. *Environ. Sci. Water Res. Technol.* **2016**, *2*, 421–442. [CrossRef]
4. Dong, S.; Li, J.; Kim, M.-H.; Park, S.-J.; Eden, J.G.; Guest, J.S.; Nguyen, T.H. Human health trade-offs in the disinfection of wastewater for landscape irrigation: Microplasma ozonation vs. chlorination. *Environ. Sci. Water Res. Technol.* **2016**, *3*, 106–118. [CrossRef]
5. Malik, S.N.; Ghosh, P.C.; Vaidya, A.N.; Mudliar, S.N. Hybrid ozonation process for industrial wastewater treatment: Principles and applications: A review. *J. Water Process Eng.* **2020**, *35*, 101193. [CrossRef]
6. Morrison, C.; Atkinson, A.; Zamyadi, A.; Kibuye, F.; McKie, M.; Hogard, S.; Mollica, P.; Jasim, S.; Wert, E.C. Critical Review and Research Needs of Ozone Applications Related to Virus Inactivation: Potential Implications for SARS-CoV-2. *Ozone Sci. Eng.* **2021**, *43*, 2–20. [CrossRef]
7. Körlü, A. Use of Ozone in the Textile Industry. In *Textile Industry and Environment*; IntechOpen: London, UK, 2019.
8. Martinelli, M.; Giovannangeli, F.; Rotunno, S.; Trombetta, C.; Montomoli, E. Water and air ozone treatment as an alternative sanitizing technology. *J. Prev. Med. Hyg.* **2017**, *58*, E48–E52. [CrossRef]
9. Kim, J.-G.; Yousef, A.E.; Dave, S. Application of Ozone for Enhancing the Microbiological Safety and Quality of Foods: A Review. *J. Food Prot.* **1999**, *62*, 1071–1087. [CrossRef]
10. Eriksson, M. Ozone Chemistry in Aqueous Solution: Ozone Decomposition and Stabilization. Doctoral Dissertation, Department of Chemistry, Royal Institute of Technology Stockholm, Stockholm, Sweden, 2005.
11. Rosenblum, J.; Ge, C.; Bohrerova, Z.; Yousef, A.; Lee, J. Ozonation as a clean technology for fresh produce industry and environment: Sanitizer efficiency and wastewater quality. *J. Appl. Microbiol.* **2012**, *113*, 837–845. [CrossRef]
12. Ward, D.B.; Tizaoui, C.; Slater, M.J. Ozone-Loaded Solvents for use in Water Treatment. *Ozone Sci. Eng.* **2003**, *25*, 485–495. [CrossRef]
13. Okada, F.; Nay, K. *Electrolysis for Ozone Water Production*; Electrolysis: London, UK, 2012.
14. Rice, R.G.; Debrum, M.; Cardis, D.; Tapp, C. The Ozone Laundry Handbook: A Comprehensive Guide for the Proper Application of Ozone in the Commercial Laundry Industry. *Ozone Sci. Eng.* **2009**, *31*, 339–347. [CrossRef]
15. Finch, G.R.; Smith, D.W.; Stiles, M.E. Dose—Response of Escherichia coli in ozone demand-free phosphate buffer. *Water Res.* **1988**, *22*, 1563–1570. [CrossRef]
16. Hunt, N.K.; Mariñas, B.J. Kinetics of Escherichia coli inactivation with ozone. *Water Res.* **1997**, *31*, 1355–1362. [CrossRef]
17. Magbanua, B.S.; Savant, G.; Truax, D.D. Combined Ozone and Ultraviolet Inactivation of Escherichia coli. *J. Environ. Sci. Health Part A* **2006**, *41*, 1043–1055. [CrossRef]
18. Sevimli, M.F.; Sarikaya, H.Z. Ozone treatment of textile effluents and dyes: Effect of applied ozone dose, pH and dye concentration. *J. Chem. Technol. Biotechnol.* **2002**, *77*, 842–850. [CrossRef]
19. Powar, A.S.; Perwuelz, A.; Behary, N.; Hoang, L.; Aussenac, T. Application of Ozone Treatment for the Decolorization of the Reactive-Dyed Fabrics in a Pilot-Scale Process—Optimization through Response Surface Methodology. *Sustainability* **2020**, *12*, 471. [CrossRef]
20. Santos, L.; Silva, E.; Oliveira, F.; Rodrigues, L.; Neves, P.; Meira, C.; Moreira, G.; Lobato, G.; Nascimento, C.; Gerhardt, M.; et al. Ozonized Water in Microbial Control: Analysis of the Stability, In Vitro Biocidal Potential, and Cytotoxicity. *Biology* **2021**, *10*, 525. [CrossRef]

21. Neral, B. Quality of the household ozone laundering. *Ind. Textila* **2018**, *69*, 304–309.
22. Palintest Ozone Meter User Manual. 2020. Available online: https://palintest.fr/application/files/7915/1845/7366/Ozone_meter_instructions.pdf (accessed on 12 May 2021).
23. Bader, H.; Hoigné, J. Determination of ozone in water by the indigo method. *Water Res.* **1981**, *15*, 449–456. [CrossRef]
24. Beloti, V.; Barros, M.A.; De Freitas, J.C.; Nero, L.A.; De Souza, J.A.; Santana, E.H.; Franco, B.D. Frequency of 2,3,5-triphenyltetrazolium chloride (TTC) non-reducing bacteria in pasteurized milk. *Rev. Microbiol.* **1999**, *30*, 137–140. [CrossRef]
25. Dip-Slides UK (2021) Dip-Slides UK (2021) BT2 Dipslide | Nutrient with TTC | High Quality. Available online: https://dip-slides.com/microbiological-dipslides/9-ttc-nutrient-ttc-nutrient-dipslides-box-of-10.html (accessed on 2 September 2021).
26. Acero, J.L.; Von Gunten, U. Influence of Carbonate on the Ozone/Hydrogen Peroxide Based Advanced Oxidation Process for Drinking Water Treatment. *Ozone Sci. Eng.* **2000**, *22*, 305–328. [CrossRef]
27. ThermoFisher Scientific Safety Data Sheet: Sodium Carbonate. Revision No. 11. Revision Date 22/12/2020. 2020. Available online: https://thermofishervn.com/images/products/MSDS/10861.pdf (accessed on 2 September 2021).
28. ThermoFisher Scientific, Safety Data Sheet: Potassium Carbonate Anhydrous. Revision No. 11. Revision Date 01/01/2021. 2021. Available online: https://www.fishersci.co.uk/chemicalProductData_uk/wercs?itemCode=10647612&lang=EN (accessed on 2 September 2021).
29. Langlais, B.; Reckhow, D.A.; Brink, D.R. *Ozone in Water Treatment: Application and Engineering*; Lewis Publishers: New York, NY, USA, 1991; p. 558.
30. Wang, Y.-H.; Chen, Q.-Y. Anodic Materials for Electrocatalytic Ozone Generation. *Int. J. Electrochem.* **2013**, *2013*, 128248. [CrossRef]
31. Tomiyasu, H.; Fukutomi, H.; Gordon, G. Kinetics and mechanism of ozone decomposition in basic aqueous solution. *Inorg. Chem.* **1985**, *24*, 2962–2966. [CrossRef]
32. Greenwood, N.N.; Earnshaw, A. *Chemistry of the Elements*; Elsevier: Amsterdam, The Netherlands, 2012; ISBN 0080501095.
33. Bulanin, K.; Bulanin, M.; Tsyganenko, A. Infrared spectra of 18O-enriched ozones in liquid oxygen solution. *Chem. Phys.* **1996**, *203*, 127–136. [CrossRef]
34. Bulanin, K.M.; Alexeev, A.V.; Bystrov, D.S.; Lavalley, J.C.; Tsyganenko, A.A. IR Study of Ozone Adsorption on $SiO_2$. *J. Phys. Chem.* **1994**, *98*, 5100–5103. [CrossRef]
35. Staehelin, J.; Buehler, R.E.; Hoigne, J. ChemInform Abstract: Ozone decomposition in water studied by pulse radiolysis. 2. hydroxyl and hydrogen tetroxide (HO4) as chain intermediates. *J. Phys. Chem.* **1984**, *88*, 5999–6004. [CrossRef]
36. Heidt, L.J.; Landi, V.R. Ozone and Ozonide Production and Stabilization in Water. *J. Chem. Phys.* **1964**, *41*, 176–178. [CrossRef]
37. Batakliev, T.; Georgiev, V.; Anachkov, M.; Rakovsky, S.; Zaikov, G.E. Ozone decomposition. *Interdiscip. Toxicol.* **2014**, *7*, 47. [CrossRef]
38. Miles, A.A.; Misra, S.S.; Irwin, J.O. The estimation of the bactericidal power of the blood. *Epidemiol. Infect.* **1938**, *38*, 732–749. [CrossRef]
39. Epelle, E.I.; Macfarlane, A.; Cusack, M.; Burns, A.; Thissera, B.; Mackay, W.; Rateb, M.E.; Yaseen, M. Bacterial and fungal disinfection via ozonation in air. *J. Microbiol. Methods* **2022**, *194*, 106431. [CrossRef]
40. US EPA Guidelines, OCSPP 810.2200: Disinfectants for Use on Hard Surfaces—Efficacy Data Recommendations. Available online: https://www.regulations.gov/document?D=EPA-HQ-OPPT-2009-0150-0021 (accessed on 2 September 2021).
41. BS EN 16616:2015-10 Chemical Disinfectants and Antiseptics—Chemical-Thermal Textile Disinfection—Test Method and Requirements. 2015. Available online: https://shop.bsigroup.com/products/chemical-disinfectants-and-antiseptics-chemical-thermal-textile-disinfection-test-method-and-requirements-phase-2-step-2/standard (accessed on 12 May 2021).

applied sciences

MDPI

*Article*

# Selective Chlorination and Extraction of Valuable Metals from Iron Precipitation Residues

**Lukas Höber *, Kerrin Witt and Stefan Steinlechner ***

Christian Doppler Laboratory for Selective Recovery of Minor Metals Using Innovative Process Concepts, Chair of Nonferrous Metallurgy, Montanuniversität Leoben, Franz Josef Strasse 18, 8700 Leoben, Austria; kerrin.witt@unileoben.ac.at

* Correspondence: lukas.hoeber@unileoben.ac.at (L.H.); stefan.steinlechner@unileoben.ac.at (S.S.)

**Featured Application: Authors are encouraged to provide a concise description of the specific application or a potential application of the work. This section is not mandatory.**

**Abstract:** Due to the aggravating situations regarding climate change, resource supply, and land consumption by the landfilling of residual materials, it is necessary to develop recycling processes that allow the recovery of valuable metals from industrial residues with significantly reduced $CO_2$ emissions. In this context, it is conceivable that processes using chlorination reactions will be of importance in the future. The simultaneous selective chlorination and evaporation of nine valuable metals was evaluated theoretically and experimentally in small-scale STA trials; then, it was tested practically on six different iron precipitation residues from the zinc and nickel industries. The metal chlorides $FeCl_3{\cdot}6H_2O$ and $MgCl_2{\cdot}6H_2O$ were identified as the most effective reactants, resulting in high extraction rates for the metals In, Ag, Zn, Pb, Au, and Bi, while lower yields are achievable for Sn, Cu, and Ni. Iron, which is predominant in volume in the residual materials, shows lower chlorination tendencies which allows the effective separation of the valuable elements of interest from the iron containing matrix.

**Keywords:** selective chlorination; jarosite recycling; multimetal recovery; $CO_2$ reduction

**Citation:** Höber, L.; Witt, K.; Steinlechner, S. Selective Chlorination and Extraction of Valuable Metals from Iron Precipitation Residues. *Appl. Sci.* **2022**, *12*, 3590. https://doi.org/10.3390/app12073590

Academic Editors: Dibyendu Sarkar, Rupali Datta, Prafulla Kumar Sahoo and Mohammad Mahmudur Rahman

Received: 9 March 2022
Accepted: 30 March 2022
Published: 1 April 2022



## 1. Introduction

Among humanity's greatest challenges for the next generations is the shift to $CO_2$-free processes for energy generation, transportation, and the production of basic materials and goods for the globalized society. A key role in this context is assigned to industry, since many processes depend on the use of carbon not only as an energy source but also as a chemical reactant. Another critical aspect of heavy industry is the continued landfilling of various residual materials. For Europe, the resource issue is essential, as the continent is already dependent on exports for many technology materials. In the long term, the global and especially the European markets are facing the absolute necessity of developing recycling processes for the recovery of valuable metals which emit significantly less greenhouse gas emissions during the process. A category of waste materials that are predominantly immobilized and disposed without recovery of the contained valuable metals are iron-containing precipitation residues such as jarosite and goethite, which are generated in the course of the hydrometallurgical production of zinc and nickel. Jarosite precipitation is by far the most important method for iron removal. The formed precipitates have the mineralogical structure $XFe_3(SO_4)_2(OH)_6$ with X being represented by monovalent cations such as $Na^+$, $K^+$, $NH_4^+$, and $H_3O^+$ or valuable elements such as $Ag^+$ or $1/2Pb^{2+}$ [1–3]. Goethite describes an iron hydroxide with the formula $FeO{\cdot}OH$ [3]. In the precipitated products of the zinc industry, significant amounts of zinc and lead can be found as well as lower concentrations of indium, silver, copper, and other valuable elements [4–7]. Jarosites from the nickel industry contain nickel [8] and arsenic. Over the past decades, a variety of methods have been developed for the processing of iron precipitation residues [9]. The only globally used process is the immobilization of jarosite in the so-called Jarofix process, in which the material is chemically stabilized by the addition of Portland cement, lime, and water and subsequently landfilled. Approaches were investigated where jarosite is mixed with clay, fly ash, and water [10,11] or marble processing rejects [12] to produce brick materials for building purposes. Other research stated the possibility of producing materials for road or dump construction by mixing jarosite with ferrous slags [13,14]. Furthermore, the residue can substitute sand in concrete mixtures [15]. Jarosite from the nickel industry was successfully tested as an alternative substitute for gypsum in cement mixtures [16]. For goethite, a concept was introduced whereby the material is mixed with blast furnace and converter slag, leading to the formation of a hard rock-like material [17]. The majority of research projects for the recovery of valuable elements or conversion of the residue into products were carried out in the field of hydrometallurgical treatment methods. The bioleaching of jarosite residues bears the potential of recovering valuable elements with low energy and chemicals consumption [18–20]. Furthermore, the materials can be converted into magnetite or hematite, which for example can be used theoretically in the pigment or iron and steel industry [21–23]. Alkaline leaching and the subsequent cyanidation of precious metals, especially silver [5,24–27], the combined treatment of sewage sludge and jarosite [28], leaching with $NH_4Cl$ to recover Pb, Cu, Cd, and Ag [6], roasting and flotation to recover silver and anglesite [29], as well as the usage of resins [30] can also be considered as relevant developments on a laboratory scale. The use of deep-eutectic solvents for the recovery of zinc was investigated to treat goethite [31]. The residual materials can also be treated in pyrometallurgical processes, whereby it was stated that a small number of zinc-producing companies use common technologies such as TSL reactors or waelz kilns to co-treat jarosite or goethite with other residues [32]. Other approaches such as the treatment of zinc plant residues in plasma furnaces [33,34], the reduction and magnetic separation to recover zinc [35], as well as the production of glass-ceramics from goethite [36,37] can be named. However, there are no generally applied methods on an industrial scale—although some zinc plants were reported to process a part of their iron precipitation residues, they remain exceptions. The chlorination of valuable elements was as well proposed in the field of recycling precipitation residues, stating that indium can be extracted as indium chloride at temperatures over 800 °C [38]. This was subsequently confirmed in small-scale trials, where electric arc furnace dust as a chlorine carrier and coal were added to jarosite waste and treated at 1100 °C, leading to the extraction of silver, indium, and lead in noteworthy extents [39]. A similar approach focused on the simultaneous extraction of zinc, lead, copper, indium, and silver by mixing jarosite with coal and $CaCl_2$ as a chlorine-bearing additive [40].

Generally, it can be summarized that processes in both categories bear significant disadvantages. Hydrometallurgical approaches are generally designed for the recovery of specific elements and often rely on narrow parameter ranges; the developed pyrometallurgical processes, on the other hand, are generally associated with the emission of greenhouse gases whereby the implementation becomes questionable in the future.

The novel recycling approach presented in this paper is based on the chlorination and extraction of valuable metals from the excess iron matrix of the precipitation residues without the addition of carbon as a reactant. The schematic process flow sheet is provided in Figure 1. The residues are mixed in their original or thermally treated (decomposed) form with suitable metal chlorides and processed at moderate temperatures of up to 1100 °C. The underlying principle of the extraction is the fact that the metal chlorides evaporate at significantly lower temperatures compared to their metallic or oxidic form. Iron shows lower thermodynamic tendencies to chloride formation than the valuable metal compounds contained in precipitation residues. Process control without carbon addition results in significantly lower specific greenhouse gas emissions. Furthermore, the formation of problematic chlorine–carbon compounds is prevented. In addition, the treatment at low temperatures does not result in the formation of liquid phases.

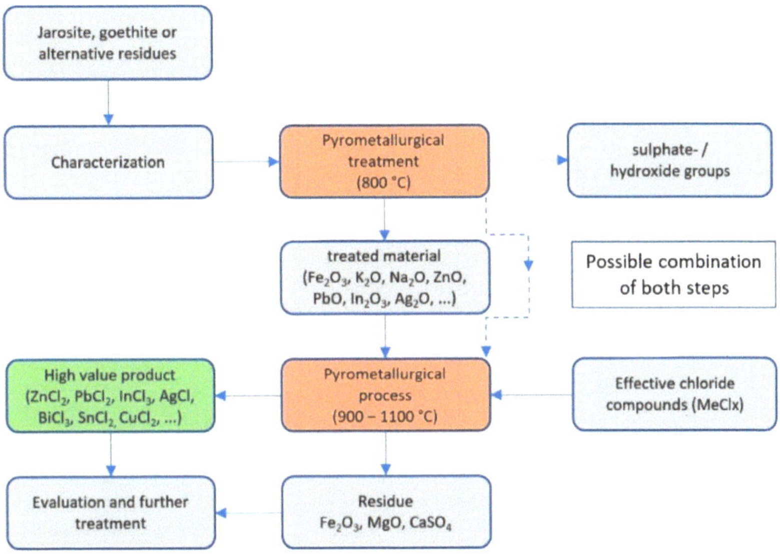


**Figure 1.** Concept flow sheet of the chlorination process for treating precipitation residues.

## 2. Theoretical Considerations and Thermodynamic Calculations

The potential of recovering valuable metals from residual materials by means of chlorination has been demonstrated repeatedly. Indium can be recovered from spent LCDs using polyvinyl chloride (PVC) [41,42] or aqueous HCl [43]; furthermore, the contained metals in CIGS–solar cells (copper–indium–gallium–selenium) can be extracted to noteworthy extents using chlorine gas or ammonium chloride [44]. Li et al. proposed the possibility of recovering Au and Ag from cyanide tailings using $CaCl_2$ [45]. Pyrite ($FeS_2$) addition positively influences the chlorination of Au as the formation of $CaSO_4$ results in the liberation of chlorine [46]. $NH_4Cl$ was stated to extract Zn via chlorination from steel mill dusts (BOS) [47]; a similar approach was developed for recovering Zn and also Pb from EAFD by treatment in an Ar-$Cl_2$ atmosphere. Comprehensive investigations were carried out in the field of chlorination of valuable elements from municipal solid waste (MSW) using HCl for Zn and Pb [48], PVC for Cu, Zn, Pb, and Cd [49]. Furthermore, metal chlorides such as NaCl, $CaCl_2$, and $MgCl_2$ were investigated, leading to noteworthy extraction of Cd, Cr, Cu, Ni, Pb, or Zn [50,51].

In the course of this work, the reactivity of various solid metal chlorides for multimetal recovery from an excess iron matrix was investigated. Metal chlorides are present in huge extents worldwide and used in various applications. The six compounds NaCl, KCl, $CaCl_2$, $MgCl_2$, $AlCl_3$, and $FeCl_3$ were investigated regarding their reactiveness for chlorinating valuable elements from residual materials. It can be stated that the chlorination of a metal of interest is highly dependent on the present form of the element chlorine itself. Hereby, three different ways for supplying chlorine by metal chlorides can be cited, being the liberation of gaseous chlorine, the liberation of gaseous hydrochloric acid, and the direct reaction of the solid, liquid, or gaseous metal chloride compound.

For example, the chlorination of Zn from ZnO using gaseous HCl proceeds according to the following reaction: $ZnO + 2HCl (g) \leftrightarrow ZnCl_2 (g) + H_2O (g)$. The demand of chlorination agent depends on the valency of the respective valuable elements. It is noteworthy in this context that all mentioned chlorides except NaCl and KCl are strongly hygroscopic (Ropp, 2013; Wiberg et al., 2007) and therefore are generally present in a partially hydrated form. Thermodynamic calculations which were carried out using the software packages FactSage (Bale et al., 2016) and HSC Chemistry V10 (Metso Outotec) revealed that NaCl and KCl are significantly more stable than the other compounds, showing no or very little tendency for reaction with valuable metal oxides.

Calculations using HSC chemistry V10 show that the chlorides $CaCl_2$, $MgCl_2$, $AlCl_3$, and $FeCl_3$ in their hydrated form ($MeCl_x{\cdot}nH_2O$) show a different behavior regarding their ability to liberate gaseous hydrochloric acid. For the theoretical interpretation of the equilibrium composition of chemical reactions in dependence of temperature, the Free Gibbs Energy (Delta-G) can be considered, whereby negative values indicate that the reaction proceeds in the given direction. For $CaCl_2$ and $MgCl_2$, there are three different types of hydrates ($nH_2O$ with n being 2, 4, or 6), while for the trivalent chlorides $AlCl_3$ and $FeCl_3$, solely the hexahydrate compounds are available in the used database. The values for Delta-G in dependence of the temperature for both the solo or multiple-step dehydration (splitting of water leaving the pure metal chloride) and the hydrolysis (splitting of HCl and $H_2O$ leading to the formation of the metal oxide) are given in Figure 2. When assuming an actual trial mixture to heat up to processing temperature, the depicted data suggest that for the three chlorides $MgCl_2{\cdot}6H_2O$, $FeCl_3{\cdot}6H_2O$, and $AlCl_3{\cdot}6H_2O$, the chemical driving force is higher for liberating gaseous hydrochloric acid than for splitting of solely the water groups (values of dashed lines are more negative than the ones of solid lines). For $MgCl_2{\cdot}6H_2O$, this accounts for all three possible steps of dehydration or hydrolysis. This effect cannot be stated for $CaCl_2{\cdot}6H_2O$, where the values for liberating water are more negative at given temperatures, indicating that the compound is unlikely to set free the highly reactive gaseous hydrochloric acid.

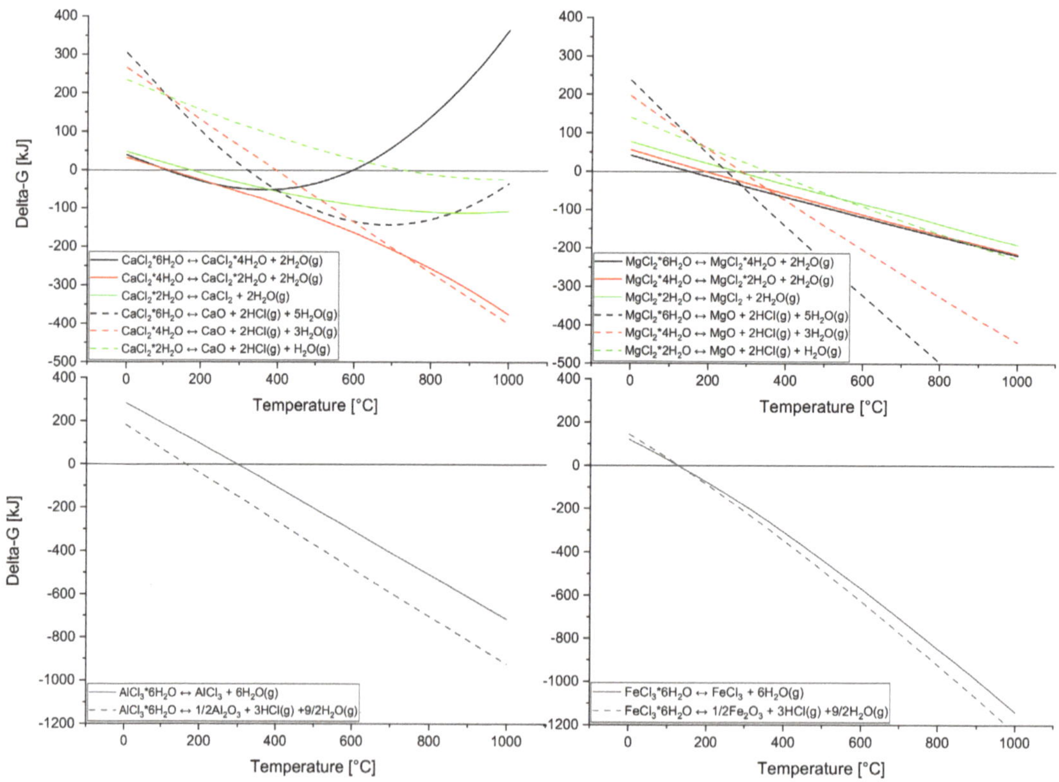


**Figure 2.** Calculated Delta-G values for hydrolysis or dehydration of the hydrated metal chlorides.

The main aspect concerning the recovery of valuable metals by selective chlorination and evaporation is the low evaporation temperature $T_{vap}$ of the formed metal chlorides compared to the elements in metallic or oxide form. Figure 3a shows the values for $T_{vap}$ of the elements of interest in various compound forms. Sulfate compounds are not included because they would generally react to oxides under the liberation of sulfur–oxygen compounds. All elements except silver evaporate completely as chlorides at temperatures below 1000 °C. In this context, it must be pointed out that all chlorides have significant vapor pressures even at lower temperatures, and extraction is effectively possible even for silver, as will be shown below. In addition to the individual values for $T_{vap}$, the range of the averaged values of all metals and oxides as well as those of the chlorides is shown. The difference between the evaporation temperatures of the compound types is significant, leading to the general conclusion that elements of value are effectively extractable in their chloridic form. Furthermore, the software package FactSage with the database FactPS was used to calculate the values for Delta-G for the reactions of the elements of interest as well as iron in the form of oxides with gaseous HCl under formation of the metal chlorides, as given in Figure 3b. Since the software considers the compounds in their most stable form at given temperatures, the slopes of the curves change due to melting and evaporation. The dashed red line shows the values of Delta-G for the reaction of $Fe_2O_3$. It can be stated that the curve is in the positive area for the majority of the temperature range. This is the theoretical basis for the selective multimetal separation from the iron matrix, which accounts for 30–40% (elemental Fe) in the iron precipitation residues, depending on the actual residue. The only other elements that show positive values are copper and nickel. The reaction of indium is in equilibrium at temperatures over 600 °C (Delta-G = 0).

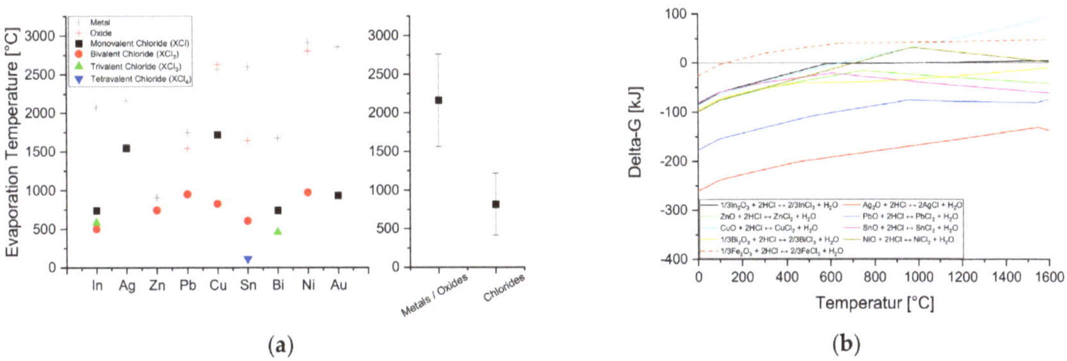


**Figure 3.** (**a**) Evaporation temperatures of the valuable metals in metallic, chloridic, and oxidic form; (**b**) Delta-G values for reactions of metal oxides with hydrochloric acid.

## 3. Materials and Methods

To confirm the reactiveness of the assessed metal chlorides as well as the potential extraction of multiple valuable elements, two different experimental approaches were conducted. The behavior of all six metal chlorides NaCl, KCl, $CaCl_2{\cdot}6H_2O$, $MgCl_2{\cdot}6H_2O$, $AlCl_3{\cdot}6H_2O$, and $FeCl_3{\cdot}6H_2O$ in their pure form was investigated, using simultaneous thermal analysis (STA). Furthermore, trials were carried out with mixtures of metal chlorides and valuable elements of interest in form of oxides and sulfates. For an evaluation of the concept with real residual materials, chlorination campaigns were executed in a muffle furnace with untreated as well as pre-treated iron precipitation residues (decomposed structure) in mixtures with specific metal chlorides, which were identified as effective.

### *3.1. STA Campaigns with Pure Substances*

As part of the fundamental research, NaCl, KCl, $CaCl_2{\cdot}6H_2O$, $MgCl_2{\cdot}6H_2O$, $AlCl_3{\cdot}6H_2O$, and $FeCl_3{\cdot}6H_2O$ were analyzed using STA regarding their behavior for liberating chlorine or hydrochloric acid. For the STA trials with mixtures, the metals silver, indium, lead, zinc, and nickel were investigated. The following oxides and sulfates in the form of pure chemicals were used for the preparation of the following mixtures: $Ag_2O$, $Ag_2SO_4$, $In_2O_3$, $In_2(SO_4)_3$, NiO, PbO, and ZnO. Based on the thermodynamic calculations performed beforehand, the chlorides $AlCl_3{\cdot}6H_2O$, $FeCl_3{\cdot}6H_2O$, and $MgCl_2{\cdot}6H_2O$ were selected as chlorination agents, respectively, for each of the metal oxides and sulfates. Furthermore, $In_2O_3$ was mixed with NaCl, KCl, and $CaCl_2{\cdot}6H_2O$ to confirm the significantly lower suitability of those chlorides (based on thermodynamic evaluations and the trials with pure chlorides). According to the chlorination reactions described in the thermodynamics section, the metal chlorides AgCl, $InCl_3$, $PbCl_2$, $ZnCl_2$, and $NiCl_2$ were expected to be formed during chlorination. In reference to the stoichiometry of the reactions, the mixtures were prepared using an excess molar amount of chlorine with a chlorination agent to metal oxide/sulfate ratio of 1.5. The STA was performed using a Netzsch 409PC analyzer with synthetic air ($N_2/O_2$ = 80/20) as purge gas and a flow rate of 0.3 L/min. Sintered $Al_2O_3$ crucibles and a sample mass of 100 mg were used; all chemicals were purchased in laboratory grade. All trials were carried out with a constant heating rate of 20 K/min from 20 °C to a maximum temperature of 1000 °C, 1200 °C, or 1400 °C. The assessment of the STA trials was conducted by evaluating the collected data (mass loss over time, temperature) and comparing the real mass of the solid residual materials in the crucible to the one expected based on theoretical calculations for each mixture.

### 3.2. Campaigns with Industrial Iron Precipitation Residues

Various iron precipitation residues were investigated and processed in defined mixtures in a muffle furnace (Nabertherm LT9/12) in a scale of 80 g. The main purposes for the trial campaign consisting of 108 experiments are at first the identification of the most effective chlorination agents for the extraction of specific metals as well as for the multi-metal recovery from various industrial residues with partly low concentrations of valuable elements. Furthermore, the influence of increasing the concentration of the respective chlorination agent as well as higher process temperatures are evaluated.

The raw materials in use were jarosite and goethite residues of the zinc industry as well as a jarosite residue originating in the nickel production. Each material was tested in its original form as well as thermally pre-treated (calcined) in a muffle furnace at 800 °C for five hours using air as atmosphere in order to decompose the mineral structures to simpler oxidic and sulfatic compounds. Chemical analyses were conducted for the untreated as well as for the pre-calcined materials at two accredited laboratories; the results are shown in Table 1.

The chlorides $AlCl_3{\cdot}6H_2O$, $FeCl_3{\cdot}6H_2O$, and $MgCl_2{\cdot}6H_2O$ were used as chlorination agents, which is premised on the thermodynamic calculations and STA trials.

The trial setup is the treatment of the mixture in an open crucible which is charged into the furnace, leading to the evaporation of a share of chloride without taking part in the targeted reactions. Furthermore, it has to be pointed out that chemical reactions do not occur ideally. Therefore, the trials were carried out with an excess of chlorine supply. The 2-fold of the total calculated stoichiometric chlorine demand for full chlorination of considered valuable elements and 5% of iron was set as the base chlorine addition. As it is of interest if the extraction of certain elements can be significantly increased by the higher addition of the respective chlorides, the experiments were carried out with 2-fold and 3-fold of the base value chlorine addition for all three chlorides.

It was assumed in the course of the calculations that the valuable element chlorides that are formed during the treatment of jarosite, goethite, calcined jarosite, and calcined goethite are AgCl, $BiCl_3$, $CuCl_2$, $InCl_3$, KCl, NaCl, $PbCl_2$, $SnCl_2$, and $ZnCl_2$. The chlorination of iron in all trials was considered to an extent of 5%, forming the trivalent compound $FeCl_3$. For the trials with nickel jarosite (original and calcined), the chlorine demand was calculated for the full chlorination of nickel and in turn 5% of Fe. The elements Na and K are considered due to their partly high concentration in precipitation residues, which would lead to significant chlorine consumption due to their affinity for chloride formation. In reference to these suppositions, the calculation of the mixture ratios for the trials was executed. Each jarosite and goethite material, untreated as well as pre-treated, was mixed with each of the mentioned chlorination agents in the three proposed mixing ratios. The resulting practically used mixtures are shown in Table S1 (Supplementary Materials). R represents the amount of residual material, and CA represents the mass of the respective chlorination agent. After the preparation of a mixture of milled dry residual material and chlorination agent, an amount of 80 g was filled into a silica crucible. The crucible was charged into a muffle furnace at trial temperature where the mass decrease was tracked via an integrated balance. Each mixture was tested at two different temperatures, 900 °C and 1100 °C, with air as atmosphere. After 30 min, the crucible was removed from the furnace to cool down at ambient air temperature, and the remaining solid residue was milled and analyzed. Test data considered are mass loss and chemical analysis of the final material.

**Table 1.** Chemical analysis of the investigated original and calcined iron precipitation residues.

| Concentration [wt %] | Jarosite | Goethite | Ni-Jarosite | Jarosite Calcined | Goethite Calcined | Ni-Jarosite Calcined |
|---|---|---|---|---|---|---|
| | | | Structural Elements | | | |
| Fe | 25.6 ± 1.8 | 29.1 ± 1.9 | 36.6 ± 0.3 | 39.9 ± 1.7 | 36.6 ± 1.3 | 61.9 ± 2.3 |
| S | 13.83 ± 0.24 | 7.16 ± 0.33 | 6.38 ± 0.08 | 7.86 ± 0.04 | 6.61 ± 0.02 | 0.19 ± 0 |
| K | 2.77 ± 0.05 | 0.2 ± 0 | 0.08 ± 0.01 | 4.03 ± 0.05 | 0.23 ± 0.05 | 0.12 ± 0 |
| Na | 1.46 ± 0.06 | <0.05 | 0.01 ± 0 | 2.24 ± 0.15 | 0.06 ± 0.01 | 0.15 ± 0.01 |
| Ca | 0.53 ± 0.02 | 4.78 ± 0.24 | 0.01 ± 0 | 0.78 ± 0.01 | 6.08 ± 0.03 | 0.01 ± 0 |
| As | 0.42 ± 0.01 | 0.19 ± 0.01 | 0.26 ± 0 | 0.63 ± 0.01 | 0.23 ± 0 | 0.45 ± 0.01 |
| Mg | 0.19 ± 0 | 0.15 ± 0 | <0.01 | 0.30 ± 0 | 0.18 ± 0 | <0.01 |
| Al | 0.37 ± 0 | 0.89 ± 0.03 | 0.19 ± 0.02 | 0.56 ± 0 | 1.10 ± 0 | 0.26 ± 0 |
| | | | Valuable Elements | | | |
| Zn | 3.57 ± 0.07 | 9.60 ± 0.21 | 0 | 5.37 ± 0.05 | 11.90 ± 0.08 | 0 |
| Pb | 1.26 ± 0 | 0.68 ± 0 | 0.06 ± 0 | 1.9 ± 0 | 0.84 ± 0 | 0.08 ± 0 |
| Ni | | | 3.08 ± 0.04 | | | 5.43 ± 0.26 |
| [ppm] | | | | | | |
| In | 591 ± 3 | 410 ± 9 | 0 | 882 ± 11 | 495 ± 9 | 0 |
| Ag | 282 ± 40 | 136 ± 32 | 0 | 403 ± 79 | 194 ± 26 | 0 |
| Cu | 2713 ± 25 | 9113 ± 67 | 47 ± 2 | 4053 ± 5 | 11,000 | 68 ± 2 |
| Sn | 1703 ± 26 | 308 ± 6 | <20 | 2617 ± 52 | 376 ± 5 | <20 |
| Bi | 941 ± 8 | 181 ± 2 | 17 ± 1 | 1440 ± 20 | 224 ± 3 | 28 ± 0 |
| [ppb] | | | | | | |
| Au | 367 ± 9 | 264 ± 13 | 73 ± 10 | 547 ± 18 | 333 ± 103 | 114 ± 8 |

Considering the chemical analyses of the used iron precipitation residues in the mixture and the solid residues of each experiment, the calculation of extraction rates $e_{El}$ for the elements of interest in percent was performed according to Equation (1):

$$e_{El} = \left(1 - \frac{x_{El(Fin)} * m_{Fin}}{x_{El(Res)} * m_{Start} * f_R}\right) * 100\% \quad (1)$$

where $m_{Start}$ is the mass of the mixture (g), $f_R$ is the respective share of precipitation residue in the mixture, and $m_{Fin}$ represents the mass of final solid residue from the chlorination treatment (g). Furthermore, $x_{El(Res)}$ represents the concentration of element El in the iron precipitation residue (wt %), and $x_{El(Fin)}$ represents the concentration of element El in the final solid residue (wt %).

Due to the high number of calculated extraction rates in trials with residues for the zinc industry and the resulting difficulty in evaluating the overall process performance, an evaluation factor was calculated that summarizes the respective extraction rates as a function of the element concentration and the current market price of the metals. The factor

(Equation (2)) is referred to as valex (value extraction) and enables the evaluation of the tests in monetary terms with regard to the simultaneous recovery of all considered valuable metals. The elements considered in calculating the valex values were In, Ag, Zn, Pb, Cu, Sn, Au, and Bi. The calculated valex factor ranges from 0 to 1 whereby it has to be stated that the amount of residue in the mixture is not considered, therefore giving only information of the extraction itself but not on the actual economic performance. The valex factor can be interpreted as the theoretic recovery of economic potential from the respective trial mixture.

$$\text{valex} = \frac{\sum x_{\text{El(Res)}} * P_{\text{El}} * e_{\text{El}}}{\sum x_{\text{El(Res)}} * P_{\text{El}}} \tag{2}$$

$P_{El}$ represents the following respective market prices as of December 2021 as given in Table 2.

**Table 2.** Market prices of the considered metals as of 29 December 2021.

| Metal | Ag [52] | Au [53] | Bi [54] | Cu [55] | In [56] | Pb [57] | Sn [58] | Zn [59] |
|---|---|---|---|---|---|---|---|---|
| Price [€/kg] | 640 | 50,900 | 6.41 | 8.5 | 390 | 2.04 | 35 | 3.16 |

## 4. Results

### 4.1. STA Trials

The objective of testing pure metal chlorides is to gather information on their potential as chlorine-liberating compounds. Figure 4a depicts the mass loss in the STA trials with NaCl and KCl showing full evaporation within the maximum temperature range. The whole amount of sample material was evaporated, indicating no release of chlorine. The highlighted peaks in the DTA signal show the melting points of the chlorides. The assumed behavior that $CaCl_2 \cdot 6H_2O$ reacts to $CaCl_2$ (no splitting of HCl) was confirmed according to Figure 4b. The theoretical remaining mass when reacting to $CaCl_2$ was calculated (black horizontal line), which covers precisely with the experimental curve. $MgCl_2 \cdot 6H_2O$ shows the opposite characteristic as the curve covers with the theoretical value for remaining MgO (evaporation of HCl). Both reactions appear to proceed stepwise. Figure 4c depicts the mass curve for the trial with $AlCl_3 \cdot 6H_2O$, which shows one continuous mass drop to the exact theoretic value of the remaining $Al_2O_3$ (full liberation of HCl). For $FeCl_3 \cdot 6H_2O$, the mass decrease appears in steps, and the final experimentally determined mass is below the theoretical value, which could be attributed to the fact that HCl is not entirely liberated, leading to the partial evaporation of $FeCl_3$ itself. This is in line with the theoretical evaluations illustrated in Figure 2, where the two curves of reactions with $FeCl_3 \cdot 6H_2O$ (hydrolysis and dehydration) are almost identical at lower temperatures. The data in Figure 4d, which summarize the trial series with $In_2O_3$ and all six chlorides, confirm that NaCl and KCl are not suitable for chlorination, as the remaining mass is very close to the calculated values for remaining $In_2O_3$ (no chlorination reaction). In a mixture of $In_2O_3$ and $CaCl_2 \cdot 6H_2O$ (green line), a first mass decrease, which is associated with the splitting of water, is followed by a stable period up to 1100 °C and a subsequent mass decrease, which is not completed at the final temperature of 1400 °C. This can be attributed to a reaction of liquid $CaCl_2$ with the present indium oxide followed by the evaporation of $InCl_3$. However, compared to the trials with $MgCl_2 \cdot 6H_2O$, $AlCl_3 \cdot 6H_2O$, and $FeCl_3 \cdot 6H_2O$, the result is poor, as they show almost full chlorination and evaporation (first two mentioned) and a precisely covering curve for iron chloride.

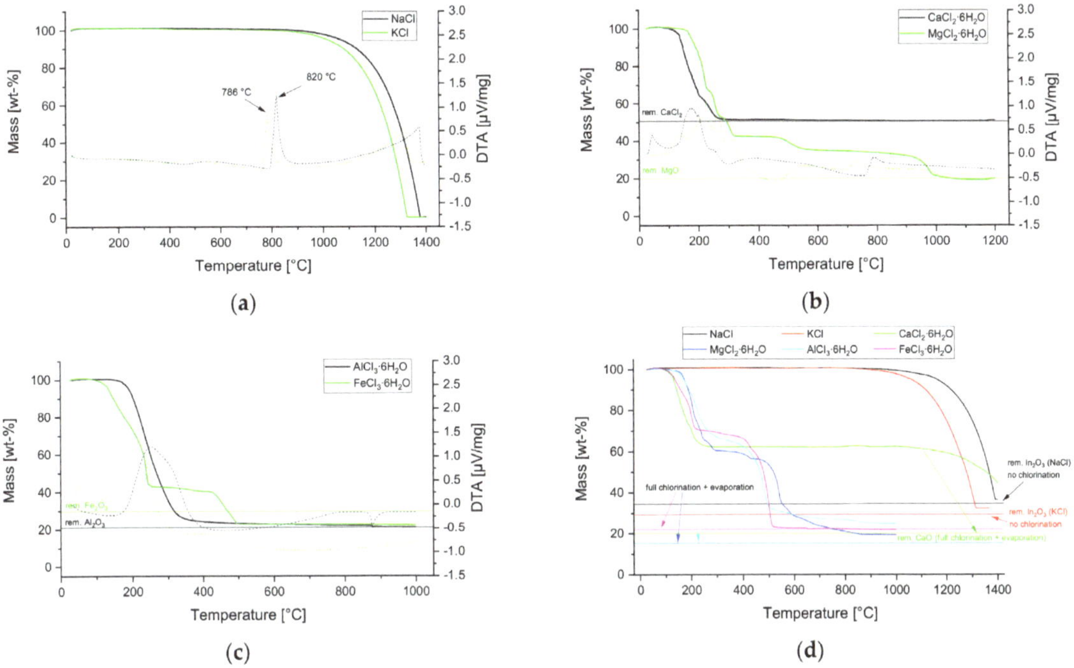


**Figure 4.** Results of STA experiments with (**a**) monovalent, (**b**) bivalent, and (**c**) trivalent metal chlorides; (**d**) comparison of thermogravimetric tests with indium oxide (1.5 times stoichiometric chlorine extent).

Figure 5a–f confirm the practical effectivity of the three utilized chlorides $MgCl_2{\cdot}6H_2O$, $AlCl_3{\cdot}6H_2O$, and $FeCl_3{\cdot}6H_2O$ as chlorination agents for further assessed valuable metal compounds. As depicted, silver chlorination (Figure 5a) proceeds effectively according to the fact that the curves cover with the theoretical remaining mass values for the chlorine carrier oxides. This is in order with the findings in Figure 3a where silver chlorination shows the highest thermodynamic driving force of all elements. As assumed, evaporation starts at elevated temperatures compared to the trials with $In_2O_3$ (Figure 4d), ZnO, and PbO (Figure 5b,c). The chlorination of NiO works insufficiently with the actual mass loss being significantly lower than the theoretical calculated values (Figure 5d). In mixtures where indium is present as sulfate, as shown in Figure 5e, it is observed that chlorination seems to be hindered, as the experimental final mass is significantly higher than the calculated values. For silver sulfate (Figure 5f), the effect is not as pronounced but still detectable compared to trials with silver oxide. It is noteworthy that in both campaigns, the three curves show mass drops in the same order with rising temperature (Fe before Al before Mg), which can be attributed to the splitting of $SO_3$ ($SO_2$ + $1/2O_2$, respectively). This is in line with findings in the literature stating decomposition temperatures of 530 °C for $Fe_2(SO_4)_3$ [60], 610 to 990 °C for $Al_2(SO_4)_3$ [61], and 1085 °C for $MgSO_4$ [62].

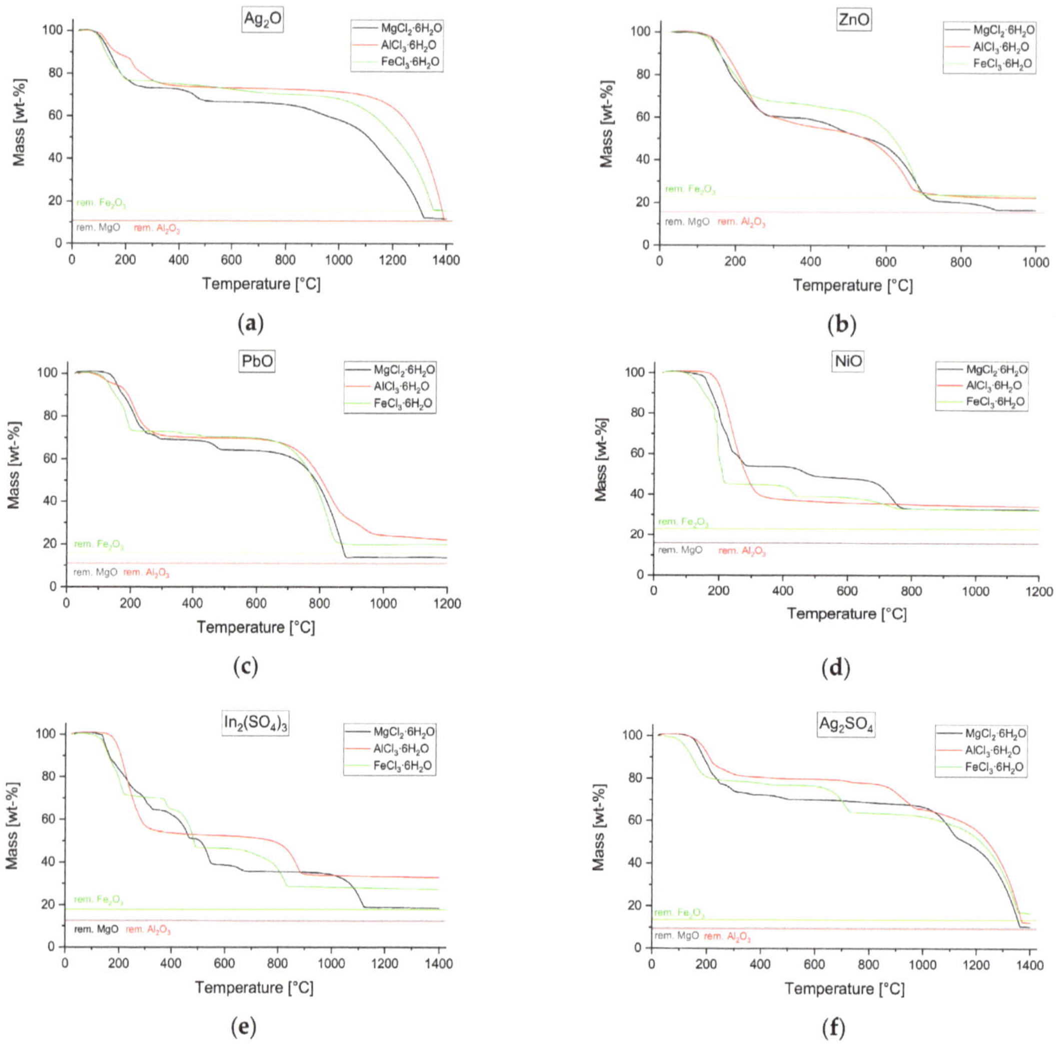


**Figure 5.** Thermogravimetric experiments with mixtures of (**a**) $Ag_2O$, (**b**) ZnO, (**c**) PbO, (**d**) NiO, (**e**) $In_2(SO_4)_3$, and (**f**) $Ag_2SO_4$ and the three investigated chlorides $MgCl_2{\cdot}6H_2O$, $AlCl_3{\cdot}6H_2O$, and $FeCl_3{\cdot}6H_2O$.

### *4.2. Campaigns with Industrial Residues in a Muffle Furnace*

The calculated valex values in Figure 6a,b show that $MgCl_2{\cdot}6H_2O$ is the most effective chlorination agent for jarosite and calcined jarosite at the base chlorine addition factor. For all reaction systems, the yield is increased with rising temperature; when using $FeCl_3{\cdot}6H_2O$, it appears to be significantly increased by higher rates in the mixture (except for calcined jarosite at 900 °C). $AlCl_3{\cdot}6H_2O$ shows the poorest outcome with very low valex values of slightly over 0.3 for the base chlorine addition. For goethite and calcined goethite (Figure 6c,d), the valex values are generally higher and less diverse with base chlorine addition; trials at 1100 °C again deliver better results compared to those at 900 °C. Yields are increased to over 90% when tripling the base addition of $FeCl_3{\cdot}6H_2O$ for both residues. In mixtures with $MgCl_2{\cdot}6H_2O$, the multimetal recovery is higher when doubling the chlorine supply and subsequently decreased (goethite) or stagnating (calcined goethite) for the three-fold addition. $AlCl_3 \cdot 6H_2O$ again appears to be the weakest chlorination agent.

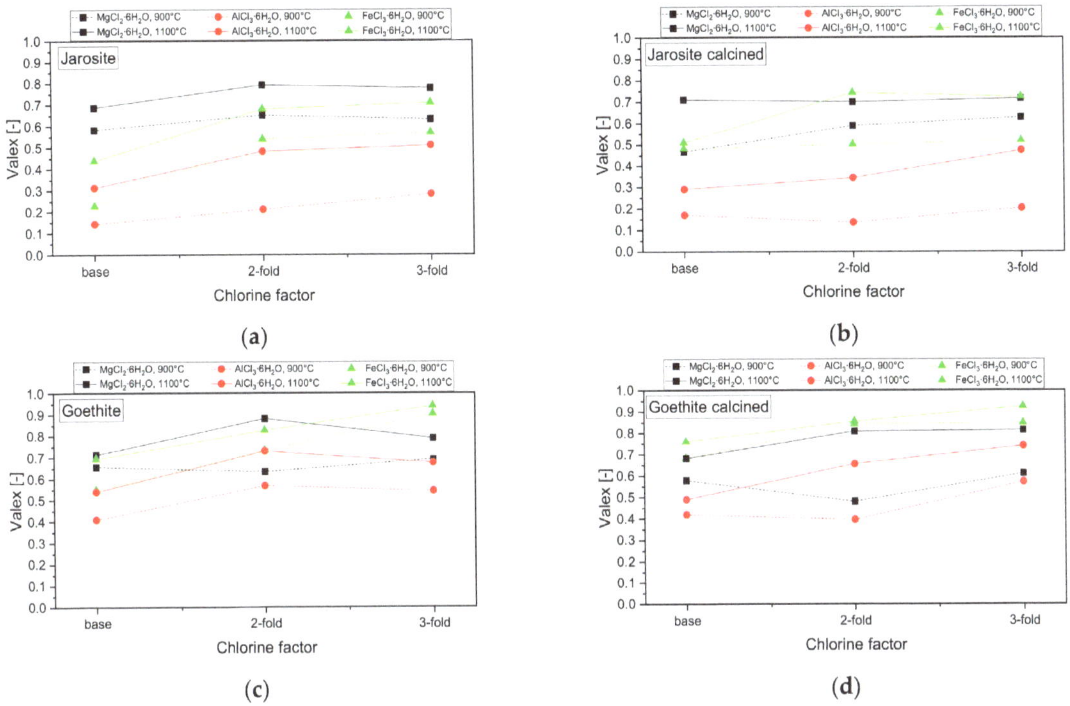


**Figure 6.** Calculated valex values of trials with precipitation residues from the zinc industry at 900 and 1100 °C. (**a**) Jarosite; (**b**) Jarosite calcined; (**c**) Goethite; (**d**) Goethite calcined.

Figure 7 shows the individual extraction rates of the elements in trials at 1100 °C. Every column is dedicated to a certain addition rate of the chloride (2Mg being the two-fold of $MgCl_2 \cdot 6H_2O$ e.g.). Some elements such as gold and bismuth are generally very easily extracted by both $MgCl_2 \cdot 6H_2O$ and $FeCl_3 \cdot 6H_2O$. Indium, zinc, and lead are preferably extracted in the presence of $MgCl_2 \cdot 6H_2O$ from all investigated residues. Silver is hard to extract from jarosite as well as calcined jarosite, but high yields are achieved with all three chlorides from goethite. Tin is almost entirely extractable with higher additions of $FeCl_3 \cdot 6H_2O$ from jarosite.

Nickel extraction (Figure 8a) via chlorination from original nickel jarosite is generally low with 35% being the highest yield when using the base addition of $MgCl_2 \cdot 6H_2O$ at 1100 °C and 38% for the three-fold rate in the mixture. The valex values when using $FeCl_3 \cdot 6H_2O$ are significantly increased by higher additions from 7% (base) to 32% at the three-fold base addition in the mixture. $AlCl_3 \cdot 6H_2O$ reveals poor results with extraction rates of up to 18%. Extraction rates from calcined materials (Figure 8b) are lower with values of 24% and 27% for two-fold and three-fold $MgCl_2 \cdot 6H_2O$ addition.

For all given trials, the extraction rates of iron and chlorine were furthermore calculated. Iron extraction is fluctuating whereby it is higher for trials with goethite materials. The average iron extraction over all trials is 15%. Hereby, the selective chlorination of the valuable elements is confirmed. As the chlorination of iron results in the generation of $FeCl_3$, which was proven to be an effective reaction agent, it is not interpreted to have a negative impact on the process performance.

In general, chlorine is entirely removed for most trials. However, for experiments with $MgCl_2$, it was observed that significant amounts remained in the material, especially for trials with jarosite and calcined jarosite from the zinc industry. When treating nickel jarosite (both original and calcined), chlorine was removed entirely in all trials.

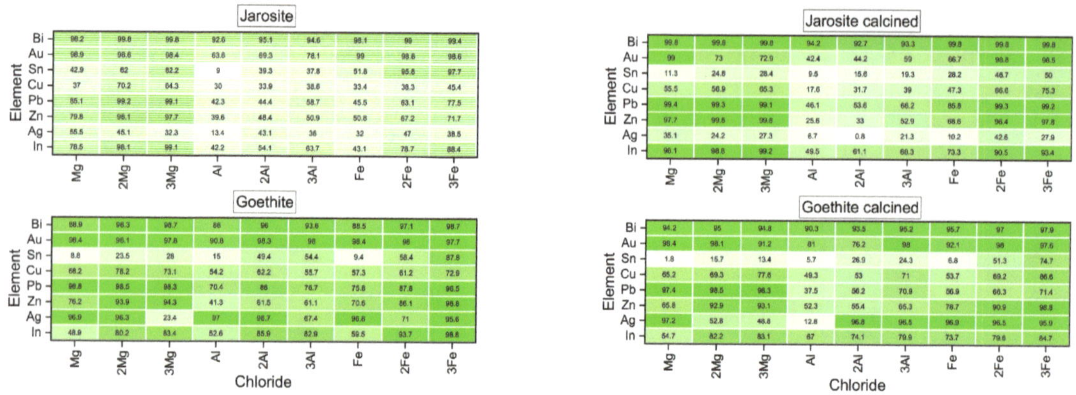


(**a**) (**b**)

**Figure 7.** Individual extraction rates in percentage in trials at 1100 °C with (**a**) jarosite and goethite from the zinc industry in original as well as in (**b**) calcined form.

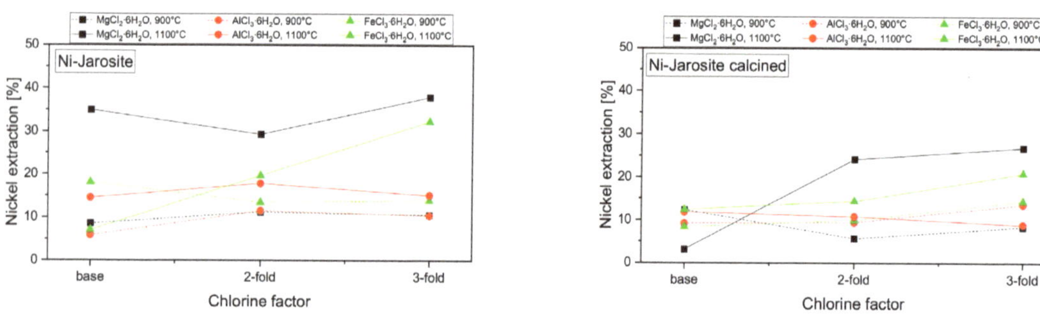


(**a**) (**b**)

**Figure 8.** Extraction rates of nickel in trials with (**a**) original and (**b**) calcined jarosite from the nickel industry.

## 5. Discussion

To cope with the drastic climate situation and security of supply for technology metals in Europe, processes are needed that make valuable metals recoverable from residues with the lowest possible $CO_2$ emissions. In the course of thermodynamic calculations and experiments using STA, as well as in comprehensive test campaigns in a resistance-heated muffle furnace, the extraction of several valuable metals via selective chlorination was demonstrated. However, the calculations and experimental investigations carried out only provide information on the feasibility on a limited scale in a small experimental setting.

In experiments with pure substances by means of STA, the three metal chlorides $MgCl_2 \cdot 6H_2O$, $AlCl_3 \cdot 6H_2O$, and $FeCl_3 \cdot 6H_2O$ that release chlorine in the form of gaseous hydrochloric acid at increasing temperatures were identified as effective for chlorination. $FeCl_3 \cdot 6H_2O$ has practically a lower tendency for hydrolysis; this finding is in agreement with thermodynamic calculations. $MgCl_2 \cdot 6H_2O$ reacts in three hydrolysis and dehydration steps to form MgO. $AlCl_3 \cdot 6H_2O$ splits off the bound crystal water and chlorine in one single decomposition step. From thermodynamic calculations, it appears that the three other chlorides NaCl, KCl, and $CaCl_2{\cdot}6H_2O$ have lower chlorine liberation tendencies. This fact was confirmed by the practical evaluations, which showed that the chlorination of $In_2O_3$ is not or is insufficiently possible with those three reactants.

In trials with pure valuable metal compounds, silver was proven to have a high chlorination tendency, but the extraction is aggravated by the high evaporation temperature. The chlorination and extraction of In, Zn, and Pb from oxides is feasible in mixtures with $MgCl_2{\cdot}6H_2O$, $AlCl_3{\cdot}6H_2O$, and $FeCl_3{\cdot}6H_2O$, Ni shows poor chlorination behavior. In the case of sulfate compounds, chlorination is more difficult as the chlorination agents react under the formation of sulfate compounds which are subsequently decomposed. In general, lower mobility of the formed intermediate phases $SO_2$ or $SO_3$ can also be mentioned as a possible reason for the lower chlorination tendency. In the tests with real residual materials, a different behavior is observed when abruptly heating the mixture by charging into the hot furnace. $AlCl_3{\cdot}6H_2O$ seems to release chlorine completely already at low temperatures due to rapid hydrolysis, where the reaction kinetics are still too low for chlorination. Only $MgCl_2{\cdot}6H_2O$ and $FeCl_3{\cdot}6H_2O$ could be identified as effective for chlorination from residual mixtures. Concerning a recycling process for iron precipitation residues, the addition of $FeCl_3{\cdot}6H_2O$ is the preferred option, since no additional impurities would be introduced.

It is worth mentioning that silver can be almost entirely recovered from goethite and calcined goethite by chlorination with $MgCl_2{\cdot}6H_2O$ and $FeCl_3{\cdot}6H_2O$ (extraction up to 97%). Zinc, lead, gold, and bismuth are extracted in high extents from all residues of the zinc industry using the two mentioned chlorides. Nickel extraction from nickel jarosite is comparably poor, yields are even lower when the material is thermally decomposed before the experiment. $MgCl_2{\cdot}6H_2O$ is the most effective chlorination agent followed by $FeCl_3{\cdot}6H_2O$. For all assessed mixtures, the yield during chlorination is significantly increased at 1100 °C compared to trials at 900 °C. Values for iron extraction are comparably low in all trials with an average yield of 15%, confirming the potential of separation of valuable elements from the excess iron matrix. Chlorine removal is effective in most trials except when $MgCl_2{\cdot}6H_2O$ is used. The experiments served to determine which chlorides are suitable in principle and whether multimetal extraction is possible while iron remains in the residue. However, in the case of a potential large-scale implementation, many aspects have to be taken into account, which were not more closely investigated in this context. For example, the aggregate used must be heated externally or by a carbon-free burner (potentially hydrogen). Furthermore, the decomposition of the precipitation residues and the addition of the hydrated chlorides generate enormous amounts of water vapor. The large amounts of gas make it difficult to filter the chloride condensates contained in low concentrations. Furthermore, it is necessary to separate the resulting condensate fractions to recover the individual valuable metals.

## 6. Conclusions

Multimetal recovery from precipitation residues using solid chlorination agents is a promising recycling option. The best values are obtained for $MgCl_2{\cdot}6H_2O$ and $FeCl_3{\cdot}6H_2O$ whereby higher chloride addition results in higher yields, especially when $FeCl_3{\cdot}6H_2O$ is used. By processing in aggregates in which the chlorine-containing atmosphere is in contact with the residual fraction for a longer time, significantly higher chlorination rates can be expected whereby the chloride additions could be significantly reduced. Furthermore, it is conceivable that the process could be carried out in the sense of combined recycling with the addition of residues that contain the chlorides in appreciable concentrations. This would result in a recycling process which, without the use of carbon as a process additive, would make it possible to recover the valuable metals contained with high economic output. Based on the investigation and positive process performance for precipitation residues as well as for the thermally decomposed residues, applicability to other residual materials is realistic.

**Supplementary Materials:** The following supporting information can be downloaded at https://www.mdpi.com/article/10.3390/app12073590/s1, Table S1: Experimental mixtures used in the trial campaigns at 900 °C and 1100 °C, base addition of the respective chlorides represents the approximate 2-fold of the stoichiometric demand of chlorine.

**Author Contributions:** Conceptualization, L.H. and K.W.; methodology, L.H.; software, L.H.; validation, L.H., K.W. and S.S.; writing—original draft preparation, L.H. and K.W.; writing—review and editing, L.H. and S.S.; visualization, L.H.; supervision, S.S.; project administration, L.H. and S.S.; funding acquisition, S.S. All authors have read and agreed to the published version of the manuscript.

**Funding:** This work was funded by the Austrian Federal Ministry for Digital and Economic Affairs, the National Foundation for Research, Technology and Development and the Christian Doppler Research Association.

**Institutional Review Board Statement:** Not applicable.

**Informed Consent Statement:** Not applicable.

**Data Availability Statement:** The data presented in this article are available on request from the corresponding author.

**Conflicts of Interest:** The authors declare no conflict of interest.

## References

1. Bayliss, P.; Kolitsch, U.; Nickel, E.H.; Pring, A. Alunite supergroup: Recommended nomenclature. *Mineral. Mag.* **2010**, *74*, 919–927. [CrossRef]
2. Dutrizac, J.E.; Jambor, J.L. Jarosites and Their Application in Hydrometallurgy. *Rev. Mineral. Geochem.* **2000**, *40*, 405–452. [CrossRef]
3. Sinclair, R.J. *The Extractive Metallurgy of Zinc*; Australasian Institute of Mining and Metallurgy: Carlton South, VIC, Australia, 2005; ISBN 9781613442166.
4. Calla-Choque, D.; Nava-Alonso, F.; Fuentes-Aceituno, J.C. Acid decomposition and thiourea leaching of silver from hazardous jarosite residues: Effect of some cations on the stability of the thiourea system. *J. Hazard. Mater.* **2016**, *317*, 440–448. [CrossRef] [PubMed]
5. González-Ibarra, A.A.; Nava-Alonso, F.; Uribe-Salas, A.; Castillo-Ventureño, E.N. Decomposition kinetics of industrial jarosite in alkaline media for the recovery of precious metals by cyanidation. *Can. Metall. Q.* **2016**, *55*, 448–454. [CrossRef]
6. Ju, S.; Zhang, Y.; Zhang, Y.; Xue, P.; Wang, Y. Clean hydrometallurgical route to recover zinc, silver, lead, copper, cadmium and iron from hazardous jarosite residues produced during zinc hydrometallurgy. *J. Hazard. Mater.* **2011**, *192*, 554–558. [CrossRef] [PubMed]
7. Orko, I.; Kangas, P.; Lundström, M.; Koukkari, P. Hydrometallurgical Processing of Jarosite Waste to Value-Added Products. In *3rd Symposium on Urban Mining and Circular Economy, SUM2016*; VTT Technical Research Centre of Finland Ltd.: Espoo, Finland, 2016.
8. Linsong, W.; Peng, Z.; Yu, F.; Sujun, L.; Yue, Y.; Li, W.; Wei, S. Recovery of metals from jarosite of hydrometallurgical nickel production by thermal treatment and leaching. *Hydrometallurgy* **2020**, *198*, 105493. [CrossRef]
9. Hoeber, L.; Steinlechner, S. A comprehensive review of processing strategies for iron precipitation residues from zinc hydrometallurgy. *Clean. Eng. Technol.* **2021**, *4*, 100214. [CrossRef]
10. Asokan, P.; Saxena, M.; Asolekar, S.R. Hazardous jarosite use in developing non-hazardous product for engineering application. *J. Hazard. Mater.* **2006**, *137*, 1589–1599. [CrossRef]
11. Asokan, P.; Saxena, M.; Asolekar, S.R. Recycling hazardous jarosite waste using coal combustion residues. *Mater. Charact.* **2010**, *61*, 1342–1355. [CrossRef]
12. Pappu, A.; Thakur, V.K.; Patidar, R.; Asolekar, S.R.; Saxena, M. Recycling marble wastes and Jarosite wastes into sustainable hybrid composite materials and validation through Response Surface Methodology. *J. Clean. Prod.* **2019**, *240*, 118249. [CrossRef]
13. Mymrin, V.; Vazquez Vaamonde, A. New construction materials from Spanish jarosite processing wastes. *Miner. Eng.* **1999**, *12*, 1399–1402. [CrossRef]
14. Mymrin, V.A.; Ponte, H.A.; Impinnisi, P.R. Potential application of acid jarosite wastes as the main component of construction materials. *Constr. Build. Mater.* **2005**, *19*, 141–146. [CrossRef]
15. Mehra, P.; Gupta, R.C.; Thomas, B.S. Properties of concrete containing jarosite as a partial substitute for fine aggregate. *J. Clean. Prod.* **2016**, *120*, 241–248. [CrossRef]
16. Katsioti, M.; Boura, P.; Agatzini, S.; Tsakiridis, P.E.; Oustadakis, P. Use of jarosite/alunite precipitate as a substitute for gypsum in Portland cement. *Cem. Concr. Compos.* **2005**, *27*, 3–9. [CrossRef]

17. Winters, J.; Vos, L.; Canoo, C. Goethite: From Residue to Secondary Building Material-Union Minière's Graveliet ® Process. In Proceedings of the Lead-Zinc 2000 Symposium which was part of the TMS Fall Extraction & Process Metallurgy Meeting, Pittsburgh, PA, USA, 22–25 October 2000; Dutrizac, J.E., Ed.; TMS: Warrendale, PA, USA, 2000; pp. 903–916; ISBN 9781118805558.
18. Castro, L.; Blázquez, M.L.; González, F.; Muñoz, J.A.; Ballester, A. Reductive leaching of jarosites by Aeromonas hydrophila. *Miner. Eng.* **2016**, *95*, 21–28. [CrossRef]
19. Castro, L.; García-Balboa, C.; González, F.; Ballester, A.; Blázquez, M.L.; Muñoz, J.A. Effectiveness of anaerobic iron bio-reduction of jarosite and the influence of humic substances. *Hydrometallurgy* **2013**, *131–132*, 29–33. [CrossRef]
20. Mäkinen, J.; Salo, M.; Hassinen, H.; Kinnunen, P. Comparison of Reductive and Oxidative Bioleaching of Jarosite for Valuable Metals Recovery. *Solid State Phenom.* **2017**, *262*, 24–27. [CrossRef]
21. Boháček, J.; Šubrt, J.; Hanslík, T.; Tláskal, J. Preparing particulate magnetites with pigment properties from suspensions of basic iron(III) sulphates with the structure of jarosite. *J. Mater. Sci.* **1993**, *28*, 2827–2832. [CrossRef]
22. Dutrizac, J.E. Converting jarosite residues into compact hematite products. *JOM* **1990**, *42*, 36–39. [CrossRef]
23. Röpenack, A.; Böhmer, W.; Rieger, H. Verfahren zur Aufarbeitung von Jarosit-Haltigen Rückständen. European Patent Application 90202618.6, 2 October 1990.
24. Cruells, M.; Roca, A.; Patiño, F.; Salinas, E.; Rivera, I. Cyanidation kinetics of argentian jarosite in alkaline media. *Hydrometallurgy* **2000**, *55*, 153–163. [CrossRef]
25. Patiño, F.; Cruells, M.; Roca, A.; Salinas, E.; Pérez, M. Kinetics of alkaline decomposition and cyanidation of argentian ammonium jarosite in lime medium. *Hydrometallurgy* **2003**, *70*, 153–161. [CrossRef]
26. Patiño, F.; Salinas, E.; Cruells, M.; Roca, A. Alkaline decomposition–cyanidation kinetics of argentian natrojarosite. *Hydrometallurgy* **1998**, *49*, 323–336. [CrossRef]
27. Patiño, F.; Viñals, J.; Roca, A.; Núñez, C. Alkaline decomposition-cyanidation kinetics of argentian plumbojarosite. *Hydrometallurgy* **1994**, *34*, 279–291. [CrossRef]
28. Hage, J.L.T.; Schuiling, R.D.; Vriend, S.P. Production of Magnetite from Sodiumjarosite under Reducing Hydrothermal Conditions. The Reduction of Fe III to Fe II with Cellulose. *Can. Metall. Q.* **1999**, *38*, 267–276. [CrossRef]
29. Han, H.; Sun, W.; Hu, Y.; Jia, B.; Tang, H. Anglesite and silver recovery from jarosite residues through roasting and sulfidization-flotation in zinc hydrometallurgy. *J. Hazard. Mater.* **2014**, *278*, 49–54. [CrossRef]
30. Riley, A.L.; Pepper, S.E.; Canner, A.J.; Brown, S.F.; Ogden, M.D. Metal recovery from jarosite waste—A resin screening study. *Sep. Sci. Technol.* **2018**, *53*, 22–35. [CrossRef]
31. Rodriguez, N.; Machiels, L.; Onghena, B.; Spooren, J.; Binnemans, K. Selective recovery of zinc from goethite residue in the zinc industry using deep-eutectic solvents. *RSC Adv.* **2020**, *10*, 7328–7335. [CrossRef]
32. Creedy, S.; Glinin, A.; Matusewicz, R.; Hughes, S.; Reuter, M. Ausmelt Technology for Treating Zinc Residues. *World Metall. Erzmetall* **2013**, *66*, 230.
33. Salminen, J.; Nyberg, J.; Imris, M.; Heegaard, B.M. Smelting Jarosite and Sulphur Residue in a Plasma Furnace. In *PBZN 2020: The 9th International Symposium on Lead and Zinc Processing*; Siegmund, A., Alam, S., Grogan, J., Kerney, U., Shibata, E., Eds.; Springer Nature: Berlin/Heidelberg, Germany, 2020; pp. 391–403; ISBN 978-3-030-37069-5.
34. Mombelli, D.; Mapelli, C.; Barella, S.; Gruttadauria, A.; Spada, E. Jarosite wastes reduction through blast furnace sludges for cast iron production. *J. Environ. Chem. Eng.* **2019**, *7*, 102966. [CrossRef]
35. Piga, L.; Stoppa, L.; Massidda, R. Recycling of industrial goethite wastes by thermal treatment. *Resour. Conserv. Recycl.* **1995**, *14*, 11–20. [CrossRef]
36. Pelino, M.; Cantalini, C.; Abbruzzese, C.; Plescia, P. Treatment and recycling of goethite waste arising from the hydrometallurgy of zinc. *Hydrometallurgy* **1996**, *40*, 25–35. [CrossRef]
37. Pelino, M.; Cantalini, C.; Boattini, P.P.; Abbruzzese, C.; Rincon, J.; Garcia Hernandez, J.E. Glass-ceramic materials obtained by recycling goethite industrial wastes. *Resour. Conserv. Recycl.* **1994**, *10*, 171–176. [CrossRef]
38. Steinlechner, S.; Antrekowitsch, J. Thermodynamic Considerations for a Pyrometallurgical Extraction of Indium and Silver from a Jarosite Residue. *Metals* **2018**, *8*, 335. [CrossRef]
39. Wegscheider, S.; Steinlechner, S.; Leuchtenmüller, M. Innovative Concept for the Recovery of Silver and Indium by a Combined Treatment of Jarosite and Electric Arc Furnace Dust. *JOM* **2017**, *69*, 388–394. [CrossRef]
40. Wang, Y.; Yang, H.; Zhang, G.; Kang, J.; Wang, C. Comprehensive recovery and recycle of jarosite residues from zinc hydrometallurgy. *Chem. Eng. J. Adv.* **2020**, *3*, 100023. [CrossRef]
41. Guan, J.; Wang, S.; Ren, H.; Guo, Y.; Yuan, H.; Yan, X.; Guo, J.; Gu, W.; Su, R.; Liang, B.; et al. Indium recovery from waste liquid crystal displays by polyvinyl chloride waste. *RSC Adv.* **2015**, *5*, 102836–102843. [CrossRef]
42. Park, K.-S.; Sato, W.; Grause, G.; Kameda, T.; Yoshioka, T. Recovery of indium from $In_2O_3$ and liquid crystal display powder via a chloride volatilization process using polyvinyl chloride. *Thermochim. Acta* **2009**, *493*, 105–108. [CrossRef]
43. Takahashi, K.; Sasaki, A.; Dodbiba, G.; Sadaki, J.; Sato, N.; Fujita, T. Recovering Indium from the Liquid Crystal Display of Discarded Cellular Phones by Means of Chloride-Induced Vaporization at Relatively Low Temperature. *Metall. Mater. Trans. A* **2009**, *40*, 891–900. [CrossRef]
44. Gustafsson, A.M.; Steenari, B.-M.; Ekberg, C. Recycling of CIGS solar cell waste materials—separation of copper, indium and gallium by high-temperature chlorination reaction with ammonium chloride. *Sep. Sci. Technol.* **2015**, *38*, 2415–2425. [CrossRef]

45. Li, H.; Ma, A.; Srinivasakannan, C.; Zhang, L.; Li, S.; Yin, S. Investigation on the recovery of gold and silver from cyanide tailings using chlorination roasting process. *J. Alloys Compd.* **2018**, *763*, 241–249. [CrossRef]
46. Qin, H.; Guo, X.; Tian, Q.; Zhang, L. Pyrite enhanced chlorination roasting and its efficacy in gold and silver recovery from gold tailing. *Sep. Purif. Technol.* **2020**, *250*, 117168. [CrossRef]
47. Jaafar, I.; Griffiths, A.J.; Hopkins, A.C.; Steer, J.M.; Griffiths, M.H.; Sapsford, D.J. An evaluation of chlorination for the removal of zinc from steelmaking dusts. *Miner. Eng.* **2011**, *24*, 1028–1030. [CrossRef]
48. Kurashima, K.; Matsuda, K.; Kumagai, S.; Kameda, T.; Saito, Y.; Yoshioka, T. A combined kinetic and thermodynamic approach for interpreting the complex interactions during chloride volatilization of heavy metals in municipal solid waste fly ash. *Waste Manag.* **2019**, *87*, 204–217. [CrossRef]
49. Wang, S.-J.; He, P.-J.; Lu, W.-T.; Shao, L.-M.; Zhang, H. Comparison of Pb, Cd, Zn, and Cu chlorination during pyrolysis and incineration. *Fuel* **2017**, *194*, 257–265. [CrossRef]
50. Nowak, B.; Frías Rocha, S.; Aschenbrenner, P.; Rechberger, H.; Winter, F. Heavy metal removal from MSW fly ash by means of chlorination and thermal treatment: Influence of the chloride type. *Chem. Eng. J.* **2012**, *179*, 178–185. [CrossRef]
51. Yu, J.; Sun, L.; Ma, C.; Qiao, Y.; Xiang, J.; Hu, S.; Yao, H. Mechanism on heavy metals vaporization from municipal solid waste fly ash by $MgCl_2 \cdot 6H_2O$. *Waste Manag.* **2016**, *49*, 124–130. [CrossRef] [PubMed]
52. Silver Price. Available online: https://www.goldpreis.de/silberpreis/ (accessed on 29 December 2021).
53. Gold Price. Available online: https://www.goldpreis.de/ (accessed on 29 December 2021).
54. Bismuth Price. Available online: https://price.metal.com/Bismuth-Selenium-Tellurium (accessed on 29 December 2021).
55. Copper Price. Available online: https://www.finanzen.at/rohstoffe/kupferpreis (accessed on 29 December 2021).
56. Indium Price. Available online: http://www.indium-preis.de/ (accessed on 29 December 2021).
57. Lead Price. Available online: https://www.finanzen.at/rohstoffe/bleipreis (accessed on 29 December 2021).
58. Tin Price. Available online: https://www.finanzen.at/rohstoffe/zinnpreis (accessed on 29 December 2021).
59. Zinc Price. Available online: https://www.finanzen.at/rohstoffe/zinkpreis (accessed on 29 December 2021).
60. Zboril, R.; Mashlan, M.; Papaefthymiou, V.; Hadjipanayis, G. Thermal decomposition of $Fe_2(SO_4)_3$: Demonstration of $Fe_2O_3$ polymorphism. *J. Radioanal. Nucl. Chem.* **2003**, *255*, 413–417. [CrossRef]
61. Pelovski, Y.; Pietkova, W.; Gruncharov, I.; Pacewska, B.; Pysiak, J. The thermal decomposition of aluminum sulfate in different gas phase environments. *Thermochim. Acta* **1992**, *205*, 219–224. [CrossRef]
62. Scheidema, M.N.; Taskinen, P. Decomposition Thermodynamics of Magnesium Sulfate. *Ind. Eng. Chem. Res.* **2011**, *50*, 9550–9556. [CrossRef]

water

MDPI

*Article*

# Open-Loop Exhaust-Gas Cleaning System: Analysis of Effects Produced on Barcelona Port Water pH

**Nelson Gustavo Díaz Delgado and Francesc Xavier Martínez De Osés ***

Department of Engineering and Nautical Sciences, UPC-Barcelona Tech, 08034 Barcelona, Spain; nelsnew@hotmail.com
* Correspondence: francesc.xavier.martinez@upc.edu

**Abstract:** The implementation of a 0.5% mass/mass sulphur cap in fuels used by ships has become a reality. Furthermore, regulation 14 of the MARPOL Convention–Annex VI (amended) establishes that the limit on fuel used by ships operating in a Sulphur Emission Control Area (SECA) cannot exceed 0.1% of mass/mass. To deal with these requirements while considering nature of the shipping business, which is the continuous carriage of cargo around the world, shipowners or shipping companies have few options for following this regulation. Apart from the use of a low-sulphur-content fuel is the use of an exhaust-gas cleaning system, also known as "scrubbers", as an alternative. The use of these systems, specifically the open-loop system, entails the discharge to the sea of residual water used during the cleaning process of exhaust gases from ship engines. The objective of this paper is to study the effect produced by discharging this residual water on the acidity level (pH) of Barcelona port water. This objective was achieved through the periodical sampling and analysis of Barcelona port water in collaboration with the IDAEA (CSIC) laboratory. We analysed the evolution of the pH results obtained so far and obtained an initial picture of the pH Barcelona port water situation in real time regarding the effect of wash-water discharges from open-loop scrubbers. Furthermore, this paper describes the implementation of a system which is going to improve the operation of open-loop exhaust-gas cleaning systems.

**Keywords:** open loop; water pollution; acidification; pH

**Citation:** Delgado, N.G.D.; Martínez De Osés, F.X. Open-Loop Exhaust-Gas Cleaning System: Analysis of Effects Produced on Barcelona Port Water pH. *Water* **2022**, *14*, 1146. https://doi.org/10.3390/w14071146

Academic Editors: Dibyendu Sarkar, Rupali Datta, Prafulla Kumar Sahoo and Mohammad Mahmudur Rahman

Received: 15 March 2022
Accepted: 31 March 2022
Published: 2 April 2022



## 1. Introduction

The MARPOL Convention [1] has implemented a new amendment which came into force on 1 January 2020, detailed in regulation 14 [2]—sulphur oxides and particulate matter, of Annex VI [3]—Prevention of Air Pollution from Ships. This amendment sets a ban on the carriage of non-compliant fuel oil for combustion purposes, for propulsion or operation onboard a ship. Non-compliant fuel means, after the entering into force of the above-mentioned amendment, a fuel with a sulphur content limit of 0.5% mass by mass. Resolution MEPC.320(74) [4], adopted on 17 May 2019, established the guidelines for ensuring the consistent implementation of the 0.5% sulphur limit under MARPOL Annex VI. These guidelines established a set of recommendations and technical considerations, as well as several requirements to be used by shipowners, administrations, shipbuilders and fuel oil suppliers regarding fuel oil availability, fuel oil tanks and fuel oil system configuration, among others. The above-mentioned guidelines also set out several recommendations to be followed by Port State Control Officers during inspections. In this context, Resolution MEPC.321(74) [5], adopted on 17 May 2019, establishes the specific guidelines for Port State Control Officers performing inspections under MARPOL Annex VI.

In this scenario, the current and available options that shipping companies have for following this specific regulation are very limited. One of them is the use of a compliant fuel whose content is less than or equal to 0.5% mass by mass. During our research, a tracking of ships calling into Barcelona port during the period between January 2020 and

January 2022 was carried out in terms of if they are using very low sulphur fuel oil (VLSFO) or an exhaust-gas cleaning system (EGCS) during their calls in port. After analysing all the obtained data, we can assert that the use of VLSFO is currently the most-chosen option among ships calling Barcelona port. The main problem with this option is that the ship depends on the availability of this fuel oil and its price in each port is different. Currently (January 2022), the price of very low sulphur fuel oil in a European port such as Rotterdam has peaked at 652.5 $/mt as showed in Figure 1. From 20 December 2021, the price of VLSFO (529 $/mt) has experienced, in general terms, a very strong increasing trend that seems to be likely to continue in the future.

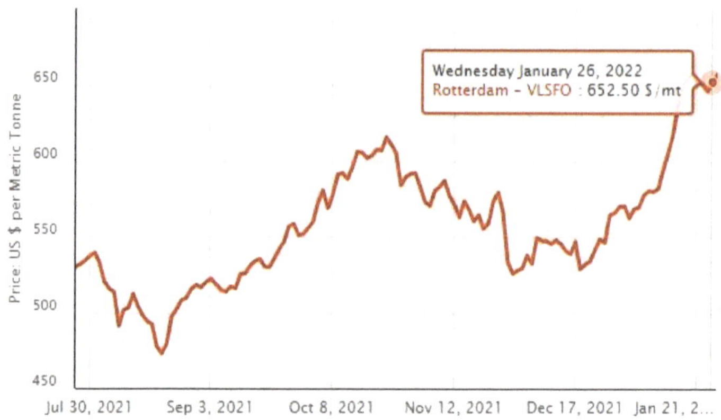


**Figure 1.** View of price evolution for VLSFO. Source: shipandbunkers.com (accessed on 14 March 2022).

Another option, one less chosen by shipping companies in the case of existing ships, is to use alternative clean fuels—for instance, natural gas. The installation of natural gas propulsion onboard an existing ship requires a great investment, and the lack of sufficient space onboard usually requires a long time in a dry dock. Ship owners mainly use this option for new ships. In Barcelona port, two ships engaged in a regular service between the Balearic Islands and Barcelona that call into the port every day are fitted with liquified natural gas engines; one of them is the Ro/Pax ship "ABEL MATUTES" from the company "BALEARIA". This ship was refitted with dual-fuel engines that can use natural gas and conventional fuels. Furthermore, the ship is fitted with two natural gas storage tanks of 178 m$^3$ each (Figure 2).

**Figure 2.** View of LNG storage tank onboard the ship "ABEL MATUTES". Source: The authors.

Finally, the third most-chosen option by shipping companies for dealing with this new regulation is to install exhaust-gas cleaning systems onboard their ships, also known as "scrubbers".

These systems represent, as stated in research regarding the costs–benefits of strategies in compliance with low sulphur [6], a great investment for shipping companies; furthermore, the profitability of this investment is strongly linked to the ship's life. As per the Bureau Veritas publication (2020) [7], the average cost of an exhaust-gas cleaning system installation is about 2 million dollars per scrubber; depending on the vessel and depending on fuel oil evolution prices, the return on investment (ROI) period for an exhaust-gas cleaning system installation is between 1–4 years [6]. It seems to make no sense to invest in this for a ship whose useful life is going to expire soon.

After the use of a compliant fuel, the third option (use of scrubbers) is what most ship owners are choosing for their ships. These systems are classified generally into two main groups, as follows: dry scrubbers and wet scrubbers. Dry scrubbers, basically, remove sulphur oxides from exhaust gases through the action of granules of hydrated lime. During the exhaust-gas cleaning process, dry scrubbers do not generate wastewater to discharge into the sea.

Wet scrubbers use water that is sprayed over the exhaust-gas cleaning system during the cleaning process to remove sulphur oxides. Wet scrubbers are also classified into three main groups: open-loop systems, closed-loop systems and hybrids.

This paper is focused on the research we are carrying out on the effects produced by the discharge of residual wash water from open-loop exhaust-gas cleaning systems into port water. We are studying the pH level of Barcelona port water and its changes, as well as its own capacity to neutralize the acidic waters from scrubber discharges.

Currently there are several studies regarding the effect that wash water discharges from open-loop scrubbers are having on seawater and/or port water parameters; we should highlight some of them, based on the aim of the research and the results obtained. All scrubbers—open-loop, closed-loop and hybrid—discharge water that is more acidic and turbid than the surrounding water. This contributes to ocean acidification and worsens water quality (Bryan Comer, Ph.D., Elise Georgeff and Liudmila Osipova, Ph.D. (2020)) [8]. The use of these scrubbers allows ships to run on heavy fuel and is potentially responsible for a general decrease in pH of at least 0.004 over the whole area of study, including the English Channel and the southern part of the North Sea, and to a decrease of 0.031 or more in coastal areas with intense shipping traffic (Dulière V., Baetens K. and Lacroix G. (2020)) [9]. The most popular type of scrubber, open loop, constantly discharges large amounts of wash water that is acidic and contains polycyclic aromatic hydrocarbons (PAHs), particulate matter, nitrates, nitrites and heavy metals including nickel, lead, copper and mercury—all of which are discharged into the aquatic environment where they can damage marine ecosystems and wildlife and worsen water quality (Liudmila Osipova, Ph.D., Elise Georgeff and Bryan Comer, Ph.D. (2021)) [10]. The average difference between inlet and outlet water pH was 2.05, meaning that on average, the outlet water was 102.05 (=110)-times more acidic than the inlet water (Anna Lunde Hermansson, Ida-Maja Hassellov, Jana Moldanova, Erik Ytreberg (2021)) [11].

This paper, as well as giving a picture of how the pH level of South Basin Barcelona port water is evolving with the use of open-loop scrubbers discharging their acidic residual wash water, also presents an alternative for avoiding these systems that contributes to acidification—thus improving their operation.

The first part of this paper talks about the current legal framework that regulates the use of exhaust-gas cleaning systems. In this section, we show a general overview of the main regulations and requirements that scrubbers must meet.

Secondly, we analyse the current situation, considering the last research on this topic, studying the availability of new technologies applied to this problem and making a compilation of main ideas from several scientific papers on this subject.

Thirdly, we explain the methodology that we have followed in this research, with an analysis of the data obtained and a study of the trends for certain water parameters.

Afterwards, we provide a summary of the preliminary results obtained, analysing the current scenario of Barcelona port water parameters and their possible evolution.

Finally, we discuss the conclusions based on the achieved results, introducing new measures for improving the operation of exhaust-gas cleaning systems.

## 2. Legal Framework

Regulation 14 [1]—sulphur oxides and particulate matter, of Annex VI [3]—Prevention of Air Pollution from Ships—MARPOL Convention [2], makes the implementation of sulphur limits on the fuels used for combustion purposes onboard ships mandatory.

Regulation 4.1 [12] of the same Annex, regarding equivalent arrangements used onboard ships in order to be in compliance with regulation 14, establishes that "The Administration of a Party may allow any fitting, material, appliance or apparatus to be fitted in a ship or other procedures, alternative fuel oils, or compliance methods used as an alternative to that required by this Annex if such fitting, material, appliance or apparatus or other procedures, alternative fuel oils, or compliance methods are at least as effective in terms of emissions reductions as that required by this Annex, including any of the standards set forth in regulations 13 and 14".

From the above, it is clear that exhaust-gas cleaning systems are effectively considered equivalent arrangements.

In addition to the previous regulation, the Resolution from the Marine Environment Protection Committee MEPC.259(68) [13] determines in Annex 10, among other things, the wash-water discharge criteria. This point states that the wash water used in the systems for the exhaust-gas cleaning process must be continuously monitored and recorded when this system is being used by ships in ports, harbours or estuaries.

The values which must be monitored are, at minimum, pH, PAH (Polycyclic Aromatic Hydrocarbons), turbidity/suspended particulate matter and temperature. For the parameters listed above, this Resolution indicates the limits which any wash-water discharge from scrubbers must comply with.

Furthermore, Resolution MEPC.259(68) [13] defines both schemes, A and B, for the approval, survey and certification of an exhaust-gas cleaning system and establishes the minimum content and all requirements for Technical Manuals for the above-mentioned schemes.

MEPC.259(68) [13] also determines the type and scope of inspection that each exhaust-gas cleaning system is subjected to, as well as the documentation set that a ship fitted with a scrubber system must have available onboard for inspection if required by authorities—for instance, during Port State Control. This documentation is set out in the above-mentioned Resolution and, depending on the scheme, is required to be as follows:

Ships fitted with an exhaust-gas cleaning system certified as per Scheme A must have onboard [9]:

1. Sulphur Oxides Emissions Compliance Plan (SECP) approved by the Administration.
2. Sulphur Oxides Emissions Compliance Certificate.
3. Technical Manual for Scheme A approved by the Administration.
4. Onboard Monitoring Manual approved by the Administration.
5. EGC Record Book, with a form approved by the Administration.

Ships fitted with an exhaust-gas cleaning system certified as per Scheme B must be provided with [9]:

1. Sulphur Oxides Emissions Compliance Plan (SECP) approved by the Administration.
2. Technical Manual for Scheme B approved by the Administration.
3. Onboard Monitoring Manual approved by the Administration.
4. EGC Record Book, with a form approved by the the Administration.

## 3. Current Situation

As we have seen before, one of the options that ship owners have to follow is Regulation 14, Annex VI, MARPOL Convention, which is to install an exhaust-gas cleaning system (scrubbers) onboard their ships. According to the DNV GL study [14], carried out in 2018 regarding the use of scrubbers for dealing with the new sulphur cap, which entered into force on 1 January 2020, around 80% of scrubbers installed onboard ships are in open-loop mode.

In the specific case of ships calling into Barcelona port with a scrubber installed onboard, as a result of the above-mentioned specific tracking performed for ships that have called into Barcelona port between January 2020 and January 2022, approximately 69% of them had the hybrid option to discharge wash water into the sea and 30% were directly open loop (continuous discharges). Therefore, only 1% of the ships calling into Barcelona port with a scrubber installed onboard were closed loop.

The main characteristic of the open-loop scrubbers is that the seawater used during the exhaust-gas cleaning process (residual wash water) is directly discharged back into the sea. The discharged wash water must follow certain water parameters (wash water criteria) established by IMO Resolution MEPC.259(68) [13] "2015 guidelines for exhaust-gas cleaning systems". Among these water parameters regulated by the IMO Resolution is the pH, which is the focus of our research.

Nowadays, there are several studies around the world looking into the impact of open-loop scrubber discharges. Most of them are focused on the effect that wash water discharges have on the marine environment, specifically on port water, harbour docks, estuaries [15] and coastal water [16]. In this paper, the acidic component of wash water discharged from the open-loop exhaust cleaning system is revealed, among other features. As stated in the paper, the exhaust gases—which are conducted through an exhaust-gas cleaning system scrubbing tower that is sprayed with seawater—produces, due to the high content of sulphur oxides in these exhaust gases, sulphurous acid ($H_2SO_3$) and sulphuric acid ($H_2SO_4$). This research was developed on the effects of wash water discharges from ships into the Antwerp harbour docks and the Scheldt estuaries (Belgium). In these scenarios, the study of the evolution of pH harbour docks and Scheldt estuaries was carried out, considering that 20% of ships calling into these docks are using open-loop scrubbers. The acidic sulphur compounds that are discharged directly into the surface water were observed and their acidifying capacity was greater than vessels operating on low-sulphur fuel (0.1% m/m). In fact, this research showed a decrease in pH of 0.015 units, directly caused by wash water discharge.

Previous research conducted in the year 2015 [12] was focused on the effects of wash water discharges from scrubbers into the North Sea and Baltic Sea, in German coastal waters. This research was focused mainly on the effects on the marine biota of these regions and determined the changes in pH values caused by wash water discharges from scrubbers. It was confirmed that these changes occurred on a small scale and that the acidification appeared faster in limited or closed systems than in the case of the North Sea and Baltic Sea.

This acidification, as concluded in this research, will first affect the phytoplankton, and then any changes in these organisms will affect other biological components:

"The direct impacts of lowering the pH on marine mammals and birds are not serious, but the impact on phytoplankton and thus on the whole food web will also become evident for higher organisms."

"A pH value reduction will have a greater effect in closed water bodies than on open sea."

Some other research regarding the buffering capacity of seawater and oceans has been worked on by authors [17], concluding that the ocean buffer capacity is being reduced due to increasing atmospheric $CO_2$ levels. According to this paper, the seawater buffer capacity will decrease due to future atmospheric $CO_2$ levels by, on average, 34% from 2000 to 2100;

additional research [18] has determined that this kind of activity may contribute to seawater buffering via dissolution, precipitation and mineral surface reactions.

In sight of the above-mentioned papers and their conclusions, we can assume that the wash-water discharges from open-loop exhaust gas-cleaning systems clearly have an acidic component and that it is changing, at least on very small scale, the seawater pH value. These small changes in water pH could endanger the marine biota, first starting to affect phytoplankton. Furthermore, these wash-water discharges are contributing to an ocean buffer capacity reduction due to their acidic component.

## 4. Data Analysis

### 4.1. Introduction

At present, and since the beginning of 2021, we are analysing the evolution of the acidity from Barcelona port water. We have chosen a specific area of the Barcelona port: the South basin. As a result of our research, we have observed that in this port area there are currently berthing ships that are using exhaust-gas cleaning systems during their calls to port—most of them operating in open-loop mode. This is the reason we have chosen this area for our study.

Inside the South Basin of Barcelona port there are mainly car carrier ships berthing. As mentioned before, most of them use open-loop exhaust-gas cleaning systems during their calls. We have to consider that there are two main scenarios regarding the use of open-loop scrubbers onboard these car carriers inside the South Basin Barcelona port; on the one hand is the moment that the ship is manoeuvring for berthing; on the other hand, is when a ship is manoeuvring on departure from the quay. In both situations, the ship is using its main and auxiliary engines at a substantial engine load (%) and the SOx emissions are greater in these moments and, consequently, more acidic water is formed.

The average car carrier ship berthing in the South Basin Barcelona port has the following characteristics:

- Type of ship: car carrier
- Main dimensions: 167 m (length) × 28 m (beam)
- Main engine power: 1 × 11,000 kW
- Auxiliary engines: 4 × 1400 kW
- Wash water flow average: 300–350 $m^3/h$

Furthermore, we must consider the particular situation of this area inside the Barcelona port, which is quite enclosed and not exposed to the open sea at all (Figures 3 and 4).

**Figure 3.** View of the South basin of Barcelona port. Source: elmercantil.com, (accessed on 14 March 2022).

**Figure 4.** View of the South basin of Barcelona port. Source: The Authors.

In order to carry out the analysis of port water, a collaboration with an independent laboratory—the Institute of Environmental Assessment and Water Research (IDAEA), which is part of the Superior Council of Scientific Research (CSIC)—has been carried out for the samples taken from South basin of Barcelona port, determining its pH level. We have been analysing up to three samples per month through the IDAEA laboratory during the first stage.

Furthermore, we have been carrying out the sampling and pH analysis of the South basin water two–three times per week through a digital device approved for measuring pH. The results obtained from this equipment have been compared and complemented with the results obtained from the Institute of Environmental Assessment and Water Research (IDAEA).

### 4.2. IDAEA pH Data

We started the sampling and pH analysis of Barcelona port South basin water with the Institute of Environmental Assessment and Water Research (IDAEA-CSIC) at the beginning of the year 2021 (Figure 5). After carefully studying the first data obtained from these analyses, we have observed that, initially, there was a clear decreasing trend in the South basin water pH levels. We are working with very small values regarding the pH changes, but the aim of this investigation lies in the early detection of a long-term change.

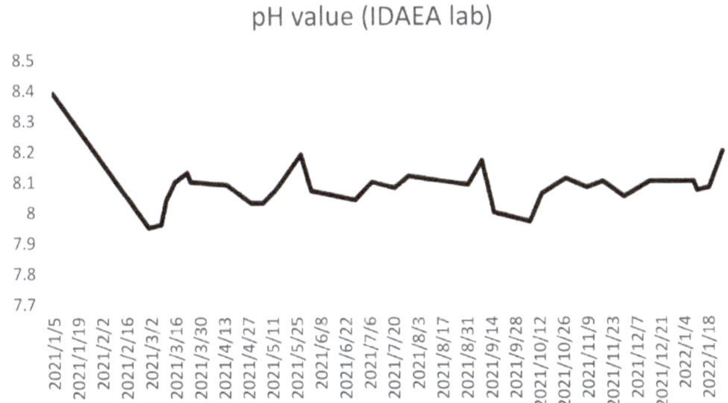


**Figure 5.** IDAEA pH results from the South basin Barcelona port water. Source: The authors.

The above-mentioned decreasing trend in the pH values continued until the first days of March, when they stabilised with a slight increasing trend.

We have also noted as that the results of the pH analysis during the second and third week of March 2021 showed a strong upward trend. From this moment, and during the month of April, we were able observe a slight decreasing tendency until the second week of May, when the pH level increased, exceeding the value of 8.18 units. From this moment there was a strong downward trend that continued until the third week of June, when it started to increase, reaching 8.12 pH units. The situation of the pH level was quite constant until 9 September 2021, when the pH level reached the value of 8.17 units. At the last analysis (27 January 2022), after several small changes, the pH level of the Barcelona port water (South Basin) was 8.2 pH units—the second highest value obtained so far from the CSIC laboratory analysis.

All these variations in the pH level were mainly due to the frequency of ships calling into the South Basin of the Barcelona port using open-loop exhaust-gas cleaning systems during the period of sampling, although they were also to a lesser extent due to the intensity, duration and frequency of precipitation in this area.

*4.3. pH Meter and Refractometer Data*

The sampling and analyses performed with the approved pH meter started at the beginning of March 2021 (Figure 6).

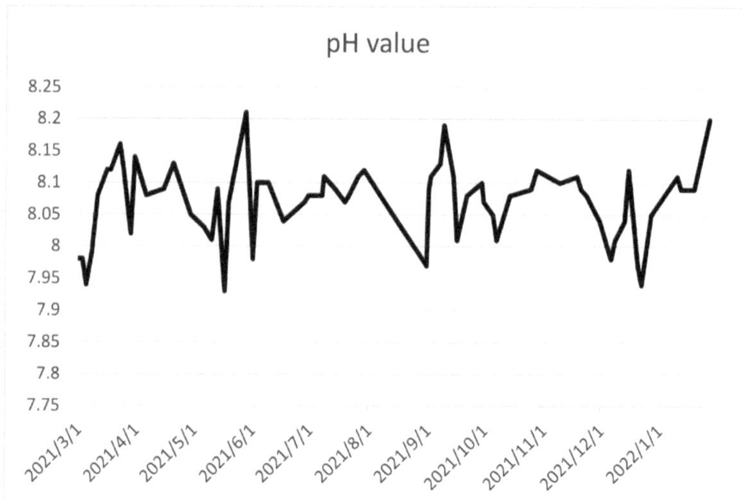


**Figure 6.** pH-meter results from the South Basin Barcelona port water. Source: The authors.

All the results obtained were really close to, and in several cases coincident with, the pH results from the IDAEA laboratory. In this way, we can assert that the results obtained from the pH meter were frequently corroborated by those obtained from the Institute of Environmental Assessment and Water Research laboratory (IDAEA-CSIC).

All the data obtained through the pH meter shows in detail how the pH level of the South basin water has been changing since the beginning of March—the moment in which the downward-trend in pH levels that had occurred since the beginning of year stabilised and began an upward trend. The use of this system helps us to see in more detail—since the pH of the water is analysed more frequently—how the pH changes from week to week.

Furthermore, we have been analysing, through the above-mentioned equipment and using a refractometer as well, other water parameters causally related to the port water basicity. We have analysed the the Oxidation-Reduction Potential (mV) (Figure 7), the salt concentration (g/L) (Figure 8) and the water density (kg/L) (Figure 9).

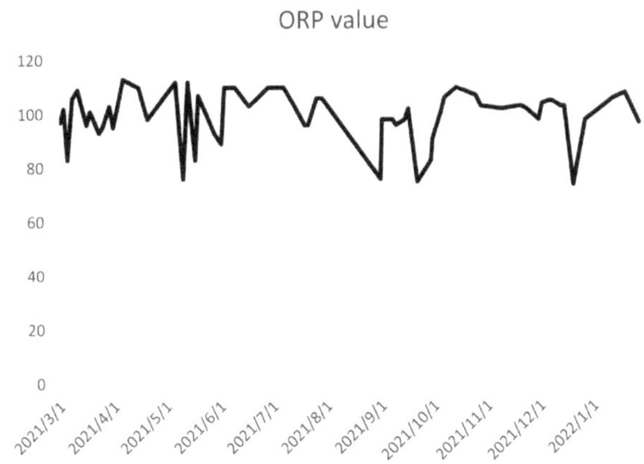


**Figure 7.** Oxidation-Reduction Potential results from the South Basin Barcelona port water. Source: The authors.

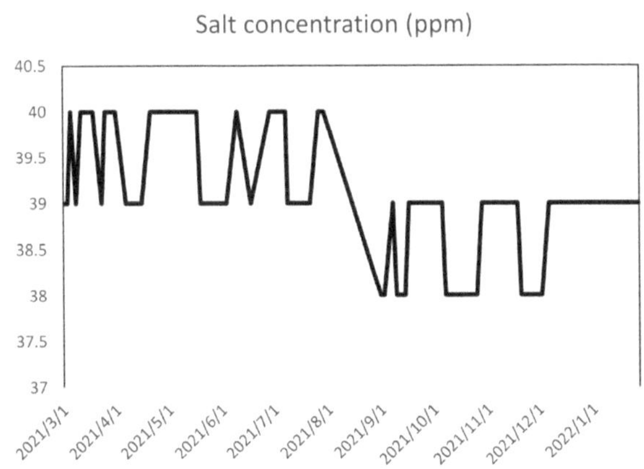


**Figure 8.** Salt concentration results from the South basin Barcelona port water. Source: The authors.

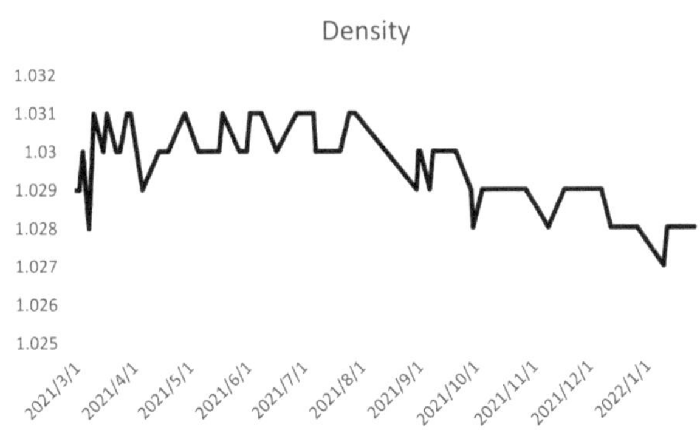


**Figure 9.** Diagram of density results from the South basin Barcelona port water. Source: The authors.

The highest pH value measured with this equipment was observed on 28 May 2021, at 8.21. Then, the pH level showed a strong downward trend at the beginning of June, until it stabilised at the last analysis on 9 June 2021 with a pH value of 8.1 units. It is at the end of July when the pH level experiments showed a clear downward trend until 31 August 2021, when it reached their third lowest value, 7.97 pH units. From this moment, we have observed a strong upward trend up to 8.19 pH units (9 September 2021). Then, the pH level showed a weaker decreasing trend until 14 October 2021, with a pH level of 8.08 units.

From this date, the pH level experienced several small increases and decreases. In the analysis carried out on 22 December 2021, the pH value was found to be 7.94 units. These continuous and slight variations may be due to the small rainfalls that took place in mid-December in Barcelona.

The last analysis performed on 27 January 2022 showed an increasing trend in port water pH, reaching the second highest value obtained so far with this specific equipment (8.2 pH units).

As mentioned before, for the case of pH levels, variations observed in the Oxidation-Reduction Potential value were mainly due to the frequency of ships calling into the South Basin of the Barcelona port using open-loop scrubbers, and due to the frequency of precipitation.

In the case of the salt concentration and port water density, all the variations observed were mainly due to the frequency, intensity and duration of precipitation, although also to a lesser extent by the frequency of ships calling into the South Basin of the Barcelona port using open-loop exhaust-gas cleaning systems.

## 5. Preliminary Results

An extract of the results obtained (Table 1) regarding the South Basin port seawater pH value from both the CSIC laboratory and our specific equipment are shown as follows.

**Table 1.** Extract of results obtained regarding the South Basin port seawater pH values from 5 January 2021 to 20 April 2021.

| Date of Analysis | Place | pH Value (CSIC) | pH Value (Specific Equipment) | ORP Value (Specific Equipment) | Density | Salt Concentration (ppm/%; Refractometer) |
|---|---|---|---|---|---|---|
| 05/01/2021 | South Basin | 8.39 | N/A | N/A | 1.029 | N/A |
| 01/03/2021 | South Basin | 7.95 | 7.98 | 97 mV(+) | 1.029 | 39,000 ppm (3.9%; 39 g/L) |
| 03/03/2021 | South Basin | N/A | 7.98 | 102 mV(+) | 1.03 | 39,000 ppm (3.9%; 39 g/L) |
| 05/03/2021 | South Basin | N/A | 7.94 | 83(+) | 1.028 | 40,000 ppm (4%; 40 g/L) |
| 08/03/2021 | South Basin | 7.96 | 7.99 | 106(+) | 1.031 | 39,000 ppm (3.9%; 39 g/L) |
| 11/03/2021 | South Basin | 8.04 | 8.08 | 109(+) | 1.03 | 40,000 ppm (4%; 40 g/L) |
| 16/03/2021 | South Basin | 8.1 | 8.12 | 96(+) | 1.031 | 40,000 ppm (4%; 40 g/L) |
| 23/03/2021 | South Basin | 8.13 | 8.16 | 93(+) | 1.03 | 39,000 ppm (3.9%; 39 g/L) |
| 25/03/2021 | South Basin | 8.1 | 8.12 | 95(+) | 1.03 | 40,000 ppm (4%; 40 g/L) |
| 29/03/2021 | South Basin | N/A | 8.02 | 103(+) | 1.031 | 40,000 ppm (4%; 40 g/L) |
| 31/03/2021 | South Basin | N/A | 8.14 | 94(+) | 1.031 | 40,000 ppm (4%; 40 g/L) |
| 06/04/2021 | South Basin | N/A | 8.08 | 113(+) | 1.029 | 39,000 ppm (3,9%; 39 g/L) |
| 15/04/2021 | South Basin | 8,09 | 8.09 | 110 | 1.03 | 39,000 ppm (3,9%; 39 g/L) |
| 20/04/2021 | South Basin | N/A | 8.13 | 98 | 1.03 | 40,000 ppm (4%; 40 g/L) |

A complete table of results obtained on the South Basin port seawater pH values from both the CSIC laboratory and our specific equipment is attached in Appendix A.

After the analysis of the obtained data, we have observed the following extremes:

The pH level recovery of South basin water is, as far as we can ascertain for the time being, very high. This is related to the basicity or alkalinity of this water, a concept that determines the capacity of solutes in aqueous systems to neutralize acid. We can observe that there was a downward trend in pH levels from the beginning of the year to the end of February. This downward trend was coincident with several ships calling during this period into the South basin. Most of these ships were using open-loop exhaust-gas cleaning systems during their calls. However, this decreasing trend was countered, stabilised and even modified upward in the short term. In fact, this pH level recovery even took place with some other ships calling into the South basin Barcelona port using exhaust-gas cleaning systems in an open-loop mode.

We can observe that the pH level in every period was causally related with the values of some parameters such as the salt concentration and density of water, among others. This is because the salt concentration of water is a clear indication of the basicity level of the water. As stated by the Arrhenius definition of acids and bases, an acid is a substance such as sulphuric acid that dissolves in water to produce H+ ions (also known as protons) and a base is a substance such as sodium hydroxide that dissolves in water to produce hydroxide ions (OH-; Chemistry libretext website) [19]. In this way, we must remember that water is an ampholyte and that it can behave like an acid or a base.

For the above-mentioned scenario, and considering our own scenario, the South basin water of the Barcelona port, depending on certain external factors such wash-water discharges from open-loop exhaust-gas cleaning systems that have an acidic component such as sulphuric acid, port water can behave like an acidic solution. However, the South basin Barcelona port water may have an extremely high alkalinity due to its high concentration of bicarbonates ($HCO_{3-}$) and carbonates ($CO_{3-2}$).

The latter would explain the great capacity to resist change in pH that the South basin Barcelona port water has. This capacity is defined as the buffering capacity of the water. This is because the above-mentioned bicarbonates and carbonates react with the released H+ from an acid, such as sulphuric acid dissociating in water, preventing the port water pH from dropping.

In all of what is mentioned above lies the importance of this investigation, checking the current situation of the pH levels of the South basin Barcelona port water and its short and long-term trends after having continuous discharges of wash water from scrubbers installed onboard the ships that are operating in open-loop mode during their calls at Barcelona port.

The sampling and analysis of the South basin Barcelona port water pH has been carried out and studied for approximately 2 years and the obtained results are being studied and introduced into a database showing the current situation of the pH levels and their short and long-term trends.

## 6. Discussion

Considering all of what is mentioned above, and despite there currently being no immediate need to reduce the acidity level of the South basin Barcelona port water after its high basicity has been ascertained, we are investigating to find a way to improve open-loop exhaust-gas cleaning systems—in particular, a way to reduce the acidity of the wash waters used during the exhaust-gas cleaning process that are discharged into the sea.

In our research we are going to propose a preventive technique that is to be applied directly to the exhaust-gas cleaning system treatment plants, where, through the implementation of a new step in the wash water treatment, this water is going to be discharged into the sea with almost the same pH as when it was collected. In this way, the current pH of the wash waters discharged from open-loop exhaust-gas cleaning systems, which is around 4–5 pH units measured on the overboard discharge point, will be increased to 7.5–8.2 pH units.

The system proposed is based on the neutralisation reaction, controlling the acidity of residual wash water. Furthermore, to prevent the generation of new residues or increases in salt concentration, an additional treatment will be performed to the wash water before is discharged.

Currently, we expect to start the testing of this proposed system on a small scale soon. Despite this, it is too early to talk about results in terms of the efficiency with which we hope to contribute to seawater acidification reduction, improving the operation of open-loop exhaust-gas cleaning systems.

## 7. Conclusions

That wash water from exhaust-gas cleaning systems has a strong acidic component is a fact. The exhaust-gas cleaning systems' manufacturer manuals consulted confirm this, as they allow a certain pH level in the wash water discharges.

The results obtained so far on the South basin port water pH values confirms that Barcelona port water has a great capacity to assume changes in pH. We have mostly noticed that, generally, the South Basin port water pH recovers to its normal values in a few days after experiencing a small pH drop. We can check the latest data from 17 May 2021, when we observed the lowest South basin port water pH value of 7.93 units. On 19 May 2021, we analysed a new South basin port water sample that was 8.07 pH units, and on 28 May 2021, a new sample was analysed with a value of 8.21 pH units. The latter defines the great capacity of the South Basin Barcelona port water for neutralizing the acidity, in this case, of wash water from scrubbers.

Keeping in mind the-above mentioned research, we can assert that seawater's buffer capacity is not unlimited. Considering that open-loop exhaust-gas cleaning systems also contribute to the acidification of seawater with wash water discharges, at the same time reducing the seawater buffer capacity, we can conclude that these wash-water discharges accelerate the process by which the seawater buffering limit is achieved.

The alkalinity of water, in general terms, is the capacity of water to deal with changes in pH—the resistance of water to becoming acidic. It is also defined as the buffering capacity of the water. This buffering capacity, in the case of seawater, is determined by the presence of certain components such as soluble hydroxide, carbonates and bicarbonates. The higher the alkalinity of the seawater, the higher is its capacity to resist changes in pH. Alkalinity is usually measured in terms of Calcium Carbonate ($CaCO_3$ ppm).

In a hypothetical case where all the buffering components in the South Basin Barcelona port water are neutralized by an acid, the concentration of H+ ions (hydrogen ions, also known as protons) in the water would be increased quickly, with a consequent drastic drop in seawater pH (acidification).

To prevent this hypothetical situation, our investigation is focused on improving, as far as possible, the operation of open-loop exhaust-gas cleaning systems by introducing a new seawater pre-treatment phase before its being used in the scrubbing process.

This new phase will allow ships using open-loop scrubbers to discharge wash water with a pH as neutral as possible into the marine ecosystem, almost at the same level as when the seawater was taken in by the system's inlet.

Our project is based on the introduction of an automatic system that continuously receives the pH data of water monitored at the outlet rack. Once the system has this data, it will compare it with the pH measurement at the inlet rack. In case a drop of pH is detected between the inlet and outlet water, the system will automatically provide an extra alkaline mixture to the tank for use during the exhaust-gas cleaning process at the scrubbing tower.

**Author Contributions:** Conceptualization, N.G.D.D. and F.X.M.D.O.; methodology, F.X.M.D.O.; validation, N.G.D.D. and F.X.M.D.O.; formal analysis, N.G.D.D.; investigation, N.G.D.D.; resources, F.X.M.D.O.; data curation, N.G.D.D.; writing—original draft preparation, N.G.D.D.; writing—review and editing, F.X.M.D.O.; supervision, F.X.M.D.O. All authors have read and agreed to the published version of the manuscript.

**Funding:** This research received no external funding and was only performed with department funds.

**Data Availability Statement:** The data presented in this study are available on request from the corresponding author.

**Conflicts of Interest:** The authors declare no conflict of interest.

## Appendix A. Results Obtained on the South Basin Port Seawater pH Value from Both the CSIC Laboratory and Our Specific Equipment

| Date of Analysis | Place | pH Value (CSIC) | pH Value (Specific Equipment) | ORP Value (Specific Equipment) | Density | Salt Concentration (ppm/%; Refractometer) |
|---|---|---|---|---|---|---|
| 05/01/2021 | South Basin | 8.39 | N/A | N/A | 1.029 | N/A |
| 01/03/2021 | South Basin | 7.95 | 7.98 | 97 mV(+) | 1.029 | 39,000 ppm (3.9%; 39 g/L) |
| 03/03/2021 | South Basin | N/A | 7.98 | 102 mV(+) | 1.03 | 39,000 ppm (3.9%; 39 g/L) |
| 05/03/2021 | South Basin | N/A | 7.94 | 83(+) | 1.028 | 40,000 ppm (4%; 40 g/L) |
| 08/03/2021 | South Basin | 7.96 | 7.99 | 106(+) | 1.031 | 39,000 ppm (3.9%; 39 g/L) |
| 11/03/2021 | South Basin | 8.04 | 8.08 | 109(+) | 1.03 | 40,000 ppm (4%; 40 g/L) |
| 16/03/2021 | South Basin | 8.1 | 8.12 | 96(+) | 1.031 | 40,000 ppm (4%; 40 g/L) |
| 23/03/2021 | South Basin | 8.13 | 8.16 | 93(+) | 1.03 | 39,000 ppm (3.9%; 39 g/L) |
| 25/03/2021 | South Basin | 8.1 | 8.12 | 95(+) | 1.03 | 40,000 ppm (4%; 40 g/L) |
| 29/03/2021 | South Basin | N/A | 8.02 | 103(+) | 1.031 | 40,000 ppm (4%; 40 g/L) |
| 31/03/2021 | South Basin | N/A | 8.14 | 94(+) | 1.031 | 40,000 ppm (4%; 40 g/L) |
| 06/04/2021 | South Basin | N/A | 8.08 | 113(+) | 1.029 | 39,000 ppm (3.9%; 39 g/L) |
| 15/04/2021 | South Basin | 8.09 | 8.09 | 110 | 1.03 | 39,000 ppm (3.9%; 39 g/L) |
| 20/04/2021 | South Basin | N/A | 8.13 | 98 | 1.03 | 40,000 ppm (4%; 40 g/L) |
| 29/04/2021 | South Basin | 8.03 | 8.05 | 106 | 1.031 | 40,000 ppm (4%; 40 g/L) |
| 06/05/2021 | South Basin | 8.03 | 8.06 | 105 | 1.03 | 40,000 ppm (4%; 40 g/L) |
| 10/05/2021 | South Basin | N/A | 8.01 | 76 | 1.03 | 40,000 ppm (4%; 40 g/L) |
| 13/05/2021 | South Basin | 8.07 | 8.09 | 112 | 1.03 | 40,000 ppm (4%; 40 g/L) |
| 17/05/2021 | South Basin | N/A | 7.93 | 83 | 1.03 | 40,000 ppm (4%; 40 g/L) |
| 19/05/2021 | South Basin | N/A | 8.07 | 107 | 1.031 | 39,000 ppm (3.9%; 39 g/L) |
| 28/05/2021 | South Basin | 8.19 | 8.21 | 93 | 1.03 | 39,000 ppm (3.9%; 39 g/L) |
| 01/06/2021 | South Basin | N/A | 7.98 | 89 | 1.03 | 39,000 ppm (3.9%; 39 g/L) |
| 03/06/2021 | South Basin | 8.07 | 8.1 | 110 | 1.031 | 39,000 ppm (3.9%; 39 g/L) |
| 09/06/2021 | South Basin | N/A | 8.1 | 110 | 1.031 | 40,000 ppm (4%; 40 g/L) |
| 28/06/2021 | South Basin | 8.04 | 8.07 | 110 | 1.03 | 40,000 ppm (4%; 40 g/L) |
| 30/06/2021 | South Basin | N/A | 8.08 | 110 | 1.031 | 40,000 ppm (4%; 40 g/L) |
| 07/07/2021 | South Basin | N/A | 8.08 | 110 | 1.031 | 40,000 ppm (4%; 40 g/L) |
| 08/07/2021 | South Basin | 8.1 | 8.11 | 109 | 1.03 | 39,000 ppm (3.9%; 39 g/L) |
| 14/07/2021 | South Basin | N/A | 8.09 | 102 | 1.03 | 39,000 ppm (3.9%; 39 g/L) |
| 19/07/2021 | South Basin | N/A | 8.07 | 96 | 1.03 | 39,000 ppm (3.9%; 39 g/L) |
| 21/07/2021 | South Basin | 8.08 | 8.08 | 96 | 1.03 | 39,000 ppm (3.9%; 39 g/L) |
| 26/07/2021 | South Basin | N/A | 8.11 | 106 | 1.031 | 40,000 ppm (4%; 40 g/L) |
| 29/07/2021 | South Basin | 8.12 | 8.12 | 106 | 1.031 | 40,000 ppm (4%; 40 g/L) |
| 31/08/2021 | South Basin | N/A | 7.97 | 76 | 1.029 | 38,000 ppm (3.8%; 38 g/L) |
| 01/09/2021 | South Basin | 8.09 | 8.1 | 98 | 1.03 | 38,000 ppm (3.8%; 38 g/L) |
| 02/09/2021 | South Basin | N/A | 8.11 | 98 | 1.03 | 38,000 ppm (3.8%; 38 g/L) |
| 07/09/2021 | South Basin | N/A | 8.13 | 98 | 1.029 | 39,000 ppm (3.9%; 39 g/L) |
| 09/09/2021 | South Basin | 8.17 | 8.19 | 96 | 1.03 | 38,000 ppm (3.8%; 38 g/L) |
| 14/09/2021 | South Basin | N/A | 8.11 | 98 | 1.03 | 38,000 ppm (3.8%; 38 g/L) |
| 16/09/2021 | South Basin | 8 | 8.01 | 102 | 1.03 | 39,000 ppm (3.9%; 39 g/L) |

| | | | | | | |
|---|---|---|---|---|---|---|
| 21/09/2021 | South Basin | N/A | 8.08 | 75 | 1.03 | 39,000 ppm (3.9%; 39 g/L) |
| 29/09/2021 | South Basin | N/A | 8.1 | 83 | 1.029 | 39,000 ppm (3.9%; 39 g/L) |
| 30/09/2021 | South Basin | N/A | 8.07 | 91 | 1.028 | 39,000 ppm (3.9%; 39 g/L) |
| 05/10/2021 | South Basin | N/A | 8.05 | 101 | 1.029 | 39,000 ppm (3.9%; 39 g/L) |
| 07/10/2021 | South Basin | 7.97 | 8.01 | 106 | 1.029 | 38,000 ppm (3.8%; 38 g/L) |
| 14/10/2021 | South Basin | 8.06 | 8.08 | 110 | 1.029 | 38,000 ppm (3.8%; 38 g/L) |
| 25/10/2021 | South Basin | N/A | 8.09 | 107 | 1.029 | 38,000 ppm (3.8%; 38 g/L) |
| 28/10/2021 | South Basin | 8.11 | 8.12 | 103 | 1.029 | 39,000 ppm (3.9%; 39 g/L) |
| 09/11/2021 | South Basin | 8.08 | 8.1 | 102 | 1.028 | 39,000 ppm (3.9%; 39 g/L) |
| 18/11/2021 | South Basin | 8.1 | 8.11 | 102 | 1.029 | 39,000 ppm (3.9%; 39 g/L) |
| 20/11/2021 | South Basin | N/A | 8.09 | 103 | 1.029 | 38,000 ppm (3.8%; 38 g/L) |
| 23/11/2021 | South Basin | N/A | 8.08 | 102 | 1.029 | 38,000 ppm (3.8%; 38 g/L) |
| 30/11/2021 | South Basin | 8.05 | 8.05 | 102 | 1.029 | 38,000 ppm (3.8%; 38 g/L) |
| 02/12/2021 | South Basin | N/A | 8.02 | 104 | 1.029 | 38,000 ppm (3.8%; 38 g/L) |
| 06/12/2021 | South Basin | N/A | 7.98 | 105 | 1.029 | 39,000 ppm (3.9%; 39 g/l) |
| 08/12/2021 | South Basin | N/A | 8.01 | 105 | 1.029 | 39,000 ppm (3.9%; 39 g/L) |
| 13/12/2021 | South Basin | N/A | 8.04 | 103 | 1.028 | 39,000 ppm (3.9%; 39 g/L) |
| 15/12/2021 | South Basin | 8.1 | 8.12 | 103 | 1.028 | 39,000 ppm (3.9%; 39 g/L) |
| 20/12/2021 | South Basin | N/A | 7.97 | 105 | 1.028 | 39,000 ppm (3.9%; 39 g/L) |
| 22/12/2021 | South Basin | N/A | 7.94 | 105 | 1.028 | 39,000 ppm (3.9%; 39 g/L) |
| 27/12/2021 | South Basin | N/A | 8.05 | 107 | 1.028 | 39,000 ppm (3.9%; 39 g/L) |
| 10/01/2022 | South Basin | 8.1 | 8.11 | 105 | 1.027 | 39,000 ppm (3.9%; 39 g/L) |
| 12/01/2022 | South Basin | 8.07 | 8.09 | 106 | 1.028 | 39,000 ppm (3.9%; 39 g/L) |
| 19/01/2022 | South Basin | 8.08 | 8.09 | 108 | 1.028 | 39,000 ppm (3.9%; 39 g/L) |
| 27/01/2022 | South Basin | 8.20 | 8.20 | 97 | 1.028 | 39,000 ppm (3.9%; 39 g/L) |

## References


1. International Maritime Organization (IMO). International Convention for the Prevention of Pollution from Ships (MARPOL). Available online: https://www.imo.org/en/About/Conventions/Pages/International-Convention-for-the-Prevention-of-Pollution-from-Ships-(MARPOL).aspx (accessed on 14 March 2022).
2. Regulation 14. Sulphur Oxides and Particulate Matter. Annex VI—Marpol Convention, Amended by Resolution MEPC.305(73), Amendments to the Annex of the Protocol of 1997 to Amend the International Convention for the Prevention of Pollution from Ships, 1973, as Modified by the Protocol of 1978 Relating Thereto. Available online: https://www.imo.org/en/OurWork/Environment/Pages/Sulphur-oxides-(SOx)----Regulation-14.aspx (accessed on 26 October 2018).
3. International Maritime Organization (IMO). International Convention for the Prevention of Pollution from Ships (MARPOL)—Annex VI, Adopted in 1997. Regulation for the Prevention of Air Pollution from Ships. Available online: https://imorules.com/MARPOL_ANNVI.html (accessed on 14 March 2022).
4. Marine Environment Protection Committee (MEPC); International Maritime Organization (IMO). Resolution MEPC.320(74), Guidelines for Consistent Implementation of the 0.50% Sulphur Limit under MARPOL Annex VI. Available online: https://wwwcdn.imo.org/localresources/en/OurWork/Environment/Documents/Resolution%20MEPC.320%2874%29.pdf (accessed on 17 May 2019).
5. Marine Environment Protection Committee (MEPC); International Maritime Organization (IMO). Resolution MEPC.321(74), Guidelines (2019) for Port State Control under MARPOL Annex VI. Available online: https://wwwcdn.imo.org/localresources/en/KnowledgeCentre/IndexofIMOResolutions/MEPCDocuments/MEPC.321(74).pdf (accessed on 17 May 2019).
6. Wu, P.-C.; Lin, C.-Y. Cost-Benefit Evaluation on Promising Strategies in Compliance with Low Sulfur Policy of IMO. *J. Mar. Sci. Eng.* **2021**, *9*, 3. [CrossRef]
7. Marine & Offshore, Bureau Veritas, Client Corner: Your Questions about Scrubbers, Answered. Available online: https://marine-offshore.bureauveritas.com/magazine/client-corner-your-questions-about-scrubbers-answered (accessed on 30 June 2020).
8. Comer, B.; Georgeff, E.; Osipova, L. Air emissions and water pollution discharges from ships with scrubbers. *Consult. Rep. Int. Counc. Clean Transp.* **2020**, 12–13. Available online: https://theicct.org/sites/default/files/publications/Air-water-pollution-scrubbers-nov2020.pdf (accessed on 14 March 2022).

9. Dulière, V.; Baetens, K.; Lacroix, G. *Potential Impact of Wash Water Effluents from Scrubbers on Water Acidification in the Southern North Sea*; Final Project Report; Royal Belgian Institute of Natural Sciences, Operational Directorate Natural Environment, Ecosystem Modelling: Bruxelles, Belgium, 2020; p. 31.
10. Osipova, L.; Georgeff, E.; Comer, B. Global scrubber washwater discharges under IMO's 2020 fuel sulfur limit. *Int. Counc. Clean Transp.* **2021**, 10–12. Available online: https://theicct.org/wp-content/uploads/2021/06/scrubber-discharges-Apr2021.pdf (accessed on 14 March 2022).
11. Hermansson, A.L.; Hassellöv, I.M.; Moldanová, J.; Ytreberg, E. Comparing emissions of polyaromatic hydrocarbons and metals from marine fuels and scrubbers. *Transp. Res. Part D Transp. Environ.* **2021**, *97*, 102912. [CrossRef]
12. International Maritime Organization (IMO). International Convention for the Prevention of Pollution from Ships (MARPOL); Regulation 4.1. Sulphur Oxides and Particulate Matter. Annex VI—Marpol Convention. Available online: https://www.marpol-annex-vi.com/emissions/sulphur-oxides-sox/ (accessed on 14 March 2022).
13. Marine Environment Protection Committee (MEPC); International Maritime Organization (IMO). Resolution MEPC.259(68), Guidelines (2015) for Exhaust Gas Cleaning Systems. Available online: https://wwwcdn.imo.org/localresources/en/KnowledgeCentre/IndexofIMOResolutions/MEPCDocuments/MEPC.259(68).pdf (accessed on 15 May 2015).
14. DNV GL webinar. Scrubber Retrofit—Meeting the Global Sulphur Cap 2020 Limits. Available online: https://www.dnv.com/maritime/webinars-and-videos/on-demand-webinars/Scrubbers-meeting-the-Global-Sulphur-Cap-2020.html (accessed on 10 October 2018).
15. Johannes, T. The Impact of Scrubbers Discharges on the Water Quality in Estuaries and Ports. Available online: https://repository.tudelft.nl/islandora/object/uuid%3A6fcfa43a-6eac-49ae-af17-1a60fafdefdc (accessed on 23 July 2020).
16. Lange, B.; Markus, T.; Helfst, L.P. Impact of Scrubbers on the Environmental Situation in Ports and Coastal Waters. Available online: https://www.umweltbundesamt.de/sites/default/files/medien/378/publikationen/texte_65_2915_impacts_of_scubbers_on_the_envoronmental_situation_in_ports_and_coastal_waters.pdf (accessed on 14 March 2022).
17. Jiang, L.Q.; Carter, B.R.; Feely, R.A.; Lauvset, S.K.; Olsen, A. *Surface Ocean pH and Buffer Capacity: Past, Present and Future;* Earth System Science Interdisciplinary Centre, University of Maryland: College Park, MA, USA, 2019.
18. Middelburg, J.J.; Soetaert, K.; Hagens, M. Ocean Alkalinity, Buffering and Biogeochemical Processes. Available online: https://agupubs.onlinelibrary.wiley.com/doi/epdf/10.1029/2019RG000681 (accessed on 9 June 2020).
19. Chemistry Libretexts, Physical_and_Theoretical_Chemistry_Textbook, the Arrhenius Theory of Acids and Bases. Available online: https://chem.libretexts.org/Bookshelves/Physical_and_Theoretical_Chemistry_Textbook_Maps/Supplemental_Moules_(Physical_and_Theoretical_Chemistry)/Acids_and_Bases/Acid/Overview_of_Acids_and_Bases (accessed on 20 December 2021).

toxics

Article

# Foliar Application of Wood Distillate Alleviates Ozone-Induced Damage in Lettuce (*Lactuca sativa* L.)

**Andrea Vannini [1,*], Riccardo Fedeli [2], Massimo Guarnieri [2] and Stefano Loppi [2,3]**

1 Department of Chemistry, Life Sciences, and Environmental Sustainability, University of Parma, Parco Area delle Scienze 11/a, 43124 Parma, Italy
2 Department of Life Sciences, University of Siena, Via PA Mattioli 4, I-53100 Siena, Italy; riccardo.fedeli@student.unisi.it (R.F.); massimo.guarnieri@unisi.it (M.G.); stefano.loppi@unisi.it (S.L.)
3 BAT Center-Interuniversity Center for Studies on Bioinspired Agro-Environmental Technology, University of Naples 'Federico II', 80138 Napoli, Italy
* Correspondence: andrea.vannini@unipr.it

**Abstract:** This study examined whether foliar applications of wood distillate (WD) have a protective effect on photosynthesis and the antioxidant power of lettuce when exposed to an ecologically relevant $O_3$ concentration. Seedlings of lettuce (*Lactuca sativa* L.) were fumigated daily with 60 ppb of $O_3$ for 30 days, five hours per day. Once per week, 50% of the fumigated plants were treated with foliar applications of 0.2% WD, while control plants were treated with water. The results clearly showed the ability of WD to protect lettuce plants from ozone-induced damage. Specifically, WD-treated plants exhibited lower damage to the photosynthetic machinery, assessed through a series of chlorophyll fluorescence parameters, a higher chlorophyll content, higher antioxidant power, as well as antioxidant molecules, i.e., caffeic acid and quercetin, and higher biomass. Counteracting the overproduction of ozone-generated reactive oxygen species (ROS) is speculated to be the main mechanism by which WD protects the plant from ozone-induced damage.

**Keywords:** bio-based product; crop resistance; horticultural plants; $O_3$; toxic effects mitigation

**Citation:** Vannini, A.; Fedeli, R.; Guarnieri, M.; Loppi, S. Foliar Application of Wood Distillate Alleviates Ozone-Induced Damage in Lettuce (*Lactuca sativa* L.). *Toxics* **2022**, *10*, 178. https://doi.org/10.3390/toxics10040178

Academic Editors: Dibyendu Sarkar, Rupali Datta, Prafulla Kumar Sahoo and Mohammad Mahmudur Rahman

Received: 15 March 2022
Accepted: 3 April 2022
Published: 5 April 2022



## 1. Introduction

Ozone ($O_3$) is a global strongly oxidizing pollutant, present at ground level following the interaction of UV light with anthropogenic gases, such as carbon monoxide (CO), nitrogen oxides ($NO_X$), methane ($CH_4$), and non-methane volatile organic compounds ($NMVOC_S$) (EPA 2021). Since the mid-20th century, global background ozone ($O_3$) concentrations have rapidly increased [1] with current average concentrations for the mid-northern latitude of ca. 30–50 ppb [2].

Plants can tolerate and adapt to $O_3$ when chronically exposed to concentrations below 20 nL/L (ca. 21 ppb; [3]). However, when ground-level exceeds the tolerance threshold generally estimated at ca. 40 ppb [4,5], $O_3$ becomes phytotoxic and capable of damaging agricultural plants and forest vegetation, and even the biodiversity of a whole ecosystem [2,6]. Following the global increase in $O_3$ background concentrations that occurred in the last 20–30 years, 2–15% reductions in the global crop yield were estimated, with a remarkable economic loss [7].

In recent years, the search for solutions to protect crop plants from $O_3$ injury has become of global interest, and among the tested methodologies, i.e., the development of $O_3$-resistant crops, the selection of resistant germplasm, the use of antiozonants and nanomaterials [8], the use of antioxidants (vitamins, phytohormones, flavonoids, and polyamines), is a very promising strategy to protect crop plants from $O_3$ phytotoxicity [9–12].

Wood distillate (WD), also known as pyroligneous acid or wood vinegar, is a bio-based liquid product obtained from the distillation of the gases produced during the pyrolysis of woody biomass for the production of green energy [13]. Such a product, still unexploited in many agricultural sectors (especially those not involved in organic farming), is now having great success in agriculture as a phytostimulant for crop plants since its use can enhance both plant productivity and endogenous defenses against pathogens [14]. WD can be produced by different feedstocks but the common feature of all WDs is the presence, to a greater or lesser extent, of antioxidant molecules such as phenols [15]. Results from a study conducted on the horticultural plant *Lactuca sativa* L. revealed that foliar application of 0.2% WD deriving from sweet chestnut (*Castanea sativa* Mill.) enhances the content of chlorophyll and biomass of this model species [16], probably owing to its high content of antioxidant compounds, such as polyphenols [17]. Foliar application of WD is also used to increase the defenses of olive (*Olea europaea* L.) trees and grapevine (*Vitis vinifera* L.) plants against pathogens [18,19], but the effectiveness of WD in contrasting the phytotoxic action of $O_3$ on crop plant metabolism is still unexplored.

For this reason, this study examines whether foliar applications of WD have a protective effect on lettuce exposed to an ecologically relevant $O_3$ concentration.

In detail, we tested if lettuce plants exposed to $O_3$ but not treated with WD have a similar response, in terms of photosynthetic system, antioxidants, and yield, as plants treated with WD.

## 2. Materials and Methods

### 2.1. Experimental

Seedlings of lettuce (*Lactuca sativa* L.) cv. Cappuccio bionda were bought from a local nursery. In the laboratory, plants were carefully removed from their growth phytocells and transplanted inside plastic pots (10 × 10 × 12 cm) using a commercial potting soil as the substrate. Seedlings were then acclimatized for one week in a climatic chamber at 20 °C, 70% RH, and 350 μM $s^{-1}$ $m^{-2}$ PAR with 15 h of photoperiod (from 5:00 a.m. to 20:00 p.m.). The whole pool of seedlings (24 plants) was then fumigated daily with 60 ppb of $O_3$ for 5 h per day (from 12:00 to 17:00). Once per week, 50% of the fumigated plants (12 plants) were treated with foliar applications of 0.2% sweet chestnut (*Castanea sativa* Mill.) WD (Biodea®), while the remaining 50% of plants were sprayed with water and used as a positive control. Foliar applications were run after the light cycle, following the method described by Vannini et al. [16]. All plants were randomly rotated every three days to minimize any possible influence of microclimatic conditions inside the climatic chamber. The experiment lasted four weeks and was replicated three times. The sweet chestnut WD was selected because it had previously been investigated for its safety, for both the environment and humans [17,20,21].

### 2.2. Photosynthetic Parameters

Since photosynthesis is considered the main target of $O_3$ phytotoxicity [22], its functionality was assessed by means of selected photosynthetic indicators: the chlorophyll fluorescence, the analysis of the fluorescence transients plotted on a logarithmic scale (OJIP transients), and the chlorophyll content.

#### 2.2.1. Chlorophyll Fluorescence Analyses

The analysis of the chlorophyll fluorescence and the analysis of the OJIP transients are key methodologies to assess the functionality of the photosynthetic machinery following exposure to $O_3$ [5,22]. The former was assessed through a number of the OJIP step fluorescence parameters, summarized in Table 1, which describes the ability of the photosynthetic system to absorb, trap, transmit, and convert the absorbed light into energy for $CO_2$ fixation [23], while the latter analyzed the $\Delta V_{OP}$ profile, calculated by the difference between the $V_{OP}$ profile of treated (Ozone + WD) and control (Ozone) samples ($\Delta V_{OP} = V_{OP}$treated (ozone + WD) − $V_{OP}$control(ozone); [24]). Prior to analysis, the seedlings were dark-adapted for 30 min under a dim green light (10 μmol photons $m^{-2}$ $s^{-1}$) and then lighted with an actinic light (3000 μmol photons $m^{-2}$ $s^{-1}$) for one second.

Fluorescence analysis was run using a plant efficiency analyzer (Handy PEA, Hansatech Ltd., Norfolk, UK). Fifteen measurements were taken for each replicate.

**Table 1.** Description of the measured photosynthetic parameters.

| Parameter | Description |
|---|---|
| $ABS/CS_0$ | Absorbance for excited cross-section |
| $TR_0/CS_0$ | Trapping flux for excited cross-section |
| $ET_0/CS_0$ | Energy transmission for excited cross-section |
| $RC/CS_0$ | Number of reaction centers for excited cross-section |
| $DI_0/CS_0$ | Heat dissipation for excited cross-section |
| $F_V/F_M$ | Photosynthetic efficiency |
| $PI_{ABS}$ | Performance index |

#### 2.2.2. Analysis of the Chlorophyll Content

The content of chlorophyll was measured by means of a chlorophyll content meter (CCM 300, Opti-Science Inc., Hudson, NH, USA), a non-destructive instrument that quantifies the amount of chlorophyll on a surface basis ($mg/m^2$) [25]. Fifteen measurements were taken for each replicate.

### 2.3. *Expression of Antioxidants*

#### 2.3.1. Total Antioxidant Power

Fresh samples (ca. 200 mg) were homogenized with 4 mL of 80% ethanol and then centrifuged at 15,000 rpm for five minutes. The supernatant (100 µL) was then added to 1 mL of a DPPH solution prepared following the protocol reported by Vannini et al. [26]. After the reaction (ca. 1 h) samples were read at 517 nm and the results were expressed as % Antiradical Activity (ARA%) following the Formulae:

$$ARA\% = 100 \times [1 - (\text{sample absorbance}/\text{control absorbance})] \quad (1)$$

where control absorbance is the absorbance of the reagents only.

#### 2.3.2. Content of Caffeic Acid and Quercetin

Given the high number of antioxidant compounds in plant leaves, caffeic acid and quercetin were selected as indicators of polyphenols and flavonoids expression [27,28], respectively. The extraction of caffeic acid and quercetin from lettuce leaves was carried out according to Tokusoglu et al. [29], with modifications. The upper part of each of the major fresh leaves (ca. 1 g) was extracted with 3 mL of 70% acetone containing 1% HCl (*v/v*). After the homogenization, 0.6 mL of pure HCl was added and the final mixture was first vigorously shaken and then left at 90 °C for two hours. Subsequently, samples were shaken and then filtered at 0.45 µm. The extracts were directly analyzed by HPLC (Perkin-Elmer series 200) coupled with a Diode Array Detector (DAD). The analysis of both caffeic acid and quercetin was run according to the method used by Kumar et al. [30], combining water (solvent A) and acetonitrile (solvent B) eluted for 21 min as a mobile phase following the gradient: 0–5 min (80% A), 5–8 min (60% A), 8–12 min (50% A), 12–17 min (40% A), 17–21 min (20% A); an Agilent C18 column (4.6 x 250 mm; particle size 5µm) was used. Runs were monitored at 280 and 325 nm for caffeic acid and quercetin, respectively. Quantifications were carried out using calibration curves of caffeic acid and quercetin (from 5–100 µg/mL), prepared by dissolving the two pure reagents (Sigma, Sant Louis, MI, USA) in the same solvent used to extract both molecules from the samples.

### 2.4. *Edible Fresh Biomass*

From each plant, visibly undamaged leaves (marketable, i.e., those without necrotic and senescent areas) were removed from shoots and then weighed on a precision balance. Results were expressed as grams on a fresh weight basis (g FW).

### 2.5. Statistical Analysis

To disentangle differences between treated (Ozone + WD) and untreated (Ozone) plants in terms of chlorophyll fluorescence and chlorophyll content, a linear mixed-effect model (LMEM) was fitted for each variable, with treatment as a fixed effect and plant as a random effect [31]. For model validation, scatterplots of the residual and fitted values were used to check for homoscedasticity, and normal probability (qqnorm) plots as well as the Shapiro–Wilk test to check for normality. Models were fitted using the restricted maximum likelihood (REML) estimation, and the significance of the models was checked with type III Anova using the Satterthwaite method [32]. For all the other parameters analyzed, a permutation t-test was used to check for differences between treated (Ozone + WD) and untreated (Ozone) plants. All calculations were run using the free R software [33]; the packages 'lme4' and 'RVAideMemoire' were used."

## 3. Results

WD-treated samples experienced better photosynthetic performances than those only fumigated with $O_3$ (Table 2). Specifically, WD-treated samples showed a higher absorbance for excited cross-section ($ABS/CS_0$; ca. 3%), energy transmission ($TR_0/CS_0$; ca. 4%), electron transport ($ET_0/CS_0$; ca. 8%), number of reaction centers ($RC/CS_0$; ca. 11%), as well as a lower energy dissipation ($DI_0/CS_0$; ca. 4%). Additionally, WD-treated plants showed a higher ($p < 0.05$) expression of the photosynthetic efficiency ($F_V/F_M$; ca. 1%), of the performance index ($PI_{Abs}$; ca. 25%), and a higher content of chlorophyll (ca. 13%). Analysis of the two fluorescence profiles confirmed the results obtained by $F_V/F_M$, while those of the $V_{OP}$ test indicated only negligible differences between treatments (Figure 1). Moreover, samples treated with WD presented a higher ($p < 0.05$) expression of the total antioxidant power (ca. 140%), as well as of both caffeic acid (ca. 400%) and quercetin (ca. 105%) (Figure 2). A higher ($p < 0.05$) biomass (+18%) was also noted (Figure 3).

**Table 2.** Expression of photosynthetic parameters (mean ± standard error) in samples of *Lactuca sativa* after 30 days of fumigation with 60 ppb of $O_3$ in combination with (Ozone + WD) or without (Ozone) weekly foliar applications of 0.2% chestnut wood distillate (WD).

| Parameter | Ozone | Ozone + WD | *p*-Value |
|---|---|---|---|
| $ABS/CS_0$ | 293 ± 2.5 a | 301 ± 2.6 b | $p < 0.05$ |
| $DI_0/CS_0$ | 51 ± 0.5 a | 49 ± 0.5 b | $p < 0.01$ |
| $TR_0/CS_0$ | 242 ± 2.0 a | 252 ± 2.1 b | $p < 0.001$ |
| $ET_0/CS_0$ | 131 ± 1.4 a | 141 ± 1.4 b | $p < 0.001$ |
| $RC/CS_0$ | 105 ± 1.1 a | 117 ± 1.1 b | $p < 0.001$ |
| $F_V/F_M$ | 0.826 ± 0.001 a | 0.837 ± 0.001 b | $p < 0.001$ |
| $PI_{ABS}$ | 2.05 ± 0.03 a | 2.56 ± 0.04 b | $p < 0.001$ |
| Chlorophyll ($mg/m^2$) | 200 ± 3.7 a | 225 ± 2.7 b | $p < 0.001$ |

Different letters indicate statistically significant ($p < 0.05$) differences between treatments. Abbreviations: absorbance for excited cross-section ($ABS/CS_0$), heat dissipation for excited cross-section ($DI_0/CS_0$), trapping flux for excited cross-section ($TR_0/CS_0$), energy transmission for excited cross-section ($ET_0/CS_0$), number of reaction centers for excited cross-section ($RC/CS_0$), photosynthetic efficiency ($F_V/F_M$), performance index ($PI_{ABS}$), chlorophyll content (Chlorophyll).

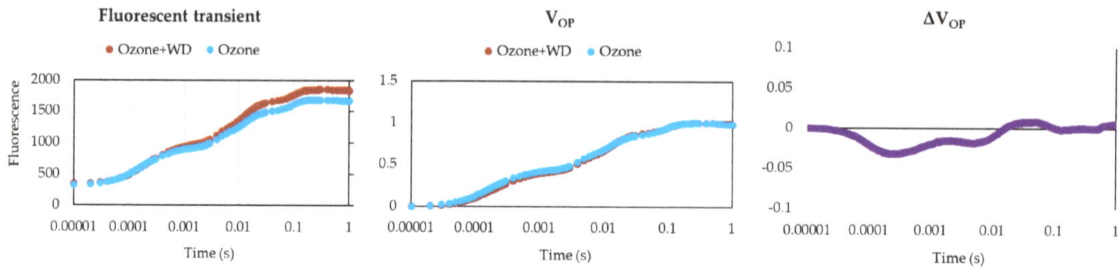


**Figure 1.** Average log-probit OJIP fluorescence curves and $V_{OP}$ analysis of samples of *Lactuca sativa* after 30 days of fumigation with 60 ppb of $O_3$ in combination with (Ozone + WD) or without (Ozone) weekly foliar applications of 0.2% chestnut wood distillate (WD).

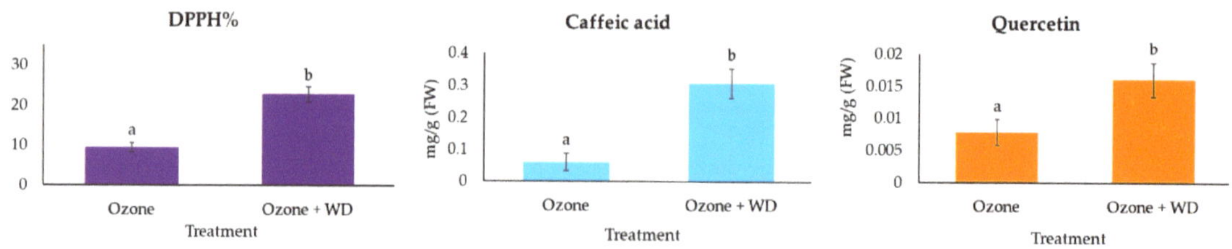


**Figure 2.** Expression of the antioxidant power (DPPH%) and concentration (mean ± standard error) of both caffeic acid and quercetin in samples of *Lactuca sativa* after 30 days of fumigation with 60 ppb of $O_3$ in combination with (Ozone + WD) or without (Ozone) weekly foliar applications of 0.2% chestnut wood distillate (WD). Different letters indicate statistically significant ($p < 0.05$) differences between treatments.

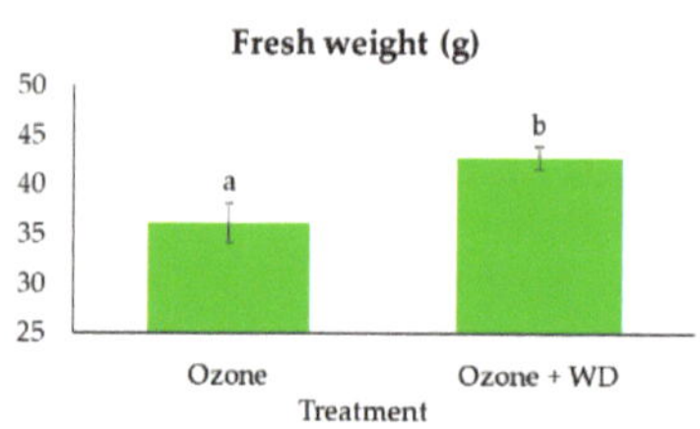


**Figure 3.** Fresh weight (mean ± standard error) of samples of *Lactuca sativa* after 30 days of fumigation with 60 ppb of $O_3$ in combination with (Ozone + WD) or without (Ozone) weekly foliar applications of 0.2% chestnut wood distillate. Different letters indicate statistically significant differences between treatments ($p$-value < 0.05).

## 4. Discussion

Lettuce is an ozone-sensitive horticultural plant [34] and $O_3$ fumigations caused a decrease in photosynthesis, chlorophyll content, and biomass of this species, consistently with other studies [35,36]. However, when 0.2% WD was sprayed on the leaves, plants showed an improvement in the photosynthetic system, a higher content of chlorophyll and antioxidant molecules, as well as higher biomass, thus suggesting the ability of WD to significantly alleviate the $O_3$-induced damage.

WD is rich in antioxidants, such as polyphenols [15,17,37], and is effectively used to increase both the chlorophyll content and biomass of crop plants, as found in some recent studies [16,38–41]. Although the mechanisms behind its effectiveness are still in need of investigation, it has been suggested that this stimulant effect on plant productivity may be due to the action of antioxidant molecules on cell division [16], in response to the activation of specific transcription genes, as previously observed by Tanase et al. [42,43].

To the best of our knowledge, information on the exact mechanism of action of WD against the $O_3$-induced damage is unknown but it is possible that antioxidant molecules play a very important role. We speculate that the pool of antioxidants could have acted in two ways: buffering the interaction between $O_3$ and the apoplast, thus reducing the overproduction of ROS, i.e., the main factor responsible for the $O_3$-induced toxicity in plants [44], and/or stimulating the synthesis of ROS scavenger molecules, such as superoxide dismutase, peroxidase, and catalase, suggested by Wang et al. [45]. In the first case, WD may have acted as a strengthener of the plant's antioxidant defenses, while in the second case it could have acted as a stimulator. However, information on which of these two processes plays the greater role has not yet been investigated.

Plants can naturally counteract the oxidative action of $O_3$ through the synthesis of antioxidants, such as reduced glutathione [46], but when insufficient or when the oxidative stress is too high, plants may experience reductions in $O_3$-sensitive antioxidant molecules, such as caffeic acid and quercetin [47–49]. However, following the application of 0.2% WD, our lettuce plants showed a higher content of these molecules compared to the untreated ones, thus suggesting a protective role of WD in preventing the oxidation of antioxidant molecules devoted to counteracting the oxidative pressure induced by $O_3$. Consistently with this assumption, WD-treated plants experienced higher antiradical scavenging activity (DPPH) than $O_3$-fumigated plants. From a nutritional point of view, both caffeic acid and quercetin have antioxidant, anti-inflammatory, and anticarcinogenic activities [50,51] and in view of the protective action on these molecules from oxidative degradation, WD can be considered not only as a tool to increase plant productivity but also to defend its nutritional content from the oxidative stress. However, caffeic acid showed higher reductions than quercetin, probably because this molecule is involved in several antioxidative roles. It is reported that this compound is also deputed to shield the photosynthetic system from oxidative stresses [52], thus explaining its higher consumption in $O_3$-fumigated plants.

Photosynthesis is the main target of $O_3$ phytotoxicity [22] and following its damage, both energy transmission mechanisms, reallocation of nutrients, and plant growth can be significantly impaired [44]. Although the photosynthetic system is naturally provided with antioxidant molecules—i.e., caffeic acid—the concentration of ROS above the threshold of the system's antioxidant defenses may damage it, and programmed cell death processes can be activated [53,54].

Ozone-induced ROS can affect photosynthesis by either damaging structurally the chloroplast, as evident by ultrastructural changes [26,55], and/or activating cascade signals which lead to stomatal closure and, in turn, generate further reductions in net photosynthesis, $CO_2$ assimilation, and plant growth [2,44]. As a result, however, reductions in carbon assimilation may induce increases in the electron transfer with the consequent increase in unwanted ROS overproduction [56]. Hence, the photosynthetic system starts a self-regulatory process to achieve protection from photo-oxidation, leading to an immediate reduction in both the content of chlorophyll and the number of reaction centers [22,57,58], i.e., the structures dedicated to intercepting light to energetically supply the photosynthetic system [59]. With this process, dissipation ($DI_0$) at the expense of absorption and transmission ($TR_0$ and $ET_0$) is favored [22] and reductions in the activity of both PSI and PSII (measured as $F_V/F_M$ and PI) can be measured, as observed for Canola (*Brassica napus* L.) and tomato (*Solanum lycopersicum* L.) plants in open top-chamber fumigations [5,60]. Nevertheless, following WD foliar applications, all the above-mentioned negative effects on the electron transport chain of the photosynthetic system were significantly alleviated. In fact, WD-treated plants experienced a higher number of active reaction centers (RC), positive values for the parameters involved in the energy absorption, trapping, and transmission from the PSII to PSI (ABS, TR, ET), as well as lower heat dissipation energy (DI). Additionally, WD-treated plants showed also a higher expression of the photosynthetic efficiency $F_V/F_M$, the maximum quantum yield of primary photochemistry, the performance index PI, the overall indicator of the PSI and PSII functionality [61], and higher content of chlorophyll. Since RCs are naturally subjected to oxidative stress forcing them to be rebuilt very quickly [53], we speculate that WD, thanks to its content of antioxidants, may have greatly reduced this turnover or increased its frequency. Higher photosynthetic efficiency also means more biomass, with a 10% increase in the ratio for every 30% increase in photosynthesis [62]. Ozone can effectively reduce photosynthesis with inevitable consequences for plant growth [44], but by shielding the photosynthetic system from oxidation, the plant's growth and biomass can also be protected.

## 5. Conclusions

The results of the present study clearly showed the ability of WD to protect lettuce plants from ozone-induced damage. Specifically, WD-treated plants exhibited lower damage to the photosynthetic machinery, a higher content of chlorophyll (ca. 13%), a higher antioxidant power (ca. 140%), as well as antioxidant molecules (i.e., caffeic acid and quercetin, ca. 400 and 105%, respectively), and higher biomass (ca. 18%). Counteracting the oxidative stress that occurred at the level of the photosynthetic system is speculated to be the main mechanism by which WD protects the plant from ozone-induced damage. Moving forward, WD can be seen not only as a natural product for stimulating plant productivity but also as a means of protecting crop plants against oxidative stress.

**Author Contributions:** Conceptualization, A.V., S.L. and R.F; methodology, A.V. and R.F.; formal analysis, A.V., R.F. and M.G.; data curation, A.V.; writing—original draft preparation, A.V. and R.F.; writing—review and editing, A.V. and S.L.; supervision, A.V. and S.L. All authors have read and agreed to the published version of the manuscript.

**Funding:** This research did not receive any specific grant from funding agencies in the public, commercial, or not-for-profit sectors.

**Institutional Review Board Statement:** Not applicable.

**Informed Consent Statement:** Not applicable.

**Data Availability Statement:** The raw data presented in this study are available on request from the corresponding author.

**Acknowledgments:** Thanks are due to Francesco Barbagli (BioDea and BioEsperia srl) for kindly providing Wood Distillate®.

**Conflicts of Interest:** The authors declare no conflict of interest.

## References

1. Cooper, O.R.; Parrish, D.D.; Ziemke, J.R.; Balashov, N.V.; Cupeiro, M.; Galbally, I.; Gilge, S.; Horowitz, L.W.; Jensen, N.R.; Lamarque, J.-F.; et al. Global distribution and trends of tropospheric ozone: An observation-based review. *Elementa Sci. Anthr.* **2014**, *2*, 000029. [CrossRef]
2. Emberson, L. Effects of ozone on agriculture, forests and grasslands. *Philos. Trans. R. Soc. A* **2020**, *378*, 20190327. [CrossRef]
3. Grulke, N.E.; Heath, R.L. Ozone effects on plants in natural ecosystems. *Plant Biol.* **2019**, *22*, 12–37. [CrossRef] [PubMed]
4. Allan, A.C.; Fluhr, R. Ozone and Reactive Oxygen Species. *eLS* **2001**. [CrossRef]
5. Maliba, B.G.; Inbaraj, P.M.; Berner, J. The Use of OJIP Fluorescence Transients to Monitor the Effect of Elevated Ozone on Biomass of Canola Plants. *Water Air Soil Pollut.* **2019**, *230*, 75. [CrossRef]
6. Feng, Z.; De Marco, A.; Anav, A.; Gualtieri, M.; Sicard, P.; Tian, H.; Fornasier, F.; Tao, F.; Guo, A.; Paoletti, E. Economic losses due to ozone impacts on human health, forest productivity and crop yield across China. *Environ. Int.* **2019**, *131*, 104966. [CrossRef] [PubMed]
7. Ainsworth, E.A. Understanding and improving global crop response to ozone pollution. *Plant J.* **2017**, *90*, 886–897. [CrossRef] [PubMed]
8. Saitanis, C.J.; Agathokleous, E. Exogenous application of chemicals for protecting plants against ambient ozone pollution: What should come next? *Curr. Opin. Environ. Sci. Health* **2021**, *19*, 100215. [CrossRef] [PubMed]
9. Didyk, N.P.; Blum, O.B. Natural antioxidants of plant origin against ozone damage of sensitive crops. *Acta Physiol. Plant.* **2010**, *33*, 25–34. [CrossRef]
10. Bellini, E.; De Tullio, M.C. Ascorbic Acid and Ozone: Novel Perspectives to Explain an Elusive Relationship. *Plants* **2019**, *8*, 122. [CrossRef] [PubMed]
11. Kittipornkul, P.; Treesubsuntorn, C.; Thiravetyan, P. Effect of exogenous catechin and salicylic acid on rice productivity under ozone stress: The role of chlorophyll contents, lipid peroxidation, and antioxidant enzymes. *Environ. Sci. Pollut. Res.* **2020**, *27*, 25774–25784. [CrossRef] [PubMed]

12. Macias-Benitez, S.; Navarro-Torre, S.; Caballero, P.; Martín, L.; Revilla, E.; Castaño, A.; Parrado, J. Biostimulant Capacity of an Enzymatic Extract From Rice Bran Against Ozone-Induced Damage in Capsicum annum. *Front. Plant Sci.* **2021**, *12*, 749422. [CrossRef] [PubMed]
13. Mathew, S.; Zakaria, Z.A. Pyroligneous acid—the smoky acidic liquid from plant biomass. *Appl. Microbiol. Biotechnol.* **2015**, *99*, 611–622. [CrossRef] [PubMed]
14. Grewal, A.; Abbey, L.; Gunupuru, L.R. Production, prospects and potential application of pyroligneous acid in agriculture. *J. Anal. Appl. Pyrolysis* **2018**, *135*, 152–159. [CrossRef]
15. Mahmud, K.N.; Hashim, N.M.; Ani, F.N.; Zakaria, Z.A. Antioxidants, Toxicity, and Nitric Oxide Inhibition Properties of Pyroligneous Acid from Palm Kernel Shell Biomass. *Waste Biomass Valorization* **2020**, *11*, 6307–6319. [CrossRef]
16. Vannini, A.; Moratelli, F.; Monaci, F.; Loppi, S. Effects of wood distillate and soy lecithin on the photosynthetic performance and growth of lettuce (*Lactuca sativa* L.). *SN Appl. Sci.* **2021**, *3*, 1–6. [CrossRef]
17. Fačkovcová, Z.; Vannini, A.; Monaci, F.; Grattacaso, M.; Paoli, L.; Loppi, S. Effects of wood distillate (pyroligneous acid) on sensitive bioindicators (lichen and moss). *Ecotoxicol. Environ. Saf.* **2020**, *204*, 111117. [CrossRef] [PubMed]
18. Biodea. Available online: https://biodea.bio/scheda-colturale/olive-trees/?lang=en (accessed on 15 January 2022).
19. Chen, Y.-H.; Li, Y.-F.; Wei, H.; Li, X.-X.; Zheng, H.-T.; Dong, X.-Y.; Xu, T.-F.; Meng, J.-F. Inhibition efficiency of wood vinegar on grey mould of table grapes. *Food Biosci.* **2020**, *38*, 100755. [CrossRef]
20. Fačkovcová, Z.; Vannini, A.; Monaci, F.; Grattacaso, M.; Paoli, L.; Loppi, S. Uptake of Trace Elements in the Water Fern *Azolla filiculoides* after Short-Term Application of Chestnut Wood Distillate (Pyroligneous Acid). *Plants* **2020**, *9*, 1179. [CrossRef]
21. Filippelli, A.; Ciccone, V.; Loppi, S.; Morbidelli, L. Characterization of the Safety Profile of Sweet Chestnut Wood Distillate Employed in Agriculture. *Safety* **2021**, *7*, 79. [CrossRef]
22. Bussotti, F.; Strasser, R.J.; Schaub, M. Photosynthetic behavior of woody species under high ozone exposure probed with the JIP-test: A review. *Environ. Pollut.* **2007**, *147*, 430–437. [CrossRef] [PubMed]
23. Strasser, R.J.; Srivastava, A.; Tsimilli-Michael, M. The fluorescence transient as a tool to characterize and screen photosynthetic samples. In *Probing Photosynthesis Mechanism, Regulation & Adaptation*, 1st ed.; Mohammad, Y., Uday, P., Eds.; CRC Press: New York, NY, USA, 2000; pp. 445–483.
24. Gomes, M.T.G.; da Luz, A.C.; dos Santos, M.R.; Batitucci, M.d.C.P.; Silva, D.M.; Falqueto, A.R. Drought tolerance of passion fruit plants assessed by the OJIP chlorophyll a fluorescence transient. *Sci. Hortic.* **2012**, *142*, 49–56. [CrossRef]
25. A Gitelson, A.; Buschmann, C.; Lichtenthaler, H.K. The Chlorophyll Fluorescence Ratio F735/F700 as an Accurate Measure of the Chlorophyll Content in Plants. *Remote Sens. Environ.* **1999**, *69*, 296–302. [CrossRef]
26. Vannini, A.; Canali, G.; Pica, M.; Nali, C.; Loppi, S. The Water Content Drives the Susceptibility of the Lichen Evernia prunastri and the Moss Brachythecium sp. to High Ozone Concentrations. *Biology* **2020**, *9*, 90. [CrossRef] [PubMed]
27. Singh, P.; Arif, Y.; Bajguz, A.; Hayat, S. The role of quercetin in plants. *Plant Physiol. Biochem.* **2021**, *166*, 10–19. [CrossRef]
28. Riaz, U.; Kharal, M.A.; Murtaza, G.; Zaman, Q.U.; Javaid, S.; Malik, H.A.; Aziz, H.; Abbas, Z. Prospective Roles and Mechanisms of Caffeic Acid in Counter Plant Stress: A Mini Review. *Pak. J. Agric. Res.* **2018**, *32*, 32. [CrossRef]
29. Tokuşoğlu, Ö.; Ünal, M.K.; Yıldırım, Z. HPLC-UV and GC-MS characterization of the fla-vonol aglycons quercetin, kaempferol, and myricetin in tomato pastes and other tomato-based products. *Acta Chromatogr.* **2003**, *13*, 196–207.
30. Kumar, N.; Bhandari, P.; Singh, B.; Gupta, A.; Kaul, V.K. Reversed phase-HPLC for rapid determination of polyphenols in flowers of rose species. *J. Sep. Sci.* **2008**, *31*, 262–267. [CrossRef]
31. Brauer, M.; Curtin, J.J. Linear mixed-effects models and the analysis of nonindependent data: A unified framework to analyze categorical and continuous independent variables that vary within-subjects and/or within-items. *Psychol. Methods* **2018**, *23*, 389–411. [CrossRef]
32. Luke, S.G. Evaluating significance in linear mixed-effects models in R. *Behav. Res. Methods* **2016**, *49*, 1494–1502. [CrossRef] [PubMed]
33. R Development Core Team. *R: A language and environment for statistical computing*; R Foundation for Statistical Computing: Vienna, Austria; Available online: https://www.R-project.org/ (accessed on 7 January 2022).
34. Mills, G.; Buse, A.; Gimeno, B.; Bermejo, V.; Holland, M.; Emberson, L.; Pleijel, H. A synthesis of AOT40-based response functions and critical levels of ozone for agricultural and horticultural crops. *Atmos. Environ.* **2007**, *41*, 2630–2643. [CrossRef]
35. Calatayud, A.; Ramirez, J.W.; Iglesias, D.J.; Barreno, E. Effects of ozone on photosynthetic CO2 exchange, chlorophyll a fluorescence and antioxidant systems in lettuce leaves. *Physiol. Plant.* **2002**, *116*, 308–316. [CrossRef]
36. Calatayud, A.; Barreno, E. Response to ozone in two lettuce varieties on chlorophyll a fluorescence, photosynthetic pigments and lipid peroxidation. *Plant Physiol. Biochem.* **2004**, *42*, 549–555. [CrossRef] [PubMed]
37. Gomez, J.P.; Velez, J.P.A.; Pinzon, M.A.; Arango, J.A.M.; Muriel, A.P. Chemical Characterization and Antiradical Properties of Pyroligneous Acid from a Preserved Bamboo, Guadua angustifolia Kunth. *Braz. Arch. Biol. Technol.* **2021**, *64*, 21190730. [CrossRef]
38. Benzon, H.R.L.; Lee, S.C. Pyroligneous Acids Enhance Phytoremediation of Heavy Metal-Contaminated Soils Using Mustard. *Commun. Soil Sci. Plant Anal.* **2017**, *48*, 2061–2073. [CrossRef]
39. Berahim, Z.; Panhwar, Q.A.; Ismail, M.R.; Saud, H.M.; Mondal, A.; Naher, U.A.; Islam, R. Rice yield improvement by foliar application of phytohormone. *J. Food Agric. Environ.* **2018**, *12*, 399–404.
40. Fedeli, R.; Vannini, A.; Guarnieri, M.; Monaci, F.; Loppi, S. Bio-Based Solutions for Agriculture: Foliar Application of Wood Distillate Alone and in Combination with Other Plant-Derived Corroborants Results in Different Effects on Lettuce (*Lactuca Sativa* L.). *Biology* **2022**, *11*, 404. [CrossRef]

41. Theerakulpisut, P.; Kanawapee, N.; Panwong, B. Seed Priming Alleviated Salt Stress Effects on Rice Seedlings by Improving Na+/K+ and Maintaining Membrane Integrity. *Int. J. Plant Biol.* **2016**, *7*, 6402. [CrossRef]
42. Tanase, C.; Bara, C.I.; Popa, V.I. Cytogenetical effect of some polyphenol compounds separated from industrial by-products on. *Cell. Chem. Technol.* **2015**, *49*, 799–805.
43. Tanase, C.; Bujor, O.-C.; Popa, V.I. Phenolic Natural Compounds and Their Influence on Physiological Processes in Plants. In *Polyphenols in Plants;* Elsevier: Amsterdam, The Netherlands, 2019; pp. 45–58. [CrossRef]
44. Wilkinson, S.; Mills, G.; Illidge, R.; Davies, W.J. How is ozone pollution reducing our food supply? *J. Exp. Bot.* **2012**, *63*, 527–536. [CrossRef]
45. Wang, Y.; Qiu, L.; Song, Q.; Wang, S.; Wang, Y.; Ge, Y. Root Proteomics Reveals the Effects of Wood Vinegar on Wheat Growth and Subsequent Tolerance to Drought Stress. *Int. J. Mol. Sci.* **2019**, *20*, 943. [CrossRef]
46. Hassan, I.A.; Haiba, N.S.; Badr, R.H.; Basahi, J.M.; Almeelbi, T.; Ismail, I.M.; Taia, W.K. Effects of ambient ozone on reactive oxygen species and antioxidant metabolites in leaves of pea (*Pisum sativum* L.) plants. *Pak. J. Bot.* **2017**, *49*, 47–55.
47. Diao, E.; Wang, J.; Li, X.; Wang, X.; Song, H.; Gao, D. Effects of ozone processing on patulin, phenolic compounds and organic acids in apple juice. *J. Food Sci. Technol.* **2019**, *56*, 957–965. [CrossRef] [PubMed]
48. Jeon, J.; Baek, S.-A.; Kim, N.S.; Sathasivam, R.; Park, J.S.; Kim, J.K.; Park, S.U. Elevated Ozone Levels Affect Metabolites and Related Biosynthetic Genes in Tartary Buckwheat. *J. Agric. Food Chem.* **2020**, *68*, 14758–14767. [CrossRef]
49. Kreczmer, B.; Dyba, B.; Barbasz, A.; Rudolphi-Szydło, E. Advantageous/Unfavorable Effect of Quercetin on the Membranes of SK-N-SH Neuroblastoma Cells. *Molecules* **2021**, *26*, 4945. [CrossRef] [PubMed]
50. Espíndola, K.M.M.; Ferreira, R.G.; Narvaez, L.E.M.; Rosario, A.C.R.S.; Da Silva, A.H.M.; Silva, A.G.B.; Vieira, A.P.O.; Monteiro, M.C. Chemical and Pharmacological Aspects of Caffeic Acid and Its Activity in Hepatocarcinoma. *Front. Oncol.* **2019**, *9*, 541. [CrossRef]
51. Salehi, B.; Machin, L.; Monzote, L.; Sharifi-Rad, J.; Ezzat, S.M.; Salem, M.A.; Merghany, R.M.; El Mahdy, N.M.; Kılıç, C.S.; Sytar, O.; et al. Therapeutic Potential of Quercetin: New Insights and Perspectives for Human Health. *ACS Omega* **2020**, *5*, 11849–11872. [CrossRef] [PubMed]
52. Hura, T.; Hura, K.; Grzesiak, S. Possible contribution of cell-wall-bound ferulic acid in drought resistance and recovery in triticale seedlings. *J. Plant Physiol.* **2009**, *166*, 1720–1733. [CrossRef]
53. Foyer, C.H. Reactive oxygen species, oxidative signaling and the regulation of photosynthesis. *Environ. Exp. Bot.* **2018**, *154*, 134–142. [CrossRef]
54. Gandin, A.; Dizengremel, P.; Jolivet, Y. Integrative role of plant mitochondria facing oxidative stress: The case of ozone. *Plant Physiol. Biochem.* **2020**, *159*, 202–210. [CrossRef] [PubMed]
55. Yamane, K.; Kawasaki, M.; Taniguchi, M.; Miyake, H. Correlation between Chloroplast Ultrastructure and Chlorophyll Fluorescence Characteristics in the Leaves of Rice (*Oryza sativa* L.) Grown under Salinity. *Plant Prod. Sci.* **2008**, *11*, 139–145. [CrossRef]
56. Lascano, H.; Melchiorre, M.N.; Luna, C.M.; Trippi, V.S. Effect of photooxidative stress induced by paraquat in two wheat cultivars with differential tolerance to water stress. *Plant Sci.* **2003**, *164*, 841–848. [CrossRef]
57. Pellegrini, E.; Cotrozzi, L.; Neri, L.; Baraldi, R.; Carrari, E.; Nali, C.; Lorenzini, G.; Paoletti, E.; Hoshika, Y. Stress markers and physiochemical responses of the Mediterranean shrub Phillyrea angustifolia under current and future drought and ozone scenarios. *Environ. Res.* **2021**, *201*, 111615. [CrossRef] [PubMed]
58. Pellegrini, E.; Francini, A.; Lorenzini, G.; Nali, C. Ecophysiological and antioxidant traits of Salvia officinalis under ozone stress. *Environ. Sci. Pollut. Res.* **2015**, *22*, 13083–13093. [CrossRef]
59. Allen, J.; Williams, J. Photosynthetic reaction centers. *FEBS Lett.* **1998**, *438*, 5–9. [CrossRef]
60. Thwe, A.A.; Kasemsap, P. Quantification of OJIP Fluorescence Transient in Tomato Plants Under Acute Ozone Stress. *Agric. Nat. Resour.* **2014**, *48*, 665–675.
61. Gottardini, E.; Cristofori, A.; Cristofolini, F.; Nali, C.; Pellegrini, E.; Bussotti, F.; Ferretti, M. Chlorophyll-related indicators are linked to visible ozone symptoms: Evidence from a field study on native Viburnum lantana L. plants in northern Italy. *Ecol. Indic.* **2014**, *39*, 65–74. [CrossRef]
62. Kirschbaum, M.U. Does Enhanced Photosynthesis Enhance Growth? Lessons Learned from CO2 Enrichment Studies. *Plant Physiol.* **2011**, *155*, 117–124. [CrossRef]

water

*Review*

# Gas Hydrate-Based Heavy Metal Ion Removal from Industrial Wastewater: A Review

**Sirisha Nallakukkala [1,2], Adeel ur Rehman [1,2,3], Dzulkarnain B. Zaini [2] and Bhajan Lal [1,2,*]**

1 Research Centre for $CO_2$ Capture (RCC$O_2$C), Universiti Teknologi PETRONAS, Bandar Seri Iskandar 32610, Perak, Malaysia; sirisha_18002191@utp.edu.my (S.N.); adeel_20000259@utp.edu.my (A.u.R.)
2 Department of Chemical Engineering, Universiti Teknologi PETRONAS, Bandar Seri Iskandar 32610, Perak, Malaysia; dzulkarnain.zaini@utp.edu.my
3 Department of Chemical Engineering, University of Karachi, Karachi 75270, Pakistan
* Correspondence: bhajan.lal@utp.edu.my; Tel.: +60-53687619 or +60-103858473

**Abstract:** Innovating methods for treating industrial wastewater containing heavy metals frequently incorporate toxicity-reduction technologies to keep up with regulatory requirements. This article reviews the latest advances, benefits, opportunities and drawbacks of several heavy metal removal treatment systems for industrial wastewater in detail. The conventional physicochemical techniques used in heavy metal removal processes with their advantages and limitations are evaluated. A particular focus is given to innovative gas hydrate-based separation of heavy metals from industrial effluent with their comparison, advantages and limitations in the direction of commercialization as well as prospective remedies. Clathrate hydrate-based removal is a potential technology for the treatment of metal-contaminated wastewater. In this work, a complete assessment of the literature is addressed based on removal efficiency, enrichment factor and water recovery, utilizing the gas hydrate approach. It is shown that gas hydrate-based treatment technology may be the way of the future for water management purposes, as the industrial treated water may be utilized for process industries, watering, irrigation and be safe to drink.

**Keywords:** heavy metals; wastewater; hydrate-based desalination; efficiency; toxicity

**Citation:** Nallakukkala, S.; Rehman, A.u.; Zaini, D.B.; Lal, B. Gas Hydrate-Based Heavy Metal Ion Removal from Industrial Wastewater: A Review. *Water* **2022**, *14*, 1171. https://doi.org/10.3390/w14071171

Academic Editors: Dibyendu Sarkar, Mohammad Mahmudur Rahman, Prafulla Kumar Sahoo and Rupali Datta

Received: 17 February 2022
Accepted: 30 March 2022
Published: 6 April 2022



## 1. Introduction

An ecosystem is the combination of biotic components, such as microbes, plants and animals, to develop a self-contained entity with their abiotic, physicochemical environment. The connection between stability and ecosystem has been extensively studied; in a way there should be more species, increased diversity gives greater resistance to change, or stress has been thoroughly invalidated, as proven by ecosystem modelling and investigations [1]. Indeed, complex ecosystems are thought to be more fragile than relatively simple dynamically stable systems, making them more vulnerable to human intervention. Because each ecosystem type reacts differently, it is impossible to accurately evaluate the impact of contaminants and heavy metals on ecosystems.

The most harmful chemical industries are those that use heavy metals such as Cd, Cu, Ni, Cr, As, Zn and Pb and dump large amounts of metal-contaminated effluent, as shown in Table 1. Due to the high solubility of heavy metals in water these metals can be absorbed by living organisms. Heavy metals can build up in significant amounts in the human body once they reach the food chain. When metals are overconsumed beyond the allowed concentration limit, they can cause serious health issues [2]. Hence, as a result, heavy metal-contaminated wastewater must be treated before being released into the environment.

Conventional treatment methods like adsorption, membrane, chemical precipitation and electrochemical techniques are used to eliminate heavy metals from inorganic effluent [3–5]. These techniques are robust but faced with major disadvantages like less selectivity, limited removal of metal, excess energy requirements and large amount of sludge generation. Many methods have been researched lately in order to advance/yield more economical and effective results in decreasing the quantity of wastewater generated and also enhancing the treated water quality. Alternative ways to water recovery are still needed, with a focus on further reducing energy demand while improving water recovery and yield.

To overcome the aforesaid limitations, novel technologies must be studied in order to provide efficient and cost-effective alternatives. In this context, clathrate hydrate-based technology has recently emerged as a viable option for treating wastewater containing dissolved minerals and heavy metals [6,7]. Gas hydrates are a non-stoichiometric crystalline structure in which the guests are encased within a crystal structure framework of host water molecules [7]. There has been a huge interest in the application of gas hydrates in desalination sector. Hydrates have three different crystal structures depending on the type of guest molecule: Structure I (methane, ethane carbon dioxide), structure II (propane, nitrogen, isobutane) and structure H. Each construction is made up of cages of various sizes and forms that are held together by van der Walls forces. Depending on their structure and formation conditions, different hydrate formers retain different numbers of water molecules. This hydration property qualifies them for water treatment applications [7]. Thus, the elimination of heavy metal ions via formation of the gas hydrate mechanism attracts substantial interest from researchers because of its ease of operation, low toxicity and good selectivity. It has the advantage of purifying the water using the application of the gas hydrate technique. The hydrate formation requires two basic raw materials, namely water and hydrate, the former which is readily available and accessible. After dissociation of hydrate, 1 $m^3$ of hydrate generates about 164 $m^3$ of gas and 0.87 $m^3$ of freshwater, which signifies a huge potential to produce treated water using gas hydrate applciation in the desalination process [7]. During this hydrate formation it eliminates all the metal ions, producing pure water. Hence this method seems to be a promising technique compared to conventional methods. This would help to address environmental problems and scarcity of water issues simultaneously.

The goal of this review paper is to summarize the conventional and novel technologies used in heavy metal treatment from industrial wastewater and evaluate their advantages and limitations based on their application. Based on this approach the conevntional treatment methods are discussed as shown in Figure 1 along with their advanatges and limitations. This paper also gives the overview of research and progress on novel technique gas hydrate-based desalination, recent innovations based on reactor design, mechanism and heavy metal ion separation, water recovery and the opportunities and challenges with the application of the gas hydrate process. The latest developments in the process of commercialization and various hydrate formers that are used to remove heavy metals and that can produce higher water recovery have been reviewed and discussed in depth.

### 1.1. Wastewater

The discharge of wastewater which is polluted with metal ions has increased significantly in a few decades as a result of urbanization, industry and agricultural activities [8,9]. In general, wastewater is divided into two types, namely industrial wastewater and domestic wastewater. Wastewater from domestic sources is comprised of solid and liquid wastes generated by non-manufacturing processes and may contain sewage, bacteria, viruses, hazardous materials, sanitary outputs, detergents, non-toxic organisms and trash [10]. However, the primary source of water contamination is untreated wastewater discharged by industries. Dyes, aromatics, pesticides, heavy metals, oil [11–16], etc., are some of the pollutants found in industrial wastewater that end up in the environment and these toxic contaminants constitute a serious risk to human health and the ecosystem [17].

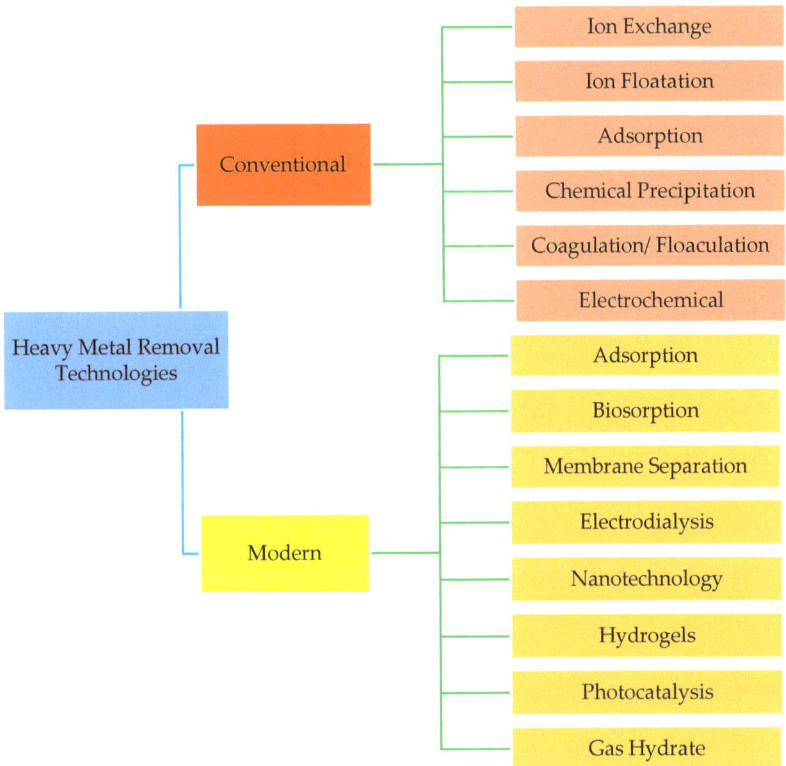


**Figure 1.** Technologies available for removal of heavy metals from industrial wastewater.

### *1.2. Heavy Metals*

A heavy metal element is defined as having a specific gravity greater than 5.0 and atomic weight between 63.5 and 200.6 g/mole [18]. The term "heavy metal" denotes a high-density element that is unsafe even in trace quantities. Heavy metals in wastewater have recently become a major environmental problem, posing considerable harm to ecosystems and human health even at very low strengths. Pollution by heavy metals is a noteworthy burden on the environment due to their non-biodegradability, accumulation, adaptability and persistence [16,17,19,20]. Heavy metals removed from dyes like amido black dye are targeted by doping cobalt where, due to the negative free energy values obtained, the elimination process has been discovered to be spontaneous. Because of the increased surface area and enhanced contact between the catalyst and the pollutant, cobalt doping resulted in up to 90% degradation [3]. Biosorption employing waste biomass materials, on the other hand, is chosen due to its low cost, simplicity, biodegradability and environmental friendliness [4]. Khan et al. [5] substantiated the need to integrate and optimise treatment approaches in order to effectively tackle the cleaning challenge.

**Table 1.** Heavy metals in wastewater from common industrial sources [16].

| Metals / Industries | Al | As | Cd | Cr | Co | Cu | Fe | Hg | Mn | Ni | Pb | Zn | References |
|---|---|---|---|---|---|---|---|---|---|---|---|---|---|
| Aviation | X | | X | X | | X | X | X | | X | | X | [17] |
| Allies, Chlorine | | X | X | X | | | | X | | | X | X | |
| Urea/Fertilizers | X | X | X | X | | X | | X | | X | X | X | |
| Glass | X | X | | X | | | X | | | X | X | | |
| Cement | X | X | | | | | | | X | X | X | X | |
| Organic chemistry | X | X | X | X | | | | X | | | X | X | |
| Paper manufacturing | | | X | X | | X | | X | | X | X | X | |
| Petroleum refiner | X | X | X | X | | X | | | | X | X | X | |
| Power plants | | | | X | | | | | | | | | |
| Steel works | | X | X | X | X | X | | X | | X | X | X | |
| Tanning | | | | X | | | | | | | | | |
| Textile mills | | | | X | | | | | | | | | |
| Pharmaceutical | | X | X | X | | X | | | | X | X | X | [21] |
| Dyes | | | X | X | | X | X | | | X | X | X | [22] |
| Engineering | | | X | X | | X | X | | | X | X | X | |
| Fine chemicals | | | X | X | | X | X | | | X | X | X | |
| Batteries | | | | | | | | X | | | | | [23] |
| Brass manufacture | | | | | | | | | | X | | | |
| Electroplating | | | X | X | | X | | | | | | X | |
| Ferromanganese alloy production | | | | | | | X | | X | | | | |
| Fungicides | | X | | | | | | | | | | | |
| Metal smelters | | X | | | | | | | | | | | |
| Mining | | | | | | | | | | X | | | |
| Nuclear fission | | | X | | | | | | | | | | |
| Pesticides | | X | X | | | X | | X | | | X | | |
| Welding | | | X | | | | | | X | | | | |

### 1.3. Various Heavy Metals and Their Effects

When heavy metals (such as $Ag^{+}$, $Hg^{2+}$, $Cd^{2+}$, $As^{3+}$ and $Pb^{2+}$) in their ionic forms mix with bioparticles in the human body, hazardous chemicals are produced; these types of interactions are also important to identify. When heavy metal concentrations exceed the tolerance limits, they become toxic and alter the cell metabolism [24]. Due to the multiple negative effects on human health and the ensuing impairment to marine life, several regulatory bodies have set acceptable limits and stringent regulations for heavy metal disposal. Furthermore, due to the severe consequences of wastewater, researchers have focused on developing improved treatment technologies [25]. The health effects of several heavy metals are summarized in Table 2 [26–28]. These heavy metals represent major dangers to the human population as well as the animals and plants that live in the water bodies. These can be absorbed into the human body, resulting in health problems such as organ damage, cancer, nervous system impairment and, in the worst-case scenario, death.

The heavy metals mentioned above can build up at any time and do so faster than they can be expelled, making them extremely dangerous to humans and the environment. Heavy metals have severe effects on aquatic environments even at low concentrations, and natural degradation of these metals does not occur. It suppresses the activity of microbes that were previously existing in waste waterways. Consequently, researchers are focusing their efforts on removing heavy metals from waste waterways at the moment.

**Table 2.** Toxicological effects of metal ion on human health.

| Metal | Route of Entry | Toxicity Effect | Disposal Limit Recommended by WHO * (mg/L) |
|---|---|---|---|
| Copper | Ingestion and inhalation | intestinal irritation, liver illness, anemia, and cancer in respiratory tract | 0.02 |
| Cadmium | Inhalation and ingestion | lung damage and limits the respiratory system | 0.06 |
| Chromium | Inhalation, ingestion, and absorption through skin | damage the lungs and limits the respiratory system | 0.05 |
| Mercury | Inhalation, ingestion, and absorption through skin | imitation of respiratory system, liver and kidney damage and loss of hearing | 0.01 (vapor) |
| Lead | Inhalation and ingestion | lungs and kidney damage | 0.15 |
| Nickel | Inhalation | lung, liver kidney damage | 0.1 |
| Zinc | Inhalation and ingestion | It causes a number of health problems, including fever, nausea, vomiting, skin irritation and anemia, despite the fact that it is required by humans at a trace amount. | 0.05 |

* World Health Organization.

## 2. Conventional Technologies for Heavy Metal Removal in Wastewater

Pollutants in wastewater are discharged in accordance to strict regulations. Due to the pollutants' inhibitory properties, a high removal strategy is required to meet the discharge protocols. Thus, the industrial sector faces several challenges in order to reduce discharge of pollutant, water usage and consumption of energy [29]. As a result, a number of treatment options to assure environmental safety were created, resulting in a noteworthy study. Each technology has its own set of benefits and drawbacks. As a result, a number of treatment options to assure environmental safety were created and every technology has its own set of benefits and drawbacks.

The conventional techniques used for wastewater treatment are precipitation, filtration, electrodialysis [30], ion exchange, supercritical fluid extraction, adsorption [31], the microbial system [32], electrochemical process, bioreactors and an advanced oxidation process [29]. These methods are robust and exciting techniques existing to remove the heavy metals. Physical, chemical and biological approaches are the three major categories that can be used to classify the methods discussed so far; each technique has a set of advantages and disadvantages, which are detailed in Table 3. A combination of procedures can be used to efficiently remove heavy metals.

**Table 3.** Merits and demerits of different metal ion removal techniques.

| S. No | Techniques | Merits | Demerits | Reference |
|---|---|---|---|---|
| 1. | Coagulation | cheap, dewatering mechanism | Production of sludge, Chemicals are utilized extensively. | [32] |
| 2. | Filtration through membrane | Heavy metals exclusion at a rapid rate demands less room. | Extremely costly, membrane fouling, and complicated procedure. | [32] |
| 3. | Adsorption | Simple operation, minimal sludge formation, and the utilization of low-cost adsorbents | Desorption cost for regenerable adsorbent is high. | [33] |
| 4. | Electrochemical treatment | Effective in eliminating metal ions with slight chemical use | The initial expenditure is significant, and a large quantity of electrical power is essential | [32] |
| 5. | Electrodialysis | Metals are separated to a higher degree. | Clogging and loss of energy | [34] |
| 6. | Ion exchange | High transformation of components | removes a little amount of metal ions, and the operational cost is significant. | [35] |
| 7. | Oxidation | No requirement for electricity | oxidation process causes rusting in the system. | [36] |

### 2.1. Coagulation/Flocculation

Precipitation of heavy metals into insoluble compounds such as sulphides, hydroxides and carbonates can now be treated with coagulation [37]. Coagulation is the destabilsation of a suspension, resulting in the formation of aggregates. On the other hand, flocculation refers to the process of causing destabilized particles to come together in order to create larger aggregates [38]. To eliminate these colloidal particles, coagulation treatment procedures are utilized to enhance the density. The efficacy of coagulation is dependent on the coagulant used, pH, dosage of the coagulant, mixing, temperature and alkalinity. Flocculants, which agglomerate the destabilized particles into larger particles, are added with the help of the agitation mechanism.

Unit operations like filtration, straining and flotation are used to separate the larger particles. By combining chitosan and mercaptoacetic acid, Chang et al. produced a novel type of macromolecule flocculant, termed mercaptoacetyl [39], that was demonstrated to eliminate turbidity and heavy metals with 98% efficiency. In a study by Sakhi et al. [40] heavy metals with their removal rates like $Cd^{2+}$ (40%), $Pb^{2+}$ (78%), $Ni^{2+}$ (62%), $Cr^{2+}$ (22%), $As^{2+}$ (81%) and $Se^{2+}$ (44%) are removed using ferric chloride as a coagulant and anionic polymer as flocculant and their respective dosage quantities and pH influenced the rate of removal. Removal percentage was found to vary with the initial concentration and dosage in the case of $Sb^{5+}$ (96%) and $Sb^{3+}$ (98%) [41] in the presence of polymeric ferric sulphate and this might be attributed to the presence of phosphate and humic acid. Colloidal particles are neutralized by coagulants, which destabilize them and allow the sludge to settle down. When colloidal particles become entrapped on a metal surface, precipitation occurs. The generation of huge amount of sludge as a result of usage of chemicals in the separation process is one of the key drawbacks of this method [42].

The sludge generated contains heavy metal elements such as nickel, cadmium, lead, chromium and zinc [43]. Recovery, recycling and reuse are all the possible options for the management of sludge [44]. Though coagulation/flocculation are robust methods in eliminating heavy metals from wastewater, they also generate by-products such as flocs, categorized as secondary pollutant, along with reusable solvents that are harmful to humans and ecosystems.

### 2.2. Ion Exchange

Ion exchange is a separation technique that effectively removes heavy metals from wastewater by substituting one type of ion for another. The ion exchange technique produces substantially less sludge compared to coagulation [45]. Ion exchange resin is used to recover or eliminate metal ions. According to Hubicki et al. [46], the chemical features of resins cause the isolation of a certain set of metal ions. There are two kinds of resins, namely synthetic and natural resins. Metal ions are substituted for the cations using both types of resins. Synthetic ion exchange resins are commonly employed in water treatment to remove a variety of unwanted dissolved particles/solids, most commonly concerning the hardness from water [45]. The resins are built on a cross-linked polymer skeleton called a matrix. Arsenic metal ion was eliminated from drinking water by means of synthetic resins [47]. The main disadvantages of synthetic resins is fouling, which happens when a metal solution with high concentration is passed through the matrix [48].

The most common ion exchanger is the cationic exchange resin, which is made up of extremely acidic and weak basic resin. Natural zeolites have been found to have the best cation exchange capability for eliminating heavy metal ions from wastewater in several investigations [49]. In the indirect elimination of chromium (Cr) metal ion, zeolites are noteworthy because they are unable to perform direct ion exchange of Cr(VI) compounds, as charge repulsions from the negatively charged framework, which limits its permeability to anion [50]. Jamil et al. [51] used two zeolites manufactured from Egyptian Kaolin to remove nickel (Ni) metal ions from industrial effluent, as well as heavy metals like Cu, Cd, Pb and Zn. They demonstrated that zeolite can eliminate metal ions up to 98%.

The anion exchange resins are best suited for treating wastewater with minimal level of pollution because of the type of functional group, composition and matrix structure. For the removal of hazardous metals like chromium (VI) and manganese, Kononova et al. [52] employed cation and anion exchangers. Ion exchange resin has the disadvantage of requiring chemical reagents for regeneration, which further leads to secondary pollution [45]. Furthermore, it adds to the operational costs and cannot be used for wastewater treatment on a large scale.

### 2.3. Flotation

Flotation is a liquid–solid separation technique in which small gas bubbles are introduced into the sample of wastewater and the heavy metals attach themselves to the bubbles, allowing them to leave the water. This is achieved by adding surfactants or collectors to aqueous solutions to remove surface-inactive ions, usually with an ion with the opposite charge to the metal ion to be removed. Once the bubbles rise to the surface, the hydrophobic particles that are concentrated are collected. Flotation contributes significantly to wastewater treatment due to its low quantity of sludge generation and high efficiency of separation. This process is most appropriate for the metals with different chemical and physical properties [53]. A few parameters, like the bubble velocity, size of the bubble and frequency of the formation of bubble, are the most vital aspects for managing the flotation method. Due to the continuous operation of flotation equipment, its drawback is that it is costly to maintain and operate [54]. Ion flotation, dissolved air flotation (DAF) and precipitate flotation are the other forms of flotation processes.

Ion flotation looks inefficient when the metal ion concentration in wastewater is low [55]. Low energy consumption, restricted volume demands, reduced sludge quantities and selective treatment were demonstrated by the ion flotation process. Microbubbles are used in the precipitation flotation process, which is essentially a chemical procedure and its precipitation takes only a few minutes [56]. The flotation methods provide a number of advantages, including quick operation, a compact process and cost effectiveness.

Synthetic chemical surfactants have been developed to have a higher capacity, to have superior selectivity and to be easy of manufacture. They are, however, constrained by cost and toxicity concerns. Biosurfactants, on the other hand, appear to be more environmentally friendly, but they have low removal rates and huge dosage amounts and are a time-consuming process. Nanoparticles have been proposed as new collectors in this area, demonstrating both synthetic and biosurfactant benefits [57].

### 2.4. Membrane Filtration

Membrane filtration is a pressure-driven technique for separating heavy metal ions [58]. Membrane filtration separates metal ions according to the concentration of solution, size, pH and pressure applied over the membrane. The filtration process can be accelerated by treating the membrane with chemical agents [50]. The membrane is made up of a porous material that has been developed to help in removing metals ions from contaminated wastewater [54]. Polymer and ceramic types are two kinds of membrane materials used for wastewater treatment. Ceramic membranes are often favored over polymer membranes for use in industrial wastewater treatment due to their chemical resilience and water-repellent nature. Despite these advantages they are faced with a major significant drawback—being fragile and expensive to manufacture [59]. Inorganic salts and organics may be extracted up to 95–99% using the RO technique [60]. The reverse osmosis process has a higher rejection rate but is faced with issues like membrane fouling and degradation.

Despite its advantages it is faced with some challenges, like membrane fouling and the fact that there is a need for membrane replacement on a frequent basis [61]. Membrane fouling is caused by dissolved organic debris (DOM) and organic matter (OM). Pretreatment is essential to lessen the impact of DOM and OM on membrane filtration, which in turn boosts the separation process' efficiency [62]. However, depending on the size of the metal ion the conventional pressure driven processes like nanofiltration, ultrafiltration, reverse osmosis and microfiltration are employed to isolate heavy metal from wastewater. Because of its low removal capabilities, the use of micro filtration in heavy metal removal has received little consideration. It has, however, been employed by changing the feed solution's membrane or chemical pre-treatment. These techniques can handle and treat enormous volumes of wastewater but, faced with the limitations of huge amounts of sludge generation, they exhibit disposal issues and post-treatment requirements. Fouling and scaling inhibition could increase membrane separation even more. Pretreatment and periodic cleaning of membranes, on the other hand, require additional costs.

### 2.5. Chemical Precipitation

Chemical precipitation is a low-cost, high-efficiency method for removing heavy metals that is used in a variety of industrial sectors. Chemicals that affect the pH of a solution cause a metal precipitate to form, in which the precipitate is prevented from dissolving in the solution [63]. Separation of phases by sedimentation results in precipitates [61]. Chemical precipitation is the most effective way to remove hazardous metals including Cu (II), Zn (II), Mn (II) and Cd (II) [45]. Tanong et al. [64] discovered that the addition of sodium carbonate ($Na_2CO_3$) aids in the exclusion of Mn and Ni metal ions by totally precipitating and raising the pH to 9.

This approach is rarely relevant for treating wastewater with less metal ion concentration, but it is relevant for large metal ion concentrations. As the insoluble metals precipitate, it results in huge amounts of sludge generation with higher content of water; this is difficult to treat and discard of, and hence is regarded as harmful waste [65]. After precipitation, the purified water could be reused or disposed of into reservoirs. To precipitate the metal ions, however, a large number of chemicals are necessary. This removal process is widely employed in various industrial applications due to its simplicity and low cost.

The sulphide precipitation and hydroxide precipitations are other methods of chemical precipitation. Hydroxide precipitation is often used because of its low cost, simplicity and changeable pH [66]. It is carried out by adding hydroxide to the agitated wastewater, resulting in the formation of insoluble metal hydroxide precipitates. However, as this approach necessitates a substantial dosage of precipitates, a high pH value is considered as a drawback. The majority of metal ions eliminated by this approach are $Zn^{2+}$, $Cu^{2+}$, $Ni^{2+}$, $Pb^{2+}$ and $Cr^{3+}$. Ain et al. could remove $Ni^{2+}$ and $Cu^{2+}$ from industrial effluent with removal efficiency of 76.66% and 100% using NaOH as precipitant [67]. In another research study by Vu et al., the authors found that the carbonation method increased $Pb^{2+}$ removal and decreased pH with a removal efficiency of 49.3% and sulphate removal of 97% [68]. There are certain downsides, such as relatively huge sludge generation, leading to dewatering, disposal difficulties and the formation of precipitation of metal hydroxide in the presence of complex agents, in addition to the need for a high dosage to achieve appropriate pH.

The sulphide participation approach outperforms the hydroxide method in terms of higher removal efficiency and low dissolved solids. This approach was found to be effective in the treatment of hazardous heavy metal ions [69]. A $Cu^{2+}$ removal of 96% was achieved in treating arsenic wastewater [70]. Lower sulphide means more concentration of zinc in the effluent, while higher sulphide means a bad odor resulting from residual sulphide. It may also produce hydrogen sulphide gas, which is noxious and odorous.

### 2.6. Electrochemical Treatment

Electrochemical (EC) treatment is a different approach that involves applying an electric current to the aqueous medium to destabilize the dissolved contaminants. These contaminants are maintained in the solution by electrical charge. These ions and other charged particles become destabilized and precipitate in a stable form when they are neutralized with ions of opposite electrical charges provided by the electrocoagulation system. Electrochemical methods are simple, quick, cheap, easy to use and environmentally friendly. By charge neutralization, these coagulants can remove contaminants or heavy metal ions from solution [71]. Electrochemical treatment is a good alternative for wastewater treatment because of its flexibility. The efficiency of electrochemical reactors is determined by the electrode materials used and the operating parameters, like mass transfer rate, current density and effluent water content [72].

Liu et al. [73] reported a removal efficiency for the metal ions $Cu^{2+}$ (97.7%), $Cd^{2+}$ (97.3%) and $Pb^{2+}$ (98.5%) using graphene oxide and carbon as an electrode. Yang et al. [74] figured that $Cr^{6+}$ reduction was faster with the existence of $K_2SO_4$ with a removal percentage of 93.7%. The usage of electrochemical treatment becomes increasingly more difficult as a result of more stringent environmental restrictions or regulations. This method works effectively for a variety of contaminants, including those that cannot be separated from their by-products. It requires a large facility as well as a steady supply of power to function. Additionally, as electrodes only last for a short period, there is an inadequate rate of mass transfer along with an increasing temperature limit with the application of this approach [75]. Electrocoagulation, electrolytic flotation, electrodeposition, electro deionization and electrodialysis are some of the electrochemical methods. The critical difficulties of using this technology are large-volume sludge production and post-treatment requirements, which are not cost effective.

### 2.7. Adsorption

Sorption is the transfer of ions from solution-to-solid phase transition. Adsorption and precipitation reactions are both included in the term "sorption" process. Adsorption has recently emerged as another treatment method for wastewater containing heavy metals. Adsorption comprises transfer of mass from liquid to solid surface, resulting in the chemical or physical process mechanism [76]. New biodegradable adsorbents [77] reliant on variety of low-cost resources, such as by products from industries, agricultural waste, natural materials or modified biopolymers have recently been produced and used to remove heavy metals from polluted wastewater. In general, sorption on solid sorbents includes three main phases: (i) Transport of the impurity from the bulk to the sorbent surface; (ii) adsorption on to the surface of a particle; and (iii) passage inside the particle sorbent [78]. As both the adsorption and desorption processes are reversible, whenever an adsorbent has been desorbed, it can be renewed and utilized for a variety of purposes. Though there are a few factors that can affect the adsorbent selection, cost effectiveness and suitability are two of the most important primary requirements [79]. The effectiveness of the adsorption process is controlled by the polarity of the adsorbent, as well as its large surface area and pore size distribution [80]. The primary limitations of this technology are the adsorbent regeneration medium and regulating the pH. Some of the key challenges of the adsorption mechanism are its capacity to remove multiple ions at once, a long retention time and adsorbent cycling stability. The heavy metal removal by conventional methods are shown in Table 4.

**Table 4.** Heavy metal ion removal from different conventional technique.

| Technique | Material Used | Metals | Removal % | Remarks | Reference |
|---|---|---|---|---|---|
| Adsorption | Modified graphene (GN) with cetyltrimethylammonium bromide | Cr | 98.2 | Low selectivity, production of waste. It is difficult to synthesize. The cost of adsorbent is too high. | [81] |
| | Activated Carbon from Prawn shell and green alga Ulva lactuca | Cr, Cd | 98, 95 | Higher quantities on larger scales are difficult to manage | [82,83] |
| Membrane Filtration | RO | Cr, Cu, Cd | 98, 99, 90 | High operational cost due to membrane fouling | [84] |
| | Ultrafiltration | Cu | 90 | | |
| Electrodialysis principles | perfluorosulfonic Nafion 117 | Co, Ni | 90, 69 | Clogging and energy loss | [85] |
| | $HNO_3$ | Cd | 70 | | [86] |

**Table 4.** *Cont.*

| Technique | Material Used | Metals | Removal % | Remarks | Reference |
|---|---|---|---|---|---|
| Chemical Precipitation | Magnesia and lime-water | Fe(III), Cr(III), Cu(II), Pb(II), Ni(II) & Cd(II) | 97 | The cost of producing high-water-content sludge and disposing of it has increased. Precipitation with lime and bisulphide, lacks specificity. When it comes to removing metal ions of low concentration, this method can be useless. | [87] |
| Ion Exchange | Magnetic ion exchange | Cr, Cu, Cd, Hg | 99.9 | Not effective when employing concentrated metal solutions since the exchange matrix is easily fouled by organic/other wastes and is the major limitation. | [88] |
| Floatation | potassium ethyl xanthate (KEtX) | Ni, Cd, Co | 98.3, 97.5 and 94.7 | High concentration of floatation agents are employed which affects the economy of the process. | [53] |

### *2.8. Gas Hydrate-Based Mechanism in Eliminating Heavy Metal Ions*

The above-mentioned conventional technologies are robust methods in treating wastewater but are faced with limitations like fouling, poor selectivity, less efficiency in metal removal, high processing costs and generation of huge amounts of sludge. New concepts have lately been examined for producing less expensive and more efficient solutions to reduce wastewater production while also enhancing the quality of treated effluent. As a result, scientists are interested in gas hydrate-based removal since it is less dangerous, easier to operate, has higher selectivity and may be used for multiple purposes other than heavy metal removal methods. This not only cleans the water by generating gas hydrate, but it also makes it easier to remove harmful metals from industrial wastes.

This technology produces hydrates by using two basic requirements, namely water and gas, where the gas molecules are enclosed inside the water cage formed by the hydrogen bond. To better understand the mechanism of the formation of hydrate in the existence of salts, it is essential to consider the hydrogen bonding of water in the bulk phase as well as at the gas–liquid interface. The gas molecules do not bind chemically to the hydrate cages, but instead interact with water cages weakly via van der Waals interactions. When 1 $m^3$ of hydrate is dissociated, it can produce up to 164 $m^3$ of gas and 0.8 $m^3$ of pure water at standard temperature and pressure. Moreover, the hydrate constitutes about 85% water and 15 mole % gas, which signifies a huge potential to produce treated water which can further be utilized for industrial or domestic usage.

Unlike ice crystals, hydrates form close to ambient conditions when a sufficient hydrate former (guest gas) and water are present. As the hydrogen-bonded water cages are so small, only small gaseous molecules and lower hydrocarbons can fit inside the cage, so it entirely depends on the type of guest component being utilized. Furthermore, the guest molecule does not interact chemically with the hydrate cages; instead, it interacts through Van der Waals forces. Hydrate-based desalination is a technique in which pure water crystallizes as a solid hydrate phase from brine solution with the help of a gaseous molecule. The ions or salt dissolved in water are surrounded by water molecules. The water molecules surround the cations and stabilize. As a result, ions in aqueous solutions are constantly hydrated, which means that they are strongly attached to water molecules via ion–dipole interactions. The amount of water molecules in the main hydration shell, which completely surrounds the ion, varies depending on the ion's radius and charge. During the formation of hydrate, all the metal ions are rejected into the effluent. Only water and gas are left in the solid hydrate, leaving significant metal ions in the effluent. This mechanism allows the elimination of salts from the wastewater, thereby producing no sludge. The freshwater produced from the hydrate provides us an opportunity to effectively eliminate metal ions from the effluent. When the solid hydrate is depressurized, it dissociates easily into liquid water and gas. Several unit operations, such as centrifugation and extrusion, can be used to separate the hydrated crystals from the brine or as a post-treatment technique [89]. Thermal stimulation or depressurization can easily disintegrate the crystals, releasing salt-free water and allowing the gas to be recovered and reused. The advantages of using the gas hydrate method over conventional technologies is high water recovery, less energy consumption, no pretreatment required, less maintenance and above all the formation of hydrate excludes all metal ions A gas hydrate-based desalination method has been suggested as being simple to install, cost effective and environmentally beneficial [90]. The formed hydrate when dissociated results in freshwater.

## 3. Overview of Gas Hydrate Technology

The lattice of the host molecule is created by hydrogen-bound water molecules, inside which the guest molecule is incorporated, is called gas hydrate and is nonstoichiometric in nature [91]. The basic requirements for gas hydrate formation are water and gas (guest molecule). The favorable conditions for formation of hydrate are low temperature and high pressure. The guest molecule spins within the water cage during hydrate formation, causing a physical change. The initial nucleation of gas hydrate crystals, followed by growth, and then dissociation of the gas hydrate, are various stages of gas hydrate process. Pressure, temperature, gas composition and the system's thermodynamic behavior conditions play a major role in determining the hydrate formation, stability and structure formation [4,92,93].

Clathrate hydrates have been identified as a potential energy resource as well as a source of problems, primarily in the area of flow assurance. On the plus side, the study discovered that gas hydrate can provide energy from a different source, with an individual hydrate unit containing significant energy can make hydrate as as sustainable energy resource. The relevance of gas hydrate as a sustainable energy resource, as well as the prospect of employing gas hydrate for storage, transportation and capture can be applied to hydrate-based applications with increased energy and environmental capabilities [94]. Hydrate-based desalination and the development of gas separation or capturing are two more prominent technical applications of hydrates. When the hydrate is formed, the dissolved impurities/salts are dispersed from the hydrate. When the hydrate is heated the gas inside is then released. This produces pure water [95]. On the other hand, the flue gas from large power plants might be used to extract carbon dioxide, thereby reducing global warming and emissions using this mechanism. This concept has advantages over present transportation techniques like liquefied natural gas (LNG) and compressed natural gas (CNG) because they do not need cryogenic temperatures or high pressure requirements [96–98]. Hydrates can be employed as a secondary refrigerant in refrigeration systems [99].

### 3.1. Gas Hydrate-Based Desalination

Desalination is the process of desalinating salty water to produce drinkable water or water beneficial for agricultural purposes [100]. Many conventional desalination processes are energy intensive, together with multi-stage flash distillation (MSF), reverse osmosis (RO) and multi-effect desalination (MED) [101]. The recovery of pure water from a saline water stream is an important factor in making any desalination process profitable or marketable. As shown in Table 5, hydrate-based desalination is a promising technology (HBD) based on water recovery and energy consumption when compared with the traditional technologies. It would be useful to practice utilizing the HBD method because it runs at a temperature greater than the freezing point of water, as shown in the comparison Table 5. Desalination systems based on indirect and direct freezing have also been developed, although they are not widely used commercially due to their high costs, though ice crystals are constituted of nearly pure water [102]. Only by developing cutting-edge desalination technologies that can deliver safe and appropriate adequate quantities of clean water while minimizing the consumption and the associated costs, as well as being environmentally benign, can this have a meaningful impact both on the environment and society [103]. The freezing or crystallization process, called hydrate-based desalination, was developed as a technique for desalinating saline water more than seven decades ago [89,90].

**Table 5.** Comparison of conventional technologies with hydrate-based technology.

| Method | Principle | T (°C) | P (MPa) | Water Recovery | Total Average Specific Energy Consumption ($kWh/m^3$ of Water) | Maintainance | Advanatge/Constraints |
|---|---|---|---|---|---|---|---|
| Distillation | Flash process | 90–120 | Less than 0.1 | 20% [3] | 23.4 [104] | Corrosion/ scaling | Used for high TDS (total dissolved solid) concentration/High energy, less water recovery |
| Reverse Osmosis | Solute diffusion | 20–35 | 5.5–7.0 | 55% | 5 [104] | Sludge generation/membrane replacement | Requires pretreatment, less water recovery, resistant to impurities |
| Hydrate desalination | Phase change | Near to 0 | 0.45–0.65 | 58.6% [105] | 0.6 [106] | No maintainance | suitable for greater TDS concentrations/higher water recovery |

In this hydrate-based desalination process, seawater acts as an electrolyte solution, with a gas hydrate former (suitable guest molecule) at promising conditions of temperature and pressure. A detailed review of the gas hydrate-based desalination developments has been discussed elsewhere [89,90,107,108]. A detailed process of hydrate formation, dissociation and separation of pure water and residual water is shown in Figure 2. As salt has no effect on the shape of the hydrate crystal, it acts as a thermodynamic hydrate inhibitor (THI) by shifting the phase equilibrium curve to a higher pressure at a given temperature condition, thereby lowering the driving force essential to form hydrates. The formed hydrate is then separated from the residual brine and further dissociated to yield pure drinking water and the guest gas (hydrate former) could be further recycled [89,90].

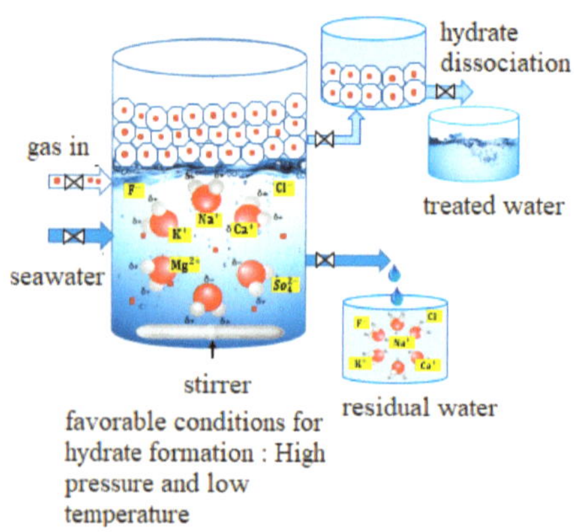


**Figure 2.** Hydrate formation, dissociation and water recovery process in hydrate-based desalination process.

The water recovery relates to the quantity of freshwater retrieved from the feed, and is estimated based on kinetics of formation of hydrate. The fraction of the volume of water converted to hydrates, and the quantity of recoverable hydrate crystals during the separation process, is estimated by hydrate formation kinetics [108]. The maximum amount of water that can be recovered is limited by the initial eutectic solution composition. Other statistics that relate to assessing the efficiency of removing salt is salt rejection. This is defined as the change in concentration of salt after hydrate formation to its initial value [89]. Scientists started researching gas hydrates and their application in the field of seawater desalination about four decades ago. Since then, numerous studies have been conducted around the world to develop appropriate hydrate formers that are environmentally friendly, stable, non-toxic, easily available and economically feasible [109]. Some of the hydrate former studies available in the literature up to today are based on propane ($C_3H_8$) [110,111], cyclopentane ($C_5H_{10}$), [89,90,112–116], carbon dioxide ($CO_2$), [117,118], refrigerants (HFC, HCFC, and CFC) [119–122] and sulphur hexafluoride ($SF_6$) [123]. Several studies have been performed in determining the phase equilibrium through a pilot scale set up [124–129] and the kinetics of hydrate formation have been boosted (by using secondary hydrate forming gases) [130–133]. Early commercialization of hydrate-based desalination was delayed based on the difficulty in separating the hydrate phase from the residual brine; however, the efficiency of salt exclusion has been examined, and ways to improve hydrate formation via subsequent treatment operations have been devised [133–136].

Following that, several improvements for hydrate separation from salt water, eliminating salt from the slurry and reducing induction time are discussed below. Despite being atmospheric guest molecules (hydrate formers) like CFC, $SF_6$, HCFC and HFC, and also because of the influence of ozone depletion, these hydrate formers are considered inapt and the flammability of cyclo-alkanes and alkanes generates threat for large-scale applications. A thermodynamically integrated technique, on the other hand, suggests that propane and ethane are good guest molecules [109]. Several hydrate formers, namely liquid, gaseous and other additives, have been analyzed to check the feasibility of the hydrate-based desalination process [122]. Moreover, after a thorough examination, cyclopentane (CP) was discovered to be an atmospheric hydrate-forming agent and it forms a suspension which requires further post-treatment [128]. The economic viability must be examined as the cost of the operation is reliant on basic variables like temperature of brine, favorable thermodynamic conditions of temperature and pressure, concentration of salt, salt mobility and yield [136–138]. Hydrate-based desalination methods are still desirable since they are environmentally useful and energy efficient in terms of use in large-scale applications.

A hybrid system with an RO-hydrate system for seawater desalination has recently been proposed, along with an optimum design strategy. Additionally, a novel desalination process based on hydrates has been developed and simulated using LNG waste cold energy as a heat sink, and the economic viability of the method has been assessed [139–141]. The outcome of substituting the external refrigeration cycle with an LNG plant is very encouraging; as a result, desalination of high strength brine solution may be performed with little energy usage, making the gas hydrate-based desalination technique commercially viable. An apparatus for multi-use desalination by gas hydrate was developed, with several unit operations and its application as a continuous process with multiple injection, separation and purging phases has been examined [142]. The desalination efficiency was determined to be greater than 80%, with a water recovery rate greater than 30%.

Utilizing $CO_2$ as a hydrate former in hydrate formation has a dual role: It provides drinkable water from salt water and also separates the gas. As a result, clathrate hydrates could be considered a viable working medium for $CO_2$ storage and desalination purposes [140–143]. Despite the fact that only limited studies have been carried out so far [144,145], additional research based on the reactor design configuration, separation of hydrate from effluent and enhancement of the kinetics of hydrate formation would undoubtedly open the door to more research in this area, particularly concerning, e.g., appropriate thermodynamics and kinetics conditions combined with minimization of cost. Recent scientific and economic breakthroughs have hastened the development of commercially viable clathrate-hydrate desalination devices. More experimental research is desirable to lower the amount of salt stuck in hydrate crystals throughout each cycle and to adjust important operational parameters based on hydrate former to improve desalination efficiency.

### 3.2. Gas Hydrate Desalination Reactor Design Innovations

The clathrate hydrate formation technique has been used to desalinate seawater worldwide. The slow kinetics and the difficulty in separating the crystal from the brine and removing salt from hydrate that has occluded the surface of the crystal have all been addressed by researchers. To separate the crystals, a conveyor belt was employed. Hydrate formation and then later dissociation was performed in different compartments using a non-stirred reactor. After dissociation of hydrate, some of the recovered water was utilized to wash the salts in the washing stage that had stuck to the surface of hydrate [146]. Ben produced hydrates at a certain depth in seawater and the hydrates rose high due to the buoyancy effect and wash water was introduced to wash the hydrate above the hydrate production zone [147].

Another breakthrough was the use of liquid propane as hydrate former to produce hydrates at the bottom of the reactor. The formed hydrate rose higher through the brine due to the buoyancy effect. It was created to flow through risers in partition plates, thereby efficiently separating hydrate crystals from saline solution. The wash process was utilized by using the produced fresh water after dissociating the hydrate [148]. To utilize the heat of hydration for decomposition of hydrate and as a solvent for formation of hydrate, a heat exchange (HE) liquid that was immiscible in water was used. To solve the problem of washing of the formed hydrate a HE liquid having a specific gravity lower than aqueous saline solution was introduced to overcome washing of hydrate. The HE liquid helped with the passage of formed hydrates from the hydrate formation zone to the dissociation section. The gas and liquid utilized in HE were recycled back to the hydrate formation zone [149]. A gas bubble was delivered by a nozzle into precooled water in another invention, resulting in the production of hydrates. Because of pressure variations between the pipeline and the mobile tank, hydrates were passed through the pipeline into the mobile tanks and then depressurized in the tank [150].

Using concentric and coaxial pipes, a high-pressure, low-temperature hydrate former was injected to an appropriate ocean depth, leading to a large investment in injecting gas to further depths. The annular zone generated hydrate slurry, which was retrieved from the surface [151]. Another invention involves injecting methane at a depth into columns and allowing hydrates to develop. Because of the buoyancy, the hydrate rose, and the heat from the hydration was dissipated into the surrounding water [152]. Atomized water is used inside the reactor to improve the interaction and surface area of gas and liquid water. Hydrate crystals are deposited as they form on a conveyor belt. Because of the porous moving surface, it was feasible to extract hydrate from feed. A roller and grinder in the production area were used to carry the washed hydrates to the dissociation zone [153]. Seawater was introduced from the top of the reactor, while gas was pumped in from below. Hydrate particles served as fluidized bed particles, thereby giving more surface area for gas–liquid contact. Fluidized hydrate particles obstruct the gas and water flow, enhancing the residence time and helping in the formation of additional hydrates [154]. The hydrate formation and dissociation took place in a single reactor, resulting in crystal formation due to buoyancy. The hydrate formed in the reactor's bottom rose upward towards the decomposition zone due to positive buoyancy [155]. It was decided to use a desalination system with a porous constraint. A thick impermeable hydrate mat was produced under this porous restriction. A thick layer of hydrate was blocking the entry of saltwater water. The hydrate close to the restraint is separated by lowering the pressure on the collection side of the constraint [156,157]. It was discovered that employing a blend of hydrochlorofluorocarbons (HFCs), hydrofluorocarbons and chlorofluorocarbon molecules might reduce the dendritic growth of the hydrate.

In another invention, ice formation over hydration was implemented to minimize the need for washing. Salts that have occluded on to the surface of crystal are washed by the melting of ice [158]. A vertical tubular reactor with upper and lower parts was employed for hydrate formation and decomposition, respectively. Stirring produced hydrate production in the reactor's lower half, and the buoyancy effect promoted easier crystal separation. The top of the reactor produced pure water for recovery [159]. It has been suggested that $CO_2$ could be sequestered by dissolving in seawater. Desalination using hydrates has been proposed as a method of concentrating brine solutions with dissolved $CO_2$ [160]. Desalination, thermal energy storage and natural gas storage have all been produced using gas hydrate modular systems [161]. A dual cylinder with a piston was employed in another design to compress the slurry, and saline water was passed through the perforations. To boost the efficiency during salt separation, concentrated saltwater from the reactor's bottom was returned to the hydrate formation region; however, maintaining hydraulic pressure requires a high energy throughput [162]. In another invention a hydrate-forming gas was pumped into a saline-water reactor utilizing a microbubble generating device for hydrate formation [163]. In order to promote hydrate nucleation in gas bubbles with diameters between $10^{-3}$ and $10^{-2}$ mm, an ultrasonic transducer was utilized to supply sonic energy. Solid material particles, such as silica gel, were introduced to the stream to increase the surface area of gas–liquid contact. A wash column was used to remove the saline solution from the interstitial pores of the crystal [164]. In a recent reactor invention [165] where hydrate formation, separation, washing and dissociation took place in a single reactor, the formation of $CO_2$ hydrate in the presence of 2 wt.% brine solution at different pressures was observed and there was a removal efficiency of 60.08% at 3.0 MPa.

The hydrate-based desalination method advances have mostly focused on resolving the fundamental problem of efficient hydrate crystal separation from the residual brine. Because hydrate nucleation is a stochastic process, researchers have tried to shorten induction time by using microbubbles, ultrasonic energy and localized supercooling by depressurizing liquid propane. While various attempts to commercialize the HyDesal process have been made, these efforts have failed due to additional technological obstacles like the difficulty in separating small crystals from effluent and the fact that a few guest molecules are toxic and they pose environmental risks. Hence, there is a need to address the challenges of energy efficiency and cost and crystal/brine separation and of course to carry out the process in a manner that fully addresses potential environmental concerns.

### 3.3. Heavy Metal Separation Mechanism Based on Gas Hydrates

New strategies for removing heavy metals from wastewater, such as separation of heavy metal by hydrate method, appear to be in high demand. The scientific relevance of this study into purification and separation using a gas hydrate approach has pricked the interest of many scholars. Parker proposed a method for producing drinkable water from highly saline sources using gas hydrate formation as early as 1942, and it has received a lot of interest over the last decade [166]. Hesse and Harrison observed a noteworthy reduction in interstitial water chlorinity when they used hydrate in deep-water sedimentary layers, and they discovered that hydrate rejects salt ions from the crystal structure, which affords the theoretical foundation for separating mixtures using hydrate-based methods [167]. Knox et al. presented a method for separating inorganic mixtures using seawater desalination, and a pilot plant was built to test this process, which produced drinking water [110]. Furthermore, Bulot et al. devised a method for purifying a solute from an aqueous solution of water and solute [168]. An overview of the metal ions removed along with their removal efficiency using gas hydrate method is listed in Table 6.

Using experimental and thermodynamic models, Ngema et al. assessed the accurate phase equilibrium data of hydrate formation in saline aqueous solutions. They measured the hydrate phase equilibrium of R134a, R410a and R507 in $MgCl_2$, NaCl and $CaCl_2$ to give the parameters for hydrate-based desalination. Despite the fact that the hydrate phase equilibrium of many guest molecules has been extensively researched, most studies only look at a single solute at low concentrations. As a result, experimental studies of phase equilibrium with mixed solutes of higher concentration are required. Gas hydrate technology could be used to develop wastewater treatment and desalination systems based on this information [121]. Huang et al. used methyl bromide, trichlorofluoromethane and 1,1-difluoroethane for separation of organic mixtures to study the concentrations of apple, orange and potato juices by gas hydrate method and found that this method removed 80% of water content [169]. When comparing hydrate desalination to reverse osmosis, Bradshaw et al. discovered that the water throughput rate and recovery are faster with hydrate-based desalination [119]. All this study suggests that hydrate-based separation methods can be applied for desalination purposes to treat wastewater.

The hydrate-based water treatment technology facilitates the solid–liquid separation more efficiently at high temperatures than conventional water freezing temperature by enriching the guest molecule with the water molecules. The guest molecule can be recycled in the system after melting the gas hydrate crystals, which is mostly freshwater, as addressed in detail in reviews elsewhere [89,90]. The following steps are followed in the hydrate-based water treatment process, as indicated in Figure 3: (i) Creation of gas hydrates, (ii) hydrate separation from effluent, (iii) post-treatment (e.g., washing, centrifuging) to improve water quality and (iv) dissociation of hydrate crystals to produce treated water and reuse gas [170].

Song et al. proposed a system for isolating heavy metals from aqueous solution built on gas hydrates, based on the aforesaid principles and mechanism. Raman spectroscopy was used to examine the separation efficiency of this approach with various R141b–effluent volume ratios, as well as the influence of a washing operation on heavy metal removal. The heavy metal ion concentration decreased from roughly 140 mg/L to less than 0.4 mg/L after hydrate-based desalination, demonstrating that the heavy metal ions had been excluded from the hydrate structure. The process employed is an intriguing method for separating heavy metal salts of different concentrations from industrial effluents using an extrapolation of hydrate-based desalination. $Cr^{3+}$, $Cu^{2+}$, $Ni^{2+}$ and $Zn^{2+}$ were removed from an aqueous solution utilizing R141b as the hydrate forming in this study [171].

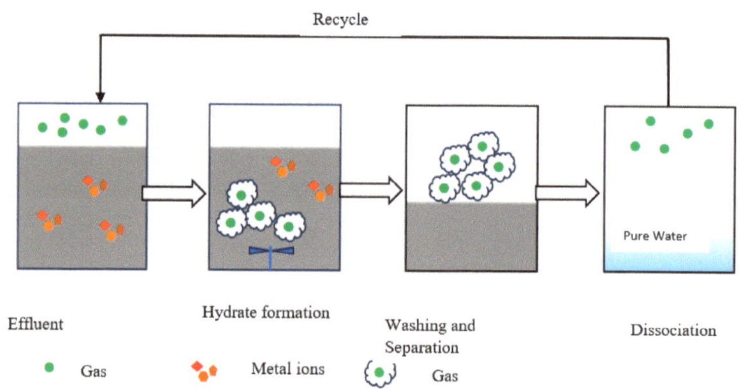


**Figure 3.** Heavy metal ion removal by gas hydrate technique.

Yang et al. investigated hydrate-based treatment looked at using hydrates to treat $Ni^{2+}$-contaminated wastewater. At a volume ratio of 3:1, cyclopentane (CP) was added to a $NiCl_2$ aqueous solution, and hydrates were produced at 2 °C with agitation at 600 rpm. Overall, the separation efficiency of the hydrate-based approach varied from 62% to 88% [172–175]. This study involved the separation of specific ions rather than series of metal ions. The separation efficiency of the method is found to be dependent on the metal ion trapped on the solid hydrate. If proper post-treatment is used, separation efficiency can be increased to up to 95% [134,176]. It is worth noting that, when the $Ni^{2+}$ concentration rises, the separation efficiency also improves.

Dong et al. devised a new hydrate-based approach for removing heavy metal ions by implementing stages of post-treatment techniques. The formation of hydrate, separation of solid from liquid, dissociation of hydrate and analysis of the effluent/residual are the four main steps in this system. They looked into a number of variables, including ion exclusion and the process for removing the hydrate from the mother liquor. Following that, a unique approach was developed to obtain high separation efficiency from high concentrations of heavy metal ions in wastewater. R141b was used as the former and copper sulphate solutions were used [177]. R141b reacts with water at temperatures less than 8.4 °C and pressures greater than 42 kPa to form sII hydrate [177,178]. Solid–liquid separation unit operations were carried out once the hydrate formation was completed. The hydrate was then melted down by adjusting the system's temperature, and separation occurred due to R141b's immiscibility with water. $Cu^{2+}$ concentrations were measured using an inductively coupled plasma optical spectrometer (ICP), and electrical conductivity variations were recorded using an electrical conductivity meter (ECM). Figure 4 shows a conceptual picture of the hydrate process for removing heavy metals from aqueous solution.

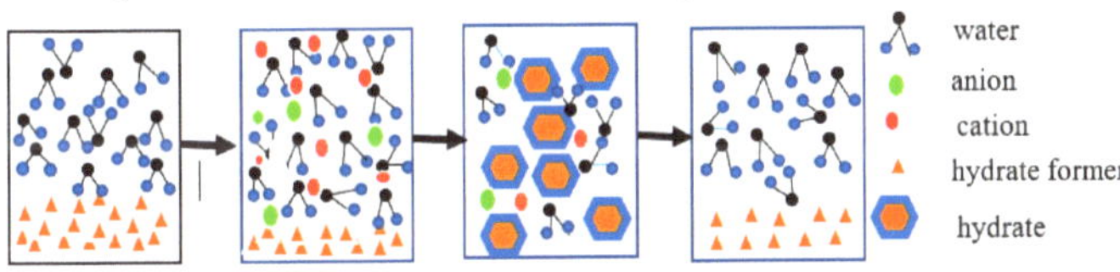


**Figure 4.** Schematic diagram for hydrate-based desalination process.

**Table 6.** Removal of heavy metals using different hydrate formers by hydrate-based desalination.

| System and Concentration | Hydrate Former | Metals Removed | Removal Efficiency | Remarks | Reference |
|---|---|---|---|---|---|
| NaCl 3.2 wt% | $CO_2$ | $K^+$, $Na^+$, $Mg^{2+}$, $B^{3+}$, $Ca^{2+}$ | $K^+$ (80.4%), $Na^+$ (78.7%), $Mg^{2+}$ (76.6%), $B^{3+}$ (73.3%), $Ca^{2+}$ (72%) | Hydraulic pressure is applied to make the hydrate into the form of pellets uisng piston which is energy intensive. 1 stage operation | [162] |
| Saline solution of 3.0 wt% | $CO_2$ | - | 20.26% | Semi batch reactor system carried out at 3.0–3.5 MPa and 6 °C. The efficiency based on individual metal ion is not discussed | [142] |
| Nacl 3.5 wt% | CP | - | (49–72%). | Post hydrate formation series of unit operations were carried to enhance efficiency. Centrifuging provided a high removal efficiency of 96%. Post-treatment is expensive as CP leaves a suspension and is also not environemnt friendly. | [178] |
| Synthetic produced water 8.9 wt% | $CO_2$<br>$CO_2$+CP<br>$CO_2$+CH | $Na^+$, $Mg^{2+}$, $K^+$, $Ca^{2+}$ | 74<br>91<br>95 | Addition of CP and CH to $CO_2$ enhance the hydrate formation temperature. Post-treatment is required as they form a suspension at interface | [179] |
| Synthetic seawater 3.5 wt% | $CO_2$<br>$CH_4$ | $Na^+$, $K^+$, $Mg^{2+}$, $Ca^{2+}$, $B^{3+}$, $Cl^-$, $SO_4{}^{2-}$ | Cations 71–94%<br>Anions 73–83%<br>68.86% | $CO_2$ based hydrate based desalaintion is better compared to $CH_4$ and might be as $CO_2$ is more soluble in water comapred to $CH_4$ | [118] |

**Table 6.** *Cont.*

| System and Concentration | Hydrate Former | Metals Removed | Removal Efficiency | Remarks | Reference |
|---|---|---|---|---|---|
| NaCl<br>4 wt%<br>at 4 °C | R 141 b | NaCl | 61.46% | Better at removing alkaline metals than it is at removing alkaline earth metals. Not environment friendly have higher global warming potential. | [121] |
| Seawater<br>3.5 wt% | Not available | $K^+$<br>$Na^+, Mg^{2+}, B^{3+}$,<br>$Ca^{2+}$ | 80.4%<br>78.7%<br>76.6%<br>73.3%<br>72% | The reverese osmosis recovery is inversely related to the gas hydrate energy consumption. With the increase in energy consumption it was founf that the efficienccy of metal ion removal increased. | [180] |
| $CuSO_4{\cdot}5H_2O$<br>Coppersulphate pentahydrate | R141b | $Cu^{2+}$ | 90.82% | Optimumratio of waterto R141b is found to be 1:5<br>Involves post-treatment techniques due to usgae of R141b and is not environment friendly | [181] |
| NaCl | R141 b | $Cr^{3+}, Cu^{2+}$,<br>$Ni^{2+}, Zn^{2+}$ | 70.02%, 71.87%,<br>71.79%, 67.82% | Lower effluent volume ratio yielded higher removal efficiency. R141B is likewise extremely flammable and when discharged into the atmosphere, this causes ozone depletion. | [170] |
| NaCl<br>3–5 wt% | CP | - | Removal efficiency increased from 50 to 79% by washing the hydrate. | Higher frequency, lower temperature (274.1 K), lower salinity could all help to form more hydrates, albeit at the expense of removal efficiency. Post-treatment of seperating the suspension is expensive | [107] |
| Produced Water<br>8.6 wt% | Compressed natural gas | $Mg^{2+}, Na^+$.<br>$Ca^{2+}$. $K^+$,<br>$HCO_3^-, Cl^-$,<br>$SO_4^{2-}$ | 79.5–84.3%<br>3 stage process | The number of water molecules in the hydrate structure decreased, resulting in powerful electrostatic interactions that cause the hydration of salt ions and ion clustering also decreased the solubility of gas | [182] |
| Brine<br>3.5 wt% | CP | - | 81%<br>3 step process, gravity seperation, flitration and washing | Because of the fine cyclopentane droplet sizes formed by the spray injection approach, more water can be turned into hydrates. | [183] |
| Seawater<br>3.4 wt% | CP | - | 63% | Washing the hydrate enhanced removal efficiency by 42%. Emulsion formed could be difficult to separate from treated water. | [184] |
| PW<br>8.6 wt% | $CO_2$ | - | 82–89.2%<br>In 3 stage process | Application of hydrate-based desalination of produced water. Removal efficiency of each metal ion is not listed. | [185] |
| Seawater<br>3.42 wt% | $CO_2$+CP | Na+, $K^+$,<br>$Mg^{2+}, Ca^{2+}$,<br>$Cl^-, SO_4^{2-}$ | 85.52%, 83.93%,<br>80.73%, 78.21%,<br>55.72%, 62.09% | A piston is used to separate the solid from liquid stream hence higher energy requirement. | [130] |
| Aqueous $NiCl_2$ solution<br>200–10,000 ppm | CP | $Ni^{2+}$ | 62–88% | Water recovery of 43% is attained. The water recovery and enrichment factory decreases with increase in concentration whereas removal efficiency increased with increased in aqueous concentration. | [171] |
| Saline solution<br>3.3 wt% + dodecane | $CH_4$+$C_2H_6$ | - | 80% | The decreased viscosity of the dodecane system allows the hydrate crystals to move through the oil layer quicker, resulting in better desalination efficiencies. | [186] |
| $CuSO_4$<br>aqueous solution | R141b | $Cu^{2+}$ | 44.79–90.82% | efficient approach was discovered to be vacuum filtration and centrifugation | [176] |

**Table 6.** *Cont.*

| System and Concentration | Hydrate Former | Metals Removed | Removal Efficiency | Remarks | Reference |
|---|---|---|---|---|---|
| PW<br>8.6 wt% | $CO_2$/<br>Natural gas | - | 73%<br>74% | $CO_2$/NG hydrate formers can be utilized to desalinate produced waters. Post-treatment/separation is not listed | [135] |
| NaCl<br>3.5 wt% | Graphite +CP | | 99.76%<br>in 4th stage | The hydrate process with graphite particles is a viable desalination technology, according to the research, with advantages such as quick nucleation, a high conversion ratio, and a fair desalting efficiency. | [187] |
| $CuSO_4$<br>3 wt% | R141b | $Cu^{2+}$ | 84% | Post-treatment by vacuum filtration combined with washing produced maximum removal efficiency Water yield and enrichment factor decreased with increase in concertation | [188] |
| NaCl<br>3 wt% | $CO_2$+$C_3H_8$<br>(90:10) | $Na^+$<br>$Cl^-$ | 87.5%<br>84% | Novel reactor design of flat bed reactor was used for hydrate-based desalination. The influence of salts on water recovery and salt rejection rate was not discussed | [177] |
| Coca cola<br>NaCl<br>15 wt% | HFC134a | $Na^+$, $Mg^{2+}$, $Ca^{2+}$, $K^+$, $B^{3+}$, $Cl^-$, $SO_4{}^{2-}$ | 75.72%<br>80% | With HFC134a in coca cola formed sII hydrate and ions did not effect the structure of hydrate | [189] |
| LiCl, LiBr, LiI | CP+graphite | $I^- > Br^- >$ $Cl^-$ | 70% | The efficiency of desalination was enhanced by increasing lithium halides, but salts restrict induction time and water recovery. | [190] |

### 3.4. Water Recovery

Water recovery denotes the volumetric efficiency of the process. It is defined as the volume of water in the initial feed that is converted into hydrate. However, in most situations, less than 100% of the hydrate crystals are recovered. As a result, after the hydrate crystal separation process, the volume of hydrate recovered from the brine is utilized to compute the amount of water recovered. The percentage of water recovered from the feed can be stated as follows:

$$\text{Water recovery\%} = \frac{\text{Volume of water converted to hydrate}}{\text{Initial volume of feed solution}} \times 100$$

Water recovery relies on the kinetics of hydrate formation and efficiency of separation. Higher water recovery necessitates the production of more hydrates and efficient separation of crystals from salt water. Addition of propane as co-guest molecule to guest molecules argon, nitrogen and carbon dioxide for hydrate-based desalination using a fixed bed reactor was studied by Nambiar et al. [191]. There was less than 2% water recovery achieved using propane with argon and nitrogen, whereas with carbon dioxide it resulted in water recovery of 41.38% due to the formation of sII structure constituting 136 water molecules. However, the amount of water that can be recovered is dependent on the wastewater concentration, hydrate former, stirring mechanism and effectiveness in separation of crystal from hydrate [89]. Babu et al. [192] used a unique cylindrical annular bed reactor design that incorporated a scraper for scraping the generated hydrate crystals on the mesh. They looked into the salt rejection rate and water recovery rate of a carbon dioxide/propane mixed gas used as a hydrate forming in porous media, and discovered that the salt rejection rate and water recovery rate could both reach 87.5% and 34.85%, respectively. The hydrate crystals that detached from the salt solution, however, remained connected to salt particles, which was unexpected.

According to Dong et al. [181], the removal effectiveness of R141b–effluent volume ratios on water recovery increased from 1:4 to 1:6. Total water content would theoretically be converted to hydrate at a ratio of 1:3.21 for R141b–effluent. As a result, there are less heavy metal ions trapped between hydrate crystal or adsorbed onto their surfaces. At ambient temperature and pressure, the dissociation of hydrate occurs, leaving lesser metal ions in the dissociated water. Therefore, an increased R141b–effluent volume ratio led to an increase in dissociated water production. Y. Yang et al. studied an innovative hydrate-based method to separate $Ni^{2+}$ from wastewater with cyclopentane at a ratio of 3:1 ($v/v$). This technique has a water recovery rate of about 43% (the average), according to the findings of this study, when using a low-concentration feeding solution. In terms of water recovery, this hydrate-based technique is on par with or even better than those used in traditional wastewater treatment. The fact that the water recovery has been relatively stable shows that this approach can treat wastewater with a wide concentration range [171].

An experimental study by Gaikwad et al. identified $As^{5+}$, $Pb^{2+}$, $Cd^{2+}$ and $Cr^{3+}$ in industrial effluents and prepared a 1000 ppm solution of each salt to imitate an industrial wastewater effluent. As the hydrate-forming gas, natural gas was chosen, cyclopentane as a liquid co-former and as the promoter, lecithin (surfactant) was selected. A maximum 30 percent water recovery has been recorded when the concentration of lecithin is increased up to 300 ppm. However, the individual metal ion removal efficiency is not discussed in this case [177]. Another study, by Nallakukkala et al. [105], found a water recovery of 66% when operating at 2.5 MPa using $CO_2$ as hydrate former by treating 2 wt% brine solution. This recovery is higher than that obtained by [191], in which the recovery was 41.38% using 10% propane and 90% $CO_2$. Ling et al. [190] evaluated the induction time, water recovery and desalination effectiveness of the LiCl, LiBr and LiI salts on the production of cyclopentane hydrates. The results showed that increasing the amount of lithium haloids improved desalination efficiency; nevertheless, these salts hinder induction time and water recovery. The inhibitory impact of halide ions on the formation of hydrates and desalination was discovered to be in the order $I^- > Br^- > Cl^-$. Water recovery by using various hydrate formers is shown in Table 7.

**Table 7.** Water recovery with removal of metal ion via gas hydrate technique.

| Metal Ion | Hydrate Former | Conc. of Former | Water Recovery | | Reference |
|---|---|---|---|---|---|
| | | | **With Additive** | **Without Additive** | |
| As | | | 28.75 | 25.72 | |
| Cd | CP | 1:6 | 24.51 | 19.76 | [177] |
| Cr | | | 23.33 | 14.97 | |
| Pb | | | 27.92 | 15.51 | |
| Cr | | | | 80 | |
| Cu | R141b | 1:4 | | 80 | [170] |
| Ni | | | | 80 | |
| Zn | | | | 80 | |
| Ni | CP | 1:3 | | 43 | [176] |
| Cu | R141b | 1:5 | | 80.83 | |
| Cu | | | | 76 | [170] |
| Ni | CP | 1:6 | | 75 | |
| Zn | | 1:2 | | 70 | [171] |
| Cr | | | | 74 | |

## 3.5. Removal Efficiency

The removal efficiency of the hydrate-based separation process is used to assess its effectiveness. A brief overview of the removal efficiencies of metal ions with washing and without washing is represented in Table 8.

$$\text{Removal Efficiency\%} = \frac{\text{Initial conc. in feed} - \text{Conc. in dissociated water}}{\text{initial conc. in feed}} \times 100$$

Song et al. investigated volume ratio of 1:6 using R141b–effluent. The removal effectiveness of hydrate crystals without washing ranged from 67.82 to 71.87%. The highly concentrated residual effluent was retained on the hydrate surface due to the porous structure of the hydrate crystal. As a result, the dissociated water still includes a significant amount of heavy metal ion, resulting in poor removal efficiency. A washing technique was employed to remove any leftover heavy metal ions, and resulted in a 19% increase in removal efficiency and a separation efficiency range of 88.01 to 90.82% [170].

To perform multi-stage desalination, Yang et al. utilized an aqueous solution with a $Ni^{2+}$ concentration of 1000 mg/L. In a single stage an efficiency of 84% was produced, whereas two stages produced an efficiency of 96%. The third-stage treatment produced efficiency of 99.2%, demonstrating that this hydrate-based technique can deliver high-quality water [171]. For separation of heavy metals during the formation and dissociation phase, Dong et al. reported that some pores inside the hydrate were interconnected and formed channels, and the hydrate former R141b was trapped in the channels inside the bulk of the aqueous solution. The hydrate decomposition shattered the small solution pockets, allowing them to pass through aqueous solutions in the hydrate slurry, thereby reducing the removal efficiency. As a result, separating the trapped solution from hydrates slurries was crucial to improving removal efficiency [192]. This experimental research used five different hydrate washing unit processes, which were referred to as: (1) No operation, (2) vacuum filtration (VF), (3) washing with a guest species followed by VF (WHVF), (4) washing with fresh water followed by VF (WFVF) and (5) VF followed by centrifugation (VCF). The highest removal efficiency was reached by utilizing VCF, which was 90.46%. When using fresh water for washing and using the vacuum filtering method, better results were obtained. Fresh water can be conserved, costs can be reduced and removal efficiency can be improved at all times with this method. This research led the researcher to the conclusion that the solid–liquid separation process has a direct impact on hydrate-based ion elimination efficacy, and that VF and centrifugation is the most successful solid–liquid separation method amongst those investigated [176].

Heavy metals with concentrations of 500 ppm, containing Zn, Cu, Ni and Cr, were utilized in a system created by Al-Hemeri et al. [193], together with CP as an additive. The removal efficiency for copper ions was as high as 92% in a CP/water solution volume ratio of (1/6) *v/v*. Due to the low strength of heavy metal ions in the discharge, the outcome in lower metal ion adherence to the hydrate crystal surface, hence the percentage removal efficiency, improved when the volume ratio was changed from 1/2 to 1/6. Heavy metal ions pollute the fresh water during hydrate dissociation by lowering the ion exclusion efficiency significantly. Increasing the water volume ratio from 0.115 to 0.192 to estimate the optimum volume, the removal efficiency was found to increase from 30.08% to 60.08% [105] when operated at 3.5 MPa. Effective desalination and kinetics of hydrate growth were observed by [176] using 6 mole % CP and there was a water recovery of 25.72% $As^{5+}$ aqueous solution of 1000 ppm; similar results were obtained using lecithin as a promoter along with 6 mole % CP. No further studies were mentioned related to removal efficiency aspect.

**Table 8.** Heavy metal ion removal efficiencies reported.

| Metal Ion | Hydrate Former | Effluent to Former Ratio | Removal Efficiency (%) With Washing | Removal Efficiency (%) Without Washing | Reference |
|---|---|---|---|---|---|
| Zn | CP | 1:6 | | 85 | [193] |
| Cu | | | | 92 | |
| Cr | | | | 50 | |
| Ni | | | | 80 | |
| Cr | R141b | 1:6 | 89.72 | 70.02 | [170] |
| Cu | | | 90.82 | 71.87 | |
| Ni | | | 89.93 | 71.79 | |
| Zn | | | 88.01 | 67.82 | |
| Cu | R141b | | | 44.7 | [176] |

**Table 8.** *Cont.*

| Metal Ion | Hydrate Former | Effluent to Former Ratio | Removal Efficiency (%) | | Reference |
|---|---|---|---|---|---|
| | | | With Washing | Without Washing | |
| Cu | R141b | 1:5 | 90.82 VF<br>87.69 WHVF<br>90.82 WFVF<br>90.46 VCF | 71.87 | [192] |
| Cu | R141b | 1:4 | 51.8 VF<br>83.80 VFC<br>87.42 VFWW | | [192] |
| Ni | Cyclopentane | 1:3 | 84 (1-stage)<br>96 (2-stage)<br>99.2 (3-stage) | | [171] |

### 3.6. Enrichment Factor (Ef)

Industrial waste has a significant environmental impact, as wastewater treatment regulations are very stringent. Wastewater reduction and zero liquid discharge (ZLD) are important goals to remove waste from a system [194]. The enrichment factor is a crucial parameter in ZLD because it determines how much wastewater is reduced and how difficult it is to produce value-added products from the residual aqueous solution. An overview of the enrichment factor obtained in removing the metal ions is shown in Table 9.

$$\text{Enrichment Factor} = \frac{\text{concentration of heavy ion in the residualeffluent.}}{\text{initial heavy metal ion cocncnetration in water solution}} \times 100$$

The influence of volume ratios on the enrichment factor was investigated by Al-Hameri et al. [195]. According to the findings of the experiments, the (Ef) dropped when the volume ratio increased from 1/2 to 1/6 because heavy metal ion concentrations in the residual effluent are a major determinant of the Ef. As a result of the lower cyclopentane/water solution volume ratio, there was a higher (Ef) and the concentrated effluent accumulates a significant percentage of heavy metal ions. The research removed heavy metals from industrial wastewater, utilizing a cyclopentane as hydrate former in the production method, in which cyclopentane was used to form hydrate due to its immiscibility, thermodynamic stability and non-toxicity. During the experiments it was noted that the (Ef) was approximately equal for all four (Ni, Cr, Zn and Cu) heavy metal ions, despite the fact that each ions' radii and charges are different. Yang et al. performed the investigation and extracted the water from the $Ni^{2+}$-contaminated wastewater using clathrate hydrate separation. The enrichment factor was found to be between 1.15 and 1.60. Like water recovery, the enrichment factor for $Ni^{2+}$ falls as the concentration of feeding solution increases. Due to the fact that water activity decreases as $Ni^{2+}$ concentration increases, it was difficult to remove water from concentrate and the feeding solution [171].

Dong et al. found that, when $Cu^{2+}$ concentration enhanced, the enrichment factor dropped, and it peaked at 2.80. This was due to the fact that less hydrate formed when the aqueous solution concentration was greater. Conversely, there was less leftover water due to increased hydrate formation. The adoption of gas hydrate technology can therefore help to minimize wastewater by reducing liquid discharges [176]. Song et al. found the enrichment factor for each heavy metal ion, an indicator of how difficult it will be to treat any remaining wastewater. They observed that the enrichment factor decreased when the R141b–effluent volume ratio was changed from 1:4 to 1:6. Even with varied ionic radii and charges, the enrichment factor was roughly the same for all four heavy metal ions (Cu, Cr, Ni, Zn) [170].

**Table 9.** Enrichment factors of metal ions.

| Metal Ion | Hydrate Former | Initial Conc. (ppm) | Enrichment Factor | Reference |
|---|---|---|---|---|
| Cr | R141b | 96.7 | 1.8501 | [170] |
| Cu | | 104.4 | 1.8247 | |
| Ni | | 97.12 | 1.805 | |
| Zn | | 93.36 | 1.8691 | |
| Ni | CP | 20 | 1.6 | [171] |
| Cu | R141b | 16.75 | 2.8 | [176] |
| Zn | | 10 | 1.4 | [178] |
| Ni | CP | 10 | 1.32 | |
| Zn | | 10 | 1.31 | |
| Cr | | 10 | 1.29 | |

### 3.7. *Gas Hydrate and Hybrid Technologies*

Another hybrid desalination process should be used as a post-treatment for the gas hydrate (GH) process to produce fresh water to improve water quality. Reverse Osmosis (RO) uses GH to make freshwater that meets design specifications, concentration and drinking water regulations [188]. From the standpoint of the RO process, the GH process is one of the pretreatment options for reducing the salinity of saltwater. Lee et al. [180] used a hybrid GH-RO simulation to find the best RO recovery approach that used the least amount of energy. The GH and RO units were connected in series in their system, and the seawater was fed into the GH system. The RO unit received the desalinated product water from GH as feed, but the concentrate was rejected. The RO unit's permeate was collected as product water, and the retentate was recycled into the GH process' feed stream. Due to their intrinsic reliance on thermal energy primarily obtained from fossil fuels, the conventional processes have significant capital and operating costs and are regarded to be highly energy intensive. The optimum RO recovery values for GH processes with salt rejections of 78, 84 and 90 percent were calculated to be 0.6, 0.8 and 0.8, respectively. According to the simulation results, the maximum permissible energy consumption of the GH process is 1.4 $kWh/m^3$ (with GH salt rejection of 78 percent) to overcome the saltwater RO process with an energy recovery device, and it can be increased to 1.9 $kWh/m^3$ when GH salt rejection is increased to 90 percent. The data revealed that by merging these procedures, energy consumption might be reduced. In another hybrid process of GH+.

Capacitive deionization (CDI) method [196] electrodes are made up of activated carbon to avoid the dissolution of binder, polyvinyl alcohol (PVA) is a binder and glutaric acid acts like a crosslinking agent. The hydrate is allowed to form and later is compacted into pellets with only a small amount of brine retained inside the pores. To boost the salt removal efficiency, the CDI approach is utilized as a post-treatment step. The hybrid desalination process could remove 82% of $Na^+$ and 100% of $K^+$, $Ca^{2+}$ and $Mg^{2+}$ ions. The researcher studied the effects of NaCl, KCl, $MgCl_2$ and $CaCl_2$ on $CO_2$ hydrate production and salt removal efficiency. These salts lowered the amount of water converted to hydrate and desalination efficiency. As each of the traditional desalination processes has its own set of advantages and demerits, researchers [194,197,198] have proposed hybrid approaches to increase desalination efficiency while cutting costs. In hyper-saline water treatment, where the RO system becomes energy intensive and expensive, such hybrid systems may be more useful.

### 3.8. *Limitations in Adopting Gas Hydrate Technique*

Maintaining low temperature and high-pressure conditions needs additional energy requirements and is the fundamental challenge in commercializing the gas hydrate-based desalination process. The kinetics of hydrate formation and dissociation is another important parameter for the commercialization potential of clathrate-based separation. Slow kinetics restrictions for hydrate-based desalination can be overcome with careful selection of a favorable guest gas/liquid and improved reactor architecture design. $CO_2$, methane, propane, cyclopentane (CP) and refrigerants like CFC were mostly used in the literature. Faster kinetics with appropriate promoters could allow the hydrate-based desalination process to be combined with a gas separation process at an industrial scale, increasing its commercial feasibility and overall efficiency.

Since the mid-nineteenth century, researchers have been looking for ideal hydrate formers that meet the criteria of non-toxicity, eco-friendliness, stability, availability and that are economically feasible [109]. Furthermore, CP is an atmospheric hydrate-forming agent and has been demonstrated by a thorough and systematic investigation, although due to its suspension formation, it requires additional post-treatment technique in desalination. Despite the fact that the hydrate-based separation method has encountered a number of technological, energy-efficiency and environmental challenges, no commercially viable systems have emerged. Because the cost of the operation is based on various factors, including the temperature of the brine, appropriate thermodynamic conditions, mobility of salt, salt concentration and yield [137,138,196], financial viability must be considered. As a result, environmentally acceptable and energy-efficient hydrate-based heavy metal exclusion procedures are still required.

A recent study [139–141] was based on utilizing LNG cold energy as a heat sink that was modelled and simulated as an energy source for a gas hydrate plant, and its economic feasibility was assessed. The results were surprising, demonstrating that the LNG heat sink can effectively replace the need for an external refrigeration arrangement, allowing for the separation of high-strength ions with minimal energy consumption, demonstrating that this type of integration can lead to the commercialization of the hydrate process. The salt separation efficiency of the process, as well as the quantity of cooling energy required for freezing and subsequent crystal formation, are used to evaluate the process efficiency of freeze desalination. As a result, the hydrate-based desalination technique, which is part of freeze desalination, is of great interest since the crystallization process runs at a significantly higher temperature than the freeze desalination process (typically above the freezing point of water), resulting in a significant reduction in energy usage for crystallization [89,90,139,140]. A hydrate-based desalination continuous module was developed, with a series of operations and separation techniques, with numerous injections, separation and purging times tested [142] with a water recovery rate of more than 30%, and with an ultimate removal efficiency of greater than 80%. As a result, the best operating parameters for the HBD process should be calculated by taking into account the greatest water yield based on the goal of beginning salinity, as well as the trade-off between salt enrichment efficiency and solid–liquid separation efficiency.

Highly saline water or produced water, for example, can have a salinity of up to 35%. Under those circumstances, the traditional desalination technologies are inefficient in dealing with such high salinity. As a result, there is a lot of potential for using the HyDesal method to treat produced water. The development of commercially effective clathrate-hydrate-based separation has been expedited by recent technological and economic developments. More research is needed, however, to improve the separation efficiency by reducing the amount of salts trapped within hydrate crystals in each cycle and optimizing the key operating factors such as the suitable hydrate formers, the use of various additives or promoters and the best operating conditions of pressure and temperature [143]. In this case the water separation is simple but the cost of pressurization is high.

### 3.9. Opportunities

However, as listed above, the limitations of hydrate-based desalination, the technology difficulties of slow hydrate formation kinetics, the difficulty of crystal separation from brine and the recovery of hydrate former, making it a continuous system and high energy consumption [198], have yet to be fully handled. From an energistics standpoint, water treatment techniques using hydrate formers with lower equilibrium formation pressure are preferable. The use of appropriate hydrate formers to increase nucleation and growth is recommended to address slow kinetics. At the plant scale, improving the hydrate formation kinetics will encourage a continuous mode of operation. Further research into benign promoters may be able to ameliorate various liquids' harmful impact on product water quality and also post-treatment.

The hydrate-based technology has proven to be competitive and cost-effective when compared to traditional methods. The years of process improvement provide a compelling rationale for promoting the use of hydrate technology in water treatment applications. However, more evaluations on the economic viability of the suggested reactor and post-treatment unit operational designs in the direction of commercialization are required. It would be interesting to investigate the appropriate integration possibilities for hydrate technology. For example, to lower the total energy consumption connected with clathrate hydrate technology, the potential refrigeration load available at LNG regasification facilities can be used for hydrate-based water reclamation. Hybrid procedures like hydrate process + RO may also be appealing for enhanced water quality and energy efficiency. As a result, future research should focus on developing sustainable solutions that combine the use of LNG cold energy, selecting appropriate gas/liquid as a guest and reforming reactor designs, as well as focusing on the combinations of various hydrate formers that are biodegradable. In addition, energy analysis and cost estimation are important variables in hydrate desalination's industrialization. It is critical to continue developing novel energy-saving technologies.

## 4. Conclusions

During the past two decades, environmental regulations have become more stringent, requiring to discharge a higher grade of treated effluent. A range of traditional technologies have been developed for the removal of heavy metals from polluted wastewater. This review summarized the numerous interactions of hydrate formers used in gas hydrate-based desalination and their natural occurrence, using gas hydrates in eliminating heavy metal ions The constant research into developing/designing new reactors, or the hydrate formers that can enhance the capability of forming more hydrates or numerous effective ways for using gas hydrates in sustainable development have been established by evaluating the feasibility of processes and enhancing their rates of formation and dissociation. In spite of the advantages of the hydrate-based mechanism for heavy metal separation methods, slow kinetics, suitable thermodynamics and separation of hydrates are still substantial barriers to overcome. More research is needed on the selection of a proper hydrate former/promoter, the cost efficient post-treatment separation techniques for higher water recovery and improvement of the efficiency of the process.

**Author Contributions:** S.N. and A.u.R.: Conceptualization, methodology, original draft and formal analysis; B.L.: resources, project administration supervision, reviewing and editing; D.B.Z.: reviewing and editing. All authors have read and agreed to the published version of the manuscript.

**Funding:** The authors would like to thank Chemical Engineering Department, Yayasan UTP cost center number 015LC0-100, Universiti Teknologi PETRONAS, Malaysia for providing adequate facilities throughout this study.

**Institutional Review Board Statement:** Not applicable.

**Informed Consent Statement:** Not applicable.

**Data Availability Statement:** Not applicable.

**Acknowledgments:** The authors would like to thank Chemical Engineering Department, $CO_2$ Research Centre Universiti Teknologi PETRONAS, Malaysia for providing adequate facilities throughout this study.

**Conflicts of Interest:** The authors declared no potential conflicts of interest with respect to the research, authorship, and/or publication of this article.

## References


1. Rosado, D.; Usero, J.; Morillo, J. Assessment of heavy metals bioavailability and toxicity toward Vibrio fischeri in sediment of the Huelva estuary. *Chemosphere* **2016**, *153*, 10–17. [CrossRef] [PubMed]
2. Azetsu-Scott, K.; Yeats, P.; Wohlgeschaffen, G.; Dalziel, J.; Niven, S.; Lee, K. Precipitation of heavy metals in produced water: Influence on contaminant transport and toxicity. *Mar. Environ. Res.* **2007**, *63*, 146–167. [CrossRef] [PubMed]
3. Ali, I.; Alharbi, O.M.; Alothman, Z.A.; Badjah, A.Y. Kinetics, thermodynamics, and modeling of amido black dye photodegradation in water using $Co/TiO_2$ nanoparticles. *Photochem. Photobiol.* **2018**, *94*, 935–941. [CrossRef] [PubMed]
4. Ali, I.; Alharbi, O.M.; ALOthman, Z.A.; Alwarthan, A.; Al-Mohaimeed, A.M. Preparation of a carboxymethylcellulose-iron composite for uptake of atorvastatin in water. *Int. J. Biol. Macromol.* **2019**, *132*, 244–253. [CrossRef] [PubMed]
5. Khan, N.A.; Ahmed, S.; Farooqi, I.H.; Ali, I.; Vambol, V.; Changani, F.; Khan, A.H. Occurrence, sources and conventional treatment techniques for various antibiotics present in hospital wastewaters: A critical review. *TrAC Trends Anal. Chem.* **2020**, *129*, 115921. [CrossRef]
6. Igunnu, E.T.; Chen, G.Z. Produced water treatment technologies. *Int. J. Low-Carbon Technol.* **2014**, *9*, 157–177. [CrossRef]
7. Sloan, E.D., Jr.; Koh, C.A. *Clathrate Hydrates of Natural Gases*, 3rd ed.; CRC Press: Boca Raton, FL, USA, 2007; pp. 116–117.
8. Gonçalves, A.L.; Pires, J.C.M.; Simões, M. A review on the use of microalgal consortia for wastewater treatment. *Algal Res.* **2017**, *24*, 403–415. [CrossRef]
9. Renuka, N.; Sood, A.; Ratha, S.K.; Prasanna, R.; Ahluwalia, A.S. Evaluation of microalgal consortia for treatment of primary treated sewage effluent and biomass production. *J. Appl. Phycol.* **2013**, *25*, 1529–1537. [CrossRef]
10. Tee, P.F.; Abdullah, M.O.; Tan, I.A.W.; Rashid, N.K.A.; Amin, M.A.M.; Nolasco-Hipolito, C.; Bujang, K. Review on hybrid energy systems for wastewater treatment and bio-energy production. *Renew. Sustain. Energy Rev.* **2016**, *54*, 235–246. [CrossRef]
11. Ngah, W.S.W.; Teong, L.C.; Hanafiah, M.A.K.M. Adsorption of dyes and heavy metal ions by chitosan composites: A review. *Carbohydr. Polym.* **2011**, *83*, 1446–1456. [CrossRef]
12. Quero, G.M.; Cassin, D.; Botter, M.; Perini, L.; Luna, G.M. Patterns of benthic bacterial diversity in coastal areas contaminated by heavy metals, polycyclic aromatic hydrocarbons (PAHs) and polychlorinated biphenyls (PCBs). *Front. Microb.* **2015**, *6*, 1053. [CrossRef] [PubMed]
13. Finizio, A.; Azimonti, G.; Villa, S. Occurrence of pesticides in surface water bodies: A critical analysis of the Italian national pesticide survey programs. *J. Environ. Monit.* **2011**, *13*, 49–57. [CrossRef] [PubMed]
14. Gupta, V.K. Application of low-cost adsorbents for dye removal—A review. *J. Environ. Manag.* **2009**, *90*, 2313–2342. [CrossRef] [PubMed]
15. Gong, Y.; Zhao, X.; Cai, Z.; O'reilly, S.E.; Hao, X.; Zhao, D. A review of oil, dispersed oil and sediment interactions in the aquatic environment: Influence on the fate, transport and remediation of oil spills. *Mar. Pollut. Bull.* **2014**, *79*, 16–33. [CrossRef] [PubMed]
16. Johnson, R.F.; Manjreker, T.G.; Halligan, J.E. Removal of oil from water surfaces by sorption on unstructured fibers. *Environ. Sci. Technol.* **1973**, *7*, 439–443. [CrossRef]
17. Xu, P.; Zeng, G.M.; Huang, D.L.; Feng, C.L.; Hu, S.; Zhao, M.H.; Liu, Z.F. Use of iron oxide nanomaterials in wastewater treatment: A review. *Sci. Total Environ.* **2012**, *424*, 1–10. [CrossRef]
18. Srivastava, N.K.; Majumder, C.B. Novel biofiltration methods for the treatment of heavy metals from industrial wastewater. *J. Hazard. Mater.* **2008**, *151*, 1–8. [CrossRef]
19. Shrestha, R.; Ban, S.; Devkota, S.; Sharma, S.; Joshi, R.; Tiwari, A.P.; Joshi, M.K. Technological Trends in Heavy Metals Removal from Industrial Wastewater: A review. *J. Environ. Chem. Eng.* **2021**, *9*, 105688. [CrossRef]
20. Tran, T.; Leu, H.; Chiu, K.; Lin, C. Electrochemical Treatment of Heavy Metal-containing Wastewater with the Removal of COD and Heavy Metal Ions. *J. Chin. Chem. Soc.* **2017**, *64*, 493–502. [CrossRef]
21. Bielen, A.; Šimatović, A.; Kosić-Vukšić, J.; Senta, I.; Ahel, M.; Babić, S.; Udiković-Kolić, N. Negative environmental impacts of antibiotic-contaminated effluents from pharmaceutical industries. *Water Res.* **2017**, *126*, 79–87. [CrossRef]
22. Lokhande, R.S.; Singare, P.U.; Pimple, D.S. Toxicity study of heavy metals pollutants in waste water effluent samples collected from Taloja industrial estate of Mumbai. *Resour. Environ.* **2011**, *1*, 13–19.
23. Alluri, H.K.; Ronda, S.R.; Settalluri, V.S.; Bondili, J.S.; Suryanarayana, V.; Venkateshwar, P. Biosorption: An eco-friendly alternative for heavy metal removal. *Afr. J. Biotechnol.* **2007**, *6*, 2924–2931.
24. Madu, P.C.; Akpaiyo, G.D.; Ikoku, P. Biosorption of $Cr^{3+}$, $Pb^{2+}$, and $Cd^{2+}$ ions from aqueous solution using modified and unmodified millet chaff. *J. Chem. Pharm. Res.* **2011**, *3*, 467–477.
25. Dhir, B.; Sharmila, P.; Saradhi, P.P. Potential of aquatic macrophytes for removing contaminants from the environment. *Crit. Rev. Environ. Sci. Technol.* **2009**, *39*, 754–781. [CrossRef]
26. Nurchi, V.M.; Crisponi, G.; Villaescusa, I. Chemical equilibria in wastewaters during toxic metal ion removal by agricultural biomass. *Coord. Chem. Rev.* **2010**, *254*, 2181–2192. [CrossRef]
27. Simate, G.S.; Maledi, N.; Ochieng, A.; Ndlovu, S.; Zhang, J.; Walubita, L.F. Coal-based adsorbents for water and wastewater treatment. *J. Environ. Chem. Eng.* **2016**, *4*, 2291–2312. [CrossRef]
28. Ali, H.; Khan, E.; Sajad, M.A. Phytoremediation of heavy metals—Concepts and applications. *Chemosphere* **2013**, *91*, 869–881. [CrossRef]
29. Guieysse, B.; Norvill, Z.N. Sequential chemical–biological processes for the treatment of industrial wastewaters: Review of recent progresses and critical assessment. *J. Hazard. Mater.* **2014**, *267*, 142–152. [CrossRef]

30. Boamah, P.O.; Huang, Y.; Hua, M.; Zhang, Q.; Wu, J.; Onumah, J. Sorption of heavy metal ions onto carboxylate chitosan derivatives—A mini-review. *Ecotoxicol. Environ. Saf.* **2015**, *116*, 113–120. [CrossRef]
31. Lesmana, S.O.; Febriana, N.; Soetaredjo, F.E.; Sunarso, J.; Ismadji, S. Studies on potential applications of biomass for the separation of heavy metals from water and wastewater. *Biochem. Eng. J.* **2009**, *44*, 19–41. [CrossRef]
32. Ahmed, M.J.K.; Ahmaruzzaman, M. A review on potential usage of industrial waste materials for binding heavy metal ions from aqueous solutions. *J. Water Proc. Eng.* **2016**, *10*, 39–47. [CrossRef]
33. Ruihua, L.; Lin, Z.; Tao, T.; Bo, L. Phosphorus removal performance of acid mine drainage from wastewater. *J. Hazard. Mater.* **2011**, *90*, 669–676. [CrossRef] [PubMed]
34. Nguyen, T.A.H.; Ngo, H.H.; Guo, W.S.; Zhang, J.; Liang, S.; Yue, Q.Y.; Nguyen, T.V. Applicability of agricultural waste and by-products for adsorptive removal of heavy metals from wastewater. *Bioresour. Technol.* **2013**, *148*, 574–585. [CrossRef] [PubMed]
35. Farooq, U.; Kozinski, J.A.; Khan, M.A.; Athar, M. Biosorption of heavy metal ions using wheat based biosorbents—A review of the recent literature. *Bioresour. Technol.* **2010**, *101*, 5043–5053. [CrossRef] [PubMed]
36. Ahmaruzzaman, M. Role of fly ash in the removal of organic pollutants from wastewater. *Energy Fuels* **2009**, *23*, 1494–1511. [CrossRef]
37. Visa, M. Synthesis and characterization of new zeolite materials obtained from fly ash for heavy metals removal in advanced wastewater treatment. *Powder Technol.* **2016**, *294*, 338–347. [CrossRef]
38. Ghernaout, D.; Al-Ghonamy, A.I.; Boucherit, A.; Ghernaout, B.; Naceur, M.W.; Messaoudene, N.A.; Elboughdiri, N.A. Brownian motion and coagulation process. *Am. J. Environ. Prot.* **2015**, *4*, 1–15. [CrossRef]
39. Chang, Q.; Zhang, M.; Wang, J. Removal of $Cu^{2+}$ and turbidity from wastewater by mercaptoacetyl chitosan. *J. Hazard. Mater.* **2009**, *169*, 621–625. [CrossRef]
40. Sakhi, D.; Rakhila, Y.; Elmchaouri, A.; Abouri, M.; Souabi, S.; Jada, A. Optimization of coagulation flocculation process for the removal of heavy metals from real textile wastewater. *Adv. Intell. Syst. Comput.* **2019**, *913*, 257–266.
41. Guo, W.; Fu, Z.; Wang, H.; Liu, S.; Wu, F.; Giesy, J.P. Removal of antimonate (Sb(V)) and antimonite (Sb(III)) from aqueous solutions by coagulation-flocculation-sedimentation (CFS): Dependence on influencing factors and insights into removal mechanisms. *Sci. Total Environ.* **2018**, *644*, 1277–1285. [CrossRef]
42. Yan, L.; Yin, H.; Zhang, S.; Leng, F.; Nan, W.; Li, H. Biosorption of inorganic and organic arsenic from aqueous solution by Acidithiobacillus ferrooxidans BY-3. *J. Hazard. Mater.* **2010**, *178*, 209–217. [CrossRef] [PubMed]
43. Łukasiewicz, E. Post-coagulation sludge management for water and wastewater treatment with focus on limiting its impact on the environment. *Econ. Environ. Stud.* **2016**, *16*, 831–841.
44. Ahmad, T.; Ahmad, K.; Alam, M. Sustainable management of water treatment sludge through 3 'R'concept. *J. Clean. Prod.* **2016**, *124*, 1–13. [CrossRef]
45. Bilal, M.; Shah, J.A.; Ashfaq, T.; Gardazi, S.M.H.; Tahir, A.A.; Pervez, A.; Haroon, H.; Mahmood, Q. Waste biomass adsorbents for copper removal from industrial wastewater—A review. *J. Hazard. Mater.* **2013**, *263*, 322–333. [CrossRef] [PubMed]
46. Hubicki, Z.; Kołodyńska, D. Selective removal of heavy metal ions from waters and waste waters using ion exchange methods. *Ion Exch. Technol.* **2012**, *7*, 193–240.
47. An, B.; Liang, Q.; Zhao, D. Removal of arsenic (V) from spent ion exchange brine using a new class of starch-bridged magnetite nanoparticles. *Water Res.* **2011**, *45*, 1961–1972. [CrossRef] [PubMed]
48. Barakat, M.A. New trends in removing heavy metals from industrial wastewater. *Arab. J. Chem.* **2011**, *4*, 361–377. [CrossRef]
49. Motsi, T.; Rowson, N.A.; Simmons, M.J.H. Adsorption of heavy metals from acid mine drainage by natural zeolite. *Int. J. Miner. Process.* **2009**, *92*, 42–48. [CrossRef]
50. Figueiredo, H.; Quintelas, C. Tailored zeolites for the removal of metal oxyanions: Overcoming intrinsic limitations of zeolites. *J. Hazard. Mater.* **2014**, *274*, 287–299. [CrossRef]
51. Ibrahim, H.S.; Jamil, T.S.; Hegazy, E.Z. Application of zeolite prepared from Egyptian kaolin for the removal of heavy metals: II. Isotherm models. *J. Hazard. Mater.* **2010**, *182*, 842–847. [CrossRef]
52. Kononova, O.N.; Bryuzgina, G.L.; Apchitaeva, O.V.; Kononov, Y.S. Ion exchange recovery of chromium (VI) and manganese (II) from aqueous solutions. *Arab. J. Chem.* **2019**, *12*, 2713–2720. [CrossRef]
53. Mahmoud, M.R.; Lazaridis, N.K.; Matis, K.A. Study of flotation conditions for cadmium (II) removal from aqueous solutions. *Process Saf. Environ. Prot.* **2015**, *94*, 203–211. [CrossRef]
54. Patil, D.S.; Chavan, S.M.; Oubagaranadin, J.U.K. A review of technologies for manganese removal from wastewaters. *J. Environ. Chem. Eng.* **2016**, *4*, 468–487. [CrossRef]
55. Hoseinian, F.S.; Rezai, B.; Kowsari, E.; Chinnappan, A.; Ramakrishna, S. Synthesis and characterization of a novel nanocollector for the removal of nickel ions from synthetic wastewater using ion flotation. *Sep. Purif. Technol.* **2020**, *240*, 116639. [CrossRef]
56. Mahne, E.J.; Pinfold, T.A. Precipitate flotation I. Removal of nickel from dilute aqueous solutions and its separation from cobalt. *J. Appl. Chem.* **1968**, *18*, 52–54. [CrossRef]
57. Peng, W.; Chang, L.; Li, P.; Han, G.; Huang, Y.; Cao, Y. An overview on the surfactants used in ion flotation. *J. Mol. Liq.* **2019**, *286*, 110955. [CrossRef]
58. Lyu, S.; Chen, W.; Zhang, W.; Fan, Y.; Jiao, W. Wastewater reclamation and reuse in China: Opportunities and challenges. *J. Environ. Sci.* **2016**, *39*, 86–96. [CrossRef]

59. Mutamim, N.S.A.; Noor, Z.Z.; Hassan, M.A.A.; Olsson, G. Application of membrane bioreactor technology in treating high strength industrial wastewater: A performance review. *Desalination* **2012**, *305*, 1–11. [CrossRef]
60. Wang, L.K.; Chen, J.P.; Hung, Y.T.; Shammas, N.K. *Membrane and Desalination Technologies*, 1st ed.; Humana Press: Totowa, NJ, USA, 2008; pp. 1–45.
61. Fu, F.; Wang, Q. Removal of heavy metal ions from wastewaters: A review. *J. Environ. Manag.* **2011**, *92*, 407–418. [CrossRef]
62. Li, W.; Zheng, P.; Guo, J.; Ji, J.; Zhang, M.; Zhang, Z.; Zhan, E.; Abbas, G. Characteristics of self-alkalization in high-rate denitrifying automatic circulation (DAC) reactor fed with methanol and sodium acetate. *Bioresour. Technol.* **2014**, *154*, 44–50. [CrossRef]
63. Hashim, M.A.; Mukhopadhyay, S.; Sahu, J.N.; Sengupta, B. Remediation technologies for heavy metal contaminated groundwater. *J. Environ. Manag.* **2011**, *92*, 2355–2388. [CrossRef] [PubMed]
64. Tanong, K.; Tran, L.H.; Mercier, G.; Blais, J.F. Recovery of Zn (II), Mn (II), Cd (II) and Ni (II) from the unsorted spent batteries using solvent extraction, electrodeposition and precipitation methods. *J. Clean. Prod.* **2017**, *148*, 233–244. [CrossRef]
65. Kuan, Y.C.; Lee, L.H.; Chern, J.M. Heavy metal extraction from PCB wastewater treatment sludge by sulfuric acid. *J. Hazard. Mater.* **2010**, *177*, 881–886. [CrossRef] [PubMed]
66. Yadav, M.; Gupta, R.; Sharma, R.K. Green and sustainable pathways for wastewater purification. In *Advances in Water Purification Techniques*; Elsevier: Amsterdam, The Netherlands, 2019; pp. 355–383.
67. Ain Zainuddin, N.; Azwan Raja Mamat, T.; Imam Maarof, H.; Wahidah Puasa, S.; Rohana Mohd Yatim, S. Removal of Nickel, Zinc and Copper from Plating Process Industrial Raw Effluent Via Hydroxide Precipitation Versus Sulphide Precipitation. *IOP Conf. Ser. Mater. Sci. Eng.* **2019**, *551*, 12122. [CrossRef]
68. Vu, H.H.T.; Gu, S.; Thriveni, T.; Khan, M.D.; Tuan, L.Q.; Ahn, J.W. Sustainable treatment for sulfate and lead removal from battery wastewater. *Sustainability* **2019**, *11*, 3497. [CrossRef]
69. Elsheikh, M.A.; Al-Hemaidi, W.K. Approach in choosing suitable technology for industrial wastewater treatment. *J. Civ. Environ. Eng.* **2012**, *2*, 5. [CrossRef]
70. Zhang, X.; Tian, J.; Hu, Y.; Han, H.; Luo, X.; Sun, W.; Zhou, H. Selective sulfide precipitation of copper ions from arsenic wastewater using monoclinic pyrrhotite. *Sci. Total Environ.* **2020**, *705*, 135816. [CrossRef]
71. Elabbas, S.; Ouazzani, N.; Mandi, L.; Berrekhis, F.; Perdicakis, M.; Pontvianne, S.; Pons, M.N.; Lapicque, F.; Leclerc, J.P. Treatment of highly concentrated tannery wastewater using electrocoagulation: Influence of the quality of aluminium used for the electrode. *J. Hazard. Mater.* **2016**, *319*, 69–77. [CrossRef]
72. de Almeida, C.C.; da Costa, P.R.F.; Melo, M.J.D.M.; dos Santos, E.V.; Martínez-Huitle, C.A. Application of electrochemical technology for water treatment of Brazilian industry effluents. *J. Mex. Chem. Soc.* **2014**, *58*, 276–286.
73. Liu, C.; Wu, T.; Hsu, P.C.; Xie, J.; Zhao, J.; Liu, K.; Cui, Y. Direct/alternating current electrochemical method for removing and recovering heavy metal from water using graphene oxide electrode. *ACS Nano* **2019**, *13*, 6431–6437. [CrossRef]
74. Yang, X.; Liu, L.; Zhang, M.; Tan, W.; Qiu, G.; Zheng, L. Improved removal capacity of magnetite for Cr (VI) by electrochemical reduction. *J. Hazard. Mater.* **2019**, *374*, 26–34. [CrossRef] [PubMed]
75. Zhang, C.; Jiang, Y.; Li, Y.; Hu, Z.; Zhou, L.; Zhou, M. Three-dimensional electrochemical process for wastewater treatment: A general review. *Chem. Eng. J.* **2013**, *228*, 455–467. [CrossRef]
76. Kurniawan, T.A.; Babel, S. A research study on Cr (VI) removal from contaminated wastewater using low-cost adsorbents and commercial activated carbon. In Proceedings of the 2nd International Conference on Energy Technology Towards a Clean Environment (RCETE), Phuket, Thailand, 12–14 February 2003; Volume 2, pp. 1110–1117.
77. Nallakukkala, S.; Lal, B.; Shaik, F. Kinetic and isothermal investigations in elimination of iron metal from aqueous mixture by using natural adsorbent. *Int. J. Environ. Sci. Technol.* **2021**, *18*, 1761–1772. [CrossRef]
78. Bose, P.; Bose, M.A.; Kumar, S. Critical evaluation of treatment strategies involving adsorption and chelation for wastewater containing copper, zinc and cyanide. *Adv. Environ. Res.* **2002**, *7*, 179–195. [CrossRef]
79. Vunain, E.; Mishra, A.K.; Mamba, B.B. Dendrimers, mesoporous silicas and chitosan-based nanosorbents for the removal of heavy-metal ions: A review. *Int. J. Biol. Macromol.* **2016**, *86*, 570–586. [CrossRef]
80. Ewecharoen, A.; Thiravetyan, P.; Wendel, E.; Bertagnolli, H. Nickel adsorption by sodium polyacrylate-grafted activated carbon. *J. Hazard. Mater.* **2009**, *171*, 335–339. [CrossRef]
81. Li, Y.; Du, Q.; Liu, T.; Peng, X.; Wang, J.; Sun, J.; Xia, L. Comparative study of methylene blue dye adsorption onto activated carbon, graphene oxide, and carbon nanotubes. *Chem. Eng. Res. Des.* **2013**, *91*, 361–368. [CrossRef]
82. El-Sikaily, A.; El Nemr, A.; Khaled, A.; Abdelwehab, O. Removal of toxic chromium from wastewater using green alga Ulva lactuca and its activated carbon. *J. Hazard. Mater.* **2007**, *148*, 216–228. [CrossRef]
83. Sathishkumar, P.; Arulkumar, M.; Palvannan, T. Utilization of agro-industrial waste Jatropha curcas pods as an activated carbon for the adsorption of reactive dye Remazol Brilliant Blue R (RBBR). *J. Clean. Prod.* **2012**, *22*, 67–75. [CrossRef]
84. Qdais, H.A.; Moussa, H. Removal of heavy metals from wastewater by membrane processes: A comparative study. *Desalination* **2004**, *164*, 105–110. [CrossRef]
85. Tzanetakis, N.; Taama, W.M.; Scott, K.; Jachuck, R.J.J.; Slade, R.S.; Varcoe, J. Comparative performance of ion exchange membranes for electrodialysis of nickel and cobalt. *Sep. Purif. Technol.* **2003**, *30*, 113–127. [CrossRef]
86. Christensen, I.V.; Pedersen, A.J.; Ottosen, L.M.; Ribeiro, A.B. Electrodialytic remediation of CCA-treated waste wood in a 2 $m^3$ pilot plant. *Sci. Total Environ.* **2006**, *364*, 45–54. [CrossRef] [PubMed]

87. BrbootI, M.M.; AbiD, B.A.; Al-ShuwaikI, N.M. Removal of heavy metals using chemicals precipitation. *Eng. Technol. J.* **2011**, *29*, 595–612.
88. Al-Enezi, G.; Hamoda, M.F.; Fawzi, N. Ion exchange extraction of heavy metals from wastewater sludges. *J. Environ. Sci. Health Part A Toxic/Hazard. Subst. Environ. Eng.* **2004**, *39*, 455–464. [CrossRef]
89. Babu, P.; Nambiar, A.; He, T.; Karimi, I.A.; Lee, J.D.; Englezos, P.; Linga, P. A review of clathrate hydrate based desalination to strengthen energy—Water nexus. *ACS Sustain. Chem. Eng.* **2018**, *6*, 8093–8107. [CrossRef]
90. Nallakukkala, S.; Lal, B. Seawater and produced water treatment via gas hydrate: Review. *J. Environ. Chem. Eng.* **2021**, *9*, 105053. [CrossRef]
91. Rehman Ur, A.; Zaini, D.B.; Lal, B. Application of Gas Hydrate Based Technique in Wastewater Treatment—A Mini Review. In Proceedings of the Third International Conference on Separation Technology 2020 (ICoST 2020), Kelantan, Malaysia, 15 August 2020; pp. 249–254.
92. Nallakukkala, S.; Kassim, Z.; Othman, N.A.; Lal, B. Advancement in Gas Hydrate Water Based Produced Water Desalination: An Overview. In Proceedings of the Third International Conference on Separation Technology 2020 (ICoST 2020), Kelantan, Malaysia, 15 August 2020.
93. Lal, B.; Nashed, O. *Chemical Additives for Gas Hydrates*, 1st ed.; Springer Publication: Cham, Switzerland, 2020; pp. 7–14.
94. Leopercio, B.C. Kinetics of Cyclopentane Hydrate Formation—An Interfacial Rheology Study. Master's Thesis, Pontifical Catholic University of Rio de Janeiro, PUC-Rio, Brazil, 2016.
95. Sangwai, J.S.; Patel, R.S.; Mekala, P.; Mech, D.; Busch, M. Desalination of seawater using gas hydrate technology-current status and future direction. In Proceedings of the 18th International Conference on Hydraulics, Water Resources, Coastal and Environmental Engineering, HYDRO, Madras, India, 4–6 December 2013.
96. Kang, S.P.; Lee, H. Recovery of $CO_2$ from flue gas using gas hydrate: Thermodynamic verification through phase equilibrium measurements. *Environ. Sci. Technol.* **2000**, *34*, 4397–4400. [CrossRef]
97. Kezirian, M.T.; Phoenix, S.L. Natural gas hydrate as a storage mechanism for safe, sustainable and economical production from offshore petroleum reserves. *Energies* **2017**, *10*, 828. [CrossRef]
98. Zhou, H. Hydrate Slurry as Cold Energy Storage and Distribution Medium: Enhancing the Performance of Refrigeration Systems. Master's Thesis, Guangzhou Institute of Energy Conversion, Guangzhou, China, 2017.
99. Aregbe, A.G. Gas hydrate—Properties, formation and benefits. *Open J. Yangtze Oil Gas* **2017**, *2*, 27–44. [CrossRef]
100. Subramani, A.; Jacangelo, J.G. Emerging desalination technologies for water treatment: A critical review. *Water Res.* **2015**, *75*, 64–187. [CrossRef]
101. Ghaffour, N.; Missimer, T.M.; Amy, G.L. Technical review and evaluation of the economics of water desalination: Current and future challenges for better water supply sustainability. *Desalination* **2013**, *309*, 197–207. [CrossRef]
102. Zheng, J.; Cheng, F.; Li, Y.; Lü, X.; Yang, M. Progress and trends in hydrate based desalination (HBD) technology: A review. *Chin. J. Chem. Eng.* **2019**, *27*, 2037–2043. [CrossRef]
103. Elimelech, M.; Phillip, W.A. The future of seawater desalination: Energy, technology, and the environment. *Science* **2011**, *333*, 712–717. [CrossRef] [PubMed]
104. Jia, X.; Klemeš, J.J.; Varbanov, P.S.; Alwi, S.R.W. Analyzing the energy consumption, GHG emission, and cost of seawater desalination in China. *Energies* **2019**, *12*, 463. [CrossRef]
105. Nallakukkala, S.; Lal, B.; Shariff, M.A. Influence of water volume on $CO_2$ hydrate-based desalination of brine solution. *Mater. Today Proc.* **2021**, *56*, 2172–2177. [CrossRef]
106. He, T.; Nair, S.K.; Babu, P.; Linga, P.; Karimi, I.A. A novel conceptual design of hydrate-based desalination (HyDesal) process by utilizing LNG cold energy. *Appl. Energy* **2018**, *222*, 13–24. [CrossRef]
107. Lv, Y.N.; Wang, S.S.; Sun, C.Y.; Gong, J.; Chen, G.J. Desalination by forming hydrate from brine in cyclopentane dispersion system. *Desalination* **2017**, *413*, 217–222. [CrossRef]
108. Xu, C.; Li, X.; Yan, K.; Ruan, X.; Chen, Z.; Xia, Z. Research progress in hydrate-based technologies and processes in China: A review. *Chin. J. Chem. Eng.* **2019**, *27*, 1998–2013. [CrossRef]
109. Sahu, P.; Krishnaswamy, S.; Ponnani, K.; Pande, N.K. A thermodynamic approach to selection of suitable hydrate formers for seawater desalination. *Desalination* **2018**, *436*, 144–151. [CrossRef]
110. Knox, W.G.; Hess, M.; Jones, G.E.; Smith, H.B. The clathrate process. *Chem. Eng. Prog.* **1961**, *57*, 66–71.
111. Ngan, Y.T.; Englezos, P. Concentration of mechanical pulp mill effluents and NaCl solutions through propane hydrate formation. *Ind. Eng. Chem. Res.* **1996**, *35*, 1894–1900. [CrossRef]
112. Nakajima, M.; Ohmura, R.; Mori, Y.H. Clathrate hydrate formation from cyclopentane-in-water emulsions. *Ind. Eng. Chem. Res.* **2008**, *47*, 8933–8939. [CrossRef]
113. Corak, D.; Barth, T.; Høiland, S.; Skodvin, T.; Larsen, R.; Skjetne, T. Effect of subcooling and amount of hydrate former on formation of cyclopentane hydrates in brine. *Desalination* **2011**, *278*, 268–274. [CrossRef]
114. Cai, L.; Pethica, B.A.; Debenedetti, P.G.; Sundaresan, S. Formation kinetics of cyclopentane–methane binary clathrate hydrate. *Chem. Eng. Sci.* **2014**, *119*, 147–157. [CrossRef]
115. Cai, L.; Pethica, B.A.; Debenedetti, P.G.; Sundaresan, S. Formation of cyclopentane methane binary clathrate hydrate in brine solutions. *Chem. Eng. Sci.* **2016**, *141*, 125–132. [CrossRef]

116. Misyura, S.Y.; Manakov, A.Y.; Morozov, V.S.; Nyashina, G.S.; Gaidukova, O.S.; Skiba, S.S.; Volkov, R.S.; Voytkov, I.S. The influence of key parameters on combustion of double gas hydrate. *J. Nat. Gas Sci. Eng.* **2020**, *80*, 103396. [CrossRef]
117. Yang, M.; Song, Y.; Jiang, L.; Liu, W.; Dou, B.; Jing, W. Effects of operating mode and pressure on hydrate-based desalination and $CO_2$ capture in porous media. *Appl. Energy* **2014**, *135*, 504–511. [CrossRef]
118. Kang, K.C.; Linga, P.; Park, K.; Choi, S.J.; Lee, J.D. Seawater desalination by gas hydrate process and removal characteristics of dissolved ions ($Na^+$, $K^+$, $Mg^{2+}$, $Ca^{2+}$, $B^{3+}$, $Cl^-$, $SO_4{}^{2-}$). *Desalination* **2014**, *353*, 84–90. [CrossRef]
119. McCormack, R.A.; Niblock, G.A. *Investigation of High Freezing Temperature, Zero Ozone, and Zero Global Warming Potential, Clathrate Formers for Desalination*; US Department of the Interior, Bureau of Reclamation, Technical Service: Denver, CO, USA, 2000.
120. Simmons, B.A.; Bradshaw, R.W.; Dedrick, D.E.; Cygan, R.T.; Greathouse, J.A.; Majzoub, E.H. *Desalination Utilizing Clathrate Hydrates (LDRD Final Report)*; National Technical Information Service: Livermore, CA, USA, 2008.
121. Karamoddin, M.; Varaminian, F. Water desalination using R141b gas hydrate formation. *Desalin. Water Treat.* **2014**, *52*, 2450–2456. [CrossRef]
122. Ngema, P.T.; Petticrew, C.; Naidoo, P.; Mohammadi, A.H.; Ramjugernath, D. Experimental measurements and thermodynamic modeling of the dissociation conditions of clathrate hydrates for (refrigerant + NaCl + water) systems. *J. Chem. Eng. Data.* **2014**, *59*, 466–475. [CrossRef]
123. Seo, Y.; Moon, D.; Lee, C.; Park, J.W.; Kim, B.S.; Lee, G.W.; Yoon, J.H. Equilibrium, kinetics, and spectroscopic studies of $SF_6$ hydrate in NaCl electrolyte solution. *Environ. Sci. Technol.* **2015**, *49*, 6045–6050. [CrossRef] [PubMed]
124. Wang, L.J.; Zhang, X.M.; Li, H.H.; Shao, L.; Zhang, D.; Jiao, L. Theory research on desalination of brackish water using gas hydrate method. *Adv. Mater. Res.* **2013**, *616*, 1202–1207. [CrossRef]
125. Lee, D.; Lee, Y.; Lee, S.; Seo, Y. Accurate measurement of phase equilibria and dissociation enthalpies of HFC-134a hydrates in the presence of NaCl for potential application in desalination. *Korean J. Chem. Eng.* **2016**, *33*, 1425–1430. [CrossRef]
126. Khan, M.N. Phase Equilibria Modeling of Inhibited Gas Hydrate Systems Including Salts: Applications in Flow Assurance, Seawater Desalination and Gas Separation. Ph.D. Thesis, Colorado School of Mines, Golden, CO, USA, 2016.
127. Zhang, Y.; Sheng, S.M.; Shen, X.D.; Zhou, X.B.; Wu, W.Z.; Wu, X.P.; Liang, D.Q. Phase equilibrium of cyclopentane + carbon dioxide binary hydrates in aqueous sodium chloride solutions. *J. Environ. Chem. Eng.* **2017**, *62*, 2461–2465. [CrossRef]
128. Ho-Van, S.; Bouillot, B.; Douzet, J.; Babakhani, S.M.; Herri, J.M. Implementing cyclopentane hydrates phase equilibrium data and simulations in brine solutions. *Ind. Eng. Chem. Res.* **2018**, *43*, 14774–14783. [CrossRef]
129. Ho-Van, S.; Bouillot, B.; Douzet, J.; Babakhani, S.M.; Herri, J.M. Cyclopentane hydrates–A candidate for desalination? *J. Environ. Chem. Eng.* **2019**, *7*, 103359. [CrossRef]
130. Lv, Q.; Li, X.; Li, G. Seawater desalination by hydrate formation and pellet production process. *Energy Procedia* **2019**, *158*, 5144–5148. [CrossRef]
131. Choi, W.; Lee, Y.; Mok, J.; Lee, S.; Lee, J.D.; Seo, Y. Thermodynamic and kinetic influences of NaCl on HFC-125a hydrates and their significance in gas hydrate-based desalination. *Chem. Eng. J.* **2019**, *358*, 598–605. [CrossRef]
132. Seo, S.D.; Hong, S.Y.; Sum, A.K.; Lee, K.H.; Lee, J.D.; Lee, B.R. Thermodynamic and kinetic analysis of gas hydrates for desalination of saturated salinity water. *Chem. Eng. J.* **2019**, *370*, 980–987. [CrossRef]
133. Han, S.; Shin, J.Y.; Rhee, Y.W.; Kang, S.P. Enhanced efficiency of salt removal from brine for cyclopentane hydrates by washing, centrifuging, and sweating. *Desalination* **2014**, *354*, 17–22. [CrossRef]
134. Karamoddin, M.; Varaminian, F. Water purification by freezing and gas hydrate processes, and removal of dissolved minerals ($Na^+$, $K^+$, $Mg^{2+}$, $Ca^{2+}$). *J. Mol. Liq.* **2016**, *223*, 1021–1031. [CrossRef]
135. Fakharian, H.; Ganji, H.; Naderifar, A.; Mofrad, H.R.; Kakavand, M. Effect of gas type and salinity on performance of produced water desalination using gas hydrates. *J. Water Reuse Desalin.* **2019**, *9*, 396–404. [CrossRef]
136. Kang, K.C.; Hong, S.Y.; Cho, S.J.; Kim, D.H.; Lee, J.D. Evaluation of desalination by nanostructured hydrate formation and pellet production process. *J. Nanosci. Nanotechnol.* **2017**, *17*, 4059–4062. [CrossRef]
137. Javanmardi, J.; Moshfeghian, M. Energy consumption and economic evaluation of water desalination by hydrate phenomenon. *Appl. Therm. Eng.* **2003**, *23*, 845–857. [CrossRef]
138. Ghalavand, Y.; Hatamipour, M.S.; Rahimi, A. A review on energy consumption of desalination processes. *Desalin. Water Treat.* **2015**, *54*, 1526–1541. [CrossRef]
139. Singh, J.; Lal, B. Prospectives on gas hydrates-based desalination. In *Gas Hydrate in Water Treatment*, 1st ed.; Lal, B., Nallakukkala, S., Eds.; John Wiley & Sons, Inc.: Hoboken, NJ, USA, 2022; Volume 1, pp. 31–53.
140. Chong, Z.R.; He, T.; Babu, P.; Zheng, J.; Linga, P. Economic evaluation of energy efficient hydrate-based desalination utilizing cold energy from liquefied natural gas (LNG). *Desalination* **2019**, *463*, 69–80. [CrossRef]
141. He, T.; Chong, Z.R.; Zheng, J.; Ju, Y.; Linga, P. LNG cold energy utilization: Prospects and challenges. *Energy* **2019**, *170*, 557–568. [CrossRef]
142. Zheng, J.; Yang, M. Experimental investigation on novel desalination system via gas hydrate. *Desalination* **2020**, *478*, 114284. [CrossRef]
143. Sarshar, M.; Sharafi, A.H. Simultaneous water desalination and $CO_2$ capturing by hydrate formation. *Desalin. Water Treat.* **2011**, *28*, 59–64. [CrossRef]
144. Babu, P.; Kumar, R.; Linga, P. Unusual behavior of propane as a co-guest during hydrate formation in silica sand: Potential application to seawater desalination and carbon dioxide capture. *Chem. Eng. Sci.* **2014**, *117*, 342–351. [CrossRef]

145. Yang, M.; Zheng, J.; Liu, W.; Liu, Y.; Song, Y. Effects of $C_3H_8$ on hydrate formation and dissociation for integrated $CO_2$ capture and desalination technology. *Energy* **2015**, *93*, 1971–1979. [CrossRef]
146. Donath, W.E. Method and Apparatus for Producing Purified Water from Aqueous Saline Solutions. U.S. Patent No. 2,904,511, 15 September 1959.
147. Buchanan, B.B. Removing Salt from Sea Water. U.S. Patent 3,027,320, 27 March 1962.
148. Walton, P.R. Continuous Saline Water Purification. U.S. Patent 3,132,096, 5 May 1964.
149. Klass, D. Hydrate Forming in Water Desalination. U.S. Patent No 3,856,492, 24 December 1974.
150. Guo, B.; Bretz, R.E.; Lee, R.L. Method and Apparatus for Generating, Transporting and Dissociating Gas Hydrates. U.S. Patent No. 5,473,904, 12 December 1995.
151. McCormack, R.A. Clathrate Freeze Desalination Apparatus and Method. U.S. Patent No. 5,553,456, 10 September 1996.
152. Max, M.D.; Pellenbarg, R.E. Desalination through Methane Hydrate. U.S. Patent No. 5,873,262, 23 February 1999.
153. Heinemann, R.F.; Huang, D.D.T.; Long, J.; Saeger, R.B. Process for Making Gas Hydrates. U.S. Patent No. 6,028,234, 22 February 2000.
154. Heinemann, R.F.; Huang, D.D.T.; Long, J.; Saeger, R.B. Method for Producing Gas Hydrates Utilizing a Fluidized Bed. U.S. Patent No. 6,180,843, 30 January 2001.
155. Max, M.D. Hydrate Formation and Growth for Hydrate-Based Desalination by Means of Enriching Water to Be Treated. U.S. Patent No. 6,890,444, 10 May 2005.
156. Max, M.D.; Korsgaard, J. Hydrate-Based Desalination with Hydrate-Elevating Density-Driven Circulation. U.S. Patent No. 6,969,467, 10 November 2005.
157. Simmons, B.A.; Bradshaw, R.W.; Dedrick, D.E.; Anderson, D.W. Complex Admixtures of Clathrate Hydrates in a Water Desalination Method. U.S. Patent 7560028B1, 14 July 2009.
158. Phelps, T.J.; Tsouris, C.; Palumbo, A.V.; Riestenberg, D.E.; McCallum, S.D. Method for Excluding Salt and Other Soluble Materials from Produced Water. U.S. Patent No. 7,569,737, 4 August 2009.
159. Li, D.; Liang, D.; Tang, C. Test Device for Desalination of Sea Water by Hydrate Method. CN Patent CN101289231B, 6 June 2010.
160. Osegovic, J.P.; Max, M.D.; Tatro, S.R. Seawater-Based Carbon Dioxide Disposal. U.S. Patent No. 8,048,309, 1 November 2011.
161. Carstens, C.; Dickinson, W.; Dickinson, W.; Myers, J. Clathrate Hydrate Modular Storage, Applications and Utilization Processes. U.S. Patent No. 7914749, 29 March 2011.
162. Park, K.N.; Hong, S.Y.; Lee, J.W.; Kang, K.C.; Lee, Y.C.; Ha, M.G.; Lee, J.D. A new apparatus for seawater desalination by gas hydrate process and removal characteristics of dissolved minerals ($Na^+$, $Mg^{2+}$, $Ca^{2+}$, $K^+$, $B^{3+}$). *Desalination* **2011**, *274*, 91–96. [CrossRef]
163. Katyal, A.A. System and Method for Hydrate-Based Desalination. U.S. Patent No. 9,643,860, 9 May 2017.
164. McCormack, R.A.; Ripmeester, J.A. Clathrate Desalination Process Using an Ultrasonic Actuator. U.S. Patent 20140223958A1, 14 August 2014.
165. Parker, A. Potable water from sea-water. *Nature* **1942**, *149*, 184–186. [CrossRef]
166. Hesse, R.; Harrison, W.E. Gas hydrates (clathrates) causing pore-water freshening and oxygen isotope fractionation in deep-water sedimentary sections of terrigenous continental margins. *Earth Planet. Sci. Lett.* **1981**, *55*, 453–462. [CrossRef]
167. Willson III, R.C.; Bulot, E.; Cooney, C. L Use of Hydrates for Aqueous Solution Treatment. U.S. Patent No. 4,678,583, 7 July 1987.
168. Huang, C.P.; Fennema, O.; Powrie, W.D. Gas hydrates in aqueous-organic systems: II. Concentration by gas hydrate formation. *Cryobiology* **1966**, *2*, 240–245. [CrossRef]
169. Truong-Lam, H.S.; Kim, S.; Seo, S.D.; Jeon, C.; Lee, J.D. Water purifying by gas hydrate: Potential applications to desalination and wastewater treatments. *Chem. Eng. Trans.* **2020**, *78*, 67–72.
170. Song, Y.; Dong, H.; Yang, L.; Yang, M.; Li, Y.; Ling, Z.; Zhao, J. Hydrate-based heavy metal separation from aqueous solution. *Sci. Rep.* **2016**, *6*, 21389. [CrossRef]
171. Yang, Y.; Zhou, H.; Li, F.; Shi, C.; Wang, S.; Ling, Z. Cyclopentane hydrate-based processes for treating heavy metal containing wastewater. *E3S Web Conf.* **2019**, *118*, 10–13. [CrossRef]
172. Li, X.S.; Yang, B.; Zhang, Y.; Li, G.; Duan, L.P.; Wang, Y.; Wu, H.J. Experimental investigation into gas production from methane hydrate in sediment by depressurization in a novel pilot-scale hydrate simulator. *Appl. Energy* **2012**, *93*, 722–732. [CrossRef]
173. Ohmura, R.; Ogawa, M.; Yasuoka, K.; Mori, Y.H. Statistical study of clathrate-hydrate nucleation in a water/hydrochlorofluorocarbon system: Search for the nature of the memory effect. *J. Phys. Chem. B* **2003**, *107*, 5289–5293. [CrossRef]
174. Murshed, M.M.; Faria, S.H.; Kuhs, W.F.; Kipfstuhl, S.; Wilhelms, F. The role of hydrochlorofluorocarbon densifiers in the formation of clathrate hydrates in deep boreholes and subglacial environments. *Ann. Glaciol.* **2007**, *47*, 109–114. [CrossRef]
175. Kato, M.; Iida, T.; Mori, Y.H. Drop formation behaviour of a hydrate-forming liquid in a water stream. *J. Fluid Mech.* **2000**, *414*, 367–378. [CrossRef]
176. Dong, H.; Zhang, L.; Ling, Z.; Zhao, J.; Song, Y. The Controlling Factors and Ion Exclusion Mechanism of Hydrate-Based Pollutant Removal. *ACS Sustain. Chem. Eng.* **2019**, *7*, 7932–7940. [CrossRef]
177. Gaikwad, N.; Nakka, R.; Khavala, V.; Bhadani, A.; Mamane, H.; Kumar, R. Gas Hydrate-Based Process for Desalination of Heavy Metal Ions from an Aqueous Solution: Kinetics and Rate of Recovery. *ACS ES&T Water* **2021**, *1*, 134–144.
178. Nallakukkala, S.; Lal, B. Waste brine management. In *Gas Hydrate in Water Treatment*, 1st ed.; Lal, B., Nallakukkala, S., Eds.; John Wiley & Sons, Inc.: Hoboken, NJ, USA, 2022; Volume 1, pp. 31–53.

179. Lee, H.; Ryu, H.; Lim, J.H.; Kim, J.O.; Dong Lee, J.; Kim, S. An optimal design approach of gas hydrate and reverse osmosis hybrid system for seawater desalination. *Desalin. Water Treat.* **2016**, *57*, 9009–9017. [CrossRef]
180. Dong, H.; Fan, Z.; Wang, B.; Xue, S.; Zhao, J.; Song, Y. Hydrate-based reduction of heavy metal ion from aqueous solution. *Energy Procedia* **2017**, *105*, 4706–4712. [CrossRef]
181. Fakharian, H.; Ganji, H.; Naderifar, A. Desalination of high salinity produced water using natural gas hydrate. *J. Taiwan Inst. Chem. Eng.* **2017**, *72*, 157–162. [CrossRef]
182. Xu, H.; Khan, M.N.; Peters, C.J.; Sloan, E.D.; Koh, C.A. Hydrate-based desalination using cyclopentane hydrates at atmospheric pressure. *J. Chem. Eng. Data* **2018**, *63*, 1081–1087. [CrossRef]
183. Han, S.; Rhee, Y.-W.; Kang, S.-P. Investigation of salt removal using cyclopentane hydrate formation and washing treatment for seawater desalination. *Desalination* **2017**, *404*, 132–137. [CrossRef]
184. Fakharian, H.; Ganji, H.; Naderifar, A. Naderifar, Saline produced water treatment using gas hydrates. *J. Environ. Chem. Eng.* **2017**, *5*, 4269–4273. [CrossRef]
185. Khan, M.N.; Peters, C.J.; Koh, C.A. Desalination using gas hydrates: The role of crystal nucleation, growth and separation. *Desalination* **2019**, *468*, 114049. [CrossRef]
186. Li, F.; Yang, Y.; Shi, C.; Yang, S.; Chen, Z.; Yang, S.; Song, Y. Hydrate-based desalination process enhanced via graphite. In *IOP Conference Series: Earth and Environmental Science*; IOP Publishing: Bristol, UK, 2019; Volume 295, p. 042028.
187. Sun, H.; Sun, L.; Zhao, Y.; Yang, S.; Zhang, L.; Dong, H.; Song, Y. A combined hydrate-based method for removing heavy metals from simulated wastewater with high concentrations. *J. Environ. Chem. Eng.* **2021**, *9*, 106633. [CrossRef]
188. Truong-Lam, H.S.; Seo, S.D.; Jeon, C.; Lee, G.P.; Lee, J.D. A gas hydrate process for high-salinity water and wastewater purification. *Desalination* **2022**, *529*, 115651. [CrossRef]
189. Ling, Z.; Shi, C.; Li, F.; Fu, Y.; Zhao, J.; Dong, H.; Song, Y. Desalination and Li+ enrichment via formation of cyclopentane hydrate. *Sep. Purif. Technol.* **2020**, *231*, 115921. [CrossRef]
190. Nambiar, A.; Babu, P.; Linga, P. Improved kinetics and water recovery with propane as co-guest gas on the hydrate-based desalination (hydesal) process. *ChemEngineering* **2019**, *3*, 31. [CrossRef]
191. Babu, P.; Nambiar, A.; Chong, Z.R.; Daraboina, N.; Albeirutty, M.; Bamaga, O.A.; Linga, P. Hydrate-based desalination (HyDesal) process employing a novel prototype design. *Chem. Eng. Sci.* **2020**, *218*, 115563. [CrossRef]
192. Al-Hemeri, S.T.; Al-Mukhtar, R.S.; Hussine, M.N. Removal of heavy metals from industrial wastewater by use of Cyclopentane-Clathrate Hydrate formation technology. In *IOP Conference Series: Materials Science and Engineering*; IOP Publishing: Istanbul, Turkey, 2020; Volume 737, p. 012178.
193. Johannsen, P.; Karlapudi, R.; Reinhold, G. High pressure reverse osmosis for wastewater minimization and zero liquid discharge applications. *Desalination* **2006**, *199*, 84–85. [CrossRef]
194. Nallakukkala, S.; Abulkhair, H.; Alsaiari, A.; Ahmad, I.; Almatrafi, E.; Bamaga, O.; Mohd Shariff, A. Suitable Binary and Ternary Thermodynamic Conditions for Hydrate Mixtures of $CH_4$, $CO_2$, and $C_3H_8$ for Gas Hydrate-Based Applications. *ACS Omega* **2022**, *7*, 10877–10889. [CrossRef]
195. Falahieh, M.M.; Bonyadi, M.; Lashanizadegan, A. A new hybrid desalination method based on the $CO_2$ gas hydrate and capacitive deionization processes. *Desalination* **2021**, *502*, 114932. [CrossRef]
196. Youssef, P.; Al-Dadah, R.; Mahmoud, S. Comparative Analysis of Desalination Technologies. *Energy Procedia* **2014**, *61*, 2604–2607. [CrossRef]
197. Yun, S.H.; Woo, D.S. Analysis of seawater desalination energy consumption based on changes in raw water characteristics and operating condition. *J. Korean Soc. Water Wastewater* **2019**, *33*, 281–289. [CrossRef]
198. Lal, B.; Nallakukkala, S. *Gas Hydrate in Water Treatment: Technological, Economic, and Industrial Aspects*, 1st ed.; John Wiley & Sons: Hoboken, NJ, USA, 2022; pp. 14–301.

applied sciences

MDPI

Article

# Impact and Challenges of Reducing Petroleum Consumption for Decarbonization

**Ryoma Matsumoto [1] and Tomohiro Tabata [2,*]**

[1] Faculty of Global Human Sciences, Kobe University, 3-11 Tsurukabuto, Nada-ku, Kobe 657-8501, Japan; ryoma7325@ezweb.ne.jp
[2] Graduate School of Human Development and Environment, Kobe University, 3-11 Tsurukabuto, Nada-ku, Kobe 657-8501, Japan
* Correspondence: tabata@people.kobe-u.ac.jp

**Abstract:** This study aimed to identify the impact of achieving the 1.5 °C target on the petroleum supply chain in Japan, and discuss the feasibility and challenges of decarbonization. First, a national material flow was established for the petroleum supply chain in Japan, including processes for crude petroleum refining, petroleum product manufacturing, plastic resin and product manufacturing, and by-product manufacturing. In particular, by-product manufacturing processes, such as hydrogen, gaseous carbon dioxide, and sulfur, were selected because they are utilized in other industries. Next, the outlook for the production of plastic resin, hydrogen, dry ice produced from carbon dioxide gas, and sulfur until 2050 was estimated for reducing petroleum consumption required to achieve the 1.5 °C target. As a result, national petroleum treatment is expected to reduce from 177,048.00 thousand kl in 2019 to 126,643.00 thousand kl in 2030 if the reduction in petroleum consumption is established. Along with this decrease, plastic resin production is expected to decrease from 10,500.00 thousand ton in 2019 to 7511.00 thousand ton by 2030. Conversely, the plastic market is expected to grow steadily, and the estimated plastic resin production in 2030 is expected to be 20,079.00 thousand ton. This result indicates that there is a large output gap between plastic supply and demand. To mitigate this gap, strongly promoting the recycling of waste plastics and making the price competitiveness of biomass plastics equal to that of petroleum-derived plastics are necessary.

**Keywords:** decarbonization; 1.5 °C target; petroleum consumption; plastics; by-products of petroleum refinery

**Citation:** Matsumoto, R.; Tabata, T. Impact and Challenges of Reducing Petroleum Consumption for Decarbonization. *Appl. Sci.* **2022**, *12*, 3738. https://doi.org/10.3390/app12083738

Academic Editor: Alberto Benato

Received: 25 February 2022
Accepted: 7 April 2022
Published: 7 April 2022



## 1. Introduction

The Paris Agreement is an international framework that focuses on reducing greenhouse gas emissions after 2020 and making efforts to limit the temperature rise from the Industrial Revolution to 2100 to less than 2 °C and to 1.5 °C [1]. The Glasgow Agreement was adopted at the 26th session of the Conference of the Parties (COP26) held in the United Kingdom in 2021. This agreement states that all countries should pursue efforts to restrict the temperature rise from before the Industrial Revolution to within 1.5 °C, while gradually reducing coal-fired power plants [2]. The background to the promotion of the "1.5 °C target" is a special report published by the Intergovernmental Panel on Climate Change (IPCC) [3] in 2018. The report states that curbing global warming to 1.5 °C can be achieved with global goals other than climate change, such as achieving sustainable development, and eradicating poverty [3]. This report also states that $CO_2$ emissions should be reduced by 45% by 2030 and reach net zero by 2050 to achieve this goal [3]. Based on this trend, the Japanese government has claimed that it will set a national goal of achieving carbon neutrality by 2050 [4]. According to the IPCC's AR5 WG1 report, 65.3% of the world's $CO_2$ emissions, as of 2010, came from fossil fuels [5]. Considering this situation, reducing fossil fuel consumption for achieving the 1.5 °C target is inevitable. From this perspective, the trend of fossil fuel reduction is rapidly progressing worldwide.

As examples of efforts to reduce fossil fuel consumption, divestment of fossil fuels such as petroleum and coal has been actively increasing from the perspectives of environmental, social, and governance investment. The international initiative "RE100" aims to cover 100% of the power consumption related to product manufacturing and business operations with renewable energy, with more than 300 companies participating [6]. Focusing on plastics, the European Union (EU) suggests the increased use of recycled plastics in its circular economy action plan [7]. The EU also applied the "Directive on single-use plastics" from July 2021, which stipulates the prohibition of the distribution of disposable plastic products such as straws and forks [8]. In Japan, the "Plastic Resource Circulation Act," which stipulates the reduction of disposable plastic product usage and the promotion of recycling of waste plastics, is scheduled to be enforced in April 2022 [9].

Welsby et al. [10] indicated that petroleum and fossil methane gas reserves and coal reserves of 60% and 90%, respectively, in 2050 will achieve the 1.5 °C target. They also argue that petroleum and fossil methane gas production needs to be reduced by 3% annually by 2050 to achieve the 1.5 °C target [10]. Many countries, such as Japan, have not clearly announced specific measures to reduce fossil fuel consumption to achieve the 1.5 °C target. However, they recognized the extreme fossil fuel reductions established by Welsby et al. [10] for achieving the 1.5 °C target. This indicates that the current socioeconomic system must be transformed.

Conversely, fossil-fuel-based industry relates to broad industries such as fuels, plastic products, fertilizers, and pharmaceuticals, and has a complicated supply chain structure. By-products, such as carbon dioxide gas and sulfur produced by petroleum refining, are also used as raw materials in other industries. For example, dry ice, which is used for maintaining low temperature when transporting refrigerated products and pharmaceuticals such as COVID-19 vaccines, is produced using carbon dioxide gas. Gaseous carbon dioxide is a by-product of petroleum refining. Quickly reducing fossil fuel consumption may affect society and the economy. Hence, rapidly proceeding with efforts to achieve the 1.5 °C target is important.

This study aims to clarify how a reduction in petroleum consumption in Japan would affect the domestic supply chain. First, the processes related to petroleum refining are organized as a material flow. Next, based on the reduction rate of Welsby et al. [10], the outlook of domestic petroleum treatment, petroleum products, and by-products required to achieve the 1.5 °C target is estimated. Then, we discuss whether there is an output gap by comparing these results with demand. This study shows the impact and challenges of reducing petroleum consumption on the Japanese petroleum supply chain. The study also intends to provide an impetus for government policymakers to consider ways to achieve the 1.5 °C target from economic and social perspectives and to examine and mitigate the obstacles to achieving the target.

Various studies discuss the social system that contributes toward achieving the Paris Agreement and the 1.5 °C target. For example, in studies focusing on energy system transformation, Liu et al. [11] indicated that $CO_2$ emissions by sectors such as industry, electricity, and transportation in China in 2050 increased by 81% compared to 2015 levels. They argued that to reduce $CO_2$ emissions, an annual acceleration rate of 2.1% is necessary to retain the temperature rise below 2 °C. Davis et al. [12] indicated that a renewable energy-based energy mix from fossil fuels can reduce greenhouse gas emissions in 2050 from 2005 levels to 90% in the case of electricity production in Alberta, Canada. Tina and Nicolosi [13] analyzed the impact of climate change on the current and future Italian electricity system. Mekonnen et al. [14] analyzed the impact of climate-change-related droughts on hydropower. Rosa and Castro [15] simulated the energy mix in Australia's electricity system for 2030. They indicated that while wind and solar energy contribute significantly to the national electricity market, more than 50% of the electricity system continues to depend on fossil fuels.

Considering studies focusing on the viewpoint of transportation, Arioli et al. [16] analyzed the path to meeting the 1.5 °C target for emerging economies and showed that achieving this requires dramatic changes in travel patterns, technology, and fuels, as well as significant enhancements to current policy approaches. Bhat and Garcia [17] examined the EU's intra-carbon emissions associated with diesel and gasoline consumption to meet road transport requirements. They argued that without significant reductions in diesel and gasoline carbon emissions, climate change targets would not be achieved. In addition, many analyses of the effects of climate change measures through the promotion of electrification of vehicles and other measures such as decarbonization in automobiles have been conducted [18–20]. Al-Enazi et al. [21] conducted a review to examine the feasibility and environmental aspects of alternative fuels to heavy fuel oil in maritime transportation.

Considering studies focusing on the viewpoint of impacts of climate change in cities and countries, Nishiura et al. [22] estimated the economic impact of achieving the 1.5 °C target in Asian countries and mitigating the economic impact of reducing greenhouse gas emissions. Alabsi et al. [23] proposed a sustainability index to evaluate climate change impacts and the effectiveness of adaptation in traditional West Asian cities. Liu et al. [24] analyzed the impact of urban sprawl on carbon emissions per capita in China. Pai et al. [25] indicated that the energy source shifts from fossil fuels to renewables to restrict temperature rise below 2 °C reduces employment in the fossil fuel industries while creating more employment in the renewables industry. Riedl [26] analyzed corporate cash flows using petroleum majors as an example. The author criticized the fact that investors still measure corporate value by operating profit margin and petroleum price against the importance of responding to climate policies for achieving the 1.5 °C target.

Considering studies that focus on national policy and innovation perspectives, Takeishi [27] indicated that reducing fossil fuel consumption would provide global policy coherence in determining future energy supply and demand scenarios. Akimoto [28] analyzed the current state of energy and $CO_2$ emissions. The author discussed the need for innovation to reduce $CO_2$ emissions. Conversely, Hansen [29] argued that countries where fossil fuels are dominated by state-owned enterprises might face political obstacles regarding climate movement because reduced fossil fuel consumption might result in financial losses.

Studies focusing on climate justice and fairness by Chapman et al. [30] have indicated that the conversion from fossil fuels to renewable-energy-based electricity improves social fairness in middle-income countries, but renewable energy shares are rising in high-income countries. They also indicated that low-income countries rely on fossil fuels for electricity, and the conversion rate to renewable energy is slow. Cronin et al. [31] indicated that while achieving the 1.5 °C target is essential from the perspectives of climate justice and fairness, economic disparities between the Global North and South might be encouraged. They argued that an interdisciplinary research agenda is necessary to conduct research that assists in establishing countries' contribution.

These studies have discussed the potential and challenges of converting fossil fuels to renewable energy sources. While many studies have discussed the importance of reducing production and consumption of petroleum in achieving the 1.5 °C target, there are few studies that have examined the potential economic and social impacts of achieving that target. This study considers the social and economic impact of achieving the 1.5 °C target on the petroleum industry and supply chain from the perspective of petroleum products and by-products. This study aimed to minimize the future impact of petroleum products and by-products on supply and demand.

This section describes the structure of this study. In Section 2, this is followed by data collection to create a material flow diagram for understanding the integrated process from petroleum refining to the manufacture of plastic products and by-products in Japan. The methodology is also described to estimate the production of plastic products and by-products if the 1.5 °C target is achieved, with 2030 and 2050 as milestones. In Section 3, a material flow diagram is created to show the current status of petroleum refining and plastic product and by-product production in Japan. Next, based on the results of the material flow diagram, the volume of plastic product and by-product production up to 2050 is estimated and compared with the results of demand projections to show the extent to which an output gap may exist. Based on these results, the impact of meeting the 1.5 °C target on Japan's economy and society is projected, and measures to mitigate this impact are discussed. Section 4 summarizes the above results and presents our conclusions.

## 2. Materials and Methods

### 2.1. Target Products and System Boundary of Material Flow

A material flow, focusing on the processes related to petroleum refining shown in Figure 1, was created to understand the trend of petroleum consumption in Japan. The material flow was created in 2019, when the latest materials related to the supply chain were available. In the material flow, the following three processes were focused on: (1) naphtha as petroleum products; (2) plastic products; and (3) hydrogen, carbon dioxide, and sulfur as petroleum by-products.

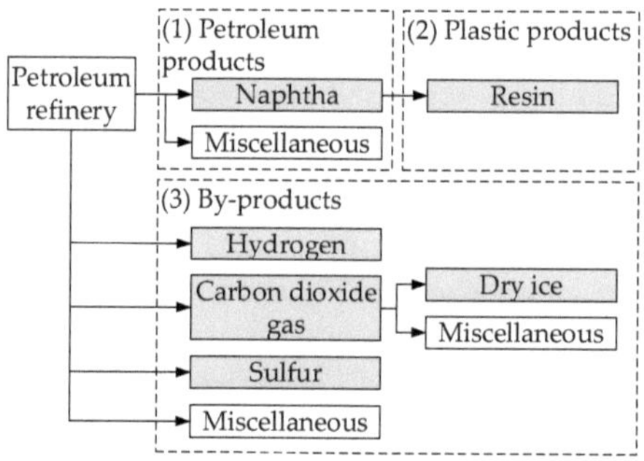


**Figure 1.** Processes related to petroleum refining.

In process (1) in Figure 1, petroleum products include naphtha, gasoline, jet fuel oil, kerosene, light oil, heavy oil, kerosene, asphalt, grease, paraffin, LPG, and LNG. Among these petroleum products, naphtha was selected because it is consumed as a product without being used as fuel, and it has the shortest life cycle until it becomes waste. In process (3) shown in Figure 1, hydrogen, carbon dioxide gas, and sulfur were selected because they were shipped. Hydrogen is a by-product of the catalytic reforming process. All the hydrogen generated by catalytic reforming is self-consumed to remove sulfur and nitrogen from petroleum products. However, because this alone is insufficient in terms of the amount of hydrogen produced, additional hydrogen is produced in hydrogen plants adjacent to petroleum refinery plants. Carbon dioxide gas is a by-product of petroleum refining. In this study, we focused on dry ice. One reason is that demand is rapidly increasing due to the transportation of the COVID-19 vaccine, while the shortage of raw materials is becoming more serious in Japan [32]. Sulfur is obtained through desulfurization during petroleum refining and used as a raw material for fertilizers and chemicals.

### 2.2. Data Collection

Data relating to the processing of petroleum products in process (1) shown in Figure 1 apply statistical data published by the Japan Agency for Natural Resources and Energy [33]. The data utilized were production, shipment, import, export, and stock. Data relating to the processing of plastic products process (2) shown in Figure 1 apply statistical data published by the Japan Plastic Waste Management Institute [34]. The utilized data were resin manufacture, plastic product manufacture and consumption, waste disposal, and recycling. Considering the data relating to by-product process (3) shown in Figure 1, there were no statistical data on hydrogen and carbon dioxide gas. Hydrogen production was applied to the survey results (14.02 billion $Nm^3$) of Mizuho Research & Technologies Ltd (Mizuho Research & Technologies Ltd, Tokyo, and Japan.) [35]. The annual domestic demand for dry ice is approximately 350.00 thousand ton, of which approximately 25.00 thousand ton is imported [36]. Hence, a difference of 325.00 thousand ton was set as the domestic dry ice production. The Japan Industrial and Medical Gases Association [37] states that 48.8% of domestic liquefied carbon dioxide production is derived from petroleum-refining plants. Therefore, this study assumed that the ratio of dry ice production from an oil refinery plant is equivalent to the production of liquefied carbon dioxide, which is estimated to be 157.30 thousand ton. Data relating to sulfur apply statistical data published by the Japan Ministry of Economy, Trade and Industry [38]. The data utilized concern sulfur production.

Data from the petroleum refinery to processes (1) to (3) are connected in a consistent manner to create a material flow.

### 2.3. Outlook of Material Flow by Meeting 1.5 °C Target

Based on the created material flow, outlook of petroleum treatment, plastic resin, and by-product production by 2050 was estimated by adopting the 3% annual petroleum production reduction rate required to achieve the 1.5 °C target [10]. These results were compared with the demand, and the social and economic impacts were discussed. Table 1 lists the material intensity coefficients of the plastic resin and the by-products. These coefficients were calculated by dividing petroleum treatment in 2019 [33] by plastic resin and by-product production in 2019 [34–38]. These coefficients were assumed to be constant until 2050.

**Table 1.** Material intensity coefficients of plastic resin and by-products.

| Products (Unit) | Material Intensity Coefficients |
|---|---|
| Plastic resin (ton/kL) | $0.59 \times 10^{-1}$ |
| Hydrogen ($Nm^3$/kL) | 79.19 |
| Dry ice (ton/kL) | $0.89 \times 10^{-3}$ |
| Sulfur (ton/kL) | $0.92 \times 10^{-2}$ |

In addition to the above target, plastic product process (2) shown in Figure 1 is discussed using the milestones in 2030, stipulated in the plastic resource recycling strategy formulated by the Japan Ministry of Environment (JMOE) [39] in 2018. These milestones set the following goals for the 3Rs (reduce, reuse, recycle) of plastics [39]:

- ✓ 25% reduction of disposable plastic by 2030,
- ✓ 60% reuse and recycling of waste plastic containers and packaging by 2030,
- ✓ 100% reuse and recycling of used plastics by 2035,
- ✓ Double the recycling of waste plastic by 2030, and
- ✓ Introducing 2000.00 thousand ton of biomass plastic by 2030.

## 3. Results and Discussion

In this section, a material flow diagram is created to understand the current status of petroleum refining and the manufacture of plastic products and by-products in Japan. This is followed by an investigation of the impact of the reduction in petroleum refining associated with the achieving of the 1.5 °C target on the demand for and supply of plastic products and by-products.

### 3.1. Material Flow of Petroleum-Related Processes

Figure 2 shows the material flow related to petroleum refining. Figure 3 shows the material flow from manufacturing to disposal and recycling of plastic products. The material flows shown in Figures 2 and 3 are linked. The total amount of petroleum received in 2019 was approximately 188,671.00 thousand kl. Approximately 177,048.00 thousand kl was used to manufacture petroleum products. As shown in process (1) in Figure 2, approximately 18,087.00 thousand kl of naphtha was manufactured. As shown in process (2) in Figure 3, approximately 10,500.00 thousand ton of plastic resin was manufactured, and approximately 10,110.00 thousand ton of plastic products was manufactured with imported and recycled resins. Approximately 7830.00 thousand ton of plastic products, which is approximately 83% of the consumption, was disposed of, and approximately 8510.00 thousand ton was waste plastic, including industrial loss. Approximately 1860.00 thousand ton of waste plastic, which is approximately 22% of waste plastic, was mechanically recycled, and approximately 910.00 thousand ton was returned to domestic plastic production as recycled resin. As shown in process (2) in Figure 3, approximately 1629.00 thousand ton of sulfur and approximately 157.00 thousand ton of dry ice were produced and 14.02 billion $Nm^3$ of hydrogen was produced, all of which was self-consumed.

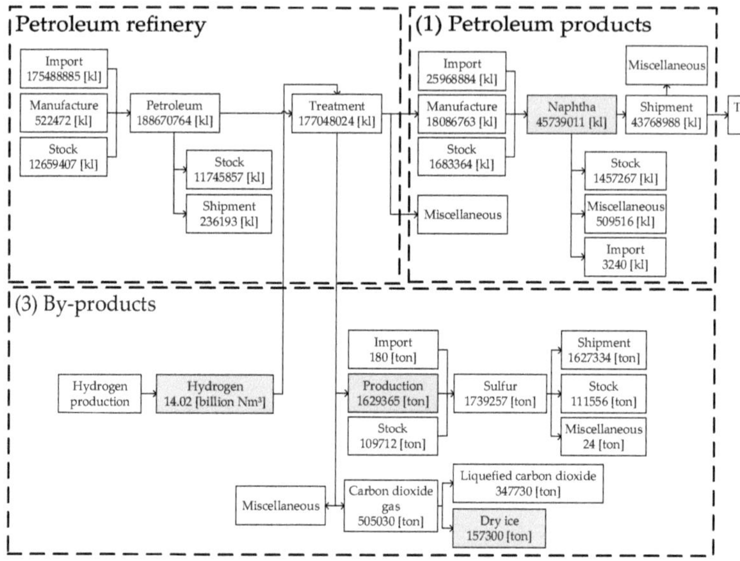


**Figure 2.** Material flow relating to petroleum refining.

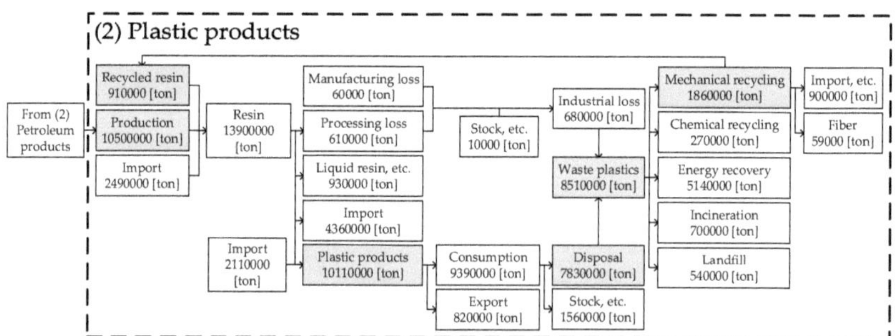


**Figure 3.** Material flow relating from manufacturing to disposal and recycling of plastic products.

### 3.2. Outlook of Naphtha, Plastic Products, and By-Products

Table 2 shows the estimation of the outlook for petroleum treatment, plastic resin, and by-product production from 2019 to 2050. Based on Welsby et al. [10], petroleum treatment needs to be reduced by approximately 61% from 2019 to 2050 for achieving the 1.5 °C target. Accordingly, the production of plastic resin and by-products is also expected to decrease, as shown in Table 2. In the following sections, the social and economic impacts of the reduced production of plastic resins and by-products are discussed.

**Table 2.** Outlook of petroleum treatment, naphtha, plastic products, and by-products.

| | 2019 | 2020 | 2030 | 2040 | 2050 |
|---|---|---|---|---|---|
| Petroleum treatment (thousand kL) | 177,048.00 | 171,737.00 | 126,643.00 | 93,389.00 | 68,868.00 |
| Plastic resin production (thousand ton) | 10,500.00 | 10,185.00 | 7511.00 | 5539.00 | 4084.00 |
| Hydrogen production (billion $Nm^3$) | 14.02 | 13.60 | 10.03 | 7.40 | 5.45 |
| Hydrogen production * (billion $Nm^3$) | 21.40 | 20.76 | 15.31 | 11.29 | 8.32 |
| Dry ice production (thousand ton) | 157.00 | 153.00 | 113.00 | 83.00 | 61.00 |
| Sulfur production (thousand ton) | 1629.00 | 1580.00 | 1165.00 | 859.00 | 634.00 |

* Maximum hydrogen production when the hydrogen production plant is operated.

### 3.2.1. Plastic Products

The plastic resin production was approximately 10,500.00 thousand ton in 2019. According to the Japan Plastic Industry Federation [40], plastic resin production has remained at the same level since 2009. This means that the domestic demand for plastic products remains high, and consumption has not been reduced. In future, the production volume should be reduced by reducing the use of disposable plastics. The global plastics market is growing steadily, with an annual growth rate of 3.4% for the projected period 2021–2028 [41]. High-performance materials such as conductive plastics are used in various high-tech products, and the demand might remain at a certain level in the future from the perspective of product weight reduction and energy saving of electronic equipment.

Figure 4 shows the outlook of the supply and demand volume of plastic resin production. The figures in Table 2 were used for the supply transition. Trends in demand were estimated assuming that the annual growth rate (3.4%) indicated above would continue until 2050. Plastic resin production is expected to reduce to approximately 7511.00 thousand ton by 2030 and approximately 4084.00 thousand ton by 2050. Conversely, the plastic resin production from petroleum refineries in 2019 was approximately 10,110.00 thousand ton from Figure 3. If the demand volume is changed at the aforementioned annual growth rate, approximately 15,168.00 thousand ton of resin production from petroleum refineries is expected to be required by 2030. In 2050, 29,602.00 thousand ton is expected to be needed. This result indicates that a large output gap might occur in the plastic resin. Japan currently relies on imports for more than half of its demand for naphtha. Covering all shortfalls of naphtha and plastic resins by imports is difficult under the international commitment to fossil fuel reduction.

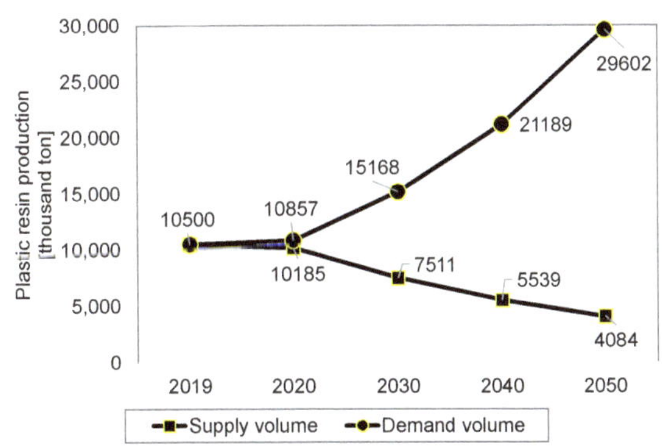


**Figure 4.** Outlook of supply and demand volume of plastic resin production.

### 3.2.2. Hydrogen

The domestic hydrogen production was approximately 14.02 billion $Nm^3$ in 2019, and this was fully self-consumed. If the hydrogen plant adjacent to the petroleum refinery was fully operational, hydrogen production would increase to approximately 16.16 billion $Nm^3$ and approximately 2.14 billion $Nm^3$ of the produced hydrogen might be supplied externally [35]. Hydrogen is used as a raw material for nitrogen fertilizer through the Haber process, as an energy source. The global fertilizer market is expanding and the annual growth rate for 2025 is expected to be 2.0% [42]. With the reduction in petroleum treatment, the external supply of hydrogen is expected to reduce to 1.53 billion $Nm^3$ in 2030 and 0.83 billion $Nm^3$ in 2050. These production volumes are expected to be used for the production of nitrogen fertilizers and as energy sources.

### 3.2.3. Dry Ice

The global plastics market is growing steadily, with an annual growth rate of 3.81% for the projected period 2016–2026 [43]. The demand for dry ice is expected to increase further, with the expansion of the food delivery market and for transportation of pharmaceuticals such as the COVID-19 vaccine. Japan currently has an output gap for dry ice, and this gap is expected to widen if the national government does not take measures. Figure 5 shows outlook of supply and demand volume of dry ice production. The figures in Table 2 were used for the supply transition. Trends in demand were estimated assuming that the annual growth rate (3.81%) indicated above would continue until 2050. Domestic dry ice production derived from petroleum by-products was approximately 157.00 thousand ton in 2019 and is expected to decrease to approximately 61.00 thousand ton in 2050. If the demand volume is changed at the aforementioned annual growth rate, approximately 237.00 thousand ton of resin production from petroleum refineries is expected to be required by 2030. In 2050, 501.00 thousand ton is expected to be needed.

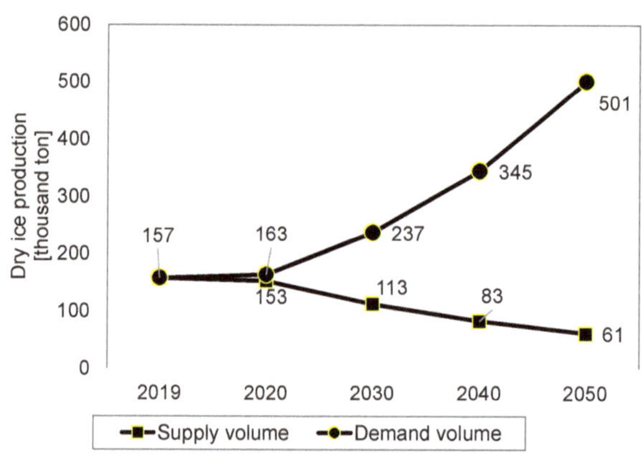


**Figure 5.** Outlook of supply and demand volume of plastic resin production.

### 3.2.4. Sulfur

Sulfur is also a raw material for fertilizers such as hydrogen, and its demand is expected to increase [42]. Figure 6 shows the outlook of the supply and demand volume of dry ice production. The figures in Table 2 were used for the supply transition. Trends in demand were estimated assuming that the annual growth rate (2.0%) indicated above would continue until 2050. Domestic sulfur production was approximately 1629.00 thousand ton in 2019 and is expected to decrease to approximately 634.00 thousand ton in 2050. If the demand volume is changed at the aforementioned annual growth rate, approximately 2026.00 thousand ton of resin production from petroleum refineries is expected to be required by 2030. In 2050, 3010.00 thousand ton is expected to be needed. Domestic sulfur production is derived entirely from desulfurization during petroleum refining [44]. Therefore, a decrease in sulfur production as a petroleum by-product directly leads to a decrease in domestic sulfur production.

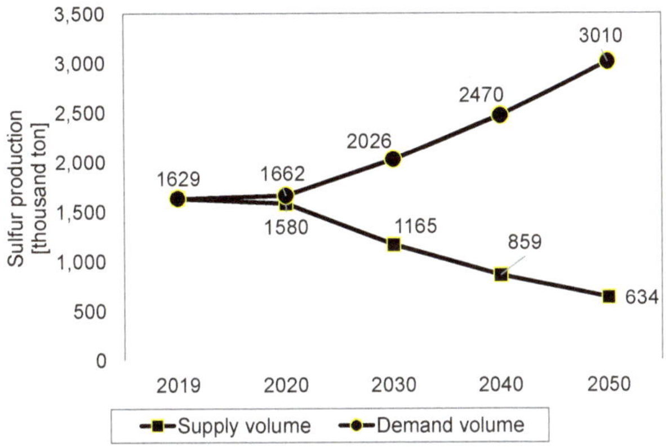


**Figure 6.** Outlook of supply and demand volume of sulfur production.

### 3.3. Discussion

This study used annual global market growth rates for each product to estimate demand for plastics, dry ice, and sulfur. The annual growth rates for these products in the Japanese market might differ from those in the world market. In addition, the growth rates might not remain the same until 2050, and the estimates are rough. This is also similar for the estimation of the supply of each of these products. Nevertheless, the fact that estimates were able to show that the output gap for these products would be large enough to achieve the 1.5 °C target is a significant contribution.

Regarding plastic products, a very large discrepancy between the supply and demand of plastic resins became evident as the 1.5 °C target was achieved. In accordance with the Plastic Resource Circulation Act, the Japanese government is required to promote the recycling of plastic waste. Figure 3 indicates that the recycling rate of plastic waste in 2019 was 85%, and most plastic waste was already being used effectively [34]. However, energy recovery, which accounts for approximately 60%, is integrated into the recycling process. By converting waste plastics that have been energy recovered, incinerated, and landfilled to mechanical recycling, the achievement of the milestone in 2030 set by JMOE [39] doubles the recycling of waste plastic from 910.00 thousand ton in 2019 to 1800.00 thousand ton in 2030. However, the national government should discuss the increased labor and infrastructure costs for transporting, sorting, and cleaning waste plastics to achieve this [45].

In addition, the introduction of 2000.00 thousand ton of biomass plastic, which JMOE [39] set as a milestone for 2030, is extremely difficult considering that the production volume was only 45.00 thousand ton in 2018 [46]. Biomass plastics are 1.5 times more expensive than polyethylene made from fossil fuels, and the price difference is an obstacle to the spread of biomass plastics [47]. For biomass plastics to have the same competitiveness as fossil-fuel-based plastics, national government should take the initiative to ensure a stable biomass supply as raw materials, reduce costs through mass production, and tax fossil-fuel-based plastics.

Regarding hydrogen, the Japanese government has positioned hydrogen as an important energy source. They have been jointly developing technologies related to the construction of fuel cell vehicles and hydrogen energy infrastructures [4]. According to the New Energy and Industrial Technology Development Organization (NEDO) [48], if hydrogen power generation becomes a reality, domestic hydrogen demand in 2030 is expected to rise to 2.2 billion $Nm^3$. According to the results of this study, 1.53 billion $Nm^3$ is expected to be available in 2030, and hydrogen production associated with petroleum refining could be one way to ensure supply against demand. However, hydrogen production declines as the 1.5 °C target is achieved, so hydrogen production is not expected to be a permanent energy source.

Regarding dry ice, one measure to solve the output gap is to increase the import volume of dry ice. However, imports not only incur higher transportation costs than domestic production but also depend on the production situation of exporting countries such as South Korea [49]. If the exporting countries were making efforts to achieve the 1.5 °C target, the production volume of the dry ice that Japan imports would also be expected to reduce accordingly. Other measures include installing new equipment to produce dry ice from low-concentration carbon dioxide, and storage equipment to stabilize the supply. However, this also incurs the cost of expanding the facilities, and the current situation cannot achieve the current demand for dry ice. This result reveals that dry ice is one of the products that have a great social and economic impact due to the reduction in petroleum consumption. The national government should recognize that securing dry ice supply is an issue for reducing petroleum consumption.

Regarding sulfur, at present, approximately 1000.00 thousand ton of produced sulfur is exported, mainly to China, and domestic demand remains at approximately 500.00 thousand ton [50]. If the production volume is decreased to the 2050 level, adjusting the export volume might be one of the measures to achieve domestic demand. However, export restrictions are a powerful diplomatic option that may be subject to retaliation from the target country and require increased attention.

## 4. Conclusions

This study discussed how the reduction in petroleum consumption required to achieve the 1.5 °C target could affect the domestic supply and demand of petroleum products and by-products. The findings of this study are described below.

(1) Data pertaining to petroleum refining, petroleum product manufacturing, and plastic product manufacturing processes in 2019 were collected to create a material flow diagram that provides a consistent picture of the current state of domestic petroleum supply and demand. The material flow also covered the production of sulfur, a petroleum by-product, as well as dry ice, one of the uses of carbon dioxide gas.

(2) The domestic production of plastic resins in 2019 was approximately 10,500.00 thousand ton, and the production in 2030 will be approximately 7511.00 thousand ton, if the reduction in petroleum consumption that contributes to achieving the 1.5 °C target is achieved. The JMOE is planning to recycle waste plastics and produce biomass plastics in its plastic resource recycling strategy. However, this requires further separation of waste plastics and reduction of production costs through mass production of biomass plastics.

(3) Hydrogen is produced and consumed during the petroleum refining process, with hydrogen produced in an adjacent hydrogen plant. Hydrogen is produced in the amount of 14.02 billion $Nm^3$, but is consumed entirely on site. At maximum operation of this hydrogen plant, there is a potential external supply of 2.14 billion $Nm^3$, which would decrease to 1.53 billion $Nm^3$ in 2030 if petroleum consumption were to decrease at the rate targeted for the reduction.

(4) Domestic production of dry ice derived from petroleum by-products in 2019 was approximately 157.00 thousand ton, which would decline to approximately 61.00 thousand ton in 2030 if the above target is achieved. There are many import and technological challenges to cover the shortfall, and it is expected to be difficult to secure supply to achieve the expected increase in demand.

(5) Domestic sulfur production from petroleum by-products in 2019 was approximately 1629.00 thousand ton. This would decrease to approximately 1165.00 thousand ton in 2030 if the above target is achieved. Since domestic demand for sulfur has remained

at approximately 500.00 thousand ton, the majority of which is exported, the future domestic demand could be achieved by adjusting the volume of exports.

This study has a limitation in that only plastic products, hydrogen, dry ice as carbon dioxide, and sulfur are the focus, but it has shown that reducing petroleum consumption might affect the supply and demand of these products. Although this study was conducted in Japan, it is not a unique case, especially in countries where petroleum refining and the production of plastics, etc. are the mainstay of their industries. It is important for other countries to calmly assess the impact on their own industries and societies to achieve the 1.5 °C target, and then to consider the impact on their own economies and societies. Measures need to be taken to minimize the impact.

Japan has adjusted to produce petroleum products that meet domestic demand by selecting the type of petroleum to be imported, before decomposing and reforming it [50]. However, perfectly matching the production of all petroleum products with demand is difficult, and the supply of naphtha is covered by imports. This situation is the same for plastic products made from naphtha and dry ice. In the future, petroleum treatment might be reduced to achieve the goal of the 1.5 °C target, and the gap in demand might be widened. The Japanese government should support the minimization of the economic and social impacts by adjusting the import volume, securing the supply volume, and funding the development of products that replace conventional petroleum products and petroleum by-products, such as biomass plastics. The milestone of plastic resource circulation announced by the JMOE [39] that this study considered, which is just a target value, involves the expansion of laws and policies regarding reuse and recycling, and reliable implementation is required. The Plastics Resource Circulation Act [9], which is scheduled to come into effect in April 2022, is expected to experience effects that might be exhibited from the perspective of reducing the petroleum consumption and substituting petroleum products. In addition to such 3R measures, national government should recognize that achieving the 1.5 °C target requires significant transition in policies and industrial and social systems.

**Author Contributions:** Conceptualization, T.T.; methodology, R.M. and T.T.; validation, R.M. and T.T.; formal analysis, R.M.; investigation, R.M.; resources, R.M.; data curation, R.M.; writing—original draft preparation, T.T.; writing—review and editing, T.T.; visualization, R.M. and T.T.; supervision, T.T.; project administration, T.T. All authors have read and agreed to the published version of the manuscript.

**Funding:** This research received no external funding.

**Institutional Review Board Statement:** Not applicable.

**Informed Consent Statement:** Not applicable.

**Data Availability Statement:** The data are obtained from [33–43].

**Acknowledgments:** This research did not receive any specific grant from funding agencies in the public, commercial, or not-for-profit sectors.

**Conflicts of Interest:** The authors declare no conflict of interest.

## References

1. The Paris Agreement. Available online: https://unfccc.int/process-and-meetings/the-paris-agreement/the-paris-agreement (accessed on 3 February 2022).
2. Glasgow Climate Pact. Available online: https://unfccc.int/documents/310475 (accessed on 3 February 2022).
3. IPCC Special Report on Global Warming of 1.5 °C. Available online: https://unfccc.int/topics/science/workstreams/cooperation-with-the-ipcc/ipcc-special-report-on-global-warming-of-15-degc (accessed on 3 February 2022).
4. Japan's 2050 Carbon Neutral Goal. Available online: https://www.meti.go.jp/english/policy/energy_environment/global_warming/roadmap/report/20201111.html (accessed on 3 February 2022).
5. Pachauri, R.K.; Allen, M.R.; Barros, V.R.; Broome, J.; Cramer, W.; Christ, R.; Church, J.A.; Clarke, L.; Dahe, Q.; Dasgupta, P.; et al. *IPCC: Climate Change 2014: Synthesis Report. Contribution of Working Groups I, II and III to the Fifth Assessment Report of the Intergovernmental Panel on Climate Change*; Core Writing Team, Pachauri, R.K., Meyer, L.A., Eds.; IPCC: Geneva, Switzerland, 2014; 151p.

6. RE100. Available online: https://www.there100.org (accessed on 3 February 2022).
7. Circular Economy Action Plan. Available online: https://ec.europa.eu/environment/strategy/circular-economy-action-plan_en (accessed on 3 February 2022).
8. Single-Use Plastics from European Comission. Available online: https://ec.europa.eu/environment/topics/plastics/single-use-plastics_en (accessed on 3 February 2022).
9. The Plastic Resource Circulation Act from Japan Mistry of Environment. Available online: https://www.env.go.jp/recycle/plastic/pdf/14.pdf (accessed on 3 February 2022).
10. Welsby, D.; Price, J.; Pye, S.; Ekins, P. Unextractable fossil fuels in a 1.5 °C world. *Nature* **2021**, *597*, 230–234. [CrossRef] [PubMed]
11. Liu, J.; Yin, M.; Xia-Hou, Q.; Wang, K.; Zou, J. Comparison of sectoral low-carbon transition pathways in China under the nationally determined contribution and 2 °C targets. *Renew. Sustain. Energy Rev.* **2021**, *149*, 111336. [CrossRef]
12. Davis, M.; Moronkeji, A.; Ahiduzzaman, M.; Kumar, A. Assessment of renewable energy transition pathways for a fossil fuel-dependent electricity-producing jurisdiction. *Energy Sustain. Dev.* **2020**, *59*, 243–261. [CrossRef]
13. Tina, G.M.; Nicolosi, C.F. Assessment of the Impacts of Climate Change on Power Systems: The Italian Case Study. *Appl. Sci.* **2021**, *11*, 11821. [CrossRef]
14. Mekonnen, T.W.; Teferi, S.T.; Kebede, F.S.; Anandarajah, G. Assessment of Impacts of Climate Change on Hydropower-Dominated Power System—The Case of Ethiopia. *Appl. Sci.* **2022**, *12*, 1954. [CrossRef]
15. Rosa, L.D.; Castro, R. Forecasting and assessment of the 2030 Australian electricity mix paths towards energy transition. *Energy* **2020**, *205*, 118020. [CrossRef]
16. Arioli, M.; Fulton, L.; Lah, O. Transportation Strategies for a 1.5 °C World: A Comparison of Four Countries. *Transp. Res. D Transp. Environ.* **2020**, *87*, 102526. [CrossRef]
17. Bhat, A.; Garcia, J.O. Sustainability and Eu Road Transport Carbon Emissions from Consumption of Diesel and Gasoline in 2000 and 2018. *Appl. Sci.* **2021**, *11*, 7601. [CrossRef]
18. Kejun, J.; Chenmin, H.; Songli, Z.; Pianpian, X.; Sha, C. Transport Scenarios for China and the Role of Electric Vehicles under Global 2 °C/1.5 °C Targets. *Energy Econ.* **2021**, *103*, 105172. [CrossRef]
19. Isik, M.; Sarica, K.; Ari, I. Driving Forces of Turkey's Transportation Sector $CO_2$ Emissions: An LMDI Approach. *Transp. Policy* **2020**, *97*, 210–219. [CrossRef]
20. Das, D.; Kalbar, P.P.; Velaga, N.R. Pathways to Decarbonize Passenger Transportation: Implications to India's Climate Budget. *J. Clean. Prod.* **2021**, *295*, 126321. [CrossRef]
21. Al-Enazi, A.; Okonkwo, E.C.; Bicer, Y.; Al-Ansari, T. A Review of Cleaner Alternative Fuels for Maritime Transportation. *Energy Rep.* **2021**, *7*, 1962–1985. [CrossRef]
22. Nishiura, O.; Fujimori, S.; Oshiro, K. Economic impact and mitigation measures of reducing greenhouse gas emission in Asian countries. *J. Jpn. Soc. Civ. Eng. Ser. G.* **2020**, *76*, I_97–I_107. [CrossRef]
23. Alabsi, A.A.N.; Wu, Y.; Koko, A.F.; Alshareem, K.M.; Hamed, R. Towards Climate Adaptation in Cities: Indicators of the Sustainable Climate-adaptive Urban Fabric of Traditional Cities in West Asia. *Appl. Sci.* **2021**, *11*, 10428. [CrossRef]
24. Liu, L.; Tang, Y.; Chen, Y.; Zhou, X.; Bedra, K.B. Urban Sprawl and Carbon Emissions Effects in City Areas Based on System Dynamics: A Case Study of Changsha City. *Appl. Sci.* **2022**, *12*, 3244. [CrossRef]
25. Pai, S.; Emmerling, J.; Drouet, L.; Zerriffi, H.; Jewell, J. Meeting well-below 2 °C target would increase energy sector jobs globally. *One Earth* **2021**, *4*, 1026–1036. [CrossRef]
26. Riedl, D. The magnitude of energy transition risk embedded in fossil fuel company valuations. *Heliyon* **2021**, *7*, e08400. [CrossRef]
27. Takeishi, R. Scenarios of energy transformation and evaluation of those expectations. *Int. Econ.* **2020**, *71*, 169–189. [CrossRef]
28. Akimoto, K. Status of long-term energy policies in terms of climate change response measures. In Proceedings of the 28th Annual Meeting of the Energy Society of Japan, Kansai University Senriyama Campus, Suita, Japan, 7–8 August 2019; p. 28. [CrossRef]
29. Hansen, T.A. Stranded assets and reduced profits: Analyzing the economic underpinnings of the fossil fuel industry's resistance to climate stabilization. *Renew. Sustain. Energy Rev.* **2022**, *158*, 112144. [CrossRef]
30. Chapman, A.; Shigetomi, Y.; Ohno, H.; McLellan, B.; Shinozaki, A. Evaluating the global impact of low-carbon energy transitions on social equity. *Environ. Innov. Soc. Transit.* **2021**, *40*, 332–347. [CrossRef]
31. Cronin, J.; Hughes, N.; Tomei, J.; Couto, L.C.; Ali, M.; Kizilcec, V.; Adewole, A.; Bisaga, I.; Broad, O.; Parikh, P.; et al. Embedding justice in the 1.5 °C transition: A transdisciplinary research agenda. *Renew. Sust. Energ. Trans.* **2021**, *1*, 100001. [CrossRef]
32. Japan's Dry Ice-Makers Scramble to Meet Vaccine-Driven Demand Surge. Available online: https://www.japantimes.co.jp/news/2021/02/15/business/corporate-business/dry-ice-demand/ (accessed on 3 February 2022).
33. 2019 Year Book of Mineral Resources and Petroleum Products Statistics (Petroleum) from Japan Agency of Natural Resources and Energy. Available online: https://www.meti.go.jp/statistics/tyo/sekiyuka/pdf/h2dhhpe2019k.pdf (accessed on 3 February 2022).
34. 2019 Year Book of Plastic Products' Manufacture, Disposal and Recycling from Japan Plastic Waste Management Institute. Available online: https://www.pwmi.or.jp/flow_pdf/flow2019.pdf (accessed on 3 February 2022).
35. Survey on Hydrogen Supply Potential as By-product Derived from Petroleum Refinery from Mizuho Research & Technologies, Ltd. Available online: https://www.nedo.go.jp/content/100937547.pdf (accessed on 3 February 2022).
36. Carbon Recycling World from Japan Carbon Recycling Fund Institute. Available online: https://carbon-recycling-fund.jp/public_relations/world/20200401 (accessed on 3 February 2022).

37. Japan Industrial and Medical Gases Association. *Outline of Raw Material Carbon Dioxide Production Flow (Revised JIGA-T-S/38/15)*; Japan Industrial and Medical Gases Association: Tokyo, Japan, 2021; 22p.
38. 2019 Current Production Survey from Japan Ministry of Economy, Trade and Industry. Available online: https://www.meti.go.jp/statistics/tyo/seidou/result/ichiran/08_seidou.html (accessed on 3 February 2022).
39. Strategy for Plastic Resource Circulation from Japan Ministry of Environment. Available online: https://www.env.go.jp/press/files/jp/111746.pdf (accessed on 3 February 2022).
40. Plastic Raw MATERIALS production (2012–2020) from Japan Plastics Industry Federation. Available online: http://www.jpif.gr.jp/english/statistics/index.html (accessed on 3 February 2022).
41. Plastic Market Size, Share & Trends Analysis Report by Product (PE, PP, PU, PVC, PET, Polystyrene, ABS, PBT, PPO, Epoxy Polymers, LCP, PC, Polyamide), by Application, by End-use, by Region, and Segment Forecasts, 2021–2028. Available online: https://www.grandviewresearch.com/industry-analysis/global-plastics-market (accessed on 3 February 2022).
42. Chemical Fertilizers Global Market Report 2021: COVID-19 Impact and Recovery to 2030. Available online: https://www.researchandmarkets.com/reports/5240331/chemical-fertilizers-global-market-report-2021 (accessed on 3 February 2022).
43. Global Dry Ice Market Bwc20107. Available online: https://www.blueweaveconsulting.com/report/global-dry-ice-market-bwc20107 (accessed on 20 February 2022).
44. Takimoto, M. Sulphur and our life. *Chem. Educ.* **2014**, *62*, 30–33. [CrossRef]
45. Tabata, T.; Hishinuma, T.; Ihara, T.; Genchi, Y. Life cycle assessment of integrated municipal solid waste management systems, taking account of climate change and landfill shortage trade-off problems. *Waste Manag. Res.* **2011**, *29*, 423–432. [CrossRef] [PubMed]
46. Roadmap for Bioplastics Introduction from Japan Ministry of Environment. Available online: https://www.env.go.jp/recycle/roadmap_for_bioplastics_introduction.html (accessed on 3 February 2022).
47. Roadmap for Bioplastics Introduction from Japan Ministry of Economy, Trade and Industry. Available online: https://www.meti.go.jp/shingikai/sankoshin/sangyo_gijutsu/haikibutsu_recycle/plastic_junkan_wg/pdf/008_s05_00.pdf (accessed on 3 February 2022).
48. NEDO Hydrogen Energy White Paper from NEDO. Available online: https://www.nedo.go.jp/library/suiso_ne_hakusyo.html (accessed on 20 February 2022).
49. Emissions from Imported Carbonated Gas from Japan Ministry of Environment. Available online: https://www.env.go.jp/earth/ondanka/ghg-mrv/methodology/material/methodology_2H3_2020.pdf (accessed on 3 February 2022).
50. Petroleum Industry 2020 from Petroleum Association of Japan. Available online: https://www.paj.gr.jp/statis/data/data/2020_data.pdf (accessed on 3 February 2022).

applied sciences

MDPI

Article

# Fixed-Bed Adsorption of Phenol onto Microporous Activated Carbon Set from Rice Husk Using Chemical Activation

**Samah B. Daffalla [1,2,*], Hilmi Mukhtar [3], Maizatul S. Shaharun [4] and Abdalhaleem A. Hassaballa [1]**

1. Department of Environment and Agricultural Natural Resources, College of Agricultural and Food Sciences, King Faisal University, P.O. Box 400, Al-Ahsa 31982, Saudi Arabia; ahassaballa@kfu.edu.sa
2. Department of Chemical Engineering, College of Engineering, University of Khartoum, Khartoum P.O. Box 321, Sudan
3. Department of Chemical Engineering, Universiti Teknologi PETRONAS, Tronoh 31750, Malaysia; hilmi_mukhtar@petronas.com
4. Department of Fundamental and Applied Sciences, Universiti Teknologi PETRONAS, Tronoh 31750, Malaysia; maizats@petronas.com.my

* Correspondence: sbalal@kfu.edu.sa

**Abstract:** In the course of this research, the potential of activated carbon from rice husk was examined as being a phenol removal medium from an aqueous solution in a fixed-bed adsorption column. The activated carbon was characterized through FESEM (Field-Emission Scanning Electron Microscopy) and BET (Brunauer–Emmett–Teller) surface area. According to the FESEM micrograph and BET surface area, RHAC (rice husk activated carbon) had a porous structure with a large surface area of 587 $m^2 \cdot g^{-1}$ and mean diameter of pores of 2.06 nm. The concentration effects on the influent phenol (100–2000 $mg \cdot L^{-1}$), rate of flow (5–10 $mL \cdot min^{-1}$), and bed depth (8.5–15.3 cm) were examined. It was found that the capacity of bed adsorption increased according to the increase in the influent concentration and bed depth. However, the capacity of bed adsorption decreased according to the increase in the feed flow rate. The regeneration of activated carbon column using 0.1 M sodium hydroxide was found to be effective with a 75% regeneration efficiency after three regeneration cycles. Data on adsorption were observed to be in line with many well-established models (i.e., Yoon–Nelson and Adams–Bohart, as well as bed depth service time models).

**Keywords:** activated carbon; breakthrough curve; column study; phenol removal; rice husk

**Citation:** Daffalla, S.B.; Mukhtar, H.; Shaharun, M.S.; Hassaballa, A.A. Fixed-Bed Adsorption of Phenol onto Microporous Activated Carbon Set from Rice Husk Using Chemical Activation. *Appl. Sci.* **2022**, *12*, 4354. https://doi.org/10.3390/app12094354

Academic Editors: Mohammad Mahmudur Rahman, Dibyendu Sarkar, Rupali Datta and Prafulla Kumar Sahoo

Received: 23 March 2022
Accepted: 20 April 2022
Published: 25 April 2022



## 1. Introduction

Phenol is an important industrial pollutant that is toxic and causes severe and long-term impacts on living things. Being exposed to reduced phenol dozes could potentially be problematic, causing conditions like dermatitis, as well as blocking in respiratory system [1–3]. Phenol is certainly essential as an environmental investigation medium as it was preferred mostly to use as the contaminant model. Moreover, much information on its devastation as well as removal exists, especially regarding treatment of wastewater [4]. Wastewater from various industrial sectors, for instance coal transformation, oil, petrochemical, pulp, pesticide and pharmaceutical, as well as industries of dye, are considered as the principal phenol and phenol derivatives sources [4–6]. Numerous practices were debated throughout previous studies with regard to phenol removal from water and wastewater, for instance microbial retrogression [7], oxidation of chemicals [8–11], extraction of solvent [12], membrane separation [4,12], and adsorption [13,14]. In most of the pre-mentioned research, adsorption has been commonly-used technology for removing water pollutant [1], and a powerful way of removing contaminants from wastewater. Essentially, the most common adsorbent used for adsorption is commercially produced activated carbon, which is a carbon-rich substance demonstrating a distinct pore composition, significant surface area, and decent structure [3,15]. The surface chemistry and pore structure of activated carbon affect the adsorption process. The porosity of carbonaceous material is considered an important factor in adsorption processes of organic compounds from aqueous solutions. It has been reported that the adsorption capacity of small molecules, such as phenol, on the inner surface of carbon correlates with the content of micropores and BET surface area, while for mesoporous activated carbons, the substituent group in the phenol and nature of the carbon also controlled the phenol adsorption [3,5,15]. Nevertheless, the expensive regeneration value of activated carbon restricts its use; in addition, the carbon's breakable nature is utilized for the adsorption of organic materials, producing carbon fines [1]. Agricultural, industrial, or municipal waste usage as adsorbents for the production of activated carbon suggest different common values in a way that makes the process more appealing. Furthermore, this kind of waste biomass can be utilized as a carbon precursor in order to study the pollution challenges associated with landfilling of such wastes, and could be a new element for biomass managing industries [3,16]. Many components have already been researched with a view to produce economical activated carbon, like banyan root [3], date pits [16], terminalia arjuna nuts [17], coconut shells [18], palm shells [19], jackfruit peel waste [20], bagasse and rice husk (RH) [21], vetiver roots [22], coconut shell [23], RH [24], sugarcane bagasse [25], Lantana camara [26], and sugar cane bagasse and sunflower seed hull [27].

Rice is among the main foods eaten worldwide. Furthermore, it produces huge waste husk quantities. RH (rice husk) is stated as being the most convenient absorbent for variable pollutants [28,29]. For every 4 tons of rice produced, a generation of 1 ton of waste RH is recorded [30]. RH consists of 32.2% cellulose, 21.4% hemicelluloses, and 21.3% lignin. The cellulose, hemicellulose, and lignin form a very stable matrix structure. The cellulose–lignin matrix burns away due to carbonization, resulting in the formation of high porosity, being full of cavities, and having almost perfect pore shapes [27]. Converting RH into activated carbon could potentially lead to the development of a particularly good adsorbent with a high removal efficiency. Zinc chloride ($ZnCl_2$) is a crucial chemical agent used as an activator. It is powerful dehydration reagent, which supports material carbonaceous decomposition and encourages the aromatization and charring of carbon, thus restraining the creation of tar and increasing the yield of carbon [31]. Numerous researchers have set activated carbon through $ZnCl_2$ activation of many carbonaceous sources [27,31–33]. Throughout this study, the phenol adsorption with rice husk activated carbon (RHAC) was examined using adsorption column studies. The crucial design parameters, like flow rate, influent phenol concentration, and bed depth, were evaluated by utilizing a fixed-bed column at laboratory scale. The phenol adsorption development curves were assessed making use of the pre-mentioned stablished models (Yoon–Nelson and Adams–Bohart, along with bed depth service time (BDST)). A model that describes the fixed bed adsorption dynamics was, hence, presented, so that, eventually, a correlation relationship between the model and the investigational figures was examined.

## 2. Materials and Methods

### 2.1. Chemicals

The chemicals employed in this study were phenol (with a purity percentage of 99.99%), sodium hydroxide (NaOH), hydrochloric acid (HCl), and $ZnCl_2$. These chemicals were handled and analytically graded by Merck, Darmstadt, Germany.

### 2.2. Preparation of Chemicals Activated Carbon

The RH utilized for setting up and carbon activation was provided by a field owned by "Pusat Benih Padi Felcra Berhad-FELCRA" Paddy Seed Company located in the state of Perak in Malaysia. The preparation entailed washing the RH many times in distilled water, in order to remove soluble impurities, 24 h drying up on an oven at 105 °C, and then milling and sieving to 125–250 µm size of particles. A chemical activation process using $ZnCl_2$ has also been applied for the raw material activation purpose. In order to decrease ash content within the sample, the RH was pending to reflux with 1 M NaOH solution for 1 h. The used $ZnCl_2$ amount was set to provide ratio of imbibition (activating agent mass: dried RH) of 1:1. A horizontal tubular furnace (model TSH17/75/450-2416-2116) was used to heat the mixture under pure $N_2$ gas at a rate of 10 °C. $min^{-1}$ to the ultimate carbonization degree of heat, then held for 1 h at a carbonization temperature of 600 °C. The ratio of imbibition and carbonization temperature were selected upon our previous published work [34]. The activated carbon was then washed with 3 M HCl at 23 ± 0.5 °C for an hour. After the acidic solution was drained, the sample was washed with distilled water several times to remove the residual chemicals. The washed sample was then dried at 105 °C for 24 h.

### 2.3. Activated Carbon Characterization

The formulated adsorbent was characterized by surface area plus the size of pore's distribution, in terms of porosity as well as morphological structure using EDX (energy dispersive X-ray) and FESEM. In order to identify the organics of the adsorbent, a detailed characterization was carried out using XRD (X-ray diffraction).

### 2.4. Adsorption Column (Fixed Bed)

Experiments concerning the column were conducted via a glass column (at lab scale) packed with RHAC with a 1.6 cm internal diameter and a 50 cm length. A supportive wool was fixed in the glass and, to ensure a good distribution of the liquid, the column was closed. Further, to obtain a particular adsorbent bed height, the column was stuffed with 5.0, 7.5, and 9.0 g of RHAC (equivalent to 8.50, 12.75, and 15.30 cm of bed depth, respectively). At certain concentrations (100, 400, 700, 1000, and 2000 mg·$L^{-1}$), phenol solution was thrust downward within the sorbent column at a convenient flow rate (5.0, 7.5, and 10.0 mL·$min^{-1}$) by a Charles Austen pump (Model CAPEX L3, Byfleet, United Kingdom). The column effluent samples were collected at consistent durations and then evaluated using a chromatography for high performance liquid (HPLC Agilent 1100 series, Waldbronn, Germanysupported with a detecting system based on a diode array (DAD). The tests were achieved at a 23 ± 1 °C without any pH adjustment. All kinetic studies of the column were repeated once more, and then mean values were considered for the analysis.

### 2.5. Fixed-Bed Column Analysis Breakthrough

The breakthrough time ($t_b$) and the exhaustible time ($t_e$) were acquired through applying a measure of the fixed-bed breakthrough curves, when $C_t/C_0$ = 0.1 and $C_t/C_0$ = 0.9, respectively. The ultimate uptake capacity ($q_{total}$, mg) for a specified concentration of the inlet and the rate of flow is equivalent to the acquired under-the-plot area for the concentration of the adsorbed phenol, and can be determined through the following formula [35,36]:

$$q_{total} = \frac{QA}{1000} = \frac{Q}{1000}\int_{t=0}^{t=t_{total}} \left(C_0 - C_{eff}\right)dt \tag{1}$$

where $Q$, $A$, and $t_{total}$ represent the flow rate (mL·$min^{-1}$), the area below the curve, as well as flow total time (min), respectively. $C_0$ is the concentration of the influent phenol, while $C_{eff}$ represents the effluent phenol concentration (mg·$L^{-1}$). The effluent volume ($V_{eff}$) is computed by Equation (2), as follows [37]:

$$V_{eff} = Q \times t_{total} \tag{2}$$

The column's equilibrium uptake ($q_{eq}$) (mg·$g^{-1}$) is described in Equation (3) as being the overall adsorbed phenol volume ($q_{total}$) per each gram of adsorbent (m) at the total flow end time [38], as follows:

$$q_{eq} = \frac{q_{total}}{m} \tag{3}$$

### 2.6. Desorption Studies

It is crucial to consider the regeneration of the adsorbent material, as it is significant in economic development. The current study was intended to accomplish phenol (adsorbate) elution from an aqueous solution through the use of 0.1 M NaOH, taking into account identical settings of a flow rate of 8 mL·min$^{-1}$, adsorbent bed height of 8.5 cm used for preloading of adsorbent, and a concentration of phenol feeding of 1000 mg·L$^{-1}$. After the elution, distilled water was conveyed via the column for bed washing purposes until the wash effluent's pH settled at nearly 7. The column was loaded once again with a phenol concentration of 1000 mg·L$^{-1}$ within conformable regeneration conditions. Studies on regeneration of the column, as well as reuse, were applied for three sequences of adsorption–desorption. The efficiency of regeneration was determined by Equation (4), as follows [39]:

$$Regeneration\ efficiency\ (\%) = \frac{(q_{e,exp})_r}{(q_{e,\ exp})_o} \times 100 \tag{4}$$

where $(q_{e,\ exp})_r$ is the adsorptive capacity of the regenerated column (mg·g$^{-1}$) and $(q_{e,\ exp})_o$ is the original capacity of the virgin adsorbent (mg·g$^{-1}$).

### 2.7. Adsorption Column Modeling

Various kinetic models with different levels of complexity and accuracy were proposed and reported; they normally predict or describe breakthrough curves of column adsorption studies accurately. In this section, three different adsorption models were studied. Each model was different from the other in terms of type of adsorption isotherm, inclusion or exclusion of chemical reaction, significant or negligible mass transfer resistance, chemical reaction kinetics, and the type of rate law used if nonequilibrium is assumed [35,40]. The governing equations for each model are presented.

#### 2.7.1. Model of Yoon–Nelson

The Yoon–Nelson model is based on the assumption that the rate of decrease in the probability of adsorption of adsorbate molecule is proportional to the probability of the adsorbate adsorption and the adsorbate breakthrough on the adsorbent. The Yoon and Nelson model is not only less complicated than other models, but also requires no detailed data concerning the characteristics of adsorbate, type of adsorbent, and the physical properties of the adsorption bed [37]. The Yoon–Nelson equation's linear expression for a sole constituent can be shown in Equation (5), as follows [37]:

$$ln\left(\frac{C_{eff}}{C_0 - C_{eff}}\right) = k_{YN}\, t - \tau\, k_{YN} \tag{5}$$

where $k_{YN}$ represents a constant for the rate (given in min$^{-1}$) in addition to $\tau$, which is the time, required for 50% adsorbate development (given in min).

#### 2.7.2. Model of Adams–Bohart

The model of Adams–Bohart is certainly an experimental formula constructed upon the supposition that the adsorption rate is relative to the capacity of the adsorbent's residual, and the adsorbate concentration. Often, such model is utilized to highlight the early phase of adsorption within the breakout curve. The expression is given as Equation (6), as follows [37]:

$$ln\left(\frac{C_{eff}}{C_0}\right) = k_{AB}\, C_0\, t - k_{AB}\, N_0\, \frac{Z}{V} \tag{6}$$

where $k_{AB}$ stands for the kinetic constant (L·mg$^{-1}$·min$^{-1}$), $N_0$ presents the concentration of saturation (mg·L$^{-1}$), $Z$ represents the column bed depth (cm), and $V$ is the linear velocity, which is determined as (flow rate/column's e section area (cm·min$^{-1}$)).

#### 2.7.3. BDST Model

The model of BDST depends physically on measuring the bed capacity at several values of breakthrough. This basic model disregards the resistant mass transfer of the intra-particles, as well as the external film resistance so that the adsorbate is adsorbed directly onto the surface of adsorbent. Considering such hypothesis, the model of BDST properly achieves and yields worthwhile modeling formulae for system parameters change. An altered version of the formula that indicates the service time at breakthrough ($t$) as being a static function concerning the process elements is the model of BDST [35,40,41]. The model of BDST is illustrated by Equation (7), as following:

$$t = \frac{N_0'}{C_0 V} Z - \frac{1}{C_0 K} \ln\left(\frac{C_0}{C_b} - 1\right) \tag{7}$$

where $t$ represents the column's service time (h), $N_0'$ represents the adsorption capacity (mg·L$^{-1}$), $V$ represents the linear velocity of flow for the bed feed (cm·h$^{-1}$), $K$ represents a constant value for adsorption rate (L·mg$^{-1}$·h$^{-1}$), and $C_b$ represents the desired concentration of solute at breakthrough (ultimate acceptable limit concentration, mg·L$^{-1}$). The BDST formula was effectively applied to characterize the adsorption of heavy metals using phosphate-treated RH [41] and copper ions on RH [35]. The formula of the Adams–Bohart equation, given in Equation (7), could be applied to determine the column's service time ($t$) of a bed depth $Z$ having the values of $N_0'$, $C_0$, and $K$, which, essentially, can be assessed for laboratory columns operated over a range of velocity values ($V$). Applying $t = 0$ while solving Equation (8) for $Z$ produces the following equation:

$$Z_0 = \frac{V}{K N_0'} \ln\left(\frac{C_0}{C_b} - 1\right) \tag{8}$$

where $Z_0$ represents the least height of column required to output the effluent concentration $C_b$, which is, moreover, recognized as the critical depth of bed. A basic form of the BDST model is given as below:

$$t = aZ + b \tag{9}$$

where

$$a = slope = \frac{N_0'}{C_0 V} \tag{10}$$

and

$$b = intercept = -\frac{1}{C_0 K} \ln\left(\frac{C_0}{C_b} - 1\right) \tag{11}$$

With the view to associate the isotherm cogency with kinetic formula more definitely, the mean percent error (APE) with a base on the differences among the investigational data, plus the model's generated data will be applied through Equation (12), as follows [42]:

$$APE\ (\%) = \frac{\sum_{i=1}^{N} \left| (A_{exp} - A_{cal}) / A_{exp} \right|}{N} \times 100 \tag{12}$$

in which $A_{exp}$ and $A_{cal}$ present the experimental, as well as determined, amounts, respectively, while $N$ is data point's number.

## 3. Results

### 3.1. Physical Morphology of the Surface

The FESEM procedure was followed in order to examine the physical morphology of the surface for RH, as well as the sample of activated carbon. Figure 1a–d summarizes the FESEM/EDX mapping of the adsorbents. There were substantial variations between the RH surface morphology and the sample of activated carbon. The morphological aspects of the RH external epidermis presented an efficient arrangement that bears a resemblance to rolling hills, and is porosity-free (Figure 1a), while the activated carbon sample's surface was completely occupied by cavities, as porous structures were formed after chemical activation (Figure 1b).

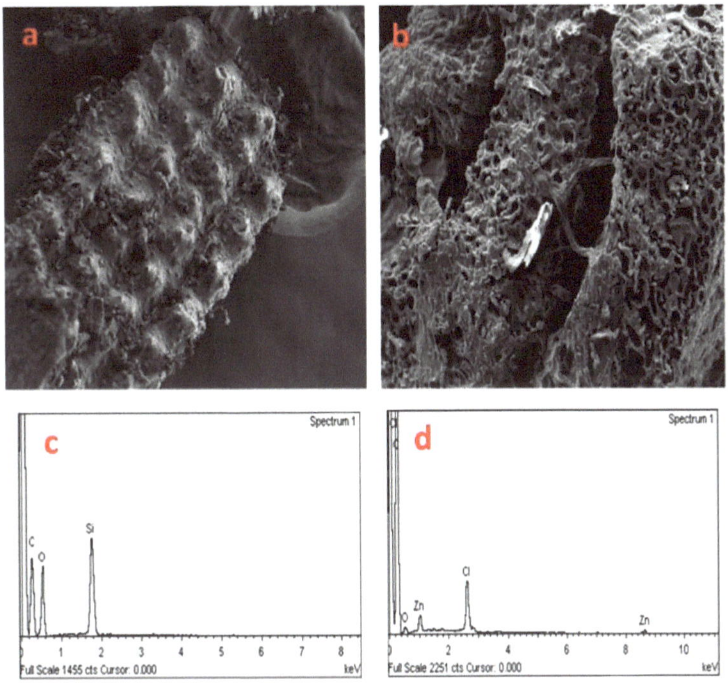


**Figure 1.** FESEM and EDX images of (**a**,**c**) rice husk (RH) and (**b**,**d**) rice husk activated carbon (RHAC) magnified 500 times.

Figure 1c,d shows the EDX images of the RH and RHAC. The EDX analysis demonstrated that C, O, Cl, and Zn were the only four elements observed on the surface of RHAC, with contents of 94.16%, 3.92%, 1.50%, and 0.41% (atomic percentage), respectively, and respective corresponding mass percentages of 88.78%, 4.93%, 4.17%, and 2.12%. The C, O, and Si contents within the RH were 58.89%, 34.98%, and 6.13% (atomic percentage), respectively, demonstrating respective equivalent mass percentages of 49.15%, 38.89%, and 11.96% (Figure 1c). To acquire a better perception of the textural properties of these adsorbents, and the size of pores, as well as surface area distribution, analyses were conducted.

### 3.2. Isotherms of Nitrogen Adsorption and Physical Properties

The surface area of BET, volume of pores, diameter of pores, and microporosity of the RH and RHAC are tabulated (Table 1). The resultant surface area produced by RH was small, whereas the RHAC had a large surface area owing to its highly developed microporous structure. The surface area of the RH increased by approximately 82.2 times from 7.14 $m^2 \cdot g^{-1}$ to 586.59 $m^2 \cdot g^{-1}$. The RHAC had high microporosity of 67.6% with a pore diameter of 2.06 nm, which produced an effective microporous adsorbent. This adsorbent's morphology also exhibited that it was porous (Figure 2).

**Table 1.** Characteristics of rice husk (RH) pore's texture and rice husk activated carbon (RHAC).

| Parameter | RH | RHAC |
|---|---|---|
| Surface area ($m^2{\cdot}g^{-1}$) | 7.14 | 586.59 |
| Micropore area ($m^2{\cdot}g^{-1}$) [a] | 1.44 | 415.80 |
| Micropore volume ($cm^3{\cdot}g^{-1}$) [a] | 0.000 | 0.204 |
| Total pore volume ($cm^3{\cdot}g^{-1}$) [b] | 0.006 | 0.301 |
| Average pore diameter (nm) [c] | 3.080 | 2.055 |
| Microporosity (%) [d] | 0.00 | 67.60 |

[a] Applying the model of Barrett–Joyner–Halenda; [b] A total pore volume for single-point adsorption; [c] Average adsorption's pore width (4V/A by BET)); [d] Microporosity given as (volume of micro-pore /total volume of pore) × 100%.

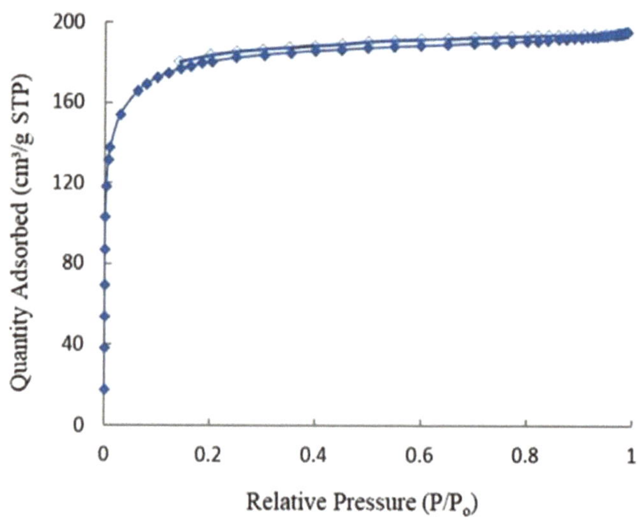


**Figure 2.** Hysteresis loops for activated carbon of rice husk.

The observed surface area expansion of the BET was a result of micro-pores presence, as demonstrated by the adsorption escalation at little relative pressures ($P/P_0$ less than 0.1), as appears in the nitrogen isotherm for the adsorption–desorption shown in Figure 2. In accordance with the International Union of Pure and Applied Chemistry classification [43], the RHAC adsorbent exhibited type I, which is a typical microporous configurations in which the ultimate adsorption is realized at relative pressures ($P/P_0$) near to unity.

### 3.3. *Identification and Structural Analysis of Rice Husk Activated Carbon*

The X-ray diffraction of the RHAC is shown in Figure 3a. From this figure it was found that the X-ray diffraction patterns of the RHAC show a broad 'hump' between 15 and 35 °θ diffraction angles, while lacking any defined peaks. This pattern is similar to the X-ray diffraction pattern of silicic acid (Figure 3b) reported by Nakbanpote et al. [44]. This suggests that the organics of this sample is characteristic of amorphous silica, which cannot be detected by X-ray diffraction. The results indicated that for the RHAC, the structure of silicon dioxide did not change from an amorphous structure to a crystalline structure, such as quartz, critobalite, or tridrimite.

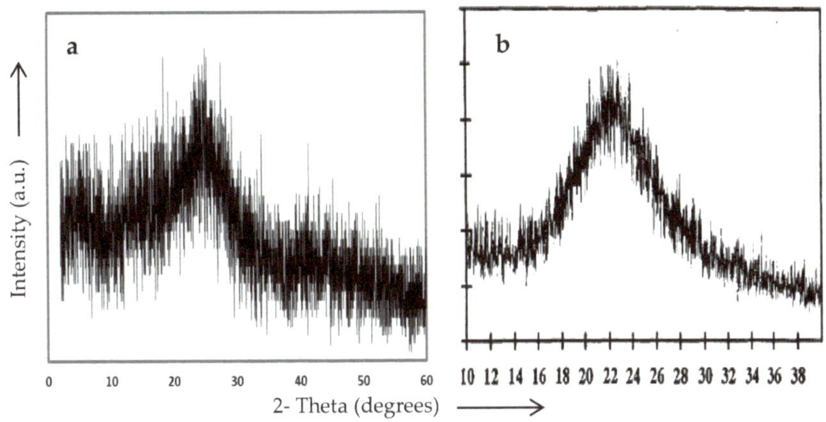


**Figure 3.** X-ray diffraction pattern of (**a**) RHAC and (**b**) silicic acid.

### 3.4. Studies of Fixed-Bed Column

#### 3.4.1. Concentration Effect of the Influent Phenol on the Curve of Breakthrough

The breakthrough curves at variable concentrations of influent phenol are presented in Figure 4. It was revealed that the process of adsorption rapidly reached the stage of saturation and the breakthrough time deteriorated according to the increase in the concentration of influent phenol. The influent phenol concentration's increase caused a shorter breakthrough along with higher adsorption capacity. Table 2 shows that the time for breakthrough decreased from 153 min to 35 min, whereas the equilibrium uptake ($q_{eq}$) increased from 17.49 mg·g$^{-1}$ to 45.75 mg·g$^{-1}$ as the inlet concentration increased from 400 mg·L$^{-1}$ to 2000 mg·L$^{-1}$. This occurred because, at lesser inlet phenol concentrations, curves of breakthrough scattered; hence, a slower breakthrough occurred. Sharper curves of breakthrough were observed once the influent concentration has boosted [45]. Additionally, a minor concentration gradient produced gentler conveyance owing to the diffusion coefficient's reduction or the coefficients of mass transfer. As the concentration of inlet elevated, the breakthrough curve's slopes and shorter times are obtained [45,46].

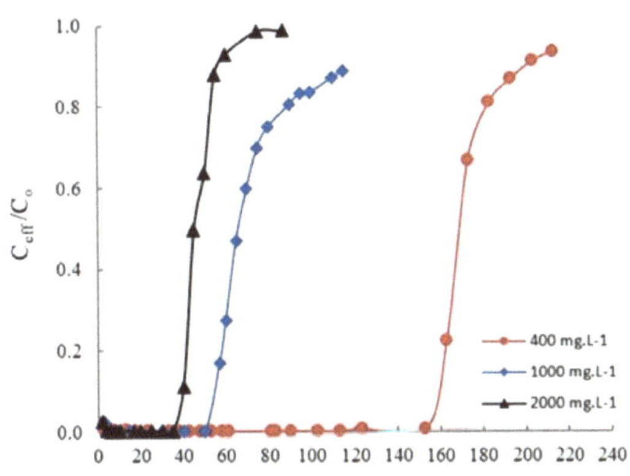


**Figure 4.** Concentration effect of the influent phenol on the curve of breakthrough of phenol adsorption of RHAC (rate of flow given as 10 mL·min$^{-1}$; bed depth given as 8.5 cm).

**Table 2.** Data on column and their acquired parameters at different influent phenol concentrations, rates of flow, and bed depths onto rice husk activated carbon (RHAC) (temperature = 23 ± 1 °C; pH = 5.58).

| $C_0$ (mg·L$^{-1}$) | $v$ (mL·min$^{-1}$) | Z (cm) | $t_b$ (min) | $t_{total}$ (min) | $q_{total}$ (mg) | $q_{eq}$ (mg·g$^{-1}$) | $V_{eff}$ (mL) |
|---|---|---|---|---|---|---|---|
| 400 | 10.0 | 8.50 | 153 | 193 | 87.45 | 17.49 | 1930 |
| 1000 | 10.0 | 8.50 | 40 | 90 | 110.85 | 22.17 | 900 |
| 2000 | 10.0 | 8.50 | 35 | 60 | 228.95 | 45.79 | 600 |
| 1000 | 7.5 | 8.50 | 60 | 130 | 135.75 | 27.15 | 975 |
| 1000 | 5.0 | 8.50 | 84 | 149 | 152.05 | 30.41 | 745 |
| 1000 | 10.0 | 12.75 | 106 | 188 | 223.28 | 29.77 | 1880 |
| 1000 | 10.0 | 15.30 | 125 | 195 | 283.41 | 31.49 | 1950 |

### 3.4.2. The Rate of Flow Effect

The curves of breakthrough at different phenol's rates of flow are demonstrated in Figure 5. Breakthrough occurred faster as the flow rate increased. The breakthrough time and $q_{eq}$ decreased from 84 min to 40 min and from 30.41 mg·g$^{-1}$ to 22.17 mg·g$^{-1}$, as the rate of flow increased from 5 mL/min to 10 mL/min$^{-1}$, respectively (Table 2). At a low influent rate of flow, phenol required more time to connect with RHAC, thereby causing a higher capacity of adsorption within the column. Nevertheless, the adsorbent at a higher linear rate of flow was saturated early; due to the minimized time of contact, a huge volume of phenol was adsorbed onto the adsorbent. Hence, there was a mis-distributed liquid into the column, which led to a decreased solute diffusivity amongst the absorbent's constituent parts [35,40]. Moreover, the reason was due to the reduction of external mass transfer as a result of an increase in the flow rate as phenol is easily transported to and from the beds surface by diffusion. Until the point of breakthrough, the treated volume was efficiently reduced as the rate of flow increased. Thus, the bed's service time has changed accordingly.

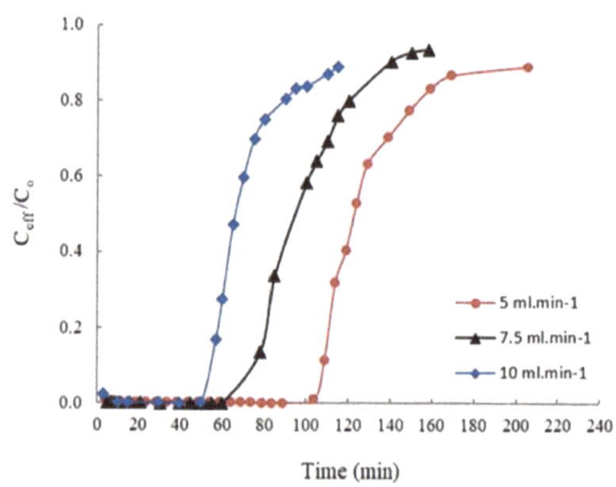


**Figure 5.** Effects of variable rates of flow on the adsorption's curve of breakthrough for RHAC (bed depth = 8.5 cm; phenol concentration at inlet = 1000 mg·L$^{-1}$).

### 3.4.3. The Bed Depth Effect

Curves of breakthrough at different bed depths are displayed in Figure 6. They demonstrated that their development usually take place faster at a reduced bed depth, and that the breakthrough slope decreases upon the increase in bed depth, which usually leads to an expanded zone of mass transfer. Table 2 shows that $q_{eq}$ was boosted from 22.17 mg·g$^{-1}$ to 31.49 mg·g$^{-1}$ as the bed depth increased from 8.5 cm to 15.3 cm, respectively. The slope of breakthrough curve declined following the increase in the bed depth, which led to an expanded zone of mass transfer. At the ultimate bed depth, a higher uptake occurred due to the adsorbent's surface area growth, leading to additional sorption sites adherence. Pollutant delayed breakthrough then led to a volumetric intensification within the processed solution [28,40].

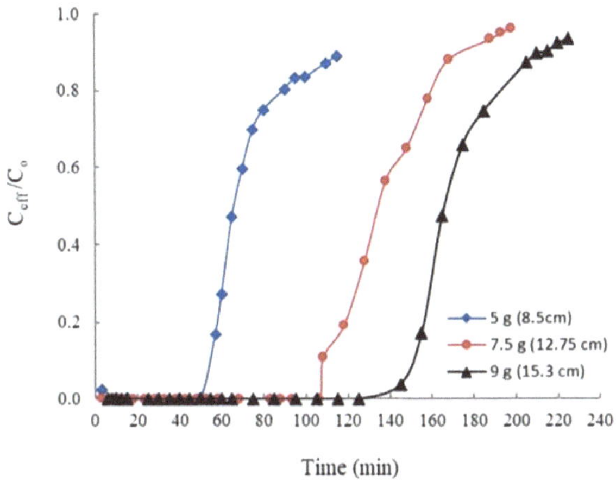


**Figure 6.** Different bed depths effect on the adsorption's curves of breakthrough onto RHAC (rate of flow of 10 mL/min; inlet phenol concentration of 1000 mg/L).

### 3.5. Study of Desorption

The desorption cycle profile of phenol is shown in Figure 7. The curves of desorption from cycles 1 to 3 exhibited a comparable asymmetrical form along with a fast increase in concentration succeeded by a flatter decrease. It was observed that desorption cycles took 2 h, after which further desorption was insignificant. In the first cycle, the eluent's total volume at 2 h was 588 mL. The phenol ultimate concentration was acquired at 7, 6, and 4 min contact times and noted as 8282, 5861, and 5562 mg·L$^{-1}$ for the first, second, and third cycle, respectively. In spite of the small volume of eluting solution, it contained a high concentration, which contributed to the processing simplicity, revival, and reuse of phenol.

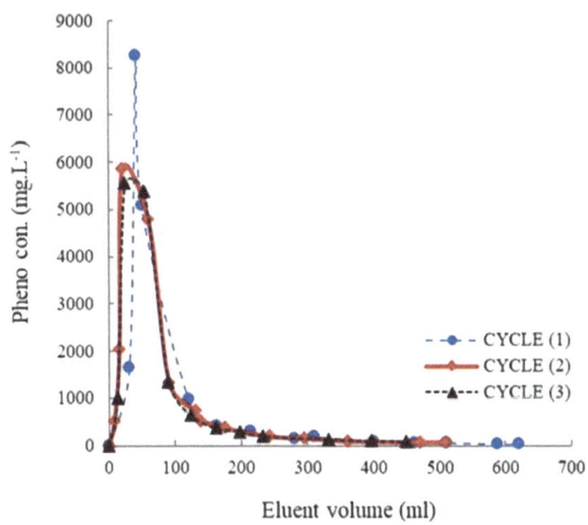


**Figure 7.** Desorption cycle profile of phenol at a rate of flow of 8 mL/min & bed depth of 8.5 cm with 0.1 M NaOH as eluent.

The regenerated RHAC was reused for further adsorption cycles. Figure 8 presents the curves of breakthrough produced for the cycles of adsorption. The breakthrough time ($t_b$) remained almost constant throughout the three cycles of sorption, which has noticeable significance for the feasible implementation of RHAC as phenol bio-sorbent. This is because RHAC entails persistent use of the bio-sorbent without an obvious lack of sorption process. Furthermore, no adsorbent swelling occurred, as specified by the constant column's bed depth [47]. The maximum uptake of phenol brought in a substantial decrease from 22.17 mg·g$^{-1}$ to 16.71 mg·g$^{-1}$ after the three cycles. The regeneration efficiency was 89.06% throughout the first cycle, then declined to 75.34% at the third cycle. The regeneration efficiency reduction occurred as some adsorbate ions were linked to the adsorbent via robust interconnection [47].

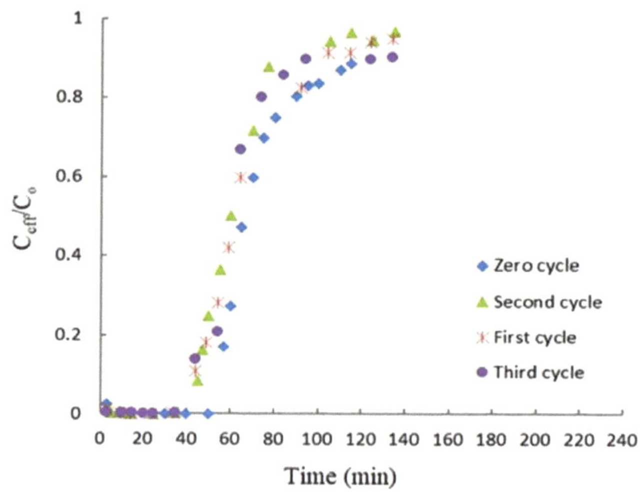


**Figure 8.** Curves of multiple breakthrough for phenol adsorption (1000 mg/L) at a constant 8 mL/min rate of flow & 8.5 cm bed depth.

### 3.6. *Curve of Breakthrough Modeling*

#### 3.6.1. Yoon–Nelson Model Application

An ordinary hypothetical model established by Yoon–Nelson was utilized to examine the phenol's breakthrough behavior within a fixed-bed column. The $k_{YN}$ (a constant for rate) and $\tau$ (required time for 50% phenol breakthrough) values were acquired using Equation (5). Such values were implemented to determine the curve of breakthrough. The $k_{YN}$ and $\tau$ values are presented in Table 3. As expected, $\tau$ decreased and $k_{YN}$ increased as the concentration in phenol inlet, along with the rate of flow, increased. Furthermore, as the bed depth increased, $\tau$ increased and $k_{YN}$ decreased. The notional curves were matched against their equivalent experimental ones in Figure 9. It can be noticed that the resulting curves of breakthrough from the experimental data are similar to the those examined by the Yoon–Nelson model at different rates of flow, primary concentrations, and bed depths. The model provided higher ($R^2 > 0.977$) and lower APE values varying from 4.09% to 12.99% (Table 3).

**Table 3.** The Yoon–Nelson model parameters at dissimilar conditions.

| $C_0$ (mg·L$^{-1}$) | $Q$ (mL·min$^{-1}$) | $Z$ (cm) | $k_{YN}$ (min$^{-1}$) | $\tau$ (min) | $R^2$ | $\tau_{exp}$ (min) | APE (%) |
|---|---|---|---|---|---|---|---|
| 1000 | 5.0 | 8.50 | 0.077 | 126.59 | 0.977 | 124 | 12.64 |
| 1000 | 7.5 | 8.50 | 0.069 | 97.81 | 0.992 | 100 | 4.89 |
| 1000 | 10.0 | 8.50 | 0.095 | 69.39 | 0.980 | 65 | 12.99 |
| 400 | 10.0 | 8.50 | 0.142 | 170.58 | 0.988 | 173 | 9.51 |
| 2000 | 10.0 | 8.50 | 0.239 | 46.67 | 0.992 | 45 | 4.73 |
| 1000 | 10.0 | 12.75 | 0.070 | 137.86 | 0.994 | 138 | 4.09 |
| 1000 | 10.0 | 15.30 | 0.077 | 170.21 | 0.989 | 165 | 11.02 |

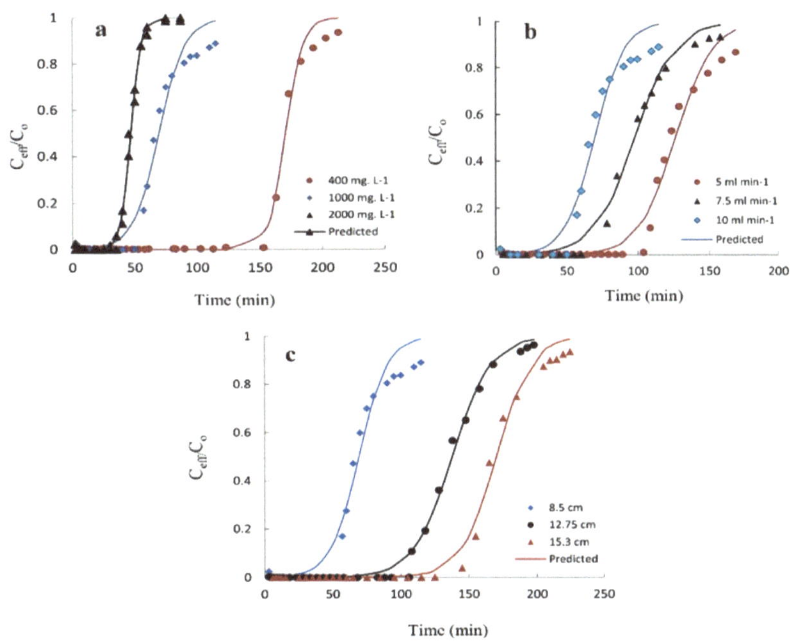


**Figure 9.** Experimental curves of breakthrough versus the predicted, attained at various (**a**) concentrations of phenol inlet, (**b**) phenol rates of flow, and (**c**) bed depths according to the model of Yoon–Nelson.

### 3.6.2. The Adams–Bohart Model Application

The sorption model of Adams–Bohart was applied to the investigational data as a way to define the primary breakthrough curve's portion. This method concentrates on the assessment of variables that describe the properties, for instance, the sorption's maximum capacity ($N_0$) and kinetic const. ($k_{AB}$) by the Adams–Bohart model. The $N_0$ and $k_{AB}$ values were determined from Equation 6, as shown in Table 4. The $k_{AB}$ value decreased in accordance with the increase in the bed depth, as well as influent concentration of phenol. However, it increased as the rate of flow increased. On the other hand, the $N_0$ value in all circumstances showed no substantial change. Moreover, the model provided a high $R^2 > 0.975$. A comparison between predicted curves against their corresponding experimental ones was conducted as presented in Figure 10. There was clearly considerable consistency between the values (experimental vs. predicted), thereby indicating that the model of Adams–Bohart was valid for concentrations up to 0.5, above which big contradictions could be observed among the plotted curves (experimental–predicted) regarding the adsorption of phenol in the RHAC column. Although the model of Adams–Bohart delivers a basic and comprehensive methodology for running and assessing sorption-column experiments, it encounters a limited validity to the level of the employed conditions [48].

**Table 4.** The Adams–Bohart model's parameters under various circumstances.

| $C_0$ (mg·L$^{-1}$) | $Q$ (mL·min$^{-1}$) | $Z$ (cm) | $k_{AB} \times 10^4$ (L·mg$^{-1}$·min$^{-1}$) | $N_0 \times 10^{-4}$ (mg·L$^{-1}$) | $R^2$ |
|---|---|---|---|---|---|
| 1000 | 5.0 | 8.50 | 1.14 | 3.808 | 0.975 |
| 1000 | 7.5 | 8.50 | 0.59 | 4.887 | 0.981 |
| 1000 | 10.0 | 8.50 | 1.31 | 4.039 | 0.982 |
| 400 | 10.0 | 8.50 | 1.68 | 5.421 | 0.979 |
| 2000 | 10.0 | 8.50 | 1.52 | 5.776 | 0.997 |
| 1000 | 10.0 | 12.75 | 0.59 | 5.796 | 0.994 |
| 1000 | 10.0 | 15.30 | 1.07 | 5.096 | 0.999 |

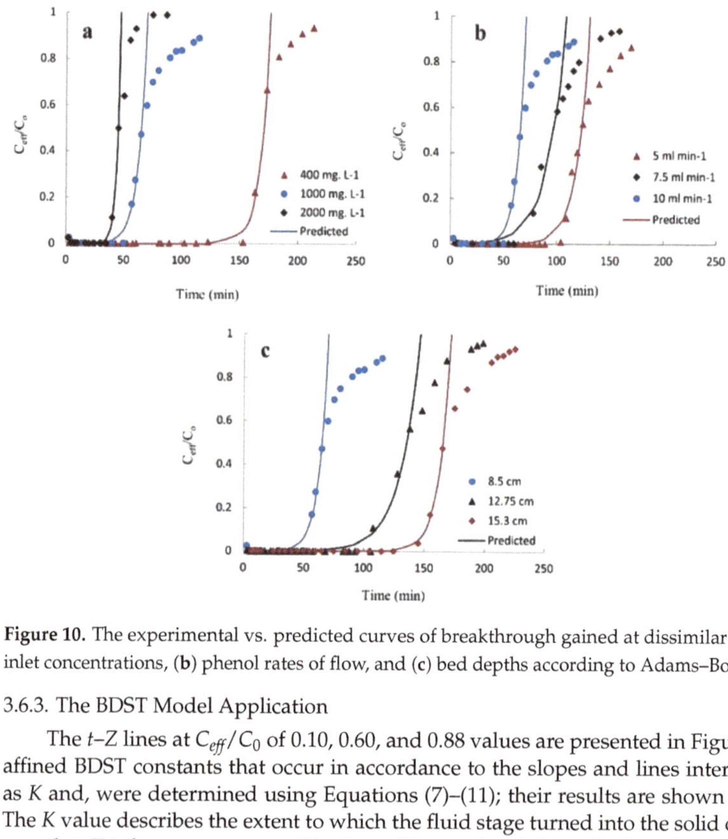


**Figure 10.** The experimental vs. predicted curves of breakthrough gained at dissimilar (**a**) pheno's inlet concentrations, (**b**) phenol rates of flow, and (**c**) bed depths according to Adams–Bohart model.

### 3.6.3. The BDST Model Application

The $t$–$Z$ lines at $C_{eff}/C_0$ of 0.10, 0.60, and 0.88 values are presented in Figure 11. The affined BDST constants that occur in accordance to the slopes and lines intercept, such as $K$ and, were determined using Equations (7)–(11); their results are shown in Table 5. The $K$ value describes the extent to which the fluid stage turned into the solid one. In the case that $K$ is large, even a small bed would avert breakthrough; though, as $K$ declines, an increasingly deeper bed is essential to avert breakthrough [49]. The capacity of BDST adsorption ($N_0'$) is a crucial constant for evaluating the efficiency of adsorbent; a powerful adsorbent is typically characterized by higher $N_0'$ values [49]. The lowest height of column ($Z_0$) essential for effluent concentration ($C_b$) of 0.1 mg·L$^{-1}$ production was determined by Equation (8). As shown in Table 5, as the $C_{eff}/C_0$ value improved, the constant ($K$) and the bed's capacity of adsorption/ a unit bed volume ($N_0'$) improved, while the minimum column height ($Z_0$) decreased. The $R^2$ values were observed between 0.9299 and 0.9975, which specified the BDST model soundness for this method. BDST model constants could be useful to increase the progression in any other rates of flow as well as concentrations with no more experimental runs.

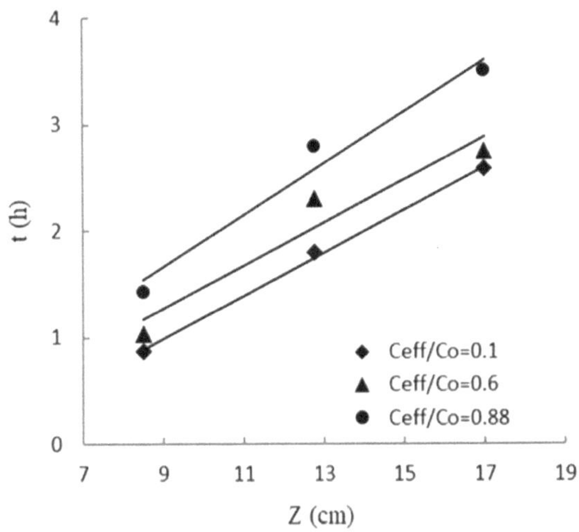


**Figure 11.** Lines of bed depth service time for 0.10, 0.60, and 0.88 breakthrough for diverse bed heights (Co given as 1000 mg/L; Q given as 10 mL/min).

**Table 5.** Service time model parameters for bed depth used for phenol adsorption ($C_0$ given as 1000 mg/L; $Q$ given as 10 mL/min).

| $C_{eff}/C_0$ | $K$ ($L{\cdot}mg^{-1}{\cdot}h^{-1}$) | $N_0' \times 10^{-3}$ ($mg{\cdot}L^{-1}$) | $Z_0$ (cm) | $R^2$ |
|---|---|---|---|---|
| 0.10 | 0.0108 | 62.048 | 4.0842 | 0.9975 |
| 0.60 | 0.0164 | 62.048 | 2.7089 | 0.9299 |
| 0.88 | 0.0171 | 74.673 | 2.1481 | 0.9665 |

## 4. Discussion

Techniques that aim at characterizing FESEM and BET were followed in order to investigate the structural features and surface areas of the RH and RHAC. Figure 1a demonstrates that the outer epidermis of the RH sample was wrapped with a uniform fluted texture. The sample of RHAC had a microporous structure with a large surface area and many adsorption active spots (Figure 1b). The analysis of EDX revealed that the RH and RHAC samples had comparable C and O elemental configurations; nonetheless, they had dissimilar ratios by weight (Figure 1c,d). The elemental composition of RH showed a higher carbon content, which makes it a convenient raw material for activated carbon preparations. Carbon represents the largest part of RHAC, whereas a complete absence of silica was observed within the RHAC, and existed merely in the RH. The carbon content of the RHAC was 94.16%, which was higher than that of the RH (49.15%), while the RHAC surface area was 586.59 $m^2{\cdot}g^{-1}$. The larger RHAC surface area observed was due to the increase in interspaces among the layers of carbon of the RHAC as a result of the $ZnCl_2$ action; microspores existed in the adsorbent [6]. The nature of the boost in the RHAC microporosity was, likewise, due to the areas left by $ZnCl_2$ during carbon activation process. This occurred as a result of the frequent washing process, which aimed to eliminate the surplus $ZnCl_2$ linked to the RHAC. Additionally, $ZnCl_2$ brings about swelling, considering cellulose as a major compound in RH; $ZnCl_2$ imposes electrolytic action to the cellulose molecular structure. The swelling action caused by $ZnCl_2$ on the sidelong can break the bonds of the molecules of cellulose, which leads to an intensification in intermicelle and intramicelle voids of RHAC [6]. Therefore, a large surface area was observed for RHAC.

Priya et al. [6] reported related findings with a greater surface area of activated carbon treated with $ZnCl_2$.

### 5. Conclusions

In this study, RHAC has been on an aqueous solution and assessed to be very powerful adsorbent used for removing phenol. The morphological features of RHAC have demonstrated a highly porous adsorbent compared with RH. RHAC demonstrated a surface area (586.6 $m^2 \cdot g^{-1}$), total volume of pores (0.3 $cm^3 \cdot g^{-1}$), and microporosity (67.6%) that greater than RH. The phenol uptake within a column of fixed-bed was reliant on the concentration of the influent phenol, the bed depth, and rate of flow. The capacity of adsorption increased as the influent concentration and bed depth increased, but reduced as the flow rate increased. The process of column sorption was improved at a higher concentration of influent phenol and minor rate of flow, in addition to greater bed depth. The models of Adams–Bohart, BDST, and Yoon–Nelson were properly utilized to predict the curves of breakthrough, thereby proving that they were appropriate for the RHAC column design. A high percentage of phenol desorption from consumed activated carbon implied that the activated carbon could be redeveloped to be used. RH can efficiently be utilized in order to equip activated carbon for treating phenol at cost effectiveness.

**Author Contributions:** Conceptualization, methodology, validation, formal analysis, investigation, resources, data curation, and visualization, S.B.D., H.M. and M.S.S.; supervision, project administration, writing—original draft preparation, and funding acquisition, S.B.D.; software, and writing—review and editing, S.B.D. and A.A.H. All authors have read and agreed to the published version of the manuscript.

**Funding:** This research was funded by the Annual Funding track by the Deanship of Scientific Research. Vice Presidency for Graduate Studies and Scientific Research, King Faisal University, Saudi Arabia [Project No. AN00048].

**Institutional Review Board Statement:** Not applicable.

**Informed Consent Statement:** Not applicable.

**Data Availability Statement:** Not applicable.

**Acknowledgments:** The authors are grateful to King Faisal University and Universiti Teknologi PETRONAS for facilitating laboratory works beside getting this work accomplished.

**Conflicts of Interest:** The authors declare no conflict of interest.

## References

1. Mishra, S.; Yadav, S.S.; Rawat, S.; Singh, J.; Koduru, J.R. Corn husk derived magnetized activated carbon for the removal of phenol and para-nitrophenol from aqueous solution: Interaction mechanism, insights on adsorbent characteristics, and isothermal, kinetic and thermodynamic properties. *J. Environ. Manag.* **2019**, *246*, 362–373. [CrossRef] [PubMed]
2. Chen, D.; Kannan, K.; Tan, H.; Zheng, Z.; Feng, Y.L.; Wu, Y.; Widelka, M. Review of the Bisphenol analogues other than BPA: Environmental occurrence, human exposure, and toxicity. *Environ. Sci. Technol.* **2016**, *50*, 5438–5453. [CrossRef] [PubMed]
3. Nirmala, G.; Murugesan, T.; Rambabu, K.; Sathiyanarayanan, K.; Show, P.L. Adsorptive removal of phenol using banyan root activated carbon. *Chem. Eng. Commun.* **2021**, *208*, 831–842. [CrossRef]
4. Busca, G.; Berardinelli, S.; Resini, C.; Arrighi, L. A short review of recent developments: Technologies for the removal of phenol from fluid streams. *J. Hazard. Mater.* **2008**, *160*, 265–288. [CrossRef]
5. Stasinakis, A.S.; Elia, I.; Petalas, A.V.; Halvadakis, C.P. Removal of total phenols from olive-mill wastewater using an agricultural by-product, olive pomace. *J. Hazard. Mater.* **2008**, *160*, 408–413. [CrossRef]
6. Priya, D.S.; Sureshkumar, M.V. Synthesis of Borassus flabellifer fruit husk activated carbon filter for phenol removal from wastewater. *Int. J. Environ. Sci. Technol.* **2020**, *17*, 829–842. [CrossRef]
7. El-Naas, M.H.; Al-Zuhair, S.; Alhaija, M.A. Removal of phenol from petroleum refinery wastewater through adsorption on date-pit activated carbon. *Chem. Eng. J.* **2010**, *162*, 997–1005. [CrossRef]
8. Aksu, Z. A review: Application of biosorption for the removal of organic pollutants. *Process Biochem.* **2005**, *40*, 997–1026. [CrossRef]

9. Cordova-Rosa, S.M.; Dams, R.I.; Cordova-Rosa, E.V.; Radetski, M.R.; Corrêa, A.X.R.; Radetski, C.M. Remediation of phenol-contaminated soil by a bacterial consortium and Acinetobacter calcoaceticus isolated from an industrial wastewater treatment plant. *J. Hazard. Mater.* **2009**, *164*, 61–66. [CrossRef]
10. Garcia-Castello, E.; Cassano, A.; Criscuoli, A.; Conidi, C.; Drioli, E. Recovery and concentration of polyphenols from olive mill wastewaters by integrated membrane system. *Water Res.* **2010**, *44*, 3883–3892. [CrossRef]
11. Jain, A.K.; Bhatnagar, A. Methylphenols removal from water by low-cost adsorbents. *J. Colloid Interface Sci.* **2002**, *251*, 39–45. [CrossRef] [PubMed]
12. Kujawski, W.; Warszawski, A.; Ratajczak, W.; Porębski, T.; Capała, W.; Ostrowska, I. Removal of phenol from wastewater by different separation techniques. *Desalination* **2004**, *163*, 287–296. [CrossRef]
13. Hameed, B.H.; Rahman, A.A. Removal of phenol from aqueous solutions by adsorption onto activated carbon prepared from biomass material. *J. Hazard. Mater.* **2008**, *160*, 576–581. [CrossRef] [PubMed]
14. Kumar, A.; Jena, H.M. Preparation and characterization of high surface area activated carbon from Fox nut (Euryale ferox) shell by chemical activation with $H_3PO_4$. *Results Phys.* **2016**, *6*, 651–658. [CrossRef]
15. Kavand, M.; Soleimani, M.; Kaghazchi, T.; Asasian, N. Competitive separation of lead, cadmium, and nickel from aqueous solutions using activated carbon: Response surface modeling, equilibrium, and thermodynamic studies. *Chem. Eng. Commun.* **2016**, *203*, 123–135. [CrossRef]
16. Krishnamoorthy, R.; Govindan, B.; Banat, F.; Sagadevan, V.; Purushothaman, M.; Show, P.L. Date pits activated carbon for divalent lead ions removal. *J. Biosci. Bioeng.* **2019**, *128*, 88–97. [CrossRef] [PubMed]
17. Mohanty, K.; Jha, M.; Meikap, B.C.; Biswas, M.N. Removal of chromium (VI) from dilute aqueous solutions by activated carbon developed from Terminalia arjuna nuts activated with zinc chloride. *Chem. Eng. Sci.* **2005**, *60*, 3049–3059. [CrossRef]
18. Azevedo, D.C.; Araújo, J.C.S.; Bastos-Neto, M.; Torres, A.E.B.; Jaguaribe, E.F.; Cavalcante, C.L. Microporous activated carbon prepared from coconut shells using chemical activation with zinc chloride. *Microporous Mesoporous Mater.* **2007**, *100*, 361–364. [CrossRef]
19. Adinata, D.; Daud, W.M.A.W.; Aroua, M.K. Preparation and characterization of activated carbon from palm shell by chemical activation with K2CO3. *Bioresour. Technol.* **2007**, *98*, 145–149. [CrossRef]
20. Prahas, D.; Kartika, Y.; Indraswati, N.; Ismadji, S.J.C.E.J. Activated carbon from jackfruit peel waste by $H_3PO_4$ chemical activation: Pore structure and surface chemistry characterization. *Chem. Eng. J.* **2008**, *140*, 32–42. [CrossRef]
21. Kalderis, D.; Bethanis, S.; Paraskeva, P.; Diamadopoulos, E. Production of activated carbon from bagasse and rice husk by a single-stage chemical activation method at low retention times. *Bioresour. Technol.* **2008**, *99*, 6809–6816. [CrossRef] [PubMed]
22. Altenor, S.; Carene, B.; Emmanuel, E.; Lambert, J.; Ehrhardt, J.J.; Gaspard, S. Adsorption studies of methylene blue and phenol onto vetiver roots activated carbon prepared by chemical activation. *J. Hazard. Mater.* **2009**, *165*, 1029–1039. [CrossRef]
23. Din, A.T.M.; Hameed, B.H.; Ahmad, A.L. Batch adsorption of phenol onto physiochemical-activated coconut shell. *J. Hazard. Materials* **2009**, *161*, 1522–1529.
24. Liou, T.H.; Wu, S.J. Characteristics of microporous/mesoporous carbons prepared from rice husk under base-and acid-treated conditions. *J. Hazard. Mater.* **2009**, *171*, 693–703. [CrossRef] [PubMed]
25. Karunarathne, H.D.S.S.; Amarasinghe, B.M.W.P.K. Fixed bed adsorption column studies for the removal of aqueous phenol from activated carbon prepared from sugarcane bagasse. *Energy Procedia* **2013**, *34*, 83–90. [CrossRef]
26. Girish, C.R.; Ramachandra Murty, V. Adsorption of phenol from aqueous solution using Lantana camara, forest waste: Kinetics, isotherm, and thermodynamic studies. *Int. Sch. Res. Not.* **2014**, *2014*, 201626.
27. Liou, T.H. Development of mesoporous structure and high adsorption capacity of biomass-based activated carbon by phosphoric acid and zinc chloride activation. *Chem. Eng. J.* **2010**, *158*, 129–142. [CrossRef]
28. Kumar, U.; Bandyopadhyay, M. Sorption of cadmium from aqueous solution using pretreated rice husk. *Bioresour. Technol.* **2006**, *97*, 104–109. [CrossRef]
29. Ngah, W.W.; Hanafiah, M.M. A review: Removal of heavy metal ions from wastewater by chemically modified plant wastes as adsorbents. *Bioresour. Technol.* **2008**, *99*, 3935–3948. [CrossRef]
30. Wong, K.K.; Lee, C.K.; Low, K.S.; Haron, M.J. Removal of Cu and Pb by tartaric acid modified rice husk from aqueous solutions. *Chemosphere.* **2003**, *50*, 23–28. [CrossRef]
31. Miao, Q.; Tang, Y.; Xu, J.; Liu, X.; Xiao, L.; Chen, Q. Activated carbon prepared from soybean straw for phenol adsorption. *J. Taiwan Inst. Chem. Eng.* **2013**, *44*, 458–465. [CrossRef]
32. Komal Kumar, N.; Shukla, V.K. Electrode material for supercapacitor using activated carbon derived from flower of *Achyranthes aspera* L. *Proc. AIP Conf.* **2020**, *2220*, 140046.
33. Joshi, S.; Bishnu, K.C. Synthesis and Characterization of Sugarcane Bagasse Based Activated Carbon: Effect of Impregnation Ratio of $ZnCl_2$. *J. Nepal Chem. Soc.* **2020**, *41*, 74–79. [CrossRef]
34. Daffalla, S.B.; Mukhtar, H.; Shaharun, M.S. Properties of activated carbon prepared from rice husk with chemical activation. *Int. J. Glob. Environ. Issues* **2012**, *12*, 107–129. [CrossRef]
35. Luo, X.; Deng, Z.; Lin, X.; Zhang, C. Fixed-bed column study for $Cu^{2+}$ removal from solution using expanding rice husk. *J. Hazard. Mater.* **2011**, *187*, 182–189. [CrossRef]
36. Aksu, Z.; Gönen, F. Biosorption of phenol by immobilized activated sludge in a continuous packed bed: Prediction of breakthrough curves. *Process Biochem.* **2004**, *39*, 599–613. [CrossRef]

37. Shi, S.L.; Lv, J.P.; Liu, Q.; Nan, F.R.; Jiao, X.Y.; Feng, J.; Xie, S.L. Application of Phragmites australis to remove phenol from aqueous solutions by chemical activation in batch and fixed-bed columns. *Environ. Sci. Pollut. Res.* **2018**, *25*, 23917–23928. [CrossRef]
38. Oguz, E.; Ersoy, M. Removal of $Cu^{2+}$ from aqueous solution by adsorption in a fixed bed column and Neural Network Modelling. *Chem. Eng. J.* **2010**, *164*, 56–62. [CrossRef]
39. Dwivedi, C.P.; Sahu, J.N.; Mohanty, C.R.; Mohan, B.R.; Meikap, B.C. Column performance of granular activated carbon packed bed for Pb (II) removal. *J. Hazard. Mater.* **2008**, *156*, 596–603. [CrossRef]
40. Han, R.; Ding, D.; Xu, Y.; Zou, W.; Wang, Y.; Li, Y.; Zou, L. Use of rice husk for the adsorption of congo red from aqueous solution in column mode. *Bioresour. Technol.* **2008**, *99*, 2938–2946. [CrossRef]
41. Mohan, S.; Sreelakshmi, G. Fixed bed column study for heavy metal removal using phosphate treated rice husk. *J. Hazard. Mater.* **2008**, *153*, 75–82. [CrossRef] [PubMed]
42. Ren, L.; Zhang, J.; Li, Y.; Zhang, C. Preparation and evaluation of cattail fiber-based activated carbon for 2, 4-dichlorophenol and 2, 4, 6-trichlorophenol removal. *Chem. Eng. J.* **2011**, *168*, 553–561. [CrossRef]
43. Pallarés, J.; González-Cencerrado, A.; Arauzo, I. Production and characterization of activated carbon from barley straw by physical activation with carbon dioxide and steam. *Biomass Bioenergy* **2018**, *115*, 64–73. [CrossRef]
44. Nakbanpote, W.; Thiravetyan, P.; Kalambaheti, C. Preconcentration of gold by rice husk ash. *Miner. Eng.* **2000**, *13*, 391–400. [CrossRef]
45. Chen, N.; Zhang, Z.; Feng, C.; Li, M.; Chen, R.; Sugiura, N. Investigations on the batch and fixed-bed column performance of fluoride adsorption by Kanuma mud. *Desalination* **2011**, *268*, 76–82. [CrossRef]
46. Ahmad, A.A.; Hameed, B.H. Fixed-bed adsorption of reactive azo dye onto granular activated carbon prepared from waste. *J. Hazard. Mater.* **2010**, *175*, 298–303. [CrossRef]
47. Sharma, N.; Kaur, K.; Kaur, S. Kinetic and equilibrium studies on the removal of $Cd^{2+}$ ions from water using polyacrylamide grafted rice (Oryza sativa) husk and (Tectona grandis) saw dust. *J. Hazard. Mater.* **2009**, *163*, 1338–1344. [CrossRef]
48. Bhaumik, M.; Setshedi, K.; Maity, A.; Onyango, M.S. Chromium (VI) removal from water using fixed bed column of polypyrrole/$Fe_3O_4$ nanocomposite. *Sep. Purif. Technol.* **2013**, *110*, 11–19. [CrossRef]
49. Al-Degs, Y.S.; Khraisheh, M.A.M.; Allen, S.J.; Ahmad, M.N. Adsorption characteristics of reactive dyes in columns of activated carbon. *J. Hazard. Mater.* **2009**, *165*, 944–949. [CrossRef]

*environments*

*Review*

# Modified Biochars and Their Effects on Soil Quality: A Review

**Abdul Hafeez [1,2,3], Taowen Pan [1,2,3], Jihui Tian [1,2,3] and Kunzheng Cai [1,2,3,*]**

1 Guangdong Provincial Key Laboratory of Eco-Circular Agriculture, South China Agricultural University, Guangzhou 510642, China; ahafeez1226@scau.edu.cn (A.H.); pantw@stu.scau.edu.cn (T.P.); jhtian@scau.edu.cn (J.T.)
2 College of Natural Resources and Environment, South China Agricultural University, Guangzhou 510642, China
3 Key Laboratory of Tropical Agro-Environment, Ministry of Agriculture, South China Agricultural University, Guangzhou 510642, China
* Correspondence: kzcai@scau.edu.cn; Tel.: +86-20-3829-7175

**Abstract:** Biochar (BC) has attracted attention due to its impacts on soil quality by enhancing soil fertility, carbon storage and contaminants immobilization. BC also induces changes in microbial community structure and enhances crop productivity in long term scenarios compared to many other organic amendments. However, information related to the role of modified BCs in altering the soil quality is still scarce. BC can be modified by using physical, chemical and microbial methods. Modified BC can change the functional groups, pore size, pore structure, surface area and chemical properties of soil, which plays a key role in changing the soil quality. The addition of modified BCs as soil amendment increased soil CEC (cation exchange capacity), EC (electron conductivity), pH, organic matter, hydraulic conductivity, soil porosity, infiltration rate, microbial activities (enzymes and community), nutrient profile and gas exchange properties, but it varies according to the soil structure and pervading environmental conditions. This study provides a basis for effective practical approaches to modifying BCs for improving soil quality.

**Keywords:** modified biochar; soil quality; physiochemical properties; microbial community; EC; CEC

**Citation:** Hafeez, A.; Pan, T.; Tian, J.; Cai, K. Modified Biochars and Their Effects on Soil Quality: A Review. *Environments* **2022**, *9*, 60. https://doi.org/10.3390/environments9050060

Academic Editors: Dibyendu Sarkar, Rupali Datta, Prafulla Kumar Sahoo and Mohammad Mahmudur Rahman

Received: 20 March 2022
Accepted: 11 May 2022
Published: 15 May 2022



## 1. Introduction

Biochar (BC) is charcoal produced from plant matter and stored in the soil as a means of removing carbon dioxide from the atmosphere. BC application as soil amendment has achieved promising results for higher crop growth and production [1–3], disposal of large scale waste biomass [4], an approved role in climate change mitigation in the long run [5], and soil biochemical property enhancement [6]. Previous studies have shown that adding BC to soil can increase saturate hydraulic conductivity and water infiltration, reduce the soil bulk density and improve the soil structure [7–9].

To improve the efficiency of BC, recent studies have focused on the use of modified BC to improve soil quality and performance for better plant growth and productivity [10–13] by playing a key role as media to provide attachment sites for microbial communities [14,15] and effectively enhance soil physicochemical properties by regulating soil pH [10], electron conductivity (EC) [16], cation exchange capacity (CEC) [14], organic matter [17], and hydraulic conductance, porosity and soil aggregates [14,15]. However, the chemical and physical properties of modified BCs greatly depend on the type of modification reagent, modification method, pyrolysis temperature, pyrolysis process and type of raw material. Generally, the properties of BC are determined by its physical structure (surface area and porosity) [18]. However, only a few studies are present to assess the effect of modified BC on soil quality [14,15,19–21], and a comprehensive review is still missing to address the effect of various modified BCs on soil quality. The main objective of this study is to expose the properties of various modified BCs and their effect on soil physiochemical properties, and their relationship with plant growth and development. The effect of modified BC on plant growth, yield attributes and soil physio-chemical properties (such as soil physical and hydraulic properties, soil organic matter, gas exchange parameters and nutrient profile, soil pH, EC and CEC and soil biological activities) are briefly discussed in this study.

## 2. Effect of Biochar on Plant Growth and Yield Attributes

BC can be used as a promising soil-amendment to increase crop growth and productivity by modulating soil properties [22–25], because of its unique characteristics such as cation exchange capacity (CEC), abundant (O)-containing functional groups, rich pore structure and large surface area [26,27]. Meta-analysis based on 153 studies and 1254 pair comparisons revealed that plant productivity varies from −31.8% to 974% under different BC conditions [23]. Moreover, they also observed that BC properties including bulk density, ash, pH, carbon content and CEC significantly modified the plant productivity by improving the nutrient uptake and soil physical structure.

Although amending soil with BC has been regarded as an effective approach for improving crop growth and productivity, it is not clear by which mechanism BC increases crop yield. Liu et al. [11] compared the relative contribution of three different BC components, such as mineral nutrient form BC ash (BA), washed BC residue (WB) and water-soluble BC extract (BE) by using *Zea mays* L. (Maize) as a model plant in two soil types Prismosol and Anthrosol. In the Primosol and Anthrosol, WB, BA and BE application increased maize biomass by 18.2%, 24.4% and 41.3%, and 27.1%, 32.7% and 41.6%, respectively. They suggested that BE had a higher plant growth promoting effect as compared to other BCs. In addition, with BE-amendment the number of maize roots, surface area, total volume and biomass were significantly improved, particularly the fine roots diameter (<0.2 mm). Intriguingly, meta-analysis of 81.3% of the dataset from 47 studies showed that BC amendment increased the root growth up to 32%, and the increment of root biomass were 12.7%, +26.9%, +66.0% and +101.6% for cereals, vegetables, grasses and trees, respectively [28]. They suggested that the rate of application and feedstocks of BC were the main factors contributing to plant growth and production. The highest positive effect was noted for gramineous (+46.2%), followed by woody plants (+25.8%) and green waste (+21.1%). Tarin et al. [29] reported that the rice BC amendment significantly increased the photosynthetic rate of *Fokienia hodginsii* (conifer tree) in all four seasons. The rice BC amendment with 80 g $kg^{-1}$ improved 36% biomass compared to control plants. Pristine and acidified rice husk-BC produced 80.5% and 110.7% root dry matter [30]. However, Bonanomi et al. [3] explored the effect of BC on the growth of lettuce (*Lactuca sativa*). They noted that BC had both a strong inhibiting and stimulating effect on plant growth. In conclusion, the application of BC effectively increased the plant biomass, photosynthetic efficiency, root diameter, surface area and root total volume by improving the soil physical structure and nutrient uptake, resulting in higher economical yield. Meta-analysis of BC application showed that BC increased the crop yield up to 10–15%, with concurrent decreases in $N_2O$ emissions [31].

## 3. Effect of Pristine Biochar on Soil Physiochemical Properties

The BC amendment effectively changed the soil CEC, bulk density, texture, pore structure and surface area [26,27]. However, according to Blanco-Canqui [32], BC generally increased the soil wet aggregate stability (3 to 32.26%), porosity (14 to 64%), soil bulk density (3 to 31%), and soil water availability (4 to 130%), but had a non-significant effect on soil penetration resistance. Additionally, soil pore distribution, size and volume differ in BC feedstock source and pyrolysis temperature. Mesopores (2–50 nm) and macropores (>50 nm) of BC are helpful for water retention and act as the sole habitat for microorganisms. Besides, the nanopores or micropores (<0.9 nm) of BC govern the sorption and chemical properties of BC [33]. Soil particles can easily interact with the fine particles of BC and form the soil aggregates by modulating the water retention and interpore shape [32,34].

During pyrolysis, many volatile compounds are released in the form of gases that increase the surface area of BC and generate honey bee like porous structures that result in higher nutrient retention and water holding capacity [35,36]. Moreover, the feedstock source of BC can also affect these properties. For instance, Weber and Quicker [37] found that the majority of BC possess biomass surface areas ranging from 100–800 $m^2$ $g^{-1}$, while BC produced from sewage sludge has a surface area of 100 $m^2$ $g^{-1}$. In another case, BC of cottonwood or aspen (populus species) had a lower surface area as compared to BC produced from maize straw [38]. Studies revealed that BC did not equally increase soil porosity, and was mostly dependent upon soil texture class and soil type. Generally, fine textured soil showed less improvement in soil porosity as compared to coarse texture soil by BC application [39].

Blanco-Canqui [32] noted that an addition of BC increased soil properties, minimized the soil water repellency, moderated soil thermal properties, altered water infiltration, and reduced the particle density and tensile strength. Tarin et al. [29] reported the effect of rice straw on the physicochemical properties of soil and suggested that BC amendment can be used to combat soil acidification and phosphorous (P) availability in P deficient soils. In another study, Sadegh-Zadeh et al. [40] found that rice BC application with 50 g $kg^{-1}$ remediates the saline-sodic soil. They noted that $Mg^{2+}$ and $Ca^{2+}$ in the BC surface could exchange $Na^{+}$ on the soil colloids, which thereby resulted in $Na^{+}$ leaching from saline-sodic soils. In addition, it is anticipated that BC is a significant source of carbon (C), and its application improves the soil aggregation properties [41]. Qayyum et al. [30] reported that apristine BC application substantially increased soil pH between 3.8 to 7.5%. In conclusion, BC soil amendment increased soil CEC, bulk density, texture, pore structure, interpose shape and surface area, soil aggregate stability and soil water availability as compared to untreated soil. Moreover, BCs application helps to combat the effect of soil acidification, P deficiency and soil salinity, and act as the sole habitat for microorganism.

## 4. Biochar Modification and Its Properties

Besides the sorption mechanism, recent developments in BC modification by loading with nanoparticles, organic functional groups, reductants, minerals, bio-materials and activation with an alkali solution in improving sorption capacity is briefly discussed below (Figure 1).

### *4.1. Biochar Physical Modification*

BC modification by using physical methods is an environmentally-friendly and cost-effective approach as compared to chemical modification. It improves the physio-chemical properties of BC including permeability and porosity with controllable measures. The common techniques in the physical modification of BC include magnetization, ball milling, microwave irradiation and steam/gas activation that improves the BC properties to enhance the soil quality. For instance, in ball milling the pristine BC is broken into small powder that increases its surface area and adsorption capacity while reducing its particle size [42]. Ball milling can be classified into chemical ball milling and physical ball milling. Chemical ball milling modifies the microporous structure and functional groups, whereas physical ball milling greatly influences the surface area and particle size of BC [43]. Magnetic BC synthesized from chemical ball milling can easily be recovered with the help of an external magnetic field [44]. The catalytic activity and surface charge of BC improve with the magnetization process, which thereby result in better environmental remediation.

Another emerging technique for BC modification is microwave irradiation, which raises the temperature of BC up to 200–300 °C by microwave heating within a short time. The microwave-modified BC showed the higher surface area and absorption capacity for many pollutants with its hydraulic functional groups. The integration of steam activation with microwave irradiation significantly improved the physiochemical properties of BC, including water holding capacity and CEC [45]. Gas/steam activation also induce porosity and increase the surface area of BC by removing the trapped residues inside the porous structure of BC due to partial combustion. This process of BC modification activates carbon dioxide and hydrogen through surface reactions [46] and, as a consequence, it shows greater adsorption capacity for nitrogen dioxide and methane over the pristine BC [47]. In conclusion, physical modifications of BCs through ball milling increases surface area, adsorption capacity and microporous structure of BCs, while microwave irradiation modification improves BCs surface area and absorption capacity for many pollutants, and gas/steam activation is important for improving BC porosity, surface area, physiochemical properties and adsorption capacity for different gasses as compared to pristine BCs.

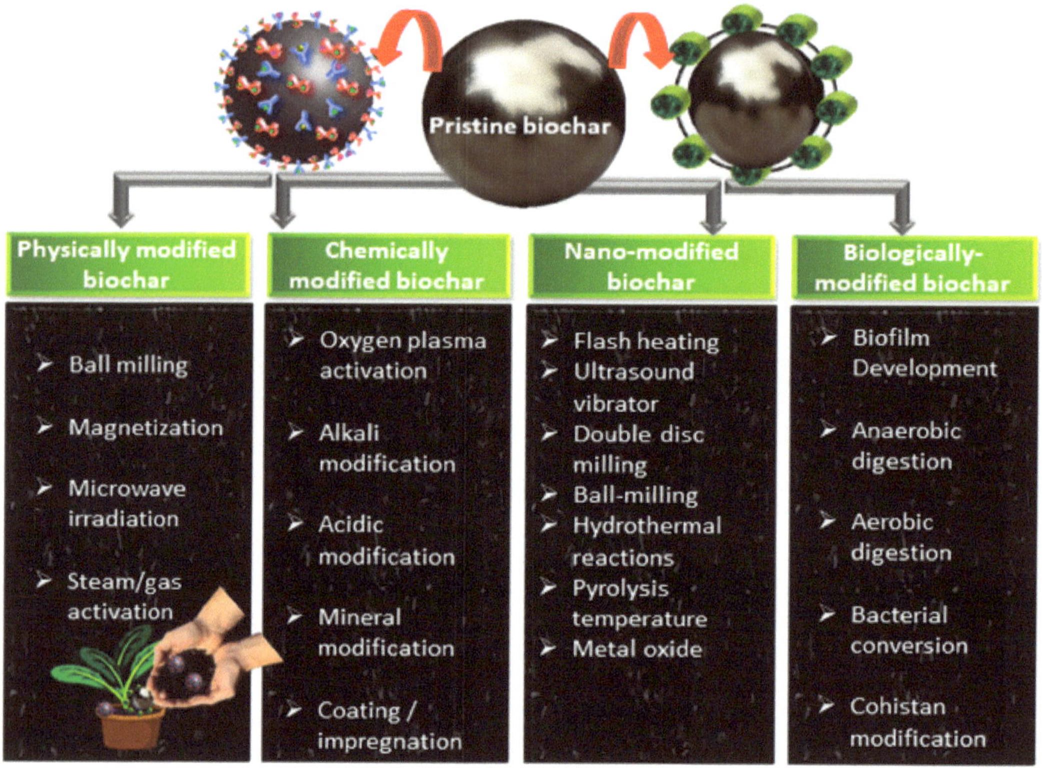


**Figure 1.** Depiction of different ways for conversion of pristine biochar to modified biochar for improving soil quality.

### *4.2. Biochar Modification with Mineral Fertilizers and Chemicals*

Biochar can be modified through reactions with different mineral fertilizers and chemicals to make them more efficient at performing specific applications by enhancing their physiochemical properties [48]. Chemical modification can be done with a range of processes, such as oxygen plasma activation, coating/impregnation, mineral coating, and acid/alkali modifications. Bamboo hardwood modified with carbon disulfide ($CS_2$) and sodium hydroxide (NaOH) at 45 °C for 8 h showed greater effectiveness for chromium (Cr) remediation. It was also observed that the sulfur-iron modified BC had more surface roughness and granular massive structure as compared to pristine BC [49]. In addition, Yin et al. [50] reported that poplar chips BC modified with 20% aluminium chloride ($AlCl_3$) at 80 °C for 48 h increased the BC yield and BET surface area (96.7%), while decreasing the carbon content (55.88) as compared to pristine. They noted that BC modified with 15% $AlCl_3$ showed optimum adsorption for nitrate, whereas the 20% $AlCl_3$ was best for phosphate adsorption.

In another study, swine manure modified BC with phosphoric acid ($H_3PO_4$) at 25 °C for 24 h showed higher mesopore, micropore volume and total pore area, while rice straw modified with the same acid showed non-significant modifications [51]. Cow manure and wheat straw BC modified with nitric acid ($HNO_3$) at 90 °C for 4 h showed negative surface charge, higher COO groups and enhanced U(VI) adsorption than in unmodified BC. The maximum U(VI) adsorption capacity of wheat straw was 40 times higher than pristine BC treatment [52]. Moreover, 3.0% cetyltrimethyl ammonium bromide (CTAB) modification with 5 g of dried Auricularia auricula dreg increased the surface area and pore diameter up to 6.1% and 16.5%, respectively. Consequently, the number of micropores and mesoporous in the unit area also increased. The adsorption quantity and rate of the modified BC were 8.0% and 6.4% higher than pristine BC [53]. In addition, NaOH (sodium hydroxide) modification in dairy manure BC showed higher oxygen-containing functional groups, ion exchange capacity and surface area, and increased the adsorption capacity of BC for Cd and Pb (lead) as compared to untreated BC. The highest sorption capacity was 68.08 and 175.53 mg $g^{-1}$ for Cd and Pb, respectively [54].

HCl+ ultra-sonication of coconut shell BC increased the microcosmic pore area (6.1%), and the average pore diameter (16.5%) structure and surface functional groups as compared to the untreated BC [55]. Moreover, potassium hydroxide (KOH) and sodium sulfide ($Na_2S$) based modification of corn straw increased BET surface area up to 59.23 and 55.58 $m^2$ $g^{-1}$, respectively, as compared to pristine BC (32.85 $m^2$ $g^{-1}$). They deduced that these types of modified BC can be effectively used for remediating mercury (Hg) pollution. Besides, Li et al. [56] used NaOH, $H_3PO_4$, $HNO_3$, $NH_4OH$ and sulphuric acid ($H_2SO_4$) to carry out treatment on coconut shell-based carbon. They noted that higher adsorption capacity (o-xylene, a hydrophobic volatile organic compound) was obtained with alkali-treated carbon as compared to acid-treated BC. The reason is that total oxygen containing functional groups diminished, and pore volume and surface area increased by alkali treatment, while it was opposite for acid application. BC activation with NaOH and KOH is quite different, KOH activation in situ showed an interaction between K and layers of the carbon crystalline, while Na did not show any interaction for carbon. In another case, Sajjadi et al. [57] noted that $H_3PO_4$, $HNO_3$, $H_2SO_4$ and HCl contains higher amounts of oxygen containing functional groups (such as carbonyl, quinone, ether, carboxylic anhydride, lactone, carboxylic acid, phenol and intercalated oxygen), while alkali modified NaOH and KOH have a greater carbon content [58].

Moreover, oxidizing agents Fe(III) and $KMnO_4$ significantly improved the specific area and pore size of BC [57]. According to Lin et al. [59], Fe-Mn modified BC increased pH, surface area, rich C content and a certain frame structure. In another case, Li et al. [47] reported that cadmium-binding BC increased surface area and increased oxygen-containing functional groups. BC modified by combining BC with nano-zero-valent iron increased the phosphorous and decreased the methane and nitrous oxide emission. Conversely, the nitrous oxide and carbon dioxide emission increased by phosphorous modified BC [60]. Moreover, *Thalia dealbata* BC modified with magnesium chloride ($MgCl_2$) showed higher surface area (110.6 $m^2$ $g^{-1}$) as compared to unmodified BC (7.1 $m^2$ $g^{-1}$) [61], mostly Mg-BC used to remediate soil from Cd and sulfamethoxazole. Additionally, oxygen gas flow through a plasma chamber that has dielectric barrier discharge under certain pressure converts plasma into oxygen ions, excited atoms, electrons and many reactive oxygen species [62], which react with BC functional groups and renders the oxygen plasma-activated BC more active as compared to chemically modified BC. The new insight of these chemical modification methods and BC properties can pave the way to improve the soil quality and sustainable agriculture. In a nutshell, the use of chemical and mineral modified BCs are most important due to their functional groups, which effectively interact with soil pollutants and improve soil quality by improving mineral uptake and soil gas exchange properties. Besides, higher adsorption capacity and greater carbon content was observed by using alkali-treated chemicals, while acid containing BCs have higher oxygen containing functional groups.

### 4.3. Biochar Modification with Nano-Particles

Nano-BC is currently used as it amalgamates the advantage of nanotechnology in BC modification. Decreasing BC particle size to the micro range (10–600 μm) led to enhancement of the available sites for adsorption, thus resulting in improved adsorption capacity [63–65], and further decreasing of BC size to nano-range (100 nm) increases its properties, such as biological effectiveness, adsorption potential, surface energy and surface to volume ratio [63,64,66]. The nano-particle size range is controlled by pyrolysis temperature or perhaps through exfoliation [66,67], flash heating [68], ultrasound vibrator [69], ball-milling [70], double disc milling [71], and hydrothermal reactions using agriculture residue as by product [72]. Metal nanoparticles increase the CEC, porosity, surface area and functional groups as compared to unmodified BCs. The impregnation can be achieved through pre-treating feedstock biomass with metal salts, supporting BC with functional nano-particles or forming composites with metal oxide nano-particles, which renders them high affinity to various pollutants [26,73]. In short, nano-modified BCs increased the adsorption capacity, biological effectiveness, functional groups efficiency, CEC, porosity, surface area, and surface to volume ratio of BCs as compared to pristine BCs.

### 4.4. Biochar Modification with Microbes

Biological modification of BC can be achieved by pre-treating the feedstock with anaerobic digestion and constructing a film on the internal and external surface of BC [43]. Digestion of waste material by aerobic and anaerobic bacteria increases the economy by producing bio-fertilizers and biofuel. BCs generated from bacterial digestion play a key role in improving hydrophobicity, CEC, and surface area, and are mostly utilized to remove heavy metals, pharmaceuticals, and pollutants from contaminated soils by developing biofilms [47,74]. BC-modified bioasphalt improves biomass utilization and enhances environmental protection [75]. In another case, Tao et al. [76] used the feasibility of combining anaerobic pyrolysis to produce biologically modified BC (corn stalk). Anaerobic ensiling improved biodegradation into non-fermented residue after 24 h of fermentation. This process's non-fermented material was pyrolyzed at 500 °C to synthesize biologically active BC. They observed that modified BC had higher oxygen-containing functional groups, mineral components and surface area as compared to pristine BC. Modified BC with anaerobic bacteria also enhanced the Cd (II) sorption up to 2.2 times as compared to unmodified BC. Moreover, chitosan (a sugar obtained from the outer skeleton of fish) modified bamboo BC, and improved the sorption of Pb on the chitosan modified BC and reduced its uptake (60%) and metal toxicity [77]. They suggested that the presence of chitosan on the BC surface increased alkalinity due to the presence of amine functional groups (weak bases), thus its amendment helps to overcome soil acidity. In addition, it also greatly lowered the C/N ratio of the BC up to 15.5 to 27.1, which make it a better choice for soil reclamation. Muhammad et al. [78] modified the wheat straw BC with soil indigenous-microbes and noted higher biosorption capacity (14.42 mg $g^{-1}$) as compared to pristine BC (6.28 mg $g^{-1}$) and wheat straw soil amendment (4.20 mg $g^{-1}$). They noted that biologically modified wheat straw BC had higher C, N, H content and surface area (6.5%), and porous morphology due to microbial degradation. In another case, a periphyton-based system comprising a BC column and a periphyton bioreactor presented a strong ability to entrap As(III) by biosorption. The calcite, -OH and -C, =O groups on the periphytic biofilm surfaces played an important role in As(III) entrapment and reduced the As toxicity and avoided microorganism poisoning [79]. Conclusively, biological modified biochar improved hydrophobicity, CEC, surface area, oxygen-containing functional groups, mineral components, and alkalinity and decreased the C/N ratio of BC as compared to pristine BC. These types of BC are mostly used to overcome heavy metal stress and other pollutants from soil.

## 5. Effect of Modified Biochar on Soil Quality

Modified BC can change the functional groups, pore size, pore structure, surface area and chemical properties of BC that play a key role in changing the soil quality by responding to its physical and hydraulic properties, nutrient profile, gas exchange properties, organic matter, soil pH, EC, CEC, and biological activities (bacteria, fungus and enzymes). The detail of soil quality with the addition of modified BC as soil amendment is briefly described below.

### 5.1. Effect of Modified Biochar on Soil Physical and Hydraulic Properties

BC as a soil amendment can influence the soil's physical and chemical properties, thus providing a means to improve soil fertility. According to Chen et al. [80], small-sized BC particles change the soil pore structure and affect soil aggregates' stability, porosity, fractal dimension, and the degree of anisotropy. Duan et al. [14] compared pristine BC with composite modified BC, particle-sized BC and acidified BC and noted that all BC modifications improved the soil water-stable aggregate contents. The soil-water aggregate content was higher with acid-modified BC at the 0–15 cm soil layer, being 1.45–1.80 and 1.59–1.96 times higher than unmodified BC and control treatments, respectively. They also found that acidified BC resulted in higher soil water holding capacity and soil infiltration rate. In another case, An et al. [15] mixed the $H_3PO_4$ and KOH modified BC of peach shell and pig manure BC with four dosages (0, 2, 3, and 8%) and noted that the $H_3PO_4$-modified BC had higher water retention as compared to KOH and pristine BCs. Instead, KOH showed reduced hydraulic functional groups on the surface of BC. They observed that pig manure BCs showed higher crack suppression intensity than functional BCs. They generally suggested a 5 to 8% BC dosage for improving water retention and reducing cracks. In conclusion, $H_3PO_4$, composite, peach shell, pig manure, particle-sized, and acidified BCs effectively improved the soil physical and hydraulic properties as compared to other modified BCs.

### 5.2. Effects of Modified Biochar on Soil Organic Matter

Soil organic matter is the component of soil that consists of plant and animal detritus at various stages of decomposition, soil microbe cells and tissues, and substances synthesized by soil microbes and play an important role in the survival of fauna and flora. The modified BC also increased the presence of the organic matter [19,20]. For instance, Li et al. [47] found that Cd-binding BC increased soil C/N ratios and soil organic carbon. Moreover, Wang et al. [81] noted higher organic matter accumulation of soil organic matter and organic carbon with the application of rice husk BC that was successively modified with NaOH, $HNO_3$ and dimethyl dithiocarbamate sodium (3% $w/w$) as compared to unmodified BC. Moreover, Moradi & Karimi [17] also noted higher soil organic matter and organic carbon with the soil amendment of Fe modified BC (2% $w/w$). Similarly, higher organic matter was also noted with the application of Fe-Mn BC (2% $w/w$) [46], S-BC (1% $w/w$), S-Fe BC (1% $w/w$) [49], iron-modified BC (3% $w/w$) [20], thiourea-modified BC (8% $w/w$), carrot pulp BC (8% $w/w$) [19], and iron-zinc oxide composite modified corn straw BC (3% $w/w$) [10] as compared to pristine BC (Table 1). In short, chemical and mineral modified BCs can play a key role in improving soil organic matter and C/N ratios.

**Table 1.** Efficacy of modified biochar in improving soil heath.

| Modified BC | BC Dose | Soil Properties | Reference |
|---|---|---|---|
| Mn oxide BC | 4% (*w*/*w*) | Increased adsorption capacity of As and pH of soil | [82] |
| Poultry manure BC | 5% (*w*/*w*) | Reduction of toxic Cr (VI) in soil | [83] |
| Sheep manure BC | 5% (*w*/*w*) | Reduction of toxic Cr (VI) in soil | [83] |
| Coconut shell BC | 5% (*w*/*w*) | Increased soil pH, CEC, bacteria, fungal and actinomyces counts, acid phosphatase, dehydrogenase, and urease while invertase was not affected. | [55] |
| S-modified rice husk BC | 5% (*w*/*w*) | Leachate total Hg concentrations decreased while increased Leachate total Hg removal | [84] |
| Magnetic BC | 2% (*w*/*w*) | Decreased Cd acid soluble fraction, and Cd reducible fraction, while no effect on Cd oxidizable fraction and Cd residual fraction were observed. Decreased Zn acid soluble fraction, and increased Zn residual fraction, while no effect on Zn oxidizable fraction and Zn reducible fraction were observed. Increased Pb acid soluble fraction, Pb reducible fraction, and Pb residual fraction, while no effect on Pb oxidizable fraction was observed. Increased Cu oxidizable fraction, Cu reducible fraction, and Cu residual fraction, while no effect on Cu acid soluble fraction was observed. | [33] |
| Fe-Mn BC | 2 wt% | Increased soil enzymes (UE, ALP/AKP, CAT, and POD) and the abundances of Proteobacteria and Firmicutes phyla. Moreover, decreased soil pH and bioavailable arsenic concentration | [59] |
| Fe-Mn BC | 2% (*w*/*w*) | Decreased pH, bioavailability of antimony and cadmium, while increased EC, available P, available K, total N, and organic matter. Moreover, UE and CAT did not affect but acid phosphatase activity was decreased | [46] |
| MgO-BC | 4.5 Mg $ha^{-1}$ | Increased available P | [12] |
| S BC | 1% (*w*/*w*) | Increased soil organic matter content and microbial community while decreased available Cd concentrations | [49] |
| S-Fe BC | 1% (*w*/*w*) | Increased soil organic matter content and microbial community while decreased available Cd concentrations | [49] |
| Fe-Mn-Ce modified BC | 2 wt% | Increased S-CAT, S-UE, S-POD and S-AKP/ALP activity. Moreover, microbial activities increased, especially Gemmatimonadaceae and Oxalobacteraceae families. | [21] |
| Multiple modified BC | 3% (*w*/*w*) | Decreased DTPA-extractable Cd, and Cu, pH, and available P while increased CEC, available K, organic matter, and dehydrogenase in soil | [85] |
| Rhamnolipid-modified BC | 2 wt% | Increased dehydrogenase activity, bacterial and fungal diversity indices, emission of $CO_2$ and $CH_4$, while reduced the emission of $N_2O$ | [86] |
| Iron-modified BC | 3% (*w*/*w*) | Decreased soil pH, available Fe, available As, available Cd, available Pb, S-CAT, and UE while increased total organic carbon | [20] |
| Thiourea-modified BC | 8% (*w*/*w*) | Increased soil pH, CEC, soil organic carbon, and EC. Decreased acid soluble fraction of Cu and reducible fraction of Cu while increased oxidizable fraction of Cu and residual fraction of Cu. Decreased acid soluble fraction of Zn while increased reducible fraction of Zn, oxidizable fraction of Zn and residual fraction of Zn | [19] |

**Table 1.** *Cont.*

| Modified BC | BC Dose | Soil Properties | Reference |
|---|---|---|---|
| Carrot pulp BC | 8% (*w*/*w*) | Increased soil pH, CEC, soil organic carbon, and EC. Decreased acid soluble fraction of Cu and reducible fraction of Cu while increased oxidizable fraction of Cu and residual fraction of Cu. Decreased acid soluble fraction of Zn while increased reducible fraction of Zn, oxidizable fraction of Zn and residual fraction of Zn | [19] |
| Fe–Mn–La-modified BC | 2 wt% | Decreased soil As concentration and increased S-CAT, S-UE, S-POD and S-AKP/ALP activity. Moreover, microbial activities increased especially γ-Proteobacteria, α-Proteobacteria, Acidobacteria, and Gemmatimonadetes. | [87] |
| Modified BC (rice husk BC and successively modified with NaOH, $HNO_3$ and dimethyl dithiocarbamate sodium) | 3% (*w*/*w*) | Increased soil pH, dissolved organic carbon, organic matter, K, Ca, Mg, Na, and available K, while reduced available P, CEC, DTPA-extractable Cd, Pb, Cu and Zn. Moreover, increased S-CAT and dehydrogenase | [81] |
| Fe modified BC | 2% (*w*/*w*) | Increased pH, soil organic carbon, microbial biomass carbon, dehydrogenase activity, Cd bound to organic matter, residual Cd and Cd bound to iron–manganese oxides while decreased Cd bound to carbonates, DTPA-extractable Cd, exchangeable Cd, and Cd mobility factor | [17] |
| *Brassica napus* BC-UV | 0.6% (*w*/*w*) | Increased soil pH and EC and decreased $CaCl_2$-extractable Cd | [16] |
| *Lolium perenne* BC-UV | 0.6% (*w*/*w*) | Increased soil pH and EC and decreased $CaCl_2$-extractable Cd | [16] |
| Iron-zinc oxide composite modified corn straw BC | 3% (*w*/*w*) | Increased pH, CEC, and dissolved organic carbon (DOC), bacterial community, i.e., Chao1 Shannon and Simpson while decreased DTPA-Cd | [10] |
| Particle size modified BC | 1% (*w*/*w*) | Soil soluble $K^+$, $Ca^{2+}$, and $Mg^{2+}$ increased while soil soluble $Na^+$ and $Na^+$ adsorption ratio decreased | [14] |
| Composite modified BC | 1% (*w*/*w*) | Soil soluble $K^+$, $Ca^{2+}$, and $Mg^{2+}$ increased while soil soluble $Na^+$ and $Na^+$ adsorption ratio decreased | [14] |

### 5.3. Gas Exchange Parameters and Nutrient Profile

Nutrient profile and gas exchange properties of soil can precisely elaborate its current quality status both for bacterial community and plant growth and development. Currently, Chen et al. [88] observed the application of nano-modified BC soil amendment reduced soil runoff, sediment and nutrient loss on sloped farmland. They reported that nano-BC (1.0%) reduced nitrate loss up to 13.6–59.8% in the sloping fields of Loess Plateau. They noted that the peak value of nitrate distribution in the soil profile moved downward, and the maximum was observed at 10 to 15 cm. In addition, Lin et al. [59] reported that Fe-Mn modified BC increased crystalline hydrous-oxide bound, soil redox-potential, amorphous-hydrous bound, while reducing the arsenic content in soil. In another case, Li et al. [47] found that Cd modified BC increased soil pH, organic carbon/total nitrogen and C/N, and decreased the nitrate nitrogen, available phosphorus, ammonium nitrogen and total nitrogen. Moreover, Qayyum et al. [30] reported that acidified BC has potential to improve phosphorous content in soil.

Additionally, Yin et al. [50] observed that $AlCl_3$-modified BC significantly improved the $PO_4^{3-}$ and $NO_3^-$ adsorption as compared to pristine BC. A 2% application of magnetic BC decreased the Zn acid-soluble fraction, and increased the Zn residual fraction, while having no effect on the Zn oxidizable fraction and Zn reducible fraction [33]. In addition, it increased the copper (Cu) oxidizable fraction, Cu reducible fraction, and Cu residual fraction, while having no effect on the Cu acid-soluble fraction. Gholami and Rahimi et al. [19] reported that thiourea-modified BC (8% *w*/*w*) decreased the acid-soluble fraction of Cu and a reducible fraction of Cu, while increasing the oxidizable fraction of Cu and a residual fraction of Cu. While a decreased acid-soluble fraction of Zn and increased reducible fraction of Zn was observed, an oxidizable fraction of Zn and a residual fraction of Zn was observed after thiourea-modified BC application. They also noted that carrot pulp BC modification decreased the acid soluble fraction of Cu and reducible fraction of Cu, while increasing the oxidizable fraction of Cu and residual fraction of Cu. Besides, a decreased acid soluble fraction of Zn and increased reducible fraction of Zn, oxidizable fraction of Zn and residual fraction of Zn was observed with carrot pulp BC modification. Moreover, Wang et al. [46] reported that available P, available K and total N decreased with soil amendment with modified Fe-Mn BC (2%). An increase in P content was found with the application of Mg-O modified and multiple modified BC as compared to pristine BC [12,85]. In another case, an iron-modified BC application (3%) reduced the availability of Fe in soil as compared to unmodified BC [20]. According to Wang et al. [81], modified rice husk BC increased the K, Ca, Mg, Na, and available K, while reducing the available P, Pb, Cu and Zn.

According to Duan et al. [14], particle size modified BC (1%) and composite modified BC (1%) increased the soil soluble $K^+$, $Ca^{2+}$, and $Mg^{2+}$, while decreasing the soil soluble $Na^+$ and $Na^+$ adsorption ratio. In addition, the treatment of 1% poultry manure BC at 550 °C changed the gas exchange properties by increasing the $CO_2$ emission from soil up to 91.4% as compared to poultry manure treatment [89] (Figure 2). Similarly, the increased emission of $CO_2$ and $CH_4$, and reduced emission of $N_2O$ was obtained with application of rhamnolipid-modified BC (2 wt%) [86] (Table 1). In brief, Fe-Mn, Cd, $AlCl_3$, magnetic, thiourea, carrot pulp, composite, Mg-O, rice husk and nano-modified BCs can improve the soil nutrient profile and prevent their leaching even in sloppy landscapes, while poultry manure and rhamnolipid modified BCs proved to be highly important for enhancing soil gas exchange properties.

### 5.4. Effects of Modified Biochar on Soil pH

Qayyum et al. [30] reported that BC acid modification with the addition of acid (1 N HCl) did not decrease the soil alkalinity. In another case, He et al. [90] modified rice straw BC with 1:1 $HNO_3/H_2SO_4$ and 15% $H_2O_2$, and used HCl-treated and unmodified BC as control to observe the acid paddy soil properties. They noted that pH buffering capacity and resistance to paddy soil acidification were effectively improved with the addition of $HNO_3/H_2SO_4$ and $H_2O_2$-modified BC. The surface functional groups were responsible for increasing the soil resistance to acidification. The generation of protonation of organic anions through dissociation of these functional groups retarded the decline in soil pH under acidification. They suggested that BC incorporation with $HNO_3/H_2SO_4$ led to higher carboxyl functional groups as compared to $H_2O_2$-modified BC, which is why it showed more soil resistance to acidification. The application of $HNO_3/H_2SO_4$-modified BC after the wet–dry cycle appeared to increase the pH of acidic paddy soil. Their work suggested that $HNO_3/H_2SO_4$-BC modification is a paramount solution to remediate acidic soil. The basic mechanism is that the weak acid functional group on the surface of BC mainly occurs in the form of organic anions under alkaline and neutral soils. Under soil acidification, these organic anions protonated with $H^+$ and converted into neutral molecules, inhibiting the soil acidification and declining the soil pH [91,92]. According to Yu et al. [82], soil pH increased with the application of Mn oxide modified BC (4%). Similar results were obtained with soil amendments of modified coconut shell BC (5%) [55], thiourea-modified BC (8%), carrot pulp BC (8%) [19], modified rice husk BC (successively modified with NaOH, $HNO_3$ and dimethyl dithiocarbamate sodium (3%) [81]. Iron-zinc oxide composite modified corn straw BC (3%) [10], Fe modified BC (2%) [17], *Brassica napus* BC-UV (0.6%) [16], *Lolium perenne* BC-UV (0.6%) [16], while decreases in pH were obtained with the use of modified Fe-Mn BC (2%) [59], Fe-Mn BC (2%) [46], multiple modified BC (3%) [85], and iron-modified BC (3%) [20] as compared to pristine BC. In short, acid modified BCs have pH buffering capacity and resistance to soil acidification due to the generation of higher protonation of organic anions, while metal-oxides and alkaline modified BCs showed a higher affinity to increased soil pH.

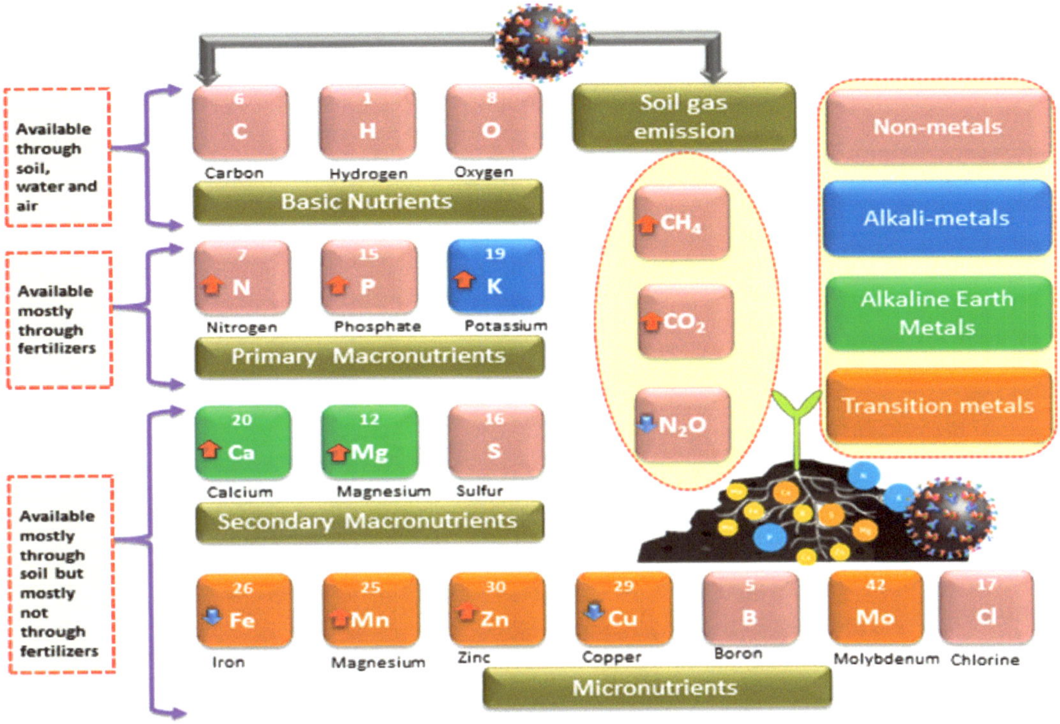


**Figure 2.** The effect of modified biochar on soil gas exchange and nutrient profile. The red color arrow represent increase, while blue arrow shows decrease in the nutrient content and gas emission.

### *5.5. Effects of Modified Biochar on Soil EC and CEC*

Qayyum et al. [30] reported that pristine BC and acidified BC had a non-significant effect on soil EC as compared to the control. The rice straw BC modified with HCl reduced the soil EC by inducing $CaCO_3$ dissolution, which added $H^+$ and $Ca^{2+}$ ions to soil, which replaced the $Na^+$ from the soil colloidal surface to facilitate the $Na^+$ leaching from the saline-sodic soil [40]. In another case, Zhang et al. [16] found that *Brassica napus* and *Lolium perenne* BC modified with UV (0.6%) showed higher soil EC as compared to pristine BC. Moreover, Gholami and Rahimi [19] also noted a higher EC with the application of modified carrot pulp BC (8%) as compared to unmodified BC. Similarly, the application of Fe-Mn modified BC also increased the soil EC [46]. In another study, thiourea-modified BC (8%) also increased soil EC as compared to untreated BC [19]. Additionally, Liu et al. [55] noted high CEC with the application of modified coconut shell BC (5% $w/w$). Similarly, an increase in EC was also found with the application of multiple modified BC (3% $w/w$) [85], thiourea-modified BC (8% $w/w$), carrot pulp BC (8% $w/w$) [19], rice modified BC [81], and iron-zinc oxide composite modified corn straw BC 3% ($w/w$) [10]. In summary, HCl, UV, Fe-Mn, thiourea and carrot pulp modified BCs can be used to increase the soil EC and CEC. However, new studies are vital to further elucidate the effect of modified BCs to enhance the soil EC and CEC.

*5.6. Effect of Modified Biochar on Soil Biological Activities*

Nano-BC application appreciably increased the microbial activity, such as the biomass of Bacteroidetes and Actinobacteria in soil, and decreased the activity of Proteobacteria, which was predominately present in contaminated soil [60]. In another study, Wu et al. [93] reported that calcium based magnetic BC significantly increased the abundance of microbial taxa and size of the bacterial population that thereby resulted in composition shift. Moreover, Lin et al. [59] reported that Fe-Mn modified BC increased the soil enzyme activity, with the exception of lower alkaline phosphatase activity and abundance of Bacteroidetes, while increasing the population of Firmicutes and Proteobacteria as compared to the control and sole addition of BC.

Moreover, Li et al. [47] found that BC aging by Cd increased microbial abundance, gram-negative bacteria, altered gram-positive/gram-negative bacteria, and decreased gram positive bacteria and microflora. In another study, Liu et al. [60] reported that nano-zero-valent iron increased the *Gemmatimanas* and *Sphingomonas* bacterial species, which resulted in higher nitrogen transformation and metabolism. They also increased the community structure of fungus composite with *Fusarium*. In addition, increased microbial activity (bacteria and fungus) was also found by the modified BC with sulfur (S) (1%) and S-Fe (1%) [49] and rhamnolipid-modified BC (2%) [86], Fe modified BC (2%) [17] and iron-zinc oxide composite modified corn straw BC (3%) [10].

In addition, Liu et al. [55] reported that coconut shell modified BC (5%) increased the bacterial and fungal community and soil enzymes including acid phosphatase, dehydrogenase, and urease, while non-significant changes were noted for invertase. Furthermore, Lin et al. [59] reported that soil amendment of Fe-Mn modified BC (2%) increased soil enzymes alkaline phosphatase/ ALP/AKP, urease (UE), peroxidase (POD), catalase (CAT) and the abundances of *Proteobacteria* and *Firmicutes* phyla. In another case, Fe-Mn-Ce modified BC (2%) increased S-CAT, S-UE, S-POD and S-AKP/ALP activity, and modified microbial activities especially increased the abundance of *Gemmatimonadaceae* and *Oxalobacteraceae* families as compared to pristine BC [21]. Similarly, Lin et al. [87] reported increased S-CAT, S-UE, S-POD and S-AKP/ALP activity and microbial population especially $\gamma$-Proteobacteria, $\alpha$-Proteobacteria, *Acidobacteria*, and *Gemmatimonadetes* with soil amendment of the Fe–Mn–La-modified BC (2 wt%). Moreover, Wen et al. [20] noted decreased activity of S-CAT, and UE with application of iron-modified BC (3%). Conversely, Wang et al. [46] reported that Fe-Mn BC (2%) had a non-significant effect on UE and CAT activities, while decreasing the phosphatase activity. In addition, dehydrogenase activity was increased with the application of multiple modified BC (3%) [85], rhamnolipid-modified BC (2%) [86], rice husk modified BC (3%) [81] and Fe modified BC (2%) [17] (Figure 3). The detailed study of soil enzymes and soil microbes changed through the application of modified-BC effectively improving the soil quality, so intensive research is direly needed to explore this field. To summarize, modified BCs effectively changed the microbial communities, increased microbial diversity and developed a microbial co-occurrence network, implied to increase soil dwelling behavior and soil ecosystem function related to nutrient- and C-cycling that results in better soil structure and quality.

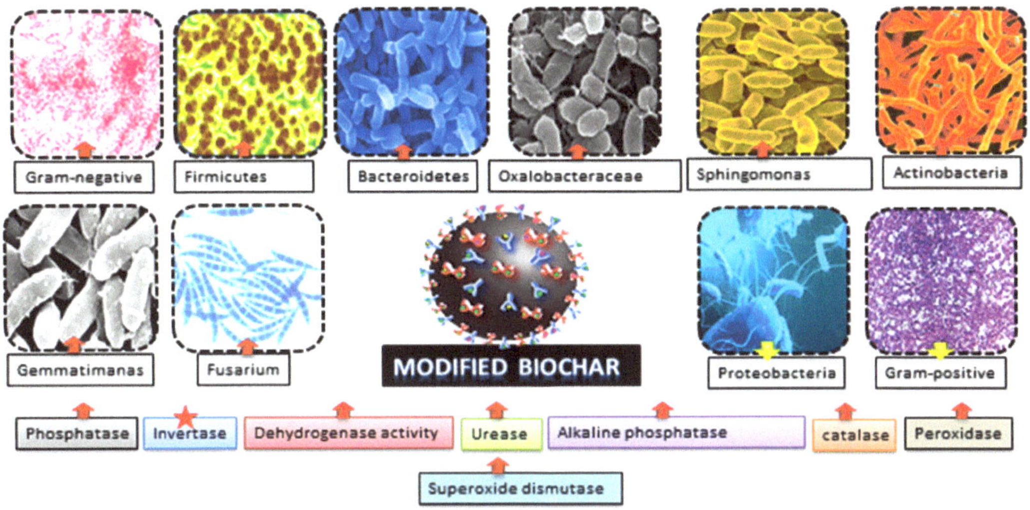


**Figure 3.** The effect of modified biochar on soil microbial community and enzymatic activity. The red arrows highlight the increase in activities, while yellow arrows show a decrease in specific activity. The star mentioned the non-significant changes.

## 6. Conclusions and Future Prospects

Modified BCs-based fertilizer addition enabled changes in the keystone taxa (algae and fungi), in the network topology structure, active participation in soil C- and nutrient-cycling, enhanced gas exchange attributes, improved soil productivity and properties including soil physical structure, enhanced soil hydraulic properties, altered soil pore structure, managed soil EC, CEC and pH, and combatted soil-based abiotic stresses. So, modified BCs soil amendments appeal from a carbon accounting and soil condition perspective. The modified BCs with mineral fertilizers, chemicals, nano-particles and biological material effectively improved the soil quality and can be used as an effective technique to promote plant growth and development. Further research is urgently required to determine how modified BCs influence the fate of the environment and to verify the proposed mechanisms involved in modifying BC properties for soil quality improvement. For instance, Li et al. [47] reported that BC modification risked the Cd desorption due to aging. The major disadvantage of ball milling modified BC is their finer size, which poses potential risk of ground water pollution. In addition, the stability of BC after pollutant absorption is also a main concern [43]. Recently, little effort has been made to explore the effect of modified BCs on soil born microbes and plants, so further studies are deliberately required to expose their environmental risk before their more intensive exploitations. Moreover, new development in the analysis of BCs properties and surface functional groups will no doubt be pivotal to uncovering the hidden dimensions of microbe interactions and soil nutrients, and pave the way for sustainable agriculture development. However, there are also some risks in using modified BCs that must be considered for developing BCs.

**Author Contributions:** Conceptualization, K.C.; structure, K.C. and A.H.; resources, A.H. and T.P.; data curation, J.T. and T.P.; writing—original draft preparation, A.H.; writing—review and editing, A.H. and J.T.; visualization, K.C. and A.H.; supervision, K.C.; project administration, K.C.; funding acquisition, K.C. All authors have read and agreed to the published version of the manuscript.

**Funding:** This study was funded by grants from the National Natural Science Foundation of China (31870420) and the Science and Technology Project of Guangdong Province No. 2021A0505030057.

**Data Availability Statement:** No new data were created or analyzed in this study. Data sharing is not applicable to this article.

**Acknowledgments:** We are thankful to anonymous reviewers for their constructive comments on the draft of the manuscript.

**Conflicts of Interest:** The authors declare no conflict of interest.

## References

1. Dickinson, D.; Balduccio, L.; Buysse, J.; Ronsse, F.; Van Huylenbroeck, G.; Prins, W. Cost-benefit analysis of using biochar to improve cereals agriculture. *Gcb Bioenergy* **2015**, *7*, 850–864. [CrossRef]
2. Changxun, G.; Zhiyong, P.; Shu'ang, P. Effect of biochar on the growth of *Poncirus trifoliata* (L.) Raf. seedlings in Gannan acidic red soil. *Soil Sci. Plant Nutr.* **2016**, *62*, 194–200. [CrossRef]
3. Bonanomi, G.; Ippolito, F.; Cesarano, G.; Nanni, B.; Lombardi, N.; Rita, A.; Saracino, A.; Scala, F. Biochar as plant growth promoter: Better off alone or mixed with organic amendments? *Front. Plant Sci.* **2017**, *8*, 1570. [CrossRef] [PubMed]
4. Jeffery, S.; Verheijen, F.G.; Kammann, C.; Abalos, D. Biochar effects on methane emissions from soils: A meta-analysis. *Soil Biol. Biochem.* **2016**, *101*, 251–258. [CrossRef]
5. Nair, V.D.; Nair, P.; Dari, B.; Freitas, A.M.; Chatterjee, N.; Pinheiro, F.M. Biochar in the agroecosystem–climate-change–sustainability nexus. *Front. Plant Sci.* **2017**, *8*, 2051. [CrossRef]
6. Hairani, A.; Osaki, M.; Watanabe, T. Effect of biochar application on mineral and microbial properties of soils growing different plant species. *Soil Sci. Plant Nutr.* **2016**, *62*, 519–525. [CrossRef]
7. Zhao, L.; Nan, H.; Kan, Y.; Xu, X.; Qiu, H.; Cao, X. Infiltration behavior of heavy metals in runoff through soil amended with biochar as bulking agent. *Environ. Pollut.* **2019**, *254*, 113114. [CrossRef]
8. Obia, A.; Mulder, J.; Martinsen, V.; Cornelissen, G.; Børresen, T. In situ effects of biochar on aggregation, water retention and porosity in light-textured tropical soils. *Soil Tillage Res.* **2016**, *155*, 35–44. [CrossRef]
9. Mahdi, Z.; El Hanandeh, A.; Yu, Q.J. Preparation, characterization and application of surface modified biochar from date seed for improved lead, copper, and nickel removal from aqueous solutions. *J. Environ. Chem. Eng.* **2019**, *7*, 103379. [CrossRef]
10. Yang, T.; Xu, Y.; Huang, Q.; Sun, Y.; Liang, X.; Wang, L.; Qin, X.; Zhao, L. An efficient biochar synthesized by iron-zinc modified corn straw for simultaneously immobilization Cd in acidic and alkaline soils. *Environ. Pollut.* **2021**, *291*, 118129. [CrossRef]
11. Liu, C.; Sun, B.; Zhang, X.; Liu, X.; Drosos, M.; Li, L.; Pan, G. The water-soluble pool in biochar dominates maize plant growth promotion under biochar amendment. *J. Plant Growth Regul.* **2021**, *40*, 1466–1476. [CrossRef]
12. Wu, L.; Wei, C.; Zhang, S.; Wang, Y.; Kuzyakov, Y.; Ding, X. MgO-modified biochar increases phosphate retention and rice yields in saline-alkaline soil. *J. Clean. Prod.* **2019**, *235*, 901–909. [CrossRef]
13. Wang, B.; Gao, B.; Fang, J. Recent advances in engineered biochar productions and applications. *Crit. Rev. Environ. Sci. Technol.* **2017**, *47*, 2158–2207. [CrossRef]
14. Duan, M.; Liu, G.; Zhou, B.; Chen, X.; Wang, Q.; Zhu, H.; Li, Z. Effects of modified biochar on water and salt distribution and water-stable macro-aggregates in saline-alkaline soil. *J. Soils Sediments* **2021**, *21*, 2192–2202. [CrossRef]
15. An, Y.; Lu, J.; Niu, R.; Li, M.; Zhao, X.; Huang, X.; Huang, H.; Garg, A.; Zhussupbekov, A. Exploring effects of novel chemical modification of biochar on soil water retention and crack suppression: Towards commercialization of production of biochar for soil remediation. *Biomass Convers. Biorefinery* **2021**, 1–14. [CrossRef]
16. Zhang, Y.; Chen, Z.; Chen, C.; Li, F.; Shen, K. Effects of UV-modified biochar derived from phytoremediation residue on Cd bioavailability and uptake in Coriandrum sativum L. in a Cd-contaminated soil. *Environ. Sci. Pollut. Res.* **2021**, *28*, 17395–17404. [CrossRef]
17. Moradi, N.; Karimi, A. Fe-Modified common reed biochar reduced cadmium (Cd) mobility and enhanced microbial activity in a contaminated calcareous soil. *J. Soil Sci. Plant Nutr.* **2021**, *21*, 329–340. [CrossRef]
18. Wani, I.; Ramola, S.; Garg, A.; Kushvaha, V. Critical review of biochar applications in geoengineering infrastructure: Moving beyond agricultural and environmental perspectives. *Biomass Convers. Biorefinery* **2021**, 1–29. [CrossRef]
19. Gholami, L.; Rahimi, G. Chemical fractionation of copper and zinc after addition of carrot pulp biochar and thiourea-modified biochar to a contaminated soil. *Environ. Technol.* **2021**, *42*, 3523–3532. [CrossRef]
20. Wen, E.; Yang, X.; Chen, H.; Shaheen, S.M.; Sarkar, B.; Xu, S.; Song, H.; Liang, Y.; Rinklebe, J.; Hou, D. Iron-modified biochar and water management regime-induced changes in plant growth, enzyme activities, and phytoavailability of arsenic, cadmium and lead in a paddy soil. *J. Hazard. Mater.* **2021**, *407*, 124344. [CrossRef]
21. Zhang, G.; Liu, X.; Gao, M.; Song, Z. Effect of Fe–Mn–Ce modified biochar composite on microbial diversity and properties of arsenic-contaminated paddy soils. *Chemosphere* **2020**, *250*, 126249. [CrossRef] [PubMed]
22. Jiang, Z.; Lian, F.; Wang, Z.; Xing, B. The role of biochars in sustainable crop production and soil resiliency. *J. Exp. Bot.* **2020**, *71*, 520–542. [CrossRef] [PubMed]
23. Dai, Y.; Zheng, H.; Jiang, Z.; Xing, B. Combined effects of biochar properties and soil conditions on plant growth: A meta-analysis. *Sci. Total Environ.* **2020**, *713*, 136635. [CrossRef]
24. Jeffery, S.; Memelink, I.; Hodgson, E.; Jones, S.; van de Voorde, T.F.; Martijn Bezemer, T.; Mommer, L.; van Groenigen, J.W. Initial biochar effects on plant productivity derive from N fertilization. *Plant Soil* **2017**, *415*, 435–448. [CrossRef]
25. Jeffery, S.; Abalos, D.; Prodana, M.; Bastos, A.C.; Van Groenigen, J.W.; Hungate, B.A.; Verheijen, F. Biochar boosts tropical but not temperate crop yields. *Environ. Res. Lett.* **2017**, *12*, 053001. [CrossRef]

26. Tan, X.F.; Liu, Y.G.; Gu, Y.L.; Xu, Y.; Zeng, G.M.; Hu, X.J.; Liu, S.B.; Wang, X.; Liu, S.M.; Li, J. Biochar-based nano-composites for the decontamination of wastewater: A review. *Bioresour. Technol.* **2016**, *212*, 318–333. [CrossRef]
27. Purakayastha, T.; Bera, T.; Bhaduri, D.; Sarkar, B.; Mandal, S.; Wade, P.; Kumari, S.; Biswas, S.; Menon, M.; Pathak, H. A review on biochar modulated soil condition improvements and nutrient dynamics concerning crop yields: Pathways to climate change mitigation and global food security. *Chemosphere* **2019**, *227*, 345–365. [CrossRef]
28. Zou, Z.; Fan, L.; Li, X.; Dong, C.; Zhang, L.; Zhang, L.; Fu, J.; Han, W.; Yan, P. Response of Plant Root Growth to Biochar Amendment: A Meta-Analysis. *Agronomy* **2021**, *11*, 2442. [CrossRef]
29. Tarin, M.W.K.; Fan, L.; Shen, L.; Lai, J.; Li, J.; Deng, Z.; Chen, L.; He, T.; Rong, J.; Zheng, Y. Rice straw biochar impact on physiological and biochemical attributes of Fokienia hodginsii in acidic soil. *Scand. J. For. Res.* **2020**, *35*, 59–68. [CrossRef]
30. Qayyum, M.F.; Haider, G.; Iqbal, M.; Hameed, S.; Ahmad, N.; ur Rehman, M.Z.; Majeed, A.; Rizwan, M.; Ali, S. Effect of alkaline and chemically engineered biochar on soil properties and phosphorus bioavailability in maize. *Chemosphere* **2021**, *266*, 128980. [CrossRef]
31. Jeffery, S. Biochar application to soil for climate change mitigation and crop production. *Asp. Appl. Biol.* **2018**, 125–132.
32. Blanco-Canqui, H. Biochar and soil physical properties. *Soil Sci. Soc. Am. J.* **2017**, *81*, 687–711. [CrossRef]
33. Lu, H.; Li, Z.; Gasco, G.; Mendez, A.; Shen, Y.; Paz-Ferreiro, J. Use of magnetic biochars for the immobilization of heavy metals in a multi-contaminated soil. *Sci. Total Environ.* **2018**, *622*, 892–899. [CrossRef] [PubMed]
34. Yao, Q.; Liu, J.; Yu, Z.; Li, Y.; Jin, J.; Liu, X.; Wang, G. Three years of biochar amendment alters soil physio-chemical properties and fungal community composition in a black soil of northeast China. *Soil Biol. Biochem.* **2017**, *110*, 56–67. [CrossRef]
35. Shakya, A.; Agarwal, T. Potential of Biochar for the Remediation of Heavy Metal Contaminated Soil. In *Biochar Applications in Agriculture and Environment Management*; Springer: Berlin/Heidelberg, Germany, 2020; pp. 77–98.
36. Rawat, J.; Saxena, J.; Sanwal, P. Biochar: A sustainable approach for improving plant growth and soil properties. In *Biochar—An Imperative Amendment for Soil and the Environment*; IntechOpen: London, UK, 2019; p. 10.
37. Weber, K.; Quicker, P. Properties of biochar. *Fuel* **2018**, *217*, 240–261. [CrossRef]
38. Zhao, B.; O'Connor, D.; Zhang, J.; Peng, T.; Shen, Z.; Tsang, D.C.; Hou, D. Effect of pyrolysis temperature, heating rate, and residence time on rapeseed stem derived biochar. *J. Clean. Prod.* **2018**, *174*, 977–987. [CrossRef]
39. Alghamdi, A.G. Biochar as a potential soil additive for improving soil physical properties—A review. *Arab. J. Geosci.* **2018**, *11*, 766. [CrossRef]
40. Sadegh-Zadeh, F.; Parichehreh, M.; Jalili, B.; Bahmanyar, M.A. Rehabilitation of calcareous saline-sodic soil by means of biochars and acidified biochars. *Land Degrad. Dev.* **2018**, *29*, 3262–3271. [CrossRef]
41. Juriga, M.; Šimanský, V. Effect of biochar on soil structure—Review. *Acta Fytotech. Zootech* **2018**, *21*, 11–19. [CrossRef]
42. Lyu, H.; Gao, B.; He, F.; Zimmerman, A.R.; Ding, C.; Huang, H.; Tang, J. Effects of ball milling on the physicochemical and sorptive properties of biochar: Experimental observations and governing mechanisms. *Environ. Pollut.* **2018**, *233*, 54–63. [CrossRef]
43. Islam, T.; Li, Y.; Cheng, H. Biochars and Engineered Biochars for Water and Soil Remediation: A Review. *Sustainability* **2021**, *13*, 9932. [CrossRef]
44. Yi, Y.; Huang, Z.; Lu, B.; Xian, J.; Tsang, E.P.; Cheng, W.; Fang, J.; Fang, Z. Magnetic biochar for environmental remediation: A review. *Bioresour. Technol.* **2020**, *298*, 122468. [CrossRef] [PubMed]
45. Lee, J.; Cho, W.-C.; Poo, K.-M.; Choi, S.; Kim, T.-N.; Son, E.-B.; Choi, Y.-J.; Kim, Y.M.; Chae, K.-J. Refractory oil wastewater treatment by dissolved air flotation, electrochemical advanced oxidation process, and magnetic biochar integrated system. *J. Water Process Eng.* **2020**, *36*, 101358. [CrossRef]
46. Wang, Y.-Y.; Ji, H.-Y.; Lyu, H.-H.; Liu, Y.-X.; He, L.-L.; You, L.-C.; Zhou, C.-H.; Yang, S.-M. Simultaneous alleviation of Sb and Cd availability in contaminated soil and accumulation in Lolium multiflorum Lam. After amendment with Fe–Mn-Modified biochar. *J. Clean. Prod.* **2019**, *231*, 556–564. [CrossRef]
47. Li, K.; Yin, G.; Xu, Q.; Yan, J.; Hseu, Z.-Y.; Zhu, L.; Lin, Q. Influence of Aged Biochar Modified by $Cd^{2+}$ on Soil Properties and Microbial Community. *Sustainability* **2020**, *12*, 4868. [CrossRef]
48. Mihoub, A.; Amin, A.E.-E.A.Z.; Motaghian, H.R.; Saeed, M.F.; Naeem, A. Citric Acid (CA)-Modified Biochar Improved Available Phosphorus Concentration and Its Half-Life in a P-Fertilized Calcareous Sandy Soil. *J. Soil Sci. Plant Nutr.* **2022**, *22*, 465–474. [CrossRef]
49. Wu, C.; Shi, L.; Xue, S.; Li, W.; Jiang, X.; Rajendran, M.; Qian, Z. Effect of sulfur-iron modified biochar on the available cadmium and bacterial community structure in contaminated soils. *Sci. Total Environ.* **2019**, *647*, 1158–1168. [CrossRef]
50. Yin, Q.; Ren, H.; Wang, R.; Zhao, Z. Evaluation of nitrate and phosphate adsorption on Al-modified biochar: Influence of Al content. *Sci. Total Environ.* **2018**, *631*, 895–903. [CrossRef]
51. Chen, T.; Luo, L.; Deng, S.; Shi, G.; Zhang, S.; Zhang, Y.; Deng, O.; Wang, L.; Zhang, J.; Wei, L. Sorption of tetracycline on H3PO4 modified biochar derived from rice straw and swine manure. *Bioresour. Technol.* **2018**, *267*, 431–437. [CrossRef]
52. Jin, J.; Li, S.; Peng, X.; Liu, W.; Zhang, C.; Yang, Y.; Han, L.; Du, Z.; Sun, K.; Wang, X. $HNO_3$ modified biochars for uranium (VI) removal from aqueous solution. *Bioresour. Technol.* **2018**, *256*, 247–253. [CrossRef]
53. Li, Y.; Wei, Y.; Huang, S.; Liu, X.; Jin, Z.; Zhang, M.; Qu, J.; Jin, Y. Biosorption of Cr (VI) onto Auricularia auricula dreg biochar modified by cationic surfactant: Characteristics and mechanism. *J. Mol. Liq.* **2018**, *269*, 824–832. [CrossRef]
54. Chen, Z.-L.; Zhang, J.-Q.; Huang, L.; Yuan, Z.-H.; Li, Z.-J.; Liu, M.-C. Removal of Cd and Pb with biochar made from dairy manure at low temperature. *J. Integr. Agric.* **2019**, *18*, 201–210. [CrossRef]

55. Liu, H.; Xu, F.; Xie, Y.; Wang, C.; Zhang, A.; Li, L.; Xu, H. Effect of modified coconut shell biochar on availability of heavy metals and biochemical characteristics of soil in multiple heavy metals contaminated soil. *Sci. Total Environ.* **2018**, *645*, 702–709. [CrossRef] [PubMed]
56. Li, L.; Liu, S.; Liu, J. Surface modification of coconut shell based activated carbon for the improvement of hydrophobic VOC removal. *J. Hazard. Mater.* **2011**, *192*, 683–690. [CrossRef]
57. Sajjadi, B.; Chen, W.-Y.; Egiebor, N.O. A comprehensive review on physical activation of biochar for energy and environmental applications. *Rev. Chem. Eng.* **2019**, *35*, 735–776. [CrossRef]
58. Wang, Y.; Zhong, B.; Shafi, M.; Ma, J.; Guo, J.; Wu, J.; Ye, Z.; Liu, D.; Jin, H. Effects of biochar on growth, and heavy metals accumulation of moso bamboo (*Phyllostachy pubescens*), soil physical properties, and heavy metals solubility in soil. *Chemosphere* **2019**, *219*, 510–516. [CrossRef]
59. Lin, L.; Li, Z.; Liu, X.; Qiu, W.; Song, Z. Effects of Fe-Mn modified biochar composite treatment on the properties of As-polluted paddy soil. *Environ. Pollut.* **2019**, *244*, 600–607. [CrossRef]
60. Liu, Z.; Tang, J.; Ren, X.; Schaeffer, S.M. Effects of phosphorus modified nZVI-biochar composite on emission of greenhouse gases and changes of microbial community in soil. *Environ. Pollut.* **2021**, *274*, 116483. [CrossRef]
61. Tao, Q.; Li, B.; Li, Q.; Han, X.; Jiang, Y.; Jupa, R.; Wang, C.; Li, T. Simultaneous remediation of sediments contaminated with sulfamethoxazole and cadmium using magnesium-modified biochar derived from Thalia dealbata. *Sci. Total Environ.* **2019**, *659*, 1448–1456. [CrossRef]
62. Gupta, R.K.; Dubey, M.; Kharel, P.; Gu, Z.; Fan, Q.H. Biochar activated by oxygen plasma for supercapacitors. *J. Power Sources* **2015**, *274*, 1300–1305. [CrossRef]
63. Taheran, M.; Naghdi, M.; Brar, S.K.; Knystautas, E.J.; Verma, M.; Ramirez, A.A.; Surampalli, R.Y.; Valéro, J.R. Adsorption study of environmentally relevant concentrations of chlortetracycline on pinewood biochar. *Sci. Total Environ.* **2016**, *571*, 772–777. [CrossRef] [PubMed]
64. Naghdi, M.; Taheran, M.; Pulicharla, R.; Rouissi, T.; Brar, S.K.; Verma, M.; Surampalli, R.Y. Pine-wood derived nanobiochar for removal of carbamazepine from aqueous media: Adsorption behavior and influential parameters. *Arab. J. Chem.* **2019**, *12*, 5292–5301. [CrossRef]
65. Lonappan, L.; Rouissi, T.; Das, R.K.; Brar, S.K.; Ramirez, A.A.; Verma, M.; Surampalli, R.Y.; Valero, J.R. Adsorption of methylene blue on biochar microparticles derived from different waste materials. *Waste Manag.* **2016**, *49*, 537–544. [CrossRef] [PubMed]
66. Sulaiman, G.M.; Mohammed, W.H.; Marzoog, T.R.; Al-Amiery, A.A.A.; Kadhum, A.A.H.; Mohamad, A.B. Green synthesis, antimicrobial and cytotoxic effects of silver nanoparticles using Eucalyptus chapmaniana leaves extract. *Asian Pac. J. Trop. Biomed.* **2013**, *3*, 58–63. [CrossRef]
67. Chen, L.; Chen, X.L.; Zhou, C.H.; Yang, H.M.; Ji, S.F.; Tong, D.S.; Zhong, Z.K.; Yu, W.H.; Chu, M.Q. Environmental-friendly montmorillonite-biochar composites: Facile production and tunable adsorption-release of ammonium and phosphate. *J. Clean. Prod.* **2017**, *156*, 648–659. [CrossRef]
68. Ramanayaka, S.; Vithanage, M.; Alessi, D.S.; Liu, W.-J.; Jayasundera, A.C.; Ok, Y.S. Nanobiochar: Production, properties, and multifunctional applications. *Environ. Sci. Nano* **2020**, *7*, 3279–3302. [CrossRef]
69. Oleszczuk, P.; Ćwikła-Bundyra, W.; Bogusz, A.; Skwarek, E.; Ok, Y.S. Characterization of nanoparticles of biochars from different biomass. *J. Anal. Appl. Pyrolysis* **2016**, *121*, 165–172. [CrossRef]
70. Fan, X.; Chang, D.W.; Chen, X.; Baek, J.-B.; Dai, L. Functionalized graphene nanoplatelets from ball milling for energy applications. *Curr. Opin. Chem. Eng.* **2016**, *11*, 52–58. [CrossRef]
71. Karinkanta, P.; Ämmälä, A.; Illikainen, M.; Niinimäki, J. Fine grinding of wood—Overview from wood breakage to applications. *Biomass Bioenergy* **2018**, *113*, 31–44. [CrossRef]
72. Guo, F.; Bao, L.; Wang, H.; Larson, S.L.; Ballard, J.H.; Knotek-Smith, H.M.; Zhang, Q.; Su, Y.; Wang, X.; Han, F. A simple method for the synthesis of biochar nanodots using hydrothermal reactor. *MethodsX* **2020**, *7*, 101022. [CrossRef]
73. Li, R.; Wang, J.J.; Zhou, B.; Zhang, Z.; Liu, S.; Lei, S.; Xiao, R. Simultaneous capture removal of phosphate, ammonium and organic substances by MgO impregnated biochar and its potential use in swine wastewater treatment. *J. Clean. Prod.* **2017**, *147*, 96–107. [CrossRef]
74. Wang, B.; Jiang, Y.-S.; Li, F.-Y.; Yang, D.-Y. Preparation of biochar by simultaneous carbonization, magnetization and activation for norfloxacin removal in water. *Bioresour. Technol.* **2017**, *233*, 159–165. [CrossRef] [PubMed]
75. Zhou, X.; Moghaddam, T.B.; Chen, M.; Wu, S.; Adhikari, S.; Xu, S.; Yang, C. Life cycle assessment of biochar modified bioasphalt derived from biomass. *ACS Sustain. Chem. Eng.* **2020**, *8*, 14568–14575. [CrossRef]
76. Tao, Q.; Li, B.; Chen, Y.; Zhao, J.; Li, Q.; Chen, Y.; Peng, Q.; Yuan, S.; Li, H.; Huang, R. An integrated method to produce fermented liquid feed and biologically modified biochar as cadmium adsorbents using corn stalks. *Waste Manag.* **2021**, *127*, 112–120. [CrossRef]
77. Zhou, Y.; Gao, B.; Zimmerman, A.R.; Fang, J.; Sun, Y.; Cao, X. Sorption of heavy metals on chitosan-modified biochars and its biological effects. *Chem. Eng. J.* **2013**, *231*, 512–518. [CrossRef]
78. Muhammad, H.; Wei, T.; Cao, G.; Yu, S.; Ren, X.; Jia, H.; Saleem, A.; Hua, L.; Guo, J.; Li, Y. Study of soil microorganisms modified wheat straw and biochar for reducing cadmium leaching potential and bioavailability. *Chemosphere* **2021**, *273*, 129644. [CrossRef]
79. Zhu, N.; Zhang, J.; Tang, J.; Zhu, Y.; Wu, Y. Arsenic removal by periphytic biofilm and its application combined with biochar. *Bioresour. Technol.* **2018**, *248*, 49–55. [CrossRef]

80. Chen, X.; Duan, M.; Zhou, B. Effects of biochar nanoparticles as a soil amendment on the structure and hydraulic characteristics of a sandy loam soil. *Soil Use Manag.* **2022**, *38*, 836–849. [CrossRef]
81. Wang, Y.; Ren, Q.; Li, T.; Zhan, W.; Zheng, K.; Liu, Y.; Chen, R. Influences of modified biochar on metal bioavailability, metal uptake by wheat seedlings (*Triticum aestivum* L.) and the soil bacterial community. *Ecotoxicol. Environ. Saf.* **2021**, *220*, 112370. [CrossRef]
82. Yu, Z.; Zhou, L.; Huang, Y.; Song, Z.; Qiu, W. Effects of a manganese oxide-modified biochar composite on adsorption of arsenic in red soil. *J. Environ. Manag.* **2015**, *163*, 155–162. [CrossRef]
83. Mandal, S.; Sarkar, B.; Bolan, N.; Ok, Y.S.; Naidu, R. Enhancement of chromate reduction in soils by surface modified biochar. *J. Environ. Manag.* **2017**, *186*, 277–284. [CrossRef] [PubMed]
84. O'Connor, D.; Peng, T.; Li, G.; Wang, S.; Duan, L.; Mulder, J.; Cornelissen, G.; Cheng, Z.; Yang, S.; Hou, D. Sulfur-modified rice husk biochar: A green method for the remediation of mercury contaminated soil. *Sci. Total Environ.* **2018**, *621*, 819–826. [CrossRef] [PubMed]
85. Wang, Y.; Zheng, K.; Zhan, W.; Huang, L.; Liu, Y.; Li, T.; Yang, Z.; Liao, Q.; Chen, R.; Zhang, C. Highly effective stabilization of Cd and Cu in two different soils and improvement of soil properties by multiple-modified biochar. *Ecotoxicol. Environ. Saf.* **2021**, *207*, 111294. [CrossRef] [PubMed]
86. Zhen, M.; Tang, J.; Li, C.; Sun, H. Rhamnolipid-modified biochar-enhanced bioremediation of crude oil-contaminated soil and mediated regulation of greenhouse gas emission in soil. *J. Soils Sediments* **2021**, *21*, 123–133. [CrossRef]
87. Lin, L.; Gao, M.; Liu, X.; Qiu, W.; Song, Z. Effect of Fe–Mn–La-modified biochar composites on arsenic volatilization in flooded paddy soil. *Environ. Sci. Pollut. Res.* **2021**, *28*, 49889–49898. [CrossRef]
88. Chen, X.; Zhou, B.; Wang, Q.; Tao, W.; Lin, H. Nano-biochar reduced soil erosion and nitrate loss in sloping fields on the Loess Plateau of China. *Catena* **2020**, *187*, 104346. [CrossRef]
89. Akanji, M.A.; Usman, A.R.; Al-Wabel, M.I. Influence of Acidified Biochar on $CO_2$–C Efflux and Micronutrient Availability in an Alkaline Sandy Soil. *Sustainability* **2021**, *13*, 5196. [CrossRef]
90. He, X.; Hong, Z.-N.; Shi, R.-Y.; Cui, J.-Q.; Lai, H.-W.; Lu, H.-L.; Xu, R.-K. The effects of $H_2O_2$-and $HNO_3/H_2SO_4$-modified biochars on the resistance of acid paddy soil to acidification. *Environ. Pollut.* **2021**, *293*, 118588. [CrossRef]
91. Shi, R.-Y.; Hong, Z.-N.; Li, J.-Y.; Jiang, J.; Kamran, M.A.; Xu, R.-K.; Qian, W. Peanut straw biochar increases the resistance of two Ultisols derived from different parent materials to acidification: A mechanism study. *J. Environ. Manag.* **2018**, *210*, 171–179. [CrossRef]
92. Shi, R.-Y.; Hong, Z.-N.; Li, J.-Y.; Jiang, J.; Baquy, M.A.-A.; Xu, R.-K.; Qian, W. Mechanisms for increasing the pH buffering capacity of an acidic Ultisol by crop residue-derived biochars. *J. Agric. Food Chem.* **2017**, *65*, 8111–8119. [CrossRef]
93. Wu, J.; Li, Z.; Huang, D.; Liu, X.; Tang, C.; Parikh, S.J.; Xu, J. A novel calcium-based magnetic biochar is effective in stabilization of arsenic and cadmium co-contamination in aerobic soils. *J. Hazard. Mater.* **2020**, *387*, 122010. [CrossRef] [PubMed]

applied sciences

MDPI

*Article*

# Thermodynamic Analysis of Air-Cycle Refrigeration Systems with Expansion Work Recovery for Compartment Air Conditioning

**Xiaoling Yang, Zhefeng Wang, Ze Zhang, Shuangtao Chen, Yu Hou and Liang Chen ***

State Key Laboratory of Multiphase Flow in Power Engineering, Xi'an Jiaotong University, Xi'an 710049, China; yxl1375055173@stu.xjtu.edu.cn (X.Y.); w1z2f3@stu.xjtu.edu.cn (Z.W.); zhangze@xjtu.edu.cn (Z.Z.); stchen.xjtu@xjtu.edu.cn (S.C.); yuhou@mail.xjtu.edu.cn (Y.H.)

* Correspondence: liangchen@mail.xjtu.edu.cn; Tel.: +86-029-8266-4921

**Abstract:** As a requirement for sustainable development, air-cycle refrigeration has received wide attention as a candidate for environmentally friendly air conditioning technology. In this study, the thermodynamic performance of air refrigeration cycles is investigated in compartment air conditioning. The effects of compressor efficiency, expander efficiency, ambient humidity, all-fresh-air supply and ambient pressure on the cycle performance are presented. The effects of compressor arrangement in the high-pressure cycle and the low-pressure cycle are compared. An open-loop high-pressure cycle has a larger COP than that of an open-loop low-pressure cycle but requires larger heat exchange. The performance of air refrigeration cycles with full fresh air is studied, and the influence of fresh air is discussed. Schemes for condensed water recirculation with wet compression are proposed, which can improve the COPs of open-loop low-pressure cycles by 44.7%, 48.8% and 48.4%. In the air conditioning of plateau trains, open-loop high-pressure cycles have slightly lower COPs, but they can supply air with elevated pressure and oxygen concentration.

**Keywords:** air cycle; reverse Brayton refrigerator; fresh air conditioner; COP

**Citation:** Yang, X.; Wang, Z.; Zhang, Z.; Chen, S.; Hou, Y.; Chen, L. Thermodynamic Analysis of Air-Cycle Refrigeration Systems with Expansion Work Recovery for Compartment Air Conditioning. *Appl. Sci.* **2022**, *12*, 5287. https://doi.org/10.3390/app12105287

Academic Editors: Mohammad Mahmudur Rahman, Dibyendu Sarkar, Rupali Datta and Prafulla Kumar Sahoo

Received: 26 April 2022
Accepted: 15 May 2022
Published: 23 May 2022



## 1. Introduction

The reverse Brayton refrigerator has been employed in space refrigeration [1,2], aircraft environmental control systems [3,4], superconducting [5,6], comparative power generation systems [7] and cryogenic refrigeration. Due to the increasing attention to environmental issues, air-cycle refrigerators based on the reverse-Brayton refrigeration cycle have received widespread attention as one of the most promising long-term solutions to achieve sustainable refrigeration [8–14].

In recent decades, one of the most pressing concerns for air conditioning has been indoor air quality, particularly with the emergence of epidemics such as the SARS virus and the novel coronavirus (SARS-CoV-2), and the fresh-air conditioner has been paid wide attention as an effective means to prevent virus accumulation in rooms. In traditional fresh-air systems, there is a return-air heat exchanger or mixer, which may pollute the fresh air. In order to improve air quality and avoid cross-infection, the all-fresh-air system has become necessary for many public buildings. This dedicated outdoor air system (DOAS), as a fresh-air system with no return air and low supplied temperature, has received much attention since its emergence [15–17]. At present, the DOAS has been used in public buildings such as hospitals and schools. The air-cycle refrigeration system offers the direct treatment of outdoor air, including the elimination of harmful substances such as viruses and pollutants, which makes it a promising candidate for a DOAS.

In high-altitude trains, the ambient pressure and density of ambient air are low, especially when the train travels from the plain to the plateau, and the variation in ambient pressure makes people uncomfortable. The problem is solved by maintaining higher indoor air pressure. High supplied-pressure is required to maintain indoor pressure, which is almost impossible to achieve in traditional air conditioning systems. In contrast, the air refrigeration system can achieve high supplied-pressure and recover the pressure energy of indoor exhaust air. In addition, the high-pressure air in an air-refrigeration cycle can be used to generate high-concentration oxygen.

In a reverse Brayton refrigerator with a co-axial configuration of the turbo-expander and compressor, the expansion work can be directly recovered into the compression process. Both Li [18] and Catalano [9] proposed a scheme to integrate a compressor and an expander in order to recover the expansion work and improve the cycle performance. Hou, Li and Zhang [10] studied open-loop air-cycle refrigeration using seawater cooling on ships and found that the efficiency of the compressor and expander has a critical influence on the system COP. Hou and Zhang [11] further studied a low-temperature open-loop reverse Brayton-cycle air refrigerator with circulating water cooling and analyzed the impact of the compressor efficiency, expander efficiency and expander inlet temperature on the system COP. With circulating water cooling, the cycle had a higher COP. However, these two studies can only be applied to situations where a water source is available and cannot be directly applied to general, compartment air conditioning. Parker [19] established a calculation model for the open-loop reverse Brayton-cycle air refrigerator by assuming the efficiency of components and studying the system performance under variable conditions.

Most of the previous studies of reverse Brayton-cycle refrigerators are mainly focused on cryogenic applications [20–26]. Although some studies [10,11,27] considered the effect of humidity, there is no comparative study of wet-air refrigeration cycles with different configurations. The all-fresh-air air-conditioner based on the reverse Brayton cycle has not been studied yet. The characteristics of the air refrigeration cycle under low ambient pressure have not been studied either. This study aimed to investigate the characteristics of the reverse Brayton-cycle air refrigerator for room temperature cooling. The influence of the humidity on the system COP is analyzed, and a scheme to improve the COP of the system by recycling condensed water is further proposed. The characteristics of the air refrigeration cycle utilized in all-fresh-air systems and low ambient pressure are studied.

## 2. Theoretical Analysis of Air Refrigeration Cycles

Air refrigeration cycles can be configured with three types of reverse Brayton cycle, including the closed cycle, the open-loop low-pressure cycle (TC cycle) and the open-loop high-pressure cycle (CT cycle), as shown in Figure 1. In the open-loop low-pressure cycle, air of ambient temperature can directly expand to low-temperature air of sub-ambient pressure, and the hot-side heat exchanger is not used. After the cooling power is consumed in a cold-side heat exchanger, the air is compressed and discharged into the outside. In the open-loop high-pressure cycle, the air after expansion is directly supplied to the room, and the cold-side heat exchanger is not used. The condensed water can be removed in the hot-side heat exchanger or after the expander in the case of high-humidity ambient air, and low-temperature, dry air is supplied to the room.

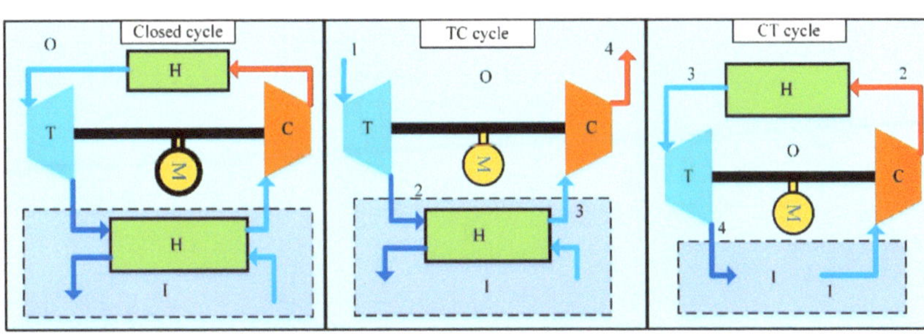


**Figure 1.** Schematic of air refrigeration cycles. T: turbo-expander; C: compressor; H: heat exchanger; I: indoor; O: outdoor.

The coefficient of performance (COP) of the cycle is defined as the ratio of the cooling capacity to the power consumption of the system. In order to simplify the calculation, the pressure drop in the pipes and the heat exchanger is ignored. The compression ratio and expansion ratio ($Er$) in the cycle are the same, and the working fluid is air which is considered an ideal gas.

In the closed cycle, the output shaft power $W_1$ and input shaft power of the cycle $W_2$ are calculated as Equations (1) and (2), respectively.

$$W_T = q_m \frac{k}{k-1} RT_{\mathrm{AC}}\left[1 - (Er)^{\frac{1-k}{k}}\right] \times \eta_T \tag{1}$$

$$W_C = q_m \frac{k}{k-1} RT_{\mathrm{CI}}\left[(Er)^{\frac{k-1}{k}} - 1\right] / \eta_C \tag{2}$$

The cooling capacity of cycle is determined by Equation (3):

$$Q_c = q_m(h_{\mathrm{CI}} - h_{EO}) = q_m(h_{\mathrm{CI}} - h_{\mathrm{AC}} + h_{AC} - h_{\mathrm{EO}}) = q_m\left[h_{\mathrm{CI}} - h_{\mathrm{AC}} + \frac{k}{k-1} RT_{\mathrm{AC}}\left(1 - Er^{\frac{1-k}{k}}\right) \times \eta_T\right] \tag{3}$$

The COP of the cycle is calculated by Equation (4):

$$\mathrm{COP}_{\mathrm{Closedcycle}} = \eta_C \frac{(T_{\mathrm{CI}} - T_{\mathrm{AC}}) + \eta_T T_{\mathrm{AC}}\left(1 - Er^{\frac{1-k}{k}}\right)}{\left[T_{\mathrm{CI}} Er^{\frac{k-1}{k}} - T_{\mathrm{CI}} - \eta_C \eta_T \eta_M T_{\mathrm{AC}}\left(1 - Er^{\frac{1-k}{k}}\right)\right]} = \eta_C \frac{\eta_T T_{\mathrm{AC}}\left(Er^{\frac{k-1}{k}} - 1\right) - (T_{\mathrm{AC}} - T_{\mathrm{CI}}) Er^{\frac{k-1}{k}}}{\left(T_{\mathrm{CI}} Er^{\frac{k-1}{k}} - \eta_T \eta_C \eta_M T_{\mathrm{AC}}\right)\left(Er^{\frac{k-1}{k}} - 1\right)} \tag{4}$$

Similarly, the COP of the open-loop low-pressure cycle and high-pressure cycle is determined by Equations (5) and (6):

$$\mathrm{COP}_{\mathrm{TCcycle}} = \eta_C \frac{(T_{\mathrm{CI}} - T_{\mathrm{O}}) + \eta_T T_{\mathrm{O}}\left(1 - Er^{\frac{1-k}{k}}\right)}{\left[T_{\mathrm{CI}} Er^{\frac{k-1}{k}} - T_{\mathrm{CI}} - \eta_C \eta_T \eta_M T_{\mathrm{O}}\left(1 - Er^{\frac{1-k}{k}}\right)\right]} = \eta_C \frac{\eta_T T_{\mathrm{O}}\left(Er^{\frac{k-1}{k}} - 1\right) - Er^{\frac{k-1}{k}}(T_{\mathrm{O}} - T_{\mathrm{CI}})}{\left(T_{\mathrm{CI}} Er^{\frac{k-1}{k}} - \eta_C \eta_T \eta_M T_{\mathrm{O}}\right)\left(Er^{\frac{k-1}{k}} - 1\right)} \tag{5}$$

$$\mathrm{COP}_{\mathrm{CTcycle}} = \eta_C \frac{(T_{\mathrm{I}} - T_{\mathrm{AC}}) + \eta_T T_{\mathrm{AC}}\left(1 - Er^{\frac{1-k}{k}}\right)}{\left[T_{\mathrm{I}} Er^{\frac{k-1}{k}} - T_{\mathrm{I}} - \eta_C \eta_T \eta_M T_{\mathrm{AC}}\left(1 - Er^{\frac{1-k}{k}}\right)\right]} = \eta_C \frac{\eta_T T_{\mathrm{AC}}\left(Er^{\frac{k-1}{k}} - 1\right) - (T_{\mathrm{AC}} - T_{\mathrm{I}}) Er^{\frac{k-1}{k}}}{\left(T_{\mathrm{I}} Er^{\frac{k-1}{k}} - \eta_C \eta_T \eta_M T_{\mathrm{AC}}\right)\left(Er^{\frac{k-1}{k}} - 1\right)} \tag{6}$$

where $T_{\mathrm{AC}}$ is the air temperature after the hot-side heat exchanger, and $T_{\mathrm{CI}}$ is the air temperature after the cold-side heat exchanger. $T_{\mathrm{I}}$ is the indoor temperature, and $T_{\mathrm{O}}$ is the outdoor temperature.

It is clear that the mechanical efficiency $\eta_{\mathrm{M}}$, compressor isentropic efficiency $\eta_{\mathrm{C}}$ and turbo-expander isentropic efficiency $\eta_{\mathrm{T}}$ determine the system performance. Moreover, the expander efficiency has a relatively greater impact, which not only affects the output work of the expander, but also determines the enthalpy drop in the expander and hence the cooling capacity of the system. Another factor affecting system performance is the expansion ratio $Er$, or the compression ratio. As the expansion ratio increases, the enthalpy drop in the expander increases, leading to an increase in the cooling capacity of the system. However, the raised expansion ratio also increases the power consumption, and the value should be optimized to obtain the maximum COP of the cycle.

By taking the partial derivative of COP with respect to $Er^{(k-1)/k}$, the optimal $Er$ can be obtained by making the derivative equal to zero. It can be found that the optimal $Er$ satisfies Equation (7):

$$Er^{\frac{k-1}{k}}{}_{\mathrm{Closedcycle,\ opt}} = \frac{\eta_T \frac{T_{\mathrm{AC}}}{T_{\mathrm{CI}}} + \sqrt{\left(\eta_T \frac{T_{\mathrm{AC}}}{T_{\mathrm{CI}}}\right)^2 - \left(1 - \frac{T_{\mathrm{AC}}}{T_{\mathrm{CI}}} + \eta_T \frac{T_{\mathrm{AC}}}{T_{\mathrm{CI}}}\right)\left[\eta_T \frac{T_{\mathrm{AC}}}{T_{\mathrm{CI}}} + \eta_C \eta_T \eta_M \frac{T_{\mathrm{AC}}}{T_{\mathrm{CI}}}\left(\frac{T_{\mathrm{AC}}}{T_{\mathrm{CI}}} - 1\right)\right]}}{1 - \frac{T_{\mathrm{AC}}}{T_{\mathrm{CI}}} + \eta_T \frac{T_{\mathrm{AC}}}{T_{\mathrm{CI}}}} \tag{7}$$

Similarly, the optimal expansion ratios of the open-loop low-pressure and high-pressure cycles can be obtained by Equations (8) and (9):

$$Er_{\text{TCcycle, opt}} = \left[ \frac{\eta_T \frac{T_O}{T_{CI}} + \sqrt{\left(\eta_T \frac{T_O}{T_{CI}}\right)^2 - \left(1 - \frac{T_O}{T_{CI}} + \eta_T \frac{T_O}{T_{CI}}\right)\left[\eta_T \frac{T_O}{T_{CI}} + \eta_C \eta_T \eta_M \frac{T_O}{T_{CI}}\left(\frac{T_O}{T_{CI}} - 1\right)\right]}}{1 - \frac{T_O}{T_{CI}} + \eta_T \frac{T_O}{T_{CI}}} \right]^{\frac{k}{k-1}} \tag{8}$$

$$Er_{\text{CTcycle, opt}} = \left[ \frac{\eta_T \frac{T_{AC}}{T_I} + \sqrt{\left(\eta_T \frac{T_{AC}}{T_I}\right)^2 - \left(1 - \frac{T_{AC}}{T_I} + \eta_T \frac{T_{AC}}{T_I}\right)\left[\eta_T \frac{T_{AC}}{T_I} + \eta_C \eta_T \eta_M \frac{T_{AC}}{T_I}\left(\frac{T_{AC}}{T_I} - 1\right)\right]}}{1 - \frac{T_{AC}}{T_I} + \eta_T \frac{T_{AC}}{T_I}} \right]^{\frac{k}{k-1}} \tag{9}$$

When $T_I$ = 300.15 K, $T_O$ = 308.15 K, $T_{AC}$ = 313.15 K, $T_{CI}$ = 295.15 K, $\eta_T$ = 0.8, $\eta_C$ = 0.75, $\eta_M$ = 0.99, it can be obtained that $Er_{\text{Closed cycle, opt}}$ = 2.27, $\text{COP}_{\text{Closed cycle, max}}$ = 0.659. For the open-loop low-pressure cycle, it can be obtained that $Er_{\text{TC cycle, opt}}$ = 1.96, $\text{COP}_{\text{TC cycle, max}}$ = 0.739. For the open-loop high-pressure cycle it can be obtained that $\text{Er}_{\text{CT cycle, opt}}$ = 1.94, $\text{COP}_{\text{CT cycle, max}}$ = 0.743. Due to the temperature difference in the heat exchanger, $T_{CI}$ is lower than the indoor temperature $T_I$, and $T_{AC}$ is higher than the outdoor temperature $T_O$, leading to a lower COP of the closed cycle than that of the open-loop cycles.

As the discharge pressure and temperature for one-stage compression is relatively high, a two-stage compression with intercooling can be employed to reduce the compression power. By integrating the expander with one of the compressors, the matching characteristics of the aerodynamic performance can be improved while the expansion work is recovered. Therefore, the TC cycle can be modified to the TC-C cycle and C-TC cycle, and the CT cycle can be modified to the C-CT cycle and CT-C cycle, as shown in Figure 2.

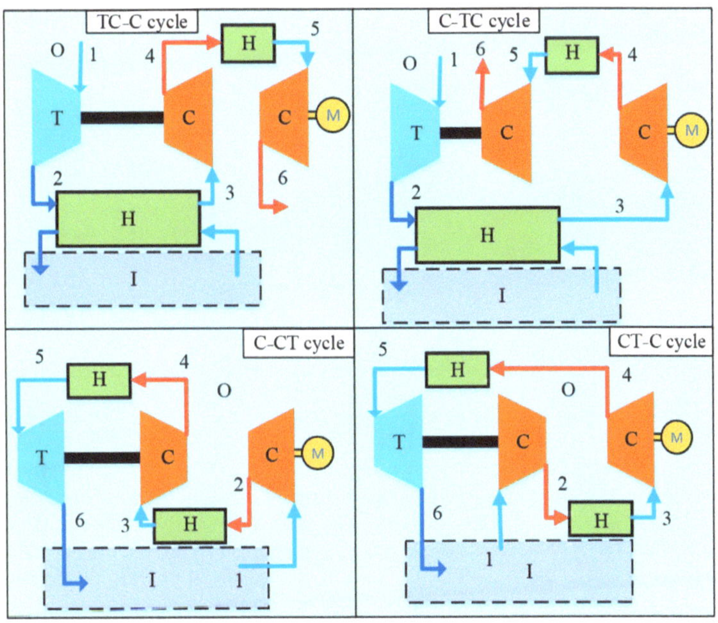


**Figure 2.** Schematic of open-loop air refrigeration cycles with two-stage compressor.

The COP of the two open-loop low-pressure refrigeration cycles (TC-C cycle and C-TC cycle) can be determined by Equations (10) and (11), respectively:

$$\text{COP}_{\text{TC-Ccycle}} = \eta_C \frac{\eta_T \frac{T_O}{T_{AC}}\left(1 - Er^{\frac{1-k}{k}}\right) - \frac{T_O - T_{CI}}{T_{AC}}}{\left(Er^{\frac{k-1}{k}} - \eta_C \eta_T \eta_M \frac{T_O}{T_{CI}}\right)\left(Er^{\frac{k-1}{k}} - 1\right)} \left[\eta_C \eta_T \eta_M \frac{T_O}{T_{CI}}\left(Er^{\frac{k-1}{k}} - 1\right) + Er^{\frac{k-1}{k}}\right] \tag{10}$$

$$\mathrm{COP}_{\mathrm{C-TCcycle}} = \eta_C \frac{\eta_T \frac{T_O}{T_{CI}}\left(1 - Er^{\frac{1-k}{k}}\right) - \frac{T_O - T_{CI}}{T_{CI}}}{\left(Er^{\frac{k-1}{k}} - \eta_C\eta_T\eta_M \frac{T_O}{T_{AC}}\right)\left(Er^{\frac{k-1}{k}} - 1\right)} \left[\eta_C\eta_T\eta_M \frac{T_O}{T_{AC}}\left(Er^{\frac{k-1}{k}} - 1\right) + Er^{\frac{k-1}{k}}\right] \tag{11}$$

The COP of the two open-loop high-pressure cycles (C-CT cycle and CT-C cycle) can be determined by Equations (12) and (13), respectively:

$$\mathrm{COP}_{\mathrm{C-CTcycle}} = \eta_C \frac{\eta_T \frac{T_{AC}}{T_I}\left(1 - Er^{\frac{1-k}{k}}\right) - \frac{T_{AC} - T_I}{T_I}}{\left(Er^{\frac{k-1}{k}} - \eta_C\eta_T\eta_M\right)\left(Er^{\frac{k-1}{k}} - 1\right)} \left[\eta_C\eta_T\eta_M\left(Er^{\frac{k-1}{k}} - 1\right) + Er^{\frac{k-1}{k}}\right] \tag{12}$$

$$\mathrm{COP}_{\mathrm{CT-Ccycle}} = \eta_C \frac{\eta_T\left(1 - Er^{\frac{1-k}{k}}\right) - \frac{T_{AC} - T_I}{T_{AC}}}{\left(Er^{\frac{k-1}{k}} - \eta_C\eta_T\eta_M \frac{T_{AC}}{T_I}\right)\left(Er^{\frac{k-1}{k}} - 1\right)} \left[\eta_C\eta_T\eta_M\left(Er^{\frac{k-1}{k}} - 1\right) + Er^{\frac{k-1}{k}}\right] \tag{13}$$

In the above case, where $T_I$ = 300.15 K, $T_O$ = 308.15 K, $T_{AC}$ = 313.15 K, $T_{CI}$ = 295.15 K, $\eta_T$ = 0.8, $\eta_C$ = 0.75, $\eta_M$ = 0.99, variations of COP with expansion ratio for various cycles are shown in Figure 3. It can be observed that the COP of the closed cycle is lower than that of the open-loop cycles due to the loss caused by the finite temperature difference in the heat exchanger. The four two-stage open-loop cycles have higher COPs, especially when the expansion ratio is larger than 1.7.

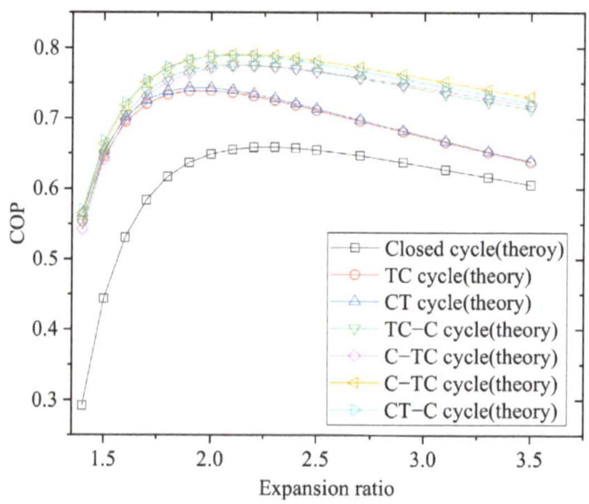


**Figure 3.** Variations in COP as a function of expansion ratio for various air refrigeration cycles.

However, in a practical cycle, the working medium in the above open-loop cycles is wet air, and the energy loss caused by the condensed water cannot be ignored. When the expansion ratio is relatively large, the air temperature after the compressor and the expander may be lower than the dew point, and the water vapor is condensed. In order to control the humidity, the air containing condensed water cannot be directly supplied to the room. As the condensed water is removed, the mass flow rate is reduced, resulting in the loss of cooling capacity. Therefore, the cycle performance in a practical air refrigerator depends on the humidity and pressure of the ambient air, in addition to the losses in expanders and compressors, which are discussed in the following section. In order to consider the effect of air temperature and absolute humidity, the enthalpy of wet air can be calculated by Equation (14):

$$h_{\mathrm{wet}} = c_{p,a}T + d\left(c_{p,v}T + L\right) \tag{14}$$

where $c_{p,a}$ is the specific heat at constant pressure of dry air, which is mostly related to temperature, and $c_{p,v}$ is the specific heat at constant pressure of water vapor. $L$ is the latent heat of the water vapor. The above data can be obtained from the NIST REFPROP database to fit the corresponding correlations; $d$ is the absolute humidity of the wet air.

The following part takes the TC cycle as an example to illustrate the thermodynamic analysis on the actual system performance of the open-loop cycle. Although the internal expansion process in the turbo-expander is too fast for the water to condense, the moist air in the downstream pipeline gradually reaches the equilibrium state as the water vapor is condensed. Therefore, the latent heat of vapor condensation needs to be considered. In the calculation, and it is assumed that no water is condensed inside the expander, and the influence of water on the isentropic efficiency of the expander is ignored. The calculation process is shown in Figure 4. In the process, the outlet temperature of the expander $T_{EO}$ and the supplied temperature $T_{sup}$ are solved iteratively, according to the change in enthalpy.

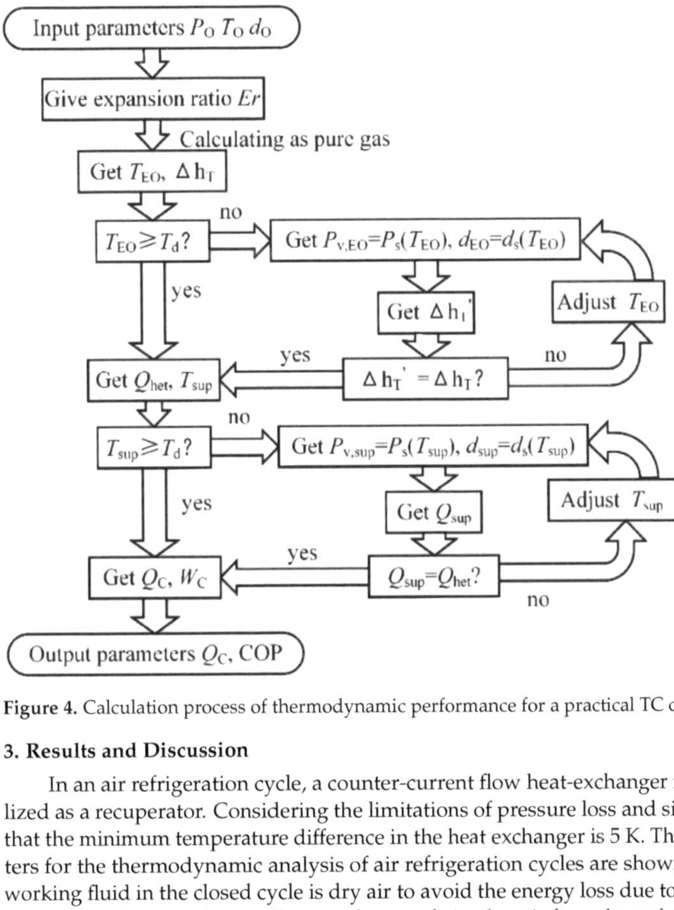


**Figure 4.** Calculation process of thermodynamic performance for a practical TC cycle.

## 3. Results and Discussion

In an air refrigeration cycle, a counter-current flow heat-exchanger is commonly utilized as a recuperator. Considering the limitations of pressure loss and size, it is assumed that the minimum temperature difference in the heat exchanger is 5 K. The design parameters for the thermodynamic analysis of air refrigeration cycles are shown in Table 1. The working fluid in the closed cycle is dry air to avoid the energy loss due to condensation of water vapor in the moist air. The open-loop cycle intakes air from the ambient environment, which contains water vapor. When the outlet temperature of the expander is lower than the dew point, the water vapor is condensed, which has an impact on the outlet temperature of the expander. The influence of the latent heat of water condensation is considered in the calculation of the outlet temperature of the expander.

**Table 1.** Ambient conditions and design parameters of air refrigeration cycles.

| Work Fluid | Ambient Pressure | Dry Bubble Temperature (Indoor) | Wet Bubble Temperature (Indoor) |
|---|---|---|---|
| Moist Air | 101.325 kPa | 300.15 K | 292.15 K |
| **Dry Bubble Temperature (Outdoor)** | **Wet Bubble Temperature (Outdoor)** | **Isentropic Efficiency (Compressor)** | **Isentropic Efficiency (Expander)** |
| 308.15 K | 297.15 K | 75% | 80% |
| **Pressure Drop (Water Separator)** | **Pressure Drop (Heat Exchanger)** | **Mechanical Efficiency** | **Temperature Difference (Heat Exchanger)** |
| 1.0 kPa | 1.0 kPa | 99% | ≥5 K |

For closed air refrigeration cycles with varied expander inlet pressure, Figure 5 shows the variations in COP and supplied-air temperature with the expansion ratio. It is apparent that the increase in operating pressure can improve cycle COP, but it has little effect on the supplied-air's temperature. As operating pressure increases, the enhancement of the cycle COP becomes less obvious. This is mainly because the system power consumption and cooling capacity mainly depend on the enthalpy difference when the pressure is beyond 0.3 MPa. Figure 5 shows that there is an optimal expansion ratio to maximize the cycle COP. When the inlet pressure of the expander is 506.625 kPa, the maximum COP of 0.652 can be reached at an expansion ratio of 2.25.

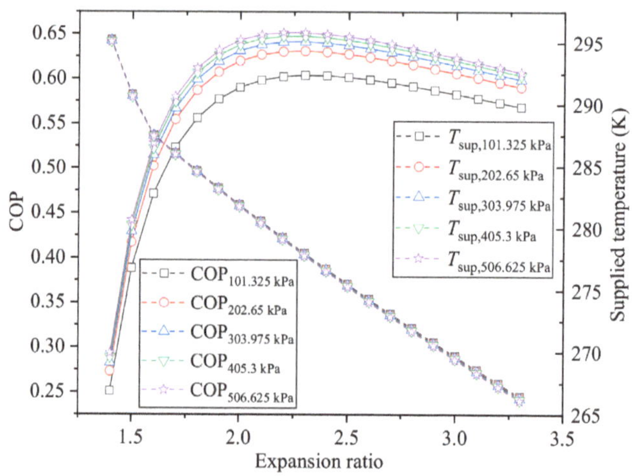


**Figure 5.** Comparisons of COP and supplied-air temperature for closed air-refrigeration cycles under different expander inlet pressures.

Figure 6a shows the comparisons of the COP and supplied-air temperature for various air refrigeration cycles. The COP of open-loop cycles is greater than closed-loop cycles, and the two-stage compression open-loop cycles, including TC-C, C-TC, C-CT and CT-C cycles have higher COPs than the one-stage compression open-loop cycles. The greatest COP for the TC cycle and CT cycle are 0.728 and 0.747, respectively, for the optimal expansion ratios of 2.10 and 2.17. Figure 6b shows the p-H diagrams for the air cycles at the respective optimal expansion ratio. In an open-loop high-pressure cycle (CT cycle), the cold air from the expander can be supplied to the compartment directly, and there is no loss of cooling capacity caused by the temperature difference in the cold-side heat exchanger. The inlet temperature of the expander in an open-loop low-pressure cycle (TC cycle) is the ambient temperature, which is generally lower than that of the CT cycle, and the cooling capacity can be increased effectively. However, the temperature difference in the cold-side heat exchanger of the TC cycle leads to a lower COP than that of the CT cycle, and the influence is more obvious when the expansion ratio is higher than the optimum value.

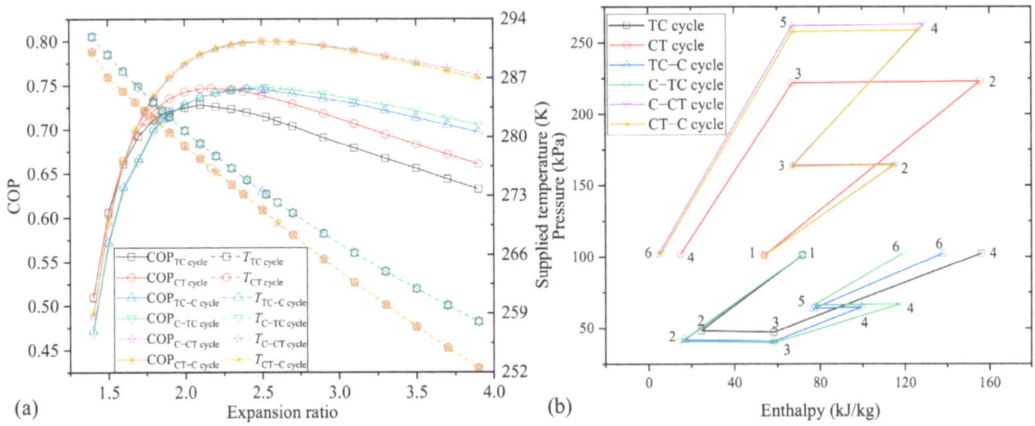

**Figure 6.** (**a**) Comparisons of COP and supplied-air temperature for different air refrigeration cycles; (**b**) p-H diagrams for each cycle at the optimal expansion ratio.

The optimal expansion ratios of the TC-C cycle, C-TC cycle, C-CT cycle and CT-C cycle are 2.41, 2.47, 2.56 and 2.52, respectively, and the corresponding maximum COPs are 0.746, 0.748, 0.800 and 0.800, respectively. When the expansion ratio is less than 1.9 for open-loop low-pressure cycles, the COP of the TC cycle is higher than that of the TC-C cycle and C-TC cycle. As the expansion ratio increases, the COP of the C-TC cycle and TC-C cycle surpasses the TC cycle, and the COP of the TC-C cycle is smaller than that of the C-TC cycle. When the supplied-air temperature is higher than about 282 K, the COP of the TC cycle is highest; otherwise, the COP of the C-TC cycle is highest. Similarly, the COP of the CT cycle is higher than that of the C-CT cycle and CT-C cycle when the expansion ratio is less than 1.6. As the expansion ratio increases, the COP of the C-CT cycle and CT-C cycle gradually exceeds the TC cycle. The CT cycle has the highest COP when the supplied-air temperature is higher than 285 K. The COP of the C-CT cycle is slightly higher than that of the CT-C under other conditions.

In addition, air refrigeration systems such as domestic air conditioners also have requirements for the supplied temperature. The supplied temperature is generally required to be greater than 283.15 K to ensure the comfort of human body, which requires a relatively low expansion ratio. This makes the open-loop cycle unable to operate at the optimal expansion ratio. At this case, the performance of CT-C cycle is the best (COP is 0.722). The COP of TC cycle is somewhat lower than that of CT-C cycle, but it has a simpler configuration with one-stage compression.

The expansion ratio should not be too large to prevent ice in the system when the supplied-air temperature is lower than 273.16 K. Open-loop low-pressure cycles can achieve the optimal pressure ratio, but their COP is lower than that of open-loop high-pressure cycles. The C-CT cycle and CT-C cycle have the highest COP of 0.798 when the supplied-air temperature is above 273.16 K. Alternatively, before being supplied to the compartment, the cold air from expander can be mixed with the indoor warm air. The temperature at expander outlet can be lower than 273.16 K in this situation, and the cooling capacity can be increased by increasing the expansion ratio.

### 3.1. *The Effect of Expander and Compressor Efficiencies*

Figure 7 reveals variations in COP and supplied-air temperature with the expansion ratio for open-loop high-pressure and open-loop low-pressure cycles. It is apparent that the

COP of the open-loop high-pressure cycles is larger than that of the open-loop low-pressure cycles, where the COP of the CT-C is the largest when the expansion ratio is greater than 1.8. When the expansion ratio is greater than 1.9 or the supplied-air temperature is lower than 281 K, the COP of the C-CT is the largest in the open low-pressure cycles. As the efficiencies of the compressor and expander increase, the COP of the open-loop cycles increases a lot, and the optimal expansion ratio decreases. The optimal expansion ratios of the TC cycle (85% and 80%), TC-C cycle (85% and 80%) and C-TC cycle (85% and 80%) are 1.96, 2.25 and 2.31, respectively, with maximum COP values of 0.974, 0.997 and 0.979, respectively.

In the open high-pressure cycles, when the expansion ratio is greater than 1.6 or the supplied temperature is lower than 286 K, the COP of the CT-C cycle is the largest. The optimal expansion ratios of the CT cycle (85% and 80%), C-CT cycle (85% and 80%) and CT-C cycle (85% and 80%) are 2.04, 2.41 and 2.35, respectively, and the corresponding maximum COPs are 0.997, 1.055 and 1.071, respectively. When the supplied-air temperature required is greater than 283.15 K, the CT-C cycle has the maximum COP of 0.969, while the TC cycle has almost the same COP (0.967) but a simpler system arrangement. When the supplied-air temperature is greater than 273.16 K, the CT-C cycle has the maximum COP of 1.069.

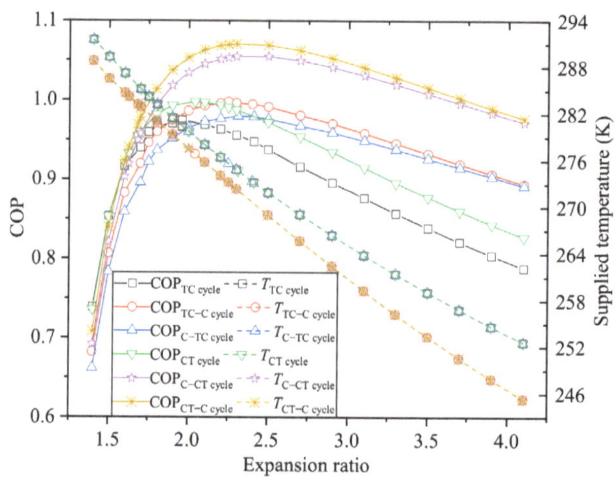


**Figure 7.** Comparison of COP and supplied-air temperature when the isentropic efficiency of expander and compressor is 85% and 80%, respectively.

### 3.2. The Effect of Humidity

Figure 8a shows variations in COP with the expansion ratio for the CT cycles with different indoor wet bulb temperatures. Figure 8b shows the p-H diagrams for the air cycles at the respective initial optimal COP. The isentropic efficiencies of the expander and compressor are 80% and 75%, respectively. In the legend, for example, 300.15 K/292.15 K denotes the dry/wet bulb temperatures. When the indoor humidity is high, there is a turning point in the curve of the COP that varies with the expansion ratio. In the zone after the turning point, the COP increases rapidly with the increase of the expansion ratio, and the temperature of after-cooler outlet is lower than the dew point. The latent heat of phase change allows the working fluid to absorb more cooling capacity in the after-cooler, which results in a decrease of absolute moisture in the air and a lower enthalpy at the expander inlet. Although the mass flow decreases slightly, the decrease in enthalpy has a greater impact on the cooling capacity of the system. Therefore, when the outlet temperature of the after-cooler is lower than the dew point, the cooling capacity of the system is greater. In addition, with the humidity increases, both the COP and the optimal expansion ratio of the cycle increase. Before the turning point, the optimal expansion ratios are 2.17, 2.20, 2.20,

2.22 and 2.24, respectively, and the corresponding maximum COPs are 0.747, 0.756, 0.766, 0.776 and 0.785, respectively.

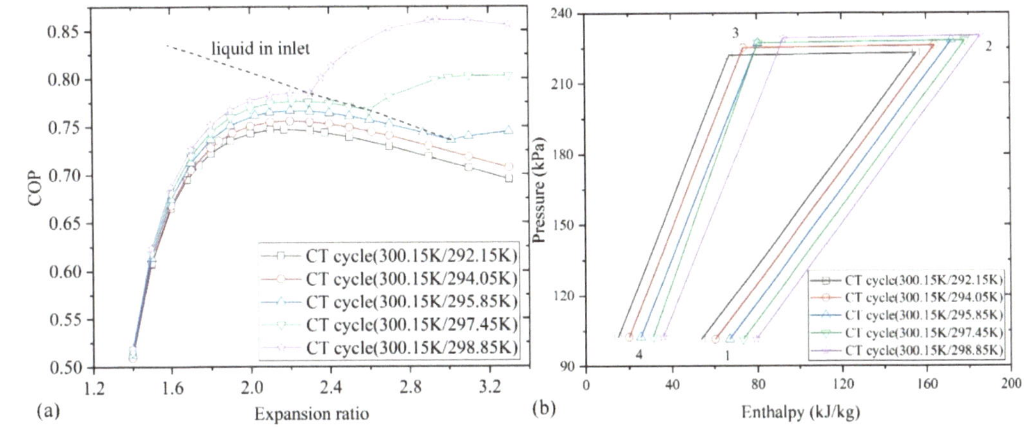


**Figure 8.** (**a**) Comparison of COP of open-loop high pressure cycle under different indoor wet bulb temperature conditions; (**b**) p-H diagrams for each cycle at the initial optimal COP.

### 3.3. The Effect of Recycling Water

The low-temperature water is directly discharged at the outlet of the air refrigeration cycle, and the cooling capacity is also wasted. If the cooling capacity of the water that separated from the separator can be recycled, the COP of the system will improve. For an open-loop low-pressure cycle, the water separated from the supplied side can be mixed with the air at the expander inlet without bringing additional wet load to the supplied side, which can reduce the temperature of the expander inlet. The water can be atomized and injected as much as possible into the inlet of the expander to cool the humid air of the expander inlet to a saturated state. In order to avoid problems such as the corrosion of parts caused by liquid at the turbo-expander inlet, the mass flow of the recirculating water should not be so large that the moist air of the expander inlet is supersaturated. The power consumption of the atomized water is ignored in the calculations. When the mass flow of the condensed water is greater than what the expander inlet requires, the remaining water can be atomized and injected into the compressor inlet, which makes a wet compression in the compressor. During the compression, the water evaporates and absorbs the compression heat, reducing the power consumption. The reduced theoretical compression work is equal to the latent heat of the water, and the effect of the increased mass flow is also calculated.

Figure 9 shows the variations in COP with the expansion ratio in open-loop low-pressure cycles with recycling water. Without wet compression, the optimal expansion ratio occurs when the mass flow of the recycling water just makes the wet air of the expander inlet reach a saturated state. When the expansion ratio is larger than the optimal expansion ratio, the temperature of the expander inlet cannot continue to decrease, reducing the effect of the recycling water on the cooling capacity of the system. The maximum COP of the TC, TC-C and C-TC cycles are 1.310, 1.297 and 1.268, respectively. Compared to the COP before recycling the water, the maximum COP of the TC, TC-C and C-TC cycles have increased by 34.5%, 30.1% and 29.5%, respectively. The optimal expansion ratio of the TC, TC-C and C-TC cycles is 1.78. At the same time, the supplied-air temperature of the system is higher than 283.15 K. When the remaining water is used for wet compression, the optimal expansion ratios of the TC, TC-C and C-TC cycles are 2.12, 2.32 and 2.35, respectively, and the corresponding maximum COP are 1.410, 1.483 and 1.453, respectively. Compared to the COPs before recycling the water, the maximum COPs of the TC, TC-C and C-TC cycles have increased by 44.7%, 48.8% and 48.4%, respectively.

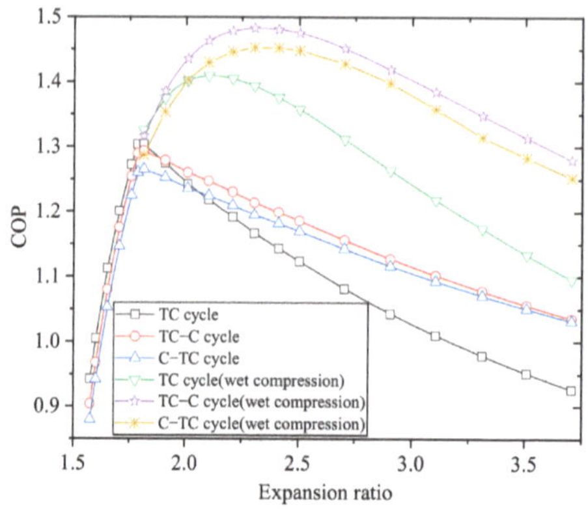


**Figure 9.** COP of air refrigerators with recycling water with isentropic efficiency of expander and compressor at 85% and 80%, respectively.

### *3.4. The Impact of Fresh Air*

A fresh-air system can significantly improve the indoor air quality in a closed indoor environment. Simultaneously, in order to inhibit the spread of bacteria and viruses, it is also necessary to use an all-fresh-air system without return air. However, the introduction of fresh air brings additional heat and wet load, thereby reducing the cooling capacity of the system. The COP of open-air cycles with the all-fresh-air system is studied.

Figure 10 shows the schematic of the open-loop air refrigeration cycles with all-fresh-air supply. The exhaust side of an open-loop low-pressure cycle intakes air indoors, while the supplied side sucks air outdoors and feeds the conditioned air indoors. The supplied side of an open-loop high-pressure cycle intakes outdoor air and feeds the conditioned air indoors, and the indoor exhaust air cools the air of the expander inlet.

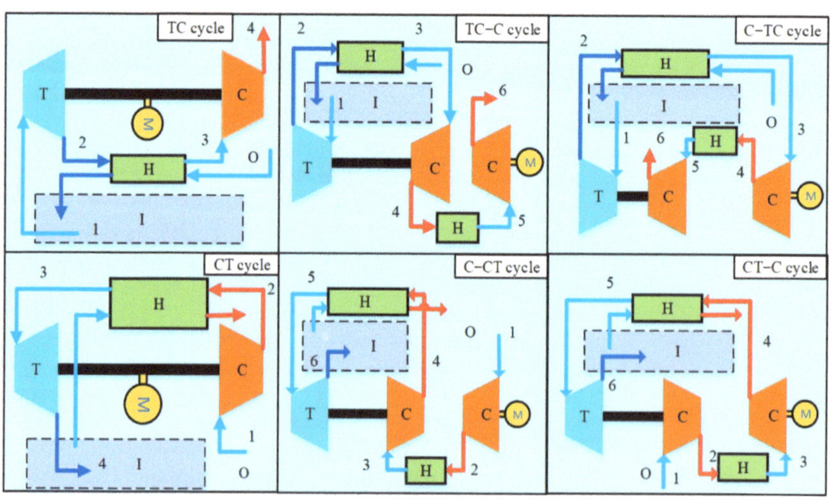


**Figure 10.** Schematic of the all-fresh-air system for open-loop air refrigeration cycles. T: turbo expander; C: compressor; I: indoor; O: outdoor.

Figure 11a reveals the variations in COP and supplied-air temperature with the expansion ratio for air refrigeration cycles with all-fresh-air systems when the isentropic efficiency of the expander and compressor is 80% and 75%, respectively. Figure 11b shows the p-H diagrams for the air cycles at the respective optimal expansion ratios. The optimal expansion ratios of the TC cycle, TC-C cycle, C-TC cycle, CT cycle, C-CT cycle and CT-C cycle are 2.25, 2.56, 2.59, 2.37, 2.76 and 2.74, respectively, and the corresponding maximum COPs are 0.620, 0.655, 0.659, 0.661, 0.739 and 0.738, respectively. Comprehensive comparison shows that when the required supplied-air temperature is greater than 283.15 K, the CT-C cycle has the best COP (0.646), which is close to that of the C-CT cycle (0.645). When the required supplied-air temperature is greater than 273.16 K, the C-CT cycle has the best COP (0.734). Adopting the all-fresh-air system, the optimal expansion ratio increases. In open-loop low-pressure cycles, the C-TC cycle has the best COP.

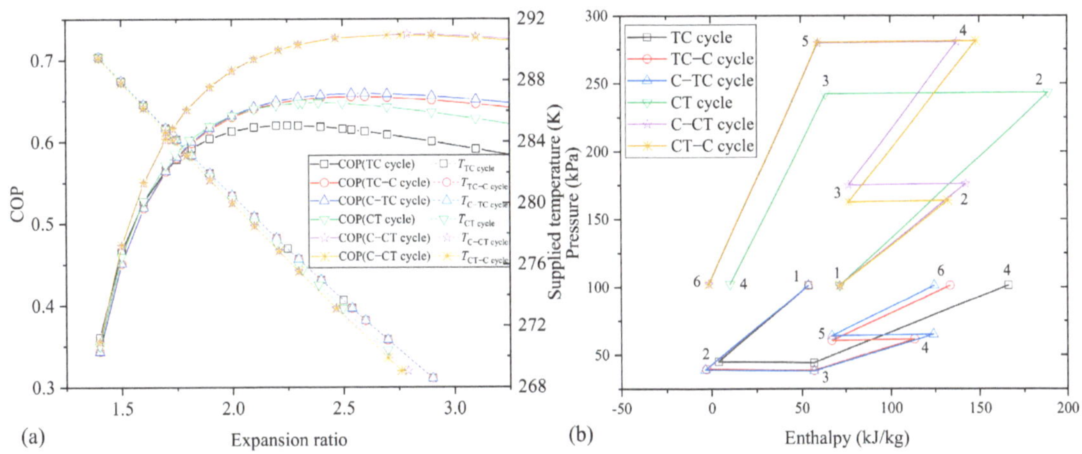


**Figure 11.** (**a**) COP and supplied-air temperature of air refrigerators with all-fresh-air systems with isentropic efficiency of expander and compressor of 80 and 75%, respectively; (**b**) p-H diagrams for each cycle at the optimal expansion ratio.

Figure 12 illustrates the variations in COP with the expansion ratio and supplied temperature for open-loop air refrigeration cycles with all-fresh-air systems when the expander and compressor isentropic efficiency is 85% and 80%. The difference from the previous example is that the COP of the TC-C cycle is greater than that of the C-TC cycle. The CT-C cycle has the best COP of the high-pressure cycles. The optimal expansion ratios of the TC cycle, TC-C cycle, C-TC cycle, CT cycle, C-CT cycle and CT-C cycle are 2.09, 2.39, 2.41, 2.23, 2.60 and 2.58, respectively, and the corresponding maximum COPs are 0.816, 0.863, 0.858, 0.872, 0.972 and 0.979, respectively. Similarly, when the supplied-air temperature is required to be greater than 283.15 K or 273.16 K, the CT-C cycle has the best COP. The TC-C cycle has the best COP in the open-loop low-pressure cycles.

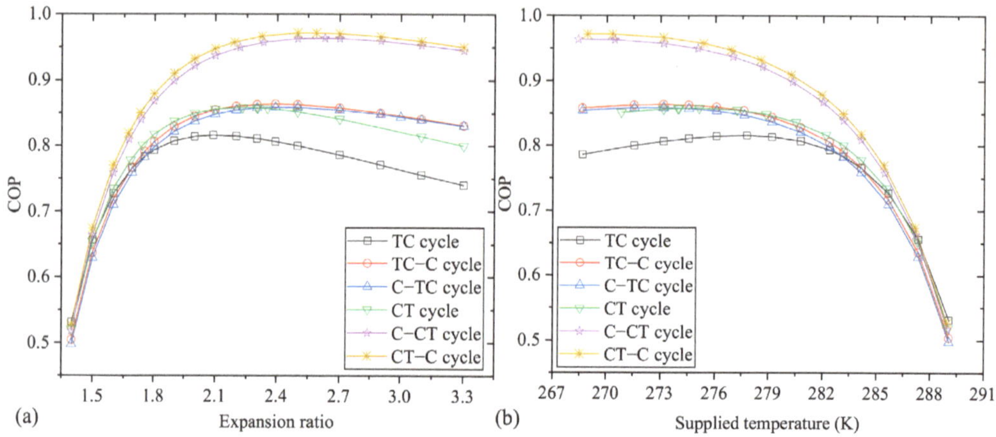


**Figure 12.** COP of air refrigerators with all-fresh-air systems with isentropic efficiency of the expander and compressor of 85% and 80%, respectively: (**a**) variable expansion ratio and (**b**) variable supplied temperature.

### 3.5. The Effect of Ambient Pressure of High-Altitude Regions

When trains are running from the plain area to the plateau area, air conditioning with elevated supply pressure and fresh air is desired in order to maintain a comfortable oxygen concentration. On the plateau train of the Qinghai–Tibet line in China, there are more than 960 km of railways where the altitude is above 4000 m and the atmospheric pressure is below 56.04 kPa, and the highest altitude is 5072 m. In that case, the respiratory and digestive systems of the human body are significantly affected and uncomfortable. As required by passenger aircraft, cabin pressure should be generally maintained around 72.84–80.9 kPa. For the conventional vapor compression air-conditioning systems, a separate fresh-air system with an air compressor is needed. Considering the sensible heat and compression work of the fresh air, the COP of the vapor compression air-conditioning system decreases significantly to a level which is comparable to that of the air-cycle refrigeration system.

For the application of an open-loop low-pressure cycle in a high-altitude train, it is necessary to employ a compressor on the supplied side to increase the pressure of outdoor fresh air. In an open-loop high-pressure cycle, an expander can be used on the cabin exhaust side to recover the pressure energy. It is assumed that the isentropic efficiencies of the expander and compressor are 85% and 80%, respectively, and the supplied air pressure is 72.84 kPa. Indoor and outdoor temperatures remain same as in previous calculations, and the absolute humidity of the indoor and outdoor air remain unchanged.

Figure 13 shows the variations in COP with the expansion ratio and supplied-air temperature for open-loop cycles when the ambient pressure is 56.04 kPa. The COP of the open-loop low-pressure cycle is better. Optimal expansion ratios of the TC, TC-C, C-TC, CT, C-CT and CT-C cycles are 2.43, 2.85, 2.93, 2.79, 3.40 and 3.40, respectively, and the corresponding maximum COP are 0.766, 0.776, 0.761, 0.688, 0.782 and 0.789, respectively. When the supplied-air temperature is greater than 283.15 K or 273.16 K, the TC cycle has the maximum COP of 0.588 or 0.763. It is obvious that the COP decreases after the atmospheric pressure is lowered. Under the same absolute humidity, the partial pressure of water vapor is lower than that under normal pressure, and the dew point of the air is lower. Under the same enthalpy drop, the temperature drop of the moist air is greater, and thus the temperature of the expander is lower. In order to increase the supplied-air temperature, it is necessary to mix the air from the expander with indoor air before supplying the air to the cabin. It is not until the supplied temperature is very low that the COP of the CT-C cycle and CT-C cycle are better. However, there is high-pressure air in the high-pressure cycles that can be used to generate high-concentration oxygen.

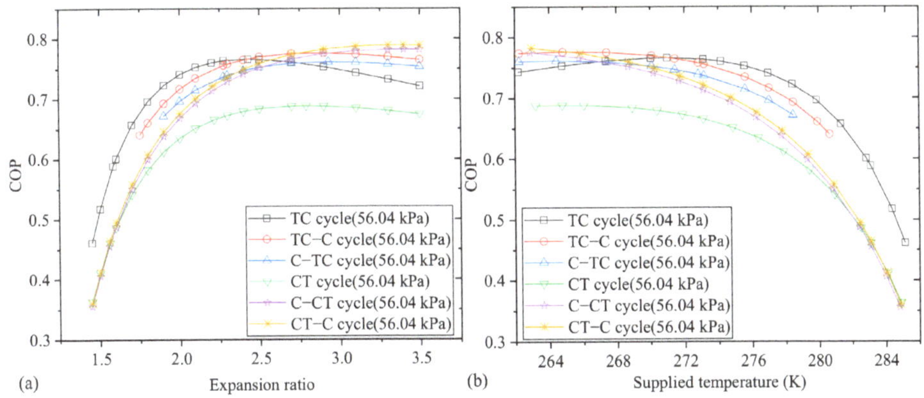


**Figure 13.** COP of air refrigerators in high altitude trains at the ambient pressure of 56.04 kPa for all fresh-air systems: (**a**) variable expansion ratio and (**b**) variable supplied temperature.

For the air-conditioning systems of plateau trains, the outside pressure also changes during the trip, especially when the train travels from the plain to the plateau. The air refrigeration cycle with the centrifugal compressors and turbo-expanders can adjust the system pressure ratio by changing the rotational speed to match the ambient pressure. Although the compression power is required to maintain the indoor pressure, the air-cycle refrigeration system can use the expander to recover the pressure energy of the indoor exhaust air, and therefore the cycle COP does not decrease too much compared to that of the plain.

## 4. Conclusions

In this study, we investigated the thermodynamic performance of three configurations of the air cycle refrigeration system for compartment air conditioning, including closed-loop cycles, open-loop low-pressure cycles and open-loop high-pressure cycles. The results show that the COP of the closed-loop cycle is generally smaller than that of the open-loop cycles because of the irreversible loss in the heat exchangers. The following conclusions can be drawn.

(1) For all the air refrigeration cycles, the optimal expansion ratio decreases with increase of the compressor and expander efficiency, but it increases with the decrease of the ambient pressure. For all-fresh-air air-conditioning, the optimal expansion ratio is increased, while the corresponding COP is reduced. Both the optimal expansion ratio and COP are increased in the two-stage compression intercooling configuration.

(2) For the normal environment, among open-loop cycles, high-pressure cycles have higher COP than low-pressure cycles. In open-loop low-pressure cycles, the COP of the TC-C cycle exceeds that of the CT-C cycle as the compressor efficiency increases, and a similar trend can also be observed in open-loop high-pressure cycles.

(3) An increase in air humidity increases the optimal expansion ratio and COP of the system. In open-loop low-pressure cycles, the condensed water can be recycled by being sprayed into the expander inlet, increasing the maximum COP of the TC, TC-C and C-TC cycles by 34.5%, 30.1% and 29.5%, respectively. Furthermore, if the remaining water is used for wet compression, the maximum COP of the TC, TC-C and C-TC cycles can be increased by 44.7%, 48.8% and 48.4%, respectively.

(4) In the air conditioning systems of plateau trains, the open-loop low-pressure cycle has the highest COP under the requirement that the supplied temperature is greater than 283.15 K. The COPs of open-loop high-pressure cycles are relatively low, but the high-pressure air in the cycle can be used to separate high-concentration oxygen.

The COP of the air refrigeration cycle at room temperature is low, and humidity can improve the COP. Future research needs to further improve the efficiency of the compressor and expander to improve the COP. Improving the efficiency of wet expansion and compression, in particular, can improve the COP.

**Author Contributions:** Conceptualization, X.Y. and L.C.; data curation, X.Y.; formal analysis, Z.Z.; investigation, X.Y.; methodology, X.Y. and Y.H.; project administration, S.C.; resources, L.C.; validation, X.Y. and Z.W.; visualization, X.Y.; writing—original draft preparation, X.Y.; writing—review and editing, X.Y. and L.C. All authors have read and agreed to the published version of the manuscript.

**Funding:** This project was supported by the National Nature Science Foundation of China (U21B2084) and the Youth Innovation Team of Shaanxi Universities.

**Data Availability Statement:** Not applicable.

**Conflicts of Interest:** The authors declare no conflict of interest.

## Nomenclature

| | |
|---|---|
| COP | coefficient of performance |
| $c_p$ | specific heat at constant pressure (J kg$^{-1}$) |
| d | humidity ratio of wet air (g kg$_a{}^{-1}$) |
| $Er$ | expansion ratio |
| $h$ | enthalpy (J kg$^{-1}$) |
| $k$ | adiabatic exponent |
| $P$ | pressure (Pa) |
| $Qc$ | cooling capacity (W) |
| $Q_{het}$ | exchanger cold duty (W) |
| $q_m$ | mass flow rate (kg s$^{-1}$) |
| $R$ | gas constant of air (J kg$^{-1}$K$^{-1}$) |
| $T$ | temperature (K) |
| $T_d$ | dew point temperature (K) |
| W | shaft power (W) |
| Greek | |
| $\eta$ | isentropic efficiency |
| $\eta_M$ | mechanical efficiency |
| $\Delta$h | the change of enthalpy (J kg$^{-1}$) |
| Subscripts | |
| a | dry air in wet air |
| AC | outlet of the after-cooler |
| C | compressor |
| CI | cold-side heat exchanger outlet |
| EO | turbo-expander outlet |
| I | indoor |
| max | maximum |
| O | outdoor |
| opt | optimal |
| s | saturated |
| sup | supplied air |
| T | turbo-expander |
| v | water vapor in wet air |

## References

1. Jedrich, N.; Zimbelman, D.; Swift, W.; Dolan, F.; Brumfield, M. Paper Session I-A—A Mechanical Cryogenic Cooler for the Hubble Space Telescope. In Proceedings of the Space Congress, Houston, TX, USA, 30 April–3 May 2002.
2. Hou, Y.; Zhao, H.L.; Chen, C.Z.; Xiong, L.Y. Developments in reverse Brayton cycle cryocooler in China. *Cryogenics* **2006**, *46*, 403–4072. [CrossRef]
3. Zhao, H.; Hou, Y.; Zhu, Y.; Chen, L.; Chen, S. Experimental study on the performance of an aircraft environmental control system. *Appl. Therm. Eng.* **2009**, *29*, 3284–3288. [CrossRef]
4. Yang, H.; Yang, C. Thermodynamic Characteristics and Order Degree of Air Cycle System. *Int. J. Refrig.* **2020**, *118*, 61–71. [CrossRef]

5. Hirai, H.; Suzuki, Y.; Hirokawa, M.; Kobayashi, H.; Kamioka, Y.; Iwakuma, M.; Shiohara, Y. Development of a turbine cryocooler for high temperature superconductor applications. *Phys. C Supercond. Its Appl.* **2009**, *469*, 1857–1861. [CrossRef]
6. Zhang, Y.; Li, Q.; Wu, J.; Li, Q.; Lu, W.; Xiong, L.; Liu, L.; Xu, X.; Sun, L.; Sun, Y.; et al. Performance analysis of a large-scale helium Brayton cryo-refrigerator with static gas bearing turboexpander. *Energy Convers. Manag.* **2015**, *90*, 207–217. [CrossRef]
7. Islam, S.; Dincer, I. Comparative performance study of an integrated air-cycle refrigeration and power generation system. *Appl. Therm. Eng.* **2016**, *106*, 493–505. [CrossRef]
8. Lewis, J.S.; Chaer, I.; Tassou, S.A. *Reviews of Alternative Refrigeration Technologies, Centre for Energy and Built Environment Research, School of Engineering and Design*; Brunel Universvity: Uxbridge, UK, 2007.
9. Catalano, L.A.; Bellis, F.D.; Amirante, R. Improved inverse joule Brayton air cycle using turbocharger units. In Proceedings of the ASME-ATI-UIT 2010 Conference on Thermal and Environmental Issues in Energy Systems, Sorrento, Italy, 16–19 May 2010.
10. Hou, S.; Li, H.; Zhang, H. An open air–vapor compression refrigeration system for air-conditioning and desalination on ship. *Desalination* **2008**, *222*, 646–655. [CrossRef]
11. Hou, S.; Zhang, H. An open reversed Brayton cycle with regeneration using moist air for deep freeze cooled by circulating water. *Int. J. Therm. Sci.* **2009**, *48*, 218–223. [CrossRef]
12. Tassou, S.A.; De-Lille, G.; Ge, Y.T. Food transport refrigeration—Approaches to reduce energy consumption and environmental impacts of road transport. *Appl. Therm. Eng.* **2009**, *29*, 1467–1477. [CrossRef]
13. Spence, S.; Doran, W.J.; Artt, D.W. Design, construction and testing of an air-cycle refrigeration system for road transport. *Int. J. Refrig.* **2004**, *27*, 503–510. [CrossRef]
14. Spence, S.; Doran, W.J.; Artt, D.W.; McCullough, G. Performance analysis of a feasible air-cycle refrigeration system for road transport. *Int. J. Refrig.* **2005**, *28*, 381–388. [CrossRef]
15. Liu, W.; Lian, Z.; Radermacher, R.; Yao, Y. Energy consumption analysis on a dedicated outdoor air system with rotary desiccant wheel. *Energy* **2007**, *32*, 1749–1760. [CrossRef] [PubMed]
16. Kim, M.H.; Jeong, J.W. Cooling performance of a 100% outdoor air system integrated with indirect and direct evaporative coolers. *Energy* **2013**, *52*, 245–257. [CrossRef]
17. Zhang, Z.Y.; Zhang, C.L.; Ge, M.C.; Yu, Y. A frost-free dedicated outdoor air system with exhaust air heat recovery. *Appl. Therm. Eng. Des. Processes Equip. Econ.* **2018**, *128*, 1041–1050. [CrossRef]
18. Li, S.; Wang, S.; Ma, Z.; Jiang, S.; Zhang, T. Using an air cycle heat pump system with a turbocharger to supply heating for full electric vehicles. *Int. J. Refrig.* **2017**, *77*, 11–19. [CrossRef]
19. Park, S.K.; Ahn, J.H.; Kim, T.S. Off-design operating characteristics of an open-cycle air refrigeration system. *Int. J. Refrig.* **2012**, *35*, 2311–2320. [CrossRef]
20. Chen, L.; Zhou, S.; Sun, F.; Wu, C. Performance of heat-transfer irreversible regenerated Brayton refrigerators. *J. Phys. D Appl. Phys.* **2001**, *34*, 830. [CrossRef]
21. Zhou, S.; Chen, L.; Sun, F.; Wu, C. Cooling Load Density Optimization of an Irreversible Simple Brayton Refrigerator. *Open Syst. Inf. Dyn.* **2002**, *9*, 325–337. [CrossRef]
22. Ust, Y. Performance analysis and optimization of irreversible air refrigeration cycles based on ecological coefficient of performance criterion. *Appl. Therm. Eng.* **2009**, *29*, 47–55. [CrossRef]
23. Giannetti, N.; Milazzo, A. Thermodynamic analysis of regenerated air-cycle refrigeration in high and low pressure configuration. *Int. J. Refrig.* **2014**, *40*, 97–110. [CrossRef]
24. Ahmadi, M.H.; Ahmadi, M.A.; Pourfayaz, F.; Bidi, M. Thermodynamic analysis and optimization for an irreversible heat pump working on reversed Brayton cycle. *Energy Convers. Manag.* **2016**, *110*, 260–267. [CrossRef]
25. Yu, H.; Yang, S.; Chen, X.; Chen, S.; Lai, T. Study on the matching performance of a low temperature reverse Brayton air refrigerator. *Energy Convers. Manag.* **2015**, *89*, 339–348.
26. Yang, S.; Chen, S.; Chen, X.; Zhang, X.; Hou, Y. Study on the coupling performance of a turboexpander compressor applied in cryogenic reverse Brayton air refrigerator. *Energy Convers. Manag.* **2016**, *122*, 386–399. [CrossRef]
27. Zhang, Z.; Liu, S.; Tian, L. Thermodynamic analysis of air cycle refrigeration system for Chinese train air conditioning. *Syst. Eng. Procedia* **2011**, *1*, 16–22. [CrossRef]

energies

MDPI

*Article*

# Simulation on the Electric and Thermal Fields of a Microwave Reactor for Ex Situ Biomass Tar Elimination

**Cheng Yang, Kanfeng Ying, Fan Yang, Huanghu Peng and Zezhou Chen ***

Department of Engineering, Huzhou University, Huzhou 313000, China; 2020388621@stu.zjhu.edu.cn (C.Y.); 2021388639@stu.zjhu.edu.cn (K.Y.); 02567@zjhu.edu.cn (F.Y.); 02472@zjhu.edu.cn (H.P.)
* Correspondence: 02643@zjhu.edu.cn

**Abstract:** Microwave treatment is an emerging technique for biomass tar elimination. The electric and thermal fields of the microwave reactor are the key to high elimination efficiency and energy utilization. In this work, we simulated the electric and thermal fields of a microwave reactor with various parameters including irradiation feed position, microwave power, silicon carbide length and flow velocity. Results show that the irradiation feed position that locates 5 mm vertically to the central plane can obtain the highest electric intensity and silicon carbide temperature (ca. 1100 K) after wave absorbing. Both the electric and thermal fields are strengthened when microwave power is increased. Extending the silicon carbide bed length will decrease the bed temperature and heating rate. A high flow velocity leads to non-uniform temperature distribution of the silicon carbide. For the purpose of achieving a high microwave energy utilization and uniform bed temperature, suitable irradiation feed position ($z_i$ = 5 mm), high microwave power ($P$ = 1000 W), short silicon carbide bed length ($l_{SiC}$ = 100 mm) and low flow velocity ($v$ = 0.02 m/s) are preferred, but the chemical kinetics of biomass tar elimination should also be considered in the practical application.

**Keywords:** microwave; biomass tar elimination; simulation; thermal field; electric field; silicon carbide

**Citation:** Yang, C.; Ying, K.; Yang, F.; Peng, H.; Chen, Z. Simulation on the Electric and Thermal Fields of a Microwave Reactor for Ex Situ Biomass Tar Elimination. *Energies* **2022**, *15*, 4143. https://doi.org/10.3390/en15114143

Academic Editors: Dibyendu Sarkar, Rupali Datta, Prafulla Kumar Sahoo and Mohammad Mahmudur Rahman

Received: 6 May 2022
Accepted: 2 June 2022
Published: 4 June 2022



## 1. Introduction

The biomass gasification that can effectively convert biological waste into synthetic gas is a promising technology for energy recovery from biomass wastes, but the process inevitably produces biomass tar as the by-product [1]. The formation of biomass tar, which usually consists of multi-ring aromatics [2], will clog the downstream pipe and cause fouling and corrosion issues [3,4], thus it becomes the most critical problem in biomass gasification. Therefore, effective tar elimination methods are necessarily required in the biomass gasification system.

The biomass tar can be removed inside the gasifier (in situ elimination) or outside the gasifier (ex situ elimination) as shown in Figure 1. In situ elimination includes the use of gasification catalysts [5,6], structurally modified gasifiers [7] and novel heating strategies [8], but it can hardly achieve a complete tar removal without secondary elimination [9]. Therefore, the ex situ elimination methods are commonly employed. The ex situ elimination methods can be further divided into physical and chemical tar removal. The physical tar removal can be realized by dry/wet gas cleaning or solid-phase adsorption/desorption, but brings drawbacks including secondary pollution, waste management cost and energy loss [10]. Compared with physical tar removal, the chemical removal methods, especially thermal or catalytic cracking [11,12], show more advantages such as high elimination efficiency and energy utilization by cracking the tar into hydrocarbon gases that can be reused and were therefore extensively studied. The widely used catalysts include nickel or noble metal-based catalysts [13,14]. These catalysts are effective in tar elimination but less attractive in industrial application because of their high cost, as well as the deactivation problem [15]. Other catalysts such as iron- or biochar-based catalysts are more economically attractive and were therefore investigated extensively, as well, in the tar elimination process [16,17].

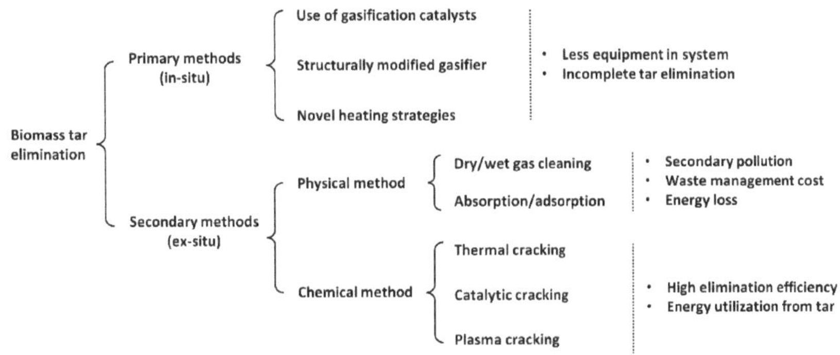


**Figure 1.** Biomass tar elimination method and their properties.

The ex situ tar elimination methods such as thermal or catalytic cracking are essentially endothermic, which makes the selection of the appropriate heating method of great importance. Microwave heating is an emerging technology in biomass treatment [18,19]. Compared to the conventional heating methods, microwave heating can penetrate the wave-absorbing materials and achieve direct energy input to the materials. It has several advantages such as rapid temperature rise with instantaneous start and stops, uniform temperature distribution throughout the volume of the materials and non-contact and selective heating [20,21]. Moreover, the hot spots and plasma generated during microwave heating can provide higher conversion efficiency and better product distribution [22]. With these advantages, the microwave-assisted elimination exhibits good performance on biomass tar removal. For example, Li et al. used the char-supported metal (Fe, Ni and Ce) catalysts as microwave absorbers and obtained the complete cracking of toluene—the model compound of biomass tar [23,24]—under microwave irradiation [25]. Similarly, Chen et al. observed a high toluene elimination efficiency (95.1%) in a microwave tube furnace with a biochar-supported Ni catalyst [26]. Zhang et al. used silicon carbide as the wave absorber in the microwave-assisted elimination of tar. They observed a maximum toluene elimination efficiency of 94.4% and found the main product to be $H_2$ [27], indicating an effective tar cracking by the microwave heating technique.

In practical industrial processes, the temperature of the biomass gasification gas is typically 450–600 °C for downdraft fixed-bed gasifiers and ca. 150 °C for updraft fixed-bed gasifiers [22]. This means the biomass tar, which is mainly composed of mono- or poly-aromatics [23,24,28], will be in the gaseous phase. Since the biomass tar is ineffective in absorbing the microwave, the microwave absorbers are required in the microwave reactor. For this reason, the microwave reactor for biomass tar elimination mainly contains a microwave oven to generate microwave irradiation, a fixed bed with a wave absorber to allow biomass tar cracking, and a quartz tube to hold the wave absorber and allow the microwave to penetrate in the absorber [21,25,26,29]. In such a microwave reactor, the biomass tar can be ex situ cracked to the hydrocarbon gases. It is obvious that the electric and thermal fields that determine the reaction conditions in the microwave reactor will significantly affect the tar cracking. Therefore, the reactor and operation parameters that would influence the reaction fields should be detailly designed to achieve a high biomass tar elimination efficiency and energy utilization.

The COMSOL Multiphysics software has been successfully used to estimate the electric field intensity, temperature distribution, heat transfer (hot and cold spots), and absorbed power during microwave heating [30,31]. In this work, we used the COMSOL Multiphysics software to simulate the electric and thermal fields of the microwave reactor under various reactor and operation parameters including irradiation feed position, microwave power, microwave absorbers bed length and flow velocity. The silicon carbide was selected as the microwave absorber in the simulation to explore its temperature distribution and heating rate under microwave irradiation. The interest of this work is to find the optimal parameters about the microwave reactor designation and the main conditions in its application for the efficient elimination of biomass tar with the microwave technique.

## 2. Simulation Methods

### 2.1. Modeling of the Microwave Reactor

The modeling of the microwave reactor was based on a real microwave reactor that was used for biomass tar elimination. Its general structure is shown in Figure 2a. The microwave was generated by a single magnetron that was located in the back of the microwave reactor and then guided to the resonant cavity through a rectangular waveguide. A quartz tube filled with wave absorber was placed in the center of the reactor. The wave absorber can absorb the microwave and be heated up to the desired temperatures. The tar was blown from the inlet and eliminated after flowing through the hot absorber bed. The simulation model was established by the COMSOL Multiphysics 5.6 software (COMSOL, Burlington, MA, USA) (Figure 2b). The resonant cavity of the microwave reactor was a multimode cavity with an inner size of 340 mm ($x$) $\times$ 330 mm ($y$) $\times$ 200 mm ($z$). In the simulation, the quartz tube ($d$ = 60 mm, $l$ = 340 mm) was filled with silicon carbide particles as the wave absorber. A $TE_{10}$ rectangular waveguide was used as the microwave irradiation feed. Its size was 100 mm ($x$) $\times$ 95 mm ($y$) $\times$ 55 mm ($z$).

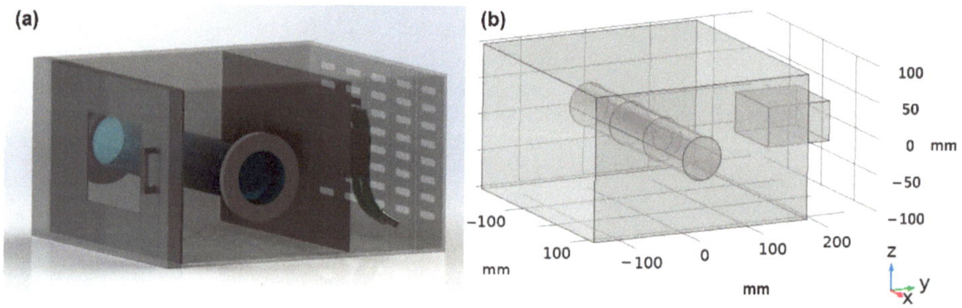


**Figure 2.** Structure of the microwave reactor for ex situ biomass tar elimination: (**a**) structural model; and (**b**) simulation model of the resonant cavity.

### 2.2. Parameter Setting

The numerical work of electromagnetic and heat transfer equations for microwave heating was computed using the COMSOL Multiphysics software, which solves the numerical models using the finite element method (FEM) [32]. Before the simulation, the model materials were first defined. Their property parameters were obtained from the COMSOL Multiphysics 5.6 built-in material library and listed in Table 1. The irradiation feed position, microwave power, silicon carbide bed length and flow velocity were the main parameters for exploration. The vertical position of the irradiation feed ($z_i$) was the main reactor parameter. Its initial location was in the center of the cavity back ($z_i$ = 0) and opposite straight to the silicon carbide bed as Figure 2b shows. The position was vertically adjusted with the $z_i$ values of 0, $\pm$5, $\pm$10 and $\pm$20 mm. The operation parameters including microwave power ($P$), silicon carbide bed length ($l_{SiC}$) and flow velocity ($v$) were adjusted in the range shown in Table 2. The electromagnetic propagation in the simulation was solved by Maxwell's equations that were subjected to certain governing equation and boundary conditions (Table 3). Other fixed simulation parameters were also listed in Table 3. Some assumptions were made in the simulation to simplify the error problem and reduce the computational time. These assumptions mainly include: (1) the dielectric and thermal conductivity properties were assumed to be independent of the temperature; (2) the SiC sample is homogeneous and isotropic; (3) the mass transfer and chemical reactions are not considered; (4) the quartz tube and the air are zero dielectric losses medium; and (5) no heat transfer was considered in the quartz tube, surrounding cavity and environment [30].

**Table 1.** Properties of the model materials.

| Item | Material | Property |
|---|---|---|
| Resonant cavity | Air | • Relative dielectric constant: 1r<br>• Relative permeability: 1<br>• $\rho$: (default) kg/m$^3$<br>• $C_p$(T): (default) J/(kg·K)<br>• K(T): (default) W/(m·K) |
| Tube | Quartz | • Thickness: 1 mm<br>• $C_p$(T): 730 J/(kg·K)<br>• K(T): 1.4 W/(m·K) |
| Carrier gas | Nitrogen (gas) | • Relative dielectric constant: 1.00058<br>• Relative permeability: 1<br>• Specific heat: 1.4<br>• $\rho$: 2 kg/m$^3$<br>• K(T): 3 W/(m·K)<br>• $C_p$(T): 3 J/(kg·K)<br>• Dynamic viscosity: 2 kg/m$^3$ |
| Wave absorber | Silicon carbide | • Relative dielectric constant: 10<br>• Relative permeability: 1<br>• Conductivity: $10^3$ S/m<br>• $C_p$(T): 1200 J/(kg·K)<br>• K(T): 450 W/(m·K) |

**Table 2.** The variable reactor and operation parameters.

| Parameters | Variable Range |
|---|---|
| $z_i$ (mm) | 0, ±5, ±10, ±15, ±20 |
| $P$ (W) | 700, 800, 900, 1000 |
| $l_{SiC}$ (mm) | 100, 110, 120, 130 |
| $v$ (m/s) | 0.02, 0.03, 0.04, 0.05 |

**Table 3.** The governing equation and boundary conditions of Maxwell's equations and other fixed simulation parameters.

| Items | Value or Expression |
|---|---|
| Governing equation * | $\nabla \times \mu_r{}^{-1}(\nabla \times E) - k_0^2(\varepsilon_r - \frac{j\sigma}{\omega\varepsilon_0})E = 0$ |
| Boundary conditions * | $\nabla^2 E + \omega^2 \mu\varepsilon E = 0$ (electric field boundary)<br>$\nabla^2 E + \omega^2 \mu\varepsilon H = 0$ (magnetic field boundary)<br>$n \times H = 0$ (thermal insulation) |
| Microwave frequency (GHz) | 2.45 |
| Wave mode | $TE_{10}$ |
| Initial value (K) | 293.15 |
| Absolute pressure (Pa) | 101,325 |
| Grid cell size (m) | 24.47 (maximum)<br>$7.3420 \times 10^{-4}$ (minimum) |
| Curvature factor | 0.6 |
| Narrow slit interval resolution | 0.5 |

* The terms in the equations include: $\mu_r$ is the relative permeability; $E$ is the electric field intensity; $H$ is the magnetic field intensity; $\omega$ is the angular frequency; $k_0$ is the wave number in free space and equals to $\omega/C_0$, in which $C_0$ is the speed of light in vacuum; $\varepsilon_r$ is the relative permittivity; $\sigma$ is the electrical conductivity; and $\varepsilon_0$ is the permittivity of vacuum ($8.85 \times 10^{-12}$ F/m).

### 2.3. Evaluation of the Electric and Thermal Fields

Distributions of the central electric intensity and temperature ($x$–$y$ plane, $z$ = 0) were obtained by the COMSOL software to evaluate the electric and thermal fields, respectively. The distribution bar with a gradient color was used to show the changes in electric intensity or temperature. The return loss is obtained to indicate the microwave reflection by the wave absorber and cavity. It is a measure of the effectiveness of power delivery from a transmission line to a load and is defined as [33]:

$$\text{Return loss} = -10\,\lg\frac{P_{ref}}{P_{in}}\ (\text{dB}) \tag{1}$$

where $P_{ref}$ and $P_{in}$ are the reflection and incident power, respectively. A higher return loss indicates less wave reflection (or a better match between the load and line), which means higher irradiation efficiency and energy utilization. The heating rate was also exported to give the temperature variation of the silicon carbide after microwave absorbing for 1200 s.

## 3. Results and Discussions

### 3.1. Effect of Irradiation Feed Position on the Simulated Electric and Thermal Fields

Figure 3a shows the variation of the electric fields with different irradiation feed positions. Due to the application of only one magnetron, the electric fields show non-uniform intensity [34,35], which is generally strong in the silicon carbide region, though it is covered by the non-conducting silicon carbide (dark blue center). As the irradiation feed position vertically moves, the electric field intensity changes accordingly. The electric intensity of the silicon carbide region becomes strong when the irradiation feed position moves from the side ($z_i = \pm 20$ mm) toward the center and becomes strongest when $z_i = \pm 5$. It is interesting to observe that when the irradiation feed position moves to the center ($z_i = 0$), the electric intensity becomes the weakest. The reason is that the microwave was reflected heavily and caused severe echo cancellation when the center of the waveguide and quartz tube was on the same plane. This can be evidenced by the microwave return loss showing that the wave reflection is the most severe (lowest return loss) when $z_i = 0$ (Figure 3b), indicating that placing the waveguide and quartz tube on the same plane results in low energy utilization. Salema and Afzal obtained a similar phenomenon that the maximum electric field intensity was not located at the center that opposites to the magnetron [30]. The reason they suggest is the larger heat losses or damped electric field at those regions, which supports the explanation about the energy loss caused by wave reflection in this simulation.

It is also reported that the sample position in the cavity greatly influences the heating characteristic [32,36] because of the high efficiency of converting electromagnetic energy into dielectric heating (typically more than 80% [37]) in conventional microwave heating systems. Therefore, the thermal fields of the reactor cavity were further investigated (Figure 4a). Since the silicon carbide is a good microwave absorber, the main high-temperature regions are centered in the silicon carbide (i.e., the middle of the quartz tube). The temperatures change with the irradiation feed position $z_i$ and reach the maximum when $z_i = \pm 5$, indicating that this irradiation feed position leads to the highest temperature of silicon carbide after wave absorbing. This is reasonable because in such a situation the silicon carbide region will have the strongest microwave intensity as indicated in Figure 3a. The temperature of the hottest region is about 1100 K, which is high enough for biomass tar cracking [38,39]. The heating rate of silicon carbide is also the highest when $z_i = \pm 5$ (Figure 4b), indicating the highest heating efficiency and energy utilization at this irradiation feed position.

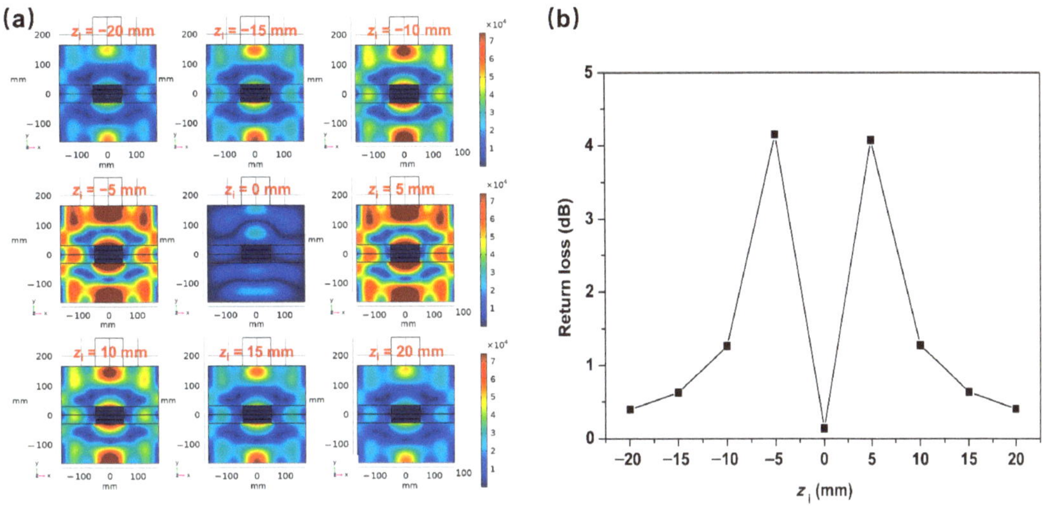


**Figure 3.** Effect of irradiation feed position on the simulated electric fields (the electric field unit: V/m; *x*–*y* plane, $z = 0$; $P = 1000$ W; $l_{SiC} = 100$ mm; $v = 0.02$ m/s): (**a**) variation of the electric fields with different irradiation feed positions; and (**b**) the microwave return loss with different irradiation feed positions.

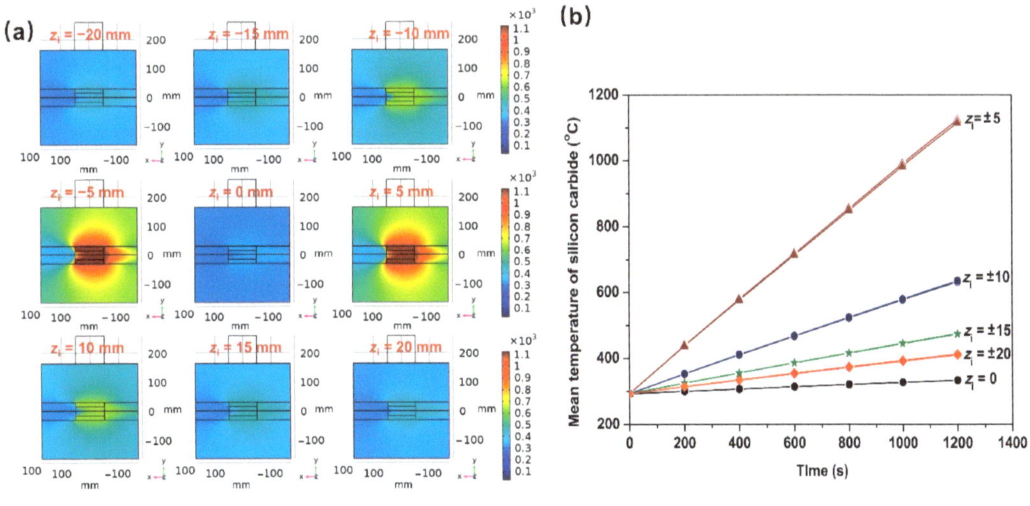


**Figure 4.** Effect of irradiation feed position on the simulated thermal fields (the temperature unit: K; *x*–*y* plane, $z = 0$; $P = 1000$ W; $l_{SiC} = 100$ mm; $v = 0.02$ m/s): (**a**) variation of the thermal fields with different irradiation feed positions; and (**b**) the heating rate of silicon carbide under microwave irradiation with different irradiation feed positions.

### *3.2. Effect of Microwave Power on the Simulated Electric and Thermal Fields*

As the input energy mainly comes from the microwave, the microwave power would have a direct impact on the electric intensity and temperature distributions. When the microwave power increases from 700 W to 1000 W, the electric field distribution changes little while the intensity strengthens a bit (Figure 5a). The microwave return loss shows almost no change with the microwave power variation (Figure 5b) because the reflection medium and environment are unchanged. The strong electric intensity caused by the high microwave power results in the high temperature of the silicon carbide bed. The central temperature of the silicon carbide bed is around 850 K at $P$ = 700 W and gradually increases to 1100 K when the microwave power rises to 1000 W (Figure 6a). Similarly, the heating rate of the silicon carbide bed also increases with the microwave power (Figure 6b). Therefore, high microwave power is positive for the bed temperature and heating rate. It is worth noting that although a high microwave power means a high energy input, it also decreases the heating time to the desired temperature. Considering the total energy that equals the power plus time, some researchers have found that the greater the microwave power is, the smaller the energy input is [32,40], which means a high microwave power can generally reduce the energy consumption.

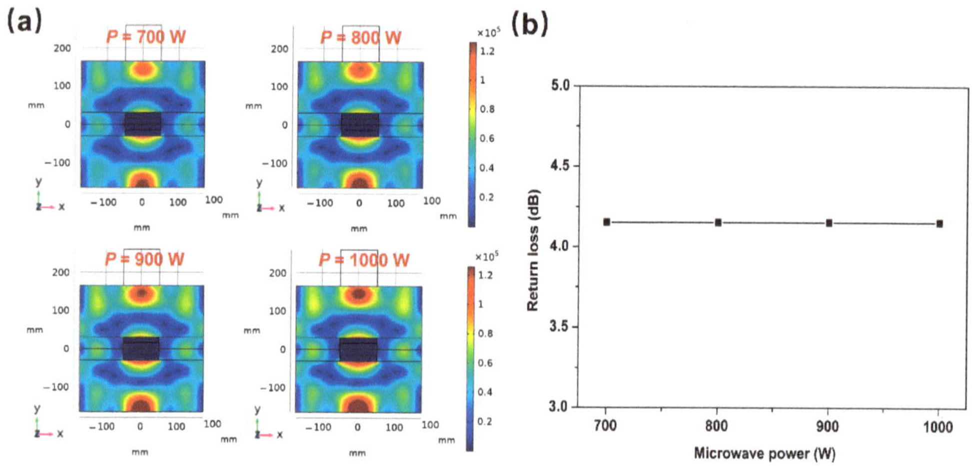


**Figure 5.** Effect of microwave power on the simulated electric fields (the electric field unit: V/m; $x$–$y$ plane, $z$ = 0; $z_i$ = 5 mm; $l_{SiC}$ = 100 mm; $v$ = 0.02 m/s): (**a**) variation of the electric fields with different microwave power; and (**b**) the microwave return loss with different microwave power.

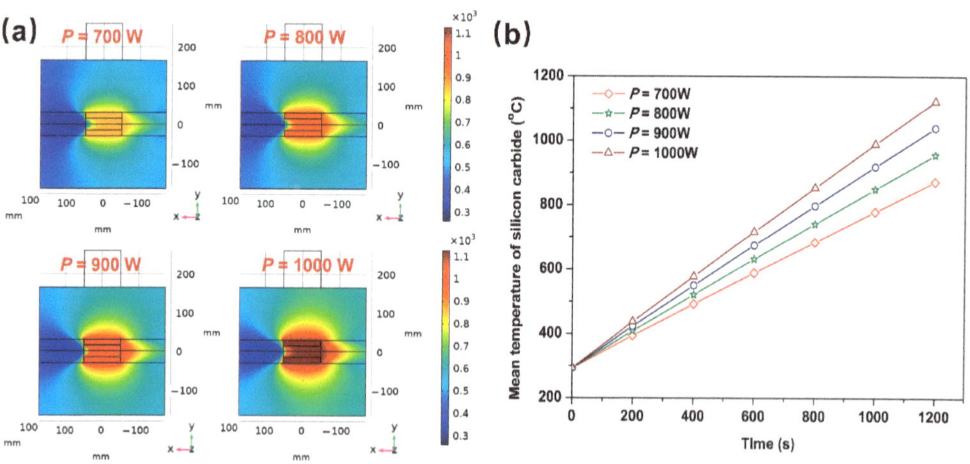


**Figure 6.** Effect of microwave power on the simulated thermal fields (the temperature unit: K; $x$–$y$ plane, $z$ = 0; $z_i$ = 5 mm; $l_{SiC}$ = 100 mm; $v$ = 0.02 m/s): (**a**) variation of the thermal fields with different microwave power; and (**b**) the heating rate of silicon carbide under microwave irradiation with different microwave power.

### 3.3. Effect of Silicon Carbide Bed Length on the Simulated Electric and Thermal Fields

As the microwave absorber, the silicon carbide provides the reaction site for tar elimination. It did not affect the microwave irradiation but influenced the wave diffraction, reflection, interference and, consequently, the electric fields as shown in Figure 7a. When the silicon carbide bed length increases from 100 mm to 130 mm, the overall electric fields, especially the strong intensity regions (red spots), become weak because the microwave reflection increases when the silicon carbide bed length gets longer (Figure 7b). Since the silicon carbide is the microwave absorber, its bed length also influences the bed temperature distribution (Figure 8a). It is observed that the temperature of the silicon carbide bed is highest when the silicon carbide bed length is 100 mm. The highest temperature can reach approximately 1100 K. Extending the bed length results in the obvious decrease of the silicon carbide bed temperature, as well as the heating rate (Figure 8b), because when the silicon carbide bed gets longer, not only does the microwave reflection become severe but, also, the side of the silicon carbide bed locates in the less-intensive electric region where the temperature is lower and less uniform than in the center [41]. These results indicate that with regard to the bed temperature for tar elimination, it is better to shorten the silicon carbide bed length. However, the silicon carbide bed length also influences the heating area, residence time (or space velocity) and, consequently, the efficiency of the biomass tar elimination. For example, extending the silicon carbide bed length increases the residence time of biomass tar on the silicon carbide bed; thus, it will promote the tar cracking if the tar cracking reactions are not fast. When the treatment capacity (or flow rate) is constant in a fixed-bed reactor, a long silicon carbide bed length may lead to a more sufficient biomass tar elimination. Therefore, the chemical kinetics of the biomass tar elimination reactions should also be considered when determining the silicon carbide bed length in the practical application.

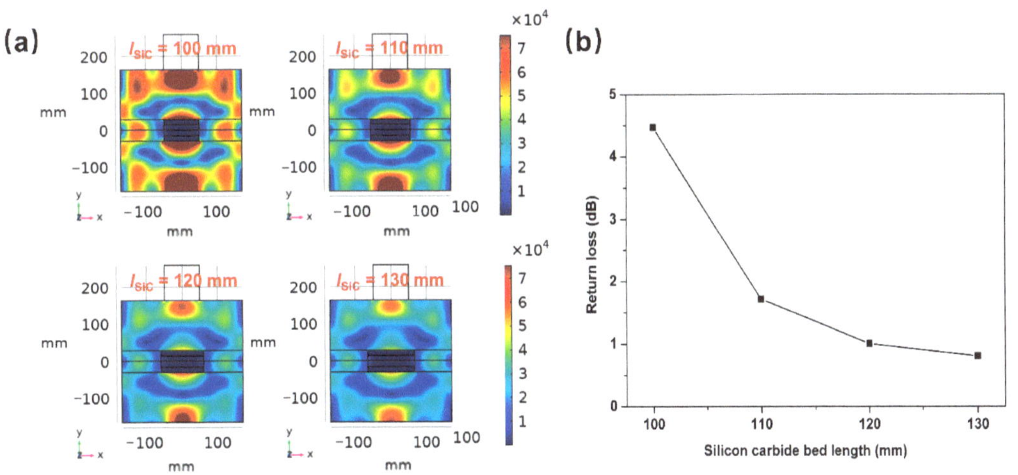


**Figure 7.** Effect of silicon carbide bed length on the simulated electric fields (the electric field unit: V/m; $x$–$y$ plane, $z$ = 0; $z_i$ = 5 mm; $P$ = 1000 W; $v$ = 0.02 m/s): (**a**) variation of the electric fields with different silicon carbide bed lengths; and (**b**) the microwave return loss with different silicon carbide bed lengths.

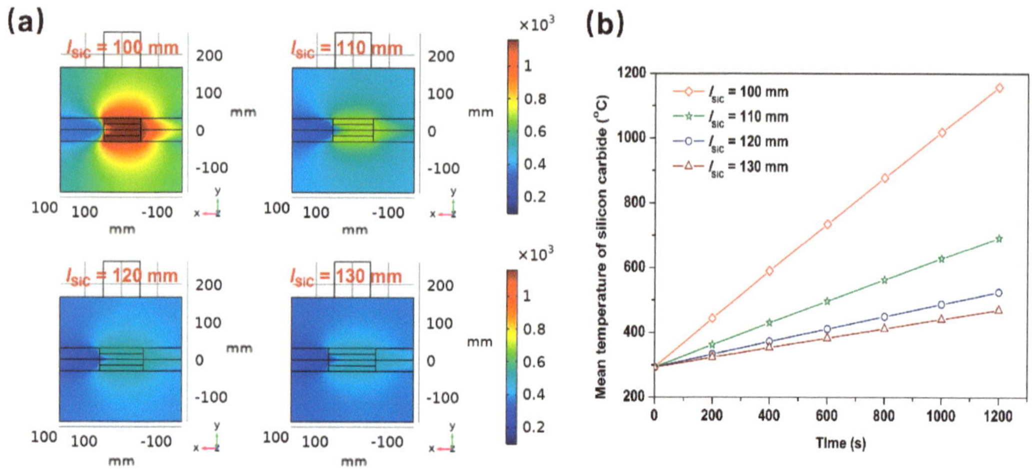


**Figure 8.** Effect of silicon carbide bed length on the simulated thermal fields (the temperature unit: K; $x$–$y$ plane, $z = 0$; $z_i = 5$ mm; $P = 1000$ W; $v = 0.02$ m/s): (**a**) variation of the thermal fields with different silicon carbide bed lengths; and (**b**) the heating rate of silicon carbide under microwave irradiation with different silicon carbide bed lengths.

Many researchers found that the sample size significantly affects the power and temperature distribution in microwave heating, as well as the heating rate within the samples [42–44]. For example, Brodie [45] reported that the thick samples generate heat just below the surface while thin samples generate heat from the center. Although the silicon carbide in this simulation is not the reactant, it does act as the sample to be heated. Its length, in general, affects the overall size and will certainly be an important factor to influence the thermal field and, consequently, the biomass tar elimination.

### 3.4. *Effect of Flow Velocity on the Simulated Thermal Fields*

The nitrogen is used as the carrier gas to blow the tar through the silicon carbide bed. It has no wave-absorbing ability and, therefore, its flow rate will not affect the electric fields under microwave irradiation. For this reason, only the thermal fields of the cavity with different flow velocities were investigated. The cold carrier gas blew from the left inlet of the quartz tube, making the hot region (especially the central hot region) of the silicon carbide shift a little to the right (Figure 9a). When the flow velocity is higher, more cold carrier gas will be blown in and take the heat of the silicon carbide away, resulting in the obvious non-uniform temperature distribution on the silicon carbide, as well as the energy loss. The heating rate of silicon carbide under microwave irradiation will also decrease a little when the flow velocity gets higher (Figure 9b). It should be noted that although a low flow velocity may be beneficial to relatively uniform temperature distribution, less energy loss and long residence time, it decreases the treatment capacity of the biomass tar. Determination of the flow velocity in a practical application for efficient tar removal should consider both the temperature requirement and the chemical reaction rate of the biomass tar elimination.

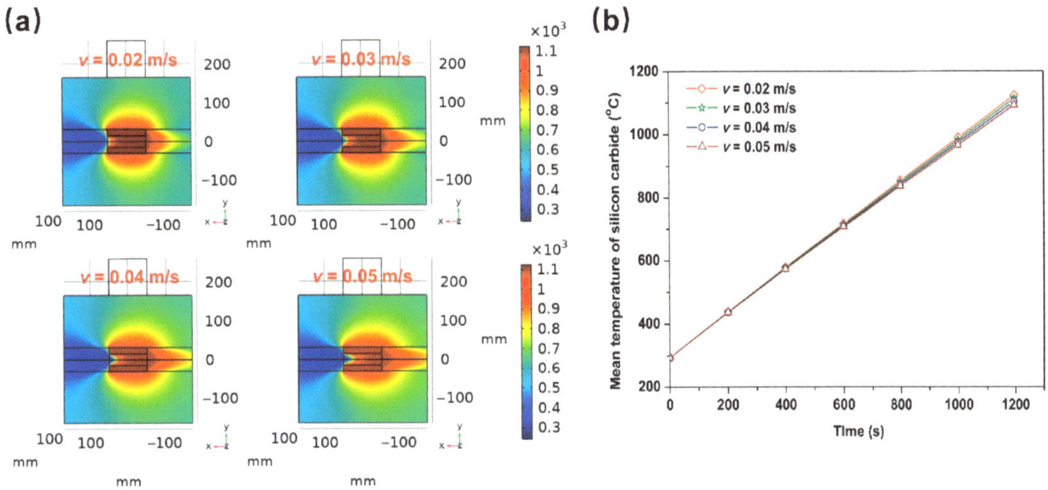


**Figure 9.** Effect of flow velocity on the simulated thermal fields (the temperature unit: K; $x$–$y$ plane, $z = 0$; $z_i = 5$ mm; $P = 1000$ W; $l_{SiC} = 100$ mm): (**a**) variation of the thermal fields with different flow velocities; and (**b**) the heating rate of silicon carbide under microwave irradiation with different flow velocities.

## 4. Conclusions

The electric and thermal fields of the microwave reactor used for ex situ biomass tar elimination were simulated under various reactor and operation parameters including irradiation feed position, microwave power, silicon carbide bed length and flow velocity. The main conclusions include:

(1) The irradiation feed position that locates 5 mm vertically to the silicon carbide plane ($z_i = 5$ mm) obtains the highest electric intensity on the silicon carbide bed. At this feed position, the silicon carbide can reach the highest temperature and fastest heating rate after wave absorbing, achieving the highest heating efficiency and energy utilization. Placing the waveguide and quartz tube on the same plane will result in the lowest energy utilization because of severe wave reflection.

(2) As the microwave power increases from 700 W to 1000 W, the temperature and heating rate of silicon carbide obviously increase, but the electric field distribution and microwave return loss changes little. The effect of silicon carbide bed length on electric fields is remarkable because of the microwave return loss. Extending the bed length from 100 mm to 130 mm will obviously decrease the bed temperature, as well as the heating rate. A high flow velocity leads to obvious non-uniform temperature distribution on silicon carbide because of the blowing of the cold carrier gas.

(3) To achieve a high microwave energy utilization and uniform bed temperature distribution, a suitable irradiation feed position ($z_i = 5$ mm), high microwave power ($P = 1000$ W), short silicon carbide bed length ($l_{SiC} = 100$ mm) and low flow velocity ($v = 0.02$ m/s) are preferred. However, determination of these parameters in a practical application of biomass tar elimination should also consider the other conditions such as chemical kinetics, treatment capacity and the temperature requirement.

**Author Contributions:** Conceptualization, C.Y. and Z.C.; methodology, C.Y. and Z.C.; data curation, K.Y. and F.Y.; formal analysis, C.Y.; validation, H.P.; funding acquisition, Z.C. and F.Y.; supervision, Z.C.; writing—original draft, C.Y.; writing—review and editing, Z.C. All authors have read and agreed to the published version of the manuscript.

**Funding:** This research was mainly funded by the Huzhou Science and Technology Project (No. 2020GZ28) and partly funded by the Zhejiang Basic Public Welfare Research Project (No. LGF22E080025) and Postdoctoral Research Project of Zhejiang Province (ZJ2020100).

**Institutional Review Board Statement:** Not applicable.

**Informed Consent Statement:** Not applicable.

**Conflicts of Interest:** The authors declared no potential conflicts of interest concerning the research, authorship and/or publication of this article.

## References

1. Han, J.; Kim, H. The reduction and control technology of tar during biomass gasification/pyrolysis: An overview. *Renew. Sustain. Energy Rev.* **2008**, *12*, 397–416. [CrossRef]
2. Li, C.; Suzuki, K. Tar property, analysis, reforming mechanism and model for biomass gasification—An overview. *Renew. Sustain. Energy Rev.* **2009**, *13*, 594–604. [CrossRef]
3. Liu, W.J.; Li, W.W.; Jiang, H.; Yu, H.Q. Fates of chemical elements in biomass during its pyrolysis. *Chem. Rev.* **2017**, *117*, 6367–6398. [CrossRef] [PubMed]
4. Guo, F.; Dong, Y.; Zhang, T.; Dong, L.; Guo, C.; Rao, Z. Experimental study on herb residue gasification in an air-blown circulating fluidized bed gasifier. Ind. Eng. *Chem. Res.* **2014**, *53*, 13264–13273. [CrossRef]
5. Zhang, X.; Chen, Z.; Cheng, L.; Xu, L.; Bi, X.; Liu, Q. Valorization of fluid petroleum coke for efficient catalytic destruction of biomass gasification tar. *J. Hazard. Mater.* **2022**, *424*, 127297. [CrossRef]
6. Li, X.; Liu, P.; Chen, W.; Wu, Y.; Lei, T.; Huang, S.; Li, Y.; Wu, S.; Wang, Z. Catalytic pyrolysis of toluene as biomass tar model component using Ni/HZSM-5 modified by $CeO_2$ and MgO promoters. *J. Anal. Appl. Pyrolysis* **2022**, *162*, 105436. [CrossRef]
7. Heidenreich, S.; Foscolo, P.U. New concepts in biomass gasification. *Prog. Energy Combust. Sci.* **2015**, *46*, 72–95. [CrossRef]
8. Janajreh, I.; Raza, S.S.; Valmundsson, A.S. Plasma gasification process: Modeling, simulation and comparison with conventional air gasification. *Energy Convers. Manag.* **2013**, *65*, 801–809. [CrossRef]
9. Devi, L.P.K.; Janssen, F.J.J.G. A review of the primary measures for tar elimination in biomass gasification processes. *Biomass Bioenergy* **2003**, *24*, 125–140. [CrossRef]
10. Zeng, X.; Ueki, Y.; Yoshiie, R.; Naruse, I.; Wang, F.; Han, Z.; Xu, G. Recent progress in tar removal by char and the applications: A comprehensive analysis. *Carbon Resour. Convers.* **2020**, *3*, 1–18. [CrossRef]
11. Anis, S.; Zainal, Z.A. Tar reduction in biomass producer gas via mechanical, catalytic and thermal methods: A review. *Renew. Sustain. Energy Rev.* **2011**, *15*, 2355–2377. [CrossRef]
12. Shen, Y.; Yoshikawa, K. Recent progresses in catalytic tar elimination during biomass gasification or pyrolysis—A review. *Renew. Sustain. Energy Rev.* **2013**, *21*, 371–392. [CrossRef]
13. Guan, G.; Kaewpanha, M.; Hao, X.; Abudula, A. Catalytic steam reforming of biomass tar: Prospects and challenges. *Renew. Sustain. Energy Rev.* **2016**, *58*, 450–461. [CrossRef]
14. Chan, F.L.; Tanksale, A. Review of recent developments in Ni-based catalysts for biomass gasification. *Renew. Sustain. Energy Rev.* **2014**, *38*, 428–438. [CrossRef]
15. Abdoulmoumine, N.; Adhikari, S.; Kulkarni, A.; Chattanathan, S. A review on biomass gasification syngas cleanup. *Appl. Energy* **2015**, *155*, 294–307. [CrossRef]
16. Cheng, L.; Wu, Z.; Zhang, Z.; Guo, C.; Ellis, N.; Bi, X.; Paul Watkinson, A.; Grace, J.R. Tar elimination from biomass gasification syngas with bauxite residue derived catalysts and gasification char. *Appl. Energy* **2020**, *258*, 114088. [CrossRef]
17. Hervy, M.; Weiss-Hortala, E.; Pham Minh, D.; Dib, H.; Villot, A.; Gérente, C.; Berhanu, S.; Chesnaud, A.; Thorel, A.; Le Coq, L.; et al. Reactivity and deactivation mechanisms of pyrolysis chars from bio-waste during catalytic cracking of tar. *Appl. Energy* **2019**, *237*, 487–499. [CrossRef]
18. Motasemi, F.; Afzal, M.T. A review on the microwave-assisted pyrolysis technique. *Renew. Sustain. Energy Rev.* **2013**, *28*, 317–330. [CrossRef]
19. Beneroso, D.; Bermúdez, J.M.; Montes-Morán, M.A.; Arenillas, A.; Menéndez, J.A. Microwave-induced cracking of pyrolytic tars coupled to microwave pyrolysis for syngas production. *Bioresour. Technol.* **2016**, *218*, 687–691. [CrossRef]
20. Muley, P.D.; Henkel, C.E.; Aguilar, G.; Klasson, K.T.; Boldor, D. Ex situ thermo-catalytic upgrading of biomass pyrolysis vapors using a traveling wave microwave reactor. *Appl. Energy* **2016**, *183*, 995–1004. [CrossRef]
21. Guo, F.; Dong, Y.; Tian, B.; Du, S.; Liang, S.; Zhou, N.; Wang, Y.; Chen, P.; Ruan, R. Applications of microwave energy in gas production and tar removal during biomass gasification. *Sustain. Energy Fuels* **2020**, *4*, 5927–5946. [CrossRef]
22. Li, J.; Tao, J.; Yan, B.; Jiao, L.; Chen, G.; Hu, J. Review of microwave-based treatments of biomass gasification tar. *Renew. Sustain. Energy Rev.* **2021**, *150*, 111510. [CrossRef]
23. Valderrama Rios, M.L.; González, A.M.; Lora, E.E.S.; Almazán del Olmo, O.A. Reduction of tar generated during biomass gasification: A review. *Biomass Bioenergy* **2018**, *108*, 345–370. [CrossRef]
24. Kostyniuk, A.; Grilc, M.; Likozar, B. Catalytic cracking of biomass-derived hydrocarbon tars or model compounds to form biobased benzene, toluene, and xylene isomer mixtures. *Ind. Eng. Chem. Res.* **2019**, *58*, 7690–7705. [CrossRef]

25. Li, J.; Tao, J.; Yan, B.; Cheng, K.; Chen, G.; Hu, J. Microwave reforming with char-supported nickel-cerium catalysts: A potential approach for thorough conversion of biomass tar model compound. *Appl. Energy* **2020**, *261*, 114375. [CrossRef]
26. Chen, G.; Li, J.; Cheng, Z.; Yan, B.; Ma, W.; Yao, J. Investigation on model compound of biomass gasification tar cracking in microwave furnace: Comparative research. *Appl. Energy* **2018**, *217*, 249–257. [CrossRef]
27. Zhang, Y.; Song, Z.; Yan, Y.; Zhao, X.; Sun, J.; Mao, Y.; Wang, W. Performance of Fe/SiC catalysts for cracking of toluene under microwave irradiation. *Int. J. Hydrogen Energy* **2018**, *43*, 7227–7236. [CrossRef]
28. Claude, V.; Courson, C.; Köhler, M.; Lambert, S.D. Overview and essentials of biomass gasification technologies and their catalytic cleaning methods. *Energy Fuels* **2016**, *30*, 8791–8814. [CrossRef]
29. Li, J.; Jiao, L.; Tao, J.; Chen, G.; Hu, J.; Yan, B.; Mansour, M.; Guo, Y.; Ye, P.; Ding, Z.; et al. Can microwave treat biomass tar? A comprehensive study based on experimental and net energy analysis. *Appl. Energy* **2020**, *272*, 115194. [CrossRef]
30. Salema, A.A.; Afzal, M.T. Numerical simulation of heating behaviour in biomass bed and pellets under multimode microwave system. *Int. J. Therm. Sci.* **2015**, *91*, 12–24. [CrossRef]
31. Zhou, R.; Yang, X.; Sun, D.; Jia, G. Multiple tube structure for heating uniformity and efficiency optimization of microwave ovens. Eur. Phys. *J. Appl. Phys.* **2015**, *69*, 20201. [CrossRef]
32. Hong, Y.; Lin, B.; Li, H.; Dai, H.; Zhu, C.; Yao, H. Three-dimensional simulation of microwave heating coal sample with varying parameters. *Appl. Therm. Eng.* **2016**, *93*, 1145–1154. [CrossRef]
33. Bird, T.S. Definition and Misuse of Return Loss [Report of the Transactions Editor-in-Chief]. *IEEE Antennas Propag.* **2009**, *51*, 166–167. [CrossRef]
34. Wang, S.; Hu, Z.; Han, Y.; Gu, Z. Effects of magnetron arrangement and power combination of microwave on drying uniformity of carrot. *Dry. Technol.* **2013**, *31*, 1206–1211. [CrossRef]
35. Wang, S.; Yang, R.; Han, Y.; Gu, Z. Effects of magnetron arrangement and power combination on temperature field uniformity of microwave drying of carrot. *Dry. Technol.* **2016**, *34*, 912–922. [CrossRef]
36. Pedreño-Molina, J.L.; Monzó-Cabrera, J.; Catalá-Civera, J.M. Sample movement optimization for uniform heating in microwave heating ovens. *Int. J. RF Microw. Comput.-Aided Eng.* **2007**, *17*, 142–152. [CrossRef]
37. Meredith, R. *Engineer's Handbook of Industrial Microwave Heating*; The Institution of Electric Engineers: London, UK, 1998.
38. Ren, J.; Cao, J.; Zhao, X.; Liu, Y. Recent progress and perspectives of catalyst design and downstream integration in biomass tar reforming. *Chem. Eng. J.* **2022**, *429*, 132316. [CrossRef]
39. Ren, J.; Liu, Y.; Zhao, X.; Cao, J. Biomass thermochemical conversion: A review on tar elimination from biomass catalytic gasification. *J. Energy Inst.* **2020**, *93*, 1083–1098. [CrossRef]
40. Samanli, S. A comparison of the results obtained from grinding in a stirred media mill lignite coal samples treated with microwave and untreated samples. *Fuel* **2011**, *90*, 659–664. [CrossRef]
41. Klinbun, W.; Rattanadecho, P.; Pakdee, W. Microwave heating of saturated packed bed using a rectangular waveguide (TE10 mode): Influence of particle size, sample dimension, frequency, and placement inside the guide. *Int. J. Heat Mass Transfer* **2011**, *54*, 1763–1774. [CrossRef]
42. Rattanadecho, P. The simulation of microwave heating of wood using a rectangular wave guide: Influence of frequency and sample size. *Chem. Eng. Sci.* **2006**, *61*, 4798–4811. [CrossRef]
43. Curet, S.; Rouaud, O.; Boillereaux, L. Effect of sample size on microwave power absorption within dielectric materials: 2D numerical results vs. closed-form expressions. *AIChE J.* **2009**, *55*, 1569–1583. [CrossRef]
44. Ayappa, K.G.; Davis, H.T.; Crapiste, G.; Davis, E.A.; Gordon, J. Microwave heating: An evaluation of power formulations. *Chem. Eng. Sci.* **1991**, *46*, 1005–1016. [CrossRef]
45. Brodie, G. The Influence of Load Geometry on Temperature Distribution During Microwave Heating. *Trans. ASABE* **2008**, *51*, 1401–1413. [CrossRef]

water

MDPI

*Article*

# Effects of Calcium on the Removal of Ammonium from Aged Landfill Leachate by Struvite Precipitation

**Hussein Rayshouni and Mahmoud Wazne** *

Civil Engineering, Lebanese American University, 301 Bassil Building, Byblos P.O. Box 36, Lebanon; hussein.rayshouni@lau.edu.lb
* Correspondence: mahmoud.wazne@lau.edu.lb

**Abstract:** Ammonium in landfill leachates is a major contributor to environmental degradation if not effectively treated. However, it could be converted to a valuable fertilizer when it is co-precipitated with phosphate and magnesium as struvite. Low-cost magnesium and phosphate sources are sought to offset the co-precipitation treatment costs, but most of the identified alternative magnesium sources have significant amounts of calcium, which may negatively impact the ammonium removal rates. In this study, the effects of calcium on ammonium removal from high-strength aged field landfill leachate as struvite were investigated. Laboratory-scale batch tests were conducted to assess the effects of the pH, $Mg^{2+}$:$NH_4^+$:$PO_4^{3-}$, and $Ca^{2+}$:$Mg^{2+}$ molar ratios on ammonium removal. Magnesium chloride salt was used as a model dissolved magnesium source, whereas different compounds derived from dolomite ($CaMg(CO_3)_2$) were used as model solid-phase magnesium sources. X-ray powder diffraction and activity ratio diagrams were used to delineate the ammonium removal mechanisms and struvite stability. The ammonium removal rate of the magnesium salt decreased from approximately 97% to 70%, upon increasing the $Ca^{2+}$:$Mg^{2+}$ molar ratio from 0 to 1.0, for the $Mg^{2+}$:$NH_4^+$:$PO_4^{3-}$ molar ratio of 1.25:1:1.25 and pH = 9.5. For similar pH values, as well as the $Mg^{2+}$:$NH_4^+$:$PO_4^{3-}$ and $Ca^{2+}$:$Mg^{2+}$ molar ratios, the ammonium removal rates by the dolomite-derived compounds reached up to 55%, which highlighted the limited availability of magnesium in solid phases, in addition to the negative impacts of calcium. The diffractometric analysis and thermodynamic calculations revealed the stable regions of struvite in the presence of competing solid phases. The new findings in this study could aid in the design of ammonium and phosphate removal and recovery systems by struvite precipitation.

**Keywords:** landfill leachate; ammonium; struvite; dolomite; activity ratio diagram

**Citation:** Rayshouni, H.; Wazne, M. Effects of Calcium on the Removal of Ammonium from Aged Landfill Leachate by Struvite Precipitation. *Water* **2022**, *14*, 1933. https://doi.org/10.3390/w14121933

Academic Editors: Dibyendu Sarkar, Rupali Datta, Prafulla Kumar Sahoo and Mohammad Mahmudur Rahman

Received: 20 April 2022
Accepted: 21 May 2022
Published: 16 June 2022



## 1. Introduction

About 40 percent of the produced solid waste is disposed of in landfills, with another 33 percent disposed of in open dumpsites [1]. The percolation of rainwater through the disposed solid waste generates leachates containing more than 200 potentially toxic compounds [2]. Similarly, the moisture embedded in the solid waste also results in the release of landfill leachates upon waste compaction. If not collected and treated, the leachates pose a severe environmental threat to the adjacent soil and water compartments. Of particular concern is dissolved ammonium, where the un-ionized form (free ammonia) exerts the most toxic effects on aquatic biota [3]. The free ammonia (FA) fraction increases at alkaline pH values such as those present at aged landfills, where ammonium concentration could reach up to 20,000 mg/L [4].

Ammonium is commonly removed from wastewater biologically, through nitrification–denitrification processes. However, the lack of sufficient electron donors in aged landfill leachates could render this process challenging. Moreover, the presence of a high ammonium concentration also exerts negative impacts on biological treatments. Ammonium concentrations below 200 mg/L can be beneficial, whereas greater concentrations may cause inhibitory effects [5]. Inhibitory effects were reported for ammonium concentrations as low as 1400 mg/L, with the free ammonia species exerting the most inhibitory effects on the methanogens [6–11].

The negative effects of a high ammonium concentration in wastewater on biological treatments can be mitigated by ammonium removal pretreatment. The most commonly used pretreatment methods include ammonia air stripping and ammonium co-precipitation as struvite [12–16]. Struvite is a mineral, with a molecular formula of $NH_4MgPO_4{\cdot}6H_2O$, that forms naturally in wastewater when magnesium, ammonium, and phosphate concentrations reach saturation level. It is also an effective slow-release fertilizer [17]. Ammonia air stripping technology suffers from the difficulty of operation and potential for ammonia release. It also requires large amounts of alkali and acids to facilitate the ammonia stripping and subsequent wastewater neutralization [18,19]. As for ammonium removal as struvite, most wastewaters contain low concentrations of magnesium and phosphate, relative to ammonium, thus requiring significant amounts of magnesium and phosphate reagents, which renders the treatment costly. To avoid the high costs of the chemical reagents, investigators searched for low-cost alternative sources of magnesium and phosphorous. Alternative phosphorous sources research focused mainly on bone wastes and waste byproducts [19–21]. On the other hand, many sources of low-cost magnesium were identified, such as wood ash, magnesium oxide production byproduct, magnesite, magnesia, dolomite, bittern, and seawater brine [22,23].

The majority of the identified alternative magnesium sources contain significant amounts of calcium, which may negatively impact struvite precipitation. Numerous studies reported on the negative impact of calcium on struvite precipitation; however, most of these studies were focused on phosphorous recovery and the effects of calcium originating from the wastewater [24–27]. The relatively low concentration of calcium in wastewater may have strong impact on the struvite crystal quality, but not necessarily the phosphate removal rates. Conversely, for studies focusing on ammonium removal from landfill leachates, where ammonium concentration could reach up to 20,000 mg/L, the calcium concentration originating from the alternative magnesium source could have significant effects on the ammonium removal rates. Nonetheless, the majority of these studies did not adequately report on the calcium effects originating from the alternative magnesium source [22,28,29]. Regarding the stoichiometric amounts of magnesium needed to precipitate struvite (where magnesium is originating from the alternative magnesium sources), the $Ca^{2+}$:$Mg^{2+}$ molar ratios in the reactors could reach up to 0.48, 0.1, 0.024, 0.098, 1.0, and 0.34 for wood ash, magnesium oxide production byproducts, magnesite, magnesia, dolomite, and bittern, respectively [23,30,31]. The corresponding potential calcium concentration in the reactors for the aforementioned $Ca^{2+}$:$Mg^{2+}$ ratio and ammonium concentration at 20,000 mg/L in the feed could reach 44,417mg/L.

One of the most promising alternative magnesium sources is bittern. Bittern is a brine residue that forms after the vaporization and crystallization of sodium chloride from sea water or lakes. The magnesium content in bittern is available in the dissolved phase (similar to the presence of magnesium in $MgCl_2$ salt), which readily enables its reaction with phosphate and ammonium to form struvite. Shin et al. [32] reported comparable phosphorus removal rates from wastewater between bittern (97%) and magnesium chloride salt (99%), but lower ammonium removal rates were achieved by bittern (72%) than by magnesium chloride (83%). The reduced efficiency of bittern for removing ammonium is most likely due to the presence of calcium in the used bittern. The use of bittern to precipitate struvite could also introduce other ions, in addition to calcium, such as chloride, potassium, sodium, and sulfate. However, the use of bittern could be hampered by accessibility or transportation costs in some localities [33]. Conversely, all of the other alternative magnesium sources are present in solid phases. The presence of magnesium in solid phases limits its availability and reactivity to form struvite. However, studies reported on the beneficial use of solid-phase alternative magnesium sources because of their high alkalinity content which could help mitigate against pH drop upon struvite precipitation [22,28].

This study investigated the effects of calcium on ammonium removal from high strength aged field landfill leachate. Laboratory-scale batch tests were conducted to assess the effects of pH, $Mg:NH_4:PO_4$ stoichiometric ratios, and $Ca^{2+}:Mg^{2+}$ molar ratios on ammonium removal. Magnesium chloride salt was used as a model dissolved magnesium source, whereas different compounds derived from dolomite ($CaMg(CO_3)_2$) were used as model solid-phase magnesium sources. X-ray powder diffraction and activity ratio diagrams were used to delineate the removal mechanisms and assess the stability of struvite in the presence of competing solid phases.

## 2. Material and Methods

### 2.1. Materials

The leachate was collected from a municipal solid waste landfill in Tripoli, Lebanon (34°27′21.539″ N, 35°50′22.559″ E). The landfill is located along the Mediterranean coastline, and it cover an area of 60,000 $m^2$. It started operations in 1980 and was planned to close in 2012; however, due to a lack of alternatives, it was still receiving an average of 450 tons/day, as of 2020 [34]. A large volume of the leachates was collected, mixed, and centrifuged at 13,000 rpm (19,685× *g*) for 30 min before characterization. The centrifuged samples were then stored in a fridge at 4 °C. The samples needed for the set experiments were allowed to reach room temperature for one hour before use. The dolomite was obtained from a local sand supplier in North Lebanon. It was first sieved using a sieve #40 to remove debris; then, it was manually grinded and sieved using sieve #200 and mechanical sieve shaker. It was then stored in plastic bags until used. Tri-sodium phosphate ($Na_3PO_4 \cdot 12H_2O$) was used as the phosphate source. Magnesium chloride ($MgCl_2$) and calcium chloride ($CaCl_2$) were used in batch experiments to assess the efficacy of the magnesium source and effects of calcium on struvite precipitation, respectively. Hydrochloric acid (HCl) and sodium hydroxide (NaOH) were used to adjust pH. All chemicals used were of analytical grade, and they were obtained from Fisher Scientific (Hampton, NH, USA).

### 2.2. Struvite Precipitation Batch Tests

For the struvite batch precipitation tests, 50 mL of centrifuged leachate was added to a beaker containing a magnetic stirring rod, which was then placed on a LabTech LMS-1003 hot plate magnetic stirrer. The phosphate, magnesium, and calcium (when applicable) sources were then added to the leachates. The reactants were mixed for 30 min, while pH was controlled at a set value ±0.05 pH units, using HCl and NaOH, and monitored using a WPA CD 500 pH meter (Cambridge, UK). After mixing for a set reaction time, the mixture was allowed to settle for 15 min, then centrifuged at 13,000 rpm (19,685× *g*) for 30 min, before analysis. Similar studies reported mixing times less than or equal to 30 min [18,19,21,28,35,36]. Resulting precipitates were dried in a desiccator and stored for analysis. All experiments were performed in triplicate at a room temperature of 20 ± 2 °C. The supernatants were analyzed for $NH_3$, $PO_4^{3-}$, $Mg^{2+}$, and $Ca^{2+}$, whereas the precipitates were analyzed for solid phases using XRPD.

### 2.3. Analytical Methods

A Hach DR3900 spectrophotometer was used for all spectrophotometry analyses procedures, and a Hach DRB200 digester block was used for the analytical procedures requiring digestion. Ammonia nitrogen $NH_3$-N was measured using the Hach salicylate method 10031. Chemical oxygen demand (COD) was measured using the Hach method 8000 with high-range digestion vials. Biochemical oxygen demand (BOD) was conducted as per Standard Method 5210B. Total organic carbon (TOC) was measured using Hach method 10128. Phosphate was measured using Hach orthophosphate method 8048; $Mg^{2+}$ and $Ca^{2+}$ and other heavy metals were measured using a PerkinElmer PinnAAcle 900H atomic absorption spectrophotometer. Chloride and sodium ions were quantified by ion chromatography, using an 882 Compact IC Plus (Metrohm, Herisau, Switzerland) equipped with a chemical suppressor module and conductivity detector. Chloride ions were detected after separation using a Metrosep A supp 5 column (250 × 4 mm), with an eluent consisting of 3.5 mmol·L$^{-1}$ sodium carbonate and 1 mmol·L$^{-1}$ sodium hydrogen carbonate, as well as an eluent flow rate of 0.7 mL·min$^{-1}$ and suppressor regenerant of 0.1 mol. L$^{-1}$ sulfuric acid. Sodium ions were detected after separation using a Metrosep C4 column (150 × 4.0 mm), with an eluent consisting of 4 µmol·L$^{-1}$ nitric acid (69%) at a flow rate of 0.9 mL·min$^{-1}$. Injection volumes of 20 µL were used for all measurements. Calcinations were performed in a Nabertherm GmbH burnout furnace L3/12. Centrifugation was performed in a TermoScientific fixed-bucket centrifuge.

Check standards were performed routinely, as well as matrix spikes, in order to check the accuracy of the instruments and analytical procedures and ensure that there were no interferences affecting the results. Three control experiments were conducted to check whether any of the ammonium removal was not due to struvite precipitation. The first experiment was conducted without adding any reagents to the leachate. In the second experiment, only dolomite was added to the leachate; in the third experiment, only $Na_3PO_4 \cdot 12H_2O$ was added. All control experiments were conducted at a reaction time of 30 min, using a pH value of 9.5 and $Mg^{2+}$:$NH_4^+$:$PO_4^{3-}$ molar ratio of 1.25:1:1.25. No ammonium removal was observed in all control experiments.

### 2.4. X-ray Powdered Diffraction

The solid phases were analyzed using a Bruker D8 Advance X-Ray diffractometer. Diffractometry was conducted at 40 kV and 40 mA using a Cu tube. Data were collected in the 2θ range of 5° to 65° using 3080 steps, with 0.2 s between each step. Qualitative analysis of the obtained patterns was performed in reference to the International Center of Diffraction Data (ICDD) database, using the DIFFRAC.EVA software Version 4.2.1.

### 2.5. Activity Ratio Diagram Modeling

Activity ratio diagrams were constructed to identify the solid phases controlling the solubility of the major ions responsible for the removal of ammonium from the leachate [37–39]. The chemical database of the geochemical speciation model, Visual Minteq, was used to plot the activity ratios of pertinent solid phases as a function of various ions [40]. The Truesdell–Jones equation was used to calculate the activity coefficients for all species, except for phosphate, due to the lack of pertinent thermodynamic data [41]. The Davies equation was used to calculate the activity coefficient of phosphate.

## 3. Results and Discussion

### 3.1. Leachate Characterization

The characterization results of the untreated landfill leachate are presented in Table 1.

**Table 1.** Characterization of the untreated leachate.

| Parameter | Value (mg/L) |
|---|---|
| pH | 8.26 |
| $BOD_5$ | 1130 |
| COD | 12,050 |
| TOC | 3400 |
| $Cl^-$ | 2676.4 |
| $NO_3^-$ | 2.5 |
| $PO_4^{3-}$ | 57.6 |
| $SO_4^{2-}$ | 47.7 |
| $Ca^{2+}$ | 58.78 |
| $K^+$ | 2483 |
| $Mg^{2+}$ | 31.82 |
| $Na^+$ | 2000 |

**Table 1.** *Cont.*

| Parameter | Value (mg/L) |
|---|---|
| $NH_4^+$ | 4584 |
| Al | 3.38 |
| As | 0.062 |
| Cd | <0.005 |
| Cr | 1.77 |
| Cu | 0.536 |
| Hg | 0.007 |
| Pb | 0.127 |
| Zn | 0.430 |

The measured $BOD_5/COD$ ratio and pH of the leachate were approximately 0.093 and 8.26, respectively, indicating that the landfill is classified as mature and in the methanogenic phase [2,42,43]. The landfill leachate had an average ammonium concentration of about 4583 mg/L, which is greater than the typical range of values of ammonium in landfill leachates of 500–2000 mg/L $NH_3$-N [2]. The free ammonia (FA) fraction in the leachates at a pH of 8.26 was calculated at approximately 6.7% or 290 mg/L, at a temperature of 20 °C [44]. This value is much greater than the concentrations of free ammonia reported to cause inhibition in biological treatments. For example, Liu et al. [5] reported a free ammonia inhibition of 40% on denitrifiers at a FA concentration of 16 mg $NH_3$/L and 80% inhibition on methanogens at a FA concentration of 40 mg $NH_3$-N/L. The characterization results also showed the presence of low concentrations of calcium and magnesium at approximately 58.8 and 32 mg/L, respectively, as compared to values reported in the literature. The concentration of phosphate at 57.6 mg/L was relatively greater that of the average value of 6 mg/L reported in landfill leachates. The concentrations of cadmium and lead at <0.005 and 0.127 mg/L, respectively, were close to the average reported values at 0.005 and 0.09 mg/L, respectively. However, the concentrations of copper and chromium at 0.536 and 1.77 mg/L, respectively, were relatively greater than the average values cited in the literature at 0.065 and 0.28 mg/L, respectively [2]. The relatively low concentrations of the heavy metals in the leachate could be due to the alkaline pH of the leachate, as metal cations have lower solubility at high pH. As for arsenic, it is expected to be present in an anionic form, where its low concentration could indicate a trace level presence or strong arsenic sorptive affinity of the solid waste constituents. The presence of chromium and copper at concentrations relatively greater than the average values reported at landfills could indicate the potential sources of contamination and lack of apparent sequesters.

### 3.2. *Ammonium Removal Using Magnesium Chloride Salt*

Batch test experiments were conducted to remove ammonium by struvite precipitation using magnesium and phosphate salt. These experiments were conducted to determine the optimal $Mg^{2+}$:$NH_4^+$:$PO_4^{3-}$ molar ratio and pH for ammonium removal from the leachate, as well as to use the test results as a benchmark for the experiments using alternative magnesium sources. All tested conditions are shown in Table 2, and the resulting concentrations of $Mg^{2+}$, $NH^4$+, $PO_4^{3-}$, and $Ca^{2+}$ are shown in Table S1.

**Table 2.** Experimental conditions tested.

| Experiment No. | $Mg^{2+}$:$NH_4^+$:$PO_4^{3-}$ Molar Ratios | $Mg^{2+}$ Source | $Ca^{2+}$:$Mg^{2+}$ Molar Ratio | $Ca^{2+}$ Source | pH |
|---|---|---|---|---|---|
| 1 | 1:1:1 | $MgCl_2$ | 0 | - | 9.5 |
| 2 | 1.25:1:1 | $MgCl_2$ | 0 | - | 9.5 |
| 3 | 1:1:1.25 | $MgCl_2$ | 0 | - | 9.5 |
| 4 | 1.25:1:1.25 | $MgCl_2$ | 0 | - | 9.5 |

**Table 2.** *Cont.*

| Experiment No. | $Mg^{2+}:NH_4^+:PO_4^{3-}$ Molar Ratios | $Mg^{2+}$ Source | $Ca^{2+}:Mg^{2+}$ Molar Ratio | $Ca^{2+}$ Source | pH |
|---|---|---|---|---|---|
| 5 | 1.25:1:1.25 | $MgCl_2$ | 0 | - | 8.5 |
| 6 | 1.25:1:1.25 | $MgCl_2$ | 0 | - | 10.5 |
| 7 | 1.25:1:1.25 | $MgCl_2$ | 0.1 | $CaCl_2$ | 9.5 |
| 8 | 1.25:1:1.25 | $MgCl_2$ | 0.2 | $CaCl_2$ | 9.5 |
| 9 | 1.25:1:1.25 | $MgCl_2$ | 0.5 | $CaCl_2$ | 9.5 |
| 10 | 1.25:1:1.25 | $MgCl_2$ | 1.0 | $CaCl_2$ | 9.5 |
| 11 | 1.25:1:1.25 | $CaMg(CO_3)_2$ | 1.0 | $CaMg(CO_3)_2$ | 9.5 |
| 12 | 1.25:1:1.25 | MgO | 1.0 | $CaCO_3$ | 9.5 |
| 13 | 1.25:1:1.25 | MgO | 1.0 | CaO | 9.5 |

### 3.2.1. Effects of Malar Ratios

The effects of the molar ratios of phosphate and magnesium to ammonium on the removal rates of ammonium were investigated for molar ratios ranging between 1 and 1.25 (Table 2). Equimolar ratios of $Mg^{2+}:NH_4^+:PO_4^{3-}$ are needed to form struvite, but the presence of other constituents in the leachates can reduce the precipitation efficiency, which results in a need for higher magnesium and phosphate molar ratios. The experiments were conducted using a pH value of 9.5.

The results of Exp.1, with $Mg^{2+}:NH_4^+:PO_4^{3-}$ molar ratios of 1:1:1, indicated an ammonium removal rate of approximately 81%, with phosphate and magnesium removal rates of 82% and 99.9%, respectively (Figure 1a). It appears that magnesium dosage was limiting, considering that it was nearly fully exhausted. It also appears that using $Mg^{2+}$ and $PO_4^{3-}$ at equal molar ratios to ammonium might not be optimal for the removal of ammonium. The incomplete ammonium removal and high residual phosphate concentration in the supernatant indicate the potential for improvement in ammonium removal by increasing the magnesium dosage. The need for molar ratios of magnesium and phosphate to ammonium that are greater than 1.0 is attributed to the formation of insoluble and, possibly, amorphous magnesium and phosphate compounds, other than struvite, such as $Mg_3(PO_4)_2$ and $Ca_3(PO_4)_2$ [19,45–48]. For example, Huang et al. [49] showed that, as the ratio of $Ca(OH)_2$ to NaOH (used to adjust pH) increased in the range of 0:1–1:1, ammonium removal efficiency decreased from 85% to 75%. Increasing magnesium dosage alone (1.25:1:1) resulted in an increase in ammonium removal rate to approximately 92%, with phosphate and magnesium removal rates at approximately 99.7% and 96%, respectively. On the other hand, increasing the phosphate dosage ($PO_4^{3-}:NH_4^+$) alone to 1.25:1 (Exp.3) resulted in an ammonium removal efficiency of approximately 83%. The corresponding removal rates of magnesium and phosphate were approximately 99.9% and 73%, respectively. This indicates that the excess amounts of added phosphate remained in the solution. When both magnesium and phosphate dosages were increased to 1.25:1, with respect to ammonium (Exp.4: 1.25:1:1.25), the ammonium removal rate increased to approximately 97%, with magnesium and phosphate removals of 99.9% and 90%, respectively. Finally, considering the concern regarding the presence of significant amounts of residual phosphate in the treated leachates, it may be prefereabe to select the treatment condition with a $Mg^{2+}:NH_4^+:PO_4^{3-}$ molar ratio of 1.25:1:1 for application in the field [50]. Even though this condition did not achieve the greatest ammonium removal rate (92%), it did achieve the greatest phosphate removal rate of about 99.7%. The residual ammonium concentration in the treated leachate may be less of a concern than the amount of residual phosphate.

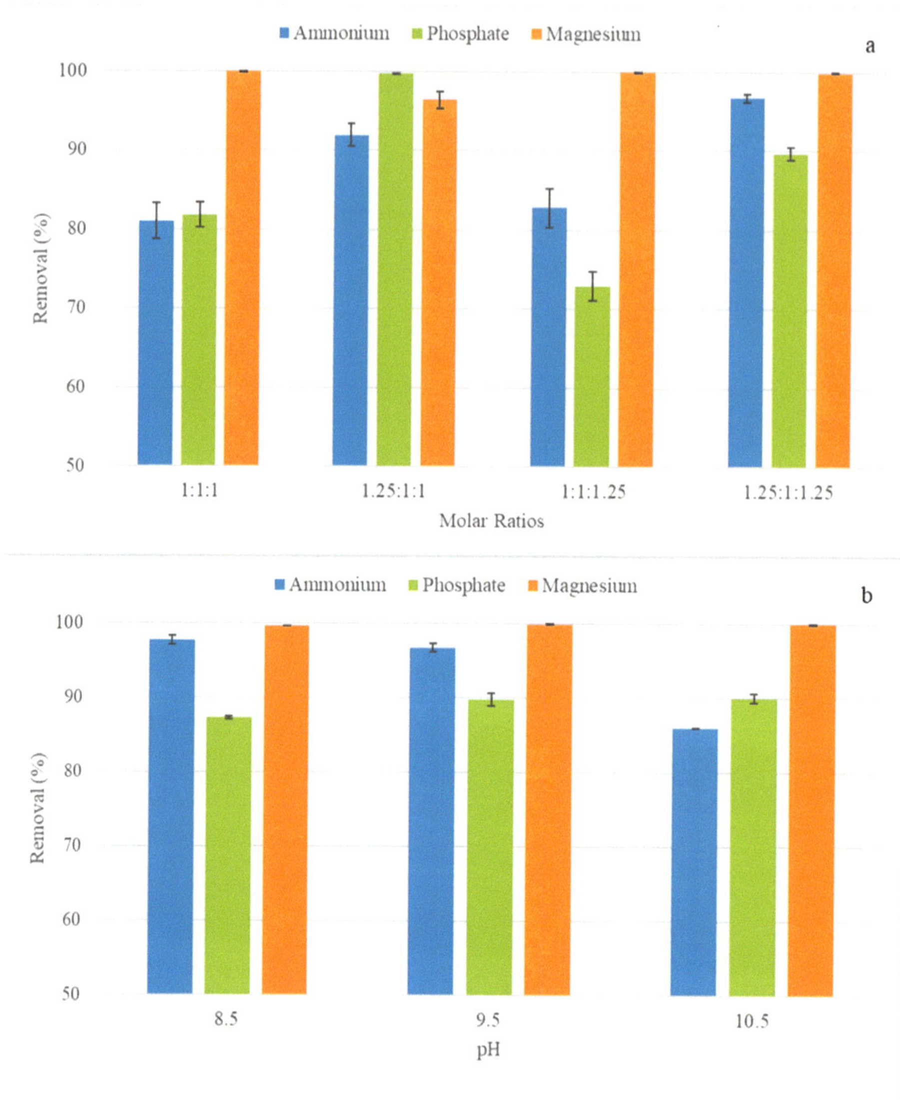


**Figure 1.** Ammonium, phosphate and magnesium removal rates vs. ($Mg^{2+}$:$NH_4{}^+$:$PO_4{}^{3-}$) molar ratios at pH = 9.5 (**a**) and vs. pH at molar ratios of 1.25:1:1.25 (**b**). The error bars represent ± one standard deviation for three replicate experiments.

3.2.2. Effects of pH

The effects of pH on ammonium removal were investigated using pH values of 8.5, 9.5, and 10.5, as well as $Mg^{2+}$:$NH_4{}^+$:$PO_4{}^{3-}$ molar ratios of 1.25:1:1.25 (Figure 1b). The results at a pH of 9.5 showed ammonium removal efficiency of approximately 97%. When pH was increased to 10.5, ammonium removal efficiency decreased to approximately 86%, whereas, when pH decreased to 8.5, ammonium removal remained roughly the same at about 98%. As pH increases (mainly above 9.5–10.0), ammonium solubility, with respect to struvite, increases, which could result in an increase in ammonium concentration in the solution [18,51]. In addition, high pH values could promote magnesium hydration, resulting in the formation of brucite $Mg(OH)_2$, which could scavenge magnesium [52]. This may explain the drop in ammonium removal efficiency at a pH of 10.5. Similarly, lower pH values increase ammonium solubility, with respect to ammonium. However, the increase in ammonium solubility upon pH change from 9.5 to 8.5 was not significant. Actually, the ammonium removal rate remained essentially the same. Other studies reported optimal ammonium removal by struvite precipitation in the pH range of 8.5–10 [48,53,54].

### *3.3. Ammonium Removal Using Dolomite-Derived Compounds*

Dolomite was pre-treated thermally to increase its activity as an alternative magnesium source. The thermal treatment of dolomite under a $CO_2$ atmosphere and 750 °C resulted in the formation of magnesium oxide (MgO) and calcite ($CaCO_3$). On the other hand, the thermal treatment under atmospheric conditions and a temperature of 950 °C resulted in the formation of MgO and lime (CaO) (Supplementary Material) [22,55,56]. In addition to the activation of the magnesium content, the thermal treatment of dolomite under the two different atmospheres also resulted in the formation of two different calcium sources, i.e., calcite and lime. Raw, $CO_2$-calcined, and air-calcined dolomite were used to investigate the removal of ammonium from the landfill leachates. The $Mg^{2+}$:$NH_4^+$:$PO_4^{3-}$ molar ratios and pH were set at 1.25:1:1.25 and 9.5, respectively. These conditions were selected to enable a performance comparison with the test results, using magnesium chloride salt under optimized ammonium removal conditions.

The batch test experiment using raw dolomite resulted in an ammonium removal rate of approximately 17% (Figure 2). When $CO_2$-calcined dolomite was used, the ammonium removal rate increased to approximately 37%, whereas, when air-calcined dolomite was used, the ammonium removal rate increased to approximately 55%. The removal rate of magnesium for all three conditions was approximately 100%. Conversely, the removal rates of phosphate were approximately 14%, 71%, and 58% for the raw dolomite, $CO_2$-calcined dolomite, and air-calcined dolomite, respectively.

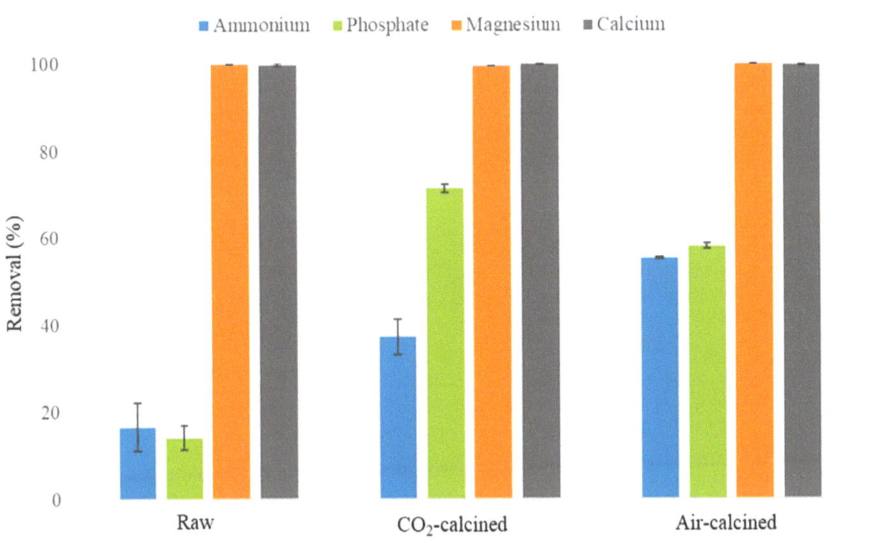


**Figure 2.** Ammonium, phosphate, magnesium, and calcium removal rates for the different treatment conditions using dolomite ($Mg^{2+}$:$NH_4^+$:$PO_4^{3-}$ = 1.25:1:1.25; pH = 9.5). The error bars represent ± one standard deviation for three replicate experiments.

It appears that magnesium was not readily available, as evidenced by the high concentration of dissolved phosphate and ammonium in the supernatant. It also appears that calcium was not readily available, as evidenced by the high concentration of phosphate in the supernatant. Greater ammonium removal rates were achieved by the air-calcined dolomite than by the $CO_2$-calcined dolomite. Conversely, greater phosphate removal rates were achieved by the $CO_2$-calcined dolomite than by the air-calcined dolomite. This could indicate that calcium was more available or active in the $CO_2$-calcined dolomite (as $CaCO_3$) than the air-calcined dolomite (as CaO), assuming that the residual phosphate was mostly removed by calcium. For field application, the air-calcined dolomite seems preferable, considering the higher ammonium removal rates and lower phosphate removal rates. Greater magnesium concentrations could be used to simultaneously increase the ammonium and phosphate removal rates. Alternatively, the supernatant could be reused in new precipitation experiments via the addition of new activated dolomite, in order to remove the residual ammonium and phosphate as struvite.

### 3.4. Effects of Calcium

In order to investigate the role of calcium in the decrease of ammonium removal rates by the dolomite-derived compounds, experiments were conducted to precipitate struvite from the landfill leachates using magnesium and phosphate salts in the presence of calcium salt. The results are shown in Figure 3. The addition of calcium, in the form of calcium salt, resulted in a decrease in the efficiency of ammonium removal. The ammonium removal rates were approximately 97%, 97%, 95%, 93%, and 70% for the calcium-to-magnesium molar ratios of 0, 0.1, 0.2, 0.5, and 1.0, respectively. Minimal effects were observed for calcium-to-magnesium ratios up to 0.5 at tested conditions. However, a significant reduction in the ammonium removal rate at 70% was observed when the calcium-to-magnesium molar ratio increased to 1.0. Greater ammonium removal rates were achieved by the magnesium salt, as compared to the corresponding removal rates by activated dolomite, for a similar $Ca^{2+}$:$Mg^{2+}$ molar ratio of one (70% versus 55%). This may indicate that dissolved alternative magnesium sources can outperform the solid-phase alternative magnesium sources for ammonium removal from wastewater, realizing that calcium was likely more available as salt than calcite or lime. The reduced efficiency of the solid-phase alternative magnesium sources could be due to the unavailability of magnesium during the co-precipitation reaction. The reduced efficiency of the solid-phase alternative magnesium sources may outweigh its added benefit, mitigating against the drop in pH upon struvite precipitation.

The formation of calcium phosphate solid phases is thermodynamically favorable under the experimental conditions, where calcium is expected to react with the added phosphates and potentially hinder the formation of struvite. However, the removal of significant amounts of ammonium indicated the formation of struvite, because struvite is the only route for ammonium removal in our system. Studies have reported that hydroxyapatite is thermodynamically stable at equilibrium, and it could uptake added phosphates as per its stoichiometry; however, that was not the case in this study, as evidenced by the high concentration of phosphate in the supernatant [26,27]. The order of the formation of the mineral phases, duration of the co-precipitation reactions experiments, and concentration of competing ions may have an impact on the resulting solid phases [25,57,58].

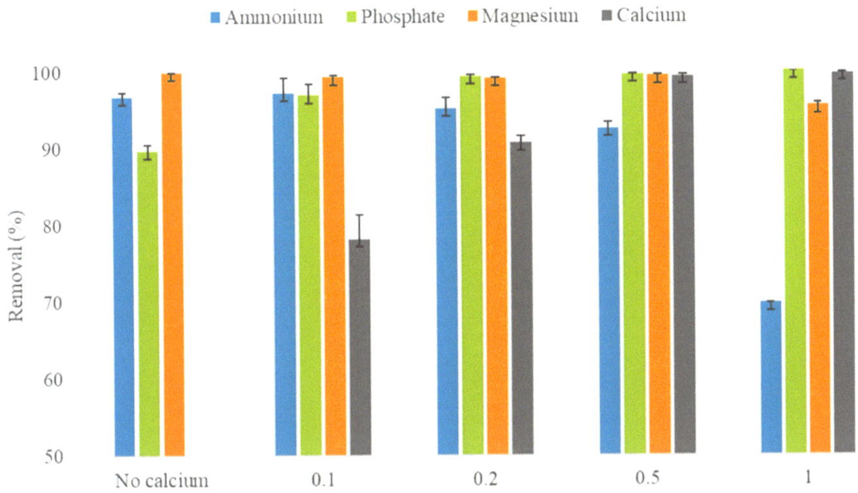


**Figure 3.** Ammonium, phosphate, magnesium, and calcium removal rates for the calcium salt source effects experiments ($Mg^{2+}$:$NH_4^+$:$PO_4^{3-}$ = 1.25:1:1.25, pH = 9.5). The error bars represent ± one standard deviation for three replicate experiments.

### 3.5. XRD Results

XRPD analysis was used to identify the solid phases that formed during the co-precipitation reactions and delineate the removal mechanisms of ammonium from the leachates. The XRPD patterns for the precipitates of the experiments using magnesium chloride salt are shown in Figure 4a. The results indicated the presence of struvite with characteristic peaks at 15.787°, 20.839°, and 30.573° (PDF 04-010-2533). Struvite is the major sequester of ammonium in the presence of magnesium and phosphate, and it is expected to be the only mechanism responsible for the removal of ammonium from the leachate. The XRPD patterns also indicated the presence of halite (NaCl), with characteristic peaks at 31.692°, 45.449°, and 56.477° (PDF 00-005-0628). Sodium is present due to the addition of sodium phosphate as the phosphate source, whereas chloride is present due to the presence of chloride in the leachates, as well as the added hydrochloric acid and calcium chloride. The absence of any crystalline calcium solid phases in the XRPD pattern, pertaining to the experiment with the added calcium chloride, is noteworthy (Figure 3a). The added calcium could be encapsulated in amorphous solid phases, as indicated by the amorphous hump in the corresponding XRPD pattern [27].

The XRPD analyses of the precipitates for the thermally-treated dolomite experiments clearly indicated the presence of struvite, with characteristic peaks at 15.787°, 20.839°, and 30.573° (PDF 04-010-2533) (Figure 4b). This indicates that ammonium was removed from the leachate by sequestration in struvite. The precipitates of the $CO_2$ calcination experiment showed characteristic peaks of calcite and periclase at 23.07°, 29.454°, 48.578° (PDF 00-024-0027), and 42.855°, 62.215° (PDF 01-071-1176), respectively, which indicates that these phases did not dissolve or hydrate completely. For the experiment with raw dolomite, XRPD analysis detected major peaks of dolomite at 30.938°, 41.127°, and 56.477° (PDF 00-036-0426), which indicated that dolomite was not significantly destabilized (XRPD patterns of raw and calcined dolomite are shown in Figure S1). No struvite peaks were identified for the raw dolomite experiments, which could be due to the presence of struvite at concentrations below the detection limit of the XRPD.

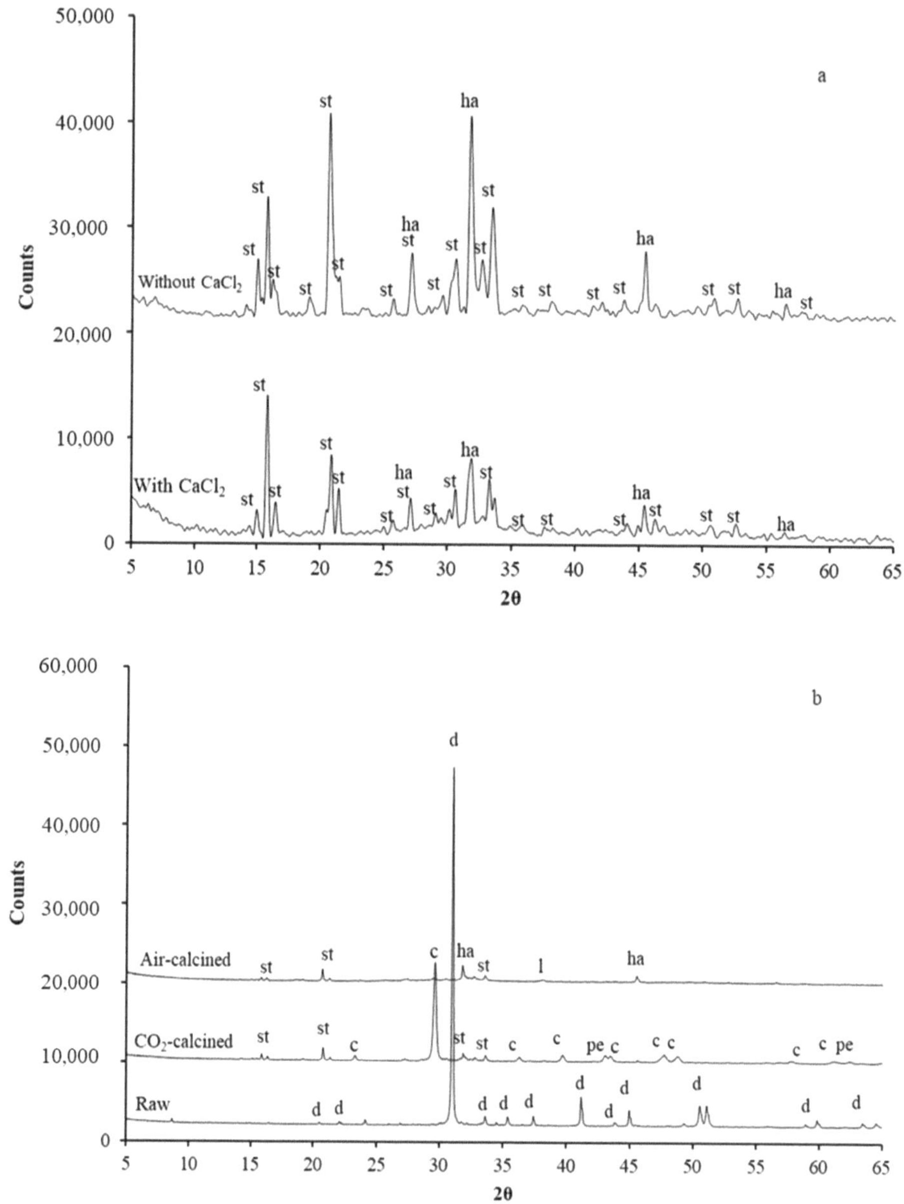


**Figure 4.** XRPD of the precipitates for the ($Mg^{2+}$:$NH_4^+$:$PO_4^{3-}$ = 1.25:1:1.25; pH = 9.5) experiments using $MgCl_2$, $Na_3PO_4.12H_2O$, and $CaCl_2$ as magnesium, phosphate, and calcium sources, respectively (**a**). And the experiments using dolomite under different pretreatment conditions (**b**). d = dolomite; c = calcite; pe = periclase; l = lime; po = portlandite; st = struvite; ha = halite.

Finally, no calcium phosphate or magnesium phosphate solid phases were identified in the XRPD analyses for all experimental conditions. This could be due to the instability of these phases, their presence at concentrations below the detection limits of the XRPD, or their presence as amorphous solid phases.

### 3.6. Activity Ratio Diagrams

Activity ratio diagrams were constructed to further delineate the ammonium removal mechanisms and investigate the stability of the solid phases controlling the solubility of ammonium and phosphate. The probable solid phases controlling the solubility of ammonium, magnesium, phosphate, and calcium are listed in Table 3. The activity ratio diagrams could qualitatively indicate the behavior of the system, especially in the presence of competing ions that could interfere with the formation of struvite and, thus, removal of ammonium. The activity of the phosphate is of particular concern, because the magnesium phosphate and calcium phosphate phases could compete with struvite for its sequestration. The thermodynamic data of the solid phases were obtained from Visual Minteq [40].

**Table 3.** Precipitation reactions used in the construction of the activity ratio solubility diagrams.

| Solid Phase | | | Log k |
|---|---|---|---|
| Struvite | = | $Mg^{2+} + NH_4^+ + PO_4^{3-}$ | −13.26 |
| $Ca_3(PO_4)_2$am1 | = | $3Ca^{2+} + 2PO_4^{3-}$ | −25.5 |
| $Ca_3(PO_4)_2$am2 | = | $3Ca^{2+} + 2PO_4^{3-}$ | −28.25 |
| $Ca_3(PO_4)_2$(beta) | = | $3Ca^{2+} + 2PO_4^{3-}$ | −28.92 |
| $Ca_4H(PO_4)_3 \cdot 3H_2O$ | = | $4Ca^{2+} + H^+ + 3PO_4^{3-} + 3H_2O$ | −47.95 |
| $CaHPO_4$ | = | $Ca^{2+} + H^+ + PO_4^{3-}$ | −19.275 |
| Hydroxyapatite + $H^+$ | = | $5Ca^{2+} + 3PO_4^{3-} + H_2O$ | −44.333 |
| $Mg_3(PO_4)_2$ | = | $3Mg^{2+} + 2PO_4^{3-}$ | −23.28 |
| $MgHPO_4.3H_2O$ | = | $Mg^{2+} + PO_4^{3-} + H+ + 3H_2O$ | −18.175 |

The activity ratios of the solid phases controlling phosphate solubility are plotted versus magnesium concentration, as shown in Figure 5a. The ammonium and calcium concentrations used in the construction of the diagram were the average values obtained in this study. The plots show that the solubility of phosphate at equilibrium and in the presence of calcium is controlled by hydroxyapatite (highest activity ratio line), independent of the concentration of magnesium. However, hydroxyapatite was not identified in the XRPD analysis, and its formation could be kinetically delayed. In the absence of hydroxyapatite, phosphate solubility is controlled by other calcium phosphate phases, except for magnesium concentrations greater than approximately$10^{-2}$ M, where phosphate solubility is controlled by struvite. If the concentration of calcium is increased by about two orders of magnitude (0.2 M), as it was at the start of the experiment for the $Ca^{2+}$:$Mg^{2+}$ ratio of 1.0, additional calcium phosphate phases activity ratio lines move above the struvite one and appear to control the solubility of the phosphate (Figure S2). The order of the activity ratio lines do not change appreciably when the ammonium concentration is concurrently increased to 0.2 M.

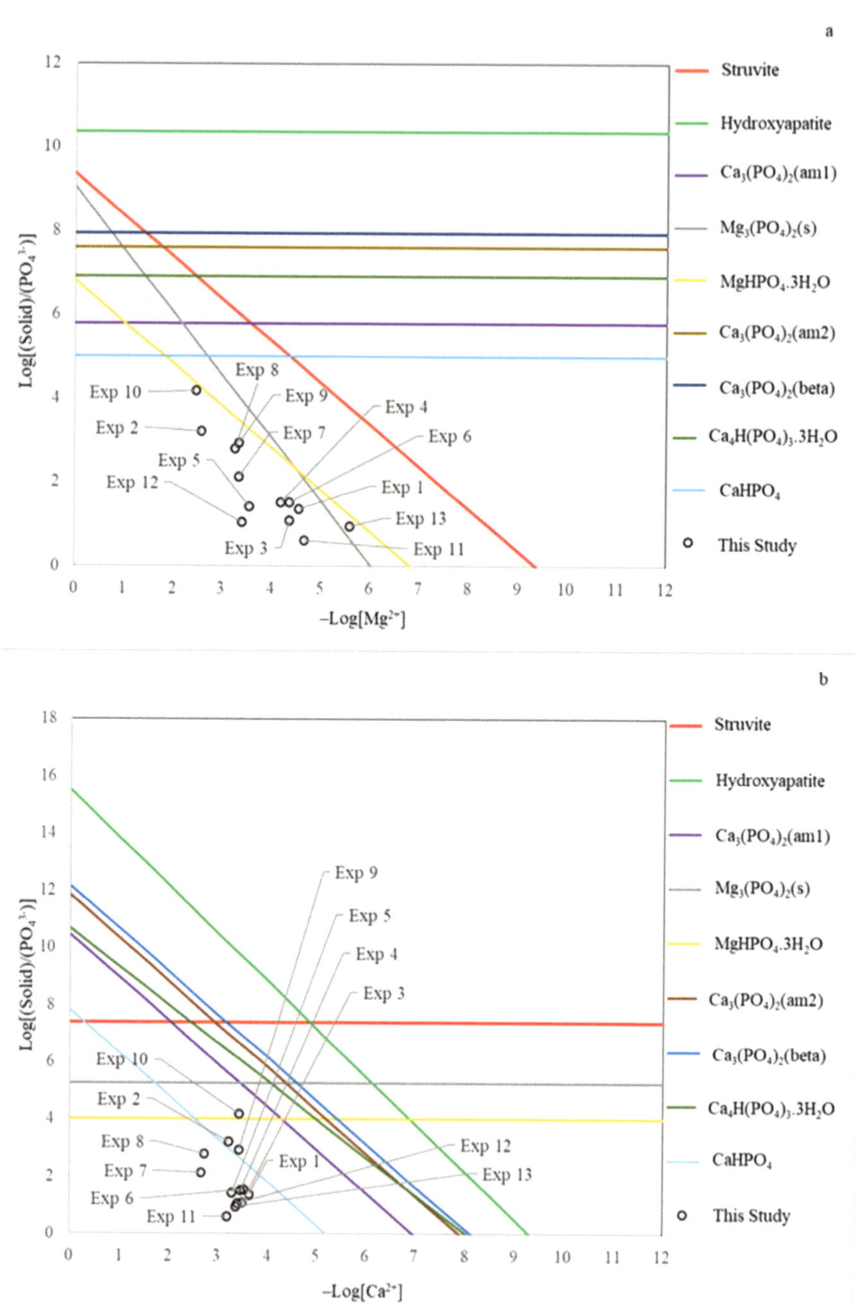


**Figure 5.** Activity ratio diagrams for: (**a**) $PO_4^{3-}$ solid phases vs. $Mg^{2+}$. pH = 9.5, $NH_4^+$ = 0.0527 M, $Ca^{2+}$ = 0.001 M; (**b**) $PO_4^{3-}$ solid phases vs. $Ca^{2+}$. pH = 9.5, $NH_4^+$ = 0.0527 M, $Mg^{2+}$ = 0.0026 M.

According to the Gay-Lussac-Ostwald (GLO) step rule, if several solid phases can potentially form with a given ion, the first solid phase to form is the one for which the activity ratio is nearest above the initial concentration. The remaining potential solid phases will form in the order of increasing activity ratio, with the rate of formation of the solid phases decreasing with increasing activity ratios. However, any solid phase can be maintained indefinitely [39]. Consequently, and as per the GLO step rule, the first phosphate solid phases that are expected to form at a magnesium concentration less than approximately $10^{-2}$ M are magnesium phosphates, $Mg_3(PO_4)_2$, and $MgHPO_4{\cdot}3H_2O$. However, phosphate solubility is ultimately controlled by calcium phosphate phases, with a struvite activity ratio line located between the magnesium phosphate and calcium phosphate activity ratio lines. If calcium concentration is exhausted or unavailable, then phosphate solubility is first controlled by magnesium phosphate solid phases and then by struvite. The $Mg_3(PO_4)_2$ and $MgHPO_4{\cdot}3H_2O$ solid phases were not identified in the XRPD analysis, but that does not necessarily mean that they did not form. It is speculated that these phases are transient ones, and they could have formed (even as amorphous ones); however, as the magnesium and phosphate concentrations decreased, due to the formation of the aforementioned magnesium phosphate phases, the equilibrium shifted towards the formation of struvite and dissolution of the magnesium phosphate phases. The significant removal of calcium could be due to the formation of amorphous calcium phosphate phases [27].

The experimental data in this study were plotted at or below the activity ratio lines corresponding to $Mg_3(PO_4)_2$ and $MgHPO_4{\cdot}3H_2O$. It appears that the presence of the data points at and below the $Mg_3(PO_4)_2$ and $MgHPO_4{\cdot}3H_2O$ activity ratio lines did not preclude the formation of struvite. The experimental data of other struvite co-precipitation studies (where calcium and magnesium concentrations were available) were also plotted within the same data cluster [19,35,59]. Liu and Wang [26] reported that struvite nucleate ahead of magnesium phosphate phases and would later transform to magnesium phosphate under excessive magnesium concentration. The later transformation of struvite to magnesium phosphate does not appear to have occurred in this study, as evidenced by the identification of struvite using XRPD analysis and significant removal rates of ammonium. However, this does not contradict the GLO rule, as any formed solid phase can be maintained indefinitely.

The concentration of ammonium was fixed during the construction of the diagram in Figure 5a as the average of the values obtained in this study. Other assumptions about ammonium concentration could be made; for example, a decrease in the concentration of ammonium by one order of magnitude will result in a downward shift in the struvite activity ratio line of one order of magnitude, placing the activity ratio line of struvite closer to that of magnesium phosphate. On the other hand, an increase in the concentration of ammonium by one order of magnitude will result in an upward shift of the struvite activity ratio line of one order of magnitude. The downward shift will move equilibrium toward the formation of struvite close to and ahead of $Mg_3(PO_4)_2$, whereas the upward shift will have no effects on the order of formation between that of the struvite and magnesium phosphate solid phases.

The plot of the activity ratio for the solid phases controlling the phosphate concentration versus the concentration of calcium are shown in Figure 5b. At equilibrium, the concentration of phosphate is expected to be controlled by struvite at calcium concentration less than approximately $10^{-5}$ M and hydroxyapatite at calcium concentrations greater than approximately $10^{-5}$ M. Assuming that the formation of hydroxyapatite could be kinetically delayed for the duration of the experiment, the practical range of calcium concentration that struvite controls phosphate solubility could be extended to less than approximately $10^{-3.3}$ M. At calcium concentrations greater than $10^{-3.3}$ M, the solubility of phosphate is first controlled by struvite and later controlled by calcium phosphate phases. If struvite nucleates first, then it has the potential to remain stable indefinitely, as per the GLO rule, which could explain its presence in the precipitates. Increasing the concentration of magnesium by about two orders of magnitude (0.2 M), as it was at the start of the experiment, will result in an upward shift of the struvite activity ratio line by two orders of magnitude, thus practically extending the range of calcium concentration, where phosphate concentration will be mainly controlled by struvite (Figure S3).

## 4. Conclusions

The experimental data showed that increasing the molar ratio of magnesium to ammonium and phosphate during the co-precipitation treatment resulted in a relatively lower ammonium removal rate and lower residual phosphate concentration in the treated leachate. This treatment option appears to be preferable to the one with a higher ammonium removal rate and higher residual phosphate concentration in the treated leachate, due to concerns regarding phosphorus discharge. The thermal treatment of dolomite significantly increased its capacity to remove ammonium from landfill leachates. However, greater ammonium removal rates were achieved by magnesium salt, as compared to the corresponding removal rates of dolomite-derived compounds for a similar $Ca^{2+}$:$Mg^{2+}$ molar ratio. This may indicate that dissolved alternative magnesium sources can outperform solid-phase alternative magnesium sources for ammonium removal from wastewater. The reduced efficiency of the solid-phase alternative magnesium sources could be due to the unavailability of magnesium during the co-precipitation reaction. Future studies could focus on the assessment of the produced precipitate as a fertilizer or soil amendment, including the impact of calcium presence. The new insights into the stability of struvite and other phosphate sequestering solid phases could be used advantageously to design better ammonium and phosphate removal and recovery systems via struvite precipitation.

**Supplementary Materials:** The following supporting information can be downloaded at: https://www.mdpi.com/article/10.3390/w14121933/s1, Figure S1: X-Ray Diffractograms of the products of the different dolomite thermal pretreatment processes. d = dolomite $MgCa(CO_3)_2$; c = calcite $CaCO_3$; pe = periclase; MgO; l = lime CaO; po = portlandite $Ca(OH)_2$; Figure S2. Activity ratio diagram for $PO_4^{3-}$ solid phases vs. $Mg^{2+}$. pH = 9.5, $NH_4^+$ = 0.0527 M, $Ca^{2+}$ = 0.2 M; Figure S3. Activity ratio diagram for $PO_4^{3-}$ solid phases vs. $Ca^{2+}$. pH = 9.5, $NH_4^+$ = 0.0527 M, $Mg^{2+}$ = 0.2 M; Table S1: The final concentrations of $Mg^{2+}$,$NH^4$+,$PO_4^{3-}$, and $Ca^{2+}$, pertaining to all batch tests in Table 2, along with the associated initial $NH_4^+$ concentration values. The results are the averages of three replicate experiments ± one standard deviation.

**Author Contributions:** Conceptualization, M.W. and H.R.; Methodology, M.W. and H.R.; Software, M.W. and H.R.; Validation, M.W. and H.R.; Formal Analysis, M.W. and H.R.; Investigation, M.W. and H.R.; Resources, M.W.; Data Curation, M.W. and H.R.; Writing—Original Draft Preparation, M.W. and H.R.; Writing—Review & Editing, M.W. and H.R.; Visualization, M.W. and H.R.; Supervision, M.W.; Project Administration, M.W.; Funding Acquisition, M.W. All authors have read and agreed to the published version of the manuscript.

**Funding:** This research received no external funding.

**Institutional Review Board Statement:** Not applicable.

**Informed Consent Statement:** Not applicable.

**Data Availability Statement:** The data presented in this study are available in article and supplementary material.

**Conflicts of Interest:** The authors declare no conflict of interest.

## References

1. Kaza, S.; Yao, L.; Bhada-Tata, P.; Van Woerden, F. *What a Waste 2.0: A Global Snapshot of Solid Waste Management to 2050*; World Bank Publications: Washington, DC, USA, 2018.
2. Kjeldsen, P.; Barlaz, M.A.; Rooker, A.P.; Baun, A.; Ledin, A.; Christensen, T.H. Present and long-term composition of MSW landfill leachate: A review. *Crit. Rev. Environ. Sci. Technol.* **2002**, *32*, 297–336. [CrossRef]
3. Shin, K.W.; Kim, S.-H.; Kim, J.-H.; Hwang, S.D.; Kang, J.-C. Toxic effects of ammonia exposure on growth performance, hematological parameters, and plasma components in rockfish, Sebastes schlegelii, during thermal stress. *Fish. Aquat. Sci.* **2016**, *19*, 1–8. [CrossRef] [PubMed]

4. Chamem, O.; Fellner, J.; Zairi, M. Ammonia inhibition of waste degradation in landfills–A possible consequence of leachate recirculation in arid climates. *Waste Manag. Res.* **2020**, *38*, 1078–1086. [CrossRef] [PubMed]
5. Liu, Y.; Ngo, H.H.; Guo, W.; Peng, L.; Wang, D.; Ni, B. The roles of free ammonia (FA) in biological wastewater treatment processes: A review. *Environ. Int.* **2019**, *123*, 10–19. [CrossRef] [PubMed]
6. Chen, Y.; Cheng, J.J.; Creamer, K.S. Inhibition of anaerobic digestion process: A review. *Bioresour. Technol.* **2008**, *99*, 4044–4064. [CrossRef]
7. Hafner, S.D.; Bisogni, J.J.; Jewell, W.J. Measurement of un-ionized ammonia in complex mixtures. *Environ. Sci. Technol.* **2006**, *40*, 1597–1602. [CrossRef]
8. Liu, T.; Sung, S. Ammonia inhibition on thermophilic aceticlastic methanogens. *Water Sci. Technol.* **2002**, *45*, 113–120. [CrossRef]
9. Procházka, J.; Dolejš, P.; Máca, J.; Dohányos, M. Stability and inhibition of anaerobic processes caused by insufficiency or excess of ammonia nitrogen. *Appl. Microbiol. Biotechnol.* **2012**, *93*, 439–447. [CrossRef]
10. Strik, D.; Domnanovich, A.; Holubar, P. A pH-based control of ammonia in biogas during anaerobic digestion of artificial pig manure and maize silage. *Process Biochem.* **2006**, *41*, 1235–1238. [CrossRef]
11. Sung, S.; Liu, T. Ammonia inhibition on thermophilic anaerobic digestion. *Chemosphere* **2003**, *53*, 43–52. [CrossRef]
12. Cheung, K.C.; Chu, L.M.; Wong, M.H. Ammonia stripping as a pretreatment for landfill leachate. *Water Air Soil Pollut.* **1997**, *94*, 209–221. [CrossRef]
13. Jurczyk, Ł.; Koc-Jurczyk, J.; Masłoń, A. Simultaneous stripping of ammonia from leachate: Experimental insights and key microbial players. *Water* **2020**, *12*, 2494. [CrossRef]
14. Kim, D.; Ryu, H.-D.; Kim, M.-S.; Kim, J.; Lee, S.-I. Enhancing struvite precipitation potential for ammonia nitrogen removal in municipal landfill leachate. *J. Hazard. Mater.* **2007**, *146*, 81–85. [CrossRef] [PubMed]
15. Warmadewanthi, I.; Zulkarnain, M.A.; Ikhlas, N.; Kurniawan, S.B.; Abdullah, S.R.S. Struvite precipitation as pretreatment method of mature landfill leachate. *Bioresour. Technol. Rep.* **2021**, *15*, 100792. [CrossRef]
16. Siciliano, A.; Limonti, C.; Curcio, G.M.; Molinari, R. Advances in struvite precipitation technologies for nutrients removal and recovery from aqueous waste and wastewater. *Sustainability* **2020**, *12*, 7538. [CrossRef]
17. Rahman, M.M.; Salleh, M.A.M.; Rashid, U.; Ahsan, A.; Hossain, M.M.; Ra, C.S. Production of slow release crystal fertilizer from wastewaters through struvite crystallization–A review. *Arab. J. Chem.* **2014**, *7*, 139–155. [CrossRef]
18. Li, X.; Zhao, Q.; Hao, X. Ammonium removal from landfill leachate by chemical precipitation. *Waste Manag.* **1999**, *19*, 409–415. [CrossRef]
19. Siciliano, A. Assessment of fertilizer potential of the struvite produced from the treatment of methanogenic landfill leachate using low-cost reagents. *Environ. Sci. Pollut. Res.* **2016**, *23*, 5949–5959. [CrossRef]
20. Darwish, M.; Aris, A.; Puteh, M.H.; Jusoh, M.; Kadir, A.A. Waste bones ash as an alternative source of P for struvite precipitation. *J. Environ. Manag.* **2017**, *203*, 861–866. [CrossRef]
21. Huang, H.; Xiao, D.; Zhang, Q.; Ding, L. Removal of ammonia from landfill leachate by struvite precipitation with the use of low-cost phosphate and magnesium sources. *J. Environ. Manag.* **2014**, *145*, 191–198. [CrossRef]
22. Chen, L.; Zhou, C.H.; Zhang, H.; Tong, D.S.; Yu, W.H.; Yang, H.M.; Chu, M.Q. Capture and recycling of ammonium by dolomite-aided struvite precipitation and thermolysis. *Chemosphere* **2017**, *187*, 302–310. [CrossRef]
23. Kataki, S.; West, H.; Clarke, M.; Baruah, D.C. Phosphorus recovery as struvite: Recent concerns for use of seed, alternative Mg source, nitrogen conservation and fertilizer potential. *Resour. Conserv. Recycl.* **2016**, *107*, 142–156. [CrossRef]
24. Enyemadze, I.; Momade, F.W.; Oduro-Kwarteng, S.; Essandoh, H. Phosphorus recovery by struvite precipitation: A review of the impact of calcium on struvite quality. *J. Water Sanit. Hyg. Dev.* **2021**, *11*, 706–718. [CrossRef]
25. Le Corre, K.S.; Valsami-Jones, E.; Hobbs, P.; Parsons, S.A. Impact of calcium on struvite crystal size, shape and purity. *J. Cryst. Growth* **2005**, *283*, 514–522. [CrossRef]
26. Liu, X.; Wang, J. Impact of calcium on struvite crystallization in the wastewater and its competition with magnesium. *Chem. Eng. J.* **2019**, *378*, 122121. [CrossRef]
27. Moragaspitiya, C.; Rajapakse, J.; Millar, G.J. Effect of Ca:Mg ratio and high ammoniacal nitrogen on characteristics of struvite precipitated from waste activated sludge digester effluent. *J. Environ. Sci.* **2019**, *86*, 65–77. [CrossRef]
28. Li, W.; Ding, X.; Liu, M.; Guo, Y.; Liu, L. Optimization of process parameters for mature landfill leachate pretreatment using MAP precipitation. *Front. Environ. Sci. Eng.* **2012**, *6*, 892–900. [CrossRef]
29. Li, X.; Zhao, Q. MAP precipitation from landfill leachate and seawater bittern waste. *Environ. Technol.* **2002**, *23*, 989–1000. [CrossRef]
30. Amrulloh, H.; Simanjuntak, W.; Situmeang, R.T.M.; Sagala, S.L.; Bramawanto, R.; Fatiqin, A.; Nahrowi, R.; Zuniati, M. Preparation of nano-magnesium oxide from Indonesia local seawater bittern using the electrochemical method. *Inorg. Nano-Met. Chem.* **2020**, *50*, 693–698. [CrossRef]
31. Bagastyo, A.Y.; Sinatria, A.Z.; Anggrainy, A.D.; Affandi, K.A.; Kartika, S.W.T.; Nurhayati, E. Resource recovery and utilization of bittern wastewater from salt production: A review of recovery technologies and their potential applications. *Environ. Technol. Rev.* **2021**, *10*, 295–322. [CrossRef]
32. Shin, H.-S.; Lee, S. Removal of nutrients in wastewater by using magnesium salts. *Environ. Technol.* **1998**, *19*, 283–290. [CrossRef]
33. Etter, B.; Tilley, E.; Khadka, R.; Udert, K.M. Low-cost struvite production using source-separated urine in Nepal. *Water Res.* **2011**, *45*, 852–862. [CrossRef] [PubMed]

34. Halwani, J.; Halwani, B.; Amine, H.; Kabbara, M.B. Waste management in Lebanon—Tripoli case study. In *Waste Management in MENA Regions*; Springer: Cham, Switzerland, 2020; pp. 223–239.
35. Camargo, C.; Guimarães, J.; Tonetti, A. Treatment of landfill leachate: Removal of ammonia by struvite formation. *Water SA* **2014**, *40*, 491–494. [CrossRef]
36. Siciliano, A.; Stillitano, M.A.; Limonti, C.; Marchio, F. Ammonium removal from landfill leachate by means of multiple recycling of struvite residues obtained through acid decomposition. *Appl. Sci.* **2016**, *6*, 375. [CrossRef]
37. Cao, X.; Ma, L.Q.; Chen, M.; Singh, S.P.; Harris, W.G. Impacts of phosphate amendments on lead biogeochemistry at a contaminated site. *Environ. Sci. Technol.* **2002**, *36*, 5296–5304. [CrossRef] [PubMed]
38. McGowen, S.; Basta, N.; Brown, G. Use of diammonium phosphate to reduce heavy metal solubility and transport in smelter-contaminated soil. *J. Environ. Qual.* **2001**, *30*, 493–500. [CrossRef]
39. Sposito, G. *The Chemistry of Soils*; Oxford Univ. Press: New York, NY, USA, 1989.
40. Gustafsson, J.P. *Visual MINTEQ 3.0 User Guide*; KTH, Department of Land and Water Recources: Stockholm, Sweden, 2011.
41. Langmuir, D. *Aqueous Environmental Geochemistry*; Prentice Hall: Hoboken, NJ, USA, 1997.
42. Luo, H.; Zeng, Y.; Cheng, Y.; He, D.; Pan, X. Recent advances in municipal landfill leachate: A review focusing on its characteristics, treatment, and toxicity assessment. *Sci. Total Environ.* **2020**, *703*, 135468. [CrossRef]
43. Teng, C.; Zhou, K.; Peng, C.; Chen, W. Characterization and treatment of landfill leachate: A review. *Water Res.* **2021**, *203*, 117525. [CrossRef]
44. Anthonisen, A.C.; Loehr, R.C.; Prakasam, T.; Srinath, E. Inhibition of nitrification by ammonia and nitrous acid. *J. Water Pollut. Control Fed.* **1976**, *48*, 835–852.
45. Çelen, I.; Türker, M. Recovery of ammonia as struvite from anaerobic digester effluents. *Environ. Technol.* **2001**, *22*, 1263–1272. [CrossRef]
46. El Diwani, G.; El Rafie, S.; El Ibiari, N.; El-Aila, H. Recovery of ammonia nitrogen from industrial wastewater treatment as struvite slow releasing fertilizer. *Desalination* **2007**, *214*, 200–214. [CrossRef]
47. Lee, S.; Weon, S.; Lee, C.; Koopman, B. Removal of nitrogen and phosphate from wastewater by addition of bittern. *Chemosphere* **2003**, *51*, 265–271. [CrossRef]
48. Siciliano, A.; Rosa, S.D. Recovery of ammonia in digestates of calf manure through a struvite precipitation process using unconventional reagents. *Environ. Technol.* **2014**, *35*, 841–850. [CrossRef] [PubMed]
49. Huang, H.; Chen, Y.; Jiang, Y.; Ding, L. Treatment of swine wastewater combined with MgO-saponification wastewater by struvite precipitation technology. *Chem. Eng. J.* **2014**, *254*, 418–425. [CrossRef]
50. Tonetti, A.L.; de Camargo, C.C.; Guimarães, J.R. Ammonia removal from landfill leachate by struvite formation: An alarming concentration of phosphorus in the treated effluent. *Water Sci. Technol.* **2016**, *74*, 2970–2977. [CrossRef] [PubMed]
51. Doyle, J.D.; Parsons, S.A. Struvite formation, control and recovery. *Water Res.* **2002**, *36*, 3925–3940. [CrossRef]
52. Maekawa, T.; Liao, C.-M.; Feng, X.-D. Nitrogen and phosphorus removal for swine wastewater using intermittent aeration batch reactor followed by ammonium crystallization process. *Water Res.* **1995**, *29*, 2643–2650. [CrossRef]
53. Korchef, A.; Saidou, H.; Amor, M.B. Phosphate recovery through struvite precipitation by CO2 removal: Effect of magnesium, phosphate and ammonium concentrations. *J. Hazard. Mater.* **2011**, *186*, 602–613. [CrossRef]
54. Kumar, R.; Pal, P. Turning hazardous waste into value-added products: Production and characterization of struvite from ammoniacal waste with new approaches. *J. Clean. Prod.* **2013**, *43*, 59–70. [CrossRef]
55. Caceres, P.; Attiogbe, E. Thermal decomposition of dolomite and the extraction of its constituents. *Miner. Eng.* **1997**, *10*, 1165–1176. [CrossRef]
56. Valverde, J.M.; Perejon, A.; Medina, S.; Perez-Maqueda, L.A. Thermal decomposition of dolomite under $CO_2$: Insights from TGA and in situ XRD analysis. *Phys. Chem. Chem. Phys.* **2015**, *17*, 30162–30176. [CrossRef]
57. Kiani, D.; Silva, M.; Sheng, Y.; Baltrusaitis, J. Experimental insights into the genesis and growth of struvite particles on low-solubility dolomite mineral surfaces. *J. Phys. Chem. C* **2019**, *123*, 25135–25145. [CrossRef]
58. Li, B.; Boiarkina, I.; Young, B.; Yu, W. Quantification and mitigation of the negative impact of calcium on struvite purity. *Adv. Powder Technol.* **2016**, *27*, 2354–2362. [CrossRef]
59. Romero-Güiza, M.; Astals, S.; Mata-Alvarez, J.; Chimenos, J.M. Feasibility of coupling anaerobic digestion and struvite precipitation in the same reactor: Evaluation of different magnesium sources. *Chem. Eng. J.* **2015**, *270*, 542–548. [CrossRef]

*energies*

MDPI

*Article*

# Predicting Mining Areas Deformations under the Condition of High Strength and Depth of Cover

**Piotr Strzałkowski**

Faculty of Mining, Safety Engineering and Industrial Automation, Silesian University of Technology, Akademicka 2A, 44-100 Gliwice, Poland; piotr.strzalkowski@polsl.pl

**Abstract:** This paper presents an analysis of mining area deformations in the rock mass consisting of high depth and strength strata deposited in the cover. The analysis of land surveying results enabled the identification of the parameters required to predict subsidence, which differed from the typical parameters for the Upper Silesian Coal Basin. The parameters of the Budryk–Knothe theory were determined based on the results of geodetic measurements. The calculations of the final state of deformations for planned mining were made using the average and characteristics for the study area parameter values. Based on experience, it is known that the range of subsidence trough depends on the mechanical properties of the rock mass. This study shows that the presence of high-strength rocks also reduces the value of the coefficient of roof control. Subsequently, calculations were made by a computer simulation of longwall mining to determine the course of indices of deformation over time. The calculations were conducted twice: on the assumption that the impact was immediate and on the assumption of the parameter values typical for the basin, and formula expressing the course of subsidence over time with the parameter values based on the measurement results. The obtained distributions of deformation indicators were diametrically opposed to each other. The results of the calculations with the parameter values appropriate for the region indicate that it is possible to carry out a planned mining operation without creating a risk to objects on the surface.

**Keywords:** mining area deformation; subsidence; rock mass

**Citation:** Strzałkowski, P. Predicting Mining Areas Deformations under the Condition of High Strength and Depth of Cover. *Energies* **2022**, *15*, 4627. https://doi.org/10.3390/en15134627

Academic Editors: Dibyendu Sarkar, Rupali Datta, Prafulla Kumar Sahoo and Mohammad Mahmudur Rahman

Received: 5 May 2022
Accepted: 22 June 2022
Published: 24 June 2022



## 1. Introduction

In Europe, hard coal extraction is being constantly reduced. For instance, in Poland the extraction of this raw material decreased from approximately 200 million Mg in the 1970s to approximately 70 million Mg currently [1]. Hard coal mining in Russia or China remains more or less at a stable level, or at least it is not being reduced so drastically [2] and amounts to approximately 3.5 billion Mg. Coal and metal ore deposits are deposited in many basins under highly urbanized areas [3–5]. This requires that mining be conducted in a way that minimizes its impact on the surface, building structures and the city infrastructure above it [2,6]. Having an appropriate mathematical formula to predict post-mining surface deformations determines the effective prevention of mining and construction damage. Therefore, we observe the development of theories that aim to predict the mining impacts on the rock mass and the surface. Among them, the following groups can be distinguished [4,7]:

1. Solutions based on adopted axioms, which may be of empirical origin. This group consists of the solutions and the method developed in the United Kingdom by the National Coal Board–NCB, [5]. This group also contains the so-called geometrical-integral theories [8–12];
2. Based on models of continuous media [13];
3. Based on the stochastic medium theory [14]. In addition, the theory of cellular automata model is also applicable [15];
4. Intermediate solutions–using models of the continuous medium and empirical axioms [16];

5. Numerical methods that are gaining popularity due to the development of digital technology [17–20]. FLAC3D software is the most frequently used. Based on the software-based calculations, Gang and Yang [3] determined a safe distance from a mining boundary to an urbanized area, which allowed extracting 1.5 million Mg of coal and avoiding damage to surface structures. It should be noted that mining was conducted at a shallow depth, which made the preparation of the data relatively easy. To a great extent, the course of the deformation process undoubtedly depends on natural and geological conditions [4]. The geological structure of the rock mass especially has an impact on the extent of deformations or the kinetics of the process. This makes it necessary to have a good understanding of the parameter values used for predictions. In these cases, using numerical methods is very helpful; however, this also has certain limitations, because the parameters of the rock strength change during extraction. Capturing these changes in a model is both difficult and time-consuming. For instance, in the work of Zhu et al. [21], the analyses were conducted to determine the impact of the depth of loose overburden on the course of deformations using FLAC3D software.

It can be concluded that although numerical methods give great opportunities to conduct comprehensive analyses, they also require a good knowledge of a series of mechanical parameters of multiple strata. Moreover, in cases of intensive mining (many workings) at great depth, the preparation of data for calculations is very laborious. This results in the widespread application of the theories from group 1 in practice. They are used in the mining industry in Poland [4,7,8], Germany [5], the United States [22,23].

In addition, predictions should also take into account the time variable [8,10,11].

One should also consider the impact of geological and mining conditions on the parameter values needed for calculations. The structure of the rock mass and its properties have a particular impact here. When using average parameter values for a given area for calculations, the obtained results might differ significantly from the actual state. In this way, the analyses of the impact of mining on built structures will be heavily error-burdened, which has a significant impact on the safety of built structures users and the structures themselves. This article is devoted to these issues.

## 2. Materials and Methods

### 2.1. Rock Mass Structure

The rock mass in the study area was built of Quaternary and Triassic overburden strata and productive carbon. The overburden was formed during the Quaternary period from clay. The thickness of these formations ranges from 1.8 m to 2.5 m. Middle Triassic strata, made of limestones and dolomites, among others, are deposited underneath. These formations reach a depth of approximately 130 m. Below, there is carbon made of sandstone interbedded with coal seams. Carbon is made of Libiąż Beds with group 100 coal beds and of Łaziska Beds with group 200 coal bed. The subject of extraction in the area under consideration was coal bed 207. Seam thickness in the area of the chosen panels ranged from 4.2 m to 5.0 m along with locally occurring interlayers of 0.10 m to 0.18 m thick clay shale. In the roof of the bed, there are clay shales as well as fine-grained and medium-grained sandstones. The angle of the bed inclination in the extraction area ranged from $3.5°$ to $5.5°$. Figure 1 shows a simplified lithological profile.

Based on the works [4,8], it is known that when strength of the layers building the rock mass increases, the range of subsidence trough also increases (the value of tan $\beta$ parameter decreases). In the case of rock mass in Upper Silesia, the value of tan $\beta$ parameter ranges from 1.5 to 2.5 depending on the strength properties of the rock mass. The structure of the rock mass also influences the course of subsidence over time. In a rock mass built of high-strength rocks, the course of subsidence is slower than in a rock mass of lower strength. The results of research conducted so far do not indicate the influence of the rock mass properties on the value of the coefficient of roof control a.

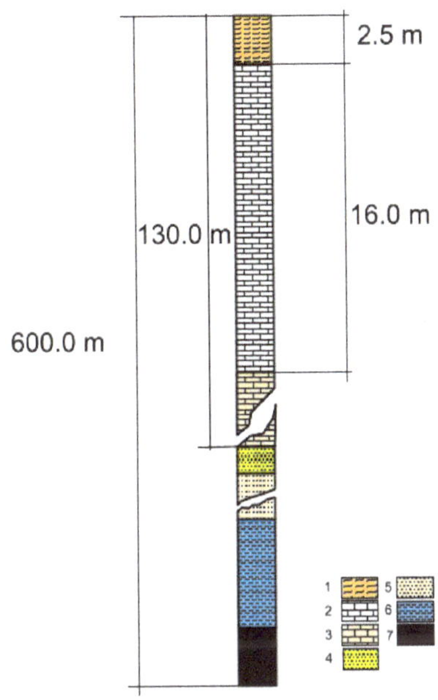


**Figure 1.** Lithological profile of the borehole in the studied area. 1—clay, 2—limestone, 3—dolomite, 4—fine-grained sandstone, 5—medium-grained sandstone, 6—clay shale, 7—coal bed.

### 2.2. Coal Mining

Mining extraction was conducted with caving of the roof of 3 m high in the 207 coal seam along two panels: 546 and 547. In order to carry out the calculations with the greatest possible accuracy, the panels were divided into lots 1–6 determined by the quarterly progress of extraction. Table 1 shows basic data on the completed extraction, along with the depth of the deposition of strata in the area of lot-H.

**Table 1.** Basic information about the extraction.

| Seam/Panel | Start | End | H, m |
|---|---|---|---|
| 546/1 | 1 December 2017 | 31 December 2017 | 597 |
| 546/2 | 1 January 2018 | 31 March 2018 | 580 |
| 546/3 | 1 April 2018 | 1 July 2018 | 565 |
| 547/4 | 1 September 2018 | 30 September 2018 | 575 |
| 547/5 | 1 October 2018 | 31 December 2018 | 565 |
| 547/6 | 1 January 2019 | 1 March 2019 | 557 |

In the analyzed region, further longwall mining is planned along the 548 panel at the height of 3 m with the roof fall at the average depth of 600 m.

Figure 2 shows the location of completed and planned extractions. It shows the division of the 546 and 547 panels into lots 1–6. Planned longwall mining along the 548 panel is shaded yellow.

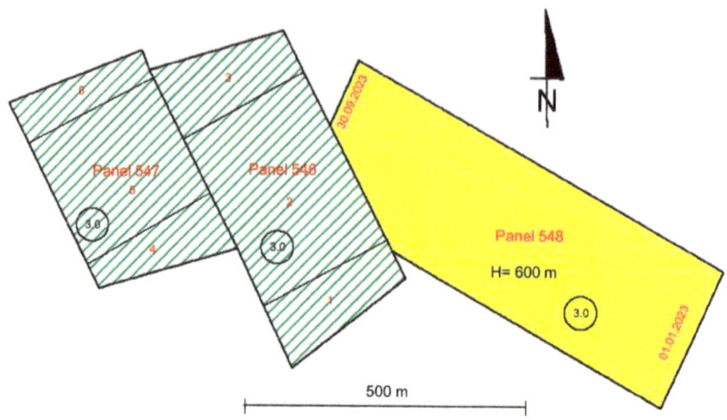


**Figure 2.** Diagram of mining location.

### 2.3. Land Surveying and Results

The impact of extraction on the surface was monitored by measurements carried out along the observation line consisting of benchmarks in the ground and wall benchmarks fixed on the building. The locations of monitoring points were set along with the actual topography on the ground surface including a street located on the ground surface above panel 546. The measurements were conducted every six months between April 2017 and July 2020. The location of the building (red mark) and measurement line are shown in Figure 3.

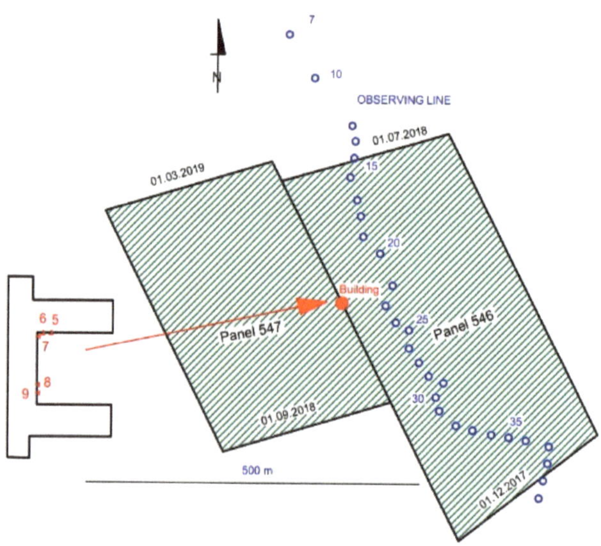


**Figure 3.** Location of the measurement line and building in relation to the extraction.

Subsidence of the line benchmarks is shown in Figure 4, and the wall benchmarks in Figure 5. It can be seen from these figures that the area has not subsided since March 2020. Some benchmarks were indeed subsidence in the last measurement, but in the previous ones they were raised, which proves measurement errors. More accurate measurement results based on the wall benchmarks confirm that the lowering process ended in March 2020.

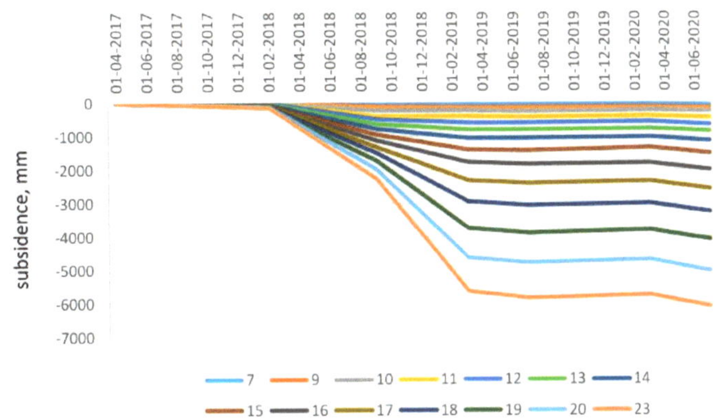


**Figure 4.** Subsidence of benchmarks in the ground over time.

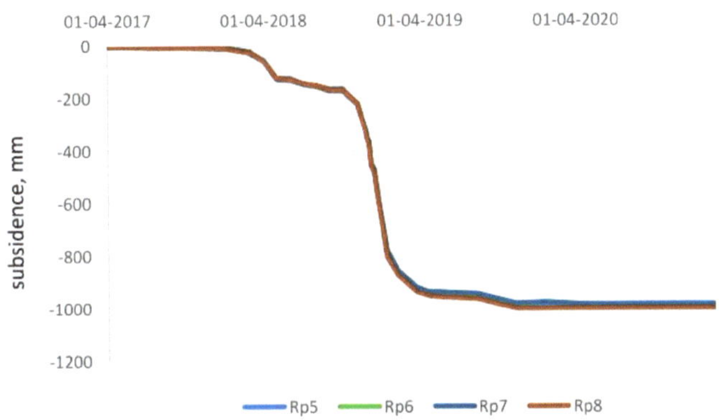


**Figure 5.** Subsidence of wall benchmarks over time.

## 2.4. Analysis Method Used

The Budryk–Knothe theory (Knothe 1983) was used for the analysis conducted in this paper due to its huge popularity in Poland [4,24,25] and in other countries [5,11,18,26,27]. Despite this popularity, it is worth providing the basic formulas for the extraction field of a rectangular shape. Longwall workings most often have a rectangular shape. Moreover, each polygon can be approximated with elementary rectangles, similar to integration by the rectangle method. Assuming a rectangular coordinate system of 0XY and point A (s, q)—Figure 6, Knothe provided a formula for subsidence of a point:

$$w(s,\ q) = \frac{w_{max}}{r^2} \int_{a-s}^{b-s} e^{\frac{-\pi\xi^2}{r^2}} d\xi \int_{c-q}^{d-q} e^{\frac{-\pi\xi\eta^2}{r^2}} d\eta \quad (1)$$

where:

$\xi = x - s$,

$\eta = y - q$.

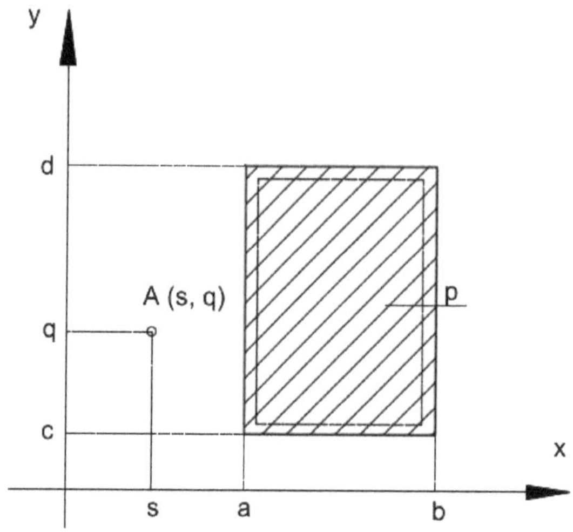


**Figure 6.** Diagram for Formula (1).

Taking into consideration the peripheral area p, the edges of mining should be shifted by its value in the directions shown in the figure, which is taken into account in the integration limits.

In Formula (1), the following parameters of the theory can be distinguished:

- Exploitation coefficient (of roof control)—a, dimensionless quantity. Its value depends on the method of liquidation of the selected part of the deposit. In the Upper Silesian Coal Basin, for caving mining the values are a = 0.7–0.9;
- Main impact radius r, expressed in meters. This parameter is used interchangeably with the parameter tan β, where β is the angle of the reach of the main impact.

Both of these parameters are related to each other:

$$\tan\beta = \frac{H}{r} \tag{2}$$

where H—depth of extraction, m;

- Extraction periphery is an additional parameter in the theory-p.

In the Upper Silesian Coal Basin (Poland), usually (routinely) the value a = 0.8 (extraction with the roof fall) and tan β = 2 are used for calculations. The value of the periphery is estimated at 0.14 r–0.20 r.

The remaining deformation indices are calculated as depression derivatives in accordance with the following dependencies.

The slopes in the X and Y directions are shown in the formula:

$$T_x = \frac{\partial w}{\partial x}, T_y = \frac{\partial w}{\partial y} \tag{3}$$

The maximum slope is shown by the formula:

$$T_{max} = |\text{grad } w(x, y)| = \sqrt{T_x^2 + T_y^2} \tag{4}$$

Vertical curves in the direction of the X and Y axes are shown by the formula:

$$K_x = \frac{\partial^2 w}{\partial x^2}, K_y = \frac{\partial^2 w}{\partial y^2} \tag{5}$$

Horizontal shifts in the direction of the X and Y axes are shown by the formula:

$$u_x = -B \cdot T_x\ ,\ u_y = -B \cdot T_y, \tag{6}$$

Horizontal deformations in the direction of the X and Y axes are shown by the formula:

$$\varepsilon_x = B \cdot \frac{\partial^2 w}{\partial x^2}, \varepsilon_y = B \cdot \frac{\partial^2 w}{\partial y^2} \tag{7}$$

where:

B = 0.32 r

Horizontal deformations in the direction of the main axes are shown by the formula:

$$\varepsilon_{1,2} = 0.5 \cdot (\varepsilon_x + \varepsilon_y) \pm \sqrt{0.25 \cdot (\varepsilon_x - \varepsilon_y)^2 + 0.5 \cdot \gamma_{xy}^2} \tag{8}$$

where:

$\gamma_{xy}$—complete differential:

$$\gamma_{xy} = \frac{\partial u_x}{\partial y} + \frac{\partial u_y}{\partial x} \tag{9}$$

Maximum horizontal deformations-εmax are understood as the maximum absolute value with ε1 and ε2. On the other hand, the values of deformation indicators are grouped into categories of mining areas, which are shown in Table 2. The categories of building structures resistance are compared to the mining area categories given in Table 2. If the area category does not exceed the category of the building resistance, then the area is not threatened with damage.

**Table 2.** Categories of mining areas depending on the values of deformation indicess.

| Category | T, mm/m | R, 1/km | ε, mm/m |
|---|---|---|---|
| 0 | $T \leq 0.5$ | $40 \leq \|R\|$ | $\|\varepsilon\| \leq 0.3$ |
| I | $0.5 < T \leq 2.5$ | $20 \leq \|R\| < 40$ | $0.3 < \|\varepsilon\| \leq 1.5$ |
| II | $2.5 < T \leq 5$ | $12 \leq \|R\| < 20$ | $1.5 < \|\varepsilon\| \leq 3.0$ |
| III | $5 < T \leq 10$ | $6 \leq \|R\| < 12$ | $3.0 < \|\varepsilon\| \leq 6.0$ |
| IV | $10 < T \leq 15$ | $4 \leq \|R\| < 6$ | $6.0 < \|\varepsilon\| \leq 9.0$ |
| V | $T > 15$ | $\|R\| < 4$ | $\|\varepsilon\| > 9$ |

To calculate the values of deformation indices at a given time t, Knothe [28] proposed an equation:

$$\frac{dw}{dt} = c\,(w_k - w(t)) \tag{10}$$

where:

w(t)—temporary subsidence,
$w_k$—asymptotic (final) subsidence,
c—subsidence rate coefficient (time), 1/year or 1/day. According to S. Knothe [28], for the Upper Silesian Coal Basin c = 0.5–7, 1/year

To facilitate the calculations, it is assumed that c = const and $w_k$ = const. Accepting the initial condition w(t = 0) = 0 provides a simple solution of Equation (10) as:

$$w(t) = w_k \cdot (1 - e^{-ct}) \tag{11}$$

Using Equation (11) is possible with the discretization of calculations–dividing the panel area into small (elementary) panels, which can be assumed to have been extracted in a short time. Then, the subsidence caused by the exploitation of elementary (rectangular) panels is summed up at the given moment t, taking into account Equation (11). In this way, the changes in final subsidence $w_k$ related to the expansion of exploitation field are taken into account.

The computer program DEFK-Win5.0 developed by Ścigała [29,30] worked on the above assumptions.

The program allows to make calculations for any number of panels in shape of any polygon. The area of the panels is approximated by elementary rectangles (similar to the integration using the rectangle method). The calculations are made using the Budryk–Knothe theory (see Formula (1) and following). Calculations can be performed for selected points on the surface or inside the rock mass. Performing calculations in a grid of points enables drawing of isoline maps of deformation indices. The program also enables identification of parameter values based on the measurement results of the final subsidence (parameters: a, tan β, p) and on the basis of values of subsidence in transient state (parameter c). The minimalization of goal function resulting from the smallest squares method is performed using the Hooke-Jeeves method.

## 3. Results

The analysis of land surveying results was conducted in order to identify the parameter values of S. Knothe theory, in order to use them to predict the state of deformation that would occur as a result of planned longwall extraction along the 548 panel. The analysis was conducted in stages:

- identification of the parameter values: a, p, tan β on the basis of final subsidence registered on the measurement line section;
- identification of parameter c value based on the lowering of the point over time.

### 3.1. Identification of Parameter Values on the Basis of Determined Subsidence

In order to identify the parameter values of the Budryk–Knothe theory based on the land surveying results conducted on the observation line, the values recorded in the last measurement were taken as the final subsidence. Computer program DEFK-Win5.0 was used for the calculations [28].

The following parameter values were obtained:

$$p = 39.5 \text{ m}, a = 0.517 \tan \beta = 1.71.$$

Table 3 shows the comparison of calculated subsidence with subsidence based on land surveying. A graphical comparison of subsidence depressions is shown in Figure 7.

**Table 3.** The comparison of calculated subsidence depressions including the peripheral areas and measurements.

| Pkt | X | Y | wrz | wt | V | VV | sumV | sumVV |
|---|---|---|---|---|---|---|---|---|
| 7 | 340.0 | 895.0 | −86.0 | −32.4 | −53.6 | 2875.2 | 53.6 | 2875.2 |
| 9 | 370.0 | 855.0 | −91.0 | −67.7 | −23.3 | 542.6 | 76.9 | 3417.8 |
| 10 | 380.0 | 830.0 | −98.0 | −100.2 | 2.2 | 4.8 | 79.1 | 3422.6 |
| 11 | 405.0 | 805.0 | −194.0 | −145.6 | −48.4 | 2344.1 | 127.5 | 5766.7 |
| 12 | 420.0 | 785.0 | −187.0 | −189.1 | 2.1 | 4.6 | 129.7 | 5771.3 |
| 13 | 435.0 | 770.0 | −216.0 | −226.4 | 10.4 | 109.0 | 140.1 | 5880.3 |

**Table 3.** *Cont.*

| Pkt | X | Y | wrz | wt | V | VV | sumV | sumVV |
|---|---|---|---|---|---|---|---|---|
| 14 | 435.0 | 740.0 | −286.0 | −311.6 | 25.6 | 657.1 | 165.7 | 6537.4 |
| 15 | 435.0 | 720.0 | −371.0 | −377.3 | 6.3 | 39.4 | 172.0 | 6576.8 |
| 16 | 435.0 | 690.0 | −469.0 | −486.3 | 17.3 | 298.1 | 189.3 | 6875.0 |
| 17 | 440.0 | 660.0 | −579.0 | −603.3 | 24.3 | 592.8 | 213.6 | 7467.8 |
| 18 | 450.0 | 635.0 | −688.0 | −700.8 | 12.8 | 164.7 | 226.5 | 7632.5 |
| 19 | 450.0 | 600.0 | −848.0 | −834.8 | −13.2 | 175.0 | 239.7 | 7807.5 |
| 20 | 475.0 | 580.0 | −931.0 | −888.9 | −42.1 | 1773.1 | 281.8 | 9580.6 |
| 23 | 485.0 | 505.0 | −1047.0 | −1060.5 | 13.5 | 181.5 | 295.3 | 9762.0 |

Standard deviation: 21.09 mm Percentage error: 2.62%. Explanations to the table: Pkt—point number; X, Y—coordinates of the point in the locally adopted Cartesian system, m; wrz—subsidence confirmed by the measurement, mm; wt—calculated subsidence, mm; V—difference between the values wt and wrz, mm; VV—square of difference v, $mm^2$; SumV—sum of difference V, mm; SumVV—sum of squares of difference VV, $mm^2$.

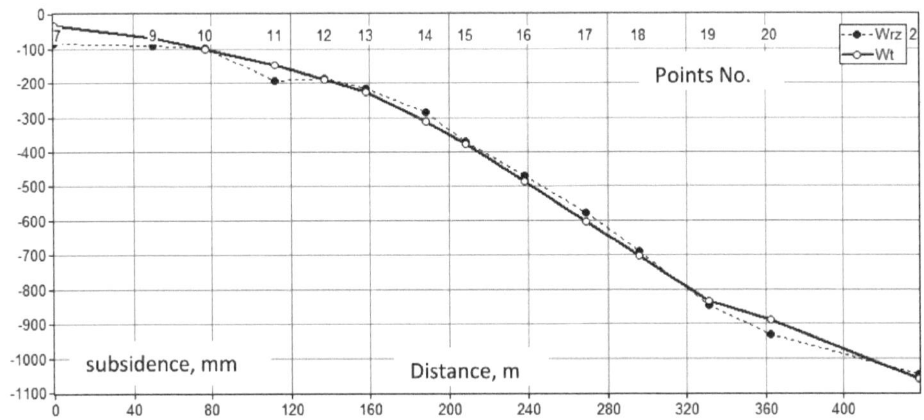


**Figure 7.** Measured and calculated final subsidence.

Obviously, it should be borne in mind that the obtained parameter values are unusual in comparison to those that are commonly used. Nevertheless, they are confirmed by the analyses carried out earlier for the same part of the basin. The results of the previous analyses are shown in Figure 8. As can be seen in the figure, with the increase in the value of the parameter tan β, the value of the parameter a also increased. The value of the parameter tan β is related to the mechanical properties of the rock mass. The higher the rock mass strength parameters, the lower the value of tan β. Thus, with subsequent extractions, the parameter value decreased. Figure 8 illustrated the analysis results—values a and tan β obtained as a result of determination based on the measurement results, the linear relationship between them as:

$$a = 0.6395 \tan \beta - 0.6338 \tag{12}$$

as well as the confidence intervals at the level of 0.95.

The value of the correlation coefficient R = 0.88 was obtained.

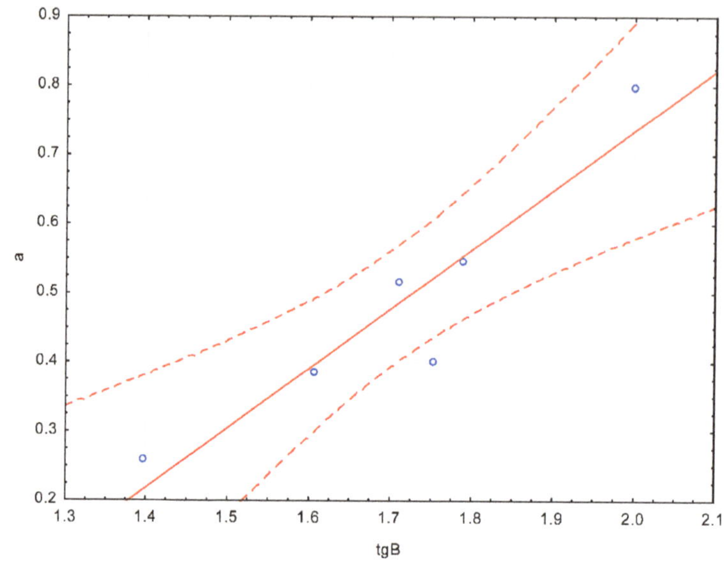


**Figure 8.** Relationship between the values of a and tan β parameters.

### 3.2. *Identification of Subsidence Rate Coefficient*

The identification of subsidence rate coefficient c was conducted based on the subsidence in the wall benchmark no. 8 stabilized on the building. The calculations were performed for one point due to the high repeatability of the measurement results obtained in points 5–8 (see Figure 5). As can be seen in Figure 5, a high repeatability of wall benchmarks subsidence was confirmed by measurements. The benchmark fixed on the building was selected for the analyses due to a higher accuracy of measurements than those carried out on the benchmarks set on the ground. Figure 9 illustrates the results of measurements with red bullet points, and the graph of calculated subsidence with a dashed line. The calculations were made using computer program DEFK-Win. In the case of the 546 panel, the obtained value was c = 0.003 1/day, and for the 547 panel c = 0.05 1/day.

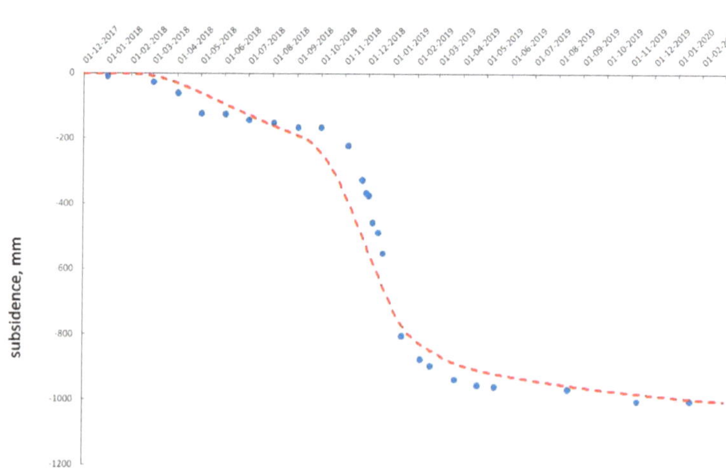


**Figure 9.** Measured and calculated subsidence for the benchmark no. 8.

### 3.3. Prediction of Ground Deformation As a Result of Planned Extraction

The calculations were made for the final state of deformations that would occur once the extraction of the 548 panel ended. The assumptions for the predictions of the impacts of the 548 panel extraction on the surface were made based on the above-mentioned analyses of the measurement results. The following parameter values were adopted:

Variant 1

- exploitation coefficient a = 0.517;
- tangent of the range of main impacts tan β = 1.71;
- extraction periphery p = 40 m.

The calculation results are presented in the form of a graphic map of subsidence isolines as well as mining areas categories resulting from maximum slopes and maximum horizontal strains. Figure 10 illustrates the boundaries of extraction of the 548 panel and isolines of predicted subsidence and areas subject to deformations within the range of individual categories. These areas are marked with appropriate colours and are illustrated with appropriate numbers.

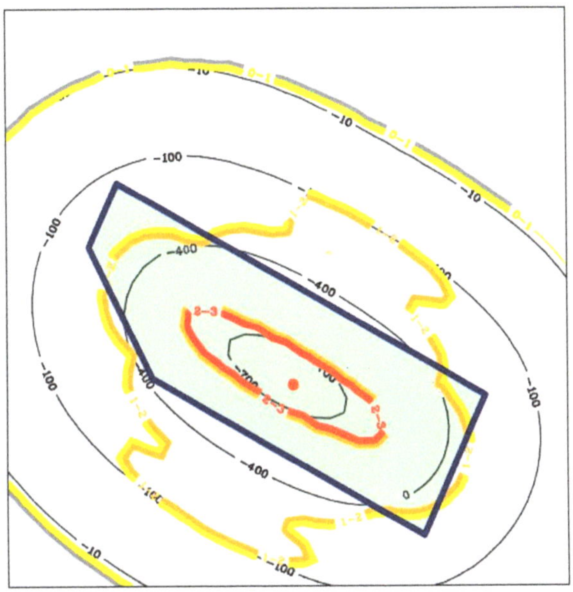


**Figure 10.** Subsidence and categories of mining areas. Variant 1.

For comparison, identical calculations were also made adopting standard, commonly accepted parameter values:

Variant 2

- exploitation coefficient a = 0.800;
- tangent of the range of main impacts tan β = 2.00;
- extraction periphery p = 40 m.

The calculation results are graphically illustrated in Figure 11, using analogous markings as in Figure 10.

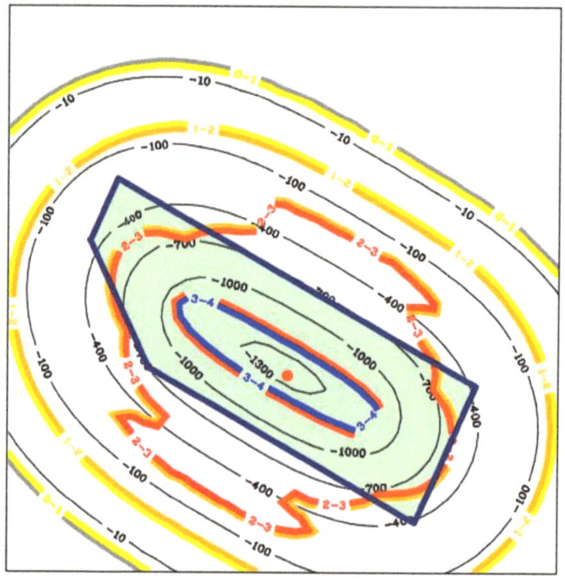


**Figure 11.** Subsidence and categories of mining areas. Variant 2.

### 3.4. Detailed Forecast for the Selected Object

Detailed calculations were carried out for a selected point above the planned extraction of the 548 panel. A computer simulation of the wall run was performed using DEK-Win 5.0 software. This point is marked in red in Figures 10 and 11. Similarly to the calculations done before, they were performed for two sets of parameters:

A variant–the calculations were made assuming a delay in the impact of mining, according to formula (11). The following parameter values were adopted:

- exploitation coefficient a = 0.517;
- tangent of the range of main impacts tan β = 1.71;
- extraction periphery p = 40 m;
- subsidence rate coefficient c = 0.05 1/day.

B variant–the calculations were made assuming immediate impact of mining, adopting standard parameter values:

- exploitation coefficient a = 0.800;
- tangent of the range of main impacts tan β = 2.00;
- extraction periphery p = 40 m;
- c -> ∞.

In other words, in the case of variant B, the subsidence at a given time t is equal to the final value of the subsidence caused by the extraction performed up to that time. In the case of variant A, the value of subsidence at time t is obviously smaller, which results from the Equation (11). In the case of variant A, the subsidence reaches the final value corresponding to the range of exploitation carried out until time t only after time T > t. Of course, the calculations according to both variants were performed with very different values of parameters a and tan β. Variant A reflects a case more similar to the actual state, both due to the inclusion of the time variable and the parameter values obtained on the basis of the measurement results.

Figure 12 shows the graphs of subsidence. Subsidence for A variant is marked with a blue line and wt symbol, whereas B variant is marked with a red line and wn symbol.

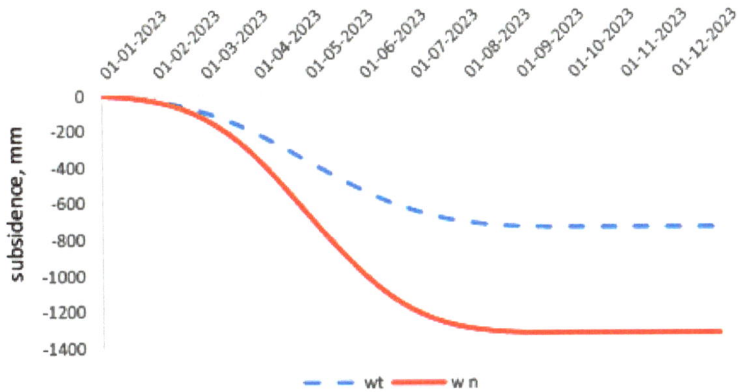


**Figure 12.** Subsidence over time for A and B variants.

Figure 13 shows the graphs of maximum slopes. The slopes for A variant are marked with a blue line and Tmax t symbol, whereas for B variant are marked with a red line and Tmax n symbol.

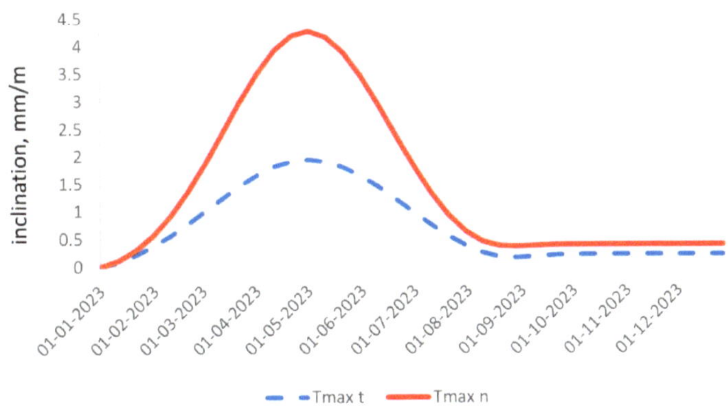


**Figure 13.** Maximum slopes over time for A and B variants.

Figure 14 illustrates the graphs of maximum horizontal strains. The strains for A variant are marked with a blue line and Emax t symbol, whereas for B variant are marked with a red line and Emax n symbol.

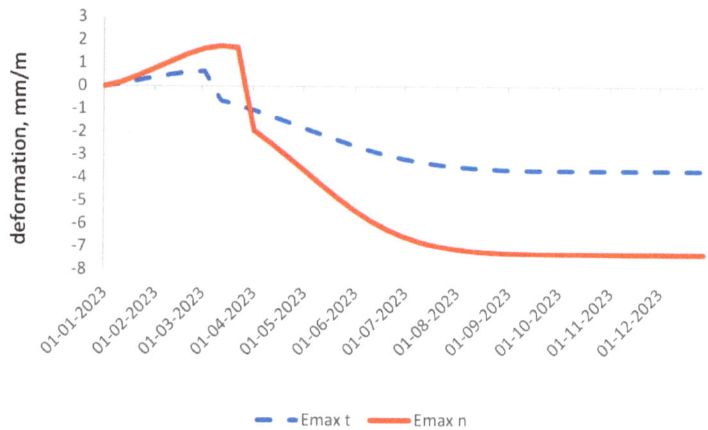


**Figure 14.** Maximum horizontal strains over time for A and B variants.

## 4. Discussion

The determination of the parameter values of the Budryk–Knothe theory based on the final trough enabled to obtain a high consistency between the calculation results and the results of subsidence measurements. This is indicated by the obtained subsidence percentage error value amounting to 2.62%—Table 3, Figure 7. It should be emphasized that the exploitation coefficient a = 0.517 is unusually low. As noted in point 2 of this paper, this value was expected to range between 0.7 and 0.9. The reason for this should be seen in the geological structure of the rock mass, i.e., high depth of strong limestone rocks of which the cover was made [31]. This corresponds with the obtained parameter value tan β = 1.71, which is characteristic of high-strength rocks [4]. In the case under consideration, not only a low value of the parameter tan β, but also of the parameter a were obtained. The correctness of the obtained results of the analyses is confirmed by the results of the previous research–Figure 8. It should be assumed that in the initial period of extraction, the parameter values a and tan β are low, and subsequently, as further extraction is carried out, they increase to the values commonly used for predictions.

The calculated subsidence of the selected point over time corresponds to the measurement results—Figure 9. Subsidence caused by the extraction of the 546 panel appeared slowly on the surface, as evidenced by the obtained value of the coefficient c = 0.003 1/day. Taking into account the above-described rock mass structure and the lack of earlier instances of mining in the immediate vicinity, such a small value is plausible. During the extraction of the 547 panel, the value c was higher and equaled c = 0.05 1/day, which should be explained by a greater degree of rock mass loosening as a result of the extraction of the 546 panel.

The following facts can be observed based on the prediction of the final state of deformations induced by the planned extraction of the 548 panel, which was conducted in the next stage of this work. Taking the parameter values obtained on the basis of the measurement results for calculations (Figure 10, variant 1) enabled the values of maximum subsidence of approximately 0.75 m and maximum deformations from the 3rd category to be obtained.

Taking the standard, commonly used parameter values for calculations (Figure 11, variant 2) resulted in obtaining much higher subsidence values equaling to 1.3 m and maximum deformations from the 4th category. As long as the occurrence of deformations in the 3rd category is accepted by local government authorities, the deformations of the 4th category mean that planned extraction under built-up areas would not be granted a permit from these authorities. Therefore, the importance of issues related to predictions of the state of deformations to truly reflect the actual state and the consequences it has, is clear.

In the paper, the need to make predictions of the deformation state, taking into account their course over time, was also considered. Such predictions are usually made for building structures of higher importance. A sample prediction was made for the object marked with a red circle in Figures 10 and 11. The calculations were carried out by making a computer simulation of the run of the workings. Similarly to the previous instance, we took the parameter values for the immediate effects (variant A) and for the delayed effects, according to Equation (11)—variant B. Here, of course, significant differences in the deformation state can also be observed. Subsidence illustrated in Figure 12 differed almost twice. Taking into account the parameters obtained based on the measurement results, the maximum slope was approximately 2 mm/m (1st category), whereas taking the typical parameter values and the immediate effects, they equaled to approximately 4.3 mm/m (2nd category)—Figure 13. Even greater differences occurred in the case of horizontal strains. Considering the parameters obtained based on the measurements results, the values of deformations were approximately +0.9 mm/m (1st category) and approximately −3.9 mm/m (3rd category), whereas considering typical parameter values and the immediate effect, they were approximately +2 mm/m (2nd category) and approximately −7.1 mm/m (4th category)—Figure 14. Of course, the course of deformation over time is very important information for construction specialists, who determine the possibility of damage occurrence to objects.

Figure 15 shows graphs of extracted and possible to extract bituminous coal resources as a result of properly conducted predictions. The resources are divided into lots 1–6 and the 548–7 panels. All the resources account for 820 500 Mg of coal.

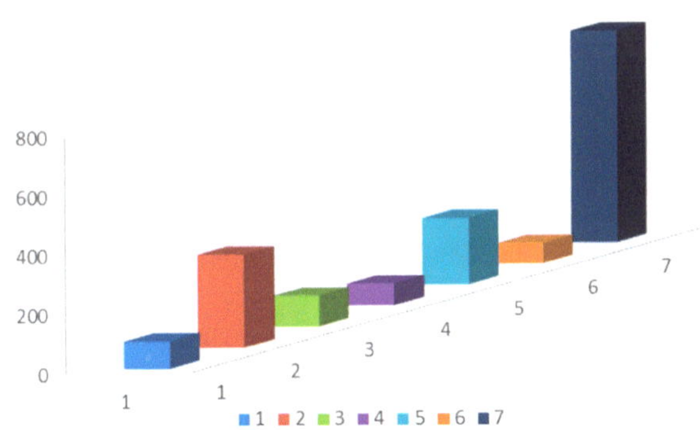


**Figure 15.** Bituminous coal resources in individual lots.

## 5. Conclusions

Based on the analyses conducted in this paper, the following conclusions and statements can be made:

1. Geological and mining conditions, including the structure of the rock mass, have a significant impact on the parameter values necessary for the preparation of mining area deformation predictions. In cases where these conditions differ significantly from the average ones in a given area, it is particularly important to identify the parameter values based on the measurement results. Taking typical parameter values significantly influences the quality of the calculation results and leads to wrong conclusions about possible damage to buildings. Consequently, it leads to wrong decision-making regarding the possibility of conducting a planned mining extraction:

2. In unusual geological conditions, when the overburden is made of strong limestone rocks, preliminary research indicated a gradual increase in the values of and tan β parameters up to the values commonly used for predictions. The previous experience has implied the influence of rock mass mechanical properties on the value of tan β parameter. Based on the research results presented in this paper it is clear that there is also a correlation between the values of the coefficient of roof control and the values of tan β, which are dependent on the rock mass structure. When analyzing the impact of planned extraction on more important objects, predictions should take into account the parameter values obtained during land surveying for specific geological and mining conditions. Such predictions should take into account a delay in the impacts, e.g., according to Equation (11).

**Funding:** The publication has been financed by Rector of Silesian Technical University (Grant no. 06/040/RGJ21/1007).

**Institutional Review Board Statement:** Not applicable.

**Informed Consent Statement:** Not applicable.

**Data Availability Statement:** Not applicable.

**Acknowledgments:** The authors thankfully acknowledge the Silesian University of Technology, Poland, for providing all the facilities to perform the research work.

**Conflicts of Interest:** The author declares no conflict of interest, no known competing financial interests or personal relationships that could influence the work.

## References

1. Strzałkowski, P.; Litwa, P. Environmental protection problems in the areas of former mines with emphasis on sinkholes: Selected examples. *Int. J. Environ. Sci. Technol.* **2021**, *18*, 771–780. [CrossRef]
2. Guo, W.; Guo, M.; Tan, Y.; Bai, E.; Zhao, G. Sustainable development of resources and the environment: Mining—Induced eco-geological environmental damage and mitigation measures—A case study in the Henan coal mining area, China. *Sustainability* **2019**, *11*, 4366. [CrossRef]
3. Li, G.; Yang, Q. Prediction of Mining Subsidence in Shallow Coal Seam. *Math. Probl. Eng.* **2020**, *2020*, 7956947. [CrossRef]
4. Strzałkowski, P. *Zarys Ochrony Terenów Górniczych*; Wyd. Pol. Śl.: Gliwice, Poland, 2015.
5. Kratzsh, H. *Mining Subsidence Engineering*; Springer: Berlin/Heidelberg, Germany; New York, NY, USA, 1983.
6. Lee, F.T.; Abel, J.F., Jr. *Subsidence from Underground Mining: Environmental Analysis and Planning Considerations*; Geological Survey Circular, 876; United States Department: Washington, DC, USA, 1983.
7. Kwiatek, J. (Ed.) *Ochrona Obiektów Budowlanych na Terenach Górniczych*; Wydawnictwo Głównego Instytutu Górnictwa: Katowice, Poland, 1997.
8. Knothe, S. *Prognozowanie Wpływów Eksploatacji Górniczej*; Wydawnictwo Śląsk: Katowice, Poland, 1984.
9. Zych, J. Metoda prognozowania wpływów eksploatacji górniczej na powierzchnię terenu uwzględniająca asymetryczny przebieg procesu deformacji. *Zesz. Nauk. Politech. Śląskiej* **1987**, *164*, 17–19.
10. Li, J.; Wang, L. Mining subsidence monitoring model based on BPM-EKTF and TLS and its application in building mining damage assessment. *Environ. Earth Sci.* **2021**, *80*, 396. [CrossRef]
11. Zhu, X.; Guo, G.; Zha, J.; Chen, T.; Fang, O.; Yang, X. Surface dynamic subsidence prediction model of solid backfill mining. *Environ. Earth Sci.* **2016**, *75*, 1007. [CrossRef]
12. Zhao, B.; Guo, Y.; Mao, X.; Zhai, D.; Zhu, D.; Huo, Y.; Sun, Z.; Wang, J. Prediction Method for Surface Subsidence of Coal Seam Mining in Loess Donga Based on the Probability Integration Model. *Energies* **2022**, *15*, 2282. [CrossRef]
13. Yu, Y.; Ma, L.; Zhang, D. Characteristics of Roof Ground Subsidence While Applying a Continuous Excavation Continuous Backfill Method in Longwall Mining. *Energies* **2020**, *13*, 95. [CrossRef]
14. Vušović, N.; Vlahović, M.; Kržanović, D. Stochastic method for prediction of subsidence due to the underground coal mining integrated with GIS, a case study. *Environ. Earth Sci.* **2021**, *80*, 67. [CrossRef]
15. Chen, Q.; Li, J.; Enke, H. Dynamic simulation for the process of mining subsidence based on cellular automata model. *Open Geosci.* **2020**, *12*, 832–839. [CrossRef]
16. Chudek, M. *Geomechanika z Podstawami Ochrony Środowiska Górniczego I Powierzchni Terenu*; Wydawnictwo Politechniki Śląskiej: Gliwice, Poland, 2002.
17. Wesołowski, M. Numerical modeling of exploitation relics and faults influence on rock mass deformations. *Arch. Min. Sci.* **2016**, *61*, 893–906. [CrossRef]

18. Guzy, A.; Witkowski, W.T. Land Subsidence Estimation for Aquifer Drainage Induced by Underground Mining. *Energies* **2021**, *14*, 4658. [CrossRef]
19. Feng, Z.; Guo, W.; Xu, F.; Yang, D.; Yang, W. Control Technology of Surface Movement Scope with Directional Hydraulic Fracturing Technology in Longwall Mining: A Case Study. *Energies* **2019**, *12*, 3480. [CrossRef]
20. Zhang, B.; Ye, J.; Zhang, Z.; Xu, L.; Xu, N. A Comprehensive Method for Subsidence Prediction on Two-Seam Longwall Mining. *Energies* **2019**, *12*, 3139. [CrossRef]
21. Zhu, X.; Zhang, W.; Wang, Z.; Wang, C.; Li, W.; Wang, C. Simulation Analysis of Influencing Factors of Subsidence Based on Mining under Huge Loose Strata: A Case Study of Heze Mining Area, China. *Geofluids* **2020**, *2020*, 6357683. [CrossRef]
22. Peng, S.S. *Coal Mine Ground Control*, 3rd ed.; West Virginia University: Morgantown, WV, USA, 2008; p. 750.
23. Li, J.; Zhang, J.X.; Huang, Y.; Zhang, Q.; Xu, J.M. An investigation of surface deformation after fully mechanized, solid back fill mining. *Int. J. Min. Sci. Technol.* **2012**, *22*, 453–457. [CrossRef]
24. Orwat, J. Causes analysis of occurrence of the terrain surface discontinuous deformations of a linear type. International Conference on Applied Sciences. *J. Phys. Conf. Ser.* **2020**, *1426*, 012016. [CrossRef]
25. Orwat, J. Mining exploitation forecasted effects caused by a hard coal extraction from a thick seam. In Proceedings of the International Conference on Applied Sciences, Hunedoara, Romania, 9–11 May 2019. [CrossRef]
26. Whittaker, B.W.; Redish, D.J. *Subsidence—Occurrence, Prediction and Control*; Elsevier: Amsterdam, The Netherlands; Elsevier: Oxford, UK; Elsevier: New York, NY, USA; Elsevier: Tokyo, Japan, 1989.
27. Han, J.; Zou, J.; Hu, C.; Yang, W. Study on Size Design of Shaft Protection Rock/Coal Pillars in Thick Soil and Thin Rock Strata. *Energies* **2019**, *12*, 2553. [CrossRef]
28. Knothe, S. Wpływ czasu na kształtowanie się niecki osiadania. *Arch. Min. Metall.* **1953**, *1*, 51–62.
29. Ścigała, R. *Komputerowe Wspomaganie Prognozowania Deformacji Górotworu i Powierzchni Wywołanych Podziemną Eksploatacją Górniczą*; Wydawnictwo Politechniki Śląskiej: Gliwice, Poland, 2008.
30. Ścigała, R. The identification of parameters of theories used for prognoses of post mining deformations by means of present software. *Arch. Min. Sci.* **2013**, *58*, 1347–1357. [CrossRef]
31. Kryzia, K.; Majcherczyk, T.; Niedbalski, Z. Variability of exploitation coefficient of Knothe theory in relation to rock mass strata type. *Arch. Min. Sci.* **2018**, *63*, 767–782. [CrossRef]

*energies*

MDPI

*Article*

# Parametric Study of Methyl Orange Removal Using Metal–Organic Frameworks Based on Factorial Experimental Design Analysis

**Abdollah Karami [1,2], Reem Shomal [1], Rana Sabouni [1,*], Mohammad H. Al-Sayah [3] and Ahmed Aidan [1]**

1 Department of Chemical Engineering, American University of Sharjah, Sharjah P.O. Box 26666, United Arab Emirates; b00029371@alumni.aus.edu (A.K.); rshomal@aus.edu (R.S.); aidan@aus.edu (A.A.)

2 Department of Chemical Engineering, McMaster University, Hamilton, ON L8S 4L8, Canada

3 Department of Chemistry, Biology and Environmental Sciences, American University of Sharjah, Sharjah P.O. Box 26666, United Arab Emirates; malsayah@aus.edu

* Correspondence: rsabouni@aus.edu

**Abstract:** Wastewater treatment plants (WWTPs) are one of the most energy-intensive industries. Every stage of wastewater treatment consumes energy, which is the primary contributor to WWTP costs. Adsorbents and process optimization are critical for energy savings. The removal of dyes from industrial wastewater by adsorption using commercially available adsorbents is inefficient. Metal–organic frameworks (MOFs) have outstanding properties that can improve separation performance over current commercial adsorbents, and thus, these materials represent a milestone in improving dye removal in water treatment methods. In this work, three types of metal–organic frameworks (Fe-BTC, Cu-BTC, and ZIF-8) have been investigated as prospective adsorbents for methyl orange removal from water in batch setups. The results showed that at 15 mg/L MO initial concentration and 100 mg dosage, Fe-BTC had the highest removal efficiency of 91%, followed by ZIF-8 (63%), and finally Cu-BTC (35%), which exhibited structural damage due to its instability in water. Fe-BTC maintained consistent adsorption capacity over a wide range of pH values. Furthermore, a $2^3$ full factorial design analysis was implemented to evaluate the conditions for maximum MO-removal efficiency. The main effects, interaction effects, analysis of variance (ANOVA), and the Pareto chart were reported. The statistical analysis demonstrated that the MOF type was the most significant factor, followed by dosage and initial concentration. The analysis indicated that the type of MOF and dosage had a positive effect on the removal efficiency, while the initial concentration had a negative effect. The two-way and three-way interactions were also found to be significant.

**Keywords:** metal–organic frameworks (MOFs); energy-saving; adsorption; methyl orange; factorial design; analysis of variance (ANOVA); pareto chart

**Citation:** Karami, A.; Shomal, R.; Sabouni, R.; Al-Sayah, M.H.; Aidan, A. Parametric Study of Methyl Orange Removal Using Metal–Organic Frameworks Based on Factorial Experimental Design Analysis. *Energies* **2022**, *15*, 4642. https://doi.org/10.3390/en15134642

Academic Editors: Dibyendu Sarkar, Rupali Datta, Prafulla Kumar Sahoo and Mohammad Mahmudur Rahman

Received: 3 May 2022
Accepted: 1 June 2022
Published: 24 June 2022



## 1. Introduction

Along with energy demand in highly urbanized and industrialized countries, population growth has led to higher demand for clean water supply and higher production of wastewater. In order to maintain the ecological status of water resources, produced wastewater requires adequate treatment before being safely discharged into the receiving water bodies [1–3]. Therefore, the number of wastewater collection and treatment plants has increased in communities [4], leading to countless plants operating throughout the world [5]. Wastewater treatment plants (WWTPs) are among the most energy-intensive industries. In 2014, the global energy demand for the water sector was estimated at 120 million tons of oil, which corresponds to 1395.6 TW-h, and projections estimate this value to double before 2040 [6–9]. This increase in energy consumption leads to significant growth in operational costs and to considerable boost in $CO_2$ emissions of wastewater treatment plants [10,11]. Therefore, the need to protect the environment from fossil fuel emissions and to have effective treatment of these industrial wastewaters before they are discharged into the environment necessitates the development of robust, economically feasible, environment-friendly, and energy-efficient solutions to be implemented for wastewater treatment industries [12].

One major component of industrial wastewater effluents is the discharge of a substantial amount of synthetic dyes due to their wide-ranging applications, including in textiles, printing, plastic, and food processing [13,14]. To prevent any potential hazards caused by these dyes to the surrounding environment of aquatic life and humans [15], industrial wastewater effluents are usually treated by several physical and/or chemical treatment processes to remove these synthetic dyes [16]. These treatments include flocculation combined with flotation, electroflocculation, membrane filtration, electrokinetic coagulation, electrochemical destruction, and ion exchange [17,18]. However, these technologies have several limitations, including low removal efficiencies, high cost, complexity of operation, high energy demand, and sensitivity to variations in wastewater influent [19]. In comparison to these techniques, however, adsorption is more effective and energy-efficient. Its low cost, simple design, easy handling, and sludge-free cleaning operations have proven that the adsorption technique is a preferred technique, especially for the removal of synthetic dyes from wastewater [18–20]. Adsorption is based on the binding of the soluble pollutant in wastewater (e.g., dye) to an insoluble adsorbent, which is then removed from the wastewater by filtration or precipitation, taking with it the adsorbed pollutant. The adsorbent can be regenerated and used for another cycle of the adsorption process.

Currently, beyond silica gel, other widely used adsorbents which rely on physical adsorption for water removal are zeolites, activated alumina, and activated carbon. Although all of them can be easily regenerated at a relatively low temperature, their main drawback is their limited adsorption capacity [20,21]. Traditionally, activated carbon (AC) is used as an adsorbent for quickly capturing dyes from wastewater and lowering their concentration in effluents. However, AC as an adsorbent generally shows very little selectivity over molecules of different sizes and has a high regeneration cost when exhausted [16]. Therefore, the development of efficient and cost-effective adsorbents is urgently needed.

Recently, metal–organic frameworks (MOFs) came into view as promising adsorbents. MOFs are porous solid materials comprised of metal clusters (blocks) connected by organic ligands via coordination bonds and have exceptional adsorbent properties that make them superior to other conventional porous solids [22]. Haque et al. (2010) were the first group to report the adsorptive removal of dyes using MOFs, where they employed chromium–benzene dicarboxylates MOFs (Cr-BDCs) for methyl orange (MO) adsorption [22]. Since then, several investigators have reported their findings on the removal of different dyes using various lab-synthesized MOFs as adsorbents [14–26]—for example, iron 1,3,5-benzenetricarboxylate (Fe-BTC) MOF, has been used in the liquid-phase separation of various molecules/compounds (organic and inorganic) [27,28]. Similarly, copper benzene-1,3,5-tricarboxylate (Cu-BTC) is a widely investigated MOF in different environmental applications, such as gas storage and adsorptive separation [29,30], and was used for methylene blue adsorption from water [31]. Additionally, 2-methylimidazole zinc salt (ZIF-8) was reported to exhibit high thermal stability up to 400 °C and excellent chemical stability in liquid-phase adsorption [32]. Recently, a new study investigated the development of new surfactant-modified carbon-coated monolith for the adsorption of cationic dyes. The results showed that the new material is suitably adsorbent for rapid and effective decolorization of methylene blue, with a high adsorption capacity of 200 mg/g [33,34].

The fact that MOFs have higher adsorption capacity, less energy-intensive regeneration conditions, and higher recyclability in comparison to other adsorbents makes them ideal energy-efficient adsorbents. However, for MOFs to be implemented for large-scale wastewater treatment plants, better assessment tools to compare MOFs with other adsorbents—emphasizing the energy-saving factor—are needed. Thus, we report herein the use of "design of experiment" (DOE) analysis to determine the optimal conditions for the adsorption process of MOFs. To achieve this objective, a well-studied system of MOF/MO is used as a model system for this study. A $2^3$ factorial design was implemented to investigate the effect of three major factors on the efficiency of the adsorption process, including (1) the type of MOF, (2) the dosage, and (3) the initial concentration on MOFs' removal efficiency of MO. In this study, three commercially available MOFs, namely Basolite® F300 (Fe-BTC) [24], C300 (Cu-BTC or HKUST-1) [25], and Z1200 (ZIF-8) [26], were used for the adsorption of azo-anionic MO dye from aqueous solutions that are widely used in the textile, printing, pharmaceutical, and food industries.

Additionally, the effect of adsorption on the structure of the MOFs and the possible adsorption mechanism was also investigated using various characterization techniques including FTIR, XRD, SEM, and TGA. Finally, the kinetics and equilibrium isotherms of adsorption were studied, and the thermodynamic parameters of MO adsorption were calculated.

We believe that this study will contribute to the efforts to better improve the process of selecting the best MOFs as promising adsorbent candidates in terms of their optimal adsorption capacity. The adsorption of methyl orange (MO) on MOFs has been contemplated as a promising advancement in wastewater treatment. Understanding the adsorption mechanism and kinetics and their effect on removal efficiency is essential for reaching the full potential of this approach. In addition, such a comparative study of MO removal and MOFs was defined by using a full factorial design method and an analysis of variance (ANOVA) statistical approach.

## 2. Materials and Methods

### 2.1. Materials

Fe-BTC, Cu-BTC, and ZIF-8 were purchased from Sigma-Aldrich under trademark Basolite® F300, Basolite® C300, and Basolite® Z1200, respectively, and were used without further modifications. Methyl orange (MO) was purchased from LabChem (Zelienople, PA, USA) as 0.1% (1000 mg/L) stock solution, and different concentrations were obtained by subsequent dilution using deionized water. For pH adjustment, 1 M HCl and NaOH aqueous solutions were used (prepared from 37% hydrochloric acid and 98% sodium hydroxide pellets, respectively). For MOF regeneration, ethanol (99.8%, Sigma-Aldrich, St. Louis, MO, USA) was used as the eluent.

### 2.2. Characterization

Several characterization analyses of the MOF samples, before and after MO adsorption, were performed. First, the crystal structure was investigated via the X-ray diffraction (XRD) measurements using a Bruker D8 ADVANCE system with a Cu tube ($\lambda$ = 1.5406 Å) and a linear detector (LYNXEYE XE). The measurements were obtained over a 2$\theta$ range of 50–60° (0.03° step size). The shapes and morphologies of the MOF samples before and after MO adsorption were examined via Scanning Electron Microscopy (SEM) and Transmission Electron Microscopy (TEM). In addition, the Fourier Transform Infrared (FTIR) spectra of the MOF samples were measured (KBr pellet technique) using PerkinElmer FTIR instrument (Waltham, MA, USA), and the measurement's scanning range was 4000 to 450 $cm^{-1}$ with a resolution of 1.0 $cm^{-1}$. The thermogravimetric analysis (TGA) was performed using a Pyris 1 TGA instrument (PerkinElmer, Waltham, MA, USA). The temperature range was 30 °C to 700 °C and the heating rate was 10 °C·$min^{-1}$. Moreover, the solid addition method was utilized to measure the surface point zero charge ($pH_{PZC}$) of each MOF. The procedure is similar to what was reported in [35,36], except for the $KNO_3$, $HNO_3$, and KOH solutions were replaced with NaCl, HCl, and NaOH solutions, respectively.

### 2.3. Adsorption Experiments

For the adsorption experiments, aqueous MO solutions were first prepared by successive dilution of the 1000 mg/L stock solution with deionized water to obtain the desired initial MO concentration. In each adsorption experiment, 100–200 mg of MOF was added to 50 mL of 5–15 mg/L MO solution at a pH of 6 at room temperature (25 °C). The mixtures were mixed thoroughly with magnetic stirring in sealed beakers and samples were taken after a predetermined time (5 to 180 min). Each sample was collected using a syringe and filtered using a membrane syringe filter (0.45 μm pore size, nylon). The MO concentration in all the samples was determined by visible light spectroscopic analysis of the solutions at the characteristic MO absorbance peak (λ = 464 nm) with a UV–Vis spectrophotometer (Evolution 60s, Thermo Scientific, Waltham, MA, USA).

At any given time (*t*), the adsorbed MO amount per unit mass of MOF ($q_t$) was calculated by Equation (1).

$$q_t = (C_o - C_t) \times \frac{V}{m} \tag{1}$$

where $C_o$, $C_t$, $V$, and $m$ are the initial MO concentration, the MO concentration in the liquid phase at time $t$, the MO solution's volume, and the MOF's dosage, respectively. Additionally, the removal efficiency (*RE*) was calculated using Equation (2).

$$RE = \frac{C_o - C_t}{C_o} \times 100 \tag{2}$$

The adsorption kinetics; isotherms; and thermodynamic analysis methodology, results, and discussion are presented in the Supplementary Information.

### 2.4. Factorial Design Analysis

DOE is a method used to plan for experiments such that the gathered experimental data can be assessed for statistical significance to reach unbiased conclusions. The basic steps of statistically designed experiments are (1) selection of factors and their corresponding levels; (2) determination of response variables; (3) selection of the type of the experimental design; (4) statistical analysis of the experimental data. In this paper, the full factorial design, which is one of the most widely used experimental designs, was chosen to investigate the effects of the aforementioned factors and their interactions. One of the main advantages of this design is the reduced number of experiments, which reduces the cost associated with the experimental work [35]. The number of experimental runs is given by $L^k$, where $k$ is the number of factors and $L$ is the number of levels in the design space (range of values over which factors are to be varied) [37]. Herein, three factors ($k$) were examined, including the type of MOF, dosage, and initial concentration. The number of levels was 2. In addition, the experiments were carried out in duplicate. Therefore, the total number of experiments was determined to be 16.

The $2^k$ factorial design is extensively utilized in adsorption experiments, where each experimental factor is investigated at two levels (high and low). In this work, the response variable was selected to be the removal efficiency of MO (*RE*), which is calculated according to Equation (2). To evaluate the statistical significance of the factors, *RE* was determined as the average of two parallel experiments. For analysis of the experimental data, Minitab Statistical Software (Version 17) was used.

## 3. Results and Discussion

### 3.1. Characterization

Figure 1 shows the measured MOF's XRD patterns before and after MO adsorption. The results demonstrate that both Cu-BTC and ZIF-8 have a crystalline structure and that the obtained patterns for the commercial MOFs are in good agreement with those reported in the literature for the equivalent lab-synthesized MOFs [32,38–44]. The main signature peaks present for Cu-BTC in the XRD patterns are at 2θ = 11.61°, 13.4°, 17.45°, 23.36°, and 29.33°. In addition, two extra peaks that are reported in the literature at 2θ = 6.7° and 9.5° were detected at low intensity. On the other hand, the signature peaks for ZIF-8 are at 2θ = 7.45°, 10.5°, 12.8°, 14.8, 16.5, and 18.1°. For the Fe-BTC sample, the broad peaks in the XRD pattern reveal that it is a material of low crystalline order, which is in agreement with what is reported in the literature for commercial Fe-BTC as well as lab-synthesized Fe-BTC [45–48]. The XRD patterns of Fe-BTC and ZIF-8 after MO adsorption show that their structure remains intact, while the XRD pattern of the Cu-BTC/MO sample revealed apparent loss of crystallinity, which is an indication that the crystalline structure of Cu-BTC was damaged after adsorption.

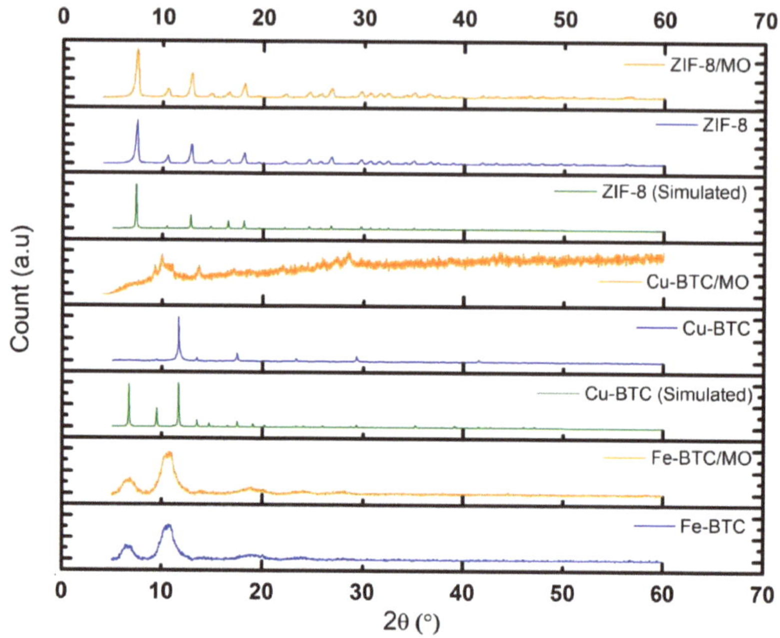


**Figure 1.** XRD patterns of MOF samples before and after MO adsorption. The simulated XRD patterns of Cu-BTC and ZIF-8 were sourced from the Cambridge Crystallographic Data Centre database.

The SEM and TEM micrographs in Figures 2 and 3, respectively, reveal the crystal morphologies of the MOF samples before and after MO adsorption. The SEM and TEM images for the fresh samples are similar to those reported in the literature for Fe-BTC [45,48,49], Cu-BTC [38,50,51], and ZIF-8 [52–54]. For Fe-BTC samples, it was difficult to obtain SEM images at higher magnification due to the magnetic character of the Fe-rich samples, whereas TEM images in Figure 3a,b clearly reveal the semi-amorphous nature of Fe-BTC. For Cu-BTC, the well-known octahedral shape can be seen in Figure 2a. However, the morphology of the after-adsorption sample shown in Figure 2b cannot be observed due to the loss of the spatial structure of the MOF. Finally, SEM images (Figure 2c,d) of the small particles of ZIF-8 and ZIF-8/MO and the TEM micrographs (Figure 3c,d) at the nm scale reveal the polyhedral shape of ZIF-8, which is retained after adsorption.

FTIR spectra of the MOF samples presented in Figure 4 are in good agreement with the literature [45,55–59]. For Fe-BTC and Cu-BTC MOFs, the measured IR spectra show absorption peaks that are quite similar. This can be explained by the similarity in nature of the two MOFs regardless of their crystalline order [45,49]. The broad peaks between 3800–3000 cm$^{-1}$ for both samples correspond to the presence of the hydroxyl group of the adsorbed water molecules. Both Fe-BTC and Cu-BTC absorb some of the moisture in the atmosphere, and this explains the presence of free water molecules in the sample before adsorption. On the other hand, water in the samples after adsorption is mainly from the adsorbed water molecule from the aqueous solution. The IR spectrum for the Fe-BTC sample shows absorption peaks at 711.6, 760.5, 1380, 1448, 1577, and 1627 cm$^{-1}$. The two peaks at 1577 cm$^{-1}$ and 1627 cm$^{-1}$ are attributed to the carboxylic (COO) asymmetric stretching, whereas the peak at 1448 cm$^{-1}$ belongs to the corresponding symmetric stretching vibration [45]. The peak at 1380 cm$^{-1}$ indicates the presence of (C-O) from the

BTC ligand [60]. In addition, the bands at 760.5 cm$^{-1}$ and 711.6 cm$^{-1}$ might be related to the Fe-O bond vibration. The IR spectrum of the Fe-BTC/MO sample shows a slight shift in some of the peaks related to the carboxylic (COO) stretching and the ligand–Fe bond vibration. Additionally, the broad band in the range of 3800–3000 cm$^{-1}$ had a narrower and less intense peak, which might be attributed to the hydroxyl group having H-bond interaction with the MO molecule. The measured IR spectra for the Cu-BTC sample, on the other hand, show absorption peaks at 729.6, 760.1, 1373, 1445, 1560, and 1620 cm$^{-1}$. The absorption peaks at 1445, 1560, and 1620 cm$^{-1}$ are related to the organic linker's COO symmetric/asymmetric vibration [45]. The presence of the C–O bond from the BTC ligand can be related to the peak at 1373 cm$^{-1}$ [60]. Additionally, the bands at 760.1 cm$^{-1}$ and 729.6 cm$^{-1}$ might be indicative of the Cu–O stretching vibration [31]. The vibrational band corresponding directly to the ligand coordinated with the centered Cu(II) ions is also visible on the IR absorption band at 487 cm$^{-1}$ [61,62]. The IR spectrum for the Cu-BTC/MO sample shows clear changes in the peaks related to the COO stretching modes in the 1700–1300 cm$^{-1}$ range, and the ligand–Cu bond vibration in the 760–730 cm$^{-1}$ range. In the IR spectrum of both Fe-BTC/MO and Cu-BTC/MO, the peak around 1120 cm$^{-1}$ can be assigned to the S-O bonds from MO, confirming the presence of MO molecules on Fe-BTC and Cu-BTC surfaces. For the ZIF-8 samples, the small peaks at 3140 cm$^{-1}$ and 2930 cm$^{-1}$ are attributed to the aliphatic/aromatic C-H stretching in the organic linker (from the methyl group and imidazole ring) [63]. Moreover, the peak at 1590 cm$^{-1}$ can be assigned to the stretching mode of C=N, whereas that observed between 1460 and 1310 cm$^{-1}$ is related to the vibration of the imidazole ring [64]. Furthermore, the peak at 1150 cm$^{-1}$ corresponds to the C-H bending in the imidazolate [64]. Finally, the peaks between 1100 and 950 cm$^{-1}$ and the two peaks at 760 and 694 cm$^{-1}$ could be related to the in-plane and out-of-plane bending of the imidazole ring, respectively [63,64]. For the ZIF-8/MO sample, the broader and more intense peak in the 3300–3600 cm$^{-1}$ range was related to the free hydroxyl group and the hydrogen-bonded hydroxyl groups confirming the presence of free water in the ZIF-8/MO sample [59]. Finally, the small peak at 1030 cm$^{-1}$ could be ascribed to the presence of MO on the ZIF-8 surface.

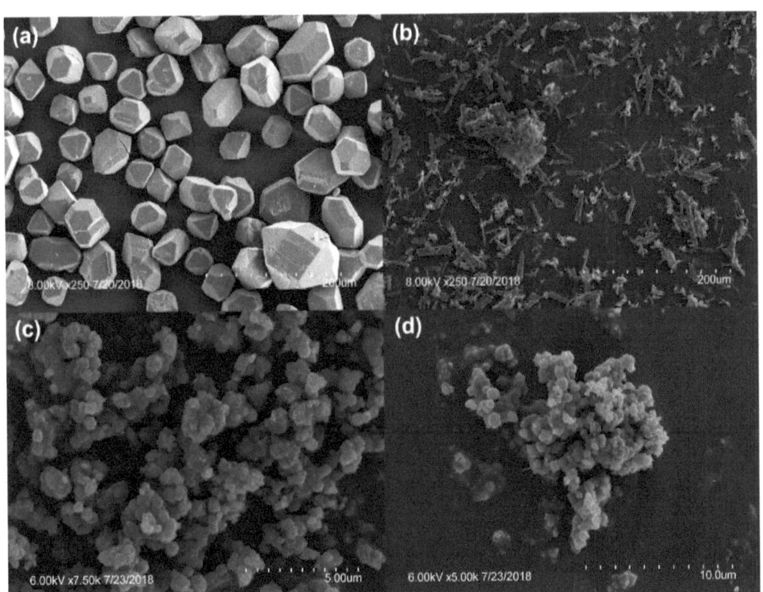


**Figure 2.** SEM micrographs of (**a**) Cu-BTC; (**b**) Cu-BTC/MO; (**c**) ZIF-8; (**d**) ZIF-8/MO.

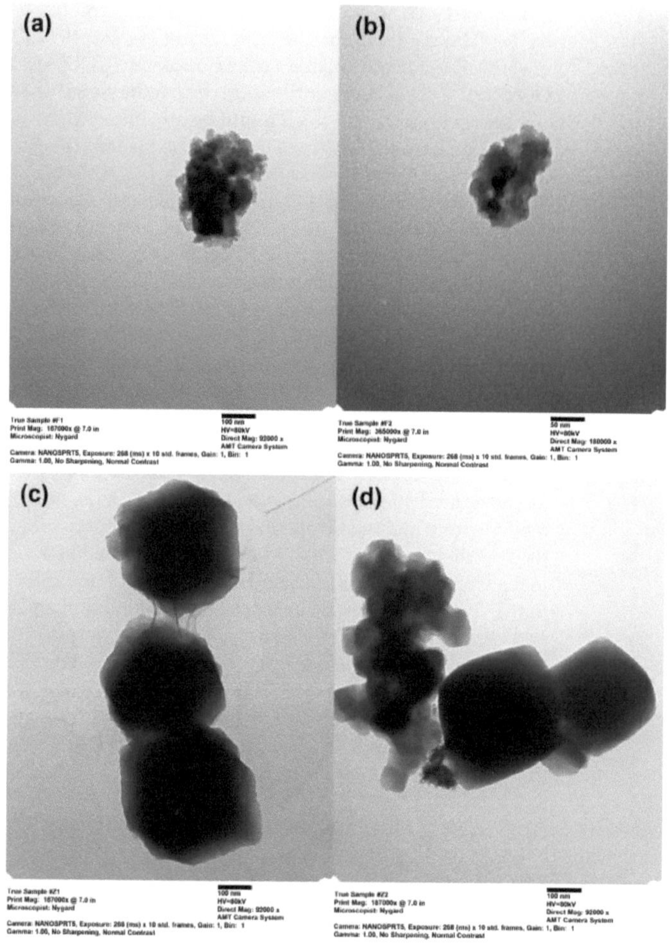

Figure 3. TEM micrographs of (a) Fe-BTC; (b) Fe-BTC/MO; (c) ZIF-8; (d) ZIF-8/MO.

TGA was used to test the stability of the MOFs before and after MO adsorption. The TGA results are presented in Figure 5, and they indicate that the overall thermal stabilities of the MOFs were comparable to those reported for the lab-synthesized MOFs [38,49,65]. For Fe-BTC MOF, the overall weight loss for the sample before adsorption was around 70%, whereas the sample after adsorption had a slightly higher weight loss of about 72%. The first drop in weight is around 100 °C, which corresponds to the evaporation of the water present in the samples. This is confirmed by the peak in derivative weight loss (DWT%) curves for both samples, where the Fe-BTC/MO sample had a slightly more intense peak, indicating higher water content in the sample, mainly from the MO aqueous solution. In addition, at 200 °C, the Fe-BTC sample lost approximately 10% of its weight, whereas the Fe-BTC/MO sample lost approximately 20%, mostly water content. Then, between 200 °C and 300 °C, both samples showed no significant weight loss. Following this, both samples continued to lose weight between 300 °C and 500 °C, where total weight loss reached around 55% at 500 °C. This loss can be attributed to the decomposition of BTC linkers [28]. Between 500 °C and 600 °C, there was no significant weight loss, followed by a sudden drop in weight above 600 °C (ca. 10% for Fe-BTC and ca. 12% for Fe-BTC/MO). This loss of weight is due to the total decomposition of the samples and the formation of iron-containing ashes [46]. For Cu-BTC MOF, the overall weight loss of the samples was around 60%. The first drop in weight is at 100 °C which corresponds to the evaporation of water present in the samples. This is confirmed by the peak in DWT% curves for both samples, where Cu-BTC/MO sample had a second peak between 150 °C and 200 °C, indicating higher water content in the sample, mainly from the MO aqueous solution. Then, above 200 °C, Cu-BTC sample showed no significant weight loss up to 360 °C after which a sudden weight loss is observed from the DWT% curve. On the other hand, the second stage of weight loss for the Cu-BTC/MO sample occurs at around 300 °C, where a broad shoulder peak in the DWT% curve is observed. In both cases, this drop in weight is attributed to the total decomposition of BTC linkers and production of CuO as the remaining product [40,41], plus the decomposition of MO in the Cu-BTC/MO sample. For ZIF-8 MOF, the overall weight loss for the sample before adsorption was around 38%, whereas the sample after adsorption had a slightly higher weight loss of about 40%. The TGA results for the ZIF-8 sample show that there was no weight loss in the temperature range of 100–450 °C, indicating that the ZIF-8 had no trapped solvent molecules inside the framework and that it was stable up to 450 °C. On the other hand, the ZIF-8/MO sample had a small weight loss (2%) at 150 °C, corresponding to the release of water adsorbed at the crystal surface. Then, both samples exhibited a gradual weight loss at around 500 °C followed by a sharp drop at 600 °C due to the structural collapse of ZIF-8 and decomposition of the organic ligand, leading to the formation of ZnO as the final product [63]. In comparison, ZIF-8 had the best overall thermal stability among the three MOFs under consideration.

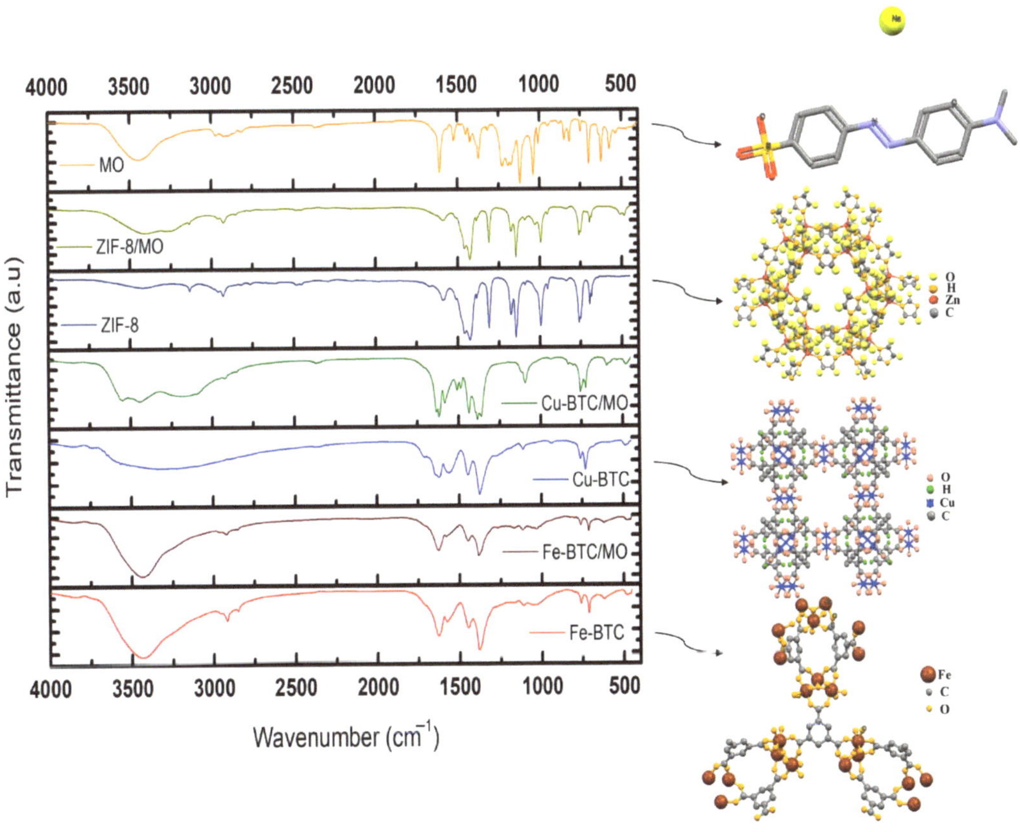


Figure 4. FTIR spectra of Fe-BTC, Cu-BTC and ZIF-8 before and after MO adsorption.

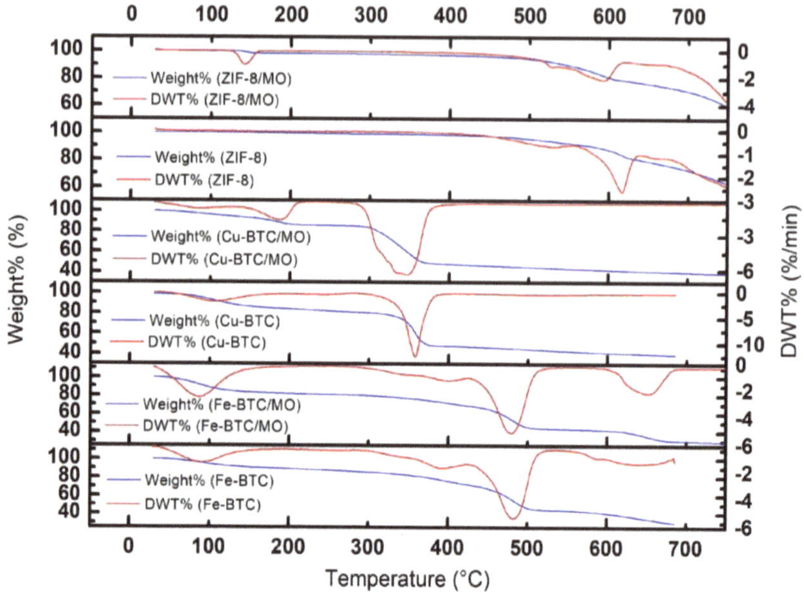


**Figure 5.** TGA profiles of the MOFs before and after MO adsorption.

### 3.2. Removal Efficiencies

Figure 6 demonstrates the removal efficiency (*RE*) as a function of time for the three adsorbents. Among the three MOFs considered, Fe-BTC had the highest maximum MO removal efficiency (91%), followed by ZIF-8 (63%) and finally Cu-BTC (35%). It can be seen that the MO adsorption was rapid in the case of Fe-BTC (80% *RE* in 10 min), and equilibrium was achieved in 60 min, while adsorption over ZIF-8 took 180 min to reach equilibrium. In contrast, the removal efficiency of Cu-BTC drops after 30 min, indicating that the structure of the adsorbent was damaged, as was confirmed by the XRD results for Cu-BTC/MO sample. Therefore, Cu-BTC cannot be counted as a good candidate for the adsorptive removal of MO from water under the investigated experimental conditions. Furthermore, although Fe-BTC had the lowest BET surface area (according to the data reported by Sigma-Aldrich), it had the highest removal efficiency, indicating that MO adsorption was not necessarily directly related to the surface area or pore volume of the MOF used; rather, it can be due to the presence of specific interactions between the MOF and MO. Additionally, Luan Tran et al. [66] studied the effect of the external surface area of ZIF-8 on dyes' adsorption capacity. According to their findings, the external surface of ZIF-8 was the only contributor to adsorption, and there was no adsorption inside the pores of the MOF. Additionally, the lower removal efficiency of ZIF-8 compared to Fe-BTC can be linked to the hydrophobic character of ZIF-8 versus the hydrophilic nature of Fe-BTC [59,67]. Therefore, these factors may be the reason for the limited MO adsorption capacity over ZIF-8 despite the larger surface area of the MOF.

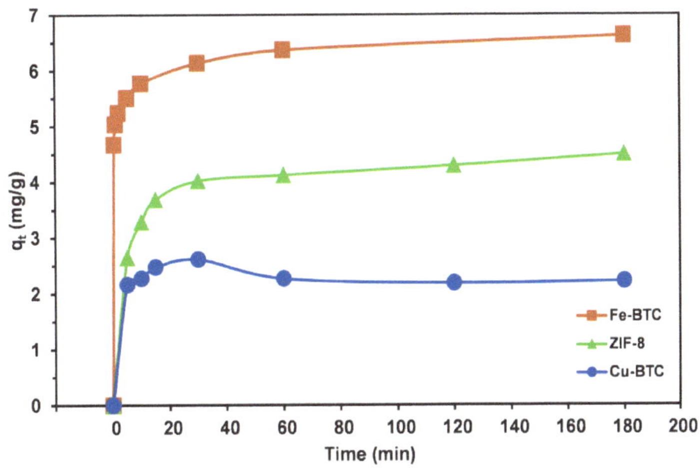


**Figure 6.** Comparison of MO removal efficiency of Fe-BTC, Cu-BTC, and ZIF-8 ($m$ = 100 mg, $C_o$ = 15 mg/L, pH = 6).

### 3.3. Initial pH ($pH_{initial}$) Effect and Adsorption Mechanism

The effect of the initial pH on MO adsorption was investigated in the 2–12 pH range. In this study, the pH effect experiments were carried out at 25 °C. The initial MO concentration was kept at 15 mg/L, and the dosage of the MOF was 150 mg. The effect of the initial pH ($pH_{initial}$) on MO adsorption is shown in Figure 7a. For Fe-BTC, the $pH_{initial}$ of the solution (in the range of 2–10) had an insignificant effect on MO equilibrium adsorption capacity ($q_e$), while in the same range, ZIF-8 had the highest adsorption capacity at $pH_{initial}$ of 2 with relatively similar capacities between 4 and 10. On the other hand, the adsorption capacity of Cu-BTC decreased as $pH_{initial}$ increased in the same range. At $pH_{initial}$ 12, the adsorption capacity of all MOFs was the lowest, suggesting the collapse of the MOFs' structures under strong alkaline conditions [68]. These findings can be linked with the results of $pH_{PZC}$ experiments shown in Figure 7b. For $pH_{initial}$ less than $pH_{PZC}$, the MOF's surface is more positively charged, which strengthens the electrostatic attraction to MO molecules that are usually in the negatively charged sulfonate form. Additionally, for $pH_{initial}$ greater than $pH_{PZC}$, the surface of the MOF is more negatively charged, which leads to strong repulsion between the MOF's surface and MO molecules (both are negatively charged) [22,68]. In the cases of Fe-BTC and ZIF-8, the $pH_{PZC}$ was around 9 and 9.5, respectively. Therefore, the drop in the adsorption capacity at pH 12 can be attributed to the electrostatic repulsion between the negative surface charge and the negatively charged MO molecule as well as to the decomposition of the MOF structure. However, for Cu-BTC, the $pH_{PZC}$ was around 4, which explains the higher adsorption capacity at $pH_{initial}$ less than 4 since the surface of Cu-BTC is more positively charged. However, it should be noted that the improvement in the amount adsorbed at $pH_{initial}$ less than 4 was not substantial, indicating that the effect of water instability on the structure of Cu-BTC was significant, as was shown previously. As a result, all other experiments were performed at the natural MO solution's pH without adjustment (pH = 6).

It is worth noting that the relatively constant $q_e$ for Fe-BTC and ZIF-8 in the pH range 2–10 could indicate the involvement of other adsorption mechanisms in addition to the electrostatic interactions, such as $\pi$–$\pi$ stacking and hydrogen bonding, between MO and the MOF. In the case of Fe-BTC, the interaction between the benzene rings in the MOF's organic linker and MO molecules is referred to as the $\pi$–$\pi$ stacking [69], while hydrogen bonds can form between MO's nitrogen or oxygen atoms and the hydrogen atoms in the benzene ring of Fe-BTC and the imidazole ring of ZIF-8. Therefore, electrostatic, $\pi$–$\pi$ stacking, and hydrogen-bond interactions are considered the primary forces involved in MO adsorption on both ZIF-8 and Fe-BTC. The details of the adsorption kinetics, isotherms, and thermodynamic analyses are discussed and presented in the supplementary information.

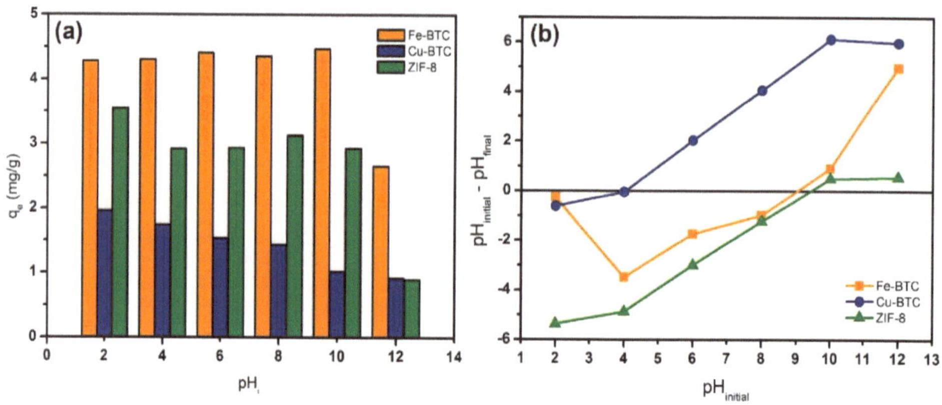


**Figure 7.** (**a**) Initial pH's (pH$_{initial}$) effect on equilibrium adsorption capacity ($q_e$); and (**b**) experimental results of (pH$_{initial}$ − pH$_{final}$) vs. pH$_{initial}$ curves used to determine pH$_{PZC}$ of the MOFs.

### 3.4. Factorial Design Analysis

The factors selected for the experimental design were the type of MOF, the dosage of MOF, and the initial concentration. For the type of MOF factor, only Fe-BTC and ZIF-8 were considered since the structure of Cu-BTC was proven to be unstable under the experimental conditions. Furthermore, the pH effect was not included since it was performed separately, and the results showed that pH effect on MO adsorption was negligible over a wide pH range (2–10).

Optimum conditions are decided by changing several factors and using different levels of these factors. Factorial designs are widely applied in experiments that are take into account several factors where it is necessary to study the interaction effect of factors on the response [20]. The factorial design describes which factor shows more impact and influences the variation of one factor on the other factors [21].

Table 1 presents the high and low levels for the $2^3$ factorial design for a total of 16 experiments. The higher level was given the "+1" designation, and the lower level was given the "−1" designation. For the type of MOF, ZIF-8 was assigned as the lower level since it had lower removal efficiency than Fe-BTC. The factorial design matrix (codified variables) along with the response values for MO removal efficiency are summarized in Table 2. Figure 8 presents a graphical representation (cube plot) of the experimental results (average *RE*) for the respective low and high levels of each factor. The following codified equation was used to explain the $2^3$ factorial design of MO removal:

$$Y = X_0 + X_1A + X_2B + X_3C + X_4AB + X_5AC + X_6BC + X_7ABC$$

where $Y$ is the predicted response (removal efficiency percentage), $X_0$ represents the global mean, $X_i$ is the regression coefficient corresponding to the main factor effects and interactions, $A$ is the MOF type, $B$ is the adsorbent dosage (mg), and $C$ is the initial dye concentration (mg/L).

Analysis of variance (ANOVA) was conducted to examine the reliability of the model describing the response as a function of the factors. The analysis was performed using Fisher's test (F-value) and *p*-values to determine the significance of the regression coefficients of the main effects and the interaction effects. In general, $p$-values < 0.05 are considered statistically significant [70]. The ANOVA results and the contribution of the main factors and their interactions with the model are shown in Table 3 as well. From the *p*-values, it can be seen that the main factors (A, B, and C), their two-way interactions (AB, AC, and BC), and their three-way interaction (ABC) are statistically significant to the removal efficiency. The coefficients and the effects of the main factor and their interactions in the model are presented in Table 4. The significance (or relative importance) of the main factors and their interactions can be determined from the absolute values in the "Effect" column. It can be seen that the order of the effects of the factors and their interactions in the model is ABC < BC < AB < AC < C < B < A. The positive values of the effects indicate that as the level of the factor increases the removal efficiency increases, while the negative values imply the opposite (i.e., increasing the level of the factor decreases removal efficiency).

**Table 1.** The factors and their corresponding levels used in the factorial design.

| Factor | Code | Low Level (−1) | High Level (+1) |
|---|---|---|---|
| Type of MOF | A | ZIF-18 | Fe-BTC |
| Dosage (mg) | B | 100 | 200 |
| Initial Concentration (mg/L) | C | 5 | 15 |

**Table 2.** The factorial design matrix and the experimental response results for MO removal efficiency.

| Runs | A | B | C | MO Removal Efficiency (%) |
|---|---|---|---|---|
| 1 | −1 | +1 | +1 | 87.3 |
| 2 | −1 | −1 | −1 | 88.9 |
| 3 | +1 | +1 | +1 | 99 |
| 4 | +1 | −1 | −1 | 96.9 |
| 5 | −1 | −1 | +1 | 61.5 |
| 6 | −1 | +1 | −1 | 97.2 |
| 7 | −1 | −1 | −1 | 88.6 |
| 8 | +1 | −1 | +1 | 90.9 |
| 9 | −1 | +1 | +1 | 86.9 |
| 10 | +1 | +1 | −1 | 100 |
| 11 | +1 | −1 | −1 | 96.8 |
| 12 | +1 | +1 | −1 | 100 |
| 13 | +1 | +1 | +1 | 98.1 |
| 14 | +1 | −1 | +1 | 90.9 |
| 15 | −1 | −1 | +1 | 66 |
| 16 | −1 | +1 | −1 | 97.2 |

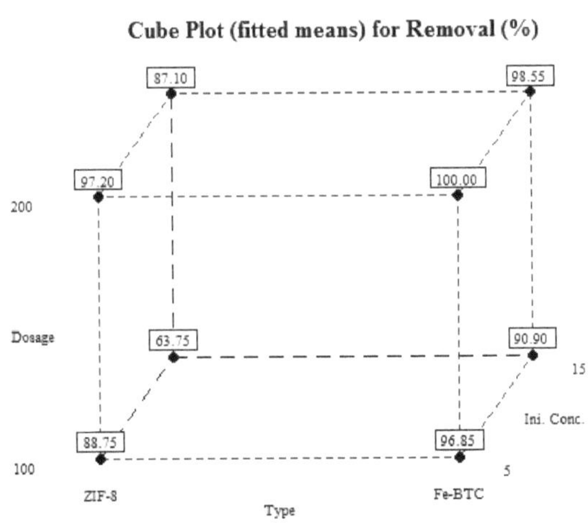


**Figure 8.** Cube plot for MO removal efficiency.

**Table 3.** Analysis of variance (ANOVA) results for *RE* calculated using Minitab.

| Source | DF | Adjusted Sum of Squares (Adj. SS) | Adjusted Mean Squares (Adj. MS) | F-Value | *p*-Value | Contribution |
|---|---|---|---|---|---|---|
| A | 1 | 612.6 | 612.6 | 459.7 | 0.000 | 31.39% |
| B | 1 | 453.7 | 453.7 | 340.5 | 0.000 | 23.25% |
| C | 1 | 451.6 | 451.6 | 338.9 | 0.000 | 23.14% |
| AB | 1 | 110.3 | 110.3 | 82.74 | 0.000 | 5.650% |
| AC | 1 | 191.8 | 191.8 | 144.0 | 0.000 | 9.830% |
| BC | 1 | 94.09 | 94.09 | 70.61 | 0.000 | 4.820% |
| ABC | 1 | 27.04 | 27.04 | 20.29 | 0.002 | 1.390% |
| Error | 8 | 10.66 | 1.332 | | | |
| Total | 7 | 1951.68 | | | | |
| Main effects | 7 | 1941.02 | 277.288 | 208.10 | 0.000 | |
| 2-way interactions | 3 | 396.16 | 132.054 | 99.10 | 0.000 | |
| 3-way interaction | 1 | 27.04 | 27.040 | 20.29 | 0.002 | |

**Table 4.** Estimated regression coefficients of the factors and their effects on MO removal.

| Term | Effect | Coefficient | Standard Error Coefficient | T-Statistic | *p*-Value |
|---|---|---|---|---|---|
| Constant | | 90.39 | 0.289 | 313.21 | 0.000 |
| A | 12.38 | 6.188 | 0.289 | 21.44 | 0.000 |
| B | 10.65 | 5.325 | 0.289 | 18.45 | 0.000 |
| C | −10.63 | −5.312 | 0.289 | −18.41 | 0.000 |
| AB | −5.250 | −2.625 | 0.289 | −9.10 | 0.000 |
| AC | 6.925 | 3.463 | 0.289 | 12.00 | 0.000 |
| BC | 4.850 | 2.425 | 0.289 | 8.40 | 0.000 |
| ABC | −2.600 | −1.300 | 0.289 | −4.50 | 0.002 |
| S | 1.154 | | | | |
| R-Sq | 99.45% | | | | |
| R-Sq (adj.) | 98.98% | | | | |

The normal probability plot can be divided into two regions: the region with percentages above 50 percent, where the factors are represented by positive coefficients (A, B, AC, and BC), and the region with percentages below 50 percent, where the factors are represented by negative coefficients (C, AB, and ABC). Furthermore, the main factors and their interactions denoted by circles are not significant, whereas the effects denoted by a square are significant. Analysis of variance is a statistical method that divides total variation into component parts, each of which corresponds to a different source of variation [20,21].

The main factors' statistical significance and their interactions were also confirmed via the Pareto chart in Figure 9b. For the 95% confidence level ($\alpha$ value of 0.05), the *t*-value was 2.31 (the vertical reference line). The values that exceed the reference line are considered significant [71]. According to Figure 9, all main factors and their interactions were statistically significant at the level of 0.05, which corroborated the results from ANOVA. Though the Pareto chart allows one to compare the absolute values of the effects of each factor and their interactions, the normal plot of standardized effects is more accurate in determining the significance and insignificance of each effect, as shown in Figure 9a. A normal probability plot could be separated into two regions: the region with percentages above 50%, where the factors are indicated by positive coefficients (A, B, AC, and BC), whereas in the region with percentages below 50%, the factors are indicated by negative coefficients (C, AB, and ABC). Moreover, the main factors and their interactions denoted as circles are not significant, while the effects shown as a square are significant. Figure 9a indicates that the main factors and their interactions were all significant. Analysis of variance is a statistical method that partitions the total variation into its component parts, each of which is associated with a different source of variation [20,21].

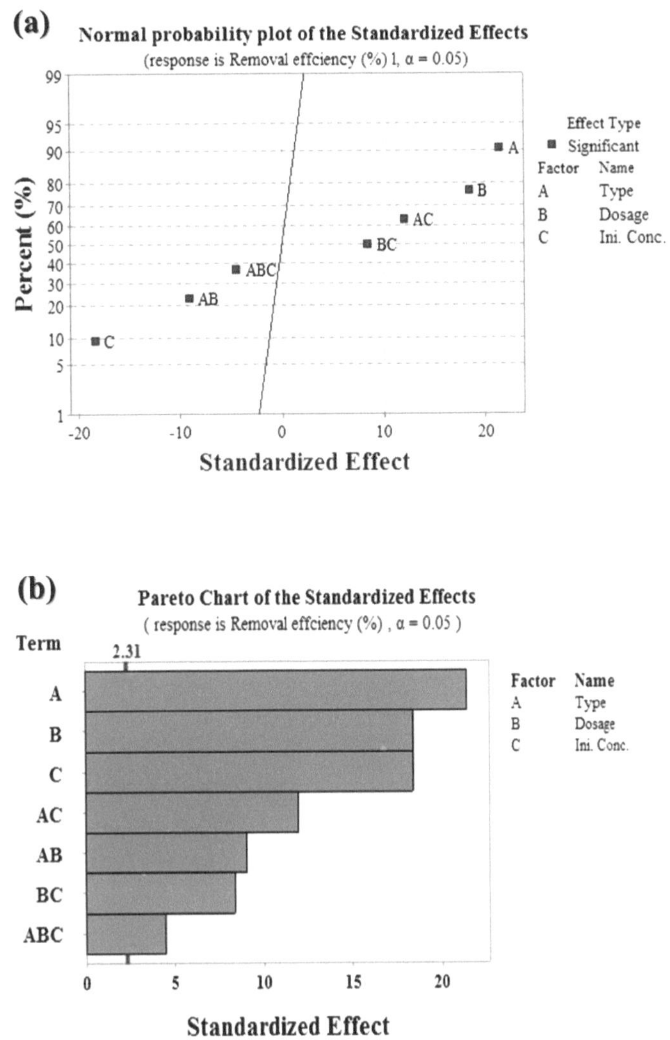


**Figure 9.** (**a**) Normal plot of the standardized effects; (**b**) Pareto chart of the standardized effects.

The main effects of each factor on the response variable (*RE*) are presented in Figure 10. The results show that changing the MOF from ZIF-8 to Fe-BTC had a positive effect on MO's removal efficiency. For the effect of MOF dosage, increasing the dosage from 100 mg to 200 mg increased the removal efficiency. As the MOF's dosage increases, the available adsorption sites increase, leading to more MO molecules being adsorbed by the MOF and thus lowering its concentration in the solution. As a result, the difference between the initial and final MO concentration increases, which increases the removal efficiency (see Equation (2)) [72–74]. On the other hand, increasing the initial concentration decreased the removal efficiency. This can be attributed to the increased amount of MO molecules in the solution—which saturates the binding sites on the MOF's surface—at a fixed MOF dosage, thus leading to a decrease in *RE* (see Equation (2)) [75–78].

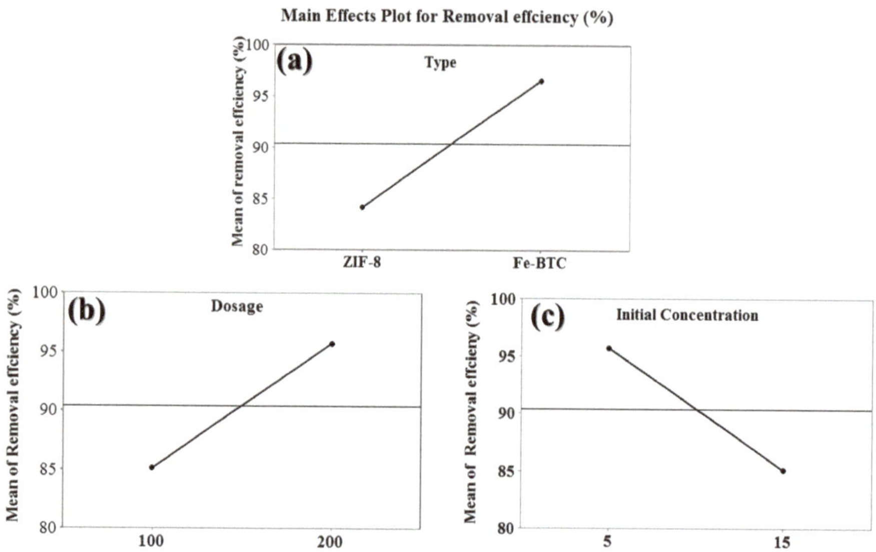


**Figure 10.** The main effects plot for MO removal efficiency of (**a**) type of MOF's, (**b**) dosage, and (**c**) initial concentration of MO.

Furthermore, an interaction effect of the different factors is considered prominent when the change in the level of one factor results in a change in the response variable (i.e., removal efficiency) that is also dependent on the level of another factor. This can be detected when the lines of the response variable are not parallel or intersect [71]. The plots, presented in Figure 11, indicate that interactions between the factors (AB, AC, and BC) were all significant, as was confirmed by the ANOVA analysis and the Pareto chart. The AB and AC plots reveal that changing the MOF type had more effect at a lower dosage and higher initial concentration. On the other hand, the BC plot showed that increasing dosage from 100 mg to 200 mg had a higher effect at higher initial concentrations. The multi-vari chart presented in Figure 12 reveals a comprehensive representation of the effects of interactions between the experimental factors. It can be seen that changing the type of MOF from ZIF-8 to Fe-BTC at 100 mg dosage and 15 mg/L MO initial concentration had the most prominent change in the removal efficiency, which is evident from the slope and length of the blue lines.

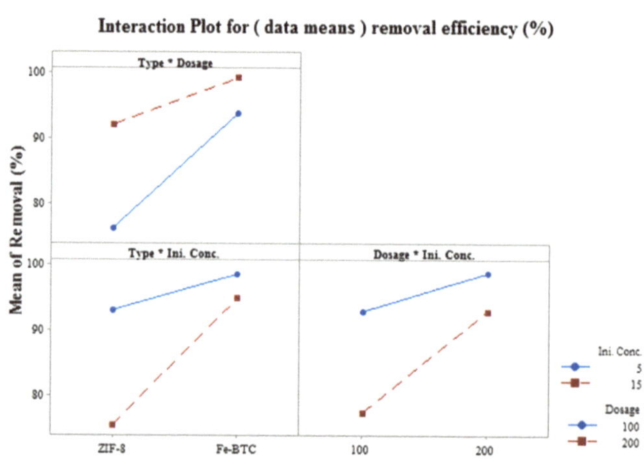


**Figure 11.** Interaction effects plot for MO removal efficiency.

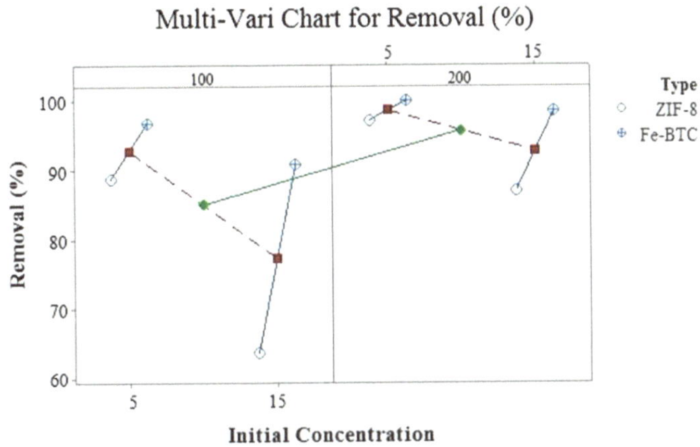


**Figure 12.** Multi-vari chart plot for MO removal efficiency. The green diamond symbols represent the mean for each level of dosage while the green line connects the dosage level means and the red square symbols correspond to the means for each level of initial concentration while the red-dashed line connects the initial concentration level means.

### 3.5. *Response Surface of Independent Variables and Their Interaction*

The last step of the statistical optimization was the analysis of the 3D response surface plots and contour plots as a function of three independent variables, which served the purpose of determining the interaction effects between three parameters while keeping the others at a fixed value. Statistical process optimization, in a given range of parameter values, allows not only for calculating the optimal condition, but also for determining the effect of the process conditions on the adsorption. A 3D graph (Figure 13) was plotted for each of initial concentration, dosage, and type of MOF. The MO removal rate reached its maximum for all MO initial concentrations and dosage loads, as shown in Figure 13A,B. However, when higher initial MO concentrations were used, the rate of MO removal decreased. Increased MOF dosage, on the other hand, increased removal efficiency. As a result, despite the dosage load, Figure 13C shows that Fe-BTC has the highest MO removals.

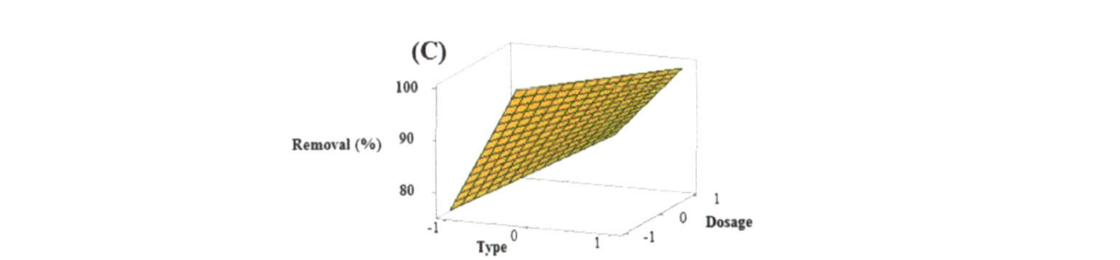


**Figure 13.** Surface graph of dye removal showing interaction of (**A**) dosage and initial concentration, (**B**) type and initial concentration, and (**C**) type and dosage.

According to Figure 14, which depicts the contour plot of the coded MO concentration and MOF dosage, the range of optimum dye removal value is related to both MO initial concentration and dosage while keeping the ZIF-8 constant (on hold). Indeed, the same trend has been demonstrated previously in the literature; the optimal dosage and concentration have a significant impact on the removal rate [76].

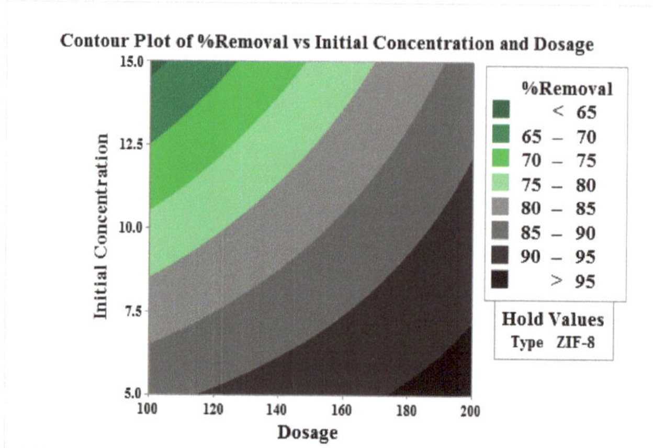


**Figure 14.** Contour plot of dye removal showing the interaction of Fe-BTC dosage and MO initial concentration.

### 3.6. Process Optimization Curve

The process optimization curve, which enables the determination of optimum conditions of independent variables for obtaining a desired value of the response, is presented in Figure 15. In the present study, the desired value of removal percentage, i.e., the goal, was to obtain a value at or near the target value of 95%, and the minimum and maximum $q_e$ were set at 10% and 100%, which means that removal percentage values lower than 61.5 and larger than 100 are not acceptable. As can be seen in Figure 15, for this target value, the predicted response of 95% was obtained at optimum conditions of the independent variables, including an initial concentration of 10.58 mg/L and an adsorbent dose of 126 g with a good desirability score. A perfect desirability score could have been obtained if uptake had achieved the ideal settings [35]. However, it is well within the acceptable range and in good accordance with the experimental response. The maximum uptake is shown to be achieved at 15 mg/L initial concentration and adsorbent dose of 200 g.

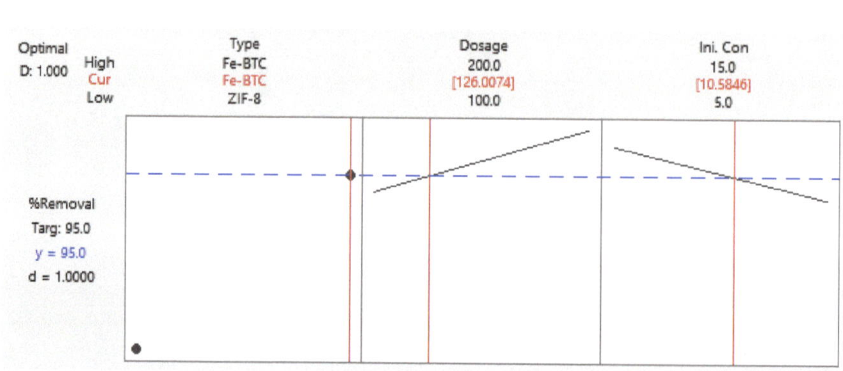


**Figure 15.** Process optimization curve for a target value of 95% removal of MO on Fe-BTC.

### 3.7. Comparison of Adsorption Capacity with Other Adsorbents

Several studies using different adsorbents were reported for the removal of methyl orange in the literature. The adsorption capacities of Fe-BTC and ZIF-8 are compared with other adsorbents in Table 5. It can be seen that the adsorption capacity of the proposed MOFs for MO is higher than most adsorbents. Thus, the value of adsorption capacity indicates that Fe-BTC is a good adsorbent for the adsorption of MO.

**Table 5.** Adsorption performance of MO with other adsorbents.

| Adsorbent Type | pH | Time (Min) | Initial Concentration (mg/L) | Adsorption Capacity (%) | References |
|---|---|---|---|---|---|
| Zirconium MOF | | | | 31 | [79] |
| chitosan/diatomite composite | 5 | 40 | 50 | 88.3 | [80] |
| Polyethyleneimine-modified persimmon tannin | 4 | 120 | 30 | 97.6 | [81] |
| MnOx-decorated MgAl | | 20 | 40 | 90.5 | [82] |
| Fly ash | 8.5 | | 60 | 66 | [83] |
| UiO-66 MOF | 5.5 | 180 | 10 | 87.4 | [84] |
| Fe-based MOF/graphene oxide | 3 | 240 | 50 | 98 | [85] |
| Fe-BTC | 6 | 60 | 15 | 99 | This work |

## 4. Conclusions

In conclusion, metal–organic frameworks with superior advantages over other conventional adsorbents in removing dyes from the aquatic environment have attracted much attention in recent years. Three MOFs (Cu-BTC, Fe-BTC, and ZIF-8) were investigated for the adsorptive removal of MO from aqueous solutions. The results showed that Fe-BTC exhibited the highest removal efficiency among the three adsorbents. Additionally, the characterization results showed that no significant changes in the internal structure of Fe-BTC and ZIF-8 after the adsorption process were detected, and they were easily regenerated by washing with ethanol, whereas the crystal structure of Cu-BTC was damaged and could not be regenerated. The maximum removal efficiencies for Fe-BTC and ZIF-8 were 91% and 63%, respectively, and both MOFs maintained their removal efficiency and adsorption capacity over a wide range of pH. Hence, the adsorption mechanism of MO on the MOFs could be due to a combination of electrostatic, $\pi$–$\pi$ stacking, and hydrogen bonding interactions. Finally, the $2^3$ factorial design utilized to study the effect of MOF type, dosage, and initial concentration on the MO removal efficiency revealed that the MOF type was the most statistically significant factor, followed by dosage and initial concentration. Both the type of MOF and dosage had a positive effect on removal efficiency, while the initial concentration had a negative effect. The two-way and three-way interactions were also found to be significant.

**Supplementary Materials:** The following supporting information can be downloaded at: https://www.mdpi.com/article/10.3390/en15134642/s1. References [86–111] are cited in the supplementary materials.

**Author Contributions:** Conceptualization, R.S. (Rana Sabouni); Data curation, R.S. (Reem Shomal); Formal analysis, A.K. and R.S. (Reem Shomal); Funding acquisition, R.S. (Rana Sabouni); Investigation, A.K. and R.S. (Reem Shomal); Methodology, A.K. and R.S. (Rana Sabouni); Project administration, R.S. (Rana Sabouni); Resources, M.H.A.-S. and A.A.; Supervision, R.S. (Rana Sabouni); Writing—original draft, A.K. and R.S. (Reem Shomal); Writing—review & editing, R.S. (Rana Sabouni) and M.H.A.-S. All authors have read and agreed to the published version of the manuscript.

**Funding:** This research was funded by [American University of Sharjah] grant number [FRG21-M-E63]. And the APC was funded by [American University of Sharjah] grant number [OAPCEN-1410-E00033].

**Institutional Review Board Statement:** Not applicable.

**Informed Consent Statement:** Not applicable.

**Data Availability Statement:** The data presented in this study are available on request from the corresponding author.

**Acknowledgments:** The authors gratefully acknowledge the financial support by the American University of Sharjah Faculty Research grant FRG21-M-E63. The work in this paper was supported in part by the Open Access Program from the American University of Sharjah (grant number: OAPCEN-1410-E00033). This paper represents the opinions of the author(s) and does not mean to represent the position or opinions of the American University of Sharjah.

**Conflicts of Interest:** The authors declare no conflict of interest.

## References

1. Katheresan, V.; Kansedo, J.; Lau, S.Y. Efficiency of various recent wastewater dye removal methods: A Review. *J. Environ. Chem. Eng.* **2018**, *6*, 4676–4697. [CrossRef]
2. Adegoke, K.A.; Bello, O.S. Dye sequestration using agricultural wastes as adsorbents. *Water Resour. Ind.* **2015**, *12*, 8–24. [CrossRef]
3. Molinos-Senante, M.; Maziotis, A. Evaluation of energy efficiency of wastewater treatment plants: The influence of the technology and aging factors. *Appl. Energy* **2022**, *310*, 118535. [CrossRef]
4. Di Fraia, S.; Massarotti, N.; Vanoli, L. A novel energy assessment of urban wastewater treatment plants. *Energy Convers. Manag.* **2018**, *163*, 304–313. [CrossRef]
5. Hernández-Chover, V.; Bellver-Domingo, Á.; Hernández-Sancho, F. Efficiency of wastewater treatment facilities: The influence of scale economies. *J. Environ. Manag.* **2018**, *228*, 77–84. [CrossRef]
6. International Energy Agency. *Water Energy Nexus—Excerpt from the World Energy Outlook 2016*; International Energy Agency: Paris, France, 2016; ISBN 978-92-64-26494-6.
7. Bodík, I.; Kubaská, M. Energy and sustainability of operation of a wastewater treatment plant. *Environ. Prot. Eng.* **2013**, *39*, 15–24. [CrossRef]
8. Wakeel, M.; Chen, B.; Hayat, T.; Alsaedi, A.; Ahmad, B. Energy consumption for water use cycles in different countries: A Review. *Appl. Energy* **2016**, *178*, 868–885. [CrossRef]
9. Ahmad, A.; Mohd-Setapar, S.H.; Chuong, C.S.; Khatoon, A.; Wani, W.A.; Kumar, R.; Rafatullah, M. Recent advances in new generation dye removal technologies: Novel search for approaches to reprocess wastewater. *RSC Adv.* **2015**, *5*, 30801–30818. [CrossRef]
10. Hao, X.; Liu, R.; Huang, X. Evaluation of the potential for operating carbon neutral wwtps in China. *Water Res.* **2015**, *87*, 424–431. [CrossRef]
11. Molinos-Senante, M.; Hernandez-Sancho, F.; Sala-Garrido, R. Benchmarking in wastewater treatment plants: A tool to save operational costs. *Clean Technol. Environ. Policy* **2013**, *16*, 149–161. [CrossRef]
12. Ananda, J. Productivity implications of the water-energy-emissions nexus: An empirical analysis of the drinking water and wastewater sector. *J. Clean. Prod.* **2018**, *196*, 1097–1105. [CrossRef]
13. Ali, H. Biodegradation of synthetic dyes—A review. *Water Air Soil Pollut.* **2010**, *213*, 251–273. [CrossRef]
14. Fatima, M.; Farooq, R.; Lindström, R.W.; Saeed, M. A review on biocatalytic decomposition of azo dyes and electrons recovery. *J. Mol. Liq.* **2017**, *246*, 275–281. [CrossRef]
15. Adeyemo, A.A.; Adeoye, I.O.; Bello, O.S. Metal organic frameworks as adsorbents for Dye adsorption: Overview, prospects and future challenges. *Toxicol. Environ. Chem.* **2012**, *94*, 1846–1863. [CrossRef]
16. Karami, A.; Sabouni, R.; Al-Sayah, M.H.; Aidan, A. Adsorption potentials of iron-based metal–organic framework for methyl orange removal: Batch and fixed-bed column studies. *Int. J. Environ. Sci. Technol.* **2021**, *18*, 3597–3612. [CrossRef]
17. Ahmad, M.A.; Alrozi, R. Optimization of rambutan peel based activated carbon preparation conditions for Remazol Brilliant Blue R Removal. *Chem. Eng. J.* **2011**, *168*, 280–285. [CrossRef]
18. Srinivasan, A.; Viraraghavan, T. Decolorization of dye wastewaters by biosorbents: A Review. *J. Environ. Manag.* **2010**, *91*, 1915–1929. [CrossRef]
19. Karami, A.; Sabouni, R.; Ghommem, M. Experimental investigation of competitive co-adsorption of naproxen and diclofenac from water by an aluminum-based metal-organic framework. *J. Mol. Liq.* **2020**, *305*, 112808. [CrossRef]
20. Konaganti, V.K.; Kota, R.; Patil, S.; Madras, G. Adsorption of anionic dyes on chitosan grafted poly(alkyl methacrylate)s. *Chem. Eng. J.* **2010**, *158*, 393–401. [CrossRef]
21. Özbay, N.; Yargıç, A.Ş.; Yarbay-Şahin, R.Z.; Önal, E. Full factorial experimental design analysis of reactive dye removal by carbon adsorption. *J. Chem.* **2013**, *2013*, 234904. [CrossRef]
22. Mtaallah, S.; Marzouk, I.; Hamrouni, B. Factorial experimental design applied to adsorption of cadmium on activated alumina. *J. Water Reuse Desalin.* **2017**, *8*, 76–85. [CrossRef]
23. Haque, E.; Lee, J.E.; Jang, I.T.; Hwang, Y.K.; Chang, J.-S.; Jegal, J.; Jhung, S.H. Adsorptive removal of methyl orange from aqueous solution with metal-organic frameworks, porous chromium-benzenedicarboxylates. *J. Hazard. Mater.* **2010**, *181*, 535–542. [CrossRef] [PubMed]
24. Ayati, A.; Shahrak, M.N.; Tanhaei, B.; Sillanpää, M. Emerging adsorptive removal of azo dye by metal–organic frameworks. *Chemosphere* **2016**, *160*, 30–44. [CrossRef] [PubMed]

25. Babu, K.F.; Kulandainathan, M.A.; Katsounaros, I.; Rassaei, L.; Burrows, A.D.; Raithby, P.R.; Marken, F. Electrocatalytic activity of BasoliteTM F300 metal-organic-framework structures. *Electrochem. Commun.* **2010**, *12*, 632–635. [CrossRef]
26. Chui, S.S.-Y.; Lo, S.M.-F.; Charmant, J.P.; Orpen, A.G.; Williams, I.D. A chemically functionalizable nanoporous material $[Cu_3 (TMA)_2 (H_2O)_3]_n$. *Science* **1999**, *283*, 1148–1150. [CrossRef]
27. Banerjee, R.; Phan, A.; Wang, B.; Knobler, C.; Furukawa, H.; O'Keeffe, M.; Yaghi, O.M. High-throughput synthesis of zeolitic imidazolate frameworks and application to $CO_2$ capture. *Science* **2008**, *319*, 939–943. [CrossRef]
28. Centrone, A.; Santiso, E.E.; Hatton, T.A. Separation of chemical reaction intermediates by metal-organic frameworks. *Small* **2011**, *7*, 2356–2364. [CrossRef]
29. Zhu, B.-J.; Yu, X.-Y.; Jia, Y.; Peng, F.-M.; Sun, B.; Zhang, M.-Y.; Luo, T.; Liu, J.-H.; Huang, X.-J. Iron and 1,3,5-benzenetricarboxylic metal–organic coordination polymers prepared by SOLVOTHERMAL method and their application in efficient as(v) removal from Aqueous Solutions. *J. Phys. Chem. C* **2012**, *116*, 8601–8607. [CrossRef]
30. Wang, Q.M.; Shen, D.; Bülow, M.; Ling Lau, M.; Deng, S.; Fitch, F.R.; Lemcoff, N.O.; Semanscin, J. Metallo-organic molecular sieve for gas separation and purification. *Microporous Mesoporous Mater.* **2002**, *55*, 217–230. [CrossRef]
31. Hamon, L.; Jolimaître, E.; Pirngruber, G.D. $CO_2$ and $CH_4$ separation by adsorption using Cu-BTC metal−organic framework. *Ind. Eng. Chem. Res.* **2010**, *49*, 7497–7503. [CrossRef]
32. Lin, S.; Song, Z.; Che, G.; Ren, A.; Li, P.; Liu, C.; Zhang, J. Adsorption behavior of metal–organic frameworks for methylene blue from aqueous solution. *Microporous Mesoporous Mater.* **2014**, *193*, 27–34. [CrossRef]
33. Park, K.S.; Ni, Z.; Cote, A.P.; Choi, J.Y.; Huang, R.; Uribe-Romo, F.J.; Chae, H.K.; O'Keeffe, M.; Yaghi, O.M. Exceptional Chemical and thermal stability of zeolitic imidazolate frameworks. *Proc. Natl. Acad. Sci. USA* **2006**, *103*, 10186–10191. [CrossRef] [PubMed]
34. Malekbala, M.R.; Khan, M.A.; Hosseini, S.; Abdullah, L.C.; Choong, T.S.Y. Adsorption/desorption of cationic dye on surfactant modified mesoporous carbon coated monolith: Equilibrium, kinetic and thermodynamic studies. *J. Ind. Eng. Chem.* **2015**, *21*, 369–377. [CrossRef]
35. Wabaidur, S.M.; Khan, M.A.; Siddiqui, M.R.; Otero, M.; Jeon, B.-H.; Alothman, Z.A.; Hakami, A.A. Oxygenated functionalities enriched mwcnts decorated with silica coated spinel ferrite—A nanocomposite for potentially rapid and efficient de-colorization of aquatic environment. *J. Mol. Liq.* **2020**, *317*, 113916. [CrossRef]
36. Fiol, N.; Villaescusa, I. Determination of Sorbent Point zero charge: Usefulness in sorption studies. *Environ. Chem. Lett.* **2008**, *7*, 79–84. [CrossRef]
37. Hosseini, S.; Khan, M.A.; Malekbala, M.R.; Cheah, W.; Choong, T.S.Y. Carbon coated monolith, a mesoporous material for the removal of methyl orange from aqueous phase: Adsorption and Desorption Studies. *Chem. Eng. J.* **2011**, *171*, 1124–1131. [CrossRef]
38. Ranjan, D.; Khan, M.A.; Otero, M.; Siddiqui, M.R.; Alshaeef, S.A. Optimization of Bromate adsorption onto Fe-CNTS nanocomposite using response surface methodology. *Water SA* **2021**, *47*, 423–429.
39. Najafi Nobar, S. Cu-BTC Synthesis, characterization and preparation for adsorption studies. *Mater. Chem. Phys.* **2018**, *213*, 343–351. [CrossRef]
40. Schlichte, K.; Kratzke, T.; Kaskel, S. Improved synthesis, thermal stability and catalytic properties of the metal-organic framework compound CU3(BTC)2. *Microporous Mesoporous Mater.* **2004**, *73*, 81–88. [CrossRef]
41. Yang, Y.; Shukla, P.; Wang, S.; Rudolph, V.; Chen, X.-M.; Zhu, Z. Significant improvement of surface area and $CO_2$ adsorption of Cu–BTC via Solvent Exchange Activation. *RSC Adv.* **2013**, *3*, 17065. [CrossRef]
42. Lin, K.-S.; Adhikari, A.K.; Ku, C.-N.; Chiang, C.-L.; Kuo, H. Synthesis and characterization of porous HKUST-1 metal organic frameworks for hydrogen storage. *Int. J. Hydrogen Energy* **2012**, *37*, 13865–13871. [CrossRef]
43. Chen, L.; Yuan, S.; Qian, J.-F.; Fan, W.; He, M.-Y.; Chen, Q.; Zhang, Z.-H. Effective adsorption separation of n-hexane/2-methylpentane in facilely synthesized zeolitic imidazolate frameworks zif-8 and zif-69. *Ind. Eng. Chem. Res.* **2016**, *55*, 10751–10757. [CrossRef]
44. Lee, Y.-R.; Jang, M.-S.; Cho, H.-Y.; Kwon, H.-J.; Kim, S.; Ahn, W.-S. Zif-8: A comparison of synthesis methods. *Chem. Eng. J.* **2015**, *271*, 276–280. [CrossRef]
45. Cho, H.-Y.; Kim, J.; Kim, S.-N.; Ahn, W.-S. High yield 1-L scale synthesis of ZIF-8 via a sonochemical route. *Microporous Mesoporous Mater.* **2013**, *169*, 180–184. [CrossRef]
46. Torres, N.; Galicia, J.; Plasencia, Y.; Cano, A.; Echevarría, F.; Desdin-Garcia, L.F.; Reguera, E. Implications of structural differences between Cu-BTC and Fe-BTC on their hydrogen storage capacity. *Colloids Surf. A Physicochem. Eng. Asp.* **2018**, *549*, 138–146. [CrossRef]
47. Martínez, F.; Leo, P.; Orcajo, G.; Díaz-García, M.; Sanchez-Sanchez, M.; Calleja, G. Sustainable fe-BTC catalyst for efficient removal of mehylene blue by advanced Fenton Oxidation. *Catal. Today* **2018**, *313*, 6–11. [CrossRef]
48. Dhakshinamoorthy, A.; Alvaro, M.; Horcajada, P.; Gibson, E.; Vishnuvarthan, M.; Vimont, A.; Grenèche, J.-M.; Serre, C.; Daturi, M.; Garcia, H. Comparison of porous iron trimesates basolite F300 and mil-100(fe) as heterogeneous catalysts for Lewis acid and oxidation reactions: Roles of structural defects and stability. *ACS Catal.* **2012**, *2*, 2060–2065. [CrossRef]
49. Shahid, S.; Nijmeijer, K. High pressure gas separation performance of mixed-matrix polymer membranes containing mesoporous fe(btc). *J. Membr. Sci.* **2014**, *459*, 33–44. [CrossRef]

50. Autie-Castro, G.; Autie, M.A.; Rodríguez-Castellón, E.; Aguirre, C.; Reguera, E. Cu-BTC and Fe-BTC Metal-organic frameworks: Role of the materials structural features on their performance for volatile hydrocarbons separation. *Colloids Surf. A Physicochem. Eng. Asp.* **2015**, *481*, 351–357. [CrossRef]
51. Dathe, H.; Peringer, E.; Roberts, V.; Jentys, A.; Lercher, J.A. Metal organic frameworks based on $Cu^{2+}$ and benzene-1,3,5-tricarboxylate as host for $SO_2$ trapping agents. *C. R. Chim.* **2005**, *8*, 753–763. [CrossRef]
52. Chowdhury, P.; Bikkina, C.; Meister, D.; Dreisbach, F.; Gumma, S. Comparison of adsorption isotherms on Cu-BTC Metal Organic Frameworks synthesized from different routes. *Microporous Mesoporous Mater.* **2009**, *117*, 406–413. [CrossRef]
53. McEwen, J.; Hayman, J.-D.; Ozgur Yazaydin, A. A comparative study of $CO_2$, $CH_4$ and $N_2$ adsorption in zif-8, zeolite-13x and BPL activated carbon. *Chem. Phys.* **2013**, *412*, 72–76. [CrossRef]
54. Yin, H.; Kim, H.; Choi, J.; Yip, A.C.K. Thermal stability of zif-8 under oxidative and Inert Environments: A practical perspective on using ZIF-8 as a catalyst support. *Chem. Eng. J.* **2015**, *278*, 293–300. [CrossRef]
55. Du, M.; Li, L.; Li, M.; Si, R. Adsorption mechanism on Metal Organic Frameworks of cu-BTC, Fe-BTC and zif-8 for $CO_2$ capture investigated by X-ray absorption fine structure. *RSC Adv.* **2016**, *6*, 62705–62716. [CrossRef]
56. Sanchez-Sanchez, M.; de Asua, I.; Ruano, D.; Diaz, K. Direct synthesis, structural features, and enhanced catalytic activity of the basolite F300-like Semiamorphous Fe-BTC Framework. *Cryst. Growth Des.* **2015**, *15*, 4498–4506. [CrossRef]
57. Dhakshinamoorthy, A.; Alvaro, M.; Chevreau, H.; Horcajada, P.; Devic, T.; Serre, C.; Garcia, H. Iron(III) metal–organic frameworks as solid Lewis acids for the isomerization of α-pinene oxide. *Catal. Sci. Technol.* **2012**, *2*, 324–330. [CrossRef]
58. Wang, Y.; Xu, Y.; Ma, H.; Xu, R.; Liu, H.; Li, D.; Tian, Z. Synthesis of zif-8 in a deep eutectic solvent using cooling-induced crystallisation. *Microporous Mesoporous Mater.* **2014**, *195*, 50–59. [CrossRef]
59. He, M.; Yao, J.; Liu, Q.; Wang, K.; Chen, F.; Wang, H. Facile synthesis of zeolitic imidazolate framework-8 from a concentrated aqueous solution. *Microporous Mesoporous Mater.* **2014**, *184*, 55–60. [CrossRef]
60. Jian, M.; Liu, B.; Zhang, G.; Liu, R.; Zhang, X. Adsorptive removal of arsenic from aqueous solution by zeolitic imidazolate framework-8 (zif-8) nanoparticles. *Colloids Surf. A Physicochem. Eng. Asp.* **2015**, *465*, 67–76. [CrossRef]
61. McKinstry, C.; Cussen, E.J.; Fletcher, A.J.; Patwardhan, S.V.; Sefcik, J. Scalable continuous production of high quality HKUST-1 via conventional and microwave heating. *Chem. Eng. J.* **2017**, *326*, 570–577. [CrossRef]
62. He, L.; Dumée, L.F.; Liu, D.; Velleman, L.; She, F.; Banos, C.; Davies, J.B.; Kong, L. Silver nanoparticles prepared by gamma irradiation across metal–organic framework templates. *RSC Adv.* **2015**, *5*, 10707–10715. [CrossRef]
63. Conde-González, J.E.; Peña-Méndez, E.M.; Rybáková, S.; Pasán, J.; Ruiz-Pérez, C.; Havel, J. Adsorption of silver nanoparticles from aqueous solution on copper-based metal organic frameworks (HKUST-1). *Chemosphere* **2016**, *150*, 659–666. [CrossRef] [PubMed]
64. Kaur, H.; Mohanta, G.C.; Gupta, V.; Kukkar, D.; Tyagi, S. Synthesis and characterization of zif-8 nanoparticles for controlled release of 6-Mercaptopurine Drug. *J. Drug Deliv. Sci. Technol.* **2017**, *41*, 106–112. [CrossRef]
65. Bustamante, E.L.; Fernández, J.L.; Zamaro, J.M. Influence of the solvent in the synthesis of zeolitic imidazolate framework-8 (zif-8) nanocrystals at room temperature. *J. Colloid Interface Sci.* **2014**, *424*, 37–43. [CrossRef]
66. Nguyen, L.T.L.; Le, K.K.A.; Phan, N.T.S. A zeolite imidazolate framework zif-8 catalyst for Friedel-Crafts acylation. *Chin. J. Catal.* **2012**, *33*, 688–696. [CrossRef]
67. Luan Tran, B.; Chin, H.-Y.; Chang, B.K.; Chiang, A.S.T. Dye adsorption in zif-8: The importance of external surface area. *Microporous Mesoporous Mater.* **2019**, *277*, 149–153. [CrossRef]
68. Seo, Y.-K.; Yoon, J.W.; Lee, J.S.; Lee, U.-H.; Hwang, Y.K.; Jun, C.-H.; Horcajada, P.; Serre, C.; Chang, J.-S. Large scale fluorine-free synthesis of hierarchically porous iron(iii) trimesate mil-100(fe) with a zeolite MTN topology. *Microporous Mesoporous Mater.* **2012**, *157*, 137–145. [CrossRef]
69. Wu, S.-C.; You, X.; Yang, C.; Cheng, J.-H. Adsorption behavior of methyl orange onto an aluminum-based metal organic framework, MIL-68(AL). *Water Sci. Technol.* **2017**, *75*, 2800–2810. [CrossRef]
70. Hasan, Z.; Jhung, S.H. Removal of hazardous organics from water using metal-organic frameworks (mofs): Plausible mechanisms for selective adsorptions. *J. Hazard. Mater.* **2015**, *283*, 329–339. [CrossRef]
71. Regti, A.; Laamari, M.R.; Stiriba, S.-E.; El Haddad, M. Use of response factorial design for process optimization of basic dye adsorption onto activated carbon derived from Persea species. *Microchem. J.* **2017**, *130*, 129–136. [CrossRef]
72. Mathialagan, T.; Viraraghavan, T. Biosorption of pentachlorophenol by fungal biomass from aqueous solutions: A factorial design analysis. *Environ. Technol.* **2005**, *26*, 571–580. [CrossRef] [PubMed]
73. Abdi, J.; Vossoughi, M.; Mahmoodi, N.M.; Alemzadeh, I. Synthesis of metal-organic framework hybrid nanocomposites based on go and CNT with high adsorption capacity for dye removal. *Chem. Eng. J.* **2017**, *326*, 1145–1158. [CrossRef]
74. Yao, Y.; Bing, H.; Feifei, X.; Xiaofeng, C. Equilibrium and kinetic studies of methyl orange adsorption on multiwalled carbon nanotubes. *Chem. Eng. J.* **2011**, *170*, 82–89. [CrossRef]
75. Senthil Kumar, P.; Ramalingam, S.; Senthamarai, C.; Niranjanaa, M.; Vijayalakshmi, P.; Sivanesan, S. Adsorption of dye from aqueous solution by cashew nut shell: Studies on equilibrium isotherm, kinetics and thermodynamics of interactions. *Desalination* **2010**, *261*, 52–60. [CrossRef]
76. Chen, C.; Zhang, M.; Guan, Q.; Li, W. Kinetic and thermodynamic studies on the adsorption of xylenol orange onto MIL-101(CR). *Chem. Eng. J.* **2012**, *183*, 60–67. [CrossRef]

77. Haque, E.; Jun, J.W.; Jhung, S.H. Adsorptive removal of methyl orange and methylene blue from aqueous solution with a metal-organic framework material, iron terephthalate (MOF-235). *J. Hazard. Mater.* **2011**, *185*, 507–511. [CrossRef] [PubMed]
78. Huo, S.-H.; Yan, X.-P. Metal–organic framework mil-100(fe) for the adsorption of Malachite Green from aqueous solution. *J. Mater. Chem.* **2012**, *22*, 7449. [CrossRef]
79. Embaby, M.S.; Elwany, S.D.; Setyaningsih, W.; Saber, M.R. The adsorptive properties of uio-66 towards organic dyes: A record adsorption capacity for the anionic dye alizarin red s. *Chin. J. Chem. Eng.* **2018**, *26*, 731–739. [CrossRef]
80. Zhao, P.; Zhang, R.; Wang, J. Adsorption of methyl orange from aqueous solution using Chitosan/diatomite composite. *Water Sci. Technol.* **2017**, *75*, 1633–1642. [CrossRef]
81. Li, X.; Wang, Z.; Ning, J.; Gao, M.; Jiang, W.; Zhou, Z.; Li, G. Preparation and characterization of a novel polyethyleneimine cation-modified persimmon tannin bioadsorbent for anionic dye adsorption. *J. Environ. Manag.* **2018**, *217*, 305–314. [CrossRef]
82. Peng, H.H.; Chen, J.; Jiang, D.Y.; Guo, X.L.; Chen, H.; Zhang, Y.X. Merging of Memory Effect and anion intercalation: Mnox-decorated mgal-LDO as a high-performance nano-adsorbent for the removal of methyl orange. *Dalton Trans.* **2016**, *45*, 10530–10538. [CrossRef] [PubMed]
83. Patra, G.; Barnwal, R.; Behera, S.K.; Meikap, B.C. Removal of dyes from aqueous solution by sorption with fly ash using a hydrocyclone. *J. Environ. Chem. Eng.* **2018**, *6*, 5204–5211. [CrossRef]
84. Ahmadijokani, F.; Mohammadkhani, R.; Ahmadipouya, S.; Shokrgozar, A.; Rezakazemi, M.; Molavi, H.; Aminabhavi, T.M.; Arjmand, M. Superior chemical stability of uio-66 metal-organic frameworks (mofs) for selective dye adsorption. *Chem. Eng. J.* **2020**, *399*, 125346. [CrossRef]
85. Tang, J.; Wang, J. Fe-based metal organic framework/graphene oxide composite as an efficient catalyst for Fenton-like degradation of methyl orange. *RSC Adv.* **2017**, *7*, 50829–50837. [CrossRef]
86. Tan, K.L.; Hameed, B.H. Insight into the adsorption kinetics models for the removal of contaminants from aqueous solutions. *J. Taiwan Inst. Chem. Eng.* **2017**, *74*, 25–48. [CrossRef]
87. Ho, Y.S.; McKay, G. Pseudo-second order model for sorption processes. *Process Biochem.* **1999**, *34*, 451–465. [CrossRef]
88. Ghaedi, M.; Ghaedi, A.M.; Negintaji, E.; Ansari, A.; Vafaei, A.; Rajabi, M. Random forest model for removal of bromophenol blue using activated carbon obtained from Astragalus bisulcatus tree. *J. Ind. Eng. Chem.* **2014**, *20*, 1793–1803. [CrossRef]
89. Tong, M.; Liu, D.; Yang, Q.; Devautour-Vinot, S.; Maurin, G.; Zhong, C. Influence of framework metal ions on the dye capture behavior of MIL-100 (Fe, Cr) MOF type solids. *J. Mater. Chem. A* **2013**, *1*, 8534. [CrossRef]
90. Ji, W.-J.; Hao, R.-Q.; Pei, W.-W.; Feng, L.; Zhai, Q.-G. Design of two isoreticular CD-biphenyltetracarboxylate frameworks for Dye adsorption, separation and photocatalytic degradation. *Dalton Trans.* **2018**, *47*, 700–707. [CrossRef]
91. Hameed, B.H.; Rahman, A.A. Removal of phenol from aqueous solutions by adsorption onto activated carbon prepared from biomass material. *J. Hazard. Mater.* **2008**, *160*, 576–581. [CrossRef]
92. Din, A.T.M.; Hameed, B.H.; Ahmad, A.L. Batch adsorption of phenol onto physiochemical-activated coconut shell. *J. Hazard. Mater.* **2009**, *161*, 1522–1529. [CrossRef]
93. Srivastava, V.C.; Swamy, M.M.; Mall, I.D.; Prasad, B.; Mishra, I.M. Adsorptive removal of phenol by bagasse fly ash and activated carbon: Equilibrium, kinetics and thermodynamics. *Colloids Surf. A Physicochem. Eng. Asp.* **2006**, *272*, 89–104. [CrossRef]
94. Teng, H.; Hsieh, C.-T. Activation Energy for Oxygen Chemisorption on Carbon at Low Temperatures. *Ind. Eng. Chem. Res.* **1999**, *38*, 292–297. [CrossRef]
95. Piasecki, W.; Rudziński, W. Application of the statistical rate theory of interfacial transport to investigate the kinetics of divalent metal ion adsorption onto the energetically heterogeneous surfaces of oxides and activated carbons. *Appl. Surf. Sci.* **2007**, *253*, 5814–5817. [CrossRef]
96. Weber, W.J.; Morris, J.C. Kinetics of Adsorption on Carbon from Solution. *J. Sanit. Eng. Div.* **1963**, *89*, 31–60. [CrossRef]
97. Mittal, A.; Malviya, A.; Kaur, D.; Mittal, J.; Kurup, L. Studies on the adsorption kinetics and isotherms for the removal and recovery of Methyl Orange from wastewaters using waste materials. *J. Hazard. Mater.* **2007**, *148*, 229–240. [CrossRef]
98. Wu, F.C.; Tseng, R.L.; Juang, R.S. Initial behavior of intraparticle diffusion model used in the description of adsorption kinetics. *Chem. Eng. J.* 2009. [CrossRef]
99. Machado, F.M.; Bergmann, C.P.; Lima, E.C.; Royer, B.; de Souza, F.E.; Jauris, I.M.; Calvete, T.; Fagan, S.B. Adsorption of Reactive Blue 4 dye from water solutions by carbon nanotubes: Experiment and theory. *Phys. Chem. Chem. Phys.* **2012**, *14*, 11139. [CrossRef]
100. Dhakshinamoorthy, A.; Alvaro, M.; Garcia, H. Aerobic oxidation of cycloalkenes catalyzed by iron metal organic framework containing N-hydroxyphthalimide. *J. Catal.* **2012**, *289*, 259–265. [CrossRef]
101. Langmuir, I. The adsorption of gases on plane surfaces of glass, mica and platinum. *J. Am. Chem. Soc.* **1918**, *40*, 1361–1403. [CrossRef]
102. Liu, Y.; Liu, Y.-J. Biosorption isotherms, kinetics and thermodynamics. *Sep. Purif. Technol.* **2008**, *61*, 229–242. [CrossRef]
103. Kousha, M.; Daneshvar, E.; Sohrabi, M.S.; Jokar, M.; Bhatnagar, A. Adsorption of acid orange II dye by raw and chemically modified brown macroalga Stoechospermum marginatum. *Chem. Eng. J.* **2012**, *192*, 67–76. [CrossRef]
104. Vijayaraghavan, K.; Padmesh, T.; Palanivelu, K.; Velan, M. Biosorption of nickel(II) ions onto Sargassum wightii: Application of two-parameter and three-parameter isotherm models. *J. Hazard. Mater.* **2006**, *133*, 304–308. [CrossRef] [PubMed]
105. Basar, C. Applicability of the various adsorption models of three dyes adsorption onto activated carbon prepared waste apricot. *J. Hazard. Mater.* **2006**, *135*, 232–241. [CrossRef]

106. Zhou, X.; Zhou, X. The unit problem in the thermodynamic calculation of adsorption using the Langmuir equation. *Chem. Eng. Commun.* **2014**, *201*, 1459–1467. [CrossRef]
107. Mahmoodi, N.M.; Najafi, F. Synthesis, amine functionalization and dye removal ability of titania/silica nano-hybrid. *Microporous Mesoporous Mater.* **2012**, *156*, 153–160. [CrossRef]
108. Shen, T.; Luo, J.; Zhang, S.; Luo, X. Hierarchically mesostructured MIL-101 metal–organic frameworks with different mineralizing agents for adsorptive removal of methyl orange and methylene blue from aqueous solution. *J. Environ. Chem. Eng.* **2015**, *3*, 1372–1383. [CrossRef]
109. Haque, E.; Lo, V.; Minett, A.I.; Harris, A.T.; Church, T.L. Dichotomous adsorption behaviour of dyes on an amino-functionalised metal–organic framework, amino-MIL-101(Al). *J. Mater. Chem. A* **2014**, *2*, 193–203. [CrossRef]
110. Angin, D. Utilization of activated carbon produced from fruit juice industry solid waste for the adsorption of Yellow 18 from aqueous solutions. *Bioresour. Technol.* **2014**, *168*, 259–266. [CrossRef]
111. Kara, M.; Yuzer, H.; Sabah, E.; Celik, M. Adsorption of cobalt from aqueous solutions onto sepiolite. *Water Res.* **2003**, *37*, 224–232. [CrossRef]

energies

MDPI

Article

# Effects of Bioliquid Recirculation on Hydrothermal Carbonization of Lignocellulosic Biomass

Sun-Ju Lee [1,2], Min-Ah Oh [2], Seung-Jin Oh [2], Na-Hyeon Cho [1], Young-Yeul Kang [1] and Jai-Young Lee [2,*]

1 Department of Resource Recirculation Research, National Institute of Environmental Research, Seo-gu, Incheon 22689, Korea; leesj1122@korea.kr (S.-J.L.); nhch0@korea.kr (N.-H.C.); kangyr@korea.kr (Y.-Y.K.)
2 School of Environmental Engineering, University of Seoul, Dongdaemun-gu, Seoul 02504, Korea; minawizz@uos.ac.kr (M.-A.O.); ohpromote1984@uos.ac.kr (S.-J.O.)
* Correspondence: leejy@uos.ac.kr; Tel.: +82-2-6490-2864; Fax: +82-2-6490-5459

**Abstract:** The characteristics of bioliquid produced through the hydrothermal carbonization (HTC) of wood wastes and the effects of recirculation on hydrochar production were analyzed. The organic acids and total organic carbon of bioliquid increased with progressive recirculation, whereas intermediate byproducts decreased. Hydrochar production by bioliquid recirculation increased mass yield, carbon content, caloric value, and energy yield of the former, while improving its quality as a solid refuse fuel. We concluded that bioliquid recirculation promoted HTC, as demonstrated by Fourier-transform infrared spectroscopy. Furthermore, contrary to predictions, a relatively constant quantity of bioliquid was generated in each step, indicating that its continuous reuse is feasible. Therefore, bioliquid recirculation can improve hydrochar production while simultaneously mitigating the environmental impact of wastewater generation. This method should be considered an important strategy toward the implementation of carbon-neutrality goals.

**Keywords:** hydrothermal carbonization; hydrochar; recirculation; bioliquid; wood waste

**Citation:** Lee, S.-J.; Oh, M.-A.; Oh, S.-J.; Cho, N.-H.; Kang, Y.-Y.; Lee, J.-Y. Effects of Bioliquid Recirculation on Hydrothermal Carbonization of Lignocellulosic Biomass. *Energies* **2022**, *15*, 4903. https://doi.org/10.3390/en15134903

Academic Editors: Dibyendu Sarkar, Rupali Datta, Prafulla Kumar Sahoo, Mohammad Mahmudur Rahman and Jaroslaw Krzywanski

Received: 22 April 2022
Accepted: 29 June 2022
Published: 4 July 2022



## 1. Introduction

In 2015, the international community launched a new climate regime to cope with the climate crisis through the Paris Agreement. South Korea established long-term low greenhouse gas emission development strategies (LEDS) becoming the 14th country in the world to enact legislation requiring the achievement of carbon neutrality by 2050. However, greenhouse gas emissions from South Korea in 2018 reached 727.6 million tons of carbon dioxide equivalent ($CO_2$eq), with the energy sector accounting for 86.9% of emissions [1]. Furthermore, $CO_2$ accounts for 91.4% of total national emissions. Thus, energy production methods that can reduce greenhouse gas emissions are necessary as a contribution toward carbon neutrality goals.

Forests cover 63.7% of Korea's land area; this proportion is the fourth highest amongst member countries of the Organization for Economic Co-operation and Development (OECD) [2]. However, as forest area decreases and growth stocks increase as a result of forest rehabilitation projects, wood wastes (e.g., roots, branches, stems) generated in processes such as construction and mountain clearing are predicted to consequently increase [3]. These wood wastes include plant biomass such as lignocellulosic biomass. Emissions from plant biomass are not included in greenhouse gas emission inventories as plants absorb $CO_2$ during growth, and the $CO_2$ generated during the burning of solid fuels produced by thermochemical conversion is also absorbed by nature [4]. As wood wastes are usually stockpiled or incinerated, measures to process them in an efficient and eco-friendly manner are required.

The hydrothermal carbonization (HTC) reaction is a thermochemical process that converts biomass into hydrochar and produces biosolid, bioliquid, and biogas. HTC is particularly useful as it can convert biomass with a high moisture content, such as lignocellulosic biomass (which consists of refractory lignin, cellulose, and hemicellulose) [5] and D-xylose, into carbon materials [6]. Previous research on hydrochar production through HTC has detailed the products and byproducts of the reaction. Acetic acid is generated first [7] and acts as a catalyst for dehydration [8,9], one of the main mechanisms of the reaction. When acetic acid is used as a catalyst in the HTC of lignocellulosic biomass, the high heating value of hydrochar increases. As a larger amount of acetic acid was added, the production of hydroxymethylfurfural (HMF), an intermediate bioliquid product, increases [10,11]. However, HTC is a process generated in the aqueous phase, and some organic materials and intermediate byproducts in the biomass are dissolved in bioliquid [12]. Therefore, bioliquid can cause environmental pollution when it is discharged as wastewater due to its high chemical oxygen demand (COD) and total organic carbon (TOC) [12,13]. For this reason, process water recirculation for the economic and ecological treatment of bioliquid has been studied—mostly for biomass from single materials such as poplar wood chips [14], orange and grape pomace [9], loblolly pine [15], sweet potato waste [12], green urban waste [16], paper [17], and laminaria [18]. Across these studies, the mass yield after recirculation always increased; however; the trends of parameters such as the carbon content and caloric value of hydrochar were not consistent. Therefore, despite inconsistencies between different types of starting biomass materials, bioliquid recirculation has been shown to form additional solid matter.

Although recirculation has previously been applied to single-source biomasses, this study aims to develop a method to produce solid fuels using a financially and ecologically sustainable process. We produce hydrochar through composite materials, wood wastes, and HTC; we suppress the generation of wastewater and recover residuals through recirculation and examine the effects of bioliquid recirculation by evaluating the characteristics of the resulting hydrochar. Overall, we sought to develop a technique that contributes to achieving carbon neutrality goals by producing solid fuels through an economic and eco-friendly process.

## 2. Materials and Methods

### 2.1. Materials

#### 2.1.1. Wood Wastes

Wood wastes (WW) composed of twigs, bark, and leaves were collected in Dongdaemun-gu, Seoul, South Korea. In addition to being used in this experiment, the samples collected were dried in a drying oven at 105 °C until a constant weight was achieved; they were then pulverized into fragments smaller than 5 mm.

#### 2.1.2. Hydrothermal Carbonization (HTC) Reactor

Hydrochar was produced using 600 mL Parr Series 5500 HP Compact Reactors (Parr Instrument Company, Moline, IL, USA). The stainless-steel reactor was composed of an internal stirrer, pressure and temperature sensors, gas inflow and outflow valves, and an internal cooling loop. The temperature, pressure and revolutions per minute were monitored and adjusted using the connected 4848 Temperature Controller.

### 2.2. Experiment

#### 2.2.1. Production of Hydrochar and Bioliquid

The mass ratio and reaction temperature of biomass and water were set to 1:5 and 220 °C, respectively [19]. In accordance with the capacity and stability of the reactor, 60 g of biomass and 300 g of distilled water were used as starting materials and reacted at 220 °C for 2 h. After the reaction, the reactor was water-cooled for 5 min, and degassed for 1 min. The head and vessel of the reactor were then separated, and the residual gas was cooled for 30 min. After filtration under reduced pressure for solid–liquid separation, the separated biosolid was dried for 12 h in a drying oven at 105 °C and defined as hydrochar H1. After the pH of the bioliquid was measured, it was marked as L1, and used for the recirculation experiment.

2.2.2. Recirculation

The same amount of biomass was used in the recirculation experiment. The total biomass (excluding the amount needed for analysis) was used in the recirculation treatment of all bioliquid. Distilled water was added to match the mass ratio of biomass to water. The subsequent hydrochar production process was performed as described above. The liquid and hydrochar produced through nine recirculation experiments were marked as L2–L10 and H2–H10, respectively, as per the relevant recirculation step.

### *2.3. Analyses*

2.3.1. Bioliquid Analysis

The bioliquid was stored in a refrigerator at 4 °C, and a 1:1000 dilution of the bioliquid was used for every analysis, excluding that of pH. The pH was measured using the K Series Multimeter K7000-PDC (iSTEK, Seoul, Republic of Korea) in accordance with Korean Standard (KS) ES 04306. The TOC was measured via combustion oxidation at 680 °C using a TOC-V Series instrument (Shimadzu, Kyoto, Japan) and an ASI-V Autosampler.

The organic acids and intermediate products were analyzed using high-performance liquid chromatography (HPLC), Ultimate 3000 (Thermo Fisher Scientific, Waltham, MA, USA), and the refractive index (RI) (ERC RefractoMax 520, DataApex, Prague, Czech Republic) was used as a detector. In the HPLC, 0.01N $H_2SO_4$ was used as the mobile phase; a 10 μL sample of this was injected into an Aminex 87H column (300 × 10 mm, Bio-Rad, Hercules, CA, USA) for analysis. The eluent flow rate was 0.5 mL/min and the oven temperature was maintained at 40 °C. A total of 7 standards were used as follows; lactic acid sodium salt (Fluka-Honeywell, Charlotte, USA), citric acid (Showa chem, Beijing, China), malic acid (Kanto chem, Tokyo, Japan), succinic acid (Sigma-Aldrich, Taufkirchen, Germany), oxalic acid (Showa chem, Beijing, China), fumaric acid (Showa chem, Beijing, China), and VOAs mixture (AccuStandard, New Haven, CT, USA).

2.3.2. Hydrochar Analysis

Fourier transform infrared spectroscopy (FT-IR) was performed using a Nicolet iS 10 FT-IR Spectrometer (Thermo Fisher Scientific, Waltham, MA, USA) to analyze changes in the chemical structure of hydrochar due to bio liquid recirculation. The mass yield of hydrochar was calculated as follows, using the mass of the input biomass ($M_b$), and the mass of the produced hydrochar ($M_h$):

$$\text{Mass Yield }(\%) = M_h/M_b \times 100 \quad (1)$$

Changes in the properties of hydrochar due to bioliquid recirculation were analyzed via industrial, elemental, and heating value analyses. For the industrial analysis, the moisture, ash, and volatile matter contents were quantified, and fixed carbon was calculated based on the values obtained, according to ASTM standards D3172, D3173, D3174 and D3175, as follows:

$$\text{Fixed carbon }(\%) = 100 - \text{moisture content }(\%) - \text{ash content }(\%) - \text{volatile matter }(\%) \quad (2)$$

The elemental analysis of hydrochar was conducted using a carbon, hydrogen, nitrogen, and sulfur (CHNS) analyzer (Thermal Fisher Scientific, Waltham, MA, USA) according to ASTM D3176. The oxygen content of the hydrochar was calculated based on the CHNS values and ash content:

$$O\,\% = 100 - (C\,\% + H\,\% + N\,\% + S\,\% + \text{Ash content }(\%)) \quad (3)$$

The high heating value (HHV) was analyzed using an ECO bomb calorimeter (DDS CALORIMETERS, South Africa) according to KS E 3707. The low heating value (LHV) was calculated based on the HHV, and results of the elemental and industrial analyses, as follows:

$$\text{LHV} = \text{HHV} - 600(9H + W) \quad (4)$$

The energy density (ED) and energy yield (EY) were calculated based on these values, as shown below, to evaluate the effects of bioliquid recirculation on the characteristics of hydrochar energy:

$$ED = \text{HHV of hydrochar/HHV of biomass} \quad (5)$$

$$EY = ED \times \text{Yield of hydrochar} \quad (6)$$

## 3. Results and Discussion

### 3.1. Characteristics of Bioliquid

#### 3.1.1. Production and pH of Bioliquid

The first HTC experiment produced 228 g of bioliquid. It was expected that the quantities produced would decrease with progressive recirculation, but similar amounts of bioliquid (in the range of 221–228 g) were produced at every step. Additionally, all of the bioliquid generated in the experiments had a yellowish-brown color that tended to darken with progressive recirculation. The pH of all bioliquid samples ranged from 3.4 to 3.6, as confirmed to be acidic as in the other studies [14–17].

#### 3.1.2. Organic Acid and Total Organic Carbon Analyses

Acetic and formic acid were detected using organic acid analysis via HPLC. Xylose accounts for approximately 20–30% by weigh of lignocellulosic biomass, and d-xylose (wood sugar) is the main structure of hemicellulose [20]. The monosaccharides in cellulose and hemicellulose are released through hydrolysis [21]. Therefore, the detected acids could be generated via the hydrolysis of hemicellulose, and the dehydration reactions of d-xylose [22,23] or lignin side chains [6,24]. In particular, they could be produced by the degradation of C5 and C6 sugars [25].

The concentration of acetic acid increased from 11.6 g/L to 35.1 g/L, rising sharply until L6 (Figure 1). After this, the increase was insignificant; however, all subsequent concentrations were higher than that of L1. Formic acid levels rose sharply to 4.1 g/L in L5 before decreasing to 4.0 g/L in L6. However, it increased significantly to 4.5 g/L in L7, showing a similar trend to that of acetic acid. Therefore, the changes in concentration of acetic and formic acid were similar, and the pH of each sample of bioliquid was affected by organic acid levels. This acidic condition can produce water-soluble intermediates by promoting the degradation of biomass into monosaccharides [26,27].

The TOC, in particular, increased sharply to 30 g/L in L6, where five cycles of recirculation had occurred, indicating that this metric was affected by increasing levels of acetic acid (Figure 1). Furthermore, $H_2O$ was generated by the decomposition of formic acid under hydrothermal conditions [6,28] (see also [29]) suggesting that a relatively constant amount of bioliquid was generated at all steps of the process.

#### 3.1.3. Intermediate Analysis of Bioliquid

In this example, 5-HMF and furfural were identified as intermediate products in the bioliquid. These are production of dehydration of hexoses [20] and xylose into pentose sugar by acid-catalyzed [30,31]. Organic acids from lignocellulosic biomass decomposition, which accumulated in the bioliquid, could act as homogeneous catalysts when hexoses and pentoses are hydrolyzed to 5-HMF and furfural [16,32,33]. Therefore, it can be considered that the generation of 5-HMF and furfural was influenced by the organic acids analyzed previously. In addition, 5-HMF and furfural can catalyze the condensation reaction of lignin [34,35].

The levels of the intermediate products decreased with the progression of bioliquid recirculation, showing a trend opposite to those of the organic acid levels (Figure 2). The 5-HMF decreased from 1.5 g/L to 0.9 g/L, while furfural decreased from 0.7 g/L to 0.2 g/L, likely as a result of their condensation reactions. Furthermore, the yellowish-brown color of every bioliquid was likely the result of furfural oxidization after contact with air as the furfural formed a yellowish-brown synthetic resin. It can also be the result of decomposition and dissolution of organic matter [24,36].

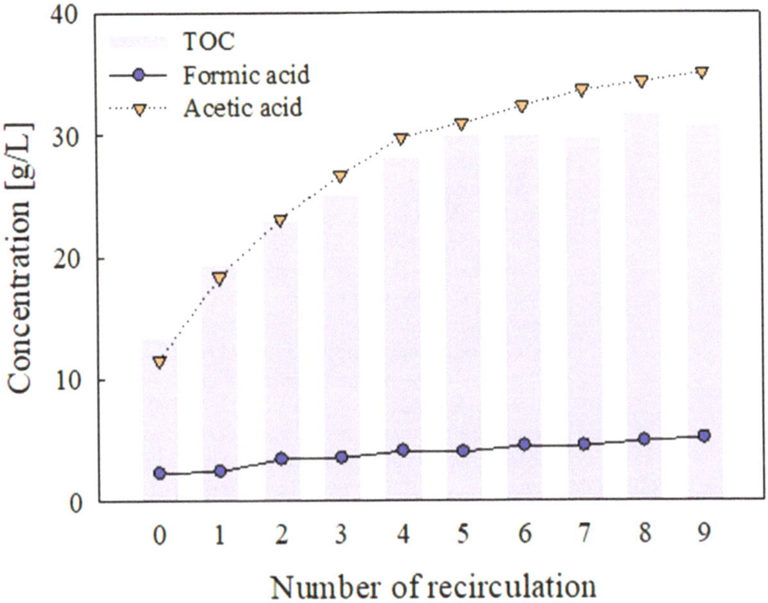


**Figure 1.** Concentration of organic acids and TOC in bioliquid.

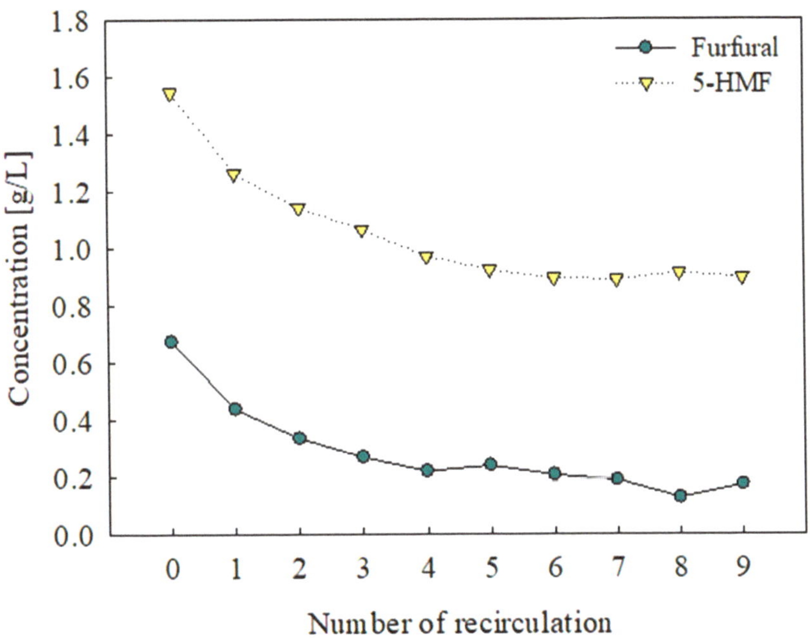


**Figure 2.** Concentration of intermediate products in bioliquid.

### 3.2. Characteristics of Hydrochar

#### 3.2.1. Fourier Transform Infrared Analysis of Hydrochar

FT-IR analysis was performed to examine the effects of bioliquid recirculation on the characteristics of hydrochar (Figure 3). A peak at 1160 $cm^{-1}$ signified the vibration characteristics of cellulose and hemicellulose in all hydrochar. This indicated the presence of undecomposed crystalline cellulose and hemicellulose. A peak at 3700 $cm^{-1}$ that was only observed in WW indicated the presence of alcohol (O-H), implying that this was what was decomposed by HTC. A peak from 3600–3200 $cm^{-1}$, which indicated an O-H bond, was discovered in every sample; however, it tended to decrease gradually with the recirculation of bioliquid. This indicated that bioliquid recirculation accelerated hydrolysis by HTC. We hypothesize that the amount of bioliquid produced remained constant because of this phenomenon. The peak decreased significantly after H6 because hydrolysis had been promoted by organic acids in the bioliquid, which were more scarce following this cycle. Due to the acidity of the initial environment, the degradation response strengthened, leading to the further decomposition of glucose. A peak at 3300 $cm^{-1}$, indicating a carboxylic acid compound, was caused by the oxidation of alcohol and the hydrolysis of ester. This peak showed a trend similar to previous peaks, decreasing significantly at H6.

In the presence of a peak from 1750–1700 $cm^{-1}$, an additional peak in the 1200–1100 $cm^{-1}$ range indicates the presence of an ester compound, which is generated by condensation polymerization after water is decomposed in acid and alcohol. The peak observed in this range showed the greatest and most distinct vibration characteristics in H1. This suggested that the ester bond was broken due to active hydrolysis during bioliquid recirculation. We thus concluded that alcohol was decomposed by HTC and that the recirculation of bioliquid accelerated the HTC mechanism.

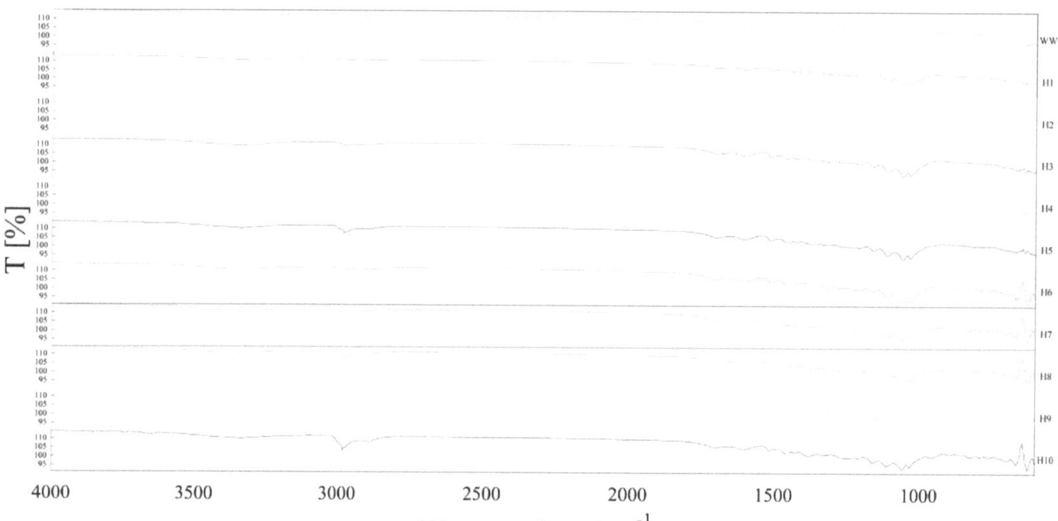


**Figure 3.** FT-IR result of biomass and hydrochar.

#### 3.2.2. Mass Yield of Hydrochar

Figure 4 shows the mass yield of hydrochar. The mass yield increased the most in the first step of bioliquid recirculation, from 65% in H1 to 69% in H2, and reached its highest, 70% in H6. The increase in mass yield in H1 to H2 is more pronounced than that in the other stages. This is because H1 is a hydrochar produced using biomass and water, whereas H2 is hydrochar produced by recirculation bioliquid. In other words, there is no material catalyzing the HTC mechanism when H1 was produced. However, when H2 is generated, the HTC mechanism is promoted by organic acids and intermediates in the bioliquid; therefore, it shows a rapid increase rate compared with other steps. The mass yield of hydrochar was lower than that of biomass because cellulose and hemicellulose are decomposed to form by-products in gaseous and liquid states [37].

The hydrochar yield did not increase consistently with the progress of recirculation, but all the values from H2–H10 were higher than the value in H1. This increase was the result of organic molecules in the bioliquid forming additional solid materials [17]. In particular, the HMF decomposed from cellulose and lignin formed a precipitate called HTC-coal via continuous polymerization and condensation reactions [6]. Therefore, the recirculation of bioliquid increased the production of hydrochar.

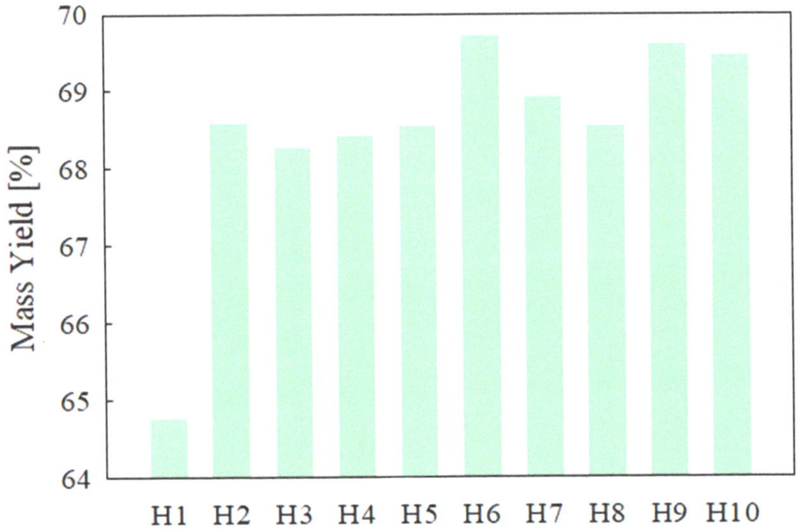


**Figure 4.** Mass yield of hydrochar.

### 3.2.3. Proximate Analysis of Hydrochar

Table 1 shows the results of the proximate analysis of biomass and hydrochar. The ash content of hydrochar generated through bioliquid recirculation increased, whereas the content of volatiles decreased. The increase in ash content can be attributed to the accumulation of inorganic components [9,38]. The fixed carbon content did not increase regularly with the progression of bioliquid recirculation, but it was higher than H1 in all other samples. In addition, H2, where bioliquid recirculation was first applied, showed the highest increase rate compared with the previous stage. This shows a similar trend to the result of the mass yield.

### 3.2.4. Changes in the Properties of Hydrochar Fuel

Table 2 shows the result of the ultimate analysis of biomass and hydrochar. The carbon content presented the largest rate of increase, rising from 55.7% in H1 to 58.4% in H2. This is similar to the result of mass yield and proximate analysis. This indicates that some of the additional solids produced by bioliquid recirculation are carbon. Additionally, all the carbon content in samples H2–H10 were higher than that in H1, reaching the highest value of 59.7% in H5. In H6, which showed the highest yield, the carbon content was 59.1%, having decreased minutely after H5.

**Table 1.** Proximate analysis of biomass and hydrochar.

| | Proximate Analysis, wt% | | | Fixed Carbon, % |
|---|---|---|---|---|
| | Moisture Content | Ash Content | Volatile Matter | |
| WW | 6.03 ± 0.16 | 1.84 ± 0.24 | 76.94 ± 0.31 | 15.19 ± 0.42 |
| H1 | 2.68 ± 0.04 | 2.35 ± 0.27 | 68.47 ± 0.04 | 26.51 ± 0.34 |
| H2 | 2.54 ± 0.14 | 2.31 ± 0.24 | 66.91 ± 0.11 | 28.24 ± 0.21 |
| H3 | 2.00 ± 0.07 | 2.37 ± 0.35 | 67.90 ± 0.19 | 27.72 ± 0.13 |
| H4 | 1.57 ± 0.04 | 2.89 ± 0.03 | 67.46 ± 1.17 | 28.08 ± 1.10 |
| H5 | 1.28 ± 0.05 | 3.09 ± 0.06 | 66.76 ± 0.09 | 28.86 ± 0.11 |
| H6 | 2.30 ± 0.02 | 2.71 ± 0.10 | 66.06 ± 0.38 | 28.93 ± 0.29 |
| H7 | 1.98 ± 0.07 | 2.85 ± 0.12 | 65.62 ± 0.27 | 29.55 ± 0.33 |
| H8 | 2.21 ± 0.07 | 2.86 ± 0.18 | 65.18 ± 0.25 | 29.74 ± 0.19 |
| H9 | 2.75 ± 0.02 | 2.88 ± 0.01 | 66.32 ± 0.45 | 28.05 ± 0.46 |
| H10 | 2.44 ± 0.01 | 3.07 ± 0.08 | 65.86 ± 0.42 | 28.63 ± 0.36 |

**Table 2.** Ultimate analysis of biomass and hydrochar.

| | Ultimate Analysis, wt% | | | | |
|---|---|---|---|---|---|
| | C | H | N | S | O |
| WW | 41.8 ± 2.13 | 6.2 ± 0.23 | 0.8 ± 0.02 | 0.05 ± 0.00 | 49.3 ± 0.06 |
| H1 | 55.7 ± 3.02 | 5.9 ± 1.41 | 0.5 ± 0.03 | 0.05 ± 0.00 | 35.5 ± 0.07 |
| H2 | 58.4 ± 0.86 | 5.9 ± 1.23 | 0.6 ± 0.03 | 0.05 ± 0.00 | 32.7 ± 0.03 |
| H3 | 59.3 ± 1.08 | 7.0 ± 1.80 | 0.6 ± 0.01 | 0.05 ± 0.00 | 30.6 ± 0.04 |
| H4 | 59.2 ± 0.24 | 7.2 ± 0.87 | 0.6 ± 0.02 | 0.05 ± 0.00 | 30.1 ± 0.01 |
| H5 | 59.7 ± 0.18 | 6.9 ± 0.95 | 0.6 ± 0.01 | 0.05 ± 0.00 | 29.7 ± 0.01 |
| H6 | 59.1 ± 0.39 | 6.7 ± 0.86 | 0.6 ± 0.02 | 0.05 ± 0.00 | 30.9 ± 0.02 |
| H7 | 58.7 ± 0.68 | 6.5 ± 1.11 | 0.6 ± 0.02 | 0.05 ± 0.00 | 31.3 ± 0.02 |
| H8 | 59.1 ± 0.74 | 6.8 ± 0.91 | 0.6 ± 0.01 | 0.05 ± 0.00 | 30.6 ± 0.03 |
| H9 | 59.0 ± 0.20 | 6.9 ± 1.11 | 0.6 ± 0.03 | 0.05 ± 0.00 | 30.6 ± 0.02 |
| H10 | 59.2 ± 0.66 | 7.5 ± 1.17 | 0.6 ± 0.01 | 0.05 ± 0.00 | 29.7 ± 0.03 |

Table 3 shows the combustion characteristics. The HHV measurements increased with bioliquid recirculation, reaching its highest value of 24.78 MJ/kg in H10. In contrast, the highest value obtained from calculating the LHV through elemental analysis was 23.04 MJ/kg in H7 and H8. This indicates that bioliquid recirculation can improve the value of hydrochar as fuel while increasing its pure carbon content. The energy densities of the biomass and hydrochar, and their energy yield, tended to increase with the progression of bioliquid recirculation. Moreover, this increase showed a similar trend to previous results. The energy yield, which was 80%, was increased to 90% after the recirculation of bioliquid, and increased continuously to 94% following the final reaction.

Overall, we found that some of the solid materials that were generated through progressive water recirculation were carbon materials, meaning that bioliquid recirculation can recover resources and increase energy recovery rates.

**Table 3.** The combustion characteristic of biomass and hydrochar.

| | HHV, MJ/kg | LHV, MJ/kg | ED | EY, % |
|---|---|---|---|---|
| WW | 18.02 ± 0.35 | 16.47 ± 0.47 | - | - |
| H1 | 22.67 ± 0.36 | 21.27 ± 0.16 | 1.26 ± 0.04 | 80.03 ± 2.27 |
| H2 | 23.24 ± 0.54 | 21.84 ± 0.32 | 1.29 ± 0.05 | 89.59 ± 3.10 |
| H3 | 23.49 ± 0.29 | 21.85 ± 0.60 | 1.30 ± 0.02 | 88.78 ± 1.34 |
| H4 | 23.63 ± 0.17 | 21.96 ± 0.06 | 1.31 ± 0.03 | 88.56 ± 2.28 |
| H5 | 23.73 ± 0.07 | 22.14 ± 0.23 | 1.32 ± 0.02 | 89.63 ± 1.48 |
| H6 | 24.52 ± 0.30 | 22.94 ± 0.11 | 1.36 ± 0.03 | 92.81 ± 2.02 |
| H7 | 24.55 ± 0.18 | 23.04 ± 0.34 | 1.36 ± 0.02 | 92.26 ± 1.10 |
| H8 | 24.63 ± 0.08 | 23.04 ± 0.28 | 1.37 ± 0.03 | 92.50 ± 1.72 |
| H9 | 24.53 ± 0.01 | 22.90 ± 0.26 | 1.36 ± 0.03 | 92.92 ± 1.74 |
| H10 | 24.78 ± 0.06 | 23.03 ± 0.25 | 1.38 ± 0.02 | 94.38 ± 1.68 |

## 4. Conclusions

To conclude, we have investigated the impact of bioliquid recirculation on solid hydrochar production by hydrothermal carbonization as a carbon-neutral wood waste processing technique. Bioliquid was generated in a relatively constant quantity for every recirculation step, and acetic acid, formic acid, 5-HMF, and furfural were generated as intermediate products. Organic acids and intermediate products promoted the HTC mechanism, thus increasing the mass yield, carbon content, heating value, and energy density of hydrochar. Our findings have a wide range of applications in the field of wood waste processing and any processing that involves hydrothermal carbonization. Therefore, bioliquid recirculation can be proposed as a method for developing an eco-friendly HTC process by recovering resources from wastewater and as a means of sustainable energy production for carbon neutrality.

**Author Contributions:** Conceptualization, S.-J.L.; Formal analysis, S.-J.O.; Investigation, S.-J.O., N.-H.C. and Y.-Y.K.; Methodology, S.-J.L.; Supervision, J.-Y.L.; Visualization, S.-J.L.; Writing—original draft, S.-J.L.; Writing—review & editing, M.-A.O. All authors have read and agreed to the published version of the manuscript.

**Funding:** This research was funded by the National Institute of Environmental Research, Republic of Korea (No. NIER-2021-03-01-006).

**Institutional Review Board Statement:** Not applicable.

**Informed Consent Statement:** Not applicable.

**Data Availability Statement:** Not applicable.

**Acknowledgments:** The author acknowledge the research support provided by the Ministry of Environment (MOE) of the Republic of Korea.

**Conflicts of Interest:** The authors declare no conflict of interest.

## References

1. Greenhouse Gas Inventory and Research Center of Korea. *National Inventory Report 2021*; Ministry of Environment: Seoul, Korea, 2021.
2. Korea Forest Service. *Forestry Statistics 2020*; Korea Forest Service: Daejeon, Korea, 2021.
3. Ministry of Environment Republic of Korea. Enforcement Rule of The Wastes Control Act. Available online: https://law.go.kr/lsSc.do?section=&menuId=1&subMenuId=15&tabMenuId=81&eventGubun=060101&query=%ED%8F%90%EA%B8%B0%EB%AC%BC%EA%B4%80%EB%A6%AC%EB%B2%95#J14551159 (accessed on 21 July 2017).
4. Intergovernmental Panel on Climate Change. *Good Practice Guidance and Uncertainty Management in National Greenhouse Gas Inventories*; Intergovernmental Panel on Climate Change: Geneva, Switzerland, 2000.
5. Anwar, Z.; Gulfraz, M.; Irshad, M. Agro-industrial lignocellulosic biomass a key to unlock the future bio-energy: A brief review. *J. Radiat. Res. Appl. Sci.* **2014**, *7*, 163–173. [CrossRef]
6. Funke, A.; Ziegler, F. Hydrothermal carbonization of biomass: A summary and discussion of chemical mechanisms for process engineering. *Biofuels Bioprod. Biorefin.* **2010**, *4*, 160–177. [CrossRef]
7. Yan, W.; Hastings, J.T.; Acharjee, T.C.; Coronella, C.J.; Vásquez, V.R. Mass and Energy Balances of Wet Torrefaction of Lignocellulosic Biomass. *Energy Fuels* **2010**, *24*, 4738–4742. [CrossRef]
8. Boutaieb, M.; Román, S.; Ledesma, B.; Sabio, E.; Guiza, M.; Ouederni, A. Towards a more efficient Hydrothermal Carbonization: Processing water recirculation under different conditions. *Waste Manag.* **2021**, *132*, 115–123. [CrossRef] [PubMed]
9. Kabadayi Catalkopru, A.; Kantarli, I.C.; Yanik, J. Effects of spent liquor recirculation in hydrothermal carbonization. *Bioresour. Technol.* **2017**, *226*, 89–93. [CrossRef] [PubMed]
10. Lynam, J.G.; Coronella, C.J.; Yan, W.; Reza, M.T.; Vasquez, V.R. Acetic acid and lithium chloride effects on hydrothermal carbonization of lignocellulosic biomass. *Bioresour. Technol.* **2011**, *102*, 6192–6199. [CrossRef]
11. Xu, C.; Leppänen, A.-S.; Eklund, P.; Holmlund, P.; Sjöholm, R.; Sundberg, K.; Willför, S. Acetylation and characterization of spruce (Picea abies) galactoglucomannans. *Carbohydr. Res.* **2010**, *345*, 810–816. [CrossRef]
12. Chen, X.; Ma, X.; Peng, X.; Lin, Y.; Wang, J.; Zheng, C. Effects of aqueous phase recirculation in hydrothermal carbonization of sweet potato waste. *Bioresour. Technol.* **2018**, *267*, 167–174. [CrossRef]
13. Becker, R.; Dorgerloh, U.; Paulke, E.; Mumme, J.; Nehls, I. Hydrothermal Carbonization of Biomass: Major Organic Components of the Aqueous Phase. *Chem. Eng. Technol.* **2014**, *37*, 511–518. [CrossRef]
14. Stemann, J.; Putschew, A.; Ziegler, F. Hydrothermal carbonization: Process water characterization and effects of water recirculation. *Bioresour. Technol.* **2013**, *143*, 139–146. [CrossRef]

15. Uddin, M.H.; Reza, M.T.; Lynam, J.G.; Coronella, C.J. Effects of water recycling in hydrothermal carbonization of loblolly pine. *Environ. Prog. Sustain. Energy* **2013**, *33*, 1309–1315. [CrossRef]
16. Köchermann, J.; Görsch, K.; Wirth, B.; Mühlenberg, J.; Klemm, M. Hydrothermal carbonization: Temperature influence on hydrochar and aqueous phase composition during process water recirculation. *J. Environ. Chem. Eng.* **2018**, *6*, 5481–5487. [CrossRef]
17. Weiner, B.; Poerschmann, J.; Wedwitschka, H.; Koehler, R.; Kopinke, F.-D. Influence of Process Water Reuse on the Hydrothermal Carbonization of Paper. *ACS Sustain. Chem. Eng.* **2014**, *2*, 2165–2171. [CrossRef]
18. Wang, F.; Wang, J.; Gu, C.; Han, Y.; Zan, S.; Wu, S. Effects of process water recirculation on solid and liquid products from hydrothermal carbonization of Laminaria. *Bioresour. Technol.* **2019**, *292*, 121996. [CrossRef] [PubMed]
19. Watanabe, A.; Morita, S.; Ozaki, Y. Study on Temperature-Dependent Changes in Hydrogen Bonds in Cellulose Iβ by Infrared Spectroscopy with Perturbation-Correlation Moving-Window Two-Dimensional Correlation Spectroscopy. *Biomacromolecules* **2006**, *7*, 3164–3170. [CrossRef] [PubMed]
20. Cheng, K.; Zheng, W.; Chen, H.; Zhang, Y.-H.P.J. Upgrade of wood sugar d-xylose to a value-added nutraceutical by in vitro metabolic engineering. *Metab. Eng.* **2019**, *52*, 1–8. [CrossRef]
21. Oliveira, P.C.; Canhaci, S.J.; Gabriel, C.B.; Moreira, C.R.; de Farias, A.M.D.; Fraga, M.A. Conversion of xylose to bioproducts on bifunctional supported platinum-group metals catalysts. *Curr. Res. Green Sustain. Chem.* **2022**, *5*, 100305. [CrossRef]
22. Świątek, K.; Gaag, S.; Klier, A.; Kruse, A.; Sauer, J.; Steinbach, D. Acid Hydrolysis of Lignocellulosic Biomass: Sugars and Furfurals Formation. *Catalysts* **2020**, *10*, 437. [CrossRef]
23. Oefner, P.J.; Lanziner, A.H.; Bonn, G.; Bobleter, O. Quantitative studies on furfural and organic acid formation during hydrothermal, acidic and alkaline degradation of D-xylose. *Monatsh. Chem.* **1992**, *123*, 547–556. [CrossRef]
24. Peterson, A.A.; Vogel, F.; Lachance, R.P.; Fröling, M.; Antal, M.J., Jr.; Tester, J.W. Thermochemical biofuel production in hydrothermal media: A review of sub- and supercritical water technologies. *Energy Environ. Sci.* **2008**, *1*, 32–65. [CrossRef]
25. Antczak, A.; Szadkowski, J.; Szadkowska, D.; Zawadzki, J. Assessment of the effectiveness of liquid hot water and steam explosion pretreatments of fast-growing poplar (*Populus trichocarpa*) wood. *Wood Sci. Technol.* **2021**, *56*, 87–109. [CrossRef]
26. Sharma, H.B.; Panigrahi, S.; Vanapalli, K.R.; Cheela, V.S.; Venna, S.; Dubey, B. Study on the process wastewater reuse and valorisation during hydrothermal co-carbonization of food and yard waste. *Sci. Total Environ.* **2022**, *806*, 150748. [CrossRef] [PubMed]
27. Picone, A.; Volpe, M.; Messineo, A. Process Water Recirculation during Hydrothermal Carbonization of Waste Biomass: Current Knowledge and Challenges. *Energies* **2021**, *14*, 2962. [CrossRef]
28. Mccollom, T.M.; Ritter, G.; Simoneit, B.R.T. Lipid synthesis under hydrothermal conditions by Fischer–Tropsch-Type reactions. *Orig. Life Evol. Biosph.* **1999**, *29*, 153–166. [CrossRef]
29. Yu, J.; Savage, P.E. Decomposition of Formic Acid under Hydrothermal Conditions. *Ind. Eng. Chem. Res.* **1998**, *37*, 2–10. [CrossRef]
30. Lange, J.-P. Lignocellulose conversion: An introduction to chemistry, process and economics. *Biofuels Bioprod. Biorefining* **2007**, *1*, 39–48. [CrossRef]
31. Schwiderski, M.; Kruse, A.; Grandl, R.; Dockendorf, D. Comparison of the influence of a Lewis acid AlCl3 and a Brønsted acid HCl on the organosolv pulping of beech wood. *Green Chem.* **2014**, *16*, 1569–1578. [CrossRef]
32. Kang, S.; Li, X.; Fan, J.; Chang, J. Characterization of Hydrochars Produced by Hydrothermal Carbonization of Lignin, Cellulose, d-Xylose, and Wood Meal. *Ind. Eng. Chem. Res.* **2012**, *51*, 9023–9031. [CrossRef]
33. Yemis, O.; Mazza, G. Acid-catalyzed conversion of xylose, xylan and straw into furfural by microwave-assisted reaction. *Bioresour. Technol.* **2011**, *102*, 7371–7378. [CrossRef]
34. Parajo, J.C.; Arbeit, D. Hydrothermal processing of lignocellulosic materials. *Eur. J. Wood Prod.* **1999**, *57*, 191–202.
35. Leng, S.; Li, W.; Han, C.; Chen, L.; Chen, J.; Fan, L.; Lu, Q.; Li, J.; Leng, L.; Zhou, W. Aqueous phase recirculation during hydrothermal carbonization of microalgae and soybean straw: A comparison study. *Bioresour. Technol.* **2020**, *298*, 122502. [CrossRef] [PubMed]
36. Xu, X.; Jiang, E. Treatment of urban sludge by hydrothermal carbonization. *Bioresour. Technol.* **2017**, *238*, 182–187. [CrossRef] [PubMed]
37. Kambo, H.S.; Minaret, J.; Dutta, A. Process Water from the Hydrothermal Carbonization of Biomass: A Waste or a Valuable Product? *Waste Biomass Valorization* **2018**, *9*, 1181–1189. [CrossRef]
38. Leng, S.; Leng, L.; Chen, L.; Chen, J.; Chen, J.; Zhou, W. The effect of aqueous phase recirculation on hydrothermal liquefaction/carbonization of biomass: A review. *Bioresour. Technol.* **2020**, *318*, 124081. [CrossRef] [PubMed]

*water*

MDPI

*Article*

# Performance Assessment of Concrete Using Discarded Membrane Filter Materials

**Sehwan Park and Junkyeong Kim ***

Safety Inspection for Infrastructure Laboratory (SIIL), Advanced Institute of Convergence Technology (AICT), 145 Gwanggyo-ro, Yeongtong-gu, Suwon-si 16229, Korea; sehwan0721@snu.ac.kr
* Correspondence: junkyeong@snu.ac.kr; Tel.: +82-31-888-9524

**Abstract:** Currently, membrane filters, which need to be replaced regularly as they get worn, are used in filtration facilities globally. The old membrane filters and housings become continuous industrial waste and are currently 100% incinerated. To solve this environmental problem, this study proposes the development of an eco-friendly concrete by mixing waste membrane resources with concrete. Through this, the environmental pollution and wastage of resources due to incineration, and the enormous amount of carbon dioxide generated during cement production, can be decreased by reducing the cement required when mixing concrete. To this end, the membrane module outer surface (acrylic butadiene styrene, ABS) and inner membrane (poly vinylidene fluoride, PVDF) were extracted from the waste membrane system and pulverized. Different mix ratios of 1%, 3%, and 5% for replacing cement were used when mixing concrete. The test specimens were then tested and compared with the reference concrete (ordinary Portland cement) specimen. It was confirmed that the compressive strength was high after 28 days in all the specimens to which ABS was added at 1%, 3%, and 5% mix ratios. Therefore, the possibility of technological development of eco-friendly concrete using waste resources from membrane filtration facilities was verified.

**Keywords:** membrane filter recycling; ordinary Portland cement (OPC); poly acrylic butadiene styrene (ABS); poly vinylidene fluoride (PVDF); eco-friendly concrete

**Citation:** Park, S.; Kim, J. Performance Assessment of Concrete Using Discarded Membrane Filter Materials. *Water* **2022**, *14*, 2167. https://doi.org/10.3390/w14142167

Academic Editors: Mohammad Mahmudur Rahman, Dibyendu Sarkar, Rupali Datta and Prafulla Kumar Sahoo

Received: 9 June 2022
Accepted: 7 July 2022
Published: 8 July 2022



## 1. Introduction

Currently, membrane filtration facilities and systems are continuously operated, with old membranes and housings being replaced when required. The old membranes and housings, which are being constantly generated, become industrial wastes that are 100% incinerated. These materials are made of plastic materials that, when incinerated, lead to environmental problems such as poor use of resources and pollution of the atmospheric environment. The issue of insufficient incineration and recycling technology of the old membrane materials (such as the separator and housing materials) is one that should be addressed. In other words, it is necessary to develop recycling technologies for each material involved in the process in order to prevent environmental pollution caused by the incineration of membrane waste.

Reusing these waste resources in concrete has the effect of reducing the cost of treating waste resources, protecting the environment from pollution, and has many advantages, such as improving some properties of concrete [1]. Therefore, research on eco-friendly concrete technology using waste resources is being actively conducted to prevent environmental pollution and reduce carbon emissions. Strength measurements and performance evaluation were performed by adding glass to concrete [2]; studies on a method using waste concrete were also conducted [3–7]. In particular, studies were actively conducted to investigate properties when using waste resources of plastic materials such as aged membranes and housings for concrete. By classifying plastics by type, strength and performance evaluations were conducted by increasing the addition ratio when mixing concrete [8–11]; when plastics were used in concrete, the compressive strength and bonding strength decreased as the plastic was added, regardless of the type of plastic [12,13], but it was confirmed that the impact resistance and ductility increased [14]. In addition, it was confirmed that round plastic particles improved the workability of concrete, whereas angled plastic particles decreased the workability of concrete [15].

However, in the existing studies, the amount of cement was generally the same, and the experiment was carried out by increasing the ratio by replacing the aggregate used when mixing concrete [16]. When 1 ton of cement is generated in the production process, about 0.7–1.0 tonnes of carbon dioxide is emitted, accounting for 7–8% of global greenhouse gas emissions [17].

In Korea, the cement industry is accountable for approximately 56.7 million tons of carbon dioxide per year [18]. To solve this emission problem, research is being conducted to reduce the amount of cement in concrete by adding admixtures such as fly ash and blast furnace slag powder [19–21].

This study investigates eco-friendly concrete technology that utilizes waste membrane resources to reduce the amount of cement in concrete. In particular, workability and strength, which are the biggest problems of concrete with plastic added, were selected as the scope of the study. To achieve this, a comparison was made between ordinary concrete, ordinary Portland cement (OPC), and concrete that replaced an amount of cement with the waste membrane material. To ease the comparison, the ratio of fine aggregates, coarse aggregates, admixtures, and w/c (water–cement ratio) were kept constant. In order to determine the strength characteristics of the generated specimens, strength measurements were performed using universal test machine (UTM) equipment, and performance evaluation was conducted by comparing the strength of ordinary concrete with concrete using different ratios of waste membrane resources.

## 2. Material Preparation for Membrane Recycled Concrete

### 2.1. Material Preparation

The cement used in the experiment was Portland cement of KS L 5201 standard, with a specific gravity of 3.15 $t/m^3$ and a fineness of 3450 $cm^2/g$. Crushed aggregates with a maximum particle diameter of 25 mm were used as the coarse aggregate, and river sand was used as the fine aggregate.

The used membrane film and housing were each composed of plastic made of poly vinylidene fluoride (PVDF) and poly acrylic butadiene styrene (ABS). As an excellent chemical-resistant material, PVDF has a wide range of applications; moreover, it has the advantage of being physiologically safe. ABS has a high heat distortion temperature and excellent chemical and oil resistance.

The outer surface of the membrane module (ABS) and the inner membrane (PVDF) were extracted from the membrane filtration facility. To use the ABS and PVDF as a substitute of cement, they should be made into fine-grained powder. As such, the extracted ABS and PVDF were cut to a certain size and then frozen with liquid nitrogen. They were then pulverized to a thickness of 0.5 mm using pulverizing equipment as shown in Figure 1.

### 2.2. Mortar Test of ABS/PVDF Mixture

Before proceeding with the concrete mixing test of the sample material, a mortar test specimen was prepared by substituting the prepared powder with a certain ratio of cement to obtain a replaceable percentage. Tables 1 and 2 present the physical and chemical properties of the prepared ABS and PVDF powder.

The mix proportions of the mortar specimens are shown in Table 3. ABS and PVDF powders were added to the mortar specimens for testing. Powders were replaced with 0%, 8%, 15%, and 25% cement. A 50 mm cube-shaped mortar test specimen was manufactured, and a compression test was conducted.

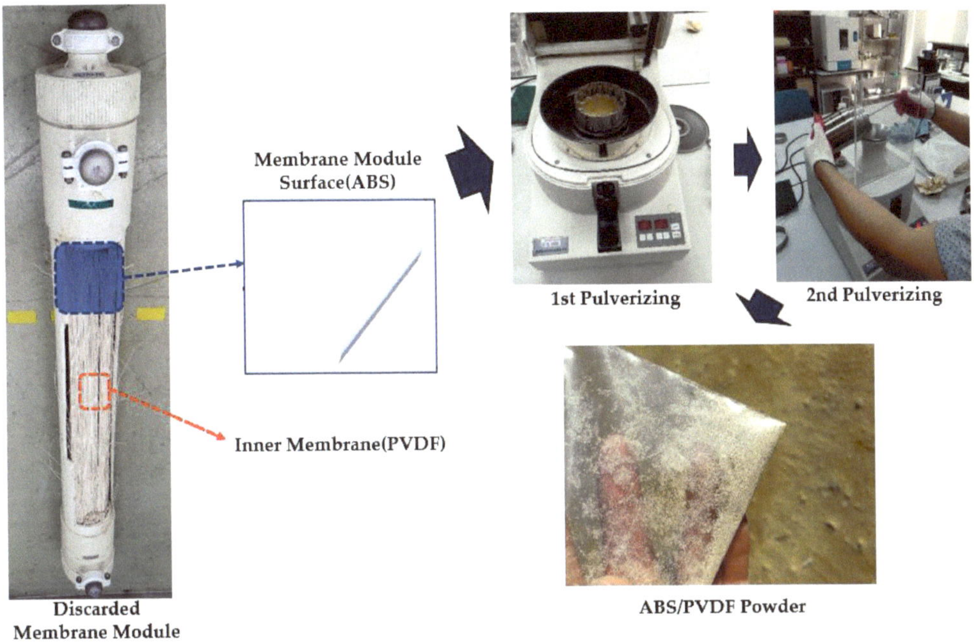


**Figure 1.** Extraction of ABS (poly acrylic butadiene styrene) and PVDF (poly vinylidene fluoride) powder from discarded membrane.

**Table 1.** Properties of ABS materials.

| Relative Density | Water Absorption Rate (%) | Tensile Strength (MPa) | Bending Strength (MPa) | Rockwell Hardness | Heat Distortion Temperature (°C) |
|---|---|---|---|---|---|
| 1.02–1.05 | 0.2–0.45 | 35–44 | 51–81 | R65–109 | 93–103 |

**Table 2.** Properties of PVDF materials.

| Relative Density | Water Absorption Rate (%) | Tensile Strength (MPa) | Bending Strength (MPa) | Rockwell Hardness | Heat Distortion Temperature (°C) |
|---|---|---|---|---|---|
| 1.78 | 0.02 | 52–54 | 70–75 | R84 | 166–170 |

**Table 3.** Mortar specimen mix proportions.

| C:S (Weight Ratio) | W/C (%) | Unit Amount (kgf/m$^3$) | | |
|---|---|---|---|---|
| | | Water (W) | Cement (C) | Fine Aggregate (S) |
| 1:2.5 | 50 | 255 | 566 | 1430 |

The results are shown in Table 4 and Figure 2. As the content of powder increased, the compressive modulus of elasticity increased by 20% compared to ordinary cement, and the compressive strength gradually decreased. However, the possibility of replacing cement by approximately 8% was confirmed.

**Table 4.** Powder-containing mortar compression test results.

| Waste Membrane Powder (%) | Modulus (GPa) | Strength (kgf/cm$^2$) | Peak Strain |
|---|---|---|---|
| 0 | 6.15 ± 0.15 | 557 ± 38.9 | 9.74 ± 1.31 |
| 8 | 7.84 ± 0.17 | 464 ± 40.6 | 8.26 ± 0.65 |
| 15 | 7.43 ± 0.18 | 314 ± 54.7 | 6.33 ± 0.55 |
| 25 | 5.39 ± 0.33 | 107 ± 22.9 | 5.00 ± 0.20 |

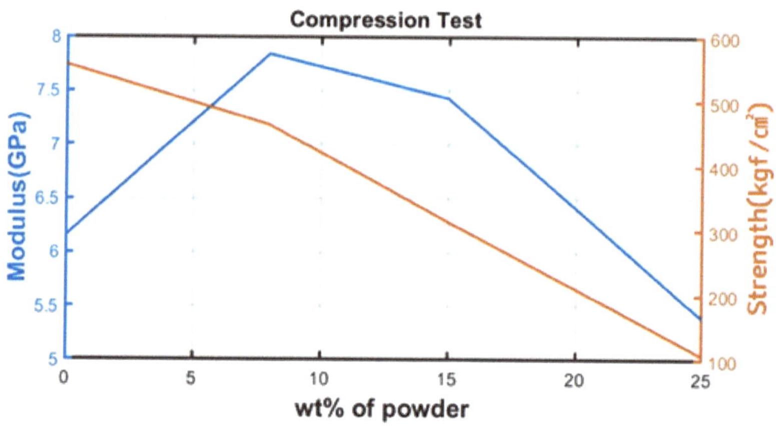


**Figure 2.** Powder-containing mortar compression test result graph.

## 3. Experimental Study

### *3.1. Preparation of OPC, ABS, and PVDF Concrete*

The specimens were mixed on site based on the sieve analysis results shown in Table 5, according to the regulations of KS F 2502. In addition, fine aggregate, coarse aggregate, and compounded water were corrected according to (*a*–*d*) in Table 5 and Equations (1)–(5).

$$S' = \frac{S - b \times (S + G)}{1 - (a + b)} \tag{1}$$

$$G' = \frac{G - a \times (S + G)}{1 - (a + b)} \tag{2}$$

**Table 5.** Aggregate sieving test results.

| | Specific Gravity | No.4 Sieve Remaining Amount | No.4 Sieve Passing Amount | Surface Water–Content Ratio |
|---|---|---|---|---|
| Fine aggregate | 2.58 | 1.0% | 99.0% | 2.7% |
| Coarse aggregate | 2.65 | 99.0% | 1.0% | −0.3% |

In Equation (1), $S'$ is the correction considering the grain size of fine aggregate, $S$ is fine aggregate, and $G$ is coarse aggregate. In (2), $G'$ is the correction considering the grain size of the coarse aggregate.

$$S'' = (1 + c) \times S \tag{3}$$

$$G'' = (1 + d) \times G' \tag{4}$$

$$W'' = W - (c \times S' + d \times G') \tag{5}$$

In (3), $S''$ is the correction considering the surface moisture of fine aggregate, and $G''$ in (4) is the correction considering the surface moisture of coarse aggregate. $W''$ in (5) is the correction considering the surface moisture of the blended water, and W is the blended water.

Based on the calibrated values, the job mix was performed, and cylindrical concrete specimens were manufactured according to the KS F 2405 standard (concrete compressive strength test method). Table 6 show the specified and job mix for each specimen; the unit of the values in the tables represents the unit amount per cube.

**Table 6.** Addition of ABS, PVDF powder (0%, 1%, 3%, 5%) in specified and job mix specimen (W: blended water, C: cement, S: fine aggregate, P: powder, G: coarse aggregate, AD: admixture).

| | Unit (kg/m$^2$) | | | | | |
|---|---|---|---|---|---|---|
| | **W** | **C** | **S** | **P** | **G** | **AD** |
| Specified mix | 167 | 383(0%)<br>379(1%)<br>372(3%)<br>364(5%) | 789 | 4(1%)<br>11(3%)<br>19(5%) | 955 | 2.68 |
| Job mix | 149 | 383(0%)<br>379(1%)<br>372(3%)<br>364(5%) | 808 | 4(1%)<br>11(3%)<br>19(5%) | 954 | 2.68 |

In specimens to which the waste membrane resource was added, cement was replaced in increments of 1%, 3%, and 5%, within 8% of the standard set in the previous experiment.

To find the applicability of ABS and PVDF concrete, the air contents test, slump test and compressive strength test were performed, as shown in Figure 3. The air contents test and slump test were performed on the day of mixing concrete, and the compressive test was performed on the 3rd, 7th, and 28th day after mixing. All the tests were performed with ABS, PVDF concrete and OPC concrete to compare with ordinary concrete.

**Figure 3.** Testing process for ABS/PVDF concrete.

### 3.2. Experimental Results

The mix for the concrete was added in the order of fine aggregate, cement, coarse aggregate, powdered ABS, and PVDF powder. After mixing, the slump and air content were checked according to the specifications of the KS F 2402 (slump test method of Portland cement concrete) and KS F 2421 (air content of hardened concrete by pressure test method) standards.

As shown in Figure 4, the air content for the OPC specimen, ABS powder specimen, and PVDF powder specimen was equal to 9%. The slump content of the OPC and ABS powder specimens was the same, at 170 mm, but the slump content of the PVDF powder specimen was measured to be approximately 100 mm. The suitable slump of concrete which is used for reinforced specimen is 60–180 mm [22]. The result of the slump test of the ABS powder specimen was suitable for standard use, but the result of the PVDF powder specimen was higher than standard slump. This was caused by the high chemical resistance of PVDF, which was derived by the fluoride.

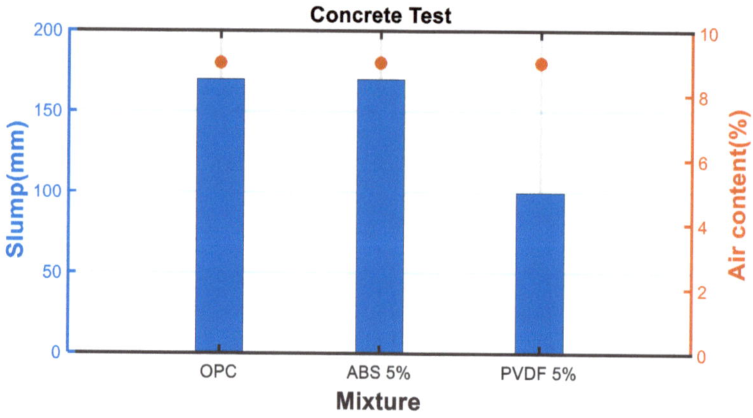


**Figure 4.** Slump and air content (Dot: air content; bar: slump).

The concrete specimens were cured in water, and strength measurements were performed according to the KS F 2405 standard using UTM equipment to measure the compressive strength of the prepared concrete by age.

Strength measurements were performed for each specimen, every 3 days, 7 days, and 28 days. The strength was measured for three specimens for each age, and the average value was derived. By comparing the average value of the compressive strength of the samples with ABS powder, PVDF powder, and the average value of the compressive strength of the OPC sample, the usability of the waste membrane was examined.

Figures 5–11 present graphs showing the strength measurement experiments for the three specimens of OPC, ABS, and PVDF. Additionally, the average strength values of the three specimens are shown together in the graphs, so OPC, ABS, and PVDF specimens can be compared using the average strength values.

By using the compressive strength tests, it was confirmed that all the specimens had a generally uniform pattern and increased with age. In addition, it was confirmed that the tendency of the average strength value and the tendency of each specimen were consistent, and thus it was verified that comparison using the average strength value between the OPC, ABS, and PVDF specimens was possible.

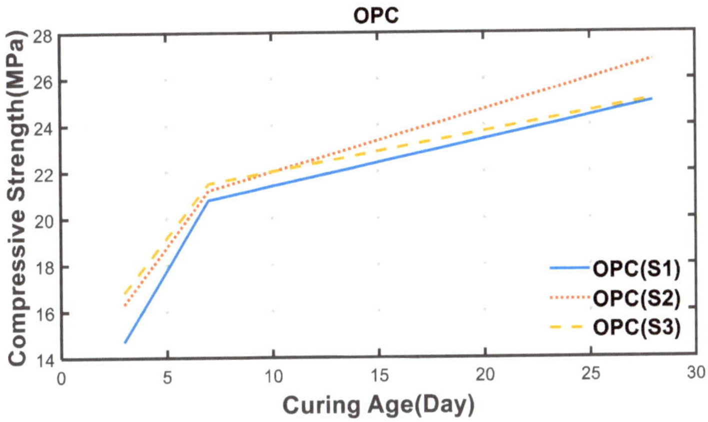


**Figure 5.** Graph of strength measurement results between OPC specimens.

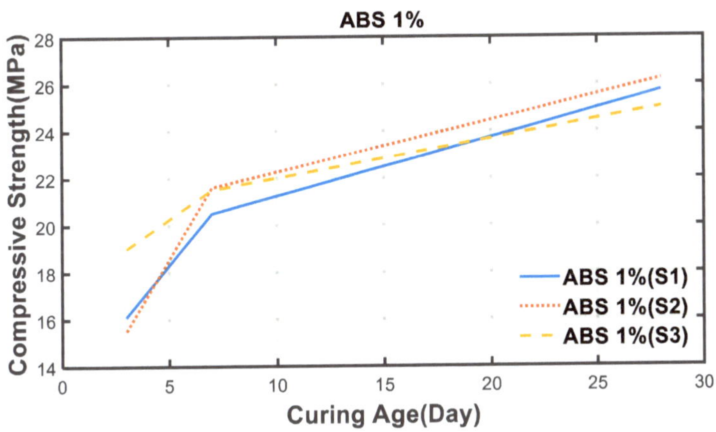


**Figure 6.** Graph of strength measurement results between ABS (1%) specimens.

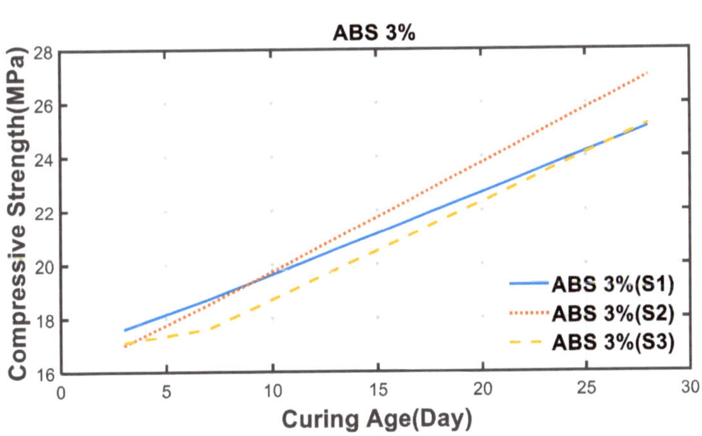


**Figure 7.** Graph of strength measurement results between ABS (3%) specimens.

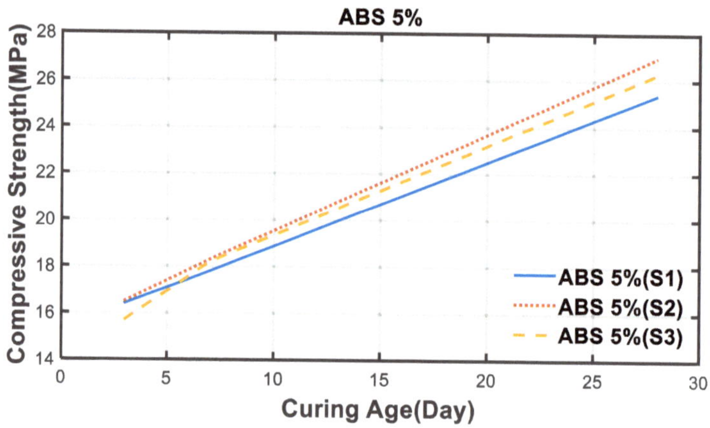


**Figure 8.** Graph of strength measurement results between ABS (5%) specimens.

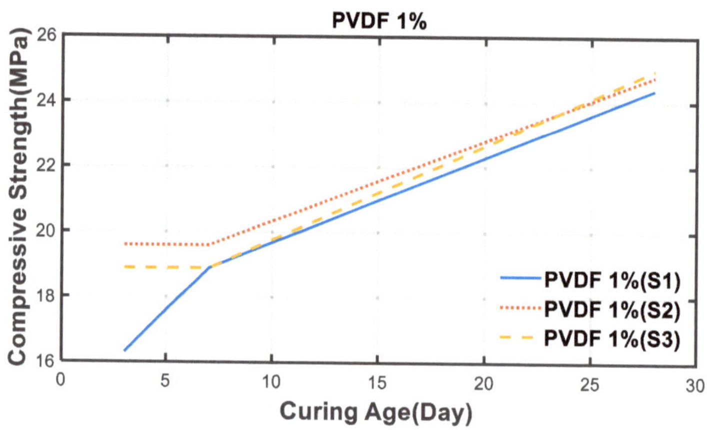


**Figure 9.** Graph of strength measurement results between PVDF (1%) specimens.

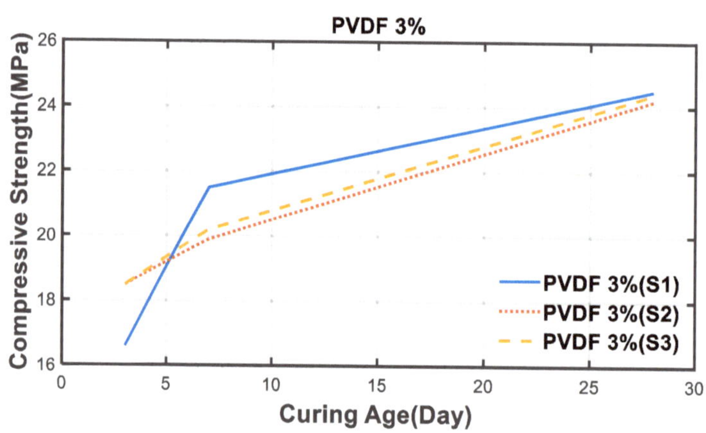


**Figure 10.** Graph of strength measurement results between PVDF (3%) specimens.

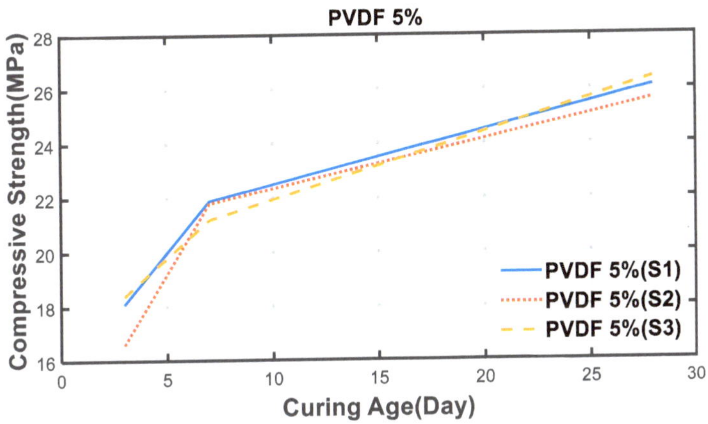


**Figure 11.** Graph of strength measurement results between PVDF (5%) specimens.

### 3.3. Experimental Discussion

Table 7 and Figure 12 show the comparison of the average strength according to the addition of the mixture and its percentage.

**Table 7.** Compressive strength by mixture addition and percentage.

| Mixture | Percentage (%) | Age (Days) | | |
|---|---|---|---|---|
| | | 3 | 7 | 28 |
| OPC | - | 15.9 MPa | 21.2 MPa | 25.6 MPa |
| ABS | 1 | 16.9 MPa | 21.2 MPa | 25.6 MPa |
| | 3 | 17.2 MPa | 18.3 MPa | 25.8 MPa |
| | 5 | 16.2 MPa | 18.1 MPa | 26.2 MPa |
| PVDF | 1 | 19.1 MPa | 19.2 MPa | 24.7 MPa |
| | 3 | 17.9 MPa | 20.5 MPa | 24.4 MPa |
| | 5 | 17.7 MPa | 21.6 MPa | 26 MPa |

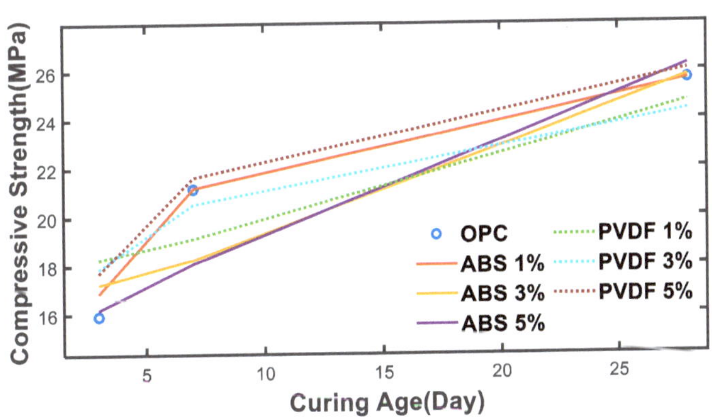


**Figure 12.** Comparison of compressive strength of specimens.

It can be seen that when ABS replaced 1% of cement, the strength of the specimen was almost the same as that of OPC concrete; however, when 3% and 5% ABS replaced the cement, it was confirmed that the strength in the first seven days was 3 MPa lower. This is suggested to be due to the replaced ABS powders interfering with the initial hydration reaction of cement. However, the strength after 28 days was high enough to use for concrete. Hence, it was confirmed that the strength of the concrete did not decrease even when ABS powder was added.

Moreover, it can be observed that the initial strengths of the PVDF specimens on the third day were all higher than that of the OPC specimen. It is assumed that the addition of PVDF powder absorbed a lot of water and resulted in high initial strength. In specimens to which 1% and 3% PVDF powder was added, it was confirmed that, except for the initial strength, the strength during the remaining days was generally low; indeed, the compressive strength on the 28th day was approximately 1 MPa lower than that of the OPC specimen. Conversely, it was confirmed that the specimen to which 5% PVDF powder was added had higher compressive strength than the OPC specimen at all ages. This phenomenon could be considered to be caused by the improved internal bonding strength due to the electric dipole of the replaced PVDF and the high moisture content of PVDF. According to the results of the experiments, the ABS and PVDF from the discarded membrane can be used as substitutes of cement to reduce carbon in regards to their compressive strength and workability.

## 4. Conclusions

This study considered reducing the amount of cement used in concrete by replacing it with discarded membrane resources. The characteristics of the age-specific strength were examined for a number of specimens with different mix ratios. A summary of this study is as follows.

1. It was found that carbon emission reduction of approximately 50 kg $CO_2/m^3$ is possible through waste membrane recycling and cement replacement. This is equivalent to 10 pine trees absorbing 5.6 kg of $CO_2$ per year.
2. To use the discarded membrane module as substitutes of cement, the outer casing (ABS) of membrane module and inner membrane (PVDF) should be made into powder form with a frozen pulverizing process.
3. The concrete specimens were made with ABS and PVDF with 1 wt%, 3 wt%, 5 wt% to substitute cement, and the air contents test and slump test were conducted. According to the test results, the workability of ABS specimens was the same as the OPC specimen; for the PVDF specimen it was lower that OPC, but it was within the standard range.
4. The compressive strength test was performed to find the applicability of ABS/PVDF concrete specimens. The compressive strengths of test specimens with ABS and PVDF were measured at the 3rd, 7th, and 28th days after pouring, and the strengths of OPC specimens were also measured for comparison with test specimens. As a result, the specimens to which 1% of ABS powder and 5% of PVDF powder was added had similar strength development to that of the OPC specimen; the numerical values were also confirmed to be almost identical.
5. Furthermore, the specimens which contained 5% of PVDF showed higher strength than OPC specimens, caused by the improved internal bonding strength due to the electric dipole of the replaced PVDF and the high moisture content of PVDF.
6. Therefore, if ABS powder is used to replace cement, it is judged that 1% replacement is possible, and in the case of PVDF powder, it is judged that it is possible to replace around 5%.

According to the results of the experiments, the ABS and PVDF from the discarded membrane can be substitutes of cement to reduce carbon in light of their compressive strength and workability.

**Author Contributions:** Conceptualization, J.K.; methodology, J.K.; data curation, S.P.; writing—original draft preparation, S.P.; writing—review and editing, J.K.; visualization, S.P.; supervision, J.K.; project administration, J.K.; funding acquisition, J.K. All authors have read and agreed to the published version of the manuscript.

**Funding:** This research and APC funded by Ministry of Oceans and Fisheries (Development of Disaster Control and Aging Management Method for Port Infrastructure) (20210603).

**Informed Consent Statement:** Not applicable.

**Data Availability Statement:** Not applicable.

**Conflicts of Interest:** The authors declare no conflict of interest.

## References

1. Hamdullah, D.N.; Hama, S.M.; Hama, S.M. Optimum Content of Waste Fine plastic aggregate Using Mini181 Software for Best Performance of Reinforced Concrete. In Proceedings of the 14th International Conference on Developments in eSystems Engineering (DeSE), Sharjah, United Arab Emirates, 7–10 December 2021; pp. 318–322.
2. Shayan, A.; Xu, A.M. Value-Added Utilisation of Waste Glass in Concrete. *Cem. Concr. Res.* **2004**, *34*, 81–89. [CrossRef]
3. Topcu, I.B.; Sengel, S. Properties of Concretes Produced with Waste Concrete Aggregate. *Cem. Concr. Res.* **2004**, *34*, 1307–1312. [CrossRef]
4. Poon, C.S.; Chan, D.X. Feasible Use of Recycled Concrete Aggregates and Crushed Clay Brick as Unbound Road Sub-Base. *Constr. Build. Mater.* **2006**, *20*, 578–585. [CrossRef]
5. Etxeberria, M.; Vazquez, E.; Mari, A.; Barra, M. Influence of Amount of Recycled Coarse Aggregates and Production Process on Properties of Recycled Aggregate Concrete. *Cem. Concr. Res.* **2007**, *37*, 735–742. [CrossRef]
6. Huang, Y.; Bird, R.N.; Heidrich, O. A Review of the Use of Recycled Solid Waste Materials in Asphalt Pavements. *Resour. Conserv. Recycl.* **2007**, *52*, 58–73. [CrossRef]
7. De Juan, M.S.; Gutierrez, P.A. Study on the Influence of Attached Mortar Content on the Properties of Recycled Concrete Aggregate. *Constr. Build. Mater.* **2009**, *23*, 872–877. [CrossRef]
8. Siddique, R.; Khatib, J.; Kaur, I. Use of Recycled Plastic in Concrete: A Review. *J. Waste Manag.* **2008**, *28*, 1835–1852. [CrossRef] [PubMed]
9. Khandelwal, V. Replacement of sand with shredded plastic in cement concrete. *Int. J. Eng. Res. Technol.* **2019**, *8*, 946–949.
10. Aleena, J.K.; Sumaiha, S.; Aswathy, A.J.; Sarath, S. An experimental study on partial replacement of M-Sand with glass powder in plastic sand brick. *Int. J. Res. Anal. Rev.* **2021**, *8*, 334–348.
11. Bahoriaa, B.V.; Parbatb, D.K.; Nagarnaik, P.B. XRD analysis of natural sand, quarry dust, waste plastic (LDPE) to be used as a fine aggregate in concrete. *Mater. Today Proc.* **2018**, *5*, 1432–1438. [CrossRef]
12. Jain, A.; Siddique, S.; Gupta, T.; Jain, S.; Sharma, R.K.; Chaudhary, S. Evaluation of concrete containing waste plastic shredded fibers: Ductility properties. *Struct. Concr.* **2021**, *22*, 566–575. [CrossRef]
13. Jain, A.; Siddique, S.; Gupta, T.; Sharma, R.K.; Chaudhary, S. Utilization of shredded waste plastic bags to improve impact and abrasion resistance of concrete. *Environ. Dev. Sustain.* **2018**, *22*, 337–362. [CrossRef]
14. Hama, S.M. Behavior of concrete incorporating waste plastic as fine aggregate subjected to compression, impact load and bond resistance. *Eur. J. Environ. Civ. Eng.* **2020**, *26*, 3372–3386. [CrossRef]
15. Hama, S.M.; Hilal, N.N. 5-Fresh properties of concrete containing plastic aggregate. In *Use of Recycled Plastics in Eco-Efficient Concrete*; Pacheco-Torgal, F., Khatib, J., Colangelo, F., Tuladhar, R., Eds.; Woodhead Publishing: Sawston, UK, 2019; pp. 85–114.
16. Kim, J.S.; Kim, K.S.; Kim, P.S. A Study Properties of concrete Recycling Cockle Shells as Fine Aggregate. *J. Korea Concr. Inst.* **2004**, *4*, 141–146.
17. Shin, K.-S.; Lee, B.-K.; Kim, G.-Y. Durability of Concrete Using Blast Furnace Slag Powder. *Mag. RCR* **2012**, *7*, 13–17.
18. Ha, J.S.; Kim, H.S.; Lee, Y.D. Properties of Strength Development of Concrete at Early Age Using High Fineness Cement and Fly Ash. *J. Korea Inst. Struct. Maint. Insp.* **2018**, *22*, 154–160.
19. Berndt, M.L. Properties of Sustainable Concrete Containing Fly Ash, Slag and Recycled Concrete Aggregate. *Constr. Build. Mater.* **2009**, *23*, 2606–2613. [CrossRef]
20. Borges, A.; Flores-Colen, I.; de Brito, J. Physical and Mechanical Performance of Cement-Based Renders with Different Contents of Fly Ash, Expanded Cork Granules and Expanded Clay. *Constr. Build. Mater.* **2018**, *191*, 535–543. [CrossRef]
21. Li, L.H.; Long, G.C.; Ma, K.L.; Ma, H.W.; Wang, W.B.; Zhang, C.; Xie, Y.J. Preparation of Green Low Strength Mixture for Foundation Reinforcement Treatment by Using Fly Ash and Waste Coal Gangue. *Materials* **2020**, *13*, 664. [CrossRef] [PubMed]
22. Ministry of Land, Infrastructure and Transport of Korea. *Building Construction Standard Specification of Korea (KSC41)*; Ministry of Land, Infrastructure and Transport of Korea: Sejong, Korea, 2021.

MDPI AG
Grosspeteranlage 5
4052 Basel
Switzerland
Tel.: +41 61 683 77 34

MDPI Books Editorial Office
E-mail: books@mdpi.com
www.mdpi.com/books

www.ingramcontent.com/pod-product-compliance
Lightning Source LLC
Chambersburg PA
CBHW041431210926
51850CB00013B/3405

* 9 7 8 3 7 2 5 8 3 0 3 9 8 *